Biosimilars of Monoclonal Antibodies

Biosimilars of Monoclonal Antibodies

A Practical Guide to Manufacturing, Preclinical, and Clinical Development

Edited by

Cheng Liu, Ph.D.
K. John Morrow, Jr., Ph.D.

WILEY

Published by John Wiley & Sons, Inc., Hoboken, New Jersey

Published simultaneously in Canada

Library of Congress Cataloging-in-Publication Data:

Names: Liu, Cheng, 1966– editor. | Morrow, John, Jr. 1938– editor.
Title: Biosimilars of monoclonal antibodies : a practical guide to manufacturing, preclinical, and clinical development / edited by Cheng Liu, K. John Morrow, Jr.
Description: Hoboken, New Jersey : John Wiley & Sons, Inc., [2017] | Includes bibliographical references and index.
Identifiers: LCCN 2016019958 (print) | LCCN 2016020707 (ebook) | ISBN 9781118662311 (cloth) | ISBN 9781118940624 (pdf) | ISBN 9781118940631 (epub)
Subjects: | MESH: Antibodies, Monoclonal | Biosimilar Pharmaceuticals
Classification: LCC QR186.7 (print) | LCC QR186.7 (ebook) | NLM QW 575.5.A6 | DDC 616.07/98–dc23
LC record available at https://lccn.loc.gov/2016019958

Printed in the United States of America

Cover image: Courtesy of the Editors

Set in 10/12pt Warnock by SPi Global, Pondicherry, India

10 9 8 7 6 5 4 3 2 1

Contents

Notes on Contributors

K. Lance Anderson focuses his law practice at Dickinson Wright PLLC on intellectual property, business and information technology related to emerging business. A registered patent attorney, he has wide-ranging experience in licensing and transactional matters, including negotiation and due diligence for licensing, mergers and acquisitions, asset transfers, and strategic planning. He often counsels on pharma and life science transactions, including research and development, in the developing anticancer and immunotherapeutic industries. His experience also includes patent prosecution, patent litigation, trade secrets, employment, corporate governance, electronic records management, and financial matters.

Lance earned his B.S. in Entomology and M.S. in Crop Science from Texas Tech University and his J.D. from the Texas Tech University School of Law.

Jonathan D. Ball, Ph.D., is a shareholder at Greenberg Traurig, LLP, with nearly 15 years of experience representing clients in intellectual property matters, with an emphasis on patent litigation and contentious Patent Office proceedings. He also has extensive experience in patent portfolio management, prosecution, and licensing and helps companies develop and implement worldwide IP strategies. He has experience in a variety of biopharmaceutical matters, including Hatch–Waxman litigation, pre-ANDA diligence, and counseling on pharmaceutical patent life cycle management.

Jonathan earned his B.A. from Vassar College, his Ph.D. in Organic Chemistry from the University of North Carolina at Chapel Hill, and his J.D. from the University of Richmond School of Law.

Maurizio Chiriva-Internati An author of 146 peer-reviewed publications, Maurizio Chiriva-Internati, Ph.D., D. Bsci., acts as the chief scientific officer at Kiromic, a biotechnology company specialized in immunotherapy, and serves as an associate professor of medicine and director of translational research at the Division of Hematology/Oncology in the Department of Internal Medicine at Texas Tech University Health Sciences Center in Lubbock, Texas. He received his Ph.D. in Immunology from the University of Nottingham and his doctorate in biology from the University of Milan. He has clinical and teaching experience in M.D., Ph.D., and M.S. school settings as well as experience as a reviewer for a variety of journals. He is an associate editor for the *Journal of Translational Medicine*, a section editor for the *International Reviews of Immunology*, and an associate editor member for both the *Open Hematology Reviews* and the *Open Hematology Journal*. For the past 14 years, he has investigated tumor antigens as therapeutic and diagnostic tools. His research is focused on developing new therapies for a variety of diseases, namely, cancers including multiple myeloma, ovarian, lung, prostate, melanoma, and breast. Using recombinant adeno-associated viruses, he is developing vaccines that direct T-cell immune responses toward unique tumor-specific or tumor-associated antigens.

Everardo Cobos, M.D., is president of Kiromic, LLC, and director of Hematologic Malignancies at Grace Health System in Lubbock, Texas. For the past 24 years, he served as a professor of medicine and oncology and chairman of the division of hematology/oncology at Texas Tech University Health Sciences Center (HSC) in Lubbock. He has also served as medical director of the Southwest Cancer Center at the University Medical Center (UMC) and as program director and founder of the Stem Cell Transplantation Program at TTUHSC/UMC. He is board certified in internal medicine, hematology, and oncology and is an expert in the treatment of hematologic malignancies, stem cell transplantation, and thrombosis and hemostasis.

Dr. Jose A. Figueroa is currently director of Solid Malignancies in the Blood Disorders and Cancer Therapeutics Center at Grace Health System in Lubbock, Texas, and vice president and chief medical officer of Kiromic, LLC, a clinically oriented biotechnology company. He earned his undergraduate and medical degrees at the University of Puerto Rico and completed his fellowship in medical oncology, hematology, and translational research at the University of Texas Health Science Center in San Antonio in 1993. He is board certified in medical oncology. He has been a recipient of the Young Oncologist Award from the American Radium Society and the American Cancer Society Research Grant and has served as medical director and director of the clinical research program of the Southwest Cancer Center/University Medical Center at Texas Tech University Health Sciences Center, School of Medicine. He is the author of many scientific abstracts, peer-reviewed publications, a book chapter on breast cancer, poetry, and a fictional novel.

João Eurico Fonseca is head of the Rheumatology Research Unit at the Instituto de Medicina Molecular and professor of rheumatology and biomedical engineering, both part of the Faculty of Medicine, University of Lisbon,

Portugal. He is also head of the Ambulatory Unit of the Rheumatology Department of the Lisbon Academic Medical Centre. He completed his M.D. and Ph.D. in Rheumatology at the University of Lisbon and has obtained a qualification in health management from the AESE School of Management and Business in Lisbon. His research focuses on bone and inflammatory joint diseases. He is also the president of the Portuguese Society of Rheumatology and a member of the editorial board of several medical journals. He has authored more than 150 scientific papers.

João Gonçalves is head of the Biopharmaceutical and Molecular Biotechnology Unit at the Institute of Innovative Medicines and professor of immunology and biotechnology, both part of the Faculty of Pharmacy, University of Lisbon, Portugal. He is also head of Antibody Engineering Laboratory at iMed—Faculdade Farmacia Universidade Lisboa. He has a pharmaceutical degree and Ph.D. in Infectious Diseases/Immunology at Harvard University and University of Lisbon and has postdoctoral experience at Centocor, Genentech, and Scripps Research Institute, La Jolla. His research focuses on therapeutic antibody discovery and development ranging from autoimmune diseases to oncology. He is a member of the Portuguese Pharmacopoeia Commission and Portuguese Commission of Medicines Evaluation. He is also a member of the editorial board of several scientific journals. He has authored more than 80 scientific papers.

Patrick G. Holder is a native of Cleveland Heights, OH. He received his Ph.D. in Chemical Biology in the lab of Matthew Francis at UC Berkeley. While there, he used bioconjugation to integrate optoelectronic materials into a scaffold of self-assembling viral proteins, including single-walled carbon nanotubes, arrays of phthalocyanines, and thermostable virus-templated polymers. As an NIH postdoctoral fellow in the lab of Daniel Nocera at MIT, in a collaboration with JoAnne Stubbe, he leveraged site-specific bioconjugation of phototriggers to perform time-resolved spectroscopy of radical transport in the enzyme ribonucleotide reductase. In 2012, his work was awarded the Energy and Environmental Science Prize from the Royal Society of Chemistry. He graduated with an A.B. in Chemistry from Columbia University, where he received the William H. Reinmuth Chemistry Scholarship. Formerly at Redwood Bioscience and currently at Bristol-Myers Squibb, he is helping advance antibody-drug conjugates from the benchtop to the clinic.

Dr. Nattamol Hosiriluck is currently a second-year resident in the Department of Internal Medicine of Texas Tech University Health Sciences Center in Lubbock, Texas. She graduated from Phramongkutkloa College of Medicine in Bangkok, Thailand, in 2012. Since then her interest has been on hematology and oncology, and she is planning to pursue her fellowship in this area.

Clarinda Islam is currently the director of Quality Assurance and Regulatory Affairs for Somru BioScience, Inc., an emerging biotechnology company based in Charlottetown, PE, Canada. In this role, she is responsible for the quality management, project management, process improvement, product development life cycle methodologies, and quality management systems. Previously she served as a senior group leader for the Immunochemistry Department at PPD (VA, USA) where she was responsible for leading a team of scientists dedicated to supporting biopharmaceutical development. Prior to joining PPD, she has held positions of increasing responsibility in the Bioanalytical Group at Covance, Inc. She has over 14 years of experience in the development of biologics in a highly regulated environment under FDA and other global regulatory guidelines.

Rafiq Islam is the senior director of Bioanalytical Services at Celerion, Inc. (NE, USA). In his current role, he is responsible for the scientific and operational leadership of both small- and large-molecule bioanalysis, which includes the development, validation, and execution of sample analysis to support pharmacokinetic, immunogenicity, bioequivalence, and biosimilar studies. Previously, he served as the scientific director for Biopharma Services at EMD Millipore. He held similar positions as the head of a bioanalytical department at Covance, Inc. (AZ, USA), and Huntingdon Life Sciences (NJ, USA). He also held several positions of increasing responsibility with CuraGen Corporation (New Haven, CT).

He has a B.S. degree in biology and M.Sc. in data mining. He has 15 years of industry experience. He is an active member of the American Association of Pharmaceutical Scientists (AAPS) and Global CRO Council (GCC).

Roy Jefferis After receiving his B.Sc. and Ph.D. in Chemistry, Jefferis moved to the Medical School (Birmingham) to study the structure and function of antibody molecules. A particular interest has been the influence of glycosylation on the expression of biologic activities, *in vitro* and *ex vivo*. Extension of these findings to glycoengineering and protein engineering of recombinant antibody therapeutics resulted in interactions with the biopharmaceutical industry. A current focus is the potential for immunogenicity of immune complexes, in contrast to aggregated forms of IgG. He is the author of more than 280 publications, with 207 of these referenced in PubMed. He is a member of the Royal College of Physicians (MRCP) and a fellow of the Royal College of Pathologists (FRCPath).

Maria de Jesus Trained as an environmental engineer in Portugal, De Jesus completed her Ph.D. in Plant Cell Culture Technology at the Swiss Federal Institute of Technology in Lausanne, Switzerland (EPFL). She joined the Laboratory of Cellular Biotechnology at the EPFL as a postdoctoral researcher and led the development of several projects as scientist and senior scientist, directed toward the establishment of manufacturing processes for recombinant proteins from CHO cells. In 2002 she initiated a self-funded, privately owned company, ExcellGene SA, acting in the role of founder and first employee. In subsequent years she established a laboratory and administrative unit of this company while simultaneously assuring the profitability and growth of the company as a CRO and CMO in the field of protein manufacture and process development for high-value therapeutic products from CHO and other cell lines. Currently she is chief operations officer, a member of the board of directors, and a significant shareholder of the company.

She has published over 40 peer-reviewed papers in the field and has lectured on process development and scale-up with animal cells in bioreactors. She is inventor of a small-scale bioreactor system, based on orbitally shaken tubes which are used in high-throughput approaches for suspension culture process development.

Dr. Weidong Jiang is a cofounder of Henlius Biopharmaceuticals, Inc. in United States. He is also senior vice president and chief scientific officer of Shanghai Henlius Biotech Co., Ltd., a joint venture formed by Henlius and Fosun Pharma in 2009 to develop antibody therapeutics. He has more than 20 years of working experience in many biotech companies, including Catalyst Biosciences, Inc.; Vasgene Therapeutics, Inc.; Applied Molecular Evolution, Inc. (subsidiary of Eli Lilly); Microcide; and ChemGenics (acquired by Millennium). He has expertise in the field of drug screening and development, particularly in research and development of antibody therapeutics and other protein biologics. He holds a Ph.D. degree from Giessen University in Germany; a master's degree from Shanghai Institute of Biochemistry and Cell Biology, Chinese Academy of Sciences; and a bachelor's degree from Zhejiang University.

Cheng Liu is the founder and CEO of Eureka Therapeutics, a California company dedicated to the development of innovative cancer immunotherapy. He worked at Chiron (now Novartis), where he championed the development of first-in-class therapeutic antibody against CSF-1, a key immune regulator, for cancer metastasis treatment. He has multiple issued US and international patents and scientific publications, including in journals of *Science* and *Nature*, in the area of therapeutic antibody discovery and engineering. In 2007, he was awarded Special US Congressional Recognition for his contributions to improving human health. He received his B.S. in Cell Biology and Genetics from Beijing University and a Ph.D. in Molecular Cell Biology from the University of California, Berkeley.

Dr. Margaret Liu is an assistant professor in the Department of Chemical and Biological Engineering at the University of Alabama. Her research areas include mammalian cell engineering, microorganism metabolic engineering and systems biology, and bioreactor process development. Specifically, her research projects focus on (i) improving the production and quality of biopharmaceuticals (e.g., biosimilars and other recombinant proteins) by host cell engineering and integrated process development via omics technologies, (ii) developing novel therapy to treat cancer and heart diseases, and (iii) producing next-generation bioenergy (e.g., biobutanol) by metabolic cell-process engineering (MCPE) using systems biology. She had worked in biopharmaceutical and biotechnology industries for six and half years before joining academia research. She was a senior scientist and team leader of Life Technologies Corporation (LTC) and a scientist of Lonza and Merck KGaA (EMD biopharmaceuticals). She has years of diverse industrial experiences in mammalian cell culture, biopharmaceutical production cell line development, omics technology, and therapeutic protein (including both innovator biologics and biosimilars) production process development in pharmaceutical and biotech companies. She accomplished her Ph.D. research under the instruction of Prof. S.T. Yang in the Department of Chemical and Biomolecular Engineering at the Ohio State University, United States. Her Ph.D. thesis work dealt with bioprocessing of high-value and biofuel products from metabolically engineered microbial mutants using a novel bioreactor production bioprocess. She earned her master's degree in bioengineering from Tianjin University and bachelor's degree in chemical engineering (major) and computer science (minor) from Shandong University, China.

Meimei Liu is a research scientist at Thermo Fisher Scientific, Inc. in Grand Island, NY. She obtained her B.S. and M.S. in Chemical Engineering from Tianjin University, China, and Ph.D. in Chemical and Biomolecular Engineering from the Ohio State University. While working at Tianjin University, she studied drug delivery and biopolymer development. While working at the Ohio State University, she investigated cell culture, cell therapy, and bioreactor process development. At Thermo Fisher Scientific, Inc., she works on cell culture and media development.

Dr. Scott Liu is a molecular biologist by training and holds a Ph.D. degree from Purdue University. As the founder, president, and CEO of Henlius Biopharmaceuticals, Inc. and Shanghai Henlius Biotech Co., Ltd., he takes overall responsibility for strategic planning and implementation, as well as operations management. Benefiting from Scott's vision in achieving quality at reasonable cost, Shanghai Henlius Biotech Co., Ltd. has adopted innovative technologies in mAb development and manufacturing and, in doing so, significantly reduced the manufacturing costs while maintaining the quality of the products. Before founding his own companies, Scott held key technical and/or leadership positions in several biopharmaceutical companies, including

Amgen, Bristol-Myers Squibb, United Biomedicals (UBI), Tanox, and Maxygen. He was the director of Quality Analytical Labs of Amgen Fremont and took responsibility for quality control operations for late-phase development and cGMP commercial manufacturing of Vectibix. Prior to Amgen, Scott held the post of associate director of Biologics QC in BMS Syracuse, where he was in charge of QC operations for late-phase development and commercial manufacturing of Orencia.

James D. Marks Dr. Marks is an internationally recognized pioneer in the field of antibody engineering and an elected member of the Institute of Medicine of the National Academies of Sciences. He has invented broadly applicable methodologies for the generation and optimization of human monoclonal antibodies. The Marks Lab at UCSF has been involved in the use of diversity libraries and display technologies to generate monoclonal antibodies (mAbs) since the earliest days of the technologies. As a graduate student in Greg Winter's Lab, he developed the first universal PCR primers which could amplify and isolate human V-gene repertoires and was the first to show that human mAbs could be directly isolated from nonimmune (naïve) phage antibody libraries. He also was the first to demonstrate the use of phage display for *in vitro* antibody affinity maturation and the direct selection of mAbs to cell surface proteins using cells. The Marks Lab has developed methods capable of generating very large nonimmune phage and phagemid single-chain Fv (scFv) antibody libraries capable of generating panels of antibodies to virtually any antigen. These libraries have been used within the Marks Lab and provided to collaborators where they have been extensively validated as a source of useful mAbs. Several antibodies generated in his lab have been tested in the clinic.

Dr. Marks serves as the chief of medical staff at San Francisco General Hospital and is professor of anesthesia and perioperative care at the University of California, San Francisco (UCSF).

Dr. Leonardo Mirandola graduated in biology and biotechnology in 2006 and received his Ph.D. in Molecular Medicine in 2009 at the University of Milano (Milan, Italy). He has published studies on intracellular signaling pathways, and his work as a Ph.D. student contributed to the molecular mechanisms by which oncogenic signals affects the response to chemokines in hematological malignancies, namely, acute leukemia and multiple myeloma. In 2010, he joined the Division of Hematology and Oncology at the Texas Tech University Health Sciences Center (TX, USA), where he focused on the discovery and validation of new immunologic targets for the development of cancer vaccines and diagnostic tests for multiple myeloma and ovarian, prostate, and breast cancer. Additionally, he leads a research and development project for the generation of new biotech drugs capable to block tumor-associated lectins, thus increasing cancer cells susceptibility to chemotherapy. His current research focuses are the design of a recombinant adeno-associated virus (rAAV) vector with modified tropism and controllable transgene expression driven by a tissue-specific promoter and the development of a new system to generate clinical-grade dendritic cells *in vitro*.

Jennifer R. Moore Meline is an associate in the Medtronic, Restorative Therapies Group Memphis office. She focuses her practice on intellectual property, with an emphasis on patent litigation, and has experience relating to a wide range of technologies, including medical devices, pharmaceuticals, biology, chemistry, and financial services. She also has experience performing prelitigation investigations, product clearance, and opinion work on behalf of clients.

Jennifer earned a B.S. in Biochemistry and English from the University of California at Los Angeles and her J.D. from Columbia Law School.

K. John Morrow, Jr. A molecular biologist, K. John Morrow, Jr. is president of Newport Biotechnology Consultants. He specializes in writing and consulting in the area of immunology, with a focus on antibody technology. He obtained his Ph.D. from the University of Washington and did postdoctoral studies in Italy at the Università degli Studi di Pavia and at the Fox Chase Cancer Center in Philadelphia. He has been employed at several universities as a faculty member including Texas Tech and Kansas University and has also worked in the private sector. For the last 10 years, he has been engaged in a number of consulting contracts as well as authoring many articles, reviews, books, and commentaries in the area of biotechnology.

Morrow has been employed since 1996 as a consultant at Meridian Bioscience, Inc. in Newtown, OH. He is presently chair of their Institutional Biosafety Committee and has performed contractual work with Meridian in the investigation and development of recombinant antibody technologies. He is also a member of the Institutional Biosafety Committee at Cincinnati Children's Hospital, Cincinnati, OH. He is a member of the Sinclair College Biotechnology Advisory Board in Dayton, OH.

He has written a total of more than 280 peer-reviewed papers, marketing reports, articles in the biotech trade press, commentaries, book chapters, and book reviews. In 2014 he published a book on epigenetics, *Cancer, Autism and Their Epigenetic Roots.*

Reena Nair is currently senior consultant in the Department of Clinical Hematology, Tata Medical Center, Kolkata. She earned her Doctor of Medicine (M.D.), postgraduate degree in internal medicine, from Goa Medical College under Bombay University in 1989. She completed her medical oncology training at Tata Memorial Hospital, Parel, Mumbai, from 1989 to 1994. Her former clinical appointments include Medical Oncology Department lecturer (1996) and professor (2008–2012) at Tata Memorial Hospital, Mumbai, India. She has been primary investigator in 35 investigator-initiated clinical trials and over 25 sponsored trials since 1994 onward. Her publications include over 95 original papers in peer-reviewed journals, both national and international. She has special research interest in lymphomas, leukemia, and breast cancer management and improving outcomes. She is also initiating collaborative clinical trials and data registries for lymphoma in India.

Dr. J. Drew Payne is currently an internal medicine resident at Texas Tech University Health Sciences Center. He completed undergraduate with honors from Texas Tech University receiving a degree in clinical laboratory sciences. He gained experience in clinical blood banking prior to receiving his doctorate of osteopathy from A.T. Still University Kirksville College of Osteopathic Medicine. He was chief resident for his third year of internal medicine training and has coauthored several peer-reviewed publications. His current interests include hematology–oncology training.

Dr. Camilo Pena is an international medical graduate currently training with the Internal Medicine Program at Texas Tech University Health Sciences Center. He graduated from the Universidad El Bosque—Escuela Colombiana de Medicina in Bogota, Colombia. He spent his last year of medical school as a foreign student in Miami, Florida, at the Jackson Memorial Hospital—University of Miami. He also became involved with the Lumen Foundation and Lumen Global, a cardiovascular research organization, where he participated in publishing 8 book chapters (*Textbook of STEMI Interventions Second Edition—2010*) and more than 10 peer-reviewed and published papers, abstracts, and posters. He then became certified by the Educational Commission for Foreign Medical Graduates (ECFMG) and obtained a residency spot with his first choice program at TTUHSC. He is currently working with the hematology and oncology division in different research projects.

Steven J. Projan Prior to joining to his industrial career, Dr. Steven J. Projan was an associate at the Public Health Research Institute, continuing his studies on plasmid replication, antibiotic resistance, and staphylococcal virulence though 1994. In 1987 he also became a senior scientist and then group leader at Applied Microbiology, Inc. (at that time an in-house biotech company at the Public Health Research Institute) working on antimicrobial enzymes and bacteriocins (small antibacterial proteins). There he developed a novel protein-based method for the prevention of bovine mastitis that was marketed to the dairy industry, and he also developed a novel protein expression system for the production of an antistaphylococcal protein that has entered clinical trials.

Dr. Projan attended MIT for his undergraduate education, receiving an S.B. degree (in the life sciences and nutrition and food science) in 1974. He then graduated with a Ph.D. from Columbia University in 1980 (also receiving M.A. and M.Phil. degrees from that institution), spending 1977–1980 at the University of Utah. His graduate work was done with Jim Wechsler and they were successful in developing the first *in vitro* system that initiated chromosomal DNA replication from the *Escherichia coli* origin of replication. He then became a postdoctoral fellow with Richard Novick at the Public Health Research Institute in New York City studying plasmid replication and virulence in *Staphylococcus aureus.*

Dr. Projan joined Wyeth in 1993 as a group leader in anti-infectives research. He became an associate director of Bacterial Genetics in January of 1997 and then director of Antibacterial Research in June of 1998. In May of 2003 he was appointed assistant vice president of Protein Technologies, and in September 2004 he was promoted to vice president and head of the newly created Department of Biological Technologies which was responsible for delivery of novel biologics to development (this group is now part of Pfizer). At Wyeth he was the biology team coleader of the Glycylcycline Discovery Team that produced tigecycline, which had been approved in the United States for the treatment of bacterial infections including those caused by multidrug-resistant strains. In 2008 he became VP and global head of Infectious Diseases at the Novartis Institute of Biomedical Research, and 2010 he assumed his current position of SVP of R&D and iMed Head for Infectious Diseases and Vaccines at MedImmune (the biologics arm of AstraZeneca).

Dr. Projan has authored over one hundred and ten papers and book chapters, several short stories, and one teleplay. He is a past chair of the Gordon Research Conference on Staphylococcal Diseases, makes an award-winning cheesecake, and served (and continues to serve) as a member of several NIH peer-reviewed study sections. He serves on five editorial boards, is a past member of the Program Committee for ECMID and ICAAC, and is a past chair of Division A (Antimicrobial Agents) of the American Society for Microbiology. Of his numerous academic appointments and honors, he has been a visiting professor at the UCLA School of Medicine and is an adjunct professor in pharmacology at Boston University. In 2004 he was elected a fellow of the American Academy of Microbiology. He has been a tireless advocate for the public and private study of microbiology and since the 1990s has sounded the alarm over declining and ineffective antibiotic research.

David Rabuka received a Ph.D. in Chemistry at the UC Berkeley as a Chevron fellow in the lab of Carolyn Bertozzi. His research included developing and applying the aldehyde protein tagging platform technology to cell surface modification. Prior to joining Bertozzi's lab, he worked at the Burnham Institute synthesizing complex glycans followed by Optimer Pharmaceuticals, which he joined as an early employee, focusing on the development of glycan- and macrolide-based antibiotics. He was CSO, president, and cofounder of Redwood Bioscience where he continued to develop novel protein conjugation methods and biotherapeutic applications such as antibody-drug conjugates. Redwood Bioscience was acquired by Catalent Pharma Solutions in October 2014, where he has continued to apply bioconjugation technologies with various collaborators and partners as a global head of R&D. He graduated with a double honors, B.S. in chemistry and biochemistry from the University of Saskatchewan, where he received the Dean's Science Award, and holds an M.S. in Chemistry from the University of Alberta. He is an author on over 35 major publications, as well as numerous book chapters and patents.

Ronald A. Rader, B.S. (Microbiology), M.L.S. (Library Science), has 35+ years' experience as a biotechnology, pharmaceutical, and chemical information specialist, author, consultant, and publisher. He is the author and publisher of BIOPHARMA: Biopharmaceutical Products in the US and European Markets, a reference source dealing with biopharmaceuticals. He is also the author/publisher of the Biosimilars/Biobetters Pipeline Database, a follow-on biologics-tracking information resource, and Biopharmaceutical Expression Systems and Genetic Engineering Technologies. Since 1988 he is the author and publisher of the *Antiviral Agents Bulletin*, a periodical specializing in antiviral/HIV drug and vaccine development. From 1994 to 2000, he authored and published the Federal Bio-Technology Transfer Directory.

Dr. Rakhshanda Layeequr Rahman is currently director of the TTUHSC Breast Center of Excellence at Amarillo, and the Interdisciplinary Breast Fellowship Program and is associate dean for faculty development at the Amarillo campus. She earned her undergraduate and medical degrees at the Aga Khan University in Pakistan and completed her fellowship in Breast Surgical Oncology at the University of Arkansas for Medical Sciences in 2004. She is board certified in surgery from the Royal College of Physicians and Surgeons of Glasgow and from the Royal College of Surgeons of Edinburgh, United Kingdom. She was also inducted as a fellow at the American College of Surgeons in 2005. She has received Texans Caring for Texans Award and Career Achievement Award by the Amarillo Women's Network. She is the Endowed Chair for Excellence in Women's Health. She specializes in the diagnosis and treatment of breast cancer and novel immunotherapeutic strategies for biologically diverse breast cancers. She is the author of numerous scientific abstracts, peer-reviewed publications, book chapters, and leadership literature.

Adair Reidy is a researcher at Kiromic, LLC, where she examines the expression of novel tumor-associated antigens for the development of immunotherapeutic treatments. She completed her B.S. in Biology at Texas Tech University and is currently pursuing an MBA with a focus in science, technology, engineering, and mathematics (STEM). While working at Kiromic and under Dr. Chiriva-Internati at the Texas Tech Health Sciences Center, she participated in various peer-reviewed projects, abstracts, posters, and patents.

Jeanene ("Gigi") Robison, MSN, RN, AOCN, works as the oncology clinical nurse specialist at the Christ Hospital, Cincinnati, OH. She earned her B.S.N. from the University of Cincinnati College of Nursing and Health in 1982 and her master's degree in oncology nursing at Case Western Reserve University in 1988. She has maintained her advanced oncology nursing certification since 1997. She has practiced in oncology nursing for 33 years and works with patients receiving chemotherapy and biotherapy drugs. She frequently presents on topics related to oncology nursing and has taught courses on chemotherapy and biotherapy since 1990.

Regis Sodoyer completed Ph.D. in Organic Chemistry at the University of Nice in 1980. Then he joined the Centre d'Immunologie INSERM/CNRS de Marseille-Luminy (CIML) in Marseille, where he directed his research to the polymorphism and structure–function relationships of HLA class I genes of the human major histocompatibility complex. He completed a second Ph.D. in Molecular Immunology in 1986. He then joined the vaccines division of Sanofi Pasteur in 1986. During this period, he was a group leader, head of the Molecular Microbiology Platform, head of the Experimental Design and Modeling Platform, and director for Technology Innovation. He recently moved to Bioaster as a biotechnology specialist, responsible for the company's training program. His fields of expertise include vaccinology, molecular biology, immunology, antibody engineering, and phage display and expression systems for the production of recombinant biotherapeutics.

Samuel D. Stimple received his B.S. degree in chemical and biomolecular engineering from the University of Notre Dame in 2012. At Notre Dame, he was an undergraduate research assistant in the laboratory of Dr. Basar Bilgicer, where his research focused on the development of a novel, small-molecule-based platform technology for the affinity chromatography purification of antibodies without the use of protein A. He is now a university fellow in the third year of his Ph.D. in Chemical and Biomolecular Engineering at the Ohio State University, where he works in the laboratory of Dr. David W. Wood. His graduate research focuses on downstream bioprocess development. Namely, the Wood lab focuses their research in the field of protein engineering, with the aim of developing protein-based biosensors and improving a platform technology for the purification of non-mAb proteins of interest using self-cleaving intein-purification tag technology. His research also includes the development of synthetic biology tools for applications in RNA and metabolic engineering.

Dr. Natallia Suvorava is currently at the last year of training in the Internal Medicine Residency Program at Texas Tech University Health Sciences Center in Lubbock, Texas, and incoming hematology oncology fellow at the University of Minnesota. She earned her undergraduate and medical degrees at the Gomel State Medical University in Belarus. She coauthored a number scientific abstracts and peer-reviewed publications.

Dr. Rashmi Verma is currently a second-year fellow in hematology and oncology at Texas Tech University, Lubbock, TX. She earned her medical degree with honors at the University College of Medical Sciences, New Delhi, India. Subsequently she did postgraduate studies at the University of Delhi in women's health and ob-gyn. She also worked as senior medical officer for 3 years at Delhi Government Hospital in 2001. She did her internal medicine residency internship at the University of South Dakota in 2009. She finished her internal medicine residency at UMDNJ, Cooper University, New Jersey, in 2012. She is board certified in internal medicine. She worked as a staff at Internal Medicine Hospitalist in 2012–2013. She started her fellowship 2013 and currently is a second-year fellow. Her interests include women's health, breast oncology, ovarian oncology, solid tumors, cancer genetics, and benign hematology. She has participated in various abstract presentations at the national level, case reviews, and peer-reviewed journals.

Adrienne R. Whitlow is a registered nurse previously employed as a critical care nurse in the Medical Intensive Care Unit at UMC Health System and is currently a pediatric nurse through Pediatric Home Health. She also serves as a second Lieutenant part-time in the Army Nurse Corps as a Medical-Surgical Nurse. She earned her B.S. degree in nursing through TTUHSC School of Nursing.

David W. Wood is an associate professor of chemical and biomolecular engineering at the Ohio State University. He received his undergraduate degree from Caltech in 1990 with a double major in chemical engineering and molecular biology. His work experience includes bioprocess development at Kelco, manufacturing at Amgen, and research at Bristol-Myers Squibb. He completed his Ph.D. in 2001 at Rensselaer Polytechnic Institute, where he was coadvised by Georges and Marlene Belfort. His primary field of research is protein engineering, especially the development of self-cleaving affinity tags for applications in recombinant protein purification. He has held faculty positions at Princeton University and the Ohio State University, where he currently resides.

He received the NSF Career Award and has two patents on intein-based technologies.

Dr. Tao Wu is a specialist in antibody research and discovery. Currently working at Boehringer Ingelheim Pharmaceuticals, Inc., he has over 10 years' experience in antibody discovery and pharmaceutical industrial development at Boehringer Ingelheim, Pfizer, and Merck, building and managing different teams and directing therapeutic antibody programs from early discovery stages into development. Currently, he is leading the antibody screening group and the cross-functional assay optimization team at Boehringer. Trained as a physician and molecular biologist, he earned his M.D. at the University of Wuhan in China and his Ph.D. in Microbiology with distinction at the University of Hong Kong. He thereafter completed postdoctoral study at the Centers for Disease Control in Atlanta, GA, and the Winship Cancer Institute at Emory University. He has published a number of peer-reviewed scientific papers in the *Journal of Immunology, Blood, Journal of Biological Chemistry, Molecular Biology of the Cell, Cancer Research,* and others.

Florian M. Wurm was trained as a biologist/molecular geneticist in Germany. He worked in industry (Behringwerke AG, Virology Department, Marburg and Genentech, Inc., Process Sciences Department, San Francisco) for 15 years during the earlier parts of his career. His work at Genentech contributed to the generation and manufacturing of several high-value products, such as Herceptin®, an antibreast cancer antibody; Pulmozyme®, a treatment for cystic fibrosis; and TNKase TPA®, a highly potent thrombolytic agent (cumulatively now sold for multibillion dollars/year globally). In 1995 he was appointed professor for biotechnology at Swiss Federal Institute of Technology, Lausanne (EPFL), where he established a research group for the field of process sciences with animal cells in bioreactors and where he gave classes in pharmaceutical biotechnology. As founder and CSO of ExcellGene (2001), he combined academic research and teaching with an entrepreneurial activity. Florian is member and past chairman of the European Society for Animal Cell Technology. He has published more than 200 papers and filed more than 30 patents, covering aspects of expression and manufacture of clinical proteins using mammalian cells in bioreactors. Some of his technology and process inventions have become globally used tools in research and in manufacturing of proteins with animal cells.

Zhinan Xia attended Nanjing College of Pharmacy in Nanjing, China, where he obtained his B.S. and M.S. degrees. He then earned his Ph.D. at the University of Kentucky College of Pharmacy in Lexington, KY. He did postdoctoral study at the Dana-Farber Cancer Institute at Harvard Medical School and subsequently worked in the private sector as principal scientist at Wyeth Discovery Research and at Pfizer Worldwide Research. He then became director of Protein engineering at Synageva Biopharma and is now head of Biotherapeutics at Moderna Therapeutics. He has worked in the fields of mAb design, Fc fusion, enzyme replacement therapy for lysosomal storage diseases, and other biotherapeutic drug development. He has published more than 35 peer-reviewed journal articles and has obtained multiple patents on therapeutic enzymes and Fc fusion proteins.

Ningning Xu is a Ph.D. candidate in the Department of Chemical and Biological Engineering at the University of Alabama. She is working on cell line development for biotherapeutic proteins and CHO cell engineering using omics technologies for antibody quality regulation. She received her bachelor's degree in the School of Chemical Engineering and Technology at Tianjin University, Tianjin, China.

Dr. Su Yan is head of Human Antibody Discovery at Eureka Therapeutics, a California company dedicated to the development of innovative cancer immunotherapy. She received her degree in medicine from Beijing University College of Medicine and a Ph.D. in Comparative Biochemistry from the University of California, Berkeley. She is coinventor of multiple issued US and international patents, including the technology on antibody ADCC enhancement. She has made major contributions in the discovery of human antibody against WT-1, an intracellular target for cancer immunotherapy, which resulted in half a dozen peer-reviewed scientific publications in the field of therapeutic antibody discovery and engineering.

Jianguo Yang holds a Ph.D. in Biotechnology from the Illinois Institute of Technology. He is an advisor and reviewer for *BioProcess International Journal* and executive director of the Sino-American Pharmaceutical Professionals Association, New England (SAPA-NE), and has several patents and numerous scientific papers.

Yang has over 20 years of experience in cell line and cell culture development and has worked at a number of global 500 pharmaceutical companies in the United States, including positions as principal scientist and senior manager at Genzyme/Sanofi, a group leader scientist at MedImmune/AstraZeneca, and as a biochemist at Abbott Laboratories. In addition he held an academic research position, working in the College of Pharmacy at the University of Illinois. Currently, she is vice president and chief scientific officer responsible for R&D in Qilu Pharmaceutical, China.

Dr. Junming Yie is a molecular biologist working in the field of metabolic diseases and drug discovery. Currently at Amgen, he has over 14-year biotech/pharmaceutical industrial experience at Amgen, Inc. and Pfizer, Inc., building groups and directing R&D projects including delivering one monoclonal antibody, three small-molecule candidates, and three recombinant protein candidates into clinical development. Trained as biochemist and molecular and cell biologist, he earned his Ph.D. with distinction in Biochemistry and Molecular Biophysics in Columbia University College of Physicians and Surgeons in New York and obtained his postdoctoral training with Dr. Bruce Spiegelman at Dana-Farber Cancer Institute and Harvard Medical School. He also holds an MBA degree from UCLA Anderson School of Management. He has published a number of high impact scientific papers in *Nature, Cell, Molecular Cell, PNAS, EMBO J*, Molecular and Cellular Biology and other journals.

Dr. Ziyang Zhong is vice president of Henlius Biotech Co., Ltd., a start-up biotechnology company located in Taiwan, with the mission to develop novel antibody therapeutics. He has more than 20 years of experiences in preclinical research and process development. Prior to joining Henlius, he worked for several start-up/medium-size biotech companies, including Scios (acquired by Johnson and Johnson), Chiron (acquired by Novartis), Abgenix, and Kosan Biosciences (acquired by BMS). He has expertise downstream purification, analytics, and bioanalytics. He earned a Ph.D. degree from Texas A&M University and a master's degree and a bachelor's degree from Sun Yat-sen University (Guangzhou, China).

Yu Zhou Department of Anesthesia and Perioperative Care, University of California San Francisco, San Francisco General Hospital, 1001 Potrero Ave., San Francisco, CA 94110, USA.

Dr. Yu (Eunice) Zhou received her B.A. and Ph.D. Molecular Biology and Biochemistry from Beijing University and is currently is an associate adjunct professor in the Department of Anesthesia and Perioperative Care, University of California, San Francisco. She has developed cancer cell internalizing antibodies and researched novel ways to identify and engineer human monoclonal antibodies to target breast cancer cells, especially cases with poor prognosis. She has also published a report of a novel method to screen phage display antibodies using both cancer cells and yeast-displayed antigens, making it possible to generate disease-associated monoclonal antibodies to virtually any antigen. Her research interests include engineered antibodies as targeted drug delivery system for cancer therapeutics and diagnosis.

Preface

With the expiration of patent protection for a range of antibody therapeutics, there is a rising expectation within the global biopharmaceutical community concerning the construction of knock-offs, so-called biosimilar antibodies, that has huge ramifications to the multibillion-dollar industry and the health of millions of patients.

Biologic agents are increasing economic demand on healthcare systems worldwide, and it is widely recognized that they represent an opportunity to increase access and reduce costs for patients and healthcare systems. A systematic approach to the characterization of the structural and functional similarities of a biosimilar molecule to the originator molecule provides specifications to guide the development of biosimilars. Identical sequence and similar host cell lines are key to development of biosimilars. A well-designed development program establishes biosimilarity based on structural, functional, preclinical, and clinical evaluation. Biosimilar endpoints are defined for each tier of characterization, and evaluation of successive tiers addresses additional relevance of biosimilarity.

Biosimilars are considered to be one of the fastest-growing sectors in the pharmaceutical industry, accelerated by the expiry of patents on multiple brands of biologics. Biosimilars in development include a large number of candidates from established and many new entrants to the biopharmaceutical industry. Nearly 20 biosimilars are already marketed in the EU and some other major market countries having implemented biosimilar approval pathways. A database tracking the biosimilars/biobetters pipeline reports nearly 700 biosimilars and 500 biobetters in development worldwide, with the great majority targeted to enter the US and other major markets where patents and other granted exclusivities are expiring. Many companies view biosimilars (and to a lesser extent, biobetters) as a more affordable, lower-risk way to enter the lucrative US and EU markets and gain credibility as a biopharmaceutical developer. With only comparative testing and trials required for biosimilar approval, this mechanism enables market entry at a cheaper and faster pace, when compared with the pursuit of innovative products receiving traditional full approvals.

Risks are also much lower, generally involving much the same active agent that has been on the market for more than two decades.

Monoclonal antibody biosimilars are expected to dominate the biosimilar field, even though it is a newcomer into the market. The value of the 10 top selling products, the majority of them antibodies and antibody-like molecules, had reached more than $50 billion in 2009, and the value of new antibodies and biosimilars is estimated to rise 7–15% per year. In the last three decades, protein yields from recombinant cell lines in bioreactors have increased 10- to 100-fold, as the result of improvements in media, bioprocess design, and cell culture process control. At the same time, the largest vessels in use for CHO cells have a working volume of about 20,000 l. This improvement of yield efficiency and production scale has lowered the production cost of making biosimilar antibodies. The strong "disposable" trend has emerged, which uses "single-use bioreactors" (SUB) and presterilized plastic bags, and dramatically reduces the capital investment in manufacturing "biosimilar" products. The huge market potential, reduced production cost, and lowering of capital investment in manufacturing facility have made the field of antibody biosimilars an attractive business.

However, there are multiple challenges facing the burgeoning monoclonal antibody biosimilars industry. As detailed by the authors in this book, the complexities and methods of manufacture create an important difference between biosimilars and conventional generic small-molecule drugs: while chemical generics can be fully characterized as identical to the originator product, biosimilars cannot. Full guaranty of similarity is only granted after equivalence parallel trials assessing PK, efficacy, safety, and immunogenicity comparing the biosimilar candidate and the originator. As the first European Medicines Agency (EMA) approval monoclonal biosimilar (infliximab) shows, biosimilars are licensed through a comparability exercise with the reference product and clinical studies to ensure equivalence of efficacy and safety profiles. Guidelines produced by the EMA detail manufacturing process requirements and the range of protein structure, isoform, aggregate, receptor binding, and biological activity assays necessary to demonstrate biological equivalence. It also outlines the required clinical and nonclinical pharmacokinetic (PK), pharmacodynamic (PD), and pharmacotoxicological evaluations necessary to assess safety and efficacy before approval. In the United States, the FDA released draft guidance for the regulatory review of biosimilars in early 2012, which is a significant step toward the first approved monoclonal antibody biosimilar in the US market.

This volume is comprised of a series of chapters dealing with this dynamic and diverse topic. It summarizes work in a number of aspects of the field to bring together the present-day state of the art. Here we review the major points laid out by the authors of this work.

Dr. Regis Sodoyer provides us with an introductory chapter describing the history of therapeutic monoclonal antibody technology. As he explains, although the power of the immune system to protect the body from disease has long been recognized, murine hybridoma fusion technology, developed in the early 1970s of the last century, has dominated drug development since its inception. Easy access to the production and engineering of murine monoclonal antibodies drove a revolution of drug discovery. The first two decades between 1975 and 1995 were marked by many challenges and the development of the field of molecular engineering. At this critical time, antibodies, still omnipresent in both diagnostic and research domains, have come to dominate the field of immunotherapy.

New technologies, such as phage display, humanized transgenic mice, and repertoire mining, have been perfected and standardized, allowing for the isolation of fully human antibodies. The natural complexity of the antibody molecules and the rapid implementation of engineering methodologies helped make them preferred candidates for the solution of complex immunotherapeutic challenges. Sodoyer updates the different antibody-derived molecules as well as a survey of the latest antibody engineering technologies. In addition, the chapter reviews the critical issue of the development of expression systems suitable for large-scale and cost-effective production of recombinant antibodies.

In Chapter 2, Dr. Zhinan Xia describes the structure, classification, and naming of therapeutic antibodies. He explains the evolution of antibody discovery, moving after the first wave of innovation based on mouse mAbs to the recombinant DNA technology providing the tools for chimeric, humanized, and fully human mAbs as well as recombinant antibody fragments and bispecific therapeutic mAbs. Over the decades, therapeutic mAbs have evolved as innovative pharmaceutical compounds, not only for the treatment of cancer but also of autoimmune and infectious diseases. A new generation of antibodies including anti-PCSK9 mAbs, now in clinical development by Amgen, Pfizer, and Regeneron (in partnership with Sanofi), respectively, may represent a class of blockbuster drugs aimed at cardiovascular disease. We are now entering a third wave of scientific advancement based on antibody architecture modification and engineering, taking advantage of new insights into Fc effector function, glycoengineering, and antibody-drug conjugation.

Among the new wave of antibody technology platforms, antibody-drug conjugates are the most prominent, with their main focus on oncology indications. Because they can offer targeted delivery of chemotherapeutic or radioactive agents, they will increase the performance of mAbs and offer a toolkit for product differentiation and life cycle management.

In Chapter 3, Dr. Yu Zhou and Dr. James Marks cover the mechanisms of action of therapeutic antibodies. Elaborating on the development of multiple antibody-based drugs directed at a single target, in the case of the autoimmune disease targeting TNFa, the differences in Ab–Ag binding properties, the ability to induce effector functions, and other factors resulted in three full-length IgG mAbs, one TNFR–Fc fusion protein, and one pegylated Fab. Meanwhile, a better understanding of antibody mechanism of action has led to the expansion of antibody drugs with increased potency and greater precision.

Therapeutic monoclonal antibodies and their targets are the topic of the fourth chapter by Dr. Jose Figueroa and his colleagues. These investigators state that neoplastic diseases are currently treated by a variety of mAbs that use diversified mechanisms against cancer-specific targets, including natural cytotoxic mechanisms as well as target-neutralizing functions, in order to produce clinical activity against various cancers. Targets presently used in mAb treatments include CD20 for patients with chronic lymphocytic leukemia and non-Hodgkin's lymphoma; CD30 for HL and ALCL; CTLA-4 and PD-1 for melanoma; HER-2 for HER-expressing breast cancer and gastroesophageal cancer; EGFR-1 for CC and squamous cell carcinoma; VEGF for CC, glioblastoma, non-small cell lung cancer, ovarian cancer, and metastatic renal cell carcinoma; and VEGFR-2 for GE cancer, gastric cancer, and NSCLC. While no specific cancer-related antigens have yet been identified, novel tumor targets, for instance, *c*ancer/*t*estis *a*ntigens (CTAs) such as SP-17, are highly restricted to cancer cells with normal expression limited to testis and placenta. The authors predict a bright future for antibody development and application. They assert that the use of new protein expression platforms and glycoengineering techniques, the design of multispecific binding domains, heterologous fusion constructs, molecular safety switches, and effective linkers will result in an explosion of more effective products with fewer and less serious side effects.

Dr. Roy Jefferis authored the fifth chapter on the topic of antibody posttranslational modifications. He explains in detail that the clinical efficacy of recombinant antibody therapeutics has resulted in their worldwide adoption as drugs of choice, for example, in oncology and inflammation. Antibodies are often referred to as "adapter molecules" as they form a bridge between a specific target and the activation of downstream molecular and cellular effector functions. Immune complexes of each of the four human IgG subclasses may bind and activate multiple host ligands with varying downstream outcomes. Recognition and activation of host ligands are dependent on interactions with the IgG-Fc region and can vary depending on the precise structure of the naturally occurring isoforms, for example, subclasses and glycoforms. Antibody therapeutics must therefore be fully characterized, structurally and functionally, as must a candidate biosimilar. Jefferis enumerates and discusses the multiple posttranslational and chemical heterogeneities that may arise within recombinant antibodies and possible consequences for mechanisms of action *in vivo*.

In Chapter 6, Drs. Ningning Xu, Meimei Liu, and Margaret Liu present the pharmacology, pharmacokinetic, and pharmacodynamic aspects of recombinant antibodies. The authors point out that pharmacological studies of anticancer mAbs focus on the interaction between a patient population and the biopharmaceutical agent. Specifically, these investigations describe the interaction between specific mAbs and deranged cancer cells as opposed to those dynamics involving normal cells. In sum, the branch of pharmacological antibody studies deals with the uses, effects, and modes of action of mAbs on the body.

Pharmacokinetics (PK) on the other hand is concerned with the biological processes that a drug undergoes in the intact organism, that is, what the body does to a drug.

The third branch of pharmacological studies is referred to as pharmacodynamics. It focuses on the therapeutic action of a drug on an organism, that is, pharmacologic response, as well as the duration and the magnitude of the response associated with the concentration of drug at an effectual site of the organism. The authors lay down a set of general principles describing *in vivo* antibody behavior to serve as a guide for initial drug development leading to future clinical applications.

In Chapter 7, Jeanne Robson outlines the clinical performance of currently approved anticancer antibodies, considering their developmental history, target biology, mechanisms of action, clinical efficacy, and FDA-approved indications.

Dr. Reena Nair discusses the development of biosimilar rituximab and her institution's clinical experience with their biosimilar in Chapter 8. Rituximab was the first monoclonal antibody approved for cancer treatment. Since its launch in 1997, it has been at the forefront of treatment for B-cell non-Hodgkin's lymphoma. Reditux^{TM} (Dr. Reddy's Laboratories Ltd.) is the first in a series of biosimilar products that dramatically expand the affordability of antibody therapy for patients in India and beyond. Regulatory requirements for biosimilars are evolving. Education of physicians and healthcare providers, patients, and payers about biosimilars will assist in informed decision making and promote acceptance of biosimilars into clinical practice. Scientific bodies should formulate practice guidelines and position statements to establish biosimilarity in efficacy, safety, comparability, and interchangeability with the reference originator biologic molecule.

In Chapter 9, Dr. S.J. Projans discusses monoclonal antibodies for infectious and cardiovascular diseases. The advent of the "age of antibiotics" in the 1940s had a major impact on extending lives, but it also brought with it a false sense of security. The overenthusiastic use of antibiotics, sometimes as a substitute for good sanitation and hygiene, has resulted in unintended consequences. We now know that drug resistance developed by these life-threatening bacteria has become a big problem for hospitals treating infectious disease. Finding new ways to prevent and treat infections caused by resistant organisms will be a major challenge in the twenty-first century. Identifying "at-risk" patients (e.g., those colonized with *Staphylococcus aureus*) and taking preemptive measures (e.g., decolonization) or perhaps developing a protective vaccine may well obviate the use of the antibiotics that have made the practice of medicine of the late twentieth and early twenty-first centuries possible. A new term of art has been coined "precision medicine," and we now have the ability to identify an infecting pathogen within 2 h. This will allow for the use of narrow-spectrum agents that, in and of themselves, cannot obviate either toxicity or the development of resistance but may well prevent cross-resistance to current antibiotics as well as ameliorate, if not prevent, horizontal gene transfer. In addition to the continued hunt for effective vaccines, currently there are numerous monoclonal antibodies in clinical development to prevent and/or treat infections caused by *Pseudomonas aeruginosa, Staphylococcus aureus, Clostridium difficile,* and others. To date, these monoclonal antibodies, mainly of human origin or design, have had an enviable safety record, and these may represent the future of antibacterial treatments.

In Chapter 10, Drs. Yie and Wu discuss monoclonal antibodies directed against markers for musculoskeletal, central nervous system, and other diseases. There are seven approved monoclonal antibodies available for the treatment of various nonmalignant conditions. They are natalizumab for autoimmune disease such as multiple sclerosis and Crohn's disease, eculizumab for rare diseases associated with the dysfunction of the complement system, ranibizumab for macular diseases, denosumab for bone diseases, and daclizumab, basiliximab, and muromonab-CD3 as immunosuppressive agents for solid organ transplantations. The approval of these seven monoclonal antibodies spanned more than two decades, reflecting the evolution of the investigative process, from murine to human/murine chimeric antibody to humanized antibody and finally a fully human antibody. In this chapter, drug development history, target biology, mechanism of action, clinical efficacy, and approved indications of each monoclonal antibody are covered thoroughly. Appropriate tables, factsheets, and sequences of the relevant biologicals round out the discussion.

In Chapter 11, the manufacture of recombinant therapeutic proteins using Chinese hamster ovary (CHO) cells in large-scale bioreactors is presented in detail by Drs. Wűrm and De Jesus. Immortalized mammalian cells in suspension culture have been used in large-scale bioreactors for the production of recombinant protein therapeutics since the mid-1980s. All recombinant antibody products with greater than $1 billion US sales/year on the market today are produced in CHO cells. Clearly they have achieved "superstar" status since more than 50% of all protein pharmaceuticals on the market are produced now in these cells. Eight of the ten top selling biologicals in the market today are derived from CHO cells. When considering products made in animal cells only, CHO cells are probably used in more than 90% of cases. Other immortalized cell lines, such as NS0, BHK, or HEK-293 cells, are used for a few individual products, but they play an insignificant role in the overall sales picture. All processes here considered involve "stable" cell lines, that is, cells that were engineered by inserting the desired gene(s) of interest into the genome (chromosomes) of the host cells.

The contribution of expression vector constructs and host cell engineering is, contrary to what numerous publications and even recent reviews are claiming, relatively minor. The reason for this surprising and counterintuitive statement is clear. While the academic world has been very active in solving problems in CHO technology by addressing them through cell engineering, the industry was bound by the conservative approaches necessary in pharmaceutical manufacturing and by the benefit of "sticking to approaches that worked." Significant investments (cell banking, testing, established operational handlings) into a relatively solid approach with a proven track record from a nonmodified cell host could not easily be abandoned just because an improving cell cycle or antiapoptotic gene might eventually provide some (minor) benefit. In fact, the backlog of knowledge concerning a given cell host provided ample opportunity to reap the benefits from modifications of media and process, with resultant dramatic improvements.

In general, the same vector components are used today as 20 years ago. Codon optimization of gene sequences coding for the desired protein of interest have improved overall expression of some proteins, but did not result in "breakthrough" yield increases. The identification and selection of the suitable clonal populations from transfected cells are facilitated by the widely accessible and cost-efficient equipment that allows high-throughput screening, including the use of flow cytometry and sorting. Some DNA elements in plasmids and novel vector/gene-transfer approaches have pushed primary expression higher. Currently marketed protein therapeutics are exclusively produced in large-volume steam-sterilizable, stainless steel bioreactors, and the majority of production processes therein can be characterized as "fed-batch" cultures, with processing times of 10–20 days. However, over the last 10 years, a strong "disposable" trend has emerged, particularly not only with innovative products that are entering the clinic but also with "biosimilar" products under development. "Single-use bioreactors" (SUB) and presterilized plastic bags, mounted into or onto containers or platforms, serve as a physical barrier between the nonsterile environment and cell culture process liquids. While stirring with marine and pitched blade impellers is still the main method for mixing of both sterilizable stainless steel and single-use bioreactors, other impeller-free approaches have emerged as well that are now applied in single-use bioreactors.

In Chapter 12, Stimple and Wood consider process development at the downstream end. They summarize several advances in antibody purification at industrial scale, with an emphasis on alternatives to conventional protein A affinity batch processing. At the molecular level, these alternatives include the use of newly developed affinity ligands, as well as reemergence of conventional chromatographic processes. The chapter also covers several novel process configurations, including the use of disposable technologies and the rapidly increasing focus on continuous processing. The general trend of these advances is toward highly flexible manufacturing environment, which will provide rapid process development as well as production on demand of antibody therapeutics in multiproduct facilities. The authors cover each of these developments, with the relative strengths, weaknesses, and potential impacts evaluated.

The impact of biosimilars and biobetters on biopharmaceutical manufacturing and contract manufacturing organizations is covered in Chapter 13 by Ronald A. Rader. He discusses the fact that biosimilars are subjected to an abbreviated but rigorous approval process involving direct comparisons of the biosimilar to its reference product, including analytical profiles and head-to-head comparative pharmacokinetics and clinical trials. Biosimilars will raise the bar for product quality expectations for all biopharmaceuticals, with regulators, healthcare professionals, patients, and the public focusing on the details of manufacturing processes and the analytical, bioprocessing, and related quality differences between products.

Dr. Rader believes that by the end of the decade, we can anticipate 10–12 entries for every successful reference product. While this may seem a chaotically competitive market, there already are many or more sources for some major selling generic drugs. The United States is projected to be the dominant biosimilars market. Many new and established companies will be moving aggressively into this market, even if this results in overcrowding and a market so fractured that it becomes impossible to realize desired profit margins. The biosimilars, biobetters, and biogenerics market will be much like the generic drug market. This includes active pharmaceutical ingredient sources and competing products originating from a range of international companies. As international competition and the available number of biosimilars and other follow-ons increase, buyers will favor the US and EU manufactured products. As a result, unlike the generic sector, development and manufacturing will be located in the industrialized countries.

Dr. Jianguo Yang's Chapter 14 covers cell line production and cell culture development for biosimilar antibody-drug manufacturing. Almost all major pharma companies are heavily committed to biosimilar product development, resulting in many active players within the industry. Likely, pharma companies with well-integrated drug R&D divisions and robust clinical trial and sales networks will come to dominate the field. Due to a wealth of current biosimilar antibody-drug candidates in various development stages, biopharmaceutical companies are increasingly interested in a platform process for mammalian cell line and cell culture development to meet productivity and quality attributes with maximum efficiency.

Cell line and cell culture processing are the most critical steps for biosimilar antibody development since both can determine productivity, quality, and cost/benefits, which are of vital importance to the bottom line. For the last several years, average productivity of mammalian cell lines has reached 3–6 g/l. These gains were achieved using fed-batch process, high-productivity cell lines, and optimized media and bioreactor platforms. These innovations are dramatically cutting costs and speeding drug development. Meanwhile, product quality is the key factor for success as many companies compete against one another in a cutthroat landscape.

Drs. Weidong Jiang, Scott Liu, and Ziyang Zhong review in Chapter 15 the analytical and bioassay methods used to characterize the finished antibody products based on predetermined specifications for biosimilar antibodies. For their identification, it is typical to have the primary sequence confirmed to the reference standard using the LC/M method. There are many options available on the market for this purpose, including quadrupole time of flight (QTOF) and Orbitrap. Both Waters and Agilent make excellent TOF instruments, while Thermo/Finnigan is the undisrupted leader in the Orbitrap technology sphere. All of these instruments offer techniques to provide both primary sequences, as well as information concerning disulfite bonds. Their purity and integrity, on the other hand, are demonstrated by conventional chromatography, such as SDS-PAGE and IEF. Similarly, HILIC is the standard method to characterize the glycan profile, which plays a key role in ADCC.

In Chapter 16, Rafiq Islam reviews the development of bioanalytical tests of biosimilar antibodies, including glycosylation pattern, ADCC/CDC activity, and FcγR binding profile. Critical features of such characterizations are the test results required to demonstrate similarity to the original product, as stated in the regulatory guidelines issued by the FDA and EMEA.

Drs. João Fonseca and João Gonçalves cover preclinical and clinical development of biosimilar antibodies in Chapter 17. They touch on a recurrent theme in this book: that biosimilars offer a highly attractive strategy for reducing medical costs and increasing accessibility to targeted biologic therapies. However, unlike small molecules, the development and production of candidate original biologics and biosimilars can be hampered by unpredictable variability that should be tackled as far as possible during the preclinical development process. Biological therapies are inherently variable, creating unavoidable differences between even subsequent batches of the same product.

Clarinda Islam covers regulatory issues in Chapter 18 providing an overview of the regulatory process with a focus on biosimilar antibody-based therapeutics. Of major concern are the challenges faced by regulatory authorities globally and at the national level as they seek to evaluate biosimilar products for safety and potency. Some of the issues include the question of how similar is similar enough and what are the implications of the observed differences for antibody performance and patient safety. This chapter will also address limitations of the regulatory process and the patients' responsibility to report safety concerns and symptoms.

Legal considerations concerning biosimilar antibodies are subjected to detailed analysis in Chapter 19 by K. Lance Anderson, Jennifer R. Moore, and Jonathan Ball. Follow-on biologics, also known as biosimilars, face unique statutory and regulatory frameworks, the understanding of which is essential for successful commercialization. Modeled after existing schemes for generic drugs, such as the Hatch–Waxman Act, biosimilars are afforded the opportunity for special treatment due to their similarity to an existing innovator product. However, due to the variable nature of a biosimilar product, whether from its origin, impurities, or method of manufacture, this process has become codified in its own legislation designed to address the additional inherent product variation. The Biologics Price Competition and Innovation Act of 2009 (BPCIA) was signed into law by President Barack Obama on March 23, 2010, as a subchapter of the Patient Protection and Affordable Care Act. The objective of the legislation was to amend the Public Health Service Act (PHS Act) to provide the FDA a framework for approving follow-on biologics. Other regulatory authorities such as the European Medicines Agency (EMA) have developed similar guidelines, although such statutory frameworks do vary.

In essence, the BPCIA provides an abbreviated approval, or licensure, pathway for biological products shown to be biosimilar to, or interchangeable with, an FDA-approved regulatory product. The FDA-licensed biological product, or reference product, may be relied upon so long as similarities can be shown between the biosimilar product and the reference product. In doing so, an abbreviated regulatory pathway is available for the biosimilar product.

In focusing on the BPCIA as it amends the PHS Act, and related FDA guidances issued periodically with regard to the applicable legislation, the basic framework of the licensure process will be appreciated in light of the reference products to which they pertain. Additionally, certain product exclusivity periods may be obtained, or must be permitted, within the biologics licensure process. As discussed further, these aspects will continue to remain of importance in coming years due to known expirations of biological products approved under Section 351(a) of the PHS Act. The exclusivities afforded may vary due to the nature of the biologic products and the sponsors involved. Further, the consideration of patent analysis and procedures critical to the BPCIA process will be discussed, along with recent guidance documents issued by the FDA, submissions of biosimilar applications, and early court cases regarding such submissions.

In Chapter 20, Dr. Cheng Liu, editor of this volume, considers antibody-dependent cell cytotoxicity enhancement technologies for next-generation therapeutic antibodies. ADCC enhancement is a key strategy for improving therapeutic antibody-drug efficacy. Recent clinical studies provided further evidence in support of the technology. It has the potential of lowering effective drug dosage, hereby benefiting patients through lower drug cost. Antibodies with ADCC enhancement are expected to eliminate variations in patient response to antibody treatments caused by genetic polymorphism and improve survival of cancer patients. The commercial value of the technology has been demonstrated by the licensing agreement between Amgen and Kyowa Hakko Kirin on an anti-CCR4 antibody, which includes $100 million up-front fee plus $420 million milestone payment and double-digit royalties. The FDA approval of obinutuzumab and its superior efficacy compared with the first-generation anti-CD20 antibody, rituximab, in leukemia was a milestone success of ADCC enhancement technology in new mAb drug discovery and development. Meanwhile, ADCC enhancement is also becoming a core technology for developing next-generation therapeutic antibody drugs with favorable clinical outcomes.

Dr. K. John Morrow, Jr., coeditor of this book, is the author of Chapter 21 which considers technologies for engineering improved antibody half-life. Biosimilar antibodies include a subcategory of "biobetters," that is, antibodies that actually represent an improvement over the original product. The loss of patent protection and technological improvements are predicted to generate a wave of biosimilar products in the coming years that will force deep inroads into the market share of the original products. The competition brought about through the introduction of biosimilars may cut costs of antibody-based therapeutics by 50% or more.

Because the IgG framework is able to sustain great variability, it lends itself to a range of engineering possibilities. Advances in cloning and engineering technologies accomplished in recent years have opened the door to improving antibody performance, in particular increasing the half-life of therapeutic mAbs. Fundamental investigations of the mechanism of half-life extension have shown that the complexing of the Fc region of the antibody with the FcRn protein is critical to the prolongation of serum half-life. This receptor protects the IgG molecule from cellular catabolism by way of a pH-dependent recycling and transcytosis pathway.

Antibody half-life can be extended by increasing the affinity of the antibody for the FcRn molecule. Extensive molecular mapping studies have revealed the mechanism of binding of the two molecules and the best mutational substitutions to the Fc region to optimize affinity. Alternatively, in some cases it may be desirable to interfere with the binding in order to shorten the half-life of the antibody. This is the strategy that is being applied in therapeutic application to various autoimmune diseases.

The FcRn molecule complexes with immunoglobulin and also with a totally unrelated molecule, albumin. This property allows for the development of stable albumin fusion molecules with notable therapeutic applications. As new biosimilars are developed, some preclinical and clinical evaluation will be required in order to establish their safety and efficacy. Since preclinical studies involve the use of mice, the recent controversy concerning the accuracy of the mouse as a model for human disease conditions should be mentioned. While there is reason for concern, the weight of the evidence indicates that the mouse is an accurate model of the human immune system, with certain caveats.

One of the most promising approaches to protecting the half-life of engineered antibodies is the use of nanotechnological devices. In its most fundamental state, the design concept is a drug-containing core in a soft sheath covered with polyethylene glycol, which attracts water molecules surrounding the nanoparticle with a liquid halo. Currently there are a number of nanotech applications that have received FDA approval, although none involving an antibody therapeutic.

Despite the problems associated with the development of biosimilars and biobetters, the future of the industry appears bright. There is an extremely robust market for biologics, now 27% of pharmaceutical sales in Europe. In the United States alone, there are 907 biologics in development, targeting more than 100 diseases. Extension of the half-life of these molecules could constitute a direct and rapid approach to improving their performance, so it is certain that this will be a major part of an R&D biobetters program.

In the final Chapter 22, Patrick G. Holder and David Rabuka discuss technologies for antibody-drug conjugation. As described in the preceding chapters, therapeutic mAbs leverage their specificity to induce cellular cytotoxicity through a variety of mechanisms. Small-molecule chemical pharmacophores can also induce cytotoxicity. An antibody-drug conjugate (ADC) is prepared by chemically joining therapeutic mAbs with these potent chemical cytotoxins. *In vivo*, a circulating ADC binds its target antigen on the cell surface and is internalized, carrying with it the chemical pharmacaphore. This “payload” can then be delivered through a variety of release mechanisms and potentiate cell death. The combination of a mAb that targets a carefully chosen antigen (Chapter 4) with a highly potent toxin can produce ADCs that significantly increase the therapeutic index of targeted treatments.

A “technology” for antibody-drug conjugation describes any of the modular components that are chemically assembled in order to produce a functional ADC. These include the location on the antibody to which drug is attached, the chemical reaction used for conjugation, the components of the linker joining mAb and cytotoxin, and the chemical functional groups that can trigger release of the cytotoxin. The exact choice of cytotoxin will depend upon the target cell and its sensitivity to a specific mode of action.

Researchers have been designing and building ADCs for over 50 years. As the construction of ADCs has changed, so have the underlying technologies. Pertinent reviews of ADC technologies have been prepared along the way; this chapter assembles current knowledge and critiques best practices.

Cheng Liu, Ph.D., and K. John Morrow, Jr. Ph.D.

1

The History of Therapeutic Monoclonal Antibodies

Regis Sodoyer

Expression Systems and Protein Engineering, Technology Research Institute Bioaster, Lyon, France

1.1 Summary

Antibodies, a main component of the immune response, have been recognized, more than a century ago, for their proven therapeutic value. The hybridoma fusion technology, proposed in the early 1970s, for the first time gave easy access to the production and engineering of murine monoclonal antibodies. The potential of these new molecules, as laboratory tools, was largely exploited during the two following decades. At present antibodies, still omnipresent in both diagnostic and research domains, have progressively come to dominate the field of immunotherapy. New technologies, such as phage display, humanized transgenic mice, and repertoire mining, have been proposed, allowing for the isolation of fully human antibodies. The natural complexity of the antibody molecules and the rapid implementation of engineering methodologies helped in making them ideal candidates for new applications and for the solution of complex immunotherapeutic challenges. The first chapter is a current update on the different antibody-derived molecules as well as a survey of the latest antibody engineering technologies. In addition the chapter reviews the critical issue of the development of expression systems suitable for large-scale and cost-effective production of recombinant antibodies.

1.2 Introduction

The historical roots of immunotherapy trace their origins to the end of the nineteenth century. In collaboration with Shibasaburō Kitasato, the first bacteriologist to succeed in cultivation of *Clostridium tetani*, Emil Adolf von Behring demonstrated the efficacy of "so-called" antitoxins to protect animals against

Biosimilars of Monoclonal Antibodies: A Practical Guide to Manufacturing, Preclinical, and Clinical Development, First Edition. Edited by Cheng Liu and K. John Morrow, Jr.

tetanus and diphtheria [1, 2]. Although not immediately recognized, the discovery was adapted to the treatment of diphtheria-stricken children and was found to significantly reduce mortality [3]. For some years serum therapy was widely adopted before being gradually replaced by active immunization. Several decades passed before the composition, physiological behavior, and chemical nature of the antitoxin components were thoroughly characterized.

These components, referred to as gamma globulins according to their electrophoretic mobility and as immunoglobulins and antibodies according to their immunological function, were at the center of multiple investigations involving a steadily increasing number of renowned scientific teams. From the early 1950s until now, the definition of antibody (Ab) became more precise, as well as the genetic aspects behind the creation of their molecular diversity. In parallel, advances in cellular biology were under way around the world, allowing a better understanding of the precise role of B cells in the immune response. Paving the way toward the advent of modern immunotherapy, hybridoma technology described by Köhler and Milstein in 1975 gave, for the first time, easy access to murine monoclonal antibodies (mAbs) [4]. mAbs were recognized as revolutionary laboratory tools from their inception, although the *in vivo* applications and therapeutic potential of these molecules were still controversial, raising some skepticism in the early 1980s. Nevertheless, significant progress in molecular biology techniques allowed mAbs to move from research to diagnostics and applications in therapy.

More recently, monoclonal and recombinant antibodies have become the focus of new technologies, such as bacteriophage or other *in vitro* display techniques, mice or larger animal transgenesis, and other technologies permitting direct access to fully human antibodies. Importantly, the pressing need for large quantities of mAbs was a major driver for the development and optimization of recombinant protein production systems. The contribution and complementarity of these different approaches will be considered in the context of large-scale industrial production of therapeutic mAbs.

1.3 New Markets for Old Antibodies, Old Markets for New Antibodies

To date, more than 40 mAbs have been approved by the US Food and Drug Administration (USFDA) or European Medicines Agency (EMA) for therapeutic applications: 5 are of murine origin, including 2 bispecific constructs, 8 chimeric, 18 humanized, and 13 fully human. Thanks to the novel technologies available [5], fully human antibodies are rapidly taking over the market, and it seems probable that even humanized molecules will be marginalized in the coming years. The composition of the early-stage antibody pipeline is indicative of this trend. In terms of revenue, substantial returns have been realized both on existing and new markets. A good illustration is provided by recent data from Lawrence and Lahteenmaki: [6] among the top 10 selling biological drugs of 2014, 5 are mAbs. Humira® (adalimumab, AbbVie Inc., North Chicago, IL, the United States), with applications to several conditions (such as rheumatoid arthritis, juvenile rheumatoid arthritis, Crohn's disease, psoriatic arthritis, psoriasis, ankylosing spondylitis, and ulcerative colitis), is number one, with a total revenue exceeding $10.5 billion. Remicade® (infliximab, Janssen Biotech, Inc., Titusville, NJ, the United States) for multiple indications (rheumatoid arthritis, psoriatic arthritis, ulcerative colitis, Crohn's disease, ankylosing spondylitis, and severe or disabling plaque psoriasis) is in second place with a total revenue over $9.2 billion. Rituxan® (rituximab, Biogen Idec Inc., Cambridge, MA, the United States, and Genentech USA, Inc., San Francisco, CA, the United States) for multiple indications (rheumatoid arthritis, chronic lymphocytic leukemia/small-cell lymphocytic lymphoma, non-Hodgkin's lymphoma, antineutrophil cytoplasmic antibody-associated vasculitis, indolent non-Hodgkin's lymphoma, and diffuse large B-cell lymphoma) is found in the sixth position with a total revenue over $7.5 billion. Avastin® (bevacizumab, Genentech USA, Inc., San Francisco, CA, the United States) and Herceptin® (trastuzumab, Genentech USA, Inc., San Francisco, CA, the United States) are, respectively, ranking seventh and ninth positions with individual revenues fluctuating between $6.8 and $7.0 billion.

Beside the success stories of some blockbusters, mAbs are set to play a role in the rapid control of emergent diseases. A striking example is the 2014 Ebola outbreak, for which more than 20 laboratories and research groups around the world, including those from Canada, Japan, Israel, Uganda, and the United States, are working simultaneously to develop therapeutic mAbs against the virus. The dire state of emergency will no doubt facilitate and shorten the approval process. A mixture of three mAbs known as ZMapp (LeafBio, Inc., San Diego, CA, the United States), never tested in humans, was exceptionally accepted despite the fact that very little is known regarding the safety and effectiveness of this treatment.

1.3.1 Intellectual Property

One possible explanation for the great commercial success of therapeutic antibodies is the proper management of intellectual property and accurate designation of strategies for protecting antibodies or antibody-derived products [7]. This is particularly meaningful if we consider the emblematic and well-known Cabilly patent filed by Genentech, one of the most ubiquitous patents in biotechnology, which covers a fundamental method for the production of therapeutic recombinant mAbs that cannot be ignored by anyone planning to commercialize an antibody [8]. Besides the uniqueness of this example, one should not underestimate the inextricably complex patent situation governing the recombinant antibody world. As a matter of illustration of this complexity, according to Sandercock and Storz [9], it might be possible to patent the precise definition of an epitope with the possibility to claim later-generation antibodies targeting the same epitope.

1.3.2 Biosimilars

A consequence of the astronomical commercial value of mAbs associated with the patent expiry date (or estimated expiry date) is their attractiveness as candidates for biosimilar drugs [10]. The European Union (EU) has been the first to establish a regulatory framework for marketing authorization application (MAA) and has named "copycat" biotherapeutic products with the term biosimilars, a term also recently adopted by the USFDA [11].

Biosimilars offer a highly attractive mechanism for reducing the cost of medical care and thus should be considered positively [12]. Nevertheless, the abbreviated approval pathway associated with their adoption requires reinforced pharmacovigilance after a biosimilar mAb is approved, in order to ensure its long-term safety and efficacy. Unfortunately, it has been observed that the studies carried out to obtain approval of the reference product are not always adequate to ensure comparability of biosimilar mAbs. A similar efficacy does not necessarily imply a similar safety profile between the innovator and biosimilar products [13]. A major challenge to be addressed is the prediction and assessment of immunogenicity of subsequent entry biologicals (SEBs) and, in particular, biosimilar products [14]. It is nevertheless important to keep in mind that the global cost of manufacturing and licensing a biosimilar product remains high, and the reduction in cost may be more limited than for a nonbiological small-molecule drug and its generic version. Vital et al. provide a good illustration for biosimilar versions of the antibody rituximab [15]. In addition to the industrial challenges of development, testing, and marketing, biosimilars and "biobetters" [16] are raising new analytic challenges concerning product characterization and immunogenicity profiling [17a, 17b].

1.3.3 Modified Antibodies, Nontherapeutic Applications

Recent technological innovations will facilitate access to alternative antibody formats including bispecific antibodies (BsAb), conjugated antibodies, and antibody fragments. The demand for these molecules is expected to rise in significance [18]. mAbs are set to play a significant role in the treatment of a wide number of indications in various therapeutic domains, although oncological therapeutics will continue to dominate the majority of applications [19, 20].

The goal of cancer therapy is to cause the direct or indirect destruction of cancer cells, by specifically targeting the tumor (Rituximab®) or the vasculature that nourishes the tumor (Avastatin®). Numerous anti-inflammatory antibodies already on the market (Remicade, Humira) have also shown the great potential of mAbs. Asthma, transplant rejection, inflammatory conditions, autoimmune disease, cardiovascular conditions [21], atherosclerosis [22], and infectious disease [23] are new domains of treatment. While treatment of infectious disease will increase in the future, full-blown success will require the development of original screening strategies to obtain broadly neutralizing antibodies [24] and/or the combination of several antibodies targeting different epitopes or stages of the pathogen life cycle [25].

1.4 Antibody Engineering: A New Approach to the Treatment of Disease

In 1975, Köhler and Milstein developed and described but did not patent a revolutionary technique for the generation of mAbs [4], based on the immortalization of mouse B cells through fusion with myeloma tumor cells. Mouse hybridomas, generated upon fusion of the two parental cell partners, became the first reliable source of mAbs, facilitating the laboratory-scale production of murine mAb (Mumab) specific for a given antigen. The first Mumab, as therapeutic agent, was unsatisfactory due to their very short half-life in serum, poor or absent activation of human effector functions, and undesirable stimulation of human anti-mouse antibody (HAMA) responses in patients when repeated administration protocols were applied [26].

These issues were addressed using genetic engineering or chemical coupling techniques but have not been entirely resolved. The different strategies chronologically developed to avoid, mask, or redirect this human immune response are "chimerization" by fusion of mouse variable regions to human constant regions [27], "humanization," [28] and "deimmunization" by removal of *in silico* predicted T-cell epitopes [29] or introducing regulatory epitopes to induce tolerance [30].

1.4.1 Chimerization: The Stone Age of Antibody Engineering

In chimeric antibodies, the murine constant regions are replaced with human equivalent regions, since the constant region significantly contributes to the molecule's immunogenicity. In addition, the presence of human constant domains guarantees a more robust interaction with human effector cells and the complement system. This strategy led to therapeutic successes such as basiliximab (Simulect®: IgG1 anti-CD25, developed by Novartis Pharmaceuticals, East Hanover, NJ, the United States) or cetuximab (Erbitux®: IgG1 anti-EGFR, developed by ImClone, LLC, New York, NY, the United States). Nevertheless, chimeric antibodies, even if perceived as less foreign and therefore less immunogenic than mouse mAbs, have been shown to induce human anti-chimeric antibody (HACA) responses. This has been seen with infliximab/Remicade [31]. The advent and increasing sophistication of humanization technologies and the direct access to fully human antibodies will likely represent the death knell for chimeric antibodies.

1.4.2 Humanization: Improvement Never Ends

In order to move forward to reduce both HAMA and HACA, antibody humanization technologies were developed and made possible by, at the least, transferring all xenogeneic complementary determining regions (CDRs) onto the framework of a human antibody. Different approaches to CDR grafting have been tested according to the human template used as a matrix: sequences with known crystalline structure [32], rearranged somatic sequences, unmodified germ line sequences [33], or consensus sequences [34].

The first humanized antibodies were constructed based on human sequences with known crystalline structures, which permits the identification of residues contributing to antigen binding. In the "best fit" strategy, the closest human sequence, usually rearranged, is used as a framework to receive the murine CDRs. Another approach for humanizing an antibody is to choose the closest human germ line sequence [33]. Indeed, human antibody genes are formed *in vivo* by rearrangements of germ line gene segments. Later in B-cell ontogeny, the hypermutation process takes place, tailoring the initial sequences to improve recognition of the specific target antigen (Ag). The body thus may be theoretically more tolerant to germ line-encoded Abs. It is likely that there will be a reduction in clinical issues if germ line sequences are used for constructing humanized Abs.

The consensus method utilizes variable light (V_L) and variable heavy (V_H) domain frameworks derived from the most common amino acid found at each position within a given human subgroup. Whatever the method, CDR grafting might not result in the complete retention of antigen-binding properties since some framework residues can directly or indirectly interact with the antigen [35] or may affect the conformation of CDR loops [32c]. In this case, the antibody must be re-engineered or back-engineered to fine-tune the structure of the antigen-binding loops and restore its original high affinity. On the other hand, humanization of a xenogeneic Ab does not necessarily abolish the immunogenicity of the molecule, since the humanized Ab can still induce response against its xenogeneic CDRs. Not all residues within the CDRs of an Ab are essential for binding to its Ag. In fact, the Ag-binding site of an Ab usually involves only 20–33% of the residues [36]. These residues have been designated as specificity-determining residues (SDRs) [37]. Therefore, a murine Ab can be humanized through restrictively grafting only its SDRs onto the human template, minimizing the immunogenicity [38].

Another humanization strategy, termed resurfacing, was proposed by Padlan and involves the replacement of solvent exposed murine framework residues in the variable regions with human residues [39]. More recently, Hanf et al. [40] have proposed humanization by redesign of CDR residues in close proximity to the acceptor framework, yielding an antibody with better binding capacity than simple CDR graft with reduced immunogenicity compared to framework redesign.

Almagro and Fransson [41] have divided the field of humanization into two main trends: rational methods based on design cycle and precise knowledge of the antibody structure and sequence information and empirical methods using large combinatorial libraries and selecting the desired variants, a so-called "rational" strategy.

In contrast, a prime example of an empirical approach is "guided selection," a process that transfers the specificity of a Mumab to novel human mAb (Humab) by creating a hybrid library of the murine heavy chain and random human light chains, subjecting them to a selection process of binding antibodies and repeating the protocol with the human light chains previously isolated and a library of human heavy chains. Adalimumab (Humira) is the first phage display-derived human antibody and was generated by "guided selection" starting from a Mumab [42].

Humaneering™ is a KaloBios Pharmaceuticals' proprietary method for converting nonhuman antibodies into engineered human antibodies. The resulting antibodies are as close to the human germ line sequences as the products of fully human antibody generation techniques. The authors assert this platform maintains epitope specificity and increases affinity.

Finally, humanization technology has taken advantage of the recent advances in bioinformatics and *in silico* modeling. Zhang et al. [43] proposed a novel antibody humanization method based on this strategy, including an epitope-scanning algorithm designed to identify antigenic residues in the framework regions (FRs) that are mutated to their human counterpart in the humanization process.

The process of humanization is not restricted to murine antibodies and can similarly be applied to antibodies from other species. Nishibori et al. [44] have described a simple method for humanizing chicken mAb by CDR grafting followed by framework fine-tuning using a chicken phage-displayed mAb, phAb4-31, as a model antibody.

Despite the reduction in the proportion of exogenous sequences, the level of immunogenicity of humanized Abs ranges from negligible to highly detrimental. The humanized anti-HER2/neu Ab Herceptin gave as low as 0.1% of human anti-human antibody (HAHA) in breast cancer patients [45], while the humanized A33 Ab elicited 49% of HAHA among colon cancer patients treated with this Ab [46]. In this case, immunogenicity might result from HAHA responses, in particular to the paratope of the antibody (anti-idiotype antibody response) [47].

To date, it is difficult to determine whether "fully human" mAbs are less immunogenic than humanized mAbs as full immunogenicity data are available only for a limited number of mAbs in each category. The answer is complex and most probably will be resolved through a case-by-case consideration and detailed analysis.

Finally, anti-allotype reactions are predicted to occur during therapy of a genetically diverse population with a single antibody reagent. Moreover, allotypic variation is not equally distributed to all antibody isotypes and might be taken into consideration before designing a therapeutic strategy and effector antibody [48].

1.4.3 Deimmunization

Biovation (the United Kingdom, www.biovation.co.uk) has developed the DeImmunization™ technology consisting of identification and removal of T helper (Th) cell epitopes from antibodies [29]. An example of a product of DeImmunization actually under clinical trials, J591 is a modified antibody binding to prostate-specific membrane antigen (PSMA) [49].

Antitope, a subsidiary of Abzema, is conducting the same kind of approach to limit the development of antidrug antibodies (ADAs), which can reduce efficacy through rapid clearance. The associated EpiScreen™ immunogenicity assessment technology is used to confirm that T-cell epitopes have been removed [50].

The company Epivax is sharing the same field and proposes *in silico* epitope discovery, immunogenicity assessment, and protein deimmunization. They include “Tregitope Analysis” based on the identification within each submitted sequence of putative regulatory T-cell epitopes. These are subregions contained within the submitted sequences, which may relate to natural regulatory T cells and which may help to dampen the immune potential of the submitted antibody sequence [51, 52].

1.5 Fully Human Antibodies, What Else?

mAbs of human origin may have greater therapeutic value; thus several methods have been developed to generate Humab: selection from human hybridomas, selection from “humanized” transgenic mice, construction of *in vitro* combinatorial libraries, or direct cloning from immunized individuals.

1.5.1 Human B-Cell Hybridoma and Immortalized B-Cell Lines

The recovery of stable human B-cell hybridoma producing high-affinity IgG mAbs has rarely been achieved, due to the lack of a suitable human myeloma cell line. The most satisfactory results were obtained using heteromyelomas (mouse–human hybrid myelomas) as fusion partners. However, the mouse–human heteromyelomas that have been used for fusion with human lymphocytes are often unstable. The situation was ameliorated by the recent development of the novel SPYMEG cell line, a human lymphocyte fusion partner, codeveloped by Naomasa Yamamoto, of Ohu University and MBL. SPYMEG was established by the cell fusion of MEG-01 with a murine myeloma cell line. The SPYMEG cell line overcomes the problem usually encountered by hybridoma cells of human origin that are prone to chromosome deletions.

As an alternative, human antibody-secreting cells (ASCs) can be immortalized by Epstein–Barr virus (EBV) infection. However, EBV-infected cells are somewhat difficult to clone and usually produce only very low yields of immunoglobulin [53]. An improved method of B-cell immortalization by EBV involving the addition of a polyclonal B-cell activator (CpG/TLR-9 agonists) has been described [54, 55]. In contrast to plasma cells, which are terminally differentiated, memory B cells retain substantial growth potential and can be immortalized by EBV [56].

The final disadvantage restricting the human B-cell hybridoma approach (and, more generally, all the technologies giving access to human antibodies) is the fact that the human circulating antibody repertoire does not generally retain specificity to "self" proteins, which represent a significant number of targets for human antibody therapeutics.

1.5.2 *Ex Vivo* Stimulation

A recent technology using a process of *ex vivo* stimulation has been proposed by Duvall et al. [57]. Naive human B cells, isolated from tonsil tissue, are immortalized. The *ex vivo* stimulation of these cells induces class switching and somatic hypermutation. The resultant human library of different IgG antibodies can then be screened against any antigen using a standard limiting dilution protocol. By eliminating immunization and humanization steps, this platform should reduce both cost and time in producing a therapeutic mAb of interest.

1.5.3 Mice and Other Animal Species Producing Human Antibodies

1.5.3.1 SCID and Other Immune-Deficient Mice

A potential source of Humab can be obtained upon transplanting a functional human immune system into immunocompromised mouse strains, such as severe combined immunodeficient (SCID), SCID-bg, Trimera, or $\gamma c^{-/-}$/RAG2$^{-/-}$ mice.

SCID mice, lacking mature T and B cells and virtually devoid of endogenous serum immunoglobulins, can be successfully reconstituted with human peripheral blood lymphocytes (PBLs). Such mice reconstituted with a competent human immune system would represent an invaluable tool for producing human immunoglobulin, after immunization with antigen. However, the use of SCID mouse can be limited by shortened life spans, spontaneous production of functional lymphocytes with aging, and residual innate immunity leading to variable levels of engraftment. Mouse natural killer (NK) cells in particular would be detrimental to engraftment of human lymphoid cells [58]. For example, the natural depletion of NK activity in SCID-bg mice facilitated engraftment of human PBL from anthrax-vaccinated (AVA) donors. Stable recombinant cell lines producing Humab were generated by hybridoma formation, and human anti-protective antigen (PA) neutralizing mAbs of high affinity were obtained [59]. Other immunodeficient strains of mice were developed, for instance, the $\gamma c^{-/-}$/RAG2$^{-/-}$ mice. In addition to T- and B-lymphocyte deficiency, $\gamma c^{-/-}$/RAG2$^{-/-}$ mice are completely deficient in NK activity [58]. In recent years, the passing fad for such technologies has been completely masked by the gain in maturity of transgenic mice expressing human antibodies.

1.5.3.2 Humanized Transgenic Mice

The availability of transgenic strains of mice expressing human Ig genes (XenoMouse®, HuMab™ Mouse, TransChromo Mouse) provides a breakthrough for isolating Humab [60]. The occurrence of rearrangements and hypermutations confirmed that the endogenous cell signaling machinery of the mouse is compatible with human immunoglobulin sequence elements [61] and would constitute a marked advantage.

XenoMouse strains from Abgenix (Fremont, CA), were engineered by functionally inactivating the murine heavy-chain and k light-chain Ig loci and incorporating megabase-sized inserts of human DNA yeast artificial chromosome (YAC) carrying Ig heavy-chain and k light-chain loci that express the vast majority of the human Ab repertoire [62]. Three different strains of XenoMouse mice have been produced, constrained to class switch from IgM to IgG1, IgG2, or IgG4. The immune repertoire of XenoMouse strains was further increased by the introduction of the entire human Igλ locus [61]. To date, results obtained in preclinical and early clinical trials with human antibody from XenoMouse mice confirm their relative lack of immunogenicity [63]. However, human antibodies from mice can be distinguished from human antibodies produced in human cells by their state of glycosylation, particularly with respect to their Galα1-3Gal residue [64]. This carbohydrate residue is widely distributed among nonprimate mammals. Anti-Gal antibodies are produced in humans throughout life, as approximately 1% of circulating Ig [65]. Thus, antibodies bearing that residue would probably be subject to an accelerated immune clearance [66].

In a parallel strategy, human minichromosomes (derived from human chromosome 2 and 14) containing the complete germ line clusters for heavy and k light chains were introduced into TransChromo mice [67]. These mice, developed by the Kirin Brewery Company (Japan), are capable of producing every subtype of fully human Ig, including IgA and IgM. However, the instability of the transchromosome carrying the Igk locus was particularly detrimental, as hybridoma production was less than 1% of that seen in wild-type mice. An instant solution to this problem was to crossbreed the Kirin TC mouse carrying the human chromosome fragment 14 (locus IgH), with the Medarex YAC transgenic mouse carrying about 50% of the Igk locus. The resulting mouse (KM) performed as well as normal mice with regard to immune responsiveness [67b].

In 2006, panitumumab, the first fully human antibody generated from transgenic mice, was approved for clinical use by the USFDA. In 2012, seven of such antibodies were approved, which clearly indicates the important contribution of humanized transgenic mice in the pipeline of novel therapeutic mAbs [68].

Over the years, models of transgenic mice have risen in sophistication. According to Murphy et al. [69] from Regeneron Pharmaceuticals, a common weakness of humanized transgenic mice is a lack of fully functional immune systems intrinsically due to the strategy used for their genetic humanization. The VelocImmune mice model efficiently produces human–mouse hybrid antibodies in which the mouse constant regions (which are rapidly convertible to fully human antibodies) are kept intact and have fully functional humoral immune systems. The rationale was that the introduced human variable region gene segments would function indistinguishably in their new genetic location, whereas the retained mouse constant regions would allow for optimal interactions and selection of the resulting antibodies within the mouse environment.

The strategy adopted by Kymab is quite similar; Kymouse™ transgenic mice have been generated upon insertion of the genetic regions corresponding to the variable genes from all three human immunoglobulin loci, including λ genes, into precise locations in the corresponding loci of mouse embryonic stem (ES) cells using a proprietary technology called sequential recombinase-mediated cassette exchange (S-RMCE) [70].

Besides the aforementioned transgenic mice, MeMo® mouse developed by Merus produces, upon immunization, human antibodies composed of a single common light chain and diversified immunoglobulin heavy chains. These antibodies serve as building blocks for the easy and direct generation of BsAb for therapy (Biclonics™). Simultaneous transfection of two cLC V_H genes gives greater than 99% pure BsAb from a single cell and might additionally provide an innovative solution to the complex problem of coexpressing a mixture of several antibodies.

1.5.3.3 Humanized Transgenic Chicken

Transgenic chickens expressing human antibodies are designed to access human targets and epitopes that have been intractable in mammalian hosts because of tolerance to conserved proteins. The major difficulty lays in the mechanism for the generation of diversity, which in the chicken is accomplished through gene conversion, a very different system from that occurring in the human and murine immune response. To resolve this issue it was necessary to confirm that a knockout chicken B-cell line lacking V_H and V_L loci was able to undergo gene conversion with inserted human V_H and V_K

regions. Gene targeting in avian primordial germ cells may provide a unique approach to antibody engineering, particularly in the field of antibody discovery [71, 72].

1.5.3.4 Humanized Transgenic Bovine: Polyclonal Revival?

Antigen-specific human polyclonal antibodies (hpAbs), produced by hyperimmunization, have proven to be useful for treating many human diseases or infections. However, yields from available transgenic mice carrying human immunoglobulin loci are too low for direct therapeutic applications. Kuroiwa et al. [73] have produced transgenic bovine antibodies after transferring a human artificial chromosome bearing the entire unrearranged human immunoglobulin heavy (hIGH) and kappa light (hIGK) chain loci to bovine fibroblasts in which two endogenous bovine IgH chain loci were inactivated. Up to 2 g/l of hIgG, either associated with human kappa light chain or with bovine kappa or lambda light chain (chimeric), can be obtained from these animals. Polyclonal sera from cattle immunized with anthrax PA proved to be highly active in an *in vitro* toxin neutralization assay and protective in an *in vivo* mouse challenge assay. The transgenic bovine platform was recently fine-tuned to further increase the proportion of fully hpAbs produced after triple genetic knockout eliminating the bovine lambda cluster [74]. A first clinical trial for pathogen-specific fully human antibodies derived from Tc bovine (currently in preparation) will assess the value of the approach. The possibility of a yearly production, of approximately 450 l of plasma per transgenic animal, makes this strategy especially appealing.

1.5.4 Antibody Display

A major step forward in the production of fully human antibodies after selection from antibody gene repertoires expressed either *in vivo* on the surface of cells or filamentous bacteriophages [75, 76] or *in vitro* took place during the early nineties. Display technologies made possible the selection of antibodies from large repertoires, and the physical linkage between genotype and phenotype enabled the recovery of the DNA encoding the selected antibody fragment.

1.5.4.1 Phage Display

The bacteriophage platform is still the most widely used and well-established technique for antibody display and library screening [77]. Large repertoires of single-chain Fv fragment (scFv) or antigen-binding fragment (Fab) antibody genes are cloned into phage or phagemid vectors as a fusion product to one of the phage coat protein genes. Expression of the fusion product and its subsequent incorporation into the mature phage coat result in the antibody display on the phage surface. During the screening procedure generally termed "biopanning," phages that display a relevant antibody will be retained on a given surface coated with antigen, while nonspecific phages will be washed away. Multiple rounds of panning are made possible and necessary to select high-affinity binders.

1.5.4.2 Naive, Immune, and Synthetic Repertoires

The different types of antibody libraries are distinguished by the source of their repertoires: naive, immune, or synthetic. Naive libraries are constructed by cloning the antibody variable domain genes from pools of nonimmunized donors [78]. Since the description of the first antibody libraries, substantial effort has been directed toward the assembly of universal large-sized repertoires. The theoretical justification for large repertoires was the expectation that they would serve to isolate antibodies with nanomolar affinities against any antigen [79].

However, the primary limiting factor is the bacterial transformation step, inherent to the process of library construction. Even if techniques based on the so-called combinatorial infection [80] have provided an elegant mean of breaking this technological barrier, the vast majority of described "naive" repertoires, with a complexity over 10^{10} clones, have been assembled through the massive and tedious accumulation of multiple small-sized subrepertoires [78b].

By definition, the complexity of a given library is the overall number of different $V_H - V_L$ combinations obtained at the DNA level. However, experience has shown a significant difference between the encoded diversity and the displayed diversity for different reasons, including toxicity upon bacterial expression, nonproper folding or assembly, competition between wild-type and protein III molecules linked to antibody fragments, and proteolysis of the displayed moiety.

Lastly, it became obvious that the major issue was not the assembly of a large repertoire but rather its maintenance over time and the prevention of the drift of its content. This reality has necessitated a move to focus on smaller or biased repertoires obtained from immunized subjects [78b]. Nevertheless, some examples of universal large-sized synthetic [79b, 81] and semisynthetic [82] repertoires are available. They are based on the use of a limited number of universal frameworks selected for their capacity to be overexpressed in *Escherichia coli*. According to the technique employed, CDRs are totally synthetic or derived from a pool of naturally expressed ones [83, 84].

1.5.4.3 Display Formats and Optimization

From their inception, display methods have been based on fusion of the C-terminus of the phage λ tail protein pV or both the N- and C-termini of the capsid D protein, which is part of the phage head. Alternative phage display systems have been described using bacteriophages T4 [85] and T7.

Because of its numerous positive features, the filamentous bacteriophage M13 became the platform of choice for antibody display. Although M13 coat protein pIII and pVIII [86] display formats have been the most instrumental in the construction of antibody libraries, other display systems have been reported, such as those based on the minor coat protein pIX [87] or pVI [88]. The value of phage display depends not only on the diversity of the library at the DNA level but also on the efficiency with which the encoded proteins are displayed on the phage surface. Extensive studies of M13 assembly and structure have enabled improvements in phage display technology [89]. Indeed, phage selections often suffer from the amplification of nonspecific binding molecules. In order to alleviate this problem, three groups have developed a system coupling the binding of a displayed peptide or protein to its target with the amplification of the displaying phage: selectively infective phage (SIP) [90], selection and amplification of phage (SAP) [91], and direct interaction rescue (DIRE) [92]. The SIP technology exploits the modular structure of the phage protein pIII, which consists of three domains: N1, N2, and CT. The N-terminal N1 domain is absolutely essential for *E. coli* infection, while the CT domain is absolutely essential for phage morphogenesis [93]. Thus, SIP consists of two components:

1) A phage particle made noninfective by replacing its N-terminal domains of pIII with an antibody fragment.
2) An "adapter" molecule in which the antigen is linked to N-terminal domains. Infectivity is restored when the displayed protein binds to the ligand. Consequently, phage propagation becomes strictly dependent on the protein–ligand interaction [94].

A vector system allowing for the display of bivalent Fabs fused to leucine zippers on phagemid virions has been reported by Lee et al. [95]. The "bivalent display" format is a way to effectively mimic the binding avidity of natural antibodies and greatly reduce the off rate for phage bound to the immobilized antigen.

1.5.4.4 Yeast Display, Bacterial Display, Anchored Periplasmic Expression (APEX)

A complementary approach is based on the display of antibody libraries on the surface of bacteria [96] or yeast cells [97]. The display of Ab on the surface of bacteria is not only an alternative expression system for the screening of binders from libraries but also opens new potential applications, such as the delivery of passive immunity to mucosal body surfaces [98] or whole-cell catalyst [99]. Unlike phage, the relatively large size of bacteria and yeast allows screening by flow cytometry [100].

A protein library-screening technology based on anchored periplasmic expression (APEX) has been developed [101]. In this method, proteins are anchored on the periplasmic face of the inner membrane of *E. coli.* After disruption of the outer membrane by tris-EDTA-lysozyme, inner membrane-anchored proteins bind fluorescent ligands, allowing screening by flow cytometry.

Baculovirus displaying antibody moieties such as scFvs has been constructed for targeting specific cell types for gene therapy protocols. But at this time the baculovirus system is not considered as an attractive choice for the construction of antibody libraries [102], and this situation will not change in the foreseeable future.

1.5.4.5 *In Vitro* Display Technologies

One limitation of the *in vivo* selection systems is the library size that could be generated and handled. The efficiency of transfer of DNA into cells often limits the library size to 10^9–10^{10} members. In addition, selection in the context of the host environment (*E. coli*) could lead to the loss of potential candidates due to their growth disadvantage or even host toxicity. The most popular *in vitro* display technologies are ribosome display [103] and mRNA display [104] based on an original idea from G. Kawasaki (US patent no. 5,643,768 and 5,658,754). These *in vitro*-based antibody selection methods have proven to be successful in the construction and selection of libraries with a high diversity and complexities (potentially up to 10^{14} members). Inherent characteristics of the *in vitro* systems could obviously be turned into advantages:

- Easily amenable to an automated process.
- The RT-PCR step between screening rounds can be performed according to error-prone conditions, thus generating an amplification of diversity [105].
- An *in vitro* process is more likely to tolerate screening steps under nonphysiological conditions such as elevated temperature or highly denaturant environment [105b].

Other potential *in vitro* systems for antibody display are covalent antibody display (CAD) and polysome display [106]. In the covalent display technology, a protein is fused to P2A, a bacteriophage DNA-nicking protein that covalently binds its own DNA and thereby is subjected to selection regimes similar to those for phage display [107].

The polysome display is a modified ribosome display method, exploiting the interaction between a tandemly fused MS2 coat protein (MSp) dimer and the RNA sequence of the corresponding specific binding motif, C-variant [106].

Finally, among the nonconventional display systems, the strategy proposed by Sepp and Griffiths [108] based on *in vitro* compartmentalization through packaging in microdroplets should be mentioned. Briefly, domain antibodies (dAbs) are *in vitro* expressed in fusion to the N-terminus of single-chain variant of phage P22 Arc repressor DNA-binding domain that links the compartmentally expressed protein molecules to their encoding PCR fragment-based genes via cognate operator sites present on the DNA.

1.5.5 Next-Generation Technologies for a Direct Access to Fully Human Monoclonal Antibodies

1.5.5.1 Single-Cell Isolation

Recent progresses in high-throughput technologies and associated platforms allow one to consider rapid and large-scale automated screening of multiple antibody-expressing cells. In combination with cloning technologies based on single-cell RT-PCR, the direct access to human antibodies is now possible. The isolated antibody sequences are then directly inserted into expression vectors for further investigation.

Trellis Bioscience, employing a 10-day human antibody platform, has screened 10 million B cells from 25 RSV-infected donors with their CellSpot technology. This approach allows the discovery of rare antibodies that neutralized both the A and B subtypes of G protein [109]. More recently, a variation of this approach was described that reveals suites of cross-clade antibodies directed to discontinuous epitopes. It takes advantage of an iterative feedback between antigen probe designs based on structure and function information. Its high-throughput and multiplexed screening method is a generally applicable strategy for efficient identification of safe, native, finely tuned antibodies with the advantage of high genetic barriers to viral escape [110].

Valneva's VIVA|Screen® technology (www.valneva.com) is a microarray-based single-cell screening proprietary technology that allows rapid high-throughput analysis and discovery of fully human therapeutic antibodies obtained directly from human donors. The development of this technology was made possible through the following:

1) Access to a very large population of healthy or diseased human donors
2) The optimization of culture and expansion of human primary memory B cells
3) The use of specifically designed microarray chips that contain 62,500 wells with size and shape optimized for a single human B lymphocyte per well

In a work by Zwick et al. [111], human neutralizing mAbs are generated from vaccinated individuals who received a booster immunization against influenza virus. Blood cells, collected at day seven post immunization, are then specifically sorted by flow cytometry for an early population of ASCs. Neutralizing mAbs against a panel of influenza subtypes were subsequently isolated from single cells and expressed as recombinant proteins.

1.5.5.2 High-Throughput Sequencing and Repertoire Mining

The rise of next-generation sequencing (NGS) technologies allowed Reddy et al. [112] to bypass the tedious screening step inherent in the isolation of antigen-specific mAbs. The strategy employs high-throughput DNA sequencing and bioinformatic analysis to mine antibody variable region (V)-gene repertoires from bone marrow plasma cells (BMPCs) of immunized mice. V-gene repertoire of BMPCs, indeed, becomes highly polarized after immunization, with the most abundant sequences represented at frequencies high enough to be sorted out of the total repertoire. The most represented variable heavy (V_H) and variable light (V_L) genes based on their relative frequencies are paired expressed as recombinant antibodies in *E. coli* or mammalian cells.

Even though extremely attractive, repertoire mining is unlikely to totally replace current screening technologies. Saggy et al. [113] have shown that phage display and repertoire mining of immune repertoires are complementary technologies that can yield different antigen-specific antibody clones.

The functional antibody repertoire of the rabbit has been investigated through NGS by Kodangattil et al. [114] as a first step toward engineering rabbit V regions to enhance their potential as therapeutic agents.

We do not have, so far, examples of repertoire mining, through NGS, from human samples leading to the isolation of a candidate therapeutic antibody, but this gap will probably be rapidly fulfilled. In addition to the discovery of novel antibodies, information gained from high-throughput sequencing (HTS) of immunoglobulin genes can be applied to detect B-cell malignancies to guide vaccine development and to understand autoimmunity. A wider application of NGS will nevertheless require the development of a standardized experimental framework that will enable the sharing and meta-analysis of sequencing data generated by different laboratories [115].

1.6 Antibody Design

1.6.1 Antibody Isotype: The Specific Case of IgG4

In the process of designing a therapeutic antibody, the rational choice for a given isotype is determined by different factors such as *in vivo* half-life, need for recruitment of effector functions, or the extent of allotypic variation. In such a context IgG4, the least abundant of the four subclasses of IgG in serum, displays unique biological properties. It can undergo heavy-chain exchange, also known as Fab-arm exchange, ending with the formation of half antibody molecules that can randomly reassociate and generate BsAb [116–118]. Its weak interaction with FcγRII and FcγRIII, and lack of complement activation, makes it relatively "noninflammatory" [119].

1.6.2 Antibody Fragments

For many therapeutic applications, antibody functions (such as cytokine inactivation, receptor blocking, or viral neutralization) do not require the recruitment of effector functions through the crystallizable fragment (Fc) portion. An antibody's therapeutic performance is to a large extent dependent on factors such as size, tissue penetration, distribution, half-life, effector functions, affinity, stability, and immunogenicity. This can explain why smaller Ab (scFv or Fab) fragments may be preferred. Consequently, one of the goals of antibody engineering has been to reduce the size to a minimum antibody fragment while still retaining both binding affinity and specificity. Another advantage of small antibody fragments is that they can be expressed in *E. coli* and yeast, dramatically reducing the cost associated with large-scale mammalian cell culture.

The Fv fragment, formed by the heterodimeric association of the two variable domains V_L and V_H of the light (L) and heavy (H) chain, respectively, can be genetically engineered into an scFv with a flexible polypeptide linker. The most commonly used linker is a flexible 15-mer peptide $(Gly_4Ser)_3$. Changing the linker length between V-domains induced oligomerization into "diabodies" or even into higher order valency antibody fragments (tribodies, tetrabodies), potentially increasing their avidity [120]. Diabodies, the best characterized of these molecules, have shown high functional affinity, greater tumor retention, and slower systemic clearance than their monovalent counterparts in preclinical studies [121]. To stabilize the association of the V_H and V_L domains, different linkage strategies have been proposed, for instance, through disulfide bridges [122] and "knob into holes" mutations [123]. An alternative format to scFv, named MoaFv, has been proposed to stabilize the Fv fragment and restore the bivalency of the antibody [124]. This MoaFv was constructed by replacing the C_{H1} and C_L domains of the Fab with heterotetrameric molybdopterin synthase (MPTS).

The fragment containing the antigen-binding (Fab) component is a heterodimer of $V_H - C_{H1}$ and $V_L - C_L$ linked together through a disulfide bond. In comparison with whole antibodies, small antibody fragments such as Fab or scFv exhibit better tissue penetration pharmacokinetics. However, Fab and scFv are monovalent and often exhibit fast off rates and poor retention time on the target.

The half-life of circulating antibody fragments (Fab, scFv) can be improved by site-specific coupling of polyethylene glycol (PEG) to the fragment, the so-called PEGylation [125]. For example, site-specific PEGylation into the hinge region has prolonged the circulating half-life of an anti-TNF, a human fragment (CDP 870, Celltech), to 14 days. Covalent attachment of PEG increases the circulating *in vivo* half-life of the antibody fragment, augmenting its apparent size above the glomerular filtration limit. PEG may have the additional benefit of increasing solubility and resistance to proteolysis and reducing immunogenicity [126]. Kitamura et al. have shown that PEG attachment to an intact Mumab reduced the HAMA response significantly when compared to the parent intact antibody [127]. The linkage of PEG to antibody fragments, done by chemical coupling, has some disadvantage in terms of global cost and the complexity of the associated downstream processing. In order to partly overcome this problem, PASylation has been proposed as an alternative. The procedure consists of a genetic fusion of conformationally disordered polypeptide chains, with expanded hydrodynamic volume, comprising the small residues Pro, Ala, and Ser (PAS) [128]. PAS polypeptides are biodegradable and do not accumulate in the host organs. Moreover, they are stable in serum and lacking toxicity or immunogenicity.

1.6.3 Bispecific Antibodies

BsAb binding two different epitopes have been used to cross-link various cells, receptors, and molecules. Numerous methods for generating BsAb have been developed: generation of hybrid hybridoma or quadroma [129] and chemical and genetic coupling of the Fab. Multimeric Ab fragments may be obtained by linking them to molecules that tend to self-associate, such as amphipathic helices [130], coiled-coil structures, and immunoglobulin constant domains [131], and molecules that are homomultimeric. Recently, a new format of BsAb has been proposed, consisting of two scFvs, one for human epidermal growth factor receptor 2 (HER2)/neu and the other for CD16, heterodimerized by a "knobs-into-holes" device from the C_{H3} domains of the human IgG1 Fc fragment [132]. Another multimerization system, based on the interaction between the prokaryotic ribonuclease barnase (110 aa) and its inhibitor barstar (89 aa), has revealed important advantages [133]. The use of such constructs is based on the exceptionally high association constant (up to $10^{14}\,M^{-1}$) of the barnase–barstar couple.

The bispecific T-cell engager (BiTE) [134] is a small-sized class of bispecific single-chain antibodies, developed for the polyclonal activation and redirection of cytotoxic T cells against pathogenic or tumorigenic target cells. They are generated from the variable domains of two distinct mAbs. A CD19-directed BiTE antibody, blinatumomab, has been approved by the USFDA in 2014 for treatment of patients with acute lymphoblastic leukemia [135].

The dual variable domain immunoglobulin (DVD-Ig™) is a very simple and elegant solution for creation of BsAb that avoids some of the issues intrinsically associated with this type of construct. In DVD-Ig proteins, each arm of the molecule contains two variable domains (VDs) linked in tandem, each of them possessing different binding specificities and therefore expressible as classical antibodies [136]. In a cancer therapeutic protocol, Gu et al. [137] have developed anti-ErbB2 dual variable domain immunoglobulin (DVD-Ig) proteins that perform the function of a combination of two anti-ErbB2 antibodies.

Fcab™ are constructs providing a new binding region within an antibody, designed from site-specific engineering of the constant (Fc) region of the antibody. BsAb are obtained by combining the natural binding of antigens via the variable (Fab) regions and introducing new antigen-binding sites into the Fc region. Fcabs maintain favorable pharmacokinetics, ability to mobilize immune effector functions, and simplicity of manufacturing [138, 139].

Merus has developed MeMo mouse producing human antibodies composed of a single common light chain and diversified immunoglobulin heavy chains. From these antibodies, BsAb for therapy (Biclonics) can be easily generated. (See description on humanized transgenic mice.)

Catumaxomab is an example of a bispecific mAb currently in use for the treatment of malignant ascites and peritoneal carcinomatosis [140, 141].

1.6.4 Conjugated Antibodies or "Armed" Antibodies

Ab fragments have been fused with a vast range of molecules to introduce different functionalities including radionuclides, toxins, or drugs for cancer cell killing, enzymes for prodrug therapy, and cytokines to stimulate the antitumor immune response. At the border between conjugated and BsAb, Doppalapudi et al. [142] have developed a new class of reagents called bispecific CovX-Bodies in which two different peptides are joined together using a branched azetidinone linker and fused to a scaffold antibody under mild conditions in a site-specific manner. This nonconventional effector has the capacity to bind both vascular endothelial growth factor (VEGF) and angiopoietin-2 (Ang2) simultaneously and shows efficacy in tumor xenograft studies.

1.6.4.1 Tumor Cell Killing

mAbs directed to tumor-associated antigens or antigens differentially expressed on the tumor vasculature have been covalently linked to a drug such as calicheamicin (Mylotarg®, Wyeth) or a radionuclide (Zevalin®). The goal was to deliver the toxic load of these antibodies directly to cancer cells, avoiding the toxicity associated with systemic administration. Immunotoxins are composed of Fv fragments of antibodies that bind to cancer cells fused to a truncated form of a very potent bacterial toxin (e.g., Pseudomonas exotoxin, ricin A chain). The antibody moiety specifically directs the toxin to cancer cells [143]. Two antibody-drug conjugates (ADCs), brentuximab vedotin (Adcetris) and ado-trastuzumab emtansine (Kadcyla), were recently approved for marketing both by the US (FDA) and the European (EMA) agencies [144–146].

1.6.4.2 Activation of Antitumoral Immune Response

Similarly, cytokine–antibody fusions (e.g., TNF-α, IFN-γ, IL-12, or IL-2) have demonstrated a capacity to initiate tumor regression in several systems [19b]. These immunocytokines allow the local activation of the antitumoral immune response, avoiding the toxicity associated with systemic cytokine administration. Cytokine–antibody fusions are being considered for the therapy of cancer and chronic inflammatory disease [147].

Other constructs such as antibody–chemokine fusions are under evaluation for tumor targeting with the antibody-based *in vivo* delivery of chemokines at disease sites [148].

Another category of antibodies with high specificity and affinity has been developed to bind specific cytokines or their receptors, inhibiting the detrimental effect of the cytokine [19a].

1.6.4.3 Antibody-Directed Enzyme Prodrug Therapy (ADEPT)

Enzymes may also be conjugated to antibodies in antibody-directed enzyme prodrug therapy [149]. ADEPT involves the pretargeting of prodrugs to tumors. An Ab–enzyme fusion protein is first administered and allowed localizing to the tumor, followed by the administration of the prodrug that is activated by the enzyme at the tumor site [150]. This approach has been applied to various tumor therapies [151]. For example, a dimeric immunoenzyme comprising an anti-CD22 diabody and *Rana pipiens* liver ribonuclease was recently generated [152]. Previous clinical trials have shown evidence of tumor response; however, the activated drug had a long half-life, which resulted in dose-limiting myelosuppression. Also, the targeting system, although giving high tumor to blood ratios of antibody conjugate (10,000 : 1), required administration of a clearing antibody in addition to the antibody–enzyme conjugate [153]. The history and latest developments of targeted enzyme prodrug therapies are reviewed by Schellmann et al. [154] and Tietze and Schmuck [155].

1.6.4.4 Pepbodies

Pepbodies are obtained by fusion of a peptide, capable of binding one or several Ab effector molecules, to an scFv or Fab. These targeting fragments are able to efficiently recruit some of the effector's functions, normally associated with the Fc fragment, such as complement activation [156]. Prototype pepbodies have been constructed by introducing C1q-binding peptides into a loop of a C region domain. These pepbodies have retained the ability to bind both the antigen and C1q.

1.6.5 Alternative Immune Repertoire

Full-length antibodies, Fab, and scFv provide an antigen-binding surface comprising six CDR loops. Some target molecules are refractory to the immune repertoire, particularly those with cavities. The specific recognition of these target molecules would require a small penetrating loop for tight binding. The natural mammalian antibody repertoire does not encode penetrating loops and only rarely has the propensity to select this type of antibody [157]. In camelid species (camels, dromedaries, llamas, and vicuña), a significant proportion of the serum immunoglobulins is constituted of antibodies naturally devoid of light chains, called heavy-chain antibodies or HCAbs [158]. HCAbs lack the first domain C_{H1} but harbor an intact variable (V) domain (VHH) encoded by different V genes. HCAbs have been found to be physically very stable and generally produced at very high levels in *E. coli*. In addition, the

CDR3 (up to 24 residues in lamas) is unusually long and especially suited for binding to the inner part of cavities [159, 160].

On the basis of these observations, Domantis (www.domantis.com, recently acquired by GSK) has developed a series of large and highly functional libraries of fully human V_H and V_L dAbs. These "camelized" human dAbs correspond to the variable regions of either the heavy (V_H) or light (V_L) chains of human antibodies.

More recently, the immunoglobulin isotype novel antigen receptor (IgNAR) was discovered in the serum of the nurse shark (*Ginglymostoma cirratum*) as a homodimeric heavy-chain complex, which also naturally lacks light chains. New antigen receptors (NARs) are composed of five constant domains and a single variable domain with two CDR-like regions [161]. The variable domains of these unique molecules are candidates for the smallest natural antibody-based immune recognition units. In contrast to camelids it is obviously difficult to design a direct immunization protocol of sharks for the recovery of antibodies. Libraries of V(NAR) domains with extensive variability in the CDR1 and CDR3 loops displayed on the surface of bacteriophage have been produced. The content of such libraries was further validated through selection against the apical membrane antigen 1 (AMA1) from *Plasmodium falciparum*. In addition, a recombinant V(NAR) with nanomolar affinities for AMA1 could be produced with strong yields in *E. coli* [162].

Regardless, the format of the antibody might be of interest with regard to its participation in the repertoires originating from different animal species. The rationale behind such investigation is that these species are not employing similar genetic strategies to create diversity and therefore exploit different spaces of diversity. For instance, in antibodies from the Australian duck-billed platypus (*Ornithorhynchus anatinus*), internal cysteine bridges in the CDR3 loop, or between the CDR2 and CDR3 loops, are found in approximately 36% of the V(H) sequences. Such cysteine bridges, also observed in cow, camel, and shark, may play a role in stabilizing long and diversified CDR3 sequences and thereby have a role in adjusting or increasing the affinity of the Ab–Ag interaction [163].

The antibody repertoire of domestic cattle (*Bos taurus*) includes a group of IgM lambda antibodies with an exceptionally long complementarity-determining region 3 of the heavy-chain (CDR3H) segments, containing multiple Cys residues [164]. Both features are susceptible to influence the 3D structure of the antibody paratope. The recombination process makes use of an atypical and specific nucleotide insertion called "conserved short nucleotide sequence" ("CSNS," 13–18 nucleotides), of non-V(H) or D(H) gene origin, at V(H)–D(H) junctions resulting in extension of CDR3H size up to 61 codons [165].

Covering the various mechanisms used among the different species to create diversity is likely to be an endless task [166], but a qualitative exploration of the structures generated to bind a given antigen is probably the most straightforward way to investigate and define new and exquisitely specific immunotherapeutic agents.

1.6.6 The Universe of Alternative Scaffolds, a Multidimension Space

The examples of VHH and NAR are indicative of the usefulness and functionality of dAbs and a clear indication that small compact structures containing distinct positions with close proximity in space, where a high degree of variability can be introduced, can perfectly mimic the binding property of an antibody. From that assumption one can easily explain the emergence of different molecular structures classified as alternative scaffolds [167].

The exploration of various immune systems throughout the animal kingdom has revealed incredible diversity. A wide range of scaffold-like structures have been described that represent analogues of adaptive immune systems in vertebrates such as the lamprey. These jawless fish naturally express tetrameric and pentameric molecules having a central core and highly flexible pairs of stalk-region "arms" with antigen-binding "hands" [168]. This ancestral structure, identified as the oldest antibody, is set to be revisited and play a role in modern biotechnology [169].

In the following section we present a sample of these newly characterized molecules to illustrate their variety, binding profiles, and possible therapeutic applications.

1.6.6.1 Ankyrin Repeat Protein

The structural characteristics of the ankyrin repeat (AR) proteins have been developed as a scaffold for the construction of libraries of novel binding molecules [170], the so-called DARPins by Molecular Partners AG. Binz et al. have designed an AR module composed of fixed framework positions and randomized potential interaction positions [171]. These monomeric soluble proteins, devoid of disulfide bonds and stable under physiological conditions, are more amenable to expression in large amount in simple and cost-effective systems. The potential of AR protein has been demonstrated as an intracellular kinase inhibitor [172] or as potential therapeutics for the treatment of Alzheimer's disease [173].

1.6.6.2 Anticalin and Duocalin

Lipocalins have also been recruited as scaffolds for the design of artificial binding proteins known as Anticalin® (Pieris Proteolab AG) [174]. Compared with antibodies, lipocalins, which exhibit biotechnological advantages due to their smaller size, are composed of a single polypeptide chain with a simple set of four hypervariable loops easily manipulable at the genetic level. Schlehuber and Skerra have also reported the construction of a functional fusion protein from two independent anticalins, a so-called duocalin [175]. The potential of anticalin formulations as therapeutic agents is currently being evaluated; some are entering clinical trials, an example is the nonrandomized first-in-human phase I trial of PRS-050, that target and antagonize vascular endothelial growth factor A (VEGF-A) [176].

1.6.6.3 Avimers

Avimers, another class of substitutes for immunoglobulin-based therapeutic proteins, are evolved from a large family of human extracellular receptor domains by *in vitro* exon shuffling and phage display, generating multidomain proteins with target binding properties. These structures have shown a good thermostability profile and resistance to proteases [177].

1.6.6.4 Adnectins

Adnectins™, a recently formulated family of therapeutic proteins derived from the 10th fibronectin type III domain, are based on a single-domain structure devoid of disulfide bonds. A PEGylated, antiangiogenic Adnectin targeting vascular endothelial growth factor (VEGF) receptor 2 was able to block the receptor's interaction with VEGF-A, VEGF-C, and VEGF-D. This molecule is currently being evaluated in phase II clinical trials for several oncology indications [178]. Another Adnectin drug is also under scrutiny as an alternative to mAbs against proprotein convertase subtilisin/kexin 9 (PCSK9) for low-density lipoprotein lowering [179].

1.6.6.5 Affibodies

Affibodies are binding molecules selected from phage display libraries and elaborated by combinatorial mutagenesis of an alpha helical domain derived from staphylococcal protein A. High-affinity affibodies have been selected against a large number of targets for use in a variety of applications, such as bioseparation, diagnostics, functional inhibition, viral targeting, and *in vivo* tumor imaging/therapy. The extremely small size (6 kDa) makes them also accessible by chemical synthesis [180].

1.6.6.6 Fynomers

Fynomers are binding proteins derived from the Fyn SH3 domain. Their specific structure and size make them very efficient in binding with high affinity and blocking enzyme activities. For instance, Fynomers inhibiting the serine protease chymase (EC = 3.4.21.39) normally expressed in the secretory granules of mast cells can have some therapeutic potential in the control of allergic reactions [181] or other diseases [182].

1.6.6.7 OBodies

The OB-fold is a natural 5-stranded β-barrel domain structure found in diverse organisms, from bacteria to mammals, binding a wide range of natural ligands, such as proteins, oligonucleotides, oligosaccharides, and other small molecules [183]. This natural scaffold has conserved similarly folded and functional concave binding face. These characteristics might be exploited *in vitro* through protein engineering and provide an alternative to antibodies in some cases [184].

1.6.6.8 Kunitz Domains

Kunitz domains are stable and small peptides (6 kDa), able to recognize specific protein structures, and also work as competitive protease inhibitors in their free form. These characteristics make them ideal candidates for designing a new class of antibody-like scaffolds [185]. DX-88 is an example of Kunitz domain-based effector molecule explored in the treatment of hereditary angioedema (HAE) [186].

1.6.6.9 WW Domains

WW domains are small β-sheet motifs naturally involved in intracellular signaling through the recognition of proline-rich or phosphorylated linear peptide sequences. Engineering the side chains exposed on its concave surface creates a nonnatural scaffold and supports modification of 25% of the natural protein to form a novel affinity reagent. Using the cell-free molecular display platform, CIS display, Patel et al. [187] have isolated binders with nanomolar affinity to VEGFR-2 capable to inhibit binding of human VEGF to its receptor.

1.6.6.10 The Future of Alternative Scaffolds

Alternative scaffolds, increasingly present in the therapeutic arsenal, may be called to replace or at least compete with full-sized antibodies for specific applications that may be unrelated to immunotherapy [188]. Although they possess numerous advantages in terms of size and easy production, they are limited by their short half-lives and intrinsic lack of capacity for recruiting effector functions [189].

1.6.7 Intrabodies, Transbodies, and Genetic Transfer

1.6.7.1 Intrabodies

Intracellular Abs, also called intrabodies [190], are usually scFv [191] or dAbs [192], expressed within the confines of the intracellular environment. During the past decades, there has been growing interest in the use of such Ab fragments, for therapeutic applications [193], targeting intracellular molecules such as tumor antigens, and in functional genomics by blocking or modulating the activity of proteins or protein domains. Intrabodies can also be designed to modulate the function of its target molecule in other ways. For example, it is possible to design an intrabody with the capacity to relocate its target molecule to another subcellular location.

However, the cytoplasmic expression of scFvs is generally limited by their instability and insolubility. The reducing environment of the cytoplasm prevents the formation of the intra-chain disulfide bridges [194], which are critical in folding almost all antibodies. Consequently, most Abs do not fold properly and are therefore inactive inside cells. A number of approaches have been devised to overcome the limitation imposed by the cellular environment.

For example, Auf der Maur et al. have proposed a "quality control" system that allows for rapid selection of stable and soluble antibody frameworks suitable for intracellular applications [194]. This approach consists of constructing randomized hypervariable loop libraries on scFv frameworks that have been preselected for their intrinsic high stability and solubility in an intracellular environment.

Intracellular antibody capture (IAC) technology [195] is an approach based on selecting intrabodies from diverse scFv phage antibody libraries, initially screened against an antigen *in vitro*, and followed by selection in a yeast *in vivo* antibody–antigen interaction assay. However, the clinical development of intrabodies has been hindered by the availability of robust gene delivery system(s) including target cell-directed gene delivery [196]. Despite this drawback, new possibilities, such as Alzheimer's disease (AD) and Parkinson's disease (PD), are currently being investigated [197]. Even if these kinds of effector molecules show some attractiveness, their future is uncertain as they are competing with other classes of small effector molecules like aptamers, anti-messengers, or siRNAs.

1.6.7.2 Transbodies

Recently, Heng and Cao have proposed to test the possibility of fusing protein transduction domains (PTDs) to scFv antibodies [198]. PTDs are short peptide sequences that enable proteins to translocate across the cell membrane and be internalized within the cytosol. This would result in a "cell-permeable" antibody or "transbody." Correct conformation, folding, and disulfide bond formation can take place prior to introduction into the target cell. Thus, the process, still within the realm of speculation, has been proposed to possess several advantages over standard intrabody technology and to be potentially more efficient than the well-established RNAi-mediated gene silencing technology.

1.6.7.3 Antibodies and Vaccination

1.6.7.3.1 Idiotypic Network

Immunization with anti-idiotype (Id) antibodies represents a conceptually sophisticated approach to active immunotherapy [199]. By definition, the paratope of an anti-idiotypic Ab might mimic the 3D structure of an Ag and thus carry its "internal image" [200]. Idiotypic-based vaccines contain neither nominal Ag nor its fragments. The concept includes the idea that the internal image of a nonpeptidic structure such as glycosaminoglycans [201] can be duplicated as well.

The anti-idiotypic approach has the potential benefit that Id vaccines may be quite limited in their undesired side or toxic effects, sometimes associated with the antigens of interest. One of the major issues of human cancer is "immune tolerance," which might be substantially blocked by an appropriate anti-Id vaccine [199b].

Mader and Kunert [202] have used the anti-idiotypic strategy to design an immunogen able to mimic the antigenic site of HIV gp41 recognized by the mAb 2F5, broadly and potently neutralizing primary HIV-1 isolates and for which the epitope was extremely difficult to be fully characterized and expressed *de novo*.

The anti-idiotype-based vaccination, considered for a long time as highly speculative, was very recently strengthened by the example of racotumomab, a murine gamma-type anti-idiotype mAb that specifically induces an antibody response against Neu-glycolyl-GM3 ganglioside (NeuGcGM3), a molecule that is overexpressed in several solid tumors [203]. In a randomized, multicenter, placebo-controlled clinical trial of switch maintenance therapy in advanced non-small cell lung cancer patients, racotumomab–alum vaccine was considered as an effective and a well-tolerated treatment option [204].

1.6.7.3.2 *Antigenized Antibodies and Troy-Bodies*

The procedures referred to as antigenization of Abs consists of the expression of foreign oligopeptides located into the hypervariable loops of an antibody molecule [205]. Immunization with antigenized antibodies is an alternative method to focus the immune response against defined epitopes of foreign antigens and a way to deliver a peptide immunogen with an extended half-life. Musselli et al. have demonstrated that the severity of experimental allergic encephalomyelitis can be greatly attenuated by active immunization with antigenized Ab expressing synthetic peptides of the TCR, which are a key regulatory element in this disease [206]. The combined expression of helper and CTL epitopes derived from two different HIV-1 isolates and inserted in both heavy and light chains of an antibody was proposed by Kumagai [207].

Troy-bodies are antigen-presenting cell (APC)-specific antibodies with T-cell epitopes inserted and "hidden" into loops of the C_{H1} domain (constant region), without disrupting immunoglobulin folding. When they bind to their target APC, the T-cell epitopes acting as clandestine passengers are released, by antigen processing, and presented in association with MHC class II molecules to $CD4^+$ T cells. Targeting of antigens to APC results in an enhanced antigen presentation and T-cell activation [156, 208].

1.6.8 Passive Immunization by Antibody Gene Transfer

Mechanisms for the control of adverse responses in the course of passive immunotherapy have been proposed by Bakker et al. [209]. Derived from established gene therapy protocols, the possibility of raising therapeutic antibodies after DNA transfer *in vivo* appears to be a workable strategy. Genes encoding fully human antibodies targeting pathogens might be complementary to existing treatments and could act synergistically in controlling infectious diseases.

Highly neutralizing antibodies have been isolated, notably following HIV immunization or infection. In contrast, the capacity of designing and producing immunogens able to elicit these antibody responses has often proven to be fruitless. A new strategy designated as vectored immunoprophylaxis (VIP), involving passive immunization by viral vector-mediated delivery of genes encoding broadly neutralizing antibodies (bnAbs) for *in vivo* expression, has recently been proposed to overcome the problem. Such a protocol conducted in mice, humanized mice, macaque, and monkeys induces expression of these mAbs at high concentrations from a single intramuscular injection and can presage an effective prophylaxis against HIV [210, 211].

1.6.9 Catalytic Antibodies (Abzymes)

Among the properties of the antibody molecule, one of the most intriguing one is the capacity to support a catalytic activity. A major and still controversial question is whether catalytic or more generally enzymatic activity of some natural antibodies represents a function or dysfunction of these molecules [212]. The literature is replete with descriptions of antibody catalytic activities frequently associated with autoimmune diseases [213] or other pathological behaviors.

Other reports have emphasized the beneficial role of catalytic antibodies, considered in that context of "defensive enzymes" [214]. Serine protease or other peptidase activities have been reported as naturally occurring catalytic antibodies [215]. Remarkably, in a number of examples, the light chain has been identified as the active component for proteolysis [216].

A variety of therapeutic functions can be envisioned for catalytic antibodies [217], including antiviral activity or prodrug activation. In the latter case the major advantage would be the lower immunogenicity of a catalyst perceived as a natural molecule by the host's immune system. The low immunogenicity combined with high stability of immunoglobulins in the bloodstream makes abzymes potent agents for detoxification [218].

In the treatment of drug addiction, catalytic antibodies are capable of breaking down cocaine to systemically inert molecules [219].

Isolating high performing catalytic antibodies is still a great challenge [220]. The major reason is that catalytic antibodies are often selected for their capacity to stabilize the transition state of a chemical reaction, which requires transition state analogues as immunogens. The chemical synthesis of stable molecules mimicking a highly ephemeral structural entity, yet similar to the transition state in terms of 3D structure, is not a trivial task. Various approaches have been evaluated in the selection of catalytic antibodies, such as immunization with electrophilic covalent reactive analogues (CRA). These are haptenic phosphonates that bind covalently to a nucleophilic residue in the active site of an abzyme [221, 222]. Other authors have used a molecular imprinting technique based on idiotypic networks. Finally, *de novo* generation of abzymes can be obtained by introduction of amino acid residues with known catalytic functions into the antigen-binding site of antibodies [223].

The other constraint is related to the construction of a screening tool for a catalytic activity, which could be very different from a standard selection based on affinity for the target. In the context of designing catalytic antibodies to treat cocaine addiction, a remarkable example of hapten design and innovative tool selection is described by Xu et al. and Meijler et al. [224].

1.6.10 Antibody Cocktails: Oligoclonal and Polyclonal Mixtures

Many therapeutic applications, such as treatment of infectious targets, require broadly neutralizing antibodies or antibody mixtures to ensure coverage of a range of infectious agents in order to limit the emergence of variants of the target pathogen. A good example is the case of post-exposure prophylaxis against rabies. The current treatment is a quick administration of human or equine polyclonal immunoglobulins, and it is clear that an mAb would hardly serve to replace these products [225, 226]. Recombinant polyclonal, oligoclonal, or simpler antibody cocktails have been proposed for future biotherapeutic formulations in order to replace available sera against envenomation [227, 228]. These alternative antibody mixtures could also be composed of small antibody fragments or antibody-related structures.

In spite of their appeal, antibody mixtures are raising new challenges in terms of bioproduction, due to requirements dictated by the FDA and the EMEA. Their rulings on this topic approve Ab mixtures as a single therapeutic only if mAbs are coproduced. Mixture of mAb can thus be obtained as separate formulations of individual mAb, single formulation of individually generated mAb, mAb mixture from a single clone, or mAb mixtures by different cell clones in single batch. Symphogen has developed an expression platform, Sympress™, to produce a mixture of full-length IgG [229]. More recently, Excelimmune has also proposed a proprietary method to express mixtures of antibodies from a single batch culture.

The transgenic MeMo mouse developed by Merus is an alternative means to obtain mAbs that can potentially be expressed clonally. Finally, transchromosome cattle producing human polyclonal sera may represent a viable competitor to the aforementioned technologies [74].

1.7 Antibody Production

Extrapolating from current trends, at least one out of three biotherapeutic agents will be an antibody or antibody-derived molecule in the next few years. The production of mAbs for therapeutic use is considered as the major driver for the development and optimization of expression systems. The vast majority of full-sized mAbs in the market are produced using mammalian cells in culture, largely Chinese hamster ovary (CHO) or CHO engineered derivatives. This platform has come to maturity, pushing the physiological boundaries to their extreme limits [230] through complex selection processes and adaptation to prohibitively expensive culture media. Whether mammalian cells retain their leading position is an open question. Nevertheless, the advent of biosimilars will encourage the establishment of new expression platforms with low cost of goods and very high volume scalability profile. Alternative production means, including plant-based systems, nonconventional unicellular eukaryotic systems, or even *in vitro* systems such as continuously fed cell-free translation technologies, may in the future mature into practical alternatives.

At the same time, protocols in which only the binding domain of the antibody is required, the alternative scaffold structures, are gaining in importance. A wide range of new, compact structures are now available and their products are being tested in clinical trials. One of the associated advantages is their production and purification in nonmammalian systems like *E. coli* and yeast [231].

The choice of the most suitable expression system is linked to the Ab format (whole-antibody, monovalent, or bivalent fragment) and the application field (therapeutic, diagnostic, experimental tools), but factors such as scale-up, coproduction of different antibodies, posttranslational modification, and regulatory issues are additional critical parameters.

Higher eukaryotic cells are covered elsewhere in this volume. Accordingly, this review will simply give rapid snapshots of other expression systems.

1.7.1 Prokaryotic Systems

Bacterial systems are highly appropriate for the expression of scFv, Fab, and artificial small-sized antibody-like structures, intracellularly within the periplasmic space or as secreted molecules.

1.7.1.1 Escherichia coli

Efficient recovery of properly folded and active Ab fragments in *E. coli* is determined largely by the composition of the primary sequence [232]. Expression of recombinant Ab is typically achieved by fusion to N-terminal signal peptides, which guide the protein to the periplasmic space of *E. coli* where chaperones (Skp, FkpA, DsbA, and DsbC) assist the folding of the Ig domains and form the correct disulfide bridges to stabilize the structure [233]. However, during accumulation of recombinant proteins in the periplasm of *E. coli*, undesired proteolytic degradations can occur. *Escherichia coli* host strains lacking the major periplasmic proteases (DegP and Prc) have been shown to increase the yield of Fab produced in the periplasm two- to threefold [234]. The yield of functional antibody fragments expressed in the bacterial periplasm depends also on the amino acid sequence, being highly variable from antibody to antibody. Alternatively, a correctly folded Fab has been produced in the cytoplasm of *E. coli* trx B gor mutants, in which redox potential of the cytoplasm is comparable to that of the mammalian endoplasmic reticulum. The expression has been improved via the coexpression of molecular chaperones such as Skp [235]. An alternative to the periplasmic and cytoplasmic expression of Abs is their secretion to culture supernatants using the α-hemolysin (HlyA) system of *E. coli* [236].

For technical reasons it was believed that the expression of full-length antibodies in the periplasmic domain of *E. coli* was not a reasonably achievable goal. Nevertheless, this assumption was controverted by Simmons et al. who described the rapid and efficient expression of complete IgGs in the periplasm of *E. coli* [187]. The binding properties of the recovered antibodies were fully retained, while, as expected, the recruitment of the effector functions linked to the Fc domain was almost completely lost.

The recurrent question of the ability to carry out glycosylation in *E. coli* was addressed by Wacker et al. [237] who established that the N-linked glycosylation pathway found in the bacterium *Campylobacter jejuni* could be transferred into *E. coli*. Recently, scientists from the International Center for Biotechnology [238] have also succeeded in transferring a complete and functional glycosylation operon from *C. jejuni* to *E. coli*. Although structural differences exist between bacteria and eukaryotic cells, this raises the possibility of expressing glycosylated antibodies or antibody fragments in *E. coli*.

As exciting as an innovative approach, we should not underestimate the difficulty of transferring complete glycosylation pathways into prokaryotic hosts and the limitation of expressing large complex molecules into the periplasm of gram-negative bacteria. Nevertheless, metabolic engineering and new technologies giving access to extensive genomic reconstruction will tentatively position *E. coli* or other bacterial hosts as candidates for economically viable glycoprotein production [239].

1.7.1.2 Alternative Prokaryotic Systems

Antibody fragments have been produced in a large number of bacterial hosts such as *Bacillus subtilis* [240], *Lactobacillus zeae* [241], *Streptomyces lividans* [242], *Staphylococcus carnosus* [243], and *Pseudomonas putida* [244]. Despite great creativity, both in terms of new host and vector design, there is to date no clear evidence that these new expression systems will replace *E. coli*, with the exception of *Pseudomonas fluorescens* [245].

As the assembly and glycosylation of full-length antibodies are still problematic in bacterial systems, mammalian cell culture or transgenic organisms are at present more suitable for production of these molecules.

1.7.2 Insect Cells and Baculovirus Expression System

The baculovirus expression system has been successfully used for the expression of functionally active antibodies [246] or antibody fragments [247]. Ready-to-use, cassette baculovirus vectors with human Fc regions have been engineered for direct transfer of specific variable regions selected by phage display and subsequent expression as a complete antibody [246b]. By its intrinsic nature, the baculovirus system is tricky for large-scale production of complex multicatenary molecules. We recommend to the reader the extensive manufacturing data provided by Protein Sciences Corporation on the production of their Flublok® vaccine [248].

1.7.3 Yeast and Filamentous Fungi

The ability to grow in chemically defined medium in the absence of animal-derived components and to secrete large amounts of recombinant protein has made yeast a system of choice to produce proteins not functionally expressed in *E. coli*, impaired by folding problems or the requirement for glycosylation.

Several antibody fragments, such as scFvs [249] or Fab [250], have been expressed at high yield in *Pichia pastoris*. This methylotrophic yeast, almost as easy to genetically manipulate as *E. coli*, possesses a true eukaryotic protein synthesis pathway [249]. The natural propensity of yeast for hypermanosylation can be detrimental for both the antibody products' immunogenicity profile and for their recruitment of Fc-mediated effector functions. To deal with this drawback, glycoengineered lines of the *Pichia* have been developed. Antibodies with specific human N-glycan structures can perform effector functions [251] as proposed by GlycoFi. Such platforms offer a credible alternative to mammalian cells and may provide more homogenous product and better efficacy, such as enhanced effector function [252]. In a recent head-to-head comparison, Zhang et al. [253] showed in preclinical evaluation that the anti-human epidermal growth factor receptor 2 (HER2) mAb produced in a glycoengineered *Pichia* was fully comparable to the original trastuzumab.

The filamentous fungi *Aspergillus awamori* [254], *Aspergillus niger* [255], and *Trichoderma reesei* (patent no. WO 92/01797) have also been used to express antibody fragment and full-length IgG.

1.7.4 Transgenic Plants: Vegetal but Not Vegetative

Antibodies extracted from plants, the so-called plantibodies, have theoretical advantages in terms of cost of production [256] and ease of scaling up to meet market demand. Moreover it is unlikely that they harbor animal-derived pathogens including prions and viruses [257]. The first functional mAb was expressed in plant in 1989 [258]. Since then, plantibodies have been successfully produced in transgenic tobacco, soybean [259], alfalfa [260], and other plants [261]. Plants have proven to be effective systems for producing functional therapeutic mAbs capable of conferring passive protection against bacterial and viral pathogens in animals [257b, 262]. To date, plants offer the only large-scale, commercially viable system for production of secretory IgA (sIgA) [263].

Particle bombardment allows the simultaneous introduction of multiple constructs, thereby expediting the recovery of transgenic lines expressing multimeric antibodies such as secretory immunoglobulin (sIgA) [264]. The availability of large amounts of secretory IgA plantibodies opens up a number of novel therapeutic opportunities for disorders of the mucosal immune system. In a human trial, the plant-derived recombinant sIgA prevented oral recolonization by *Streptococcus mutans* [261].

Plant-based production, however, suffers from numerous drawbacks: one is related to the time necessary to establish and screen transgenic plants, and another is more related to social perception of vegetal GMOs in certain countries. A tentative solution to these issues is the advent of transient expression means such as "agroinfiltration" [265], in which plants are just a substrate and never become transgenic. These systems are generally based on target gene delivery into genetically unmodified plant host via viral or bacterial vectors; viral vectors often use plant RNA viruses, where nonessential sequences are replaced with the gene of interest [266]. Above all, the real advantage of transient expression is the very short time frame from vector design to large-scale production.

Researchers developing plant-based systems need to find acceptable levels of homogeneity, authenticity, and absence of immunogenicity for antibodies destined for human patients. These concerns have been addressed by targeted gene replacement, for instance, in the moss *Physcomitrella*, in which strains with nonimmunogenic humanized glycan patterns were created [155]. The implementation of large-scale production and downstream processing under good manufacturing practice conditions is also still challenging [267].

The field of vaccine production has suggested that transgenic plants can accommodate the "edible vaccine" concept. Such a concept has sparked interest and often controversy for varying reasons. Similarly in the specific domain of passive immunization, it is also conceivable that the production of antibodies in edible tissues from a variety of plants, such as potato, tobacco, maize, rice, tomato, and pea, would allow for a unique, convenient, needle-less, oral passive immunization at the gastric mucosal surface [268].

In targeted applications such as immediate treatment of bioterrorism agents, there are important applications for plant-derived biotherapeutics, such as antibodies against ricin toxin [269, 270]. The outcome of therapy using the antibody cocktail ZMapp against the recent Ebola outbreak will definitely influence the current perception on plantibodies.

1.7.5 Transgenic Animal-Based Production Systems

The mammary gland has generally been considered the tissue of choice to express valuable recombinant protein since milk is easily collected in large volumes. Among the transgenic farm livestock species used so far, cattle, goats, sheep, pigs, and rabbits are candidates for the expression of tens to hundreds of grams of genetically engineered milk proteins [271]. Transgenic farm animals (goats, cows) could solve many of the problems associated with either microbial or animal cell expression systems: improper folding of complex proteins and lack of adequate posttranslational modifications [272]. Several reports describe expression of antibodies in the milk of transgenic animals [273, 274] despite limitations by the relatively long interval from birth to first lactation encountered with domestic livestock, the discontinuous nature of the lactation cycle, and substantial material investments required to produce transgenic dairy animals [275].

The transgenic production of antibodies in egg white has been reported to take approximately 18 months [276]. The cost of generating nonpurified material from transgenic chickens is calculated at $\$0.50\,g^{-1}$. However, this cost does not take into account production under current good manufacturing practice (cGMP) conditions.

Kaneka Corporation, in Japan, has developed transgenic chickens using a retroviral vector encoding an expression cassette for a fusion protein of the extracellular domain of the human tumor necrosis factor (TNF) receptor 2 and Fc region of human IgG1 (TNFR/Fc). This hybrid antibody derivative is proposed as a therapy for inflammatory diseases such as rheumatoid arthritis [277].

Insect, plant, and yeast cells are attractive as possible hosts for the production of recombinant antibodies as they represent inexpensive and versatile expression systems. The principal limitations to their more widespread employment are that the N-glycans generated differ from those of human glycoproteins due to the presence of some host glycoepitopes on expressed glycoproteins, which may elicit immune responses in humans and poor pharmacological activity *in vivo* due to the rapid clearance from the circulatory system of glycoproteins with nonhuman glycans.

1.7.6 Cell-Free Translation, *"In Vitro Veritas"*

Since the early 1950s it has been shown that disrupted cells are still capable of synthesizing proteins. Based on this observation, "cell-free translation" systems have been developed and used to obtain minute amounts of labeled proteins for identification or characterization. Their productivity has improved greatly over the last two decades [278]. An important step was to move from qualitative to semiquantitative or even quantitative use of the technique and to further explore the so-called continuous-flow cell-free translation (CFCF) or continuous-exchange cell-free translation (CECF). These new generation systems, able to produce milligram amounts of a desired product, are suitable for automation.

Recent improvements include the possibility of expressing proteins with disulfide bonds, through cotranslation or addition of disulfide isomerase and/or chaperones to assist with protein folding. Moreover, cell-free expression opens the way to the synthesis of unanticipated new molecules such as polypeptides containing nonnatural amino acids. For instance, modified amino acids artificially linked to suppressor tRNAs can be incorporated, through recognition of amber codons, at specific sites within a sequence.

Thus, scFv [279] and Fab [280] have been reported to have been synthesized using *in vitro* systems. However, functional Abs with antigen-binding activity are obtained only if disulfide formation and rearrangement are allowed to take place during the translation reaction (e.g., with addition of protein disulfide isomerase (PDI) or chaperones) [281]. The potential for scale-up of cell-free transcription/translation systems has recently been exploited to reach the milligram range using wheat germ extracts (Cell-free Sciences, Japan) [282] or eventually a much larger scale (grams to kilograms) with the *E. coli*-based open cell-free synthesis (OCFS) proposed by Sutro Biopharma [283]. The OCFS system was further evolved to identify chaperone proteins that improve the folding and assembly of trastuzumab. Through the system, the disulfide isomerase DsbC and the prolyl isomerase FkpA were identified as important positive effectors of IgG folding [284].

1.7.7 Glycosylation Pattern and Expression Systems

All the aforementioned expression systems are not equivalent for the post-translational modifications of antibodies, in particular glycosylation. Although the processing of the initial Glc2Man9GlcNac2 oligosaccharide to Man8GlcNAc2 in the endoplasmic reticulum shows significant similarities between plants, insects, yeasts, and mammals, very different processing events occur in the Golgi compartments [285]. Glycosylation profoundly affects the effector functions of IgG antibodies [66]. In particular, an optimal effector function is dependent upon the correct carbohydrate structure at Asn 297 of the Fc region of the two heavy chains. This oligosaccharide is buried between the CH2 domains, forming extensive contacts with amino acid residues within CH2. The terminal β-1,4-galactose residues are in close contact with the protein backbone. It has been suggested that the Gal residues are critical for antibody performance, not only in effector functions but also in correct antibody folding [66]. Differences in glycan usage between human and expression hosts (mammalian cells, insect cells, plants, yeasts) have caused some concerns that recombinant mAbs could be immunogenic or even allergenic. Consequently, several groups investigated the possibility of humanizing glycosylation pathways in various expression systems to produce humanlike glycoproteins [285, 286].

Glycosylation engineering of mammalian cell lines by stable transfection with genes encoding terminal human glycosyltransferases allows production of products with tailored glycosylation in high yields [287]. Umana et al. showed that the efficacy of antibodies could be improved by enhancing the potency of their natural immune effector functions [288]. Stable overexpression of *N*-acetylglucosaminyltransferase III, an enzyme not naturally expressed in CHO and NSO cells, in recombinant antibody-producing cells generate IgGs with high levels of bisected, nonfucosylated oligosaccharides in the Fc region [288]. On the other hand, the CHO cell line has been engineered to express the missing enzyme α-2,6-sialylytransferase [289], and feeding with galactose has proven to improve galactosylation of produced antibodies. In order to obtain highly sialylated products as similar as possible to human proteins, Mai et al. [290] have designed ST6Gal minigenes to optimize the ST6GalI sialyltransferase activity and used them to engineer ST6(+)CHO cells.

Insect cells, like other eukaryotic cells, modify many of their proteins by N-glycosylation. However, the endogenous insect cell N-glycan processing machinery generally does not produce complex, terminally sialylated N-glycans such as those found in mammalian systems [286]. Similarly, glycoengineering has been used to improve baculovirus–insect cell production systems [291].

In plants, N-linked glycans may be processed to include potentially antigenic and/or allergenic β(1,2)-xylose residues attached to the α-linked mannose of the glycan core and α(1,3)-fucose residues linked to the proximal Glc-NAc [292]. However, recent studies suggest that these glycans have limited immunogenicity. Even if plant glycan patterns may not represent a problem in terms of human health, they may affect conformational epitopes or clearance of plant-derived antibodies. In an attempt to "humanize" glycans on antibodies produced in plants, the human β-1,4-galactosyltransferase was stably expressed in tobacco plants [293].

Most efforts to humanize yeast glycosylation pathways [294] have focused on the deletion of specific yeast genes involved in hypermanosylation (α-1,6-mannosyltransferase Och1p) and the introduction of genes catalyzing the synthesis, transport, and addition of human sugars (α-1,2-mannosidase, galactosyltransferases) [295]. This approach has been successfully used in *P. pastoris* [296]. In addition to specific glycoengineering of host biosynthetic pathways, complementary techniques such as *in vitro* chemoenzymatic glycosylation remodeling and chemoselective and site-specific glycosylation of proteins should be mentioned [297].

1.8 Recombinant Antibodies: No Limits...

Recombinant mAbs have become the major success story of the pharmaceutical industry. This predominance has come about through a synergism between the continuing high demand from the market and the progressive maturation and improvement of antibody engineering technology. In the decades between their discovery and today, the industry has been faced over and over again with daunting challenges. These include unanticipated side effects (including death), vast overruns on production costs, and failure to bring about long-term responses in cancer patients. The industry has addressed these issues on all levels, and now approval of new antibodies by the appropriate regulatory agencies is becoming routine, although less frequent than would be wished. However, until recently, there were no biosimilars (or biobetters) approved by regulatory agencies. This picture is beginning to change, with important implications for affordable and more effective antibody therapies.

For scientists, clinicians, and leaders of the pharmaceutical industry, mAbs are an exciting, sometimes frustrating technology. With so many options opened through the advances in recombinant antibody technology, we believe that the introduction of new therapies is limited only by the innovative spirit of their inventors. Continuously reinvented, recombinant antibody technology is ready to meet the medical and industrial needs of a generation of challenges.

Acknowledgments

No external funding was used to assist in the preparation of this manuscript. The author has no conflict of interest directly relevant to the content of this review.

Very special thanks to Bettina Werle, Charlotte Mignon, Adrien Lugari, Laurent Beloeil, and Patrick Lécine for critical reading and helpful discussions.

References

1. von Behring, E., Untersuchungen ueber das Zustandekommen der Diphtherie-Immunitat bei Thieren. *Dtsch Med Wochenschr* 1890, 16, 1145–1148.
2. von Behring, E.; Kitasato, S., Ueber das Zustandekommen der Diptherie-Immunitat und des Tetanus-Immunitat bei Thieren. *Dtsch Med Wochenschr* 1890, 16, 1113–1114.
3. Grundbacher, F. J., Behring's discovery of diphtheria and tetanus antitoxins. *Immunol Today* 1992, 13 (5), 188–190.
4. Köhler, G.; Milstein, C., Continuous cultures of fused cells secreting antibody of predefined specificity. *Nature* 1975, 256, 495–497.

5. Laffly, E.; Sodoyer, R., Monoclonal and recombinant antibodies, 30 years after. *Hum Antibodies* 2005, 14 (1–2), 33–55.
6. Lawrence, S.; Lahteenmaki, R., Public biotech 2013—the numbers. *Nat Biotechnol* 2014, 32 (7), 626–632.
7. Storz, U., Intellectual property protection: strategies for antibody inventions. *MAbs* 2011, 3 (3), 310–317.
8. Waltz, E., Genentech's Cabilly victory. *Nat Biotechnol* 2009, 27 (4), 307.
9. Sandercock, C. G.; Storz, U., Antibody specification beyond the target: claiming a later-generation therapeutic antibody by its target epitope. *Nat Biotechnol* 2012, 30 (7), 615–618.
10. Ledford, H., "Biosimilar" drugs poised to penetrate market. *Nature* 2010, 468 (7320), 18–19.
11. Tsiftsoglou, A. S.; Ruiz, S.; Schneider, C. K., Development and regulation of biosimilars: current status and future challenges. *BioDrugs* 2013, 27 (3), 203–211.
12. Cortes, J.; Curigliano, G.; Dieras, V., Expert perspectives on biosimilar monoclonal antibodies in breast cancer. *Breast Cancer Res Treat* 2014, 144 (2), 233–239.
13. Calvo, B.; Zuniga, L., Therapeutic monoclonal antibodies: strategies and challenges for biosimilars development. *Curr Med Chem* 2012, 19 (26), 4445–4450.
14. Wadhwa, M.; Thorpe, R., The challenges of immunogenicity in developing biosimilar products. *IDrugs* 2009, 12 (7), 440–444.
15. Vital, E. M.; Kay, J.; Emery, P., Rituximab biosimilars. *Expert Opin Biol Ther* 2013, 13 (7), 1049–1062.
16. Beck, A.; Sanglier-Cianferani, S.; Van Dorsselaer, A., Biosimilar, biobetter, and next generation antibody characterization by mass spectrometry. *Anal Chem* 2012, 84 (11), 4637–4646.
17. (a) Barbosa, M. D.; Kumar, S.; Loughrey, H.; Singh, S. K., Biosimilars and biobetters as tools for understanding and mitigating the immunogenicity of biotherapeutics. *Drug Discov Today* 2012, 17 (23–24), 1282–1288;
(b) Ebbers, H. C.; van Meer, P. J.; Moors, E. H.; Mantel-Teeuwisse, A. K.; Leufkens, H. G.; Schellekens, H., Measures of biosimilarity in monoclonal antibodies in oncology: the case of bevacizumab. *Drug Discov Today* 2013, 18 (17–18), 872–879.
18. (a) Pavlou, A. K.; Belsey, M. J., The therapeutic antibodies market to 2008. *Eur J Pharm Biopharm* 2005, 59 (3), 389–396;
(b) Nieri, P.; Donadio, E.; Rossi, S.; Adinolfi, B.; Podesta, A., Antibodies for therapeutic uses and the evolution of biotechniques. *Curr Med Chem* 2009, 16 (6), 753–779.
19. (a) Helguera, G.; Penichet, M. L., Antibody-cytokine fusion proteins for the therapy of cancer. *Methods Mol Med* 2005, 109, 347–374;

(b) Dela Cruz, J. S.; Huang, T. H.; Penichet, M. L.; Morrison, S. L., Antibody-cytokine fusion proteins: innovative weapons in the war against cancer. *Clin Exp Med* 2004, 4 (2), 57–64;
(c) Stern, M.; Herrmann, R., Overview of monoclonal antibodies in cancer therapy: present and promise. *Crit Rev Oncol Hematol* 2005, 54, 11–29.

20. Ribatti, D., From the discovery of monoclonal antibodies to their therapeutic application: an historical reappraisal. *Immunol Lett* 2014, 161 (1), 96–99.

21. (a) Jacquemin, M.; Saint-Remy, J. M., The use of antibodies to coagulation factors for anticoagulant therapy. *Curr Med Chem* 2004, 11, 2291–2296;
(b) Nigam, A.; Kopecky, S. L., Therapeutic potential of monoclonal antibodies in myocardial reperfusion injury. *Am J Cardiovasc Drugs* 2002, 2, 367–376.

22. Kuhnast, S.; van der Hoorn, J. W.; Pieterman, E. J.; van den Hoek, A. M.; Sasiela, W. J.; Gusarova, V.; Peyman, A.; Schafer, H. L.; Schwahn, U.; Jukema, J. W.; Princen, H. M., Alirocumab inhibits atherosclerosis, improves the plaque morphology, and enhances the effects of a statin. *J Lipid Res* 2014, 55 (10), 2103–2112.

23. Casadevall, A.; Dadachova, E.; Pirofski, L. A., Passive antibody therapy for infectious diseases. *Nat Rev Microbiol* 2004, 2 (9), 695–703.

24. Rudicell, R. S.; Kwon, Y. D.; Ko, S. Y.; Pegu, A.; Louder, M. K.; Georgiev, I. S.; Wu, X.; Zhu, J.; Boyington, J. C.; Chen, X.; Shi, W.; Yang, Z. Y.; Doria-Rose, N. A.; McKee, K.; O'Dell, S.; Schmidt, S. D.; Chuang, G. Y.; Druz, A.; Soto, C.; Yang, Y.; Zhang, B.; Zhou, T.; Todd, J. P.; Lloyd, K. E.; Eudailey, J.; Roberts, K. E.; Donald, B. R.; Bailer, R. T.; Ledgerwood, J.; Mullikin, J. C.; Shapiro, L.; Koup, R. A.; Graham, B. S.; Nason, M. C.; Connors, M.; Haynes, B. F.; Rao, S. S.; Roederer, M.; Kwong, P. D.; Mascola, J. R.; Nabel, G. J., Enhanced potency of a broadly neutralizing HIV-1 antibody *in vitro* improves protection against lentiviral infection *in vivo*. *J Virol* 2014, 88 (21), 12669–12682.

25. Moog, C.; Dereuddre-Bosquet, N.; Teillaud, J. L.; Biedma, M. E.; Holl, V.; Van Ham, G.; Heyndrickx, L.; Van Dorsselaer, A.; Katinger, D.; Vcelar, B.; Zolla-Pazner, S.; Mangeot, I.; Kelly, C.; Shattock, R. J.; Le Grand, R., Protective effect of vaginal application of neutralizing and nonneutralizing inhibitory antibodies against vaginal SHIV challenge in macaques. *Mucosal Immunol* 2014, 7 (1), 46–56.

26. Schroff, R. W.; Foon, K. F.; Beatty, S. M.; Oldham, R. K.; Morgan, A. C. J., Human anti-murine immunoglobulin responses in patients receiving monoclonal antibody therapy. *Cancer Res* 1985, 45, 879–885.

27. Morrison, S. L.; Johnson, M. J.; Herzenberg, L. A.; Oi, V. T., Chimeric human antibody molecules mouse antigen-binding domains with human constant region domains. *Proc Natl Acad Sci U S A* 1984, 81, 6851–6855.

28. Jones, P. T.; Dear, P. H.; Foote, J.; Neuberger, M.; Winter, G., Replacing the complementary-determining regions in a human antibody with those from a mouse. *Nature* 1986, 321, 522–525.

29. (a) Roque-Navarro, L.; Mateo, C.; Lombardero, J.; Mustelier, G.; Fernandez, A.; Sosa, K.; Morrison, S. L.; Perez, R., Humanization of predicted T-cell epitopes reduces the immunogenicity of chimeric antibodies: new evidence supporting a simple method. *Hybrid Hybridomics* 2003, 22, 245–257;
(b) Mateo, C.; Lombardero, J.; Moreno, E.; Bombino, G.; Coloma, J.; Wims, L.; Morrison, S. L.; Perez, R., Removal of amphipathic epitopes from genetically engineered antibodies: production of modified immunoglobulins with reduced immunogenicity. *Hybridoma* 2000, 19, 463–471.

30. Yu, C. Y.; Ng, G.; Liao, P., Therapeutic antibodies in stroke. *Transl Stroke Res* 2013, 4 (5), 477–483.

31. Baert, F.; Noman, M.; Vermeire, S.; Van Assche, G.; D'Haens, G.; Carbonez, A.; Rutgeerts, P., Influence of immunogenicity on the long-term efficacy of infliximab in Crohn's disease. *N Engl J Med* 2003, 348 (7), 601–608.

32. (a) Yazaki, P. J.; Sherman, M. A.; Shively, J. E.; Ikle, D.; Williams, L. E.; Wong, J. Y.; Colcher, D.; Wu, A. M.; Raubitschek, A. A., Humanization of the anti-CEA T84.66 antibody based on crystal structure data. *Protein Eng Des Sel* 2004, 17, 481–489;
(b) Riechmann, L.; Clark, H.; Waldmann, H.; Winter, G., Reshaping human antibodies for therapy. *Nature* 1988, 332, 323–327;
(c) Foote, J.; Winter, G., Antibody framework residues affecting the conformation of the hypervariable loops. *J Mol Biol* 1992, 224, 487–499.

33. Tan, P.; Mitchell, D. A.; Buss, T. N.; Holmes, M. A.; Anasetti, C.; Foote, J., "Superhumanized" antibodies: reduction of immunogenic potential by complementarity-determining region grafting with human germline sequences: application to an anti-CD28. *J Immunol* 2002, 169, 1119–1125.

34. Carter, P.; Presta, L.; Gorman, C. M.; Ridgway, J. B. B.; Henner, D.; Wong, W. L. T.; Rowland, A. M.; Kotts, C.; Carver, M. E.; Shepard, H. M., Humanization of an anti-p185HER2 antibody for human cancer therapy. *Proc Natl Acad Sci U S A* 1992, 89, 4285–4289.

35. Mian, I. S.; Bradwell, A. R.; Olson, A. J., Structure, function and properties of antibody binding sites. *J Mol Biol* 1991, 217, 133–151.

36. Padlan, E., Anatomy of the antibody molecule. *Mol Immunol* 1994, 31, 169–217.

37. Padlan, E.; Abergel, C.; Tipper, J. P., Identification of specificity-determining residues in antibodies. *FASEB J* 1995, 9, 133–139.

38. Gonzales, N. R.; Padlan, E.; De Pascalis, R.; Schuck, P.; Schlom, J.; Kashmiri, S. V., SDR grafting of a murine antibody using multiple human germline templates to minimize its immunogenicity. *Mol Immunol* 2004, 41, 863–872.

39. (a) Roguska, M. A.; Pedersen, J. T.; Keddy, C. A.; Henry, A. H.; Searle, S. J.; Lambert, J. M.; Goldmacher, V. S.; Blattler, W. A.; Rees, A. R.; Guild, B. C., Humanization of murine monoclonal antibodies through variable domain resurfacing. *Proc Natl Acad Sci U S A* 1994, 91, 969–973;
(b) Padlan, E., A possible procedure for reducing the immunogenicity of

antibody variable domains while preserving their ligand-binding properties. *Mol Immunol* 1991, 28, 489–498;
(c) Delagrave, S.; Catalan, J.; Sweet, C.; Drabik, G.; Henry, A.; Rees, A.; Monath, T. P.; Guirakhoo, F., Effect of humanization by variable domain resurfacing on the antiviral activity of a single-chain antibody against respiratory syncitial virus. *Protein Eng* 1999, 12, 357–362.

40. Hanf, K. J.; Arndt, J. W.; Chen, L. L.; Jarpe, M.; Boriack-Sjodin, P. A.; Li, Y.; van Vlijmen, H. W.; Pepinsky, R. B.; Simon, K. J.; Lugovskoy, A., Antibody humanization by redesign of complementarity-determining region residues proximate to the acceptor framework. *Methods* 2014, 65 (1), 68–76.

41. Almagro, J. C.; Fransson, J., Humanization of antibodies. *Front Biosci* 2008, 13, 1619–1633.

42. (a) Jespers, L. S.; Roberts, A.; Mahler, S. M.; Winter, G.; Hoogenboom, H. R., Guiding the selection of human antibodies from phage display repertoires to a single epitope of an antigen. *Biotechnology (N Y)* 1994, 12 (9), 899–903;
(b) Baca, M.; Presta, L. G.; O'Connor, S.; Wells, J. A., Antibody humanization using monovalent phage display. *J Biol Chem* 1997, 272, 10678–10684.

43. Zhang, D.; Chen, C. F.; Zhao, B. B.; Gong, L. L.; Jin, W. J.; Liu, J. J.; Wang, J. F.; Wang, T. T.; Yuan, X. H.; He, Y. W., A novel antibody humanization method based on epitopes scanning and molecular dynamics simulation. *PLoS One* 2013, 8 (11), e80636.

44. Nishibori, N.; Horiuchi, H.; Furusawa, S.; Matsuda, H., Humanization of chicken monoclonal antibody using phage-display system. *Mol Immunol* 2006, 43 (6), 634–642.

45. Cobleigh, M. A.; Vogel, C. L.; Tripathy, D.; Robert, N. J.; Scholl, S.; Fehrenbacher, L.; Wolter, J. M.; Paton, V.; Shak, S.; Lieberman, G.; Slamon, D. J., Multinational study of the efficacy and safety of humanized anti-HER2 monoclonal antibody in women who have HER2-overexpressing metastatic breast cancer that has progressed after chemotherapy for metastatic disease. *J Clin Oncol* 1999, 17, 2639–2648.

46. Ritter, G. L.; Cohen, L. S.; Williams, C. J.; Richards, E. C.; Old, L. J.; Welt, S., Serological analysis of human anti-human antibody responses in colon cancer patients treated with repeated doses of humanized monoclonal antibody A33. *Cancer Res* 2001, 61, 6851.

47. Clark, M., Antibody humanization: a case of the "Empror's new clothes"? *Immunol Today* 2000, 21, 397–402.

48. Salfeld, J. G., Isotype selection in antibody engineering. *Nat Biotechnol* 2007, 25 (12), 1369–1372.

49. Galsky, M. D.; Eisenberger, M.; Moore-Cooper, S.; Kelly, W. K.; Slovin, S. F.; DeLaCruz, A.; Lee, Y.; Webb, I. J.; Scher, H. I., Phase I trial of the prostate-specific membrane antigen-directed immunoconjugate MLN2704 in patients with progressive metastatic castration-resistant prostate cancer. *J Clin Oncol* 2008, 26 (13), 2147–2154.

50. Baker, M. P.; Reynolds, H. M.; Lumicisi, B.; Bryson, C. J., Immunogenicity of protein therapeutics: the key causes, consequences and challenges. *Self Nonself* 2010, 1 (4), 314–322.
51. Scott, D. W.; De Groot, A. S., Can we prevent immunogenicity of human protein drugs? *Ann Rheum Dis* 2010, 69 (Suppl 1), i72–i76.
52. Su, Y.; Rossi, R.; De Groot, A. S.; Scott, D. W., Regulatory T cell epitopes (Tregitopes) in IgG induce tolerance *in vivo* and lack immunogenicity *per se. J Leukoc Biol* 2013, 94 (2), 377–383.
53. Karpas, A.; Dremucheva, A.; Czepulkowski, B. H., A human myeloma cell line suitable for the generation of human monoclonal antibodies. *Proc Natl Acad Sci U S A* 2001, 98, 1799–1804.
54. Traggiai, E.; Becker, S.; Subbarao, K.; Kolesnikova, L.; Uematsu, Y.; Gismondi, M. R.; Murphy, B. R.; Rappuoli, R.; Lanzavecchia, A., An efficient method to make human monoclonal antibodies from memory B cells: potent neutralization of SARS coronavirus. *Nat Med* 2004, 10, 871–875.
55. Aryan, Z.; Holgate, S. T.; Radzioch, D.; Rezaei, N., A new era of targeting the ancient gatekeepers of the immune system: toll-like agonists in the treatment of allergic rhinitis and asthma. *Int Arch Allergy Immunol* 2014, 164 (1), 46–63.
56. Lanzavecchia, A.; Corti, D.; Sallusto, F., Human monoclonal antibodies by immortalization of memory B cells. *Curr Opin Biotechnol* 2007, 18 (6), 523–528.
57. Duvall, M.; Bradley, N.; Fiorini, R. N., A novel platform to produce human monoclonal antibodies: the next generation of therapeutic human monoclonal antibodies discovery. *MAbs* 2011, 3 (2), 203–208.
58. Goldman, J. P.; Blundell, M. P.; Lopes, L.; Kinnon, C.; Di Santo, J.; Thrasher, A. J., Enhanced human cell engraftment in mice deficient in RAG2 and the common cytokine receptor γ chain. *Br J Haematol* 1998, 103, 335–342.
59. Sawada-Hirai, R.; Jiang, I.; Wang, F.; Sun, S. M.; Nedellec, R.; Ruther, P.; Alvarez, A.; Millis, D.; Morrow, P. R.; Kang, A. S., Human anti-anthrax protective antigen neutralizing monoclonal antibodies derived from donors vaccinated with anthrax vaccine adsorbed. *J Immune Based Ther Vaccines* 2004, 2, 5.
60. Davis, C. G.; Jia, X. C.; Feng, X.; Haak-Frendscho, M., Production of human antibodies from transgenic mice. *Methods Mol Biol* 2004, 248, 191–200.
61. Kellermann, S.-A.; Green, L., Antibody discovery: the use of transgenic mice to generate human monoclonal antibodies for therapeutics. *Curr Opin Biotechnol* 2002, 13, 593–597.
62. (a) Jakobovits, A., Production of fully human antibodies by transgenic mice. *Curr Opin Biotechnol* 1995, 6, 561–566;
(b) He, Y.; Honnen, W. J.; Krachmarov, C. P.; Burkhart, M.; Kayman, S. C.; Corvalan, J.; Pinter, A., Efficient isolation of novel human monoclonal antibodies with neutralizing activity against HIV-1 from transgenic mice expressing human Ig loci. *J Immunol* 2002, 169, 595–605.

63. (a) Rowinsky, E. K.; Schwartz, G. H.; Gollob, J. A.; Thompson, J. A.; Vogelzang, N. J.; Figlin, R.; Bukowski, R.; Haas, N.; Lockbaum, P.; Li, Y. P.; Arends, R.; Foon, K. A.; Schwab, G.; Dutcher, J., Safety, pharmacokinetics, and activity of ABX-EGF, a fully human anti-epidermal growth factor receptor monoclonal antibody in patients with metastatic renal cell cancer. *J Clin Oncol* 2004, 22 (15), 3003–3015;
(b)Foon, K. A.; Yang, X. D.; Weiner, L. M.; Belldegrun, A. S.; Figlin, R. A.; Crawford, J.; Rowinsky, E. K.; Dutcher, J. P.; Vogelzang, N. J.; Gollub, J.; Thompson, J. A.; Schwartz, G.; Bukowski, R. M.; Roskos, L. K.; Schwab, G. M., Preclinical and clinical evaluations of ABX-EGF, a fully human anti-epidermal growth factor receptor antibody. *Int J Radiat Oncol Biol Phys* 2004, 58 (3), 984–990.

64. Lund, J.; Takahashi, N.; Nakagawa, H.; Goodall, M.; Bentley, T.; Hindley, S. A.; Tyler, R.; Jefferis, R., Control of IgG/Fc glycosylation: a comparison of oligosaccharides from chimeric human/mouse and mouse subclass immunoglobulin Gs. *Mol Immunol* 1993, 30, 741–748.

65. Galili, U.; Rachmilewitz, E. A.; Peleg, A.; Flechner, I., A unique human IgG antibody with anti-α-galactosyl specificity. *J Exp Med* 1984, 160, 1519–1531.

66. Wright, A.; Morrison, S. L., Effect of glycosylation on antibody function: implications for genetic engineering. *Trends Biotechnol* 1997, 15, 26–32.

67. (a) Tomizuka, K.; Shinohara, T.; Yoshida, H.; Uejima, H.; Ohguma, H.; Tanaka, S.; Sato, K.; Oshimura, M.; Ishida, I., Double trans-chromosomic mice: maintenance of two individual human chromosome fragments containing Ig heavy and kappa loci and expression of fully human antibodies. *Proc Natl Acad Sci U S A* 2000, 97, 722–727;
(b) Ishida, I.; Tomizuka, K.; Yoshida, H.; Tahara, T.; Takahashi, N.; Ohguma, A.; Tanaka, S.; Umehashi, M.; Maeda, H.; Nozaki, C.; Halk, E.; Lonberg, N., Production of human monoclonal and polyclonal antibodies in TransChromo animals. *Cloning Stem Cells* 2002, 4, 91–102.

68. Lee, E. C.; Owen, M., The application of transgenic mice for therapeutic antibody discovery. *Methods Mol Biol* 2012, 901, 137–148.

69. Murphy, A. J.; Macdonald, L. E.; Stevens, S.; Karow, M.; Dore, A. T.; Pobursky, K.; Huang, T. T.; Poueymirou, W. T.; Esau, L.; Meola, M.; Mikulka, W.; Krueger, P.; Fairhurst, J.; Valenzuela, D. M.; Papadopoulos, N.; Yancopoulos, G. D., Mice with megabase humanization of their immunoglobulin genes generate antibodies as efficiently as normal mice. *Proc Natl Acad Sci U S A* 2014, 111 (14), 5153–5158.

70. Lee, E. C.; Liang, Q.; Ali, H.; Bayliss, L.; Beasley, A.; Bloomfield-Gerdes, T.; Bonoli, L.; Brown, R.; Campbell, J.; Carpenter, A.; Chalk, S.; Davis, A.; England, N.; Fane-Dremucheva, A.; Franz, B.; Germaschewski, V.; Holmes, H.; Holmes, S.; Kirby, I.; Kosmac, M.; Legent, A.; Lui, H.; Manin, A.; O'Leary, S.; Paterson, J.; Sciarrillo, R.; Speak, A.; Spensberger, D.; Tuffery, L.; Waddell, N.; Wang, W.; Wells, S.; Wong, V.; Wood, A.; Owen, M. J.; Friedrich, G. A.; Bradley, A., Complete humanization of the mouse immunoglobulin loci

enables efficient therapeutic antibody discovery. *Nat Biotechnol* 2014, 32 (4), 356–363.
71. Schusser, B.; Collarini, E. J.; Yi, H.; Izquierdo, S. M.; Fesler, J.; Pedersen, D.; Klasing, K. C.; Kaspers, B.; Harriman, W. D.; van de Lavoir, M. C.; Etches, R. J.; Leighton, P. A., Immunoglobulin knockout chickens via efficient homologous recombination in primordial germ cells. *Proc Natl Acad Sci U S A* 2013, 110 (50), 20170–20175.
72. Schusser, B.; Yi, H.; Collarini, E. J.; Izquierdo, S. M.; Harriman, W. D.; Etches, R. J.; Leighton, P. A., Harnessing gene conversion in chicken B cells to create a human antibody sequence repertoire. *PLoS One* 2013, 8 (11), e80108.
73. Kuroiwa, Y.; Kasinathan, P.; Sathiyaseelan, T.; Jiao, J. A.; Matsushita, H.; Sathiyaseelan, J.; Wu, H.; Mellquist, J.; Hammitt, M.; Koster, J.; Kamoda, S.; Tachibana, K.; Ishida, I.; Robl, J. M., Antigen-specific human polyclonal antibodies from hyperimmunized cattle. *Nat Biotechnol* 2009, 27 (2), 173–181.
74. Matsushita, H.; Sano, A.; Wu, H.; Jiao, J. A.; Kasinathan, P.; Sullivan, E. J.; Wang, Z.; Kuroiwa, Y., Triple immunoglobulin gene knockout transchromosomic cattle: bovine lambda cluster deletion and its effect on fully human polyclonal antibody production. *PLoS One* 2014, 9 (3), e90383.
75. Huls, G. A.; Hienjen, I.; Cuomo, M.; Koningsberger, J.; Wiegman, L.; Boel, E.; van der Vuurst de Vries, A.; Loyson, S.; Helfrich, W.; van Berge Henegouwen, G.; van Meijer, M.; de Kruif, J.; Logtenberg, T., A recombinant, fully human monoclonal antibody with antitumor activity constructed from phage-displayed antibody fragments. *Nat Biotechnol* 1999, 17, 276–281.
76. Smith, G. P., Filamentous fusion phage: novel expression vectors that display cloned antigens on the virion surface. *Science* 1985, 228, 1315–1317.
77. McCafferty, J.; Griffiths, A. D.; Winter, G.; Chiswell, D. J., Phage antibodies: filamentous phage displaying antibody variable domains. *Nature* 1990, 348, 552–554.
78. (a) Marks, J. D.; Hoogenboom, H. R.; Bonnert, T. P.; McCafferty, J.; Griffiths, A. D.; Winter, G., By-passing immunization: human antibodies from V-gene libraries displayed on phage. *J Mol Biol* 1991, 222, 581–597;
(b) Vaughan, T. J.; Williams, A. J.; Pritchard, K.; Osbourn, J. K.; Pope, A. R.; Earnshaw, J. C.; McCafferty, J.; Hodits, R. A.; Wilton, J.; Johnson, K. J., Human antibodies with sub-nanomolar affinities isolated from a large non-immunized phage display library. *Nat Biotechnol* 1996, 14, 309–314.
79. (a) Perelson, A. S.; Oster, G. F., Theoretical studies of clonal selection: minimal antibody repertoire size and reliability of self-non-self discrimination. *J Theor Biol* 1979, 81 (4), 645–670;
(b) Griffiths, A. D.; Williams, S. C.; Hartley, O.; Tomlinson, I. M.; Waterhouse, P.; Crosby, W. L.; Kontermann, R. E.; Jones, P. T.; Low, N. M.; Allison, T. J.; Prospero, T. D.; Hoogenboom, H. R.; Nissim, A.; Cox, J. P. L.; Harrison, J. L.;

Zaccolo, M.; Gherardi, E.; Winter, G., Isolation of high affinity human antibodies directly from large synthetic repertoires. *EMBO J* 1994, 13, 3245–3260.

80. (a) Waterhouse, P.; Griffiths, A. D.; Johnson, K. S.; Winter, G., Combinatorial infection and *in vivo* recombination: a strategy for making large phage antibody repertoires. *Nucleic Acids Res* 1993, 21, 2265–2266;
(b) Geoffroy, F.; Sodoyer, R.; Aujame, L., A new phage display system to construct multicombinatorial libraries of very large antibody repertoires. *Gene* 1994, 151, 109–113.

81. Knappik, A.; Ge, L.; Honegger, A.; Pack, P.; Fischer, M.; Wellnhofer, G.; Hoess, A.; Wolle, J.; Pluckthun, A.; Virnekas, B., Fully synthetic human combinatorial antibody libraries (HuCAL) based on modular consensus frameworks and CDRs randomized with trinucleotides. *J Mol Biol* 2000, 296, 57–86.

82. Carlsson, R.; Soderlind, E., n-CoDeR concept: unique types of antibodies for diagnostic use and therapy. *Expert Rev Mol Diagn* 2001, 1, 102–108.

83. Rothe, C.; Urlinger, S.; Lohning, C.; Prassler, J.; Stark, Y.; Jager, U.; Hubner, B.; Bardroff, M.; Pradel, I.; Boss, M.; Bittlingmaier, R.; Bataa, T.; Frisch, C.; Brocks, B.; Honegger, A.; Urban, M., The human combinatorial antibody library HuCAL GOLD combines diversification of all six CDRs according to the natural immune system with a novel display method for efficient selection of high-affinity antibodies. *J Mol Biol* 2008, 376 (4), 1182–1200.

84. Tiller, T.; Schuster, I.; Deppe, D.; Siegers, K.; Strohner, R.; Herrmann, T.; Berenguer, M.; Poujol, D.; Stehle, J.; Stark, Y.; Hessling, M.; Daubert, D.; Felderer, K.; Kaden, S.; Kolln, J.; Enzelberger, M.; Urlinger, S., A fully synthetic human Fab antibody library based on fixed VH/VL framework pairings with favorable biophysical properties. *MAbs* 2013, 5 (3), 445–470.

85. Ren, Z.; Black, L. W., Phage T4 SOC and HOC display of biologically active, full-length proteins on the viral capsid. *Gene* 1998, 215, 439–444.

86. Huse, W. D.; Stinchcombe, T. J.; Glaser, S. M.; Starr, L.; MacLean, M.; Hellstrom, K. E.; Hellstrom, I.; Yelton, D. E., Application of a filamentous phage pVIII fusion protein system suitable for efficient production, screening and mutagenesis of F(ab) antibody fragments. *J Immunol* 1992, 149, 3914–3920.

87. Gao, C.; Mao, S.; Kaufmann, G.; Wirshing, P.; Lerner, R. A.; Janda, K. D., A method for the generation of combinatorial antibody libraries using pIX phage display. *Proc Natl Acad Sci U S A* 2002, 99, 12612–12616.

88. Jespers, L. S.; Messens, J. H.; De Keyser, A.; Eeckhout, D.; Van den Brande, I.; Gansemans, Y. G.; Lauwereys, M. J.; Vlasuk, G. P.; Stanssens, P. E., Surface expression and ligand-based selection of cDNAs fused to filamentous phage gene VI. *Biotechnology (N Y)* 1995, 13, 378–382.

89. Sidhu, S. S., Engineering M13 for phage display. *Biomol Eng* 2001, 18, 57–63.

90. Krebber, C.; Spada, S.; Desplancq, D.; Plückthun, A., Co-selection of cognate antibody-antigen pairs by selectively infective phages. *FEBS Lett* 1995, 377, 227–231.
91. Duenas, M.; Borrebaeck, C. A., Clonal selection and amplification of phage displayed antibodies by linking antigen recognition and phage replication. *Biotechnology (N Y)* 1994, 12, 999–1002.
92. Gramatikoff, K.; Georgiev, O.; Schaffner, W., Direct interaction rescue, a novel filamentous phage technique to study protein-protein interactions. *Nucleic Acids Res* 1994, 22, 5761–5762.
93. Jung, S.; Arndt, K. M.; Muller, K. M.; Pluckthun, A., Selectively infective phage (SIP) technology: scope and limitations. *J Immunol Methods* 1999, 9, 93–104.
94. Krebber, C.; Spada, S.; Desplancq, D.; Krebber, A.; Ge, L.; Plückthun, A., Selectively-infective phage (SIP): a mechanistic dissection of a novel *in vivo* selection for protein-ligand interactions. *J Mol Biol* 1997, 268, 607–618.
95. Lee, C. V.; Sidhu, S. S.; Fuh, G., Bivalent antibody phage display mimics natural immunoglobulin. *J Immunol Methods* 2004, 284, 119–132.
96. Samuelson, P.; Gunneriusson, E.; Nygren, P.-A.; Stahl, S., Display of proteins on bacteria. *J Biotechnol* 2002, 96, 129–154.
97. Van Deventer, J. A.; Wittrup, K. D., Yeast surface display for antibody isolation: library construction, library screening, and affinity maturation. *Methods Mol Biol* 2014, 1131, 151–181.
98. Cano, F.; Plotnicky-Gilquin, H.; Nguyen, T. N.; Liljeqvist, S.; Samuelson, P.; Bonnefoy, J. Y.; Stahl, S.; Robert, A., Partial protection to respiratory syncytial virus (RSV) elicited in mice by intranasal immunization using live staphylococci with surface-displayed RSV-peptides. *Vaccine* 2000, 18, 2743–2752.
99. Latteman, C. T.; Maurer, J.; Gerland, E.; Meyer, T. F., Autodisplay: functional display of active beta-lactamase on the surface of *Escherichia coli* by the AIDA-I autotransporter. *J Bacteriol* 2000, 182, 3726–3733.
100. Feldhaus, M.; Siegel, R., Flow cytometric screening of yeast surface display libraries. *Methods Mol Biol* 2004, 263, 311–332.
101. Harvey, B. R.; Georgiou, G.; Hayhurst, A.; Jeong, K. J.; Iverson, B. L.; Rogers, G. K., Anchored periplasmic expression, a versatile technology for the isolation of high-affinity antibodies from *Escherichia coli*-expressed libraries. *Proc Natl Acad Sci U S A* 2004, 101, 9193–9198.
102. Mottershead, D. G.; Alfthan, K.; Ojala, K.; Takkinen, K.; Oker-Blom, C., Baculoviral display of functional scFv and synthetic IgG-binding domains. *Biochem Biophys Res Commun* 2000, 275, 84–90.
103. Groves, M. A.; Osbourn, J. K., Applications of ribosome display to antibody drug discovery. *Expert Opin Biol Ther* 2005, 5, 125–135.
104. Hanes, J.; Jermutus, L.; Pluckthun, A., Selecting and evolving functional proteins *in vitro* by ribosome display. *Adv Protein Chem* 2000, 328, 404–430.

105. (a) Hanes, J.; Schaffitzel, C.; Knappik, A.; Pluckthun, A., Picomolar affinity antibodies from a fully synthetic naïve library selected and evolved by ribosome display. *Nat Biotechnol* 2000, 18, 1287–1292;
(b) Jermutus, L.; Honegger, A.; Schwesinger, F.; Hanes, J.; Pluckthun, A., Tailoring *in vitro* evolution for protein affinity or stability. *Proc Natl Acad Sci U S A* 2001, 98, 75–80.

106. Sawata, S. Y.; Taira, K., Development of an advanced polysome display system dependent on a specific protein-RNA motif interaction. *Nucleic Acids Res* 2001 (1), 99–100.

107. Reiersen, H.; Lobersli, I.; Loset, G. A.; Hvattum, E.; Simonsen, B.; Stacy, J. E.; McGregor, D.; Fitzgerald, K.; Welschof, M.; Brekke, O. H.; Marvik, O. J., Covalent antibody display- an *in vitro* antibody DNA library selection system. *Nucleic Acids Res* 2005, 33, e10.

108. Sepp, A.; Griffiths, A., Cell-free selection of domain antibodies by *in vitro* compartmentalization. *Methods Mol Biol* 2012, 911, 183–198.

109. Collarini, E. J.; Lee, F. E.; Foord, O.; Park, M.; Sperinde, G.; Wu, H.; Harriman, W. D.; Carroll, S. F.; Ellsworth, S. L.; Anderson, L. J.; Tripp, R. A.; Walsh, E. E.; Keyt, B. A.; Kauvar, L. M., Potent high-affinity antibodies for treatment and prophylaxis of respiratory syncytial virus derived from B cells of infected patients. *J Immunol* 2009, 183 (10), 6338–6345.

110. McCutcheon, K. M.; Gray, J.; Chen, N. Y.; Liu, K.; Park, M.; Ellsworth, S.; Tripp, R. A.; Tompkins, S. M.; Johnson, S. K.; Samet, S.; Pereira, L.; Kauvar, L. M., Multiplexed screening of natural humoral immunity identifies antibodies at fine specificity for complex and dynamic viral targets. *MAbs* 2014, 6 (2), 460–473.

111. Zwick, M. B.; Gach, J. S.; Burton, D. R., A welcome burst of human antibodies. *Nat Biotechnol* 2008, 26 (8), 886–887.

112. Reddy, S. T.; Ge, X.; Miklos, A. E.; Hughes, R. A.; Kang, S. H.; Hoi, K. H.; Chrysostomou, C.; Hunicke-Smith, S. P.; Iverson, B. L.; Tucker, P. W.; Ellington, A. D.; Georgiou, G., Monoclonal antibodies isolated without screening by analyzing the variable-gene repertoire of plasma cells. *Nat Biotechnol* 2010, 28 (9), 965–969.

113. Saggy, I.; Wine, Y.; Shefet-Carasso, L.; Nahary, L.; Georgiou, G.; Benhar, I., Antibody isolation from immunized animals: comparison of phage display and antibody discovery via V gene repertoire mining. *Protein Eng Des Sel* 2012, 25 (10), 539–549.

114. Kodangattil, S.; Huard, C.; Ross, C.; Li, J.; Gao, H.; Mascioni, A.; Hodawadekar, S.; Naik, S.; Min-debartolo, J.; Visintin, A.; Almagro, J. C., The functional repertoire of rabbit antibodies and antibody discovery via next-generation sequencing. *MAbs* 2014, 6 (3), 628–636.

115. Georgiou, G.; Ippolito, G. C.; Beausang, J.; Busse, C. E.; Wardemann, H.; Quake, S. R., The promise and challenge of high-throughput sequencing of the antibody repertoire. *Nat Biotechnol* 2014, 32 (2), 158–168.

116. Aalberse, R. C.; van der Gaag, R.; van Leeuwen, J., Serologic aspects of IgG4 antibodies. I. Prolonged immunization results in an IgG4-restricted response. *J Immunol* 1983, 130 (2), 722–726.

117. Aalberse, R. C.; Schuurman, J., IgG4 breaking the rules. *Immunology* 2002, 105 (1), 9–19.

118. Rispens, T.; Davies, A. M.; Ooijevaar-de Heer, P.; Absalah, S.; Bende, O.; Sutton, B. J.; Vidarsson, G.; Aalberse, R. C., Dynamics of inter-heavy chain interactions in human immunoglobulin G (IgG) subclasses studied by kinetic Fab arm exchange. *J Biol Chem* 2014, 289 (9), 6098–6109.

119. van der Neut Kolfschoten, M.; Schuurman, J.; Losen, M.; Bleeker, W. K.; Martinez-Martinez, P.; Vermeulen, E.; den Bleker, T. H.; Wiegman, L.; Vink, T.; Aarden, L. A.; De Baets, M. H.; van de Winkel, J. G.; Aalberse, R. C.; Parren, P. W., Anti-inflammatory activity of human IgG4 antibodies by dynamic Fab arm exchange. *Science* 2007, 317 (5844), 1554–1557.

120. Iliades, P.; Kortt, A. A.; Hudson, P. J., Triabodies: single chain Fv fragments without a linker form trivalent trimers. *FEBS Lett* 1997, 409, 437–441.

121. Todorovska, A.; Roovers, R. C.; Dolezal, O.; Kortt, A. A.; Hoogenboom, H. R.; Hudson, P. J., Design and application of diabodies, triabodies and tetrabodies for cancer targeting. *J Immunol Methods* 2001, 248, 47–66.

122. (a) Brinkmann, U.; Reiter, Y.; Jung, S. H.; Lee, B.; Pastan, I., A recombinant immunotoxin containing a disulfide-stabilized fv fragment. *Proc Natl Acad Sci U S A* 1993, 90, 7538–7542;
(b) Reiter, Y.; Brinkmann, U.; Lee, B.; Pastan, I., Engineering antibody Fv fragments for cancer detection and therapy: disulfide-stabilized Fv fragments. *Nat Biotechnol* 1996, 14, 1239–1245.

123. Ridgway, J. B.; Presta, L. G.; Carter, P., "Knobs-into-holes" engineering of antibody CH3 domains for heavy chain heterodimerization. *Protein Eng* 1996, 9, 617–621.

124. Petrov, K.; Dion, M.; Hoffmann, L.; Dintinger, T.; Defontaine, A.; Tellier, C., Bivalent Fv antibody fragments obtained by substituting the constant domains of Fab fragment with heterotetrameric molybdopterin synthase. *J Mol Biol* 2004, 341, 1039–1048.

125. Chapman, A. P.; Antoniw, P.; Spitali, M.; West, S.; Stephen, S.; King, D. J., Therapeutic antibody fragments with prolonged *in vivo* half-lives. *Nat Biotechnol* 1999, 17, 780–783.

126. Chapman, A. P., PEGylated antibodies and antibody fragments for improved therapy: a review. *Adv Drug Deliv Rev* 2002, 54, 531–545.

127. Kitamura, K.; Takahashi, T.; Takashina, K.; Yamaguchi, T.; Noguchi, A.; Tsurumi, H.; Toyokuni, T.; Hakomori, S., Polyethylene glycol modification of the monoclonal antibody A7 enhances its tumor localization. *Biochem Biophys Res Commun* 1990, 171 (3), 1387–1394.

128. Schlapschy, M.; Binder, U.; Borger, C.; Theobald, I.; Wachinger, K.; Kisling, S.; Haller, D.; Skerra, A., PASylation: a biological alternative to PEGylation for extending the plasma half-life of pharmaceutically active proteins. *Protein Eng Des Sel* 2013, 26 (8), 489–501.

129. Suresh, M. R.; Cuello, A. C.; Milstein, C., Bispecific monoclonal antibodies from hybrid hybridomas. *Methods Enzymol* 1986, 121, 210–228.

130. de Kruif, J.; Logtenberg, T., Leucine Zipper dimerized bivalent and bispecific scFv antibodies from a semi-synthetic antibody phage display library. *J Biol Chem* 1996, 271, 7630–7634.

131. Hu, S.; Shively, L.; Raubitschek, A.; Sherman, M.; Williams, L. E.; Wong, J. Y.; Shively, J. E.; Wu, A. M., Minibody: a novel engineered anti-carcinoembryonic antigen antibody fragment (single-chain Fv-CH3) which exhibits rapid, high level targeting of xenografts. *Cancer Res* 1996, 56, 3055–3061.

132. Xie, Z.; Guo, N.; Yu, M.; Hu, M.; Shen, B., A new format of bispecific antibody: highly efficient heterodimerization, expression and tumor cell lysis. *J Immunol Methods* 2005, 296, 95–101.

133. Deyev, S. M.; Waibel, R.; Lebedenko, E. N.; Shubiger, A. P.; Pluckthun, A., Design of multivalent complexes using the barnase*barstar module. *Nat Biotechnol* 2003, 21, 1486–1492.

134. Wolf, E.; Hofmeister, R.; Kufer, P.; Schlereth, B.; Baeuerle, P. A., BiTEs: bispecific antibody constructs with unique anti-tumor activity. *Drug Discov Today* 2005, 10 (18), 1237–1244.

135. Walter, R. B., Biting back: BiTE antibodies as a promising therapy for acute myeloid leukemia. *Expert Rev Hematol* 2014, 7 (3), 317–319.

136. DiGiammarino, E.; Ghayur, T.; Liu, J., Design and generation of DVD-Ig molecules for dual-specific targeting. *Methods Mol Biol* 2012, 899, 145–156.

137. Gu, J.; Yang, J.; Chang, Q.; Lu, X.; Wang, J.; Chen, M.; Ghayur, T.; Gu, J., Identification of anti-ErbB2 dual variable domain immunoglobulin (DVD-Ig) proteins with unique activities. *PLoS One* 2014, 9 (5), e97292.

138. Wozniak-Knopp, G.; Bartl, S.; Bauer, A.; Mostageer, M.; Woisetschlager, M.; Antes, B.; Ettl, K.; Kainer, M.; Weberhofer, G.; Wiederkum, S.; Himmler, G.; Mudde, G. C.; Ruker, F., Introducing antigen-binding sites in structural loops of immunoglobulin constant domains: Fc fragments with engineered HER2/neu-binding sites and antibody properties. *Protein Eng Des Sel* 2010, 23 (4), 289–297.

139. Woisetschlager, M.; Antes, B.; Borrowdale, R.; Wiederkum, S.; Kainer, M.; Steinkellner, H.; Wozniak-Knopp, G.; Moulder, K.; Ruker, F.; Mudde, G. C., *In vivo* and *in vitro* activity of an immunoglobulin Fc fragment (Fcab) with engineered Her-2/neu binding sites. *Biotechnol J* 2014, 9 (6), 844–851.

140. Strohlein, M. A.; Heiss, M. M., The trifunctional antibody catumaxomab in treatment of malignant ascites and peritoneal carcinomatosis. *Future Oncol* 2010, 6 (9), 1387–1394.

141. Pilanci, K. N.; Ordu, C.; Akpinar, H.; Balci, C.; Bassulu, N.; Koksal, U. I.; Elbuken, F.; Okutur, K.; Bulbul, G.; Saglam, S.; Demir, G., Dramatic response

to catumaxomab treatment for malign ascites related to renal cell carcinoma with sarcomotoid differentiation. *Am J Ther* 2014.

142. Doppalapudi, V. R.; Huang, J.; Liu, D.; Jin, P.; Liu, B.; Li, L.; Desharnais, J.; Hagen, C.; Levin, N. J.; Shields, M. J.; Parish, M.; Murphy, R. E.; Del Rosario, J.; Oates, B. D.; Lai, J. Y.; Matin, M. J.; Ainekulu, Z.; Bhat, A.; Bradshaw, C. W.; Woodnutt, G.; Lerner, R. A.; Lappe, R. W., Chemical generation of bispecific antibodies. *Proc Natl Acad Sci U S A* 2010, 107 (52), 22611–22616.
143. Reiter, Y.; Pastan, I., Recombinant Fv immunotoxins and Fv fragments as novel agents for cancer therapy and diagnosis. *Trends Biotechnol* 1998, 16, 513–520.
144. Beck, A.; Reichert, J. M., Antibody-drug conjugates: present and future. *MAbs* 2014, 6 (1), 15–17.
145. Corrigan, P. A.; Cicci, T. A.; Auten, J. J.; Lowe, D. K., Ado-trastuzumab emtansine: a HER2-positive targeted antibody-drug conjugate. *Ann Pharmacother* 2014, 48 (11), 1484–1493.
146. Newland, A. M.; Li, J. X.; Wasco, L. E.; Aziz, M. T.; Lowe, D. K., Brentuximab vedotin: a CD30-directed antibody-cytotoxic drug conjugate. *Pharmacotherapy* 2013, 33 (1), 93–104.
147. Hemmerle, T.; Doll, F.; Neri, D., Antibody-based delivery of IL4 to the neovasculature cures mice with arthritis. *Proc Natl Acad Sci U S A* 2014, 111 (33), 12008–12012.
148. Hess, C.; Neri, D., Evaluation of antibody-chemokine fusion proteins for tumor-targeting applications. *Exp Biol Med (Maywood)* 2014, 239 (7), 842–852.
149. Xu, G.; McLeod, H. L., Strategies for enzyme/prodrug cancer therapy. *Clin Cancer Res* 2001, 7, 3314–3324.
150. Sharma, S. K.; Pedley, R. B.; Bhatia, J.; Boxer, G. M.; El-Emir, E.; Qureshi, U.; Tolner, B.; Lowe, H.; Michael, N. P.; Minton, N.; Begent, R. H.; Chester, K. A., Sustained tumor regression of human colorectal cancer xenografts using a multifunctional mannosylated fusion protein in antibody-directed enzyme prodrug therapy. *Clin Cancer Res* 2005, 11, 814–825.
151. (a) Henry, M. D.; Wen, S.; Silva, M. D.; Chandra, S.; Milton, M.; Worland, P. J., A prostate-specific membrane antigen-targeted monoclonal antibody-chemotherapeutic conjugate designed for the treatment of prostate cancer. *Cancer Res* 2004, 64, 7995–8001;
(b) Arditti, F. D.; Rabinkov, A.; Miron, T.; Reisner, Y.; Berrebi, A.; Wilchek, M.; Mirelman, D., Apoptotic killing of B-chronic lymphocytic leukemia tumor cells by allicin generated in situ using a rituximab-alliinase conjugate. *Mol Cancer Ther* 2005, 4 (2), 325–331;
(c) Law, C. L.; Cerveny, C.; Gordon, K. A.; Klussman, K.; Mixan, B. J.; Chace, D. F.; Meyer, D. L.; Doronina, S. O.; Siegall, C. B.; Francisco, J. A.; Senter, P. D.; Wahl, A. F., Efficient elimination of B-lineage lymphomas by anti-CD20-auristatin conjugates. *Clin Cancer Res* 2004, 10, 7842–7851.

152. Krauss, J.; Arndt, K. M.; Vu, B. K.; Newton, D. L.; Seeber, S.; Rybak, S. M., Efficient killing of CD22+ tumor cells by a humanized diabody-RNase fusion protein. *Biochem Biophys Res Commun* 2005, 331, 595–602.
153. Francis, R. J.; Sharma, S. K.; Springer, C.; Green, A. J.; Hope-Stone, L. D.; Sena, L.; Martin, J.; Adamson, K. L.; Robbins, A.; Gumbrell, L.; O'Malley, D.; Tsiompanou, E.; Shahbakhti, H.; Webley, S.; Hochhauser, D.; Hilson, A. J.; Blakey, D.; Begent, R. H., A phase I trial of antibody directed enzyme prodrug therapy (ADEPT) in patients with advanced colorectal carcinoma or other CEA producing tumours. *Br J Cancer* 2002, 87, 600–607.
154. Schellmann, N.; Deckert, P. M.; Bachran, D.; Fuchs, H.; Bachran, C., Targeted enzyme prodrug therapies. *Mini Rev Med Chem* 2010, 10 (10), 887–904.
155. Tietze, L. F.; Schmuck, K., Prodrugs for targeted tumor therapies: recent developments in ADEPT, GDEPT and PMT. *Curr Pharm Des* 2011, 17 (32), 3527–3547.
156. Lunde, E.; Lauvrak, V.; Rasmussen, I. B.; Schjetne, K. W.; Thompson, K. M.; Michaelsen, T. E.; Brekke, O. H.; Sollid, L. M.; Bogen, B.; Sandlie, I., Troybodies and pepbodies. *Biochem Soc Trans* 2002, 30, 500–506.
157. Hudson, P. J.; Souriau, C., Engineered antibodies. *Nat Med* 2003, 9, 129–134.
158. Hamers-Casterman, C.; Atarhouch, T.; Muyldermans, S.; Robinson, G.; Hamers, G.; Songa, E. B.; Bendahman, N.; Hamers, R., Naturally occurring antibodies devoid of light chains. *Nature* 1993, 363, 446–448.
159. Holt, L. J.; Herring, C.; Jespers, L. S.; Woolven, B. P.; Tomlinson, I. M., Domain antibodies: proteins for therapy. *Trends Biotechnol* 2003, 21, 484–490.
160. Vincke, C.; Muyldermans, S., Introduction to heavy chain antibodies and derived Nanobodies. *Methods Mol Biol* 2012, 911, 15–26.
161. (a) Nuttall, S. D.; Krishnan, U. V.; Hattarki, M.; De Gori, R.; Irving, R. A.; Hudson, P. J., Isolation of the new antigen receptor from wobbegong sharks, and use as a scaffold for the display of protein loop libraries. *Mol Immunol* 2001, 38, 313–326;
(b) Nuttall, S. D.; Krishnan, U. V.; Doughty, L.; Pearson, K.; Ryan, M. T.; Hoogenraad, N. J.; Hattarki, M.; Carmichael, J. A.; Irving, R. A.; Hudson, P. J., Isolation and characterization of an IgNAR variable domain specific for the human mitochondrial translocase receptor Tom70. *Eur J Biochem* 2003, 270, 3543–3554.
162. Nuttall, S. D.; Humberstone, K. S.; Krishnan, U. V.; Carmichael, J. A.; Doughty, L.; Hattarki, M.; Coley, A. M.; Casey, J. L.; Anders, R. F.; Foley, M.; Irving, R. A.; Hudson, P. J., Selection and affinity maturation of IgNAR variable domains targeting *Plasmodium falciparum* AMA1. *Proteins* 2004, 55, 187–197.
163. Johansson, J.; Aveskogh, M.; Munday, B.; Hellman, L., Heavy chain V region diversity in the duck-billed platypus (*Ornithorhynchus anatinus*): long and

highly variable complementarity-determining region 3 compensates for limited germline diversity. *J Immunol* 2002, 168 (10), 5155–5162.

164. Saini, S. S.; Farrugia, W.; Ramsland, P. A.; Kaushik, A. K., Bovine IgM antibodies with exceptionally long complementarity-determining region 3 of the heavy chain share unique structural properties conferring restricted VH + Vlambda pairings. *Int Immunol* 2003, 15 (7), 845–853.
165. Koti, M.; Kataeva, G.; Kaushik, A. K., Novel atypical nucleotide insertions specifically at VH-DH junction generate exceptionally long CDR3H in cattle antibodies. *Mol Immunol* 2010, 47 (11–12), 2119–2128.
166. Sun, Y.; Wei, Z.; Li, N.; Zhao, Y., A comparative overview of immunoglobulin genes and the generation of their diversity in tetrapods. *Dev Comp Immunol* 2013, 39 (1–2), 103–109.
167. Boersma, Y. L.; Pluckthun, A., DARPins and other repeat protein scaffolds: advances in engineering and applications. *Curr Opin Biotechnol* 2011, 22 (6), 849–857.
168. Herrin, B. R.; Alder, M. N.; Roux, K. H.; Sina, C.; Ehrhardt, G. R.; Boydston, J. A.; Turnbough, C. L., Jr.; Cooper, M. D., Structure and specificity of lamprey monoclonal antibodies. *Proc Natl Acad Sci U S A* 2008, 105 (6), 2040–2045.
169. Pancer, Z.; Mariuzza, R. A., The oldest antibodies newly discovered. *Nat Biotechnol* 2008, 26 (4), 402–403.
170. Kohl, A.; Binz, H. K.; Forrer, P.; Stumpp, M. T.; Plückthun, A.; Grütter, M. G., Designed to be stable: crystal structure of a consensus ankyrin repeat protein. *Proc Natl Acad Sci U S A* 2003, 100, 1700–1705.
171. Binz, H. K.; Stumpp, M. T.; Forrer, P.; Amstutz, H.; Pluckthun, A., Designing repeat proteins: well-expressed, soluble and stable proteins from combinatorial libraries of consensus ankyrin repeat proteins. *J Mol Biol* 2003, 332, 489–503.
172. Amstutz, P.; Binz, H. K.; Parizek, P.; Stumpp, M. T.; Kohl, A.; Grutter, G.; Forrer, P.; Pluckthun, A., Intracellular kinase inhibitors selected from combinatorial libraries of designed ankyrin repeat proteins. *J Biol Chem* 2005, 280, 24715–24722.
173. Hanenberg, M.; McAfoose, J.; Kulic, L.; Welt, T.; Wirth, F.; Parizek, P.; Strobel, L.; Cattepoel, S.; Spani, C.; Derungs, R.; Maier, M.; Pluckthun, A.; Nitsch, R. M., Abeta-specific DARPins as a novel class of potential therapeutics for Alzheimer's disease. *J Biol Chem* 2014, 289 (39), 27080–27089.
174. (a) Schlehuber, S.; Skerra, A., Tuning ligand affinity, specificity, and folding stability of an engineered lipocalin variant - a so-called "anticalin" - using a molecular random approach. *Biophys Chem* 2002, 96, 213–228;
(b) Schlehuber, S.; Skerra, A., Lipocalins in drug discovery: from natural ligand-binding proteins to "anticalins". *Drug Discov Today* 2005, 10 (1), 23–33.

175. Schlehuber, S.; Skerra, A., Duocalins: engineered ligand-binding proteins with dual specificity derived from the lipocalin fold. *Biol Chem* 2001, 382, 1335–1342.

176. Mross, K.; Richly, H.; Fischer, R.; Scharr, D.; Buchert, M.; Stern, A.; Gille, H.; Audoly, L. P.; Scheulen, M. E., First-in-human phase I study of PRS-050 (Angiocal), an Anticalin targeting and antagonizing VEGF-A, in patients with advanced solid tumors. *PLoS One* 2013, 8 (12), e83232.

177. Silverman, J.; Liu, Q.; Bakker, A.; To, W.; Duguay, A.; Alba, B. M.; Smith, R.; Rivas, A.; Li, P.; Le, H.; Whitehorn, E.; Moore, K. W.; Swimmer, C.; Perlroth, V.; Vogt, M.; Kolkman, J.; Stemmer, W. P., Multivalent avimer proteins evolved by exon shuffling of a family of human receptor domains. *Nat Biotechnol* 2005, 23 (12), 1556–1561.

178. Lipovsek, D., Adnectins: engineered target-binding protein therapeutics. *Protein Eng Des Sel* 2011, 24 (1–2), 3–9.

179. Mitchell, T.; Chao, G.; Sitkoff, D.; Lo, F.; Monshizadegan, H.; Meyers, D.; Low, S.; Russo, K.; DiBella, R.; Denhez, F.; Gao, M.; Myers, J.; Duke, G.; Witmer, M.; Miao, B.; Ho, S. P.; Khan, J.; Parker, R. A., Pharmacologic profile of the Adnectin BMS-962476, a small protein biologic alternative to PCSK9 antibodies for low-density lipoprotein lowering. *J Pharmacol Exp Ther* 2014, 350 (2), 412–424.

180. Feldwisch, J.; Tolmachev, V., Engineering of affibody molecules for therapy and diagnostics. *Methods Mol Biol* 2012, 899, 103–126.

181. Schlatter, D.; Brack, S.; Banner, D. W.; Batey, S.; Benz, J.; Bertschinger, J.; Huber, W.; Joseph, C.; Rufer, A.; van der Klooster, A.; Weber, M.; Grabulovski, D.; Hennig, M., Generation, characterization and structural data of chymase binding proteins based on the human Fyn kinase SH3 domain. *MAbs* 2012, 4 (4), 497–508.

182. Banner, D. W.; Gsell, B.; Benz, J.; Bertschinger, J.; Burger, D.; Brack, S.; Cuppuleri, S.; Debulpaep, M.; Gast, A.; Grabulovski, D.; Hennig, M.; Hilpert, H.; Huber, W.; Kuglstatter, A.; Kusznir, E.; Laeremans, T.; Matile, H.; Miscenic, C.; Rufer, A. C.; Schlatter, D.; Steyaert, J.; Stihle, M.; Thoma, R.; Weber, M.; Ruf, A., Mapping the conformational space accessible to BACE2 using surface mutants and cocrystals with Fab fragments, Fynomers and Xaperones. Acta crystallographica. Section D. *Acta Crystallogr D Biol Crystallogr* 2013, 69 (Pt 6), 1124–1137.

183. Arcus, V., OB-fold domains: a snapshot of the evolution of sequence, structure and function. *Curr Opin Struct Biol* 2002, 12 (6), 794–801.

184. Steemson, J. D.; Baake, M.; Rakonjac, J.; Arcus, V. L.; Liddament, M. T., Tracking molecular recognition at the atomic level with a new protein scaffold based on the OB-fold. *PLoS One* 2014, 9 (1), e86050.

185. Hosse, R. J.; Rothe, A.; Power, B. E., A new generation of protein display scaffolds for molecular recognition. *Protein Sci* 2006, 15 (1), 14–27.

186. Williams, A.; Baird, L. G., DX-88 and HAE: a developmental perspective. *Transfus Apher Sci* 2003, 29 (3), 255–258.

187. Simmons, L. C.; Reilly, D.; Klimowski, L.; Raju, T. S.; Meng, G.; Sims, P.; Hong, K.; Shields, R. L.; Damico, L. A.; Rancatore, P.; Yansura, D. G., Expression of full-length immunoglobulins in *Escherichia coli*: rapid and efficient production of aglycosylated antibodies. *J Immunol Methods* 2002, 263, 133–147.

188. de Marco, A., Biotechnological applications of recombinant single-domain antibody fragments. *Microb Cell Fact* 2011, 10, 44.

189. Nelson, A. L., Antibody fragments: hope and hype. *MAbs* 2010, 2 (1), 77–83.

190. (a) Popkov, M.; Jendreyko, N.; McGavern, D. B.; Rader, C.; Barbas, C. F. I., Targeting tumor angiogenesis with adenovirus-delivered anti-Tie2 intrabody. *Cancer Res* 2005, 65, 972–981;
(b) Stocks, M., Intrabodies as drug discovery tools and therapeutics. *Curr Opin Chem Biol* 2005, 9 (4), 359–365.

191. Auf der Maur, A.; Zahnd, C.; Fischer, F.; Spinelli, S.; Honegger, A.; Cambillau, C.; Escher, D.; Pluckthun, A.; Barberis, A., Direct *in vivo* screening of intrabody libraries constructed on a highly stable single-chain framework. *J Biol Chem* 2002, 277, 45075–45085.

192. Tanaka, T.; Lobato, M. N.; Rabbitts, T. H., Single domain intracellular antibodies: a minimal fragment for direct *in vivo* selection of antigen-specific intrabodies. *J Mol Biol* 2003, 331, 1109–1120.

193. Jendreyko, N.; Popkov, M.; Rader, C.; Barbas, C. F., 3rd, Phenotypic knockout of VEGF-R2 and Tie-2 with an intradiabody reduces tumor growth and angiogenesis *in vivo*. *Proc Natl Acad Sci U S A* 2005, 102 (23), 8293–8298.

194. Auf der Maur, A.; Tissot, K.; Barberis, A., Antigen-independent selection of intracellular stable antibody frameworks. *Methods* 2004, 34, 215–224.

195. Visintin, M.; Quondam, M.; Cattaneo, A., The intracellular antibody capture technology: towards the high-throughput selection of functional intracellular antibodies for target validation. *Methods* 2004, 34 (2), 200–214.

196. Lo, A. S.; Zhu, Q.; Marasco, W. A., Intracellular antibodies (intrabodies) and their therapeutic potential. *Handb Exp Pharmacol* 2008 (181), 343–373.

197. Cardinale, A.; Merlo, D.; Giunchedi, P.; Biocca, S., Therapeutic application of intrabodies against age-related neurodegenerative disorders. *Curr Pharm Des* 2014, 20 (38), 6028–6036.

198. Heng, B. C.; Cao, T., Making cell-permeable antibodies (Transbody) through fusion transduction domains (PTD) with single chain variable fragment (scFv) antibodies: potential advantages over antibodies expressed within the intracellular environment (Intrabody). *Med Hypotheses* 2005, 64, 1105–1108.

199. (a) Bendandi, M., The role of idiotype vaccines in the treatment of human B-cell malignancies. *Expert Rev Vaccines* 2004, 3, 163–170;
(b) Bhattacharya-Chatterjee, M.; Chatterjee, S. K.; Foon, K. A., Anti-idiotype vaccine against cancer. *Immunol Lett* 2000, 74, 51–58.

200. Fields, B. A.; Goldbaum, F. A.; Ysern, X.; Poljak, R. J.; Mariuzza, R. A., Molecular basis of antigen mimicry by an anti-idiotope. *Nature* 1995, 374, 739–742.
201. Brito, V.; Mellal, K.; Portelance, S. G.; Perez, A.; Soto, Y.; deBlois, D.; Ong, H.; Marleau, S.; Vazquez, A. M., Induction of anti-anti-idiotype antibodies against sulfated glycosaminoglycans reduces atherosclerosis in apolipoprotein E-deficient mice. *Arterioscler Thromb Vasc Biol* 2012, 32 (12), 2847–2854.
202. Mader, A.; Kunert, R., Humanization strategies for an anti-idiotypic antibody mimicking HIV-1 gp41. *Protein Eng Des Sel* 2010, 23 (12), 947–954.
203. Gajdosik, Z., Racotumomab—a novel anti-idiotype monoclonal antibody vaccine for the treatment of cancer. *Drugs Today (Barc)* 2014, 50 (4), 301–307.
204. Alfonso, S.; Valdes-Zayas, A.; Santiesteban, E. R.; Flores, Y. I.; Areces, F.; Hernandez, M.; Viada, C. E.; Mendoza, I. C.; Guerra, P. P.; Garcia, E.; Ortiz, R. A.; de la Torre, A. V.; Cepeda, M.; Perez, K.; Chong, E.; Hernandez, A. M.; Toledo, D.; Gonzalez, Z.; Mazorra, Z.; Crombet, T.; Perez, R.; Vazquez, A. M.; Macias, A. E., A randomized, multicenter, placebo-controlled clinical trial of racotumomab-alum vaccine as switch maintenance therapy in advanced non-small cell lung cancer patients. *Clin Cancer Res* 2014, 20 (14), 3660–3671.
205. Zanetti, M., Antigenized antibodies. *Nature* 1992, 355, 476–477.
206. Musselli, C.; Daverio-Zanetti, S.; Zanetti, M., Antigenized antibodies expressing Vbeta8.2 TCR peptides immunize against rat experimental allergic encephalomyelitis. *J Immune Based Ther Vaccines* 2004, 2, 9.
207. Kumagai, Y., Epitope-grafted and antigenized antibodies can be used for versatile vaccination strategies to induce epitope-specific immune responses. *J Nippon Med Sch* 2011, 78 (2), 66–67.
208. Lunde, E.; Rasmussen, I. B.; Eidem, J. K.; Gregers, T. F.; Western, K. H.; Bogen, B.; Sandlie, I., "Troy-bodies": antibodies as vector proteins for T cell epitopes. *Biomol Eng* 2001, 18, 109–116.
209. Bakker, J. M.; Bleeker, W. K.; Parren, P. W., Therapeutic antibody gene transfer: an active approach to passive immunity. *Mol Ther* 2004, 10, 411–416.
210. Balazs, A. B.; Chen, J.; Hong, C. M.; Rao, D. S.; Yang, L.; Baltimore, D., Antibody-based protection against HIV infection by vectored immunoprophylaxis. *Nature* 2012, 481 (7379), 81–84.
211. Yang, L.; Wang, P., Passive immunization against HIV/AIDS by antibody gene transfer. *Viruses* 2014, 6 (2), 428–447.
212. Nevinsky, G. A.; Kanyshkova, T. G.; Buneva, V. N., Natural catalytic antibodies (abzymes) in normalcy and pathology. *Biochemistry* 2000, 65, 1245–1255.

213. Lacroix-Desmazes, S.; Wootla, B.; Delignat, S.; Dasgupta, S.; Nagaraja, V.; Kazatchkine, M. D.; Kaveri, S. V., Pathophysiology of catalytic antibodies. *Immunol Lett* 2006, 103 (1), 3–7.

214. Paul, S.; Nishiyama, Y.; Planque, S.; Karle, S.; Taguchi, H.; Hanson, C.; Weksler, M. E., Antibodies as defensive enzymes. *Springer Semin Immunopathol* 2005, 26, 485–503.

215. Lacroix-Desmazes, S.; Bayry, J.; Kaveri, S. V.; Hayon-Sonsino, D.; Thorenoor, N.; Charpentier, J.; Luyt, C. E.; Mira, J. P.; Nagaraja, V.; Kazatchkine, M. D.; Dhainaut, J. F.; Mallet, V. O., High levels of catalytic antibodies correlate with favorable outcome in sepsis. *Proc Natl Acad Sci U S A* 2005, 102, 4109–4113.

216. Mitsuda, Y.; Tsuruhata, K.; Hifumi, E.; Takagi, M.; Uda, T., Investigation of active form of catalytic antibody light chain 41S-2-L. *Immunol Lett* 2005, 96, 63–71.

217. Padiolleau-Lefevre, S.; Ben Naya, R.; Shahsavarian, M. A.; Friboulet, A.; Avalle, B., Catalytic antibodies and their applications in biotechnology: state of the art. *Biotechnol Lett* 2014, 36 (7), 1369–1379.

218. Kurkova, I. N.; Smirnov, I. V.; Belogurov, A. A., Jr.; Ponomarenko, N. A.; Gabibov, A. G., Creation of catalytic antibodies metabolizing organophosphate compounds. *Biochemistry (Mosc)* 2012, 77 (10), 1139–1146.

219. (a) Larsen, N. A.; de Prada, P.; Deng, S. X.; Mittal, A.; Braskett, M.; Zhu, X.; Wilson, I. A.; Landry, D. W., Crystallographic and biochemical analysis of cocaine-degrading antibody 15A10. *Biochemistry* 2004, 43, 8067–8076; (b) Homayoun, P.; Mandal, T.; Landry, D.; Komiskey, H., Controlled release of anti-cocaine catalytic antibody from biodegradable polymer microspheres. *J Pharm Pharmacol* 2003, 55, 933–938.

220. Mahendra, A.; Sharma, M.; Rao, D. N.; Peyron, I.; Planchais, C.; Dimitrov, J. D.; Kaveri, S. V.; Lacroix-Desmazes, S., Antibody-mediated catalysis: induction and therapeutic relevance. *Autoimmun Rev* 2013, 12 (6), 648–652.

221. Paul, S.; Planque, S.; Zhou, Y. X.; Taguchi, H.; Bhatia, G.; Karle, S.; Hanson, C.; Nishiyama, Y., Specific HIV gp120-cleaving antibodies induced by covalently reactive analog of gp120. *J Biol Chem* 2003, 278 (22), 20429–20435.

222. Li, J. W.; Xia, L.; Su, Y.; Liu, H.; Xia, X.; Lu, Q.; Yang, C.; Reheman, K., Molecular imprint of enzyme active site by camel nanobodies: rapid and efficient approach to produce abzymes with alliinase activity. *J Biol Chem* 2012, 287 (17), 13713–13721.

223. Fletcher, M. C.; Kuderova, A.; Cygler, M.; Lee, J. S., Creation of a ribonuclease abzyme through site-directed mutagenesis. *Nat Biotechnol* 1998, 16 (11), 1065–1067.

224. (a) Xu, Y.; Yamamoto, N.; Janda, K. D., Catalytic antibodies: hapten design strategies and screening methods. *Bioorg Med Chem* 2004, 12, 5247–5268;

(b) Meijler, M. M.; Kaufmann, G. F.; Qi, L.; Mee, J. M.; Coyle, A. R.; Moss, J. A.; Wirshing, P.; Matsuhita, M.; Janda, K. D., Fluorescent cocaine probes: a tool for the selection and engineering of therapeutic antibodies. *J Am Chem Soc* 2005, 127, 2477–2484.

225. Bakker, A. B.; Python, C.; Kissling, C. J.; Pandya, P.; Marissen, W. E.; Brink, M. F.; Lagerwerf, F.; Worst, S.; van Corven, E.; Kostense, S.; Hartmann, K.; Weverling, G. J.; Uytdehaag, F.; Herzog, C.; Briggs, D. J.; Rupprecht, C. E.; Grimaldi, R.; Goudsmit, J., First administration to humans of a monoclonal antibody cocktail against rabies virus: safety, tolerability, and neutralizing activity. *Vaccine* 2008, 26 (47), 5922–5927.

226. Nagarajan, T.; Rupprecht, C. E.; Dessain, S. K.; Rangarajan, P. N.; Thiagarajan, D.; Srinivasan, V. A., Human monoclonal antibody and vaccine approaches to prevent human rabies. *Curr Top Microbiol Immunol* 2008, 317, 67–101.

227. Muzard, J.; Billiald, P.; Goyffon, M.; Aubrey, N., [Recombinant antibodies: a new application in scorpion envenomation?]. *Bull Soc Pathol Exot* 2005, 98 (5), 383–385.

228. Bouhaouala-Zahar, B.; Ben Abderrazek, R.; Hmila, I.; Abidi, N.; Muyldermans, S.; El Ayeb, M., Immunological aspects of scorpion toxins: current status and perspectives. *Inflamm Allergy Drug Targets* 2011, 10 (5), 358–368.

229. Rasmussen, S. K.; Naested, H.; Muller, C.; Tolstrup, A. B.; Frandsen, T. P., Recombinant antibody mixtures: production strategies and cost considerations. *Arch Biochem Biophys* 2012, 526 (2), 139–145.

230. Glassy, M. C.; Gupta, R., Technical and ethical limitations in making human monoclonal antibodies (an overview). *Methods Mol Biol* 2014, 1060, 9–36.

231. (a) Jeong, K. J.; Jang, S. H.; Velmurugan, N., Recombinant antibodies: engineering and production in yeast and bacterial hosts. *Biotechnol J* 2011, 6 (1), 16–27;
(b) Spadiut, O.; Capone, S.; Krainer, F.; Glieder, A.; Herwig, C., Microbials for the production of monoclonal antibodies and antibody fragments. *Trends Biotechnol* 2014, 32 (1), 54–60.

232. Fernandez, L. A., Prokaryotic expression of antibodies and affibodies. *Curr Opin Biotechnol* 2004, 15, 364–373.

233. Yoon, S. H.; Kim, S. K.; Kim, J. F., Secretory production of recombinant proteins in *Escherichia coli*. *Recent Pat Biotechnol* 2010, 4 (1), 23–29.

234. Chen, C.; Snedecor, B.; Nishihara, J. C.; Joly, J. C.; McFarland, N.; Andersen, D. C.; Battersby, J. E.; Champion, K. M., High-level accumulation of a recombinant antibody fragment in the periplasm of *Escherichia coli* requires a triple-mutant (degP prc spr) host strain. *Biotechnol Bioeng* 2004, 85, 463–474.

235. Levy, R.; Weiss, R.; Chen, G.; Iverson, B. L.; Georgiou, G., Production of correctly folded Fab antibody fragment in the cytoplasm of *Escherichia coli* trxB gor mutants via the coexpression of molecular chaperones. *Protein Expr Purif* 2001, 23 (2), 338–347.

236. Fernandez, L. A.; Sola, I.; Enjuanes, L.; de Lorenzo, V., Specific secretion of active single-chain Fv antibodies into the supernatants of *Escherichia coli* cultures by use of the hemolysin system. *Appl Environ Microbiol* 2000, 66, 5024–5029.
237. Wacker, M.; Linton, D.; Hitchen, P. G.; Nita-Lazar, M.; Haslam, S. M.; North, S. J.; Panico, M.; Morris, H. R.; Dell, A.; Wren, B. W.; Aebi, M., N-linked glycosylation in *Campylobacter jejuni* and its functional transfer into *E. coli. Science* 2002, 298, 1790–1793.
238. Srichaisupakit, A.; Ohashi, T.; Fujiyama, K., Identification of a protein glycosylation operon from *Campylobacter jejuni* JCM 2013 and its heterologous expression in *Escherichia coli. J Biosci Bioeng* 2014, 118 (3), 256–262.
239. Jaffe, S. R.; Strutton, B.; Levarski, Z.; Pandhal, J.; Wright, P. C., *Escherichia coli* as a glycoprotein production host: recent developments and challenges. *Curr Opin Biotechnol* 2014, 30c, 205–210.
240. Wu, S.-C.; Yeung, J. C.; Duan, Y.; Ye, R.; Szarka, S. J.; Habibi, H. R.; Wong, S.-L., Functional production and characterization of a fibrin-specific single-chain antibody fragment from *Bacillus subtilis*: effects of molecular chaperones and a wall-bound protease on antibody fragment production. *Appl Environ Microbiol* 2002, 68, 3261–3269.
241. Krugger, C.; Hu, Y.; Pan, Q.; Marcotte, H.; Hultberg, A.; Delwar, D.; Van Dalen, P. J.; Pouwels, P. H.; Leer, R. J.; Kelly, C. G.; Van Dollenweerd, C.; Ma, J.; Hammarstrom, L., In situ delivery of passive immunity by lactobacilli producing single-chain antibodies. *Nat Biotechnol* 2002, 20, 702–706.
242. Ueda, Y.; Tsumoto, K.; Watanabe, K.; Kumagai, I., Synthesis and expression of a DNA encoding the Fv domain of an anti-lysozyme monoclonal antibody, HyHEL10, in *Streptomyces lividans. Gene* 1993, 129, 129–134.
243. Pschorr, J.; Bieseler, B.; Fritz, H. J., Production of the immunoglobulin variable domain REiv via a fusion protein synthetized and secreted by *Staphylococcus carnosus. Biol Chem Hoppe Seyler* 1994, 375, 271–280.
244. Dammeyer, T.; Steinwand, M.; Kruger, S. C.; Dubel, S.; Hust, M.; Timmis, K. N., Efficient production of soluble recombinant single chain Fv fragments by a *Pseudomonas putida* strain KT2440 cell factory. *Microb Cell Fact* 2011, 10, 11.
245. Retallack, D. M.; Schneider, J. C.; Mitchell, J.; Chew, L.; Liu, H., Transport of heterologous proteins to the periplasmic space of *Pseudomonas fluorescens* using a variety of native signal sequences. *Biotechnol Lett* 2007, 29 (10), 1483–1491.
246. (a) Cao, J. Y.; Liang, M. F.; Meng, Q. L.; Wang, X. F.; Xu, Y. G.; Guo, K. Q.; Zhan, M. Y.; Bi, S. L.; Li, D. X., [Baculovirus expression of two human recombinant neutralizing IgG monoclonal antibodies to hepatitis A virus]. *Zhonghua Shi Yan He Lin Chuang Bing Du Xue Za Zhi* 2004, 18 (1), 20–23; (b) Liang, M.; Dubel, S.; Li, D.; Queitsch, I.; Li, W.; Bautz, E. K., Baculovirus expression cassette vectors for rapid production of complete human IgG from phage display selected antibody fragments. *J Immunol Methods* 2001, 247 (1–2), 119–130;

(c) Guttieri, M. C.; Liang, M., Human antibody production using insect-cell expression systems. *Methods Mol Biol* 2004, 248, 269–299.

247. Kurasawa, J. H.; Shestopal, S. A.; Jha, N. K.; Ovanesov, M. V.; Lee, T. K.; Sarafanov, A. G., Insect cell-based expression and characterization of a single-chain variable antibody fragment directed against blood coagulation factor VIII. *Protein Expr Purif* 2013, 88 (2), 201–206.

248. Buckland, B.; Boulanger, R.; Fino, M.; Srivastava, I.; Holtz, K.; Khramtsov, N.; McPherson, C.; Meghrous, J.; Kubera, P.; Cox, M. M., Technology transfer and scale-up of the Flublok® recombinant hemagglutinin (HA) influenza vaccine manufacturing process. *Vaccine* 2014, 32 (42), 5496–5502.

249. Fischer, R.; Drossard, J.; Emans, N.; Commandeur, U.; Hellwig, S., Towards molecular farming in the future: *Pichia pastoris*-based production of single-chain antibody fragments. *Biotechnol Appl Biochem* 1999, 30, 117–120.

250. Ning, D.; Junjian, X.; Xunzhang, W.; Wenying, C.; Qing, Z.; Kuanyuan, S.; Guirong, R.; Xiangrong, R.; Qingxin, L.; Zhouyao, Y., Expression, purification, and characterization of humanized anti-HBs Fab fragment. *J Biochem* 2003, 134, 813–817.

251. Li, H.; Sethuraman, N.; Stadheim, T. A.; Zha, D.; Prinz, B.; Ballew, N.; Bobrowicz, P.; Choi, B. K.; Cook, W. J.; Cukan, M.; Houston-Cummings, N. R.; Davidson, R.; Gong, B.; Hamilton, S. R.; Hoopes, J. P.; Jiang, Y.; Kim, N.; Mansfield, R.; Nett, J. H.; Rios, S.; Strawbridge, R.; Wildt, S.; Gerngross, T. U., Optimization of humanized IgGs in glycoengineered *Pichia pastoris*. *Nat Biotechnol* 2006, 24 (2), 210–215.

252. Zha, D., Glycoengineered Pichia-based expression of monoclonal antibodies. *Methods Mol Biol* 2013, 988, 31–43.

253. Zhang, N.; Liu, L.; Dumitru, C. D.; Cummings, N. R.; Cukan, M.; Jiang, Y.; Li, Y.; Li, F.; Mitchell, T.; Mallem, M. R.; Ou, Y.; Patel, R. N.; Vo, K.; Wang, H.; Burnina, I.; Choi, B. K.; Huber, H. E.; Stadheim, T. A.; Zha, D., Glycoengineered Pichia produced anti-HER2 is comparable to trastuzumab in preclinical study. *MAbs* 2011, 3 (3), 289–298.

254. Sotiriadis, A.; Keshavarz, T.; Keshavarz-Moore, E., Factors affecting the production of a single-chain antibody fragment by *Aspergillus awamori* in a stirred tank reactor. *Biotechnol Prog* 2001, 17, 618–623.

255. Ward, M.; Lin, C.; Victoria, D. C.; Fox, J. A.; Wong, D. L.; Meerman, H. J.; Pucci, J. P.; Fong, R. B.; Heng, M. H.; Tsurushita, N.; Gieswein, C.; Park, M.; Wang, H., Characterization of humanized antibodies secreted by *Aspergillus niger*. *Appl Environ Microbiol* 2004, 70, 2567–2576.

256. Hefferon, K. L., Plant virus expression vectors set the stage as production platforms for biopharmaceutical proteins. *Virology* 2012, 433 (1), 1–6.

257. (a) Schillberg, S.; Fischer, R.; Emans, N., Molecular farming of recombinant antibodies in plant. *Cell Mol Life Sci* 2003, 60, 433–445;
(b) Stoger, E.; Sack, M.; Fischer, R.; Christou, P., Plantibodies: applications, advantages and bottlenecks. *Curr Opin Biotechnol* 2002, 13, 161–166.

258. Hiatt, A.; Cafferkey, R.; Bowdish, K., Production of antibodies in transgenic plants. *Nature* 1989, 342, 76–78.
259. Zeitlin, L.; Olmsted, S. S.; Moench, T. R.; Co, M. S.; Martinell, B. J.; Paradkar, V. M.; Russell, D. R.; Queen, C.; Cone, R. A.; Whaley, K. J., A humanized monoclonal antibody produced in transgenic plants for immunoprotection of the vagina against genital herpes. *Nat Biotechnol* 1998, 16, 1361–1364.
260. Khoudi, H.; Laberge, S.; Ferullo, J.-M.; Bazin, R.; Darveau, A.; Castongay, Y.; Allard, G.; Lemiex, R.; Vezina, L.-P., Production of a diagnostic monoclonal antibody in perennial alfalfa plants. *Biotechnol Bioeng* 1999, 64, 135–143.
261. Ma, J.; Hikmat, B. Y.; Wycoff, K.; Vine, N. D.; Charlelegue, D.; Yu, L.; Hein, M. B.; Lehner, T., Characterization of a recombinant plant monoclonal secretory antibody and preventive immunotherapy in humans. *Nat Med* 1998, 4, 601–606.
262. Ma, J.; Drake, P. M.; Charlelegue, D.; Obregon, P.; Prada, A., Antibody processing and engineering in plants, and new strategies for vaccine production. *Vaccine* 2005, 23, 1814–1818.
263. Larrick, J. W.; Yu, L.; Naftzger, C.; Jaiswal, S.; Wycoff, K., Production of secretory IgA antibodies in plants. *Biomol Eng* 2001, 18, 87–94.
264. Peeters, K.; De Wilde, C.; De Jaeger, G.; Angenon, G.; Depicker, A., Production of antibodies and antibody fragments in plants. *Vaccine* 2001, 19, 2756–2761.
265. Rodriguez, M.; Ramirez, N. I.; Ayala, M.; Freyre, F.; Perez, L.; Triguero, A.; Mateo, C.; Selman-Housein, G.; Gavilondo, J. V.; Pujol, M., Transient expression in tobacco leaves of an aglycosylated recombinant antibody against the epidermal growth factor receptor. *Biotechnol Bioeng* 2005, 89 (2), 188–194.
266. Yusibov, V.; Streatfield, S. J.; Kushnir, N.; Roy, G.; Padmanaban, A., Hybrid viral vectors for vaccine and antibody production in plants. *Curr Pharm Des* 2013, 19 (31), 5574–5586.
267. Stoger, E.; Sack, M.; Nicholson, L.; Fischer, R.; Christou, P., Recent progress in plantibody technology. *Curr Pharm Des* 2005, 11 (19), 2439–2457.
268. Virdi, V.; Depicker, A., Role of plant expression systems in antibody production for passive immunization. *Int J Dev Biol* 2013, 57 (6–8), 587–593.
269. Sully, E. K.; Whaley, K. J.; Bohorova, N.; Bohorov, O.; Goodman, C.; Kim do, H.; Pauly, M. H.; Velasco, J.; Hiatt, E.; Morton, J.; Swope, K.; Roy, C. J.; Zeitlin, L.; Mantis, N. J., Chimeric plantibody passively protects mice against aerosolized ricin challenge. *Clin Vaccine Immunol* 2014, 21 (5), 777–782.
270. Olinger, G. G., Jr.; Pettitt, J.; Kim, D.; Working, C.; Bohorov, O.; Bratcher, B.; Hiatt, E.; Hume, S. D.; Johnson, A. K.; Morton, J.; Pauly, M.; Whaley, K. J.; Lear, C. M.; Biggins, J. E.; Scully, C.; Hensley, L.; Zeitlin, L., Delayed treatment of Ebola virus infection with plant-derived monoclonal antibodies provides protection in rhesus macaques. *Proc Natl Acad Sci U S A* 2012, 109 (44), 18030–18035.
271. Samiec, M.; Skrzyszowska, M., Transgenic mammalian species, generated by somatic cell cloning, in biomedicine, biopharmaceutical industry and human nutrition/dietetics—recent achievements. *Pol J Vet Sci* 2011, 14 (2), 317–328.

272. Pollock, D. P.; Kutzko, J. P.; Birck-Wilson, E.; Williams, J. L.; Echelard, Y.; Meade, H. M., Transgenic milk as a method for the production of recombinant antibodies. *J Immunol Methods* 1999, 231, 147–157.

273. Limonta, J.; Pedraza, A.; Rodriguez, A.; Freyre, F. M.; Barral, A. M.; Castro, F. O.; Lleonart, R.; Gracia, C. A.; Gavilondo, J. V.; de la Fuente, J., Production of active anti-CD6 mouse/human chimeric antibodies in the milk of transgenic mice. *Immunotechnology* 1995, 1, 107–113.

274. Houdebine, L. M., Production of pharmaceutical proteins by transgenic animals. *Comp Immunol Microbiol Infect Dis* 2009, 32 (2), 107–121.

275. Dick, M. K.; Lacroix, D.; Pothier, F.; Sirard, M. A., Making recombinant proteins in animals—different systems, different applications. *Trends Biotechnol* 2003, 21, 394–399.

276. Chadd, H. E.; Chamow, S. M., Therapeutic antibody expression technology. *Curr Opin Biotechnol* 2001, 12 (2), 188–194.

277. Kyogoku, K.; Yoshida, K.; Watanabe, H.; Yamashita, T.; Kawabe, Y.; Motono, M.; Nishijima, K.; Kamihira, M.; Iijima, S., Production of recombinant tumor necrosis factor receptor/Fc fusion protein by genetically manipulated chickens. *J Biosci Bioeng* 2008, 105 (5), 454–459.

278. Shirokov, V. A.; Kommer, A.; Kolb, V. A.; Spirin, A. S., Continuous-exchange protein-synthesizing systems. *Methods Mol Biol* 2007, 375, 19–55.

279. Merk, H.; Stiege, W.; Tsumoto, K.; Kumagai, I.; Erdmann, V. A., Cell-free expression of two single-chain monoclonal antibodies against lysozyme: effect of domain arrangement on the expression. *J Biochem* 1999, 125, 328–333.

280. Jiang, X.; Ookubo, Y.; Fuji, Y.; Nakano, H.; Yamane, T., Expression of a Fab fragment of catalytic antibody 6D9 in an *Escherichia coli in vitro* coupled transcription/translation system. *FEBS Lett* 2002, 514, 290–294.

281. (a) Ryabova, L. A.; Desplancq, D.; Spirin, A. S.; Pluckthun, A., Functional antibody production using cell-free translation: effects of protein disulfide isomerase and chaperones. *Nat Biotechnol* 1997, 15, 79–84;
(b) Ying, B. W.; Taguchi, H.; Ueda, H.; Ueda, T., Chaperone-assisted folding of a single-chain antibody in a reconstituted translation system. *Biochem Biophys Res Commun* 2004, 320, 1359–1364.

282. Takai, K.; Sawasaki, T.; Endo, Y., Practical cell-free protein synthesis system using purified wheat embryos. *Nat Protoc* 2010, 5 (2), 227–238.

283. Yin, G.; Garces, E. D.; Yang, J.; Zhang, J.; Tran, C.; Steiner, A. R.; Roos, C.; Bajad, S.; Hudak, S.; Penta, K.; Zawada, J.; Pollitt, S.; Murray, C. J., Aglycosylated antibodies and antibody fragments produced in a scalable *in vitro* transcription-translation system. *MAbs* 2012, 4 (2), 217–225.

284. Groff, D.; Armstrong, S.; Rivers, P. J.; Zhang, J.; Yang, J.; Green, E.; Rozzelle, J.; Liang, S.; Kittle, J. D., Jr.; Steiner, A. R.; Baliga, R.; Thanos, C. D.; Hallam, T. J.; Sato, A. K.; Yam, A. Y., Engineering toward a bacterial "endoplasmic reticulum" for the rapid expression of immunoglobulin proteins. *MAbs* 2014, 6 (3), 671–678.

285. Betenbaugh, M. J.; Tomiya, N.; Narang, S.; Hsu, J. T. A.; Lee, Y. C., Biosynthesis of human-type N-glycans in heterologous systems. *Curr Opin Struct Biol* 2004, 14, 601–606.

286. Hollister, J.; Grabenhorst, E.; Nimtz, M.; Conradt, H.; Jarvis, D. L., Engineering the protein N-glycosylation pathway in insect cells for production of biantennary complex N-glycans. *Biochemistry* 2002, 41, 15093–15104.

287. Grabenhorst, E.; Schlenke, P.; Pohl, S.; Nimtz, M.; Conradt, H. S., Genetic engineering of recombinant glycoproteins and the glycosylation pathway in mammalian host cells. *Glycoconj J* 1999, 16, 81–97.

288. Umana, P.; Jean-Mairet, J.; Moudry, R.; Amstutz, H.; Bailey, J. E., Engineered glycoforms of an antineuroblastoma IgG1 with optimized antibody-dependant cellular cytotoxic activity. *Nat Biotechnol* 1999, 17, 176–180.

289. Bragonzi, A.; Distefano, G.; Buckberry, L. D.; Acerbis, G.; Foglieni, C.; Lamotte, D.; Campi, G.; Marc, A.; Soria, M. R.; Jenkins, N.; Monaco, L., A new chinese hamster ovary cell line expressing a2,6-sialyltransferase used as universal host for the production of human-like sialylated recombinant glycoproteins. *Biochim Biophys Acta* 2000, 1474, 273–282.

290. El Mai, N.; Donadio-Andrei, S.; Iss, C.; Calabro, V.; Ronin, C., Engineering a human-like glycosylation to produce therapeutic glycoproteins based on 6-linked sialylation in CHO cells. *Methods Mol Biol* 2013, 988, 19–29.

291. Chang, G.-D.; Chen, C.-J.; Lin, C.-Y.; Chen, H.-C.; Chen, H., Improvement of glycosylation in insect cells with mammalian glycosyltransferases. *J Biotechnol* 2003, 102, 61–71.

292. Ko, K.; Tekoah, Y.; Rudd, P. M.; Harvey, D. J.; Dwek, R. A.; Spitsin, S.; Hanlon, C. A.; Ruppprecht, C.; Dietzschold, B.; Golovkin, M.; Koprowski, H., Function and glycosylation of plant-derived antiviral monoclonal antibody. *Proc Natl Acad Sci U S A* 2003, 100, 8013–8018.

293. Bakker, H.; Bardor, M.; Molthoff, J. W.; Gomord, V.; Elbers, I.; Stevens, L. H.; Jordi, W.; Lommen, A.; Faye, L.; Lerouge, P.; Bosch, D., Galactose-extended glycans of antibodies produced by transgenic plants. *Proc Natl Acad Sci U S A* 2001, 98, 2899–2904.

294. Wildt, S.; Gerngross, T. U., The humanization of N-glycosylation pathways in yeast. *Nat Rev Microbiol* 2005, 3, 119–128.

295. Bretthauer, R. K., Genetic engineering of *Pichia pastoris* to humanize N-glycosylation of proteins. *Trends Biotechnol* 2003, 21, 459–462.

296. (a) Vervecken, W.; Kaigorodov, V.; Callewaert, N.; Geysens, S.; De Vusser, K.; Contreras, R., *In vivo* synthesis of mammalian-like, hybrid-type N-glycans in *Pichia pastoris*. *Appl Environ Microbiol* 2004, 70, 2639–2646;
(b) Bobrowicz, P.; Davidson, R. C.; Li, H.; Potgieter, T. I.; Nett, J. H.; Hamilton, S. R.; Stadheim, T. A.; Miele, R. G.; Bobrowics, B.; Mitchell, T.; Rausch, S.; Renfer, E.; Wildt, S., Engineering of an artificial glycosylation pathway blocked in core oligosaccharide assembly in the yeast *Pichia pastoris*: production of complex humanized glycoproteins with terminal galactose. *Glycobiology* 2004, 14, 757–766.

297. Wang, L. X.; Lomino, J. V., Emerging technologies for making glycan-defined glycoproteins. *ACS Chem Biol* 2012, 7 (1), 110–122.

2

Structure, Classification, and Naming of Therapeutic Monoclonal Antibodies

Zhinan Xia

Biotherapeutics, Moderna Therapeutics, Cambridge, MA, USA

2.1 Summary

The introduction almost 40 years ago of monoclonal antibodies generated by hybridoma technology has fueled a revolution, initially in the field of biotechnology and now disseminated throughout the entire pharmaceutical industry. After the first wave of innovation based on mouse mAbs, recombinant DNA technology provided the tools for constructing chimeric, humanized, and fully human mAbs as well as recombinant antibody fragments and bispecific therapeutic mAbs.

Over the decades, therapeutic mAbs have evolved as innovative pharmaceutical compounds, not only for the treatment of cancer but also of autoimmune and infectious diseases. A new generation of antibodies including anti-PCSK9 mAbs, now in clinical development by Amgen, Pfizer, and Regeneron (Sanofi), may represent a blockbuster drug aimed at cardiovascular disease. We are now entering a third wave of scientific advancement based on antibody architecture modification and engineering, taking advantage of new insights into Fc effector function, glycoengineering, and antibody-drug conjugation.

The ADCs are the most prominent of the new antibody technology platforms, with their main focus on oncology indications. Because ADCs can offer targeted delivery of chemotherapeutic or radioactive agents, they will increase the performance of mAbs and offer a toolkit for product differentiation and life-cycle management.

Biosimilars of Monoclonal Antibodies: A Practical Guide to Manufacturing, Preclinical, and Clinical Development, First Edition. Edited by Cheng Liu and K. John Morrow, Jr.

2.2 Introduction

Therapeutic monoclonal antibodies (mAbs) have become the cornerstone in the clinical practice of modern biotherapeutics, especially in oncology, inflammation, and autoimmune diseases. They are the most dominant class of biologics in current drug development pipelines. The success of mAbs as therapeutics in clinics and drug development processes stems from their ability to recognize target antigens with exquisite specificity and high affinity. Other beneficial attributes as drug molecules include their solubility, stability, long-circulation half-life, and manufacturability.

In the 1970s, the seminal invention of hybridoma mAb generation technology by Köhler and Milstein allowed scientists to make large amounts of homogeneous, target-specific mAbs [1]. These mAbs, as reagent tools, have been instrumental in life science and biomedical research. At the same time, the advent of recombinant DNA technologies and antibody phage display allowed drug developers to significantly reduce the murine amino acid sequence content in therapeutic mAbs and therefore the risk of immunogenicity. Technological breakthroughs in these areas combined with the advancement of modern serum-free suspension mammalian cell culture systems have made it possible to manufacture therapeutic mAbs on an industrial scale and use them widely as protein drugs in the clinical sphere.

Throughout the progression of therapeutic mAb drug development, there have been four major mAb types developed: murine, chimeric, humanized, and human. Initial therapeutic antibodies were murine mAbs (suffix -omab, International Nonproprietary Names (INN), and the United States Adopted Names (USAN) usually end with -omab), which were largely unsuccessful. It has since been shown that these antibodies have a short half-life *in vivo* due to immune complex formation, limited penetration into tumor sites, and inadequate recruitment of host effector functions (due to the dissimilarity between murine and human immune systems) [2]. Major shortcomings associated with murine mAbs included immunogenicity (from mild allergic reactions to severe anaphylactic shock) or reduced stimulation of cytotoxicity and the formation of immune complexes after repeated administration.

To overcome these difficulties, chimeric (suffix -ximab) and humanized mAbs (suffix -zumab) have been developed from murine mAbs in a modern therapeutic antibody context. Murine mAbs were engineered to remove immunogenic content and to increase their immunological efficiency [2]. The advancement was initially achieved by the production of chimeric mAbs and further improved with the introduction of humanized mAbs. Chimeric mAbs are composed of murine variable regions fused onto human antibody constant regions and contain approximately 65% human sequences. It significantly reduces immunogenicity and increases the Fc effector function as well as the serum half-life.

Humanized antibodies are produced by grafting murine complementarity determining regions (CDRs) of amino acid domains into human antibodies.

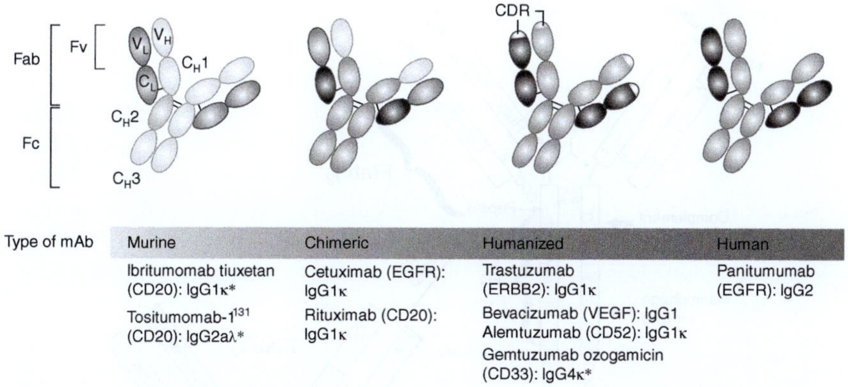

Figure 2.1 Classification of therapeutic mAb according to its increasing content of human sequence. ©2006 Nature Publishing Group. Nature Reviews/Cancer. (*See insert for color representation of the figure.*)

It results in a molecule of approximately 95% human origin. However the result is often mAbs with much lower binding affinity to target antigens than the parental murine mAbs following initial humanization, with reported decreases in affinity of up to several hundredfold. Increases in antibody–antigen binding strength have been achieved by introducing back mutations into the initial humanized CDRs, using techniques such as chain shuffling, randomization of CDRs, and generation of antibody libraries with mutations within the CDRs by error-prone PCR, *Escherichia coli* mutator strains, and site-specific mutagenesis [3–5]. Complete human mAbs (100% human sequence, suffix -umab) coding sequences can be obtained from single human B cell PCR, phage display of human B cell repertoires, or human Ig loci transgenic mouse immunized with the desired antigen. The heavy and light chains of human IgG proteins are expressed in structural polymorphic (allotypic) forms. Human IgG allotype has been considered as one of the many factors that can contribute to immunogenicity [6]. The general scheme of a monoclonal antibody development program is described by Chapman et al. [7]. The schematic summary of classification of therapeutic mAbs according to the content of human sequence is shown in Figure 2.1.

2.3 Antibody Structure

The basic structural unit of most mammalian antibodies is a glycoprotein (MW ~150,000 Da) comprising four polypeptide chains: two light chains and two heavy chains, which are connected by disulfide bonds (Figure 2.2). Each light chain has a molecular weight of approximately 25,000 Da and is composed of two domains: one variable domain (V_L) and one constant domain

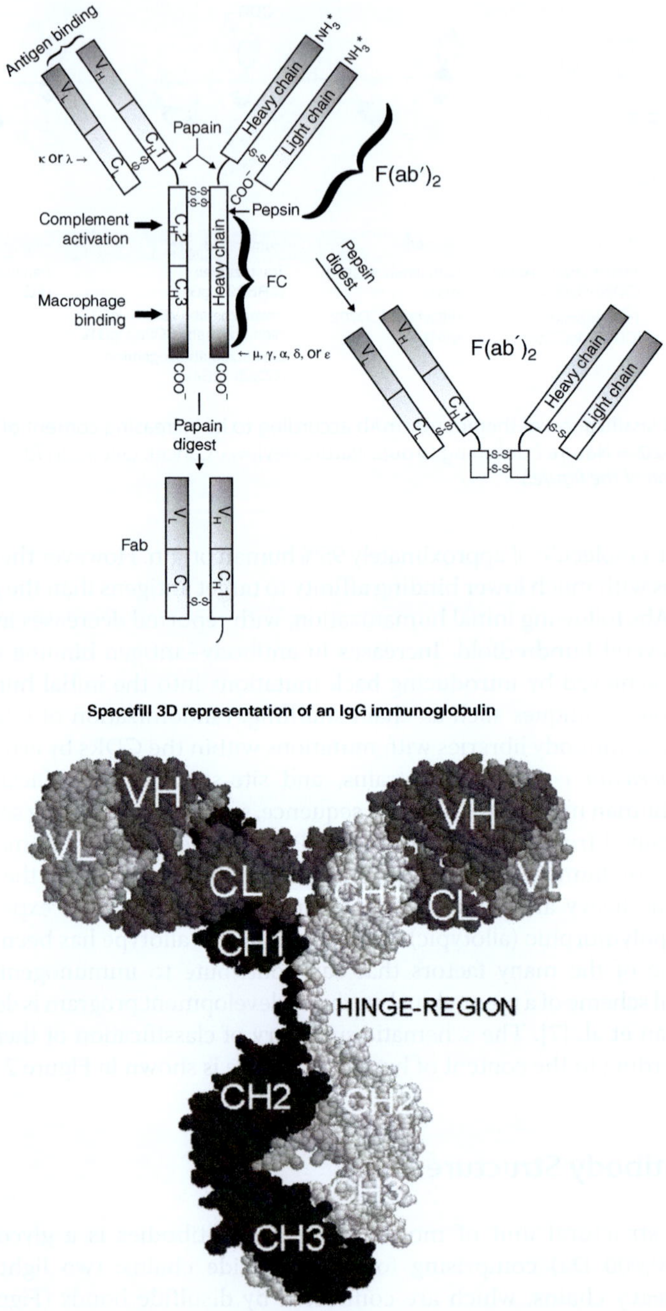

Figure 2.2 General structure of antibody. (*See insert for color representation of the figure.*)

(C_L). There are two types of light chains: lambda (λ) and kappa (k). In humans, 60% of the light chains are k, and 40% are λ, whereas in mice, 95% of the light chains are k and only 5% are λ. A single natural monoclonal antibody molecule contains either k or λ light chains, that is, two k or two λ light chains but never both k and λ light chains in the same antibody.

Each heavy chain has a molecular weight of approximately 50,000 Da and two regions: the constant region and the variable region. The constant region is identical in all antibodies of the same isotype but differs in antibodies of different isotypes. The heavy and light chains contain a number of homologous sections consisting of similar but not identical groups of amino acid sequences. These homologous units consist of about 110 amino acids and are referred to as the immunoglobulin (Ig) domains. The heavy chain contains one variable domain (V_H) and either three or four constant domains (C_H1, C_H2, C_H3, and C_H4, depending on the antibody class or isotype). The linker region between the C_H1 and C_H2 domains is called the hinge, and it permits flexibility between the two Fab arms of the Y-shaped antibody molecule, allowing them to open and close to accommodate binding to two antigenic determinants separated by a prescribed distance. Heavy chains γ, α, and δ have a constant region composed of three tandem Ig domains and a hinge region for added flexibility; heavy chains μ and ε have a constant region composed of four tandem Ig domains. The variable region of the heavy chain differs in antibodies produced by different B cells but is the same for all antibodies produced by a single B cell or B cell clone. The variable region of each heavy chain is approximately 110 amino acids long and is composed of a single Ig domain [8].

Each light chain has two successive domains: one constant domain and one variable domain. The approximate length of a light chain is 211–217 amino acids. Each antibody contains two identical light chains. Only one type of light chain, k or λ, is present per antibody in mammals. Other types of light chains, such as the iota chain, are found in nonmammalian vertebrates such as sharks (Chondrichthyes) and bony fishes (Teleostei).

The heavy chain also serves to determine the functional activity of the antibody molecule. There are five antibody classes—IgG, IgA, IgM, IgE, and IgD—which are distinguished by their heavy chains γ, α, μ, ε, and δ, respectively (Figure 2.3). The IgD, IgE, and IgG antibody classes are each made up of a single structural unit, whereas IgA antibodies may contain either one or two units and IgM antibodies consist of five disulfide-linked structural units. IgG antibodies are further divided into four subclasses (often referred to as isotypes) although the nomenclature differs slightly depending on the species producing the antibody (Table 2.1 and Figure 2.3).

In mammals, IgG is the major serum antibody. But in the avian class, the major serum antibody is called IgY (also found in egg yolk), which is quite different from mammalian IgG. The Fc portion of the IgY molecule does not

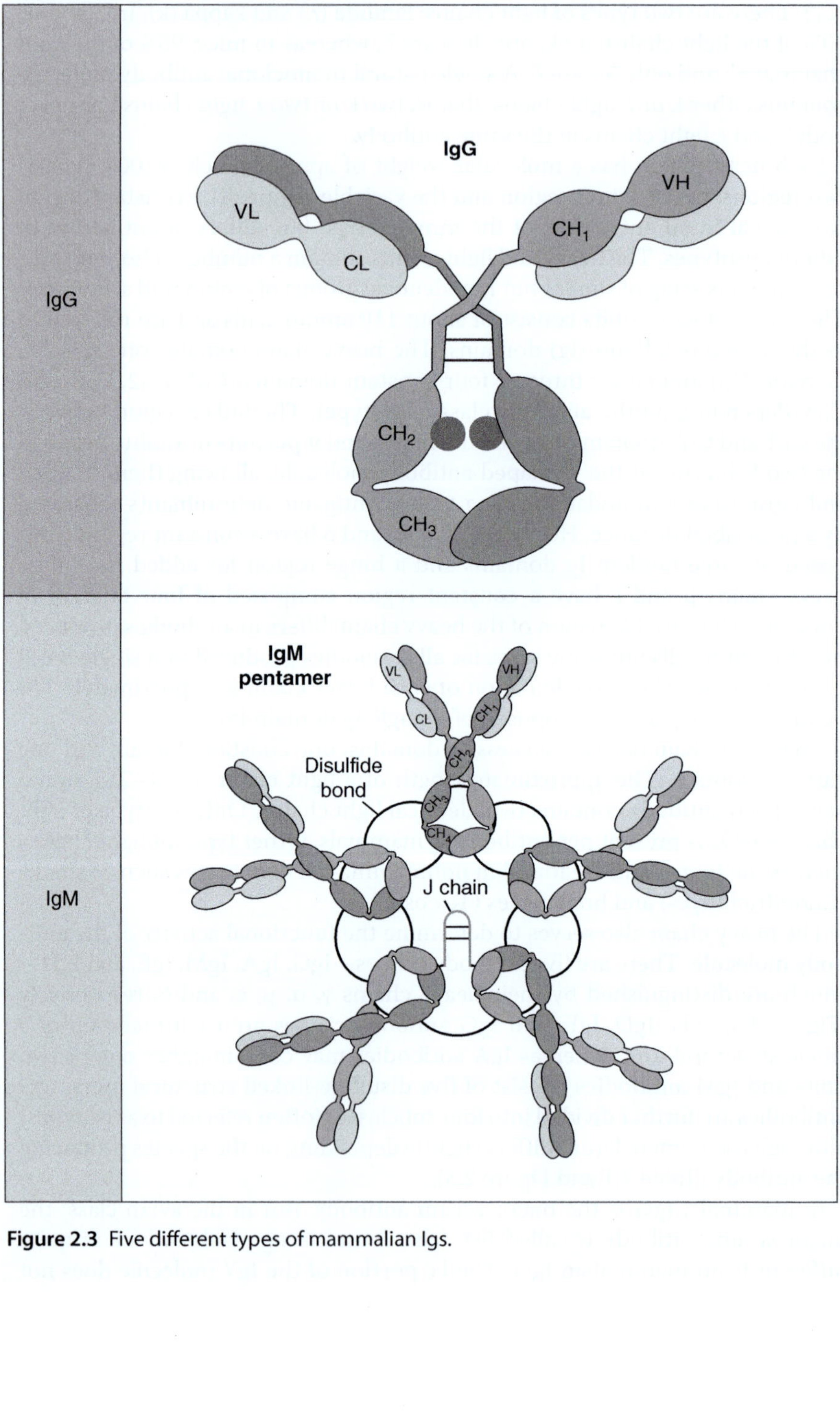

Figure 2.3 Five different types of mammalian Igs.

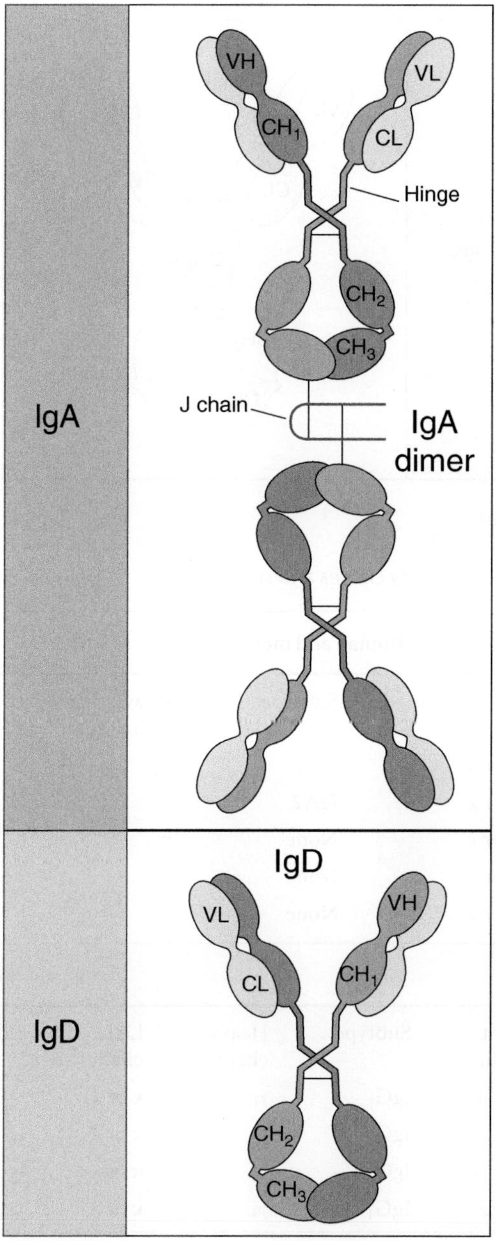

Figure 2.3 (Continued)

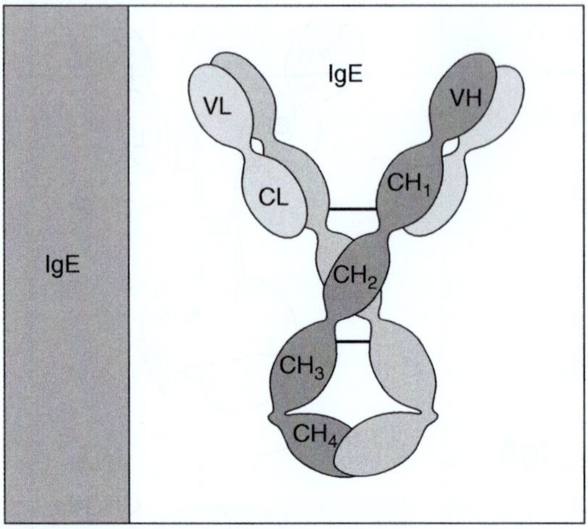

Figure 2.3 (Continued)

Table 2.1 Overview of antibody classes and subclasses.

Antibody	**Human and mouse**					
	Light chain	Subtype	Heavy chain			
IgA	κ or λ	IgA_1	α_1			
	κ or λ	IgA2	α_2			
IgE	κ or λ	None	ε			
IgD	κ or λ	None	δ			
IgM	κ or λ	None	μ			
Antibody	**Human**			**Mouse**		
IgG	Light chain	Subtype	Heavy chain	Light chain	Subtype	Heavy chain
	κ or λ	IgG_1	γ_1	κ or λ	IgG_1	γ_1
	κ or λ	IgG_2	γ_2	κ or λ	IgG_{2a}	γ_{2a}
	κ or λ	IgG_3	γ_3	κ or λ	IgG_{2b}	γ_{2b}
	κ or λ	IgG_4	γ_4	κ or λ	IgG_3	γ_3

bind protein A or protein G; therefore it requires alternative purification platforms. However, in some older literature and even on some commercial life sciences product websites, it is often confused with IgG.

Structure/function studies on IgGs have been aided by the discovery that the proteolytic enzymes pepsin and papain cleave the molecule into precise fragments with specific biological properties. Treatment of an IgG molecule with pepsin generates the $F(ab')_2$ fragment, which broadly encompasses the two Fab regions linked by the hinge region (Figure 2.2). Because the $F(ab')_2$ molecule is bivalent, it is capable of precipitating an antigen. Papain cleaves the IgG molecule in the hinge region between the C_H1 and C_H2 domains to yield two identical Fab fragments, which retain their antigen-binding ability, and one nonantigen-binding fragment—the (Fc) region. The Fc region is glycosylated and has many effector functions (e.g., binding complement, binding to cell receptors on macrophages, and monocytes), and it serves to distinguish one class of antibody from another [9].

Some parts of the antibody molecule have the same functions. The arms of the Y, for example, contain the sites that can bind to antigens (in general, identical) and, therefore, recognize specific foreign objects. This region of the antibody is called the Fab (fragment, antigen-binding) region. It is composed of one constant and one variable domain from each heavy and light chain of the antibody. The paratope is shaped at the amino terminal end of the antibody monomer by the variable domains from the heavy and light chains. The variable domain is also referred to as the F_V region and is the most important region for binding to antigens. To be specific, variable loops of β-strands, three each on the light (V_L) and heavy (V_H) chains, are responsible for binding to the antigen. scFv, short for single-chain Fv, is recognized as V_L and V_H linked by a flexible linker. These loops are often called the CDRs. The structures of these CDRs have been clustered and classified by Chothia et al. [10] and more recently by North et al. [11]. In the framework of the immune network theory, CDRs are also called idiotypes. According to immune network theory, the adaptive immune system is regulated by interactions between idiotypes. The base of the Y plays a role in modulating immune cell activity. This region is called the Fc (Fragment, crystallizable) region and is composed of two heavy chains that contribute two or three constant domains depending on the class of the antibody. Thus, the Fc region ensures that each antibody generates an appropriate immune response for a given antigen by binding to a specific class of Fc receptors, or to other immune molecules, such as complement proteins. By doing this, it mediates different physiological effects including recognition of opsonized particles, lysis of cells, and degranulation of mast cells, basophils, and eosinophils [12].

2.4 Classification of Antibodies

As noted previously, distinct heavy chains differ in size and composition; α, δ, and γ contain approximately 450 amino acids, whereas μ and ε have approximately 550 amino acids (see Table 2.1).

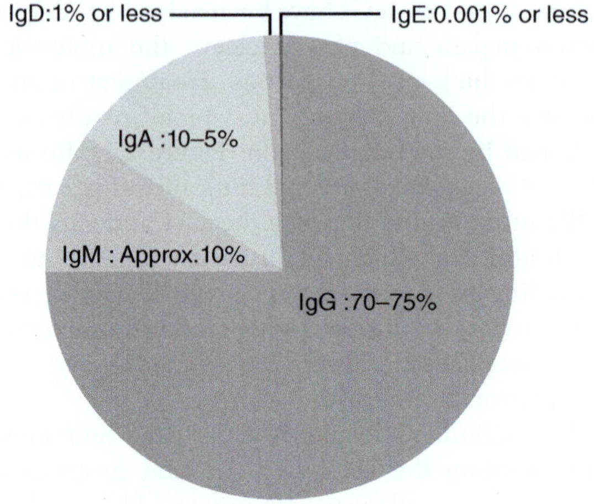

Figure 2.4 Population distribution of five types of antibodies in blood (%).

The general structure of five types of mammalian antibody is shown in Figures 2.2 and 2.3, and their physiological functions described in the following text.

IgG is the main type of antibody in human blood (comprise 70–75% of total serum Ig. Figure 2.4) and is the only antibody capable of crossing the placenta to provide passive immunity to the fetus. It protects a newborn until a week after birth. Widely distributed throughout the blood and tissue, it is derived from an IgY-like ancestor. First appearing in amphibians and able to undergo the class switch, secreted IgM is made up of five antibodies (pentamer) with very high avidity to antigen. IgM is a primordial Ig and has a key role in the immediate immune response to foreign antigens. It expresses on the surface of B cells as monomer in the earliest immune response. It possesses the ability to eliminate pathogens in the early stages of B cell-mediated (humoral) immunity before there is sufficient IgG buildup. IgM comprises about 10% of total serum Ig.

IgA first appears in reptiles. Secreted IgA is made up of two antibodies. It is found in mucosal areas, such as gut, respiratory, and urogenital tract, and acts to prevent colonization by pathogens. IgA is also found in saliva, tears, and breast milk in which it fulfills the important function to protect the gastrointestinal tract of newborns from bacterial and viral infection (maternal immunity). Ten to fifteen percentage of circulating Ig is IgA.

IgD is the homologue of IgW in fish. It is a primordial Ig that shows great variability over the course of evolution. It functions mainly as an antigen receptor on B cells that have not been exposed to antigens. The secreted form is associated with granulocytes and inflammation. It has been shown to activate basophils and mast cells to produce antimicrobial factors. IgD comprise only 1% or less of the total serum Ig. IgE is derived from an IgY-like ancestor.

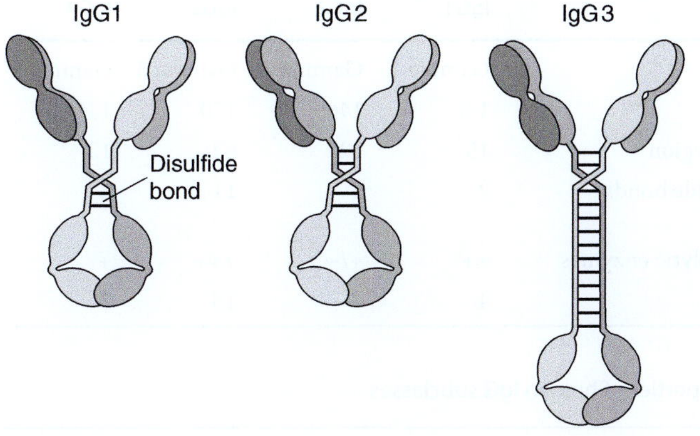

Figure 2.5 Subtypes of IgG human antibodies.

It is believed to be responsible for immune reactions against parasites, binds to allergens and triggers histamine release from mast cells and basophils, and is involved in allergy and hypersensitivity reaction. It is known to protect against parasitic worms. Vast majority of IgE is found on cell surface of mast cells and basophils but very little free IgE in the circulation of blood.

2.5 IgG Subtype

Most therapeutic mAbs belong to the IgG type, particularly the IgG1 subtype. There are four known IgG subclasses. As shown in Figure 2.5, major structural differences between IgG subclasses are at the hinge region. Physiochemical and biological properties of human IgG subclasses are listed and compared in Tables 2.2 and 2.3.

2.6 Nomenclature of Therapeutic mAbs

Drugs from the same therapeutic or chemical class are usually given names with the same stem. Stems are mostly placed word-finally, but in some cases word-initial stems are used. They are collected in a publication informally known as the *Stem Book*. Examples are as follows [13]:

- *-anib* for angiogenesis inhibitors (e.g., pazopanib)
- *-anserin* for serotonin receptor antagonists, especially 5-HT_2 antagonists (e.g., ritanserin and mianserin)
- *-arit* for antiarthritic agents (e.g., lobenzarit)

Table 2.2 Physiochemical properties of human IgG subclasses.

	IgG1	**IgG2**	**IgG3**	**IgG4**
Heavy chain type	Gamma 1	Gamma 2	Gamma 3	Gamma 4
Molecular mass (kDa)	146	146	170	146
Amino acids in hinge region	15	12	62	12
Interheavy chain disulfide bonds (in hinge region)	2	4	11	2
Susceptibility to proteolytic enzymes	++	+/−	+++	+
Number of allotypes	4	1	13	0

Table 2.3 Biological properties of human IgG subclasses.

	IgG1	**IgG2**	**IgG3**	**IgG4**
Adult serum level range (g/l) (mean, g/l)	4.9–11.4 (6.98)	1.5–6.4 (3.8)	0.20–1.10 (0.51)	0.08–1.40 (0.56)
Proportion of total IgG (%)	43–75	16–48	1.7–7.5	0.8–11.7
Half-life (days)	21	21	7	21
Placental transfer	+	+	+	+
Antibody response to				
Proteins	++	+/−	++	+/−
Polysaccharides	+	++	(−)	(−)
Allergens	+	(−)	(−)	++
Complement activation				
C1q binding	++	+	+++	−
C1q binding, high epitope density enhancement alternative pathway	−	+	−	+/−
Binding to Fcγ *receptors*				
Fcγ RI (CD64:monocytes, macrophages, neutrophils, dendritic cells)	++	−	+++	+
Fcγ RII (CD32:monocytes, macrophages, neutrophils, eosinophils, platelets, B cells, dendritic cells, endothelial cells)	++	(a)	+++	−
Fcγ RIIa-H131	++	+++	+++	−
Fcγ RIIa-R131	++	−	++	−
Fcγ RIII (CD16:neutrophils, eosinophils, macrophages, NK cells, subsets of T cells)	++	−	++	−
Fcγ RIIIb-NA1	+++	−	+++	−
Fcγ RIIIb-NA2	++	−	++	−
Binding to staphylococcal protein A	++	++	(b)	+
Binding to streptococcal protein G	++	++	++	++
(a): Fcγ RII allotype dependent				
(b): IgG3 allotype dependent				

- *-ase* for enzymes (e.g., asparaginase)
- *-azepam* for benzodiazepines (e.g., diazepam and oxazepam)
- *-caine* for local anesthetics (e.g., procaine or cocaine)
- *-cain-* for class I antiarrhythmics (e.g., procainamide)
- *-coxib* for COX-2 inhibitors, a type of anti-inflammatory drugs (e.g., celecoxib)
- *-mab* for monoclonal antibodies (e.g., infliximab)
- *-navir* for antiretroviral protease inhibitors (e.g., darunavir)
- *-olol* for beta-blockers (e.g., atenolol)
- *-pril* for ACE inhibitors (e.g., captopril)
- *-sartan* for angiotensin II receptor antagonists (e.g., losartan)
- *-tinib* for tyrosine kinase inhibitors (e.g., imatinib)
- *-vastatin* for HMG-CoA reductase inhibitors, a group of cholesterol-lowering agents (e.g., simvastatin)
- *-vir* for antivirals (e.g., aciclovir or ritonavir)
- *arte-* for artemisinin antimalarials (e.g., artemether)
- *cef-* for cefalosporins (e.g., cefalexin)
- *io-* for iodine-containing radiopharmaceuticals (e.g., iobenguane)

The nomenclature of therapeutic mAbs is a naming scheme for assigning generic or nonproprietary names to a group of medicines that belongs to the mAb class. This scheme is used for both the World Health Organization's INN and the USAN. In general, word stems are used to identify classes of drugs, in most cases placed word-finally. All mAb therapeutic names end with the stem -mab. Unlike most other pharmaceuticals, mab nomenclature uses different preceding word parts depending on structure and function. These are called substems and sometimes called infixes. This nomenclature is also used for fragments of mAbs, such as antigen-binding fragments and single-chain variable fragments. Most therapeutic mAb drugs have two types of names: generic or nonproprietary names, the most important of which are the INNs, and trade names, also called brand names. A therapeutic mAb drug may have more than one brand name. For example, anti-CD20 therapeutic mAb rituximab is marketed in the United States as brand name Rituxan while sold in the European Union as MabThera. Generic names for therapeutic mAb drugs are currently constructed out of affixes and stems that classify the drugs into different categories and also separate drugs within the categories. A marketed therapeutic mAb drug might also have a company code or a compound code. A firm that is developing a therapeutic mAb often gives the drug a company code used by people within the company to identify the drug under development. For example, CDP870 is UCB's company code for Cimzia. Many of these codes, although not all, have prefixes that correspond to the company name.

2.6.1 Stem -mab

The stem *-mab* is used for monoclonal antibodies as well as for their fragments, as long as at least one variable domain (the domain that contains the target binding structure) is included. This is the case for antigen-binding fragments and single-chain variable fragments, among other artificial proteins. Other antibody parts (such as the Fc regions) and antibody mimetics use different naming schemes. The general scheme of a generic name for a mAb drug is as follows.

Prefix–target–source–suffix

For example, rituximab is the generic name for anti-CD20 mAb drug Rituxan. Its name is composed of prefix ri-, plus target stem -tu-, plus source stem -xi-, plus suffix -mab. This shows that the drug is a chimeric mAb acting against tumors. Table 2.4 lists the commonly used source and target substems in generic names of mAbs.

2.6.2 Substem for Origin/Source

The substem preceding the stem denotes the animal from which the antibody is obtained. The first monoclonal antibodies were produced in mice (substem *-o-*, yielding the ending *-omab*; usually *Mus musculus*, the house mouse) or other nonhuman organisms. Neither INN nor USAN has ever been requested for antibodies from rats (theoretically *-a-*), hamsters (*-e-*), and primates (*-i-*).

Rat/mouse hybrid antibodies can be engineered with binding sites for two different antigens. These drugs, termed *trifunctional antibodies*, have the substem *-axo-*.

2.6.3 Substem for Target

The substem preceding the source of the antibody refers to the medicine's target. Examples of targets are tumors, organ systems like the circulatory system, or infectious agents like bacteria or viruses. The term *target* does not imply what sort of action the antibody exerts. Therapeutic, prophylactic, and diagnostic agents are not distinguished by this nomenclature.

In the naming scheme as originally developed, these substems mostly consist of a consonant, a vowel, and then another consonant. The final letter may be dropped if the resulting name would be difficult to pronounce otherwise. Examples include *-ci*(*r*)- for the circulatory system, *-li*(*m*)- for the immune system (*lim* stands for lymphocyte), and *-ne*(*r*)- for the nervous system. The final letter is usually omitted if the following source substem begins with a

Table 2.4 Commonly used source and target substems in generic names of mAbs.

Prefix	Target		Source		Suffix
Variable	*-o(s)-*	Bone	*-u-*	Human	*-mab*
	-vi(r)-	Viral	*-o-*	Mouse	
	-ba(c)-	Bacterial	*-a-*	Rat	
	-li(m)-	Immune	*-e-*	Hamster	
	-le(s)-	Infectious lesions	*-i-*	Primate	
	-ci(r)-	Cardiovascular	*-xi-*	Chimeric	
	-mu(l)-	Musculoskeletal	*-zu-*	Humanized	
	-ki(n)-	Interleukin	*-axo-*	Rat/murine hybrid	
	-co(l)-	Colonic tumor			
	-me(l)-	Melanoma			
	-ma(r)-	Mammary tumor			
	-go(t)-	Testicular tumor			
	-go(v)-	Ovarian tumor			
	-pr(o)-	Prostate tumor			
	-tu(m)-	Miscellaneous tumor			
	-neu(r)-	Nervous system			
	-tox(a)-	Toxin as target			

consonant (such as *-zu-* or *-xi-*), but not all target substems are used in their shortened form. *-mul-*, for example, is never reduced to *-mu-* because no chimeric or humanized antibodies targeting the musculoskeletal system ever received an INN. Combination of target and source substems results in endings like *-limumab* (immune system, human) or *-ciximab* (circulatory system, chimeric, consonant *r* dropped).

New and shorter target substems were adopted in 2009. They mostly consist of a consonant plus a vowel that is omitted if the source substem begins with a vowel. For example, human antibodies targeting the immune system receive names ending in *-lumab* instead of the old *-limumab*. Some endings like *-ciximab* remain unchanged. The old system employed seven different substems for tumor targets, depending on the type of tumor. Because many antibodies are investigated for several tumor types, the new convention only has *-t(u)-*.

Table 2.5 Therapeutic mAbs approved or in review in the European Union or United States.

International nonproprietary name	Brand name	Target; format	Indication first approved or reviewed	First EU approval year	First US approval year
Brodalumab		IL-17R; human IgG2	[In review in Japan]	NA	NA
Sarilumab	(Pending)	IL-6R; human IgG1	Rheumatoid arthritis	NA	In review
Begelomab	Begedina	CD26; murine IgG2b	Graft-versus-host disease	In review	NA
Reslizumab	(Pending)	IL-5; humanized IgG4	Asthma	In review	In review
Obiltoxaximab	Anthim	Protective antigen of B. anthracis exotoxin; chimeric IgG1	Prevention of inhalational anthrax	NA	In review
Ixekizumab	(Pending)	IL-17a; humanized IgG4	Psoriasis	In review	NA
Necitumumab	Portrazza	EGFR; human IgG1	Nonsmall cell lung cancer	In review	2015
Elotuzumab	Empliciti	SLAMF7; humanized IgG1	Multiple myeloma	In review	2015
Daratumumab	Darzalex	CD38; human IgG1	Multiple myeloma	In review	2015
Mepolizumab	Nucala	IL-5; humanized IgG1	Severe eosinophilic asthma	2015	2015
Idarucizumab	Praxbind	Dabigatran; humanized Fab	Reversal of dabigatran-induced anticoagulation	2015	2015
Alirocumab	Praluent	PCSK9; human IgG1	High cholesterol	2015	2015
Evolocumab	Repatha	PCSK9; human IgG2	High cholesterol	2015	2015
Dinutuximab	Unituxin	GD2; chimeric IgG1	Neuroblastoma	2015	2015
Secukinumab	Cosentyx	IL-17a; human IgG1	Psoriasis	2015	2015
Nivolumab	Opdivo	PD1; human IgG4	Melanoma, nonsmall cell lung cancer	2015	2014

Blinatumomab	Blincyto	CD19, CD3; murine bispecific tandem scFv	Acute lymphoblastic leukemia	2015	2014
Pembrolizumab	Keytruda	PD1; humanized IgG4	Melanoma	2015	2014
Ramucirumab	Cyramza	VEGFR2; human IgG1	Gastric cancer	2014	2014
Vedolizumab	Entyvio	α4β7 integrin; humanized IgG1	Ulcerative colitis, Crohn's disease	2014	2014
Siltuximab	Sylvant	IL-6; chimeric IgG1	Castleman's disease	2014	2014
Obinutuzumab	Gazyva	CD20; humanized IgG1; Glycoengineered	Chronic lymphocytic leukemia	2014	2013
Ado-trastuzumab emtansine	Kadcyla	HER2; humanized IgG1; immunoconjugate	Breast cancer	2013	2013
Raxibacumab	(Pending)	B. anthracis PA; human IgG1	Anthrax infection	NA	2012
Pertuzumab	Perjeta	HER2; humanized IgG1	Breast cancer	2013	2012
Brentuximab vedotin	Adcetris	CD30; chimeric IgG1; immunoconjugate	Hodgkin's lymphoma, systemic anaplastic large-cell lymphoma	2012	2011
Belimumab	Benlysta	BLyS; human IgG1	Systemic lupus erythematosus	2011	2011
Ipilimumab	Yervoy	CTLA-4; human IgG1	Metastatic melanoma	2011	2011
Denosumab	Prolia	RANK-L; human IgG2	Bone loss	2010	2010
Tocilizumab	RoActemra, Actemra	IL-6R; humanized IgG1	Rheumatoid arthritis	2009	2010
Ofatumumab	Arzerra	CD20; human IgG1	Chronic lymphocytic leukemia	2010	2009
Canakinumab	Ilaris	IL-1β; human IgG1	Muckle–Wells syndrome	2009	2009
Golimumab	Simponi	TNF; human IgG1	Rheumatoid and psoriatic arthritis, ankylosing spondylitis	2009	2009

(Continued)

Table 2.5 (Continued)

International nonproprietary name	Brand name	Target; format	Indication first approved or reviewed	First EU approval year	First US approval year
Ustekinumab	Stelara	IL-12/23; human IgG1	Psoriasis	2009	2009
Certolizumab pegol	Cimzia	TNF; humanized Fab, pegylated	Crohn's disease	2009	2008
Catumaxomab	Removab	EPCAM/CD3;rat/mouse bispecific mAb	Malignant ascites	2009	NA
Eculizumab	Soliris	C5; humanized IgG2/4	Paroxysmal nocturnal hemoglobinuria	2007	2007
Ranibizumab	Lucentis	VEGF; humanized IgG1 Fab	Macular degeneration	2007	2006
Panitumumab	Vectibix	EGFR; human IgG2	Colorectal cancer	2007	2006
Natalizumab	Tysabri	a4 Integrin; humanized IgG4	Multiple sclerosis	2006	2004
Bevacizumab	Avastin	VEGF; humanized IgG1	Colorectal cancer	2005	2004
Cetuximab	Erbitux	EGFR; chimeric IgG1	Colorectal cancer	2004	2004
Efalizumab	Raptiva	CD11a; humanized IgG1	Psoriasis	2004[a]	2003[a]
Omalizumab	Xolair	IgE; humanized IgG1	Asthma	2005	2003
Tositumomab-I131	Bexxar	CD20; murine IgG2a	Non-Hodgkin lymphoma	NA	2003[a]
Ibritumomab tiuxetan	Zevalin	CD20; murine IgG1	Non-Hodgkin lymphoma	2004	2002
Adalimumab	Humira	TNF; human IgG1	Rheumatoid arthritis	2003	2002

Alemtuzumab	MabCampath, Campath-1H; Lemtrada	CD52; humanized IgG1	Chronic myeloid leukemia[a]; multiple sclerosis	2001[a]; 2013	2001[a]; 2014
Gemtuzumab ozogamicin	Mylotarg	CD33; humanized IgG4	Acute myeloid leukemia	NA	2000[a]
Trastuzumab	Herceptin	HER2; humanized IgG1	Breast cancer	2000	1998
Infliximab	Remicade	TNF; chimeric IgG1	Crohn's disease	1999	1998
Palivizumab	Synagis	RSV; humanized IgG1	Prevention of respiratory syncytial virus infection	1999	1998
Basiliximab	Simulect	IL-2R; chimeric IgG1	Prevention of kidney transplant rejection	1998	1998
Daclizumab	Zenapax; Zinbryta	IL-2R; humanized IgG1	Prevention of kidney transplant rejection; multiple sclerosis	1999[a]; in review	1997[a]; in review
Rituximab	MabThera, Rituxan	CD20; chimeric IgG1	Non-Hodgkin lymphoma	1998	1997
Abciximab	Reopro	GPIIb/IIIa; chimeric IgG1 Fab	Prevention of blood clots in angioplasty	1995[b]	1994
Muromonab-CD3	Orthoclone Okt3	CD3; murine IgG2a	Reversal of kidney transplant rejection	1986[b]	1986[a]

Source: Janice M. Reichert, PhD, Reichert Biotechnology Consulting LLC; table updated December 14, 2015.

Note: Information current as of 14 December, 2015.

NA, not approved or in review in the European Union; information on review status in the United States not available.

[a] Withdrawn or marketing discontinued for the first approved indication.

[b] Country-specific approval.

2.6.4 Prefix

The prefix carries no special meaning. It is unique for each medicine and contribute to a well-sounding name. This means that antibodies with the same source and target substems are only distinguished by their prefix. Even antibodies targeting exactly the same structure are differently prefixed, such as the *ada*limumab and *go*limumab, both of which are TNF inhibitors but differ in their chemical structure.

2.6.5 Additional Words

A second word following the name of the antibody indicates that another substance is attached, which is done for several reasons. An antibody can be PEGylated (attached to molecules of polyethylene glycol) to slow down its degradation by enzymes and to decrease its immunogenicity; [14] this is shown by the word *pegol* as in *alacizumab pegol* [15].

A cytotoxic agent can be linked to an antitumor antibody for drug targeting purposes. The word *vedotin,* for example, stands for monomethyl auristatin E that is toxic by itself but predominantly affects cancer cells if used in conjugates like glembatumumab vedotin [16].

A chelator for binding a radioisotope can be attached. Pendetide, a derivative of pentetic acid, is used, for example, in capromab pendetide to chelate indium-111 [17]. If the drug contains a radioisotope, the name of the isotope precedes the name of the antibody. Consequently, *indium* (^{111}In) *capromab pendetide* is the name for the aforementioned example including indium-111 [17].

2.7 List of Therapeutic mAbs on Market or in Review in the European Union and the United States

Table 2.5 lists all mAbs (both INN and brand names listed) so approved in the United States and the European Union as of November 2015.

References

1. Köhler G, Milstein C (1975). Continuous cultures of fused cells secreting antibody of predefined specificity. *Nature* 256 (5517): 495–497.
2. Stern M, Herrmann R (2005). Overview of monoclonal antibodies in cancer therapy: present and promise. *Crit Rev Oncol Hematol* 54 (1): 11–29.
3. Hudson PJ, Souriau C (2003). Engineered antibodies. *Nat Med* 9 (1): 129–134.

4. Carter P, Presta L, Gorman CM, et al. (1992). Humanization of an anti-p185HER2 antibody for human cancer therapy. *Proc Natl Acad Sci U S A* 89 (10): 4285–4289.
5. Presta LG, Lahr SJ, Shields RL, et al. (1993). Humanization of an antibody directed against IgE. *J Immunol* 151 (5): 2623–2632.
6. Jefferis R, Marie-Paule L (2009). Human immunoglobulin allotypes. *MAbs* 1 (4): 332–338.
7. Chapman K, Pullen N, Lee C, et al. (2009). Preclinical development of monoclonal antibodies. *MAbs* 1 (5): 505–516.
8 Chothia C, Lesk AM, Tramontano A, et al. (1989). Conformations of immunoglobulin hypervariable regions. *Nature* 342 (6252): 877–883.
9. Woof J, Burton D (2004). Human antibody-Fc receptor interactions illuminated by crystal structures. *Nat Rev Immunol* 4 (2): 89–99.
10. Al-Lazikani B, Lesk AM, Chothia C (1997). Standard conformations for the canonical structures of immunoglobulins. *J Mol Biol* 273 (4): 927–948.
11. North B, Lehmann A, Dunbrack RL (2010). A new clustering of antibody CDR loop conformations. *J Mol Biol* 406 (2): 228–256.
12. Heyman B (1996). Complement and Fc-receptors in regulation of the antibody response. *Immunol Lett* 54 (2–3): 195–199.
13. World Health Organization (2011). The use of stems in the selection of International Nonproprietary Names (INN) for pharmaceutical substances (PDF). WHO/EMP/QSM/2011.3. Available at http://www.who.int/medicines/services/inn/StemBook_2011_Final.pdf. Archived from the original on October 26, 2013. Retrieved March 21, 2014.
14. Veronese F, Pasut G (2005). PEGylation, successful approach to drug delivery. *Drug Discov Today* 10 (21): 1451–1458.
15. World Health Organization (2008). International Nonproprietary Names for Pharmaceutical Substances (INN) (PDF). *WHO Drug Information* 22 (3): 221. Available at http://www.who.int/medicines/services/inn/INN_2008_list60.pdf. Retrieved June 8, 2010.
16. National Cancer Institute. Drug Dictionary: Glembatumumab vedotin. NCI drug Dictionary. Available at http://www.cancer.gov/publications/dictionaries (accessed on May 15, 2015).
17. World Health Organization (2009). International Nonproprietary Names for Pharmaceutical Substances (INN). *WHO Drug Information* 15 (1). Available at http://www.who.int/medicines/publications/druginformation/innlists/RL61final.pdf. Retrieved February 13, 2011.

3

Mechanism of Action for Therapeutic Antibodies

Yu Zhou and James D. Marks

Department of Anesthesia, San Francisco General Hospital, University of California, San Francisco, San Francisco, CA, USA

3.1 Introduction

The complexity of cancer and other diseases calls for the development of multiple antibody (Ab)-based drugs for even a single disease target. In the case of targeting tumor necrosis factor (TNF) for autoimmune diseases, five drugs (three full-length IgG mAbs, one TNFR-Fc fusion protein, and one pegylated Fab) were developed with different biological activities. The differences include Ab–antigen (Ag) binding properties, the ability to induce effector functions, and other factors. Meanwhile, a better understanding of Ab-based mechanisms of action (MOAs) has led to the expansion of pharmaceuticals with increased potency and greater precision.

Ab-based drugs work via multiple mechanisms, recapitulating some of the many mechanisms observed in the immune system. The ability to engineer novel Ab molecules has greatly expanded their range. Common MOAs proposed for existing Ab drugs include (i) disruption of ligand–receptor interaction; (ii) target cell elimination via antibody-dependent cellular cytotoxicity (ADCC), complement-dependent cytotoxicity (CDC), and antibody-dependent cellular phagocytosis (ADCP); (iii) engagement of cytotoxic T cell by bispecific Abs; (iv) receptor downregulation by enhanced internalization and degradation; and (v) targeted drug delivery. Here, we review these mechanisms using a number of Ab drug examples.

Biosimilars of Monoclonal Antibodies: A Practical Guide to Manufacturing, Preclinical, and Clinical Development, First Edition. Edited by Cheng Liu and K. John Morrow, Jr.

3.2 Blockade of Ligand–Receptor Interaction

The ligand–receptor interaction is the initial step in signaling transduction for a number of cellular events, including cell proliferation, adhesion, and migration. Some of these are associated with cancer progression. The blockage of the ligand–receptor interaction is thought to interrupt the chain of transmission. Antibodies (Abs) that bind either side of the ligand–receptor interface can disrupt the binding, interfering with signal transduction, and thus halting disease progression.

3.2.1 Ligand-Binding Abs

Among the approved therapeutic mAbs, the ligand targets include growth factors (VEGF), cytokines (TNF-a, IL-1b, IL-12/23, IL-6, IL-17a, RANKL, BAFF), soluble immune components (C5, IgE), as well as viral and bacterial targets (RSV and *Bacillus anthracis* PA) (see Table 3.1). To some ligand targets (TNF, VEGF), a variety of technologies were employed to develop multiple mAbs in different formats [11, 13]. The main mechanism for ligand-binding mAbs is blocking the interaction between the ligand and the cognate receptor, which results in the inhibition of the ligand-stimulated cellular signaling. While the MOAs for these types of mAbs are similar, the subsequent effects vary due to a number of intrinsic properties that include the mAb-binding epitope, affinity, and serum half-life, etc. (Table 3.1). In the following text we discuss three exemplary cases of how these mAb properties affect the clinical efficacy.

3.2.1.1 TNF mAbs

Multiple anti-TNF alpha Abs have been identified using mouse hybridoma technology and phage Ab display. Anti-TNF alpha Ab drugs are used clinically to treat various inflammatory and autoimmune diseases (Table 3.2). Infliximab has a slow off rate that contributes to the forming of stable Ab–TNF complex, while adalimumab binds directly at the TNFR and TNF interface with greater area than infliximab [42]. The epitopes of four TNF mAbs are similar but not identical, and the differences may contribute to the downstream signaling variations [44].

3.2.1.2 VEGF mAbs

Bevacizumab (Avastin, Genentech) is the first antiangiogenesis Ab drug developed and approved for treatment of colorectal cancer in 2004 [46]. Overexpression of VEGF stimulates neovasculature formation and facilitates tumor growth and metastasis via VEGF and VEGFR signaling activation on endothelial cells [47]. Signaling blockade by bevacizumab dampens the angiogenesis signal and starves the tumor cells leading to tumor shrinkage [12]. Novel VEGF mAbs that bind the epitope of the VEGFR–VEGF interface

Table 3.1 Antibodies against soluble ligands.

Ab name (brand name)	Targets	Ab format	Proposed MOA	References
Palivizumab (Synagis)	RSV	Humanized IgG1	Blocks binding of virus protein F to cell surface receptor; inhibits its entry into the cells	[1, 2]
Infliximab (Remicade)	TNF	Chimeric IgG1	Binds soluble and transmembrane TNF and blocks TNF binding to TNFRs; induces apoptosis of activated T cell and macrophage	[3–5]
Adalimumab (Humira)	TNF	Human IgG1	Binds soluble and transmembrane TNF and blocks TNF binding to TNFRs; induces apoptosis of activated T cell and macrophage	[6–8]
Omalizumab (Xolair)	IgE	Humanized IgG1	Binds soluble and membrane-bound IgE, blocks IgE binding to FcεRI, to reduce release of allergic response mediators, and downregulates FcεRI on mast cells and basophils	[9–11]
Bevacizumab (Avastin)	VEGF	Humanized IgG1	Inhibits angiogenesis by blocking VEGF-A signaling	[12]
Ranibizumab (Lucentis)	VEGF	Humanized IgG1 Fab	Blocks VEGF-A signaling but with faster clearance	[13]
Eculizumab (Soliris)	C5	Human IgG2/4	Binds C5, inhibits its cleavage to C5a and C5b, and prevents the generation of the terminal membrane attack complex C5b–C9	[14, 15]
Certolizumab pegol (Cimzia)	TNF	Humanized –Fab	Binds soluble and transmembrane TNF, blocks TNF binding to TNFRs; lack of placenta transfer	[16–21]
Golimumab (Simponi)	TNF	Human IgG1	Binds soluble and transmembrane TNF, blocks TNF binding to TNFRs; induces apoptosis of activated T cell and macrophage	[22]
Canakinumab (Ilaris)	IL1b	Human IgG1	Antagonizes proinflammatory signaling via IL1b	[23]

(Continued)

Table 3.1 (Continued)

Ab name (brand name)	Targets	Ab format	Proposed MOA	References
Ustekinumab (Stelara)	IL 12/23	Human IgG1	Blocks IL12/23 binding to their receptors, suppresses Th1 and Th17 signaling, and inhibits T-cell-mediated inflammatory responses	[24–27]
Denosumab (Prolia)	RANKL	Human IgG2	Blocks RANKL binding to RANK and inhibits RANK activation, thus reduces osteoclastogenesis	[28, 29]
Belimumab (Benlysta)	BLyS (BAFF)	Human IgG1	Blocks BLys binding to its receptors (TACI, BCMA, and BR-3), inhibits BLys-induced B-cell proliferation, and neutralizes BLys bioactivity *in vitro* and *in vivo*	[30–32]
Raxibacumab (Anthrax)	*Bacillus anthracis* PA	Human IgG1	Binds the protective antigen (PA) and neutralizes the anthrax toxin	[33–35]
Siltuximab (Sylvant)	IL-6	Chimeric IgG1	Neutralizes IL-6 activity by blocking the binding to IL6R	[36–38]
Secukinumab (Cosentyx)	IL-17a	Human IgG1	Blocks IL-17a binding to its receptors and inhibits proinflammatory signaling via IL-17a	[39]

Table 3.2 Properties of TNF mAbs.

	Epitope	Format	Affinity	Immune-related effect	Disease indications	References
Infliximab	C-D, E-F loops	Chimeric IgG1	117 pM	ADCC, CDC	RA, PA, PP, AS, Crohn's, UC	[40, 41]
Adalimumab	TNFR epitope	Human IgG1	115 pM	ADCC, CDC	RA, PA, PP, AS, Crohn's, PJIA	[42, 43]
Certolizumab pegol	TNFR–TNF Interface	Humanized Fab	89 pM	None	RA, Crohn's	[16, 44, 45]
Golimumab	TNFR–TNF Interface	Human IgG1	18 pM	ADCC, CDC	RA, PA, AS	[22]

Website: http://grantarthritis.com/Humira-Enbrel-Remicade-Simponi-Cimzia.

cross-react with both mouse and human VEGF and may block the VEGF and VEGFR interaction more effectively than Avastin [48, 49].

In the case of both TNF and VEGF mAbs, different Ab binding properties result in distinct biological activities, suggesting that a better understanding of their molecular biology clarified the signaling blockade necessary for the design of mAbs with the desired therapeutic function.

3.2.1.3 RANKL mAb

Denosumab (Prolia) is a fully human mAb against RANKL approved by the FDA to treat osteoporosis due to various causes. Denosumab mimics some of the properties of osteoprotegerin, a natural anti-RANKL protein that inhibits binding of RANK to its receptor. Denosumab binds RANKL at the binding site where both RANK and osteoprotegerin interact with the receptor [50]. However, Denosumab is specific to RANKL, while osteoprotegerin binds TRAIL and other TNF family members as well. RANKL is expressed as two forms: membrane bound and soluble protein. In order to avoid Ab-mediated depletion of osteoblasts expressing RANKL on the surface, denosumab was engineered as an IgG2, which has very low FcR binding and relatively low levels of complement activation. Denosumab has a long serum half-life (25.4 days) and can be administered every 6 months by subcutaneous injection, which has made it more convenient for osteoporosis treatment [29, 51].

3.2.2 Ab Binding to Cell Surface Receptors

Most of the Ab-based therapeutics for oncological applications target cell surface receptors (Table 3.3). The first two mAbs approved for cancer therapy are rituximab against CD20, which is overexpressed on non-Hodgkin's lymphoma B cells [111], and trastuzumab against HER2 overexpressed on malignant mammary epithelial cells in about 25% of metastatic breast cancers [64].

The actions of Abs targeting cell surface receptor are threefold: (i) inhibit receptor signaling by blocking the receptor binding to the ligand; (ii) downregulate the receptor surface display; and (iii) engage immune effector functions including antibody-dependent cellular cytotoxicity (ADCC), complement-dependent cytotoxicity (CDC), and ADCP.

While some therapeutic mAbs depend on the disruption of receptor–ligand interaction to be effective, such as anti-EGFR Abs, cetuximab, and panitumumab, [78, 79] they also elicit effector function thought to contribute significantly to the antitumor efficacy [79]. However, in some cases, the effector functions are not desired. For instances, the checkpoint mAbs nivolumab and pembrolizumab that target PD1 are both human IgG4 type, which lack complement activation and relatively lower FcR binding compared to IgG1 [112]. The anti-PD1 mAbs primarily interfere with the signaling between PD1 and PD-L1 and, thus, shut off the inhibition of cytotoxic T-cell activity and

Table 3.3 MAbs to cell surface receptors.

Ab name	Targets	Ab format	Proposed MOA	References
Muromonab-CD3 (Orthoclone Okt3)	CD3	Murine IgG2a	Induction of T-cell apoptosis, effector cell-mediated depletion, and Ab-mediated CD3-TCR downregulation by internalization and metabolism	[52–57]
Abciximab (Reopro)	GPIIβ/IIIα	Chimeric IgG1 Fab	Blockade of platelets adhesion and short plasma half-life but stays effective up to 48 h	[58]
Rituximab (MabThera, Rituxan)	CD20	Chimeric IgG1	Sensitizes cells to chemotherapy; induces apoptosis, ADCC, ADCP, and CDC	[59–61]
Basiliximab (Simulect)	IL2R (CD25)	Chimeric IgG1	Saturates receptor, blocks IL2R(CD25) activation of T cell, and prevents T cell from replication and from activating the B cells (production of Abs to transplant)	[62]
Daclizumab (Zenapax)	IL2R (CD25)	Humanized IgG1	Saturate receptor, blocks IL2R(CD25) activation of T cell, prevents T cell from replication and from activating the B cells (production of Abs to transplant)	[63]
Trastuzumab (Herceptin)	HER2	Humanized IgG1	HER2 downregulation, inhibition of dimerization, G1 arrest and induction of p27, prevention of HER2 cleavage, induction of immune mechanism, and inhibition of angiogenesis by decrease VEGF production	[64–67]
Gemtuzumab ozogamicin (Mylotarg)	CD33	Humanized IgG1	Ab-drug conjugate delivers extremely toxic DNA-disrupting small molecule to CD33-positive relapsed AML, withdrawn in 2010 due to the increase of patient death and lack of additive benefit	[68–70]

(Continued)

Table 3.3 (Continued)

Ab name	Targets	Ab format	Proposed MOA	References
Alemtuzumab (MabCampath, Campath-1H)	CD52	Humanized IgG1	ADCC	[71–73]
Tositumomab-I131 (Bexxar)	CD20	Murine IgG2a	CD20-directed radiotherapeutic drug, withdrawn in 2014 due to low sales	[74]
Efalizumab (Raptiva)	CD11α	Humanized IgG1	Blocks ICAM1 binding to LFA-1 a chain, withdrawn in 2009 due to fatal brain infection	[75, 76]
Cetuximab	EGFR	Chimeric IgG1	Blocks EGFR activation by EGF, shuts off the downstream signaling, and ADCC	[77–79]
Ibritumomab tiuxetan (Zevalin)	CD20	Murine IgG1	CD20-targeted radiation therapy: rituximab with tiuxetan as a linker to conjugate ^{111}In for imaging and ^{90}Y for radiation therapy with enhanced efficacy for rituximab-resistant patients	[80, 81]
Natalizumab (Tysabri)	α4 Integrin	Humanized IgG4	Blocks α4β1 and α4β7 binding to VCAM1 and MdCAM1; inhibits leukocyte adhesion and entry into the central nervous system and the intestine	[82, 83]
Panitumumab (Vectibix)	EGFR	Human IgG2	Blocks EGFR activation by EGF and shuts off the downstream signaling	[84]
Catumaxomab (Removab)	EPCAM/ CD3	Rat/mouse bispecific mAb	Enhances immune responses by directing a T cell and antigen-presenting cell to a tumor cell to link innate and adaptive immunity	[85, 86]
Tocilizumab (RoActemra, Actemra)	IL6R-α	Humanized IgG1	Blocks IL6 signaling through membrane-bound IL6R and soluble IL6R/GP130 and inhibits IL6-induced T and B cell activation.	[87–89]

Ofatumumab (Arzerra)	CD20	Human IgG1	Enhanced CDC activity compared to Rituximab	[90]
Ipilimumab (Yervoy)	CTLA-4	Human IgG1	Blocks CTLA-4-inhibiting signal to CTL, allowing CTLs to destroy tumor cells effectively	[91]
Brentuximab vedotin (Adcetris)	CD30	Chimeric IgG1, immunoconjugate	Targets CD30-positive lymphoma cells and delivers antimicrotubule drug intracellularly to destruct the cancer cells	[92, 93]
Pertuzumab (Perjeta)	HER2	Humanized IgG1	Blocks homo- and heterodimerization between HER2 and other EGF-receptor members	[94, 95]
Ado-trastuzumab emtansine (Kadcyla)	HER2	Humanized IgG1, immunoconjugate	Targets HER2-positive breast cancer cell and delivers toxic drug intracellularly	[96–98]
Vedolizumab (Entyvio)	α4β7 integrin	Humanized IgG1	Blocks α4β7 binding to MdCAM1; inhibits leukocyte adhesion, migration, and entry into the intestine, with reduced PML side effect compared to natalizumab	[99, 100]
Ramucirumab (Cyramza)	VEGFR2	Human IgG1	Blocks VEGF signaling to inhibit tumor angiogenesis	[101]
Obinutuzumab (Gazyva)	CD20	Humanized IgG1-glycoengineered	Targets CD20-positive lymphocyte with enhanced ADCC function	[102–104]
Pembrolizumab (Keytruda)	PD1	Humanized IgG4	Enhances CTL function by blocking PD1 and PD-L1 signaling	[105]
Nivolumab (Opdivo)	PD1	Human IgG4	Enhances CTL function by blocking PD1 and PD-L1 signaling	[106]
Dinutuximab (Unituxin)	GD2	Chimeric IgG1	Targets glycolipid GD2-postitive tumor cell and induces ADCC activity	[107–109]
Blinatumomab (Blincyto)	CD19/CD3	Murine bispecific tandem scFv	Directs a T cell to a malignant B cell to achieve anticancer effect	[110]

enhance the cellular antitumor effect. Obviously, the responses to anti-PD1 mAbs are correlated with the tumor PD-L1 level [113].

3.3 Target Depletion via ADCC and CDC

The therapeutic activity of a large proportion of mAbs results from the effector functions engaged by the Ab Fc region binding to Fc receptors and complement [59, 114]. The effector functions include ADCC, ADCP, and CDC.

For soluble ligands, mAb binding forms immune complexes that can be eliminated by effector cells; for cell surface receptors, the mAb-labeled Ag-expressing cells can be attacked by effector cells leading to cell destruction and depletion, which exceed the signaling inhibition effect *in vivo* (Table 3.3) [114]. The evidence of Fc receptor contribution to efficacy has been demonstrated by Fc knockout mice.

3.3.1 CD20 mAbs

The first CD20 chimeric Ab, rituximab, was approved by the FDA for treatment of non-Hodgkin's lymphoma. Its clinical efficacy correlated with the FCGR3A genotype [59]. Since approval of rituximab, two more CD20 Abs ofatumumab and obinutuzumab were approved for treatment of chronic lymphocytic leukemia (CLL) [119], and two CD20 drug conjugates tositumomab-I131 and ibritumomab tiuxetan were developed to treat non-Hodgkin's lymphoma [74, 80]. While all the anti-CD20 Ab therapeutics target CD20 and kill B cells, the MOAs are not the same. Although rituximab can inhibit the growth of CD20 positive B cells by G1 arrest and induction of apoptosis [60, 120], the efficacy for the naked IgG Abs predominantly resulted from Ab-engaged effector function including CDC [111], ADCP [121], and ADCC [61, 114]. Both ofatumumab and obinutuzumab were superior to rituximab when combined with chemotherapy [103, 119], by different mechanisms. Ofatumumab recognizes the smaller extracellular loop of CD20 that elicits exceptionally stronger CDC as compared to rituximab [90, 122, 123]. It is important to note that obinutuzumab is glycoengineered and nonfucosylated to enhance the CD16A binding and natural killer cell activity [102, 104].

Ab–radionuclide conjugates have shown additive anticancer effects compared to the naked IgGs. Ibritumomab tiuxetan is supplied as a single dosage kit, which can be linked to indium-111 for imaging and yttrium-90 for immune radiotherapy. Compared to rituximab, ibritumomab tiuxetan showed higher overall response rate (80 vs. 56%) and complete response rate (34 vs. 20%) and was effective for patients that are rituximab resistant [81, 124]. However, another Ab–radionuclide conjugate, tositumomab, was discontinued in February 2014, suggesting that a better understanding of Ab MOAs and the advent of new technology may advance the development of more potent and less toxic therapeutic drugs and thus shift the paradigm for Ab field.

3.3.2 HER2 mAbs

Herceptin was the only targeted therapy for treatment of metastatic breast cancer with HER2 overexpression for 14 years (1998–2012). The MOAs include signaling intervention leading to receptor downregulation and G1 arrest [64, 125], destruction of tumor cells by ADCC activity [67, 114], and adaptive immunity [126].

Unlike Herceptin that recognizes the domain 4 of HER2, another HER2 mAb, pertuzumab binds to domain 2 preventing both hetero- and homodimerization of HER2 with EGFR [95, 127]. Blocking receptor tyrosine kinase dimerization directly leads to signaling shutoff and thus inhibits cell proliferation, which is a consequence of signaling activation.

3.3.3 CD52 mAbs

Alemtuzumab is an approved mAb for CLL. It is engineered for maximal lympholytic activity and reduced immunogenic response. As such it is another example of a human IgG1 that takes advantage of ADCC to exert its *in vivo* antitumor efficacy [71, 112, 128]. CD52 is present on mature lymphocytes but not on stem cells, allowing specific mAb directed depletion of leukemic cells in chronic myeloid leukemia to achieve optimal clinical efficacy [72].

3.3.4 Engineered Fc for Enhanced Effector Function

Molecular modifications of the IgG constant domains have been shown to enhance ADCC function of existing therapeutic Abs [129, 130]. The modifications include defucosylation of Asn 297 in the Fc domain [130] and point mutations of the residues that are involved in the Fc and Fc receptor binding [129]. Obinutuzumab is the first approved Fc-glycoengineered type II anti-CD20 Ab with enhanced direct and immune effector cell-mediated cytotoxicity [131]. Although the obinutuzumab epitope overlaps that of rituximab, it does not induce CD20 redistribution into lipid rafts as do other type I mAbs in order to engage a strong CDC. Rather its MOA is through the triggering of an optimal ADCC response and a robust programmed cell death (PCD) response with the modified elbow hinge region of the IgG [102, 104, 131]. It is possible that the existing therapeutic mAbs could be substantially improved using engineered Fc regions, thereby improving clinical benefit without investing more on discovery of new mAbs.

3.4 Engaging Cytotoxic T Cell Through the Use of Bispecific Abs

Among many antitumor effectors, cytotoxic T cells play an essential role, and their numbers and functionality within tumors can reflect the clinical outcome in colorectal cancers [132]. However, antitumor T-cell responses are limited by immune escape mechanisms of cancer cells [133]. Monoclonal Abs targeting T-cell receptors can mimic the effect of Ag recognition, induce specific T-cell proliferation, and release cytotoxic factors, such as IL-2 and INFγ [134, 135]. Such T-cell-engaging mAbs can only lyse cells that express Fcγ receptors, which are not present on tumor cells. To direct the T cells to tumor cells, bispecific T-cell engagers (BiTEs) are designed to carry two single-chain Fv (scFv) Abs. One scFv binds the cell surface Ag overexpressed on the target tumor cells, and the other binds CD3, part of the T-cell receptor. The bispecific molecule is expected to bridge between the tumor cell and T cell leading to T-cell activation and destruction of the tumor cell on the other end of the bridge [86]. The first BiTE targeting CD19 and CD3, namely, blinatumomab, was approved to treat acute lymphoblastic leukemia (ALL) in the United States in December 2014 [110]. Another bispecific full-length IgG Ab with two Fab arms targeting EpCAM and CD19, respectively, was approved to treat ascites-based cancers in Europe in 2009 [85].

3.5 Receptor Downregulation by Enhanced Internalization and Degradation

Despite the fact that human IgG1 can recruit immune cells to destroy tumor cells, targeting EGFR with either cetuximab (IgG1) or panitumumab (IgG2), the monotherapy had only a moderate response rate, 11% [77] and 8% [84, 136], respectively. A synergistic downregulation of receptor tyrosine kinases by combining Ab-binding nonoverlapping epitopes leading to a more efficient shutoff of EGFR signaling was observed for both anti-EGFR and HER2 [137]. This approach may represent a novel strategy to Ab therapy. Symphogen has developed two Ab combinations for different diseases. The most promising EGFR two mAb-combo, Sym004, is in phase 3 clinical trial for metastatic colorectal cancer [138]. To achieve strong synergy-enhancing receptor internalization and downregulation, not only are the two mAbs designed to bind distinct epitopes, but they are also configured for the optimal epitopes.

With an extensive group of such antitumor Ab combinations in the pipeline, the mechanism and critical factors can be studied effectively and this information employed in the design of more potent Ab combinations [139, 140].

3.6 Targeted Drug Delivery

Antibody-drug conjugate (ADC) and Ab immunoliposomal drugs (MM-302). Two ADC drugs (anti-CD30 monomethyl auristatin) and T-DM1 (ado-trastuzumab emtansine) have been approved for clinical use [98]. The MOA includes four steps: Ab binding to the target, Ab–Ag complex internalization, drug release inside the cell, and the drug acting on essential cellular

Table 3.4 Armed mAbs.

Ab name	Trade name	Target	Ab format	Mechanism of action	References
Gemtuzumab ozogamicin	Mylotarg	CD33	Humanized IgG1-immunoconjugate	Ab-drug conjugate delivers extremely toxic DNA-disrupting small molecule to CD33-positive relapsed AML	[68]
Brentuximab vedotin	Adcetris	CD30	Immunoconjugate	Ab targeting; microtubule polymerization inhibition	[115, 116]
Trastuzumab emtansine	Kadcyla	HER2	Immunoconjugate	Ab targeting; microtubule polymerization inhibition	[97, 98]
MM302	Phase II clinical trial	HER2	Immunoliposome	Ab targeting; DNA replication inhibition	[117, 118]

functions, which include DNA and microtubule disruption by calicheamicin and taxane derivatives and DNA synthesis interruption by doxorubicin (Table 3.4). The efficiency of targeted drug delivery depends on the internalization of Ab–Ag complex and the intracellular breakdown of the ADC, which can be affected by the Ag endocytic property and the Ab–Ag binding properties.

In the case of the anti-CD33 Ab, the unconjugated Ab did not have *in vivo* efficacy despite its inhibiting activity of tumor cell proliferation *in vitro* [141]. However, the Ab-toxic-drug conjugate coupled with the endocytic property of CD33 resulted in therapeutic efficacy [68].

In contrast to the situation of ADCs, which can carry only a few drug molecules per Ab, each immunoliposome encapsulates over 10,000 drug molecules, allowing the use of less potent chemotherapeutic drugs [117]. Like the other ADCs, endocytic activity is a prerequisite for intracellular drug delivery although the immunoliposome can be deposited at the tumor lesion by enhanced permeability and a retention effect [142]. The immunoliposome may have an additional advantage over other ADCs, as it employs use of the same targeting particle to encapsulate both toxic drugs for therapy and radioisotopes for tumor visualization [143].

3.7 Summary

Ab drugs have offered clinical benefit for a number of cancers for almost two decades. With better understanding of the MOAs, more potent Ab drugs can be developed more rapidly, with less cost for discovery and more precisely tailored for various patient backgrounds.

References

1. Groothuis, J. R., E. A. Simoes and V. G. Hemming (1995). "Respiratory syncytial virus (RSV) infection in preterm infants and the protective effects of RSV immune globulin (RSVIG). Respiratory Syncytial Virus Immune Globulin Study Group." *Pediatrics* 95(4): 463–467.
2. Johnson, S., C. Oliver, G. A. Prince, V. G. Hemming, D. S. Pfarr, S. C. Wang, M. Dormitzer, J. O'Grady, S. Koenig, J. K. Tamura, R. Woods, G. Bansal, D. Couchenour, E. Tsao, W. C. Hall and J. F. Young (1997). "Development of a humanized monoclonal antibody (MEDI-493) with potent in vitro and in vivo activity against respiratory syncytial virus." *J Infect Dis* 176(5): 1215–1224.
3. Knight, D. M., H. Trinh, J. Le, S. Siegel, D. Shealy, M. McDonough, B. Scallon, M. A. Moore, J. Vilcek, P. Daddona et al. (1993). "Construction and initial characterization of a mouse-human chimeric anti-TNF antibody." *Mol Immunol* 30(16): 1443–1453.

4. Scallon, B. J., M. A. Moore, H. Trinh, D. M. Knight and J. Ghrayeb (1995). "Chimeric anti-TNF-alpha monoclonal antibody cA2 binds recombinant transmembrane TNF-alpha and activates immune effector functions." *Cytokine* 7(3): 251–259.
5. Siegel, S. A., D. J. Shealy, M. T. Nakada, J. Le, D. S. Woulfe, L. Probert, G. Kollias, J. Ghrayeb, J. Vilcek and P. E. Daddona (1995). "The mouse/human chimeric monoclonal antibody cA2 neutralizes TNF in vitro and protects transgenic mice from cachexia and TNF lethality in vivo." *Cytokine* 7(1): 15–25.
6. Furst, D. E., M. H. Schiff, R. M. Fleischmann, V. Strand, C. A. Birbara, D. Compagnone, S. A. Fischkoff and E. K. Chartash (2003). "Adalimumab, a fully human anti tumor necrosis factor-alpha monoclonal antibody, and concomitant standard antirheumatic therapy for the treatment of rheumatoid arthritis: results of STAR (Safety Trial of Adalimumab in Rheumatoid Arthritis)." *J Rheumatol* 30(12): 2563–2571.
7. Mease, P. J. (2007). "Adalimumab in the treatment of arthritis." *Ther Clin Risk Manag* 3(1): 133–148.
8. Salfeld, J., Z. Kayamakcalan, D. Tracey, A. Roberts and R. Kamen (1998). "Generation of fully human anti-TNF antibody D2E7." *Arthritis Rheum* 41(Suppl): S57.
9. Chang, T. W. (2000). "The pharmacological basis of anti-IgE therapy." *Nat Biotechnol* 18(2): 157–162.
10. MacGlashan, D. W., Jr., B. S. Bochner, D. C. Adelman, P. M. Jardieu, A. Togias, J. McKenzie-White, S. A. Sterbinsky, R. G. Hamilton and L. M. Lichtenstein (1997). "Down-regulation of Fc(epsilon)RI expression on human basophils during in vivo treatment of atopic patients with anti-IgE antibody." *J Immunol* 158(3): 1438–1445.
11. Presta, L. G., S. J. Lahr, R. L. Shields, J. P. Porter, C. M. Gorman, B. M. Fendly and P. M. Jardieu (1993). "Humanization of an antibody directed against IgE." *J Immunol* 151(5): 2623–2632.
12. Ferrara, N., K. J. Hillan, H. P. Gerber and W. Novotny (2004). "Discovery and development of bevacizumab, an anti-VEGF antibody for treating cancer." *Nat Rev Drug Discov* 3(5): 391–400.
13. Ferrara, N., L. Damico, N. Shams, H. Lowman and R. Kim (2006). "Development of ranibizumab, an anti-vascular endothelial growth factor antigen binding fragment, as therapy for neovascular age-related macular degeneration." *Retina* 26(8): 859–870.
14. Hillmen, P., N. S. Young, J. Schubert, R. A. Brodsky, G. Socie, P. Muus, A. Roth, J. Szer, M. O. Elebute, R. Nakamura, P. Browne, A. M. Risitano, A. Hill, H. Schrezenmeier, C. L. Fu, J. Maciejewski, S. A. Rollins, C. F. Mojcik, R. P. Rother and L. Luzzatto (2006). "The complement inhibitor eculizumab in paroxysmal nocturnal hemoglobinuria." *N Engl J Med* 355(12): 1233–1243.

15. Rother, R. P., S. A. Rollins, C. F. Mojcik, R. A. Brodsky and L. Bell (2007). "Discovery and development of the complement inhibitor eculizumab for the treatment of paroxysmal nocturnal hemoglobinuria." *Nat Biotechnol* 25(11): 1256–1264.
16. Fleischmann, R. and D. Shealy (2003). "Developing a new generation of TNFalpha antagonists for the treatment of rheumatoid arthritis." *Mol Interv* 3(6): 310–318.
17. Goel, N. and S. Stephens (2010). "Certolizumab pegol." *MAbs* 2(2): 137–147.
18. Nesbitt, A., G. Fossati, M. Bergin, P. Stephens, S. Stephens, R. Foulkes, D. Brown, M. Robinson and T. Bourne (2007). "Mechanism of action of certolizumab pegol (CDP870): in vitro comparison with other anti-tumor necrosis factor alpha agents." *Inflamm Bowel Dis* 13(11): 1323–1332.
19. Sandborn, W. J., B. G. Feagan, S. Stoinov, P. J. Honiball, P. Rutgeerts, D. Mason, R. Bloomfield, S. Schreiber and P. S. Investigators (2007). "Certolizumab pegol for the treatment of Crohn's disease." *N Engl J Med* 357(3): 228–238.
20. Schreiber, S., M. Khaliq-Kareemi, I. C. Lawrance, O. O. Thomsen, S. B. Hanauer, J. McColm, R. Bloomfield, W. J. Sandborn and P. S. Investigators (2007). "Maintenance therapy with certolizumab pegol for Crohn's disease." *N Engl J Med* 357(3): 239–250.
21. Winter, T. A., W. J. Sandborn, W. J. de Villiers and S. Schreiber (2007). "Treatment of Crohn's disease with certolizumab pegol." *Expert Rev Clin Immunol* 3(5): 683–694.
22. Shealy, D. J., A. Cai, K. Staquet, A. Baker, E. R. Lacy, L. Johns, O. Vafa, G. Gunn, 3rd, S. Tam, S. Sague, D. Wang, M. Brigham-Burke, P. Dalmonte, E. Emmell, B. Pikounis, P. J. Bugelski, H. Zhou, B. J. Scallon and J. Giles-Komar (2010). "Characterization of golimumab, a human monoclonal antibody specific for human tumor necrosis factor alpha." *MAbs* 2(4): 428–439.
23. Lachmann, H. J., I. Kone-Paut, J. B. Kuemmerle-Deschner, K. S. Leslie, E. Hachulla, P. Quartier, X. Gitton, A. Widmer, N. Patel, P. N. Hawkins and Canakinumab in CAPS Study Group (2009). "Use of canakinumab in the cryopyrin-associated periodic syndrome." *N Engl J Med* 360(23): 2416–2425.
24. Koutruba, N., J. Emer and M. Lebwohl (2010). "Review of ustekinumab, an interleukin-12 and interleukin-23 inhibitor used for the treatment of plaque psoriasis." *Ther Clin Risk Manag* 6: 123–141.
25. Leonardi, C. L., A. B. Kimball, K. A. Papp, N. Yeilding, C. Guzzo, Y. Wang, S. Li, L. T. Dooley, K. B. Gordon and PHOENIX 1 Study Investigators (2008). "Efficacy and safety of ustekinumab, a human interleukin-12/23 monoclonal antibody, in patients with psoriasis: 76-week results from a randomised, double-blind, placebo-controlled trial (PHOENIX 1)." *Lancet* 371(9625): 1665–1674.
26. Papp, K. A., R. G. Langley, M. Lebwohl, G. G. Krueger, P. Szapary, N. Yeilding, C. Guzzo, M. C. Hsu, Y. Wang, S. Li, L. T. Dooley, K. Reich and PHOENIX 1 Study Investigators (2008). "Efficacy and safety of ustekinumab, a human interleukin-12/23 monoclonal antibody, in patients with psoriasis: 52-week results from a randomised, double-blind, placebo-controlled trial (PHOENIX 2)." *Lancet* 371(9625): 1675–1684.

27. Sandborn, W. J., B. G. Feagan, R. N. Fedorak, E. Scherl, M. R. Fleisher, S. Katz, J. Johanns, M. Blank, P. Rutgeerts and Ustekinumab Crohn's Disease Study Group (2008). "A randomized trial of Ustekinumab, a human interleukin-12/23 monoclonal antibody, in patients with moderate-to-severe Crohn's disease." *Gastroenterology* 135(4): 1130–1141.
28. Kostenuik, P. J., H. Q. Nguyen, J. McCabe, K. S. Warmington, C. Kurahara, N. Sun, C. Chen, L. Li, R. C. Cattley, G. Van, S. Scully, R. Elliott, M. Grisanti, S. Morony, H. L. Tan, F. Asuncion, X. Li, M. S. Ominsky, M. Stolina, D. Dwyer, W. C. Dougall, N. Hawkins, W. J. Boyle, W. S. Simonet and J. K. Sullivan (2009). "Denosumab, a fully human monoclonal antibody to RANKL, inhibits bone resorption and increases BMD in knock-in mice that express chimeric (murine/human) RANKL." *J Bone Miner Res* 24(2): 182–195.
29. McClung, M. R., E. M. Lewiecki, S. B. Cohen, M. A. Bolognese, G. C. Woodson, A. H. Moffett, M. Peacock, P. D. Miller, S. N. Lederman, C. H. Chesnut, D. Lain, A. J. Kivitz, D. L. Holloway, C. Zhang, M. C. Peterson, P. J. Bekker and AMG 162 Bone Loss Study Group (2006). "Denosumab in postmenopausal women with low bone mineral density." *N Engl J Med* 354(8): 821–831.
30. Baker, K. P., B. M. Edwards, S. H. Main, G. H. Choi, R. E. Wager, W. G. Halpern, P. B. Lappin, T. Riccobene, D. Abramian, L. Sekut, B. Sturm, C. Poortman, R. R. Minter, C. L. Dobson, E. Williams, S. Carmen, R. Smith, V. Roschke, D. M. Hilbert, T. J. Vaughan and V. R. Albert (2003). "Generation and characterization of LymphoStat-B, a human monoclonal antibody that antagonizes the bioactivities of B lymphocyte stimulator." *Arthritis Rheum* 48(11): 3253–3265.
31. Furie, R., M. Petri, O. Zamani, R. Cervera, D. J. Wallace, D. Tegzova, J. Sanchez-Guerrero, A. Schwarting, J. T. Merrill, W. W. Chatham, W. Stohl, E. M. Ginzler, D. R. Hough, Z. J. Zhong, W. Freimuth, R. F. van Vollenhoven and B.-S. Group (2011). "A phase III, randomized, placebo-controlled study of belimumab, a monoclonal antibody that inhibits B lymphocyte stimulator, in patients with systemic lupus erythematosus." *Arthritis Rheum* 63(12): 3918–3930.
32. Stohl, W. and D. M. Hilbert (2012). "The discovery and development of belimumab: the anti-BLyS-lupus connection." *Nat Biotechnol* 30(1): 69–77.
33. Chen, Z., M. Moayeri and R. Purcell (2011). "Monoclonal antibody therapies against anthrax." *Toxins (Basel)* 3(8): 1004–1019.
34. Froude, J. W., 2nd, P. Thullier and T. Pelat (2011). "Antibodies against anthrax: mechanisms of action and clinical applications." *Toxins (Basel)* 3(11): 1433–1452.
35. Migone, T. S., G. M. Subramanian, J. Zhong, L. M. Healey, A. Corey, M. Devalaraja, L. Lo, S. Ullrich, J. Zimmerman, A. Chen, M. Lewis, G. Meister, K. Gillum, D. Sanford, J. Mott and S. D. Bolmer (2009). "Raxibacumab for the treatment of inhalational anthrax." *N Engl J Med* 361(2): 135–144.

36. van Rhee, F., L. Fayad, P. Voorhees, R. Furman, S. Lonial, H. Borghaei, L. Sokol, J. Crawford, M. Cornfeld, M. Qi, X. Qin, J. Herring, C. Casper and R. Kurzrock (2010). "Siltuximab, a novel anti-interleukin-6 monoclonal antibody, for Castleman's disease." *J Clin Oncol* 28(23): 3701–3708.
37. van Zaanen, H. C., H. M. Lokhorst, L. A. Aarden, H. J. Rensink, S. O. Warnaar, J. van der Lelie and M. H. van Oers (1998). "Chimaeric anti-interleukin 6 monoclonal antibodies in the treatment of advanced multiple myeloma: a phase I dose-escalating study." *Br J Haematol* 102(3): 783–790.
38. van Zaanen, H. C., H. M. Lokhorst, L. A. Aarden, H. J. Rensink, S. O. Warnaar and M. H. Van Oers (1998). "Blocking interleukin-6 activity with chimeric anti-IL6 monoclonal antibodies in multiple myeloma: effects on soluble IL6 receptor and soluble gp130." *Leuk Lymphoma* 31(5–6): 551–558.
39. Patel, D. D., D. M. Lee, F. Kolbinger and C. Antoni (2013). "Effect of IL-17A blockade with secukinumab in autoimmune diseases." *Ann Rheum Dis* 72(Suppl 2): ii116–ii123.
40. Liang, S., J. Dai, S. Hou, L. Su, D. Zhang, H. Guo, S. Hu, H. Wang, Z. Rao, Y. Guo and Z. Lou (2013). "Structural basis for treating tumor necrosis factor alpha (TNFalpha)-associated diseases with the therapeutic antibody infliximab." *J Biol Chem* 288(19): 13799–13807.
41. Scallon, B., A. Cai, N. Solowski, A. Rosenberg, X. Y. Song, D. Shealy and C. Wagner (2002). "Binding and functional comparisons of two types of tumor necrosis factor antagonists." *J Pharmacol Exp Ther* 301(2): 418–426.
42. Hu, S., S. Liang, H. Guo, D. Zhang, H. Li, X. Wang, W. Yang, W. Qian, S. Hou, H. Wang, Y. Guo and Z. Lou (2013). "Comparison of the inhibition mechanisms of adalimumab and infliximab in treating tumor necrosis factor alpha-associated diseases from a molecular view." *J Biol Chem* 288(38): 27059–27067.
43. Kaymakcalan, Z., P. Sakorafas, S. Bose, S. Scesney, L. Xiong, D. K. Hanzatian, J. Salfeld and E. H. Sasso (2009). "Comparisons of affinities, avidities, and complement activation of adalimumab, infliximab, and etanercept in binding to soluble and membrane tumor necrosis factor." *Clin Immunol* 131(2): 308–316.
44. Henry, A. I., G. Haiping and A. M. Nesbitt (2011). Mapping the certolizumab pegol epitope on TNF and comparison with infliximab, adalimumab and etanercept. *J Transl Med* 9(Suppl 2): 43.
45. Pasut, G. (2014). "Pegylation of biological molecules and potential benefits: pharmacological properties of certolizumab pegol." *BioDrugs* 28(Suppl 1): S15–S23.
46. Hurwitz, H., L. Fehrenbacher, W. Novotny, T. Cartwright, J. Hainsworth, W. Heim, J. Berlin, A. Baron, S. Griffing, E. Holmgren, N. Ferrara, G. Fyfe, B. Rogers, R. Ross and F. Kabbinavar (2004). "Bevacizumab plus irinotecan, fluorouracil, and leucovorin for metastatic colorectal cancer." *N Engl J Med* 350(23): 2335–2342.

47. Ellis, L. M. (2006). "Mechanisms of action of bevacizumab as a component of therapy for metastatic colorectal cancer." *Semin Oncol* 33(5 Suppl 10): S1–S7.

48. Fuh, G., P. Wu, W. C. Liang, M. Ultsch, C. V. Lee, B. Moffat and C. Wiesmann (2006). "Structure-function studies of two synthetic anti-vascular endothelial growth factor Fabs and comparison with the Avastin Fab." *J Biol Chem* 281(10): 6625–6631.

49. Liang, W. C., X. Wu, F. V. Peale, C. V. Lee, Y. G. Meng, J. Gutierrez, L. Fu, A. K. Malik, H. P. Gerber, N. Ferrara and G. Fuh (2006). "Cross-species vascular endothelial growth factor (VEGF)-blocking antibodies completely inhibit the growth of human tumor xenografts and measure the contribution of stromal VEGF." *J Biol Chem* 281(2): 951–961.

50. Schieferdecker, A., M. Voigt, K. Riecken, F. Braig, T. Schinke, S. Loges, C. Bokemeyer, B. Fehse and M. Binder (2014). "Denosumab mimics the natural decoy receptor osteoprotegerin by interacting with its major binding site on RANKL." *Oncotarget* 5(16): 6647–6653.

51. Kostenuik, P. J. (2005). "Osteoprotegerin and RANKL regulate bone resorption, density, geometry and strength." *Curr Opin Pharmacol* 5(6): 618–625.

52. Chatenoud, L., M. F. Baudrihaye, N. Chkoff, H. Kreis, G. Goldstein and J. F. Bach (1986). "Restriction of the human in vivo immune response against the mouse monoclonal antibody OKT3." *J Immunol* 137(3): 830–838.

53. Chatenoud, L., M. F. Baudrihaye, H. Kreis, G. Goldstein, J. Schindler and J. F. Bach (1982). "Human in vivo antigenic modulation induced by the anti-T cell OKT3 monoclonal antibody." *Eur J Immunol* 12(11): 979–982.

54. Herold, K. C., W. Hagopian, J. A. Auger, E. Poumian-Ruiz, L. Taylor, D. Donaldson, S. E. Gitelman, D. M. Harlan, D. Xu, R. A. Zivin and J. A. Bluestone (2002). "Anti-CD3 monoclonal antibody in new-onset type 1 diabetes mellitus." *N Engl J Med* 346(22): 1692–1698.

55. Janssen, O., S. Wesselborg and D. Kabelitz (1992). "Immunosuppression by OKT3—induction of programmed cell death (apoptosis) as a possible mechanism of action." *Transplantation* 53(1): 233–234.

56. Kung, P., G. Goldstein, E. L. Reinherz and S. F. Schlossman (1979). "Monoclonal antibodies defining distinctive human T cell surface antigens." *Science* 206(4416): 347–349.

57. Wesselborg, S., O. Janssen and D. Kabelitz (1993). "Induction of activation-driven death (apoptosis) in activated but not resting peripheral blood T cells." *J Immunol* 150(10): 4338–4345.

58. Tam, S. H., P. M. Sassoli, R. E. Jordan and M. T. Nakada (1998). "Abciximab (ReoPro, chimeric 7E3 Fab) demonstrates equivalent affinity and functional blockade of glycoprotein IIb/IIIa and alpha(v)beta3 integrins." *Circulation* 98(11): 1085–1091.

59. Cartron, G., L. Dacheux, G. Salles, P. Solal-Celigny, P. Bardos, P. Colombat and H. Watier (2002). "Therapeutic activity of humanized anti-CD20 monoclonal antibody and polymorphism in IgG Fc receptor FcgammaRIIIa gene." *Blood* 99(3): 754–758.

60. Taji, H., Y. Kagami, Y. Okada, M. Andou, Y. Nishi, H. Saito, M. Seto and Y. Morishima (1998). "Growth inhibition of CD20-positive B lymphoma cell lines by IDEC-C2B8 anti-CD20 monoclonal antibody." *Jpn J Cancer Res* 89(7): 748–756.
61. Weiner, G. J. (2010). "Rituximab: mechanism of action." *Semin Hematol* 47(2): 115–123.
62. Nashan, B., R. Moore, P. Amlot, A. G. Schmidt, K. Abeywickrama and J. P. Soulillou (1997). "Randomised trial of basiliximab versus placebo for control of acute cellular rejection in renal allograft recipients. CHIB 201 International Study Group." *Lancet* 350(9086): 1193–1198.
63. Vincenti, F., R. Kirkman, S. Light, G. Bumgardner, M. Pescovitz, P. Halloran, J. Neylan, A. Wilkinson, H. Ekberg, R. Gaston, L. Backman and J. Burdick (1998). "Interleukin-2-receptor blockade with daclizumab to prevent acute rejection in renal transplantation. Daclizumab Triple Therapy Study Group." *N Engl J Med* 338(3): 161–165.
64. Baselga, J. and J. Albanell (2001). "Mechanism of action of anti-HER2 monoclonal antibodies." *Ann Oncol* 12(Suppl 1): S35–S41.
65. Baselga, J., L. Norton, J. Albanell, Y. M. Kim and J. Mendelsohn (1998). "Recombinant humanized anti-HER2 antibody (Herceptin) enhances the antitumor activity of paclitaxel and doxorubicin against HER2/neu overexpressing human breast cancer xenografts." *Cancer Res* 58(13): 2825–2831.
66. Fendly, B. M., M. Winget, R. M. Hudziak, M. T. Lipari, M. A. Napier and A. Ullrich (1990). "Characterization of murine monoclonal antibodies reactive to either the human epidermal growth factor receptor or HER2/neu gene product." *Cancer Res* 50(5): 1550–1558.
67. Lewis, G. D., I. Figari, B. Fendly, W. L. Wong, P. Carter, C. Gorman and H. M. Shepard (1993). "Differential responses of human tumor cell lines to anti-p185HER2 monoclonal antibodies." *Cancer Immunol Immunother* 37(4): 255–263.
68. Bernstein, I. D. (2000). "Monoclonal antibodies to the myeloid stem cells: therapeutic implications of CMA-676, a humanized anti-CD33 antibody calicheamicin conjugate." *Leukemia* 14(3): 474–475.
69. Bross, P. F., J. Beitz, G. Chen, X. H. Chen, E. Duffy, L. Kieffer, S. Roy, R. Sridhara, A. Rahman, G. Williams and R. Pazdur (2001). "Approval summary: gemtuzumab ozogamicin in relapsed acute myeloid leukemia." *Clin Cancer Res* 7(6): 1490–1496.
70. Laszlo, G. S., E. H. Estey and R. B. Walter (2014). "The past and future of CD33 as therapeutic target in acute myeloid leukemia." *Blood Rev* 28(4): 143–153.
71. Greenwood, J., M. Clark and H. Waldmann (1993). "Structural motifs involved in human IgG antibody effector functions." *Eur J Immunol* 23(5): 1098–1104.
72. Keating, M. J., S. O'Brien, D. Kontoyiannis, W. Plunkett, C. Koller, M. Beran, S. Lerner and H. Kantarjian (2002). "Results of first salvage therapy for patients refractory to a fludarabine regimen in chronic lymphocytic leukemia." *Leuk Lymphoma* 43(9): 1755–1762.

73. Waldmann, H. and G. Hale (2005). "CAMPATH: from concept to clinic." *Philos Trans R Soc Lond B Biol Sci* 360(1461): 1707–1711.
74. Kaminski, M. S., A. D. Zelenetz, O. W. Press, M. Saleh, J. Leonard, L. Fehrenbacher, T. A. Lister, R. J. Stagg, G. F. Tidmarsh, S. Kroll, R. L. Wahl, S. J. Knox and J. M. Vose (2001). "Pivotal study of iodine I 131 tositumomab for chemotherapy-refractory low-grade or transformed low-grade B-cell non-Hodgkin's lymphomas." *J Clin Oncol* 19(19): 3918–3928.
75. Papp, K., R. Bissonnette, J. G. Krueger, W. Carey, D. Gratton, W. P. Gulliver, H. Lui, C. W. Lynde, A. Magee, D. Minier, J. P. Ouellet, P. Patel, J. Shapiro, N. H. Shear, S. Kramer, P. Walicke, R. Bauer, R. L. Dedrick, S. S. Kim, M. White and M. R. Garovoy (2001). "The treatment of moderate to severe psoriasis with a new anti-CD11a monoclonal antibody." *J Am Acad Dermatol* 45(5): 665–674.
76. Werther, W. A., T. N. Gonzalez, S. J. O'Connor, S. McCabe, B. Chan, T. Hotaling, M. Champe, J. A. Fox, P. M. Jardieu, P. W. Berman and L. G. Presta (1996). "Humanization of an anti-lymphocyte function-associated antigen (LFA)-1 monoclonal antibody and reengineering of the humanized antibody for binding to rhesus LFA-1." *J Immunol* 157(11): 4986–4995.
77. Cunningham, D., Y. Humblet, S. Siena, D. Khayat, H. Bleiberg, A. Santoro, D. Bets, M. Mueser, A. Harstrick, C. Verslype, I. Chau and E. Van Cutsem (2004). "Cetuximab monotherapy and cetuximab plus irinotecan in irinotecan-refractory metastatic colorectal cancer." *N Engl J Med* 351(4): 337–345.
78. Fan, Z., H. Masui, I. Altas and J. Mendelsohn (1993). "Blockade of epidermal growth factor receptor function by bivalent and monovalent fragments of 225 anti-epidermal growth factor receptor monoclonal antibodies." *Cancer Res* 53(18): 4322–4328.
79. Goldstein, N. I., M. Prewett, K. Zuklys, P. Rockwell and J. Mendelsohn (1995). "Biological efficacy of a chimeric antibody to the epidermal growth factor receptor in a human tumor xenograft model." *Clin Cancer Res* 1(11): 1311–1318.
80. Witzig, T. E., I. W. Flinn, L. I. Gordon, C. Emmanouilides, M. S. Czuczman, M. N. Saleh, L. Cripe, G. Wiseman, T. Olejnik, P. S. Multani and C. A. White (2002). "Treatment with ibritumomab tiuxetan radioimmunotherapy in patients with rituximab-refractory follicular non-Hodgkin's lymphoma." *J Clin Oncol* 20(15): 3262–3269.
81. Witzig, T. E., L. I. Gordon, F. Cabanillas, M. S. Czuczman, C. Emmanouilides, R. Joyce, B. L. Pohlman, N. L. Bartlett, G. A. Wiseman, N. Padre, A. J. Grillo-Lopez, P. Multani and C. A. White (2002). "Randomized controlled trial of yttrium-90-labeled ibritumomab tiuxetan radioimmunotherapy versus rituximab immunotherapy for patients with relapsed or refractory low-grade, follicular, or transformed B-cell non-Hodgkin's lymphoma." *J Clin Oncol* 20(10): 2453–2463.
82. Miller, D. H., O. A. Khan, W. A. Sheremata, L. D. Blumhardt, G. P. Rice, M. A. Libonati, A. J. Willmer-Hulme, C. M. Dalton, K. A. Miszkiel, P. W. O'Connor

and International Natalizumab Multiple Sclerosis Trial Group (2003). "A controlled trial of natalizumab for relapsing multiple sclerosis." *N Engl J Med* 348(1): 15–23.

83. von Andrian, U. H. and B. Engelhardt (2003). "Alpha4 integrins as therapeutic targets in autoimmune disease." *N Engl J Med* 348(1): 68–72.
84. Van Cutsem, E., M. Peeters, S. Siena, Y. Humblet, A. Hendlisz, B. Neyns, J. L. Canon, J. L. Van Laethem, J. Maurel, G. Richardson, M. Wolf and R. G. Amado (2007). "Open-label phase III trial of panitumumab plus best supportive care compared with best supportive care alone in patients with chemotherapy-refractory metastatic colorectal cancer." *J Clin Oncol* 25(13): 1658–1664.
85. Burges, A., P. Wimberger, C. Kumper, V. Gorbounova, H. Sommer, B. Schmalfeldt, J. Pfisterer, M. Lichinitser, A. Makhson, V. Moiseyenko, A. Lahr, E. Schulze, M. Jager, M. A. Strohlein, M. M. Heiss, T. Gottwald, H. Lindhofer and R. Kimmig (2007). "Effective relief of malignant ascites in patients with advanced ovarian cancer by a trifunctional anti-EpCAM x anti-CD3 antibody: a phase I/II study." *Clin Cancer Res* 13(13): 3899–3905.
86. Zeidler, R., G. Reisbach, B. Wollenberg, S. Lang, S. Chaubal, B. Schmitt and H. Lindhofer (1999). "Simultaneous activation of T cells and accessory cells by a new class of intact bispecific antibody results in efficient tumor cell killing." *J Immunol* 163(3): 1246–1252.
87. Betts, B. C., E. T. St Angelo, M. Kennedy and J. W. Young (2011). "Anti-IL6-receptor-alpha (tocilizumab) does not inhibit human monocyte-derived dendritic cell maturation or alloreactive T-cell responses." *Blood* 118(19): 5340–5343.
88. Naka, T., N. Nishimoto and T. Kishimoto (2002). "The paradigm of IL-6: from basic science to medicine." *Arthritis Res* 4(Suppl 3): S233–S242.
89. Sebba, A. (2008). "Tocilizumab: the first interleukin-6-receptor inhibitor." *Am J Health Syst Pharm* 65(15): 1413–1418.
90. Teeling, J. L., R. R. French, M. S. Cragg, J. van den Brakel, M. Pluyter, H. Huang, C. Chan, P. W. Parren, C. E. Hack, M. Dechant, T. Valerius, J. G. van de Winkel and M. J. Glennie (2004). "Characterization of new human CD20 monoclonal antibodies with potent cytolytic activity against non-Hodgkin lymphomas." *Blood* 104(6): 1793–1800.
91. Phan, G. Q., J. C. Yang, R. M. Sherry, P. Hwu, S. L. Topalian, D. J. Schwartzentruber, N. P. Restifo, L. R. Haworth, C. A. Seipp, L. J. Freezer, K. E. Morton, S. A. Mavroukakis, P. H. Duray, S. M. Steinberg, J. P. Allison, T. A. Davis and S. A. Rosenberg (2003). "Cancer regression and autoimmunity induced by cytotoxic T lymphocyte-associated antigen 4 blockade in patients with metastatic melanoma." *Proc Natl Acad Sci U S A* 100(14): 8372–8377.
92. Francisco, J. A., C. G. Cerveny, D. L. Meyer, B. J. Mixan, K. Klussman, D. F. Chace, S. X. Rejniak, K. A. Gordon, R. DeBlanc, B. E. Toki, C. L. Law, S. O. Doronina, C. B. Siegall, P. D. Senter and A. F. Wahl (2003). "cAC10-vcMMAE, an anti-CD30-monomethyl auristatin E conjugate with potent and selective antitumor activity." *Blood* 102(4): 1458–1465.

93. Younes, A., N. L. Bartlett, J. P. Leonard, D. A. Kennedy, C. M. Lynch, E. L. Sievers and A. Forero-Torres (2010). "Brentuximab vedotin (SGN-35) for relapsed CD30-positive lymphomas." *N Engl J Med* 363(19): 1812–1821.
94. Badache, A. and N. E. Hynes (2004). "A new therapeutic antibody masks ErbB2 to its partners." *Cancer Cell* 5(4): 299–301.
95. Franklin, M. C., K. D. Carey, F. F. Vajdos, D. J. Leahy, A. M. de Vos and M. X. Sliwkowski (2004). "Insights into ErbB signaling from the structure of the ErbB2-pertuzumab complex." *Cancer Cell* 5(4): 317–328.
96. Isakoff, S. J. and J. Baselga (2011). "Trastuzumab-DM1: building a chemotherapy-free road in the treatment of human epidermal growth factor receptor 2-positive breast cancer." *J Clin Oncol* 29(4): 351–354.
97. Junttila, T. T., G. Li, K. Parsons, G. L. Phillips and M. X. Sliwkowski (2011). "Trastuzumab-DM1 (T-DM1) retains all the mechanisms of action of trastuzumab and efficiently inhibits growth of lapatinib insensitive breast cancer." *Breast Cancer Res Treat* 128(2): 347–356.
98. Lewis Phillips, G. D., G. Li, D. L. Dugger, L. M. Crocker, K. L. Parsons, E. Mai, W. A. Blattler, J. M. Lambert, R. V. Chari, R. J. Lutz, W. L. Wong, F. S. Jacobson, H. Koeppen, R. H. Schwall, S. R. Kenkare-Mitra, S. D. Spencer and M. X. Sliwkowski (2008). "Targeting HER2-positive breast cancer with trastuzumab-DM1, an antibody-cytotoxic drug conjugate." *Cancer Res* 68(22): 9280–9290.
99. Feagan, B. G., P. Rutgeerts, B. E. Sands, S. Hanauer, J. F. Colombel, W. J. Sandborn, G. Van Assche, J. Axler, H. J. Kim, S. Danese, I. Fox, C. Milch, S. Sankoh, T. Wyant, J. Xu, A. Parikh and G. S. Group (2013). "Vedolizumab as induction and maintenance therapy for ulcerative colitis." *N Engl J Med* 369(8): 699–710.
100. Sandborn, W. J., B. G. Feagan, P. Rutgeerts, S. Hanauer, J. F. Colombel, B. E. Sands, M. Lukas, R. N. Fedorak, S. Lee, B. Bressler, I. Fox, M. Rosario, S. Sankoh, J. Xu, K. Stephens, C. Milch, A. Parikh and GEMINI 2 Study Group (2013). "Vedolizumab as induction and maintenance therapy for Crohn's disease." *N Engl J Med* 369(8): 711–721.
101. Franklin, M. C., E. C. Navarro, Y. Wang, S. Patel, P. Singh, Y. Zhang, K. Persaud, A. Bari, H. Griffith, L. Shen, P. Balderes and P. Kussie (2011). "The structural basis for the function of two anti-VEGF receptor 2 antibodies." *Structure* 19(8): 1097–1107.
102. Alduaij, W., A. Ivanov, J. Honeychurch, E. J. Cheadle, S. Potluri, S. H. Lim, K. Shimada, C. H. Chan, A. Tutt, S. A. Beers, M. J. Glennie, M. S. Cragg and T. M. Illidge (2011). "Novel type II anti-CD20 monoclonal antibody (GA101) evokes homotypic adhesion and actin-dependent, lysosome-mediated cell death in B-cell malignancies." *Blood* 117(17): 4519–4529.
103. Goede, V., K. Fischer, R. Busch, A. Engelke, B. Eichhorst, C. M. Wendtner, T. Chagorova, J. de la Serna, M. S. Dilhuydy, T. Illmer, S. Opat, C. J. Owen, O. Samoylova, K. A. Kreuzer, S. Stilgenbauer, H. Dohner, A. W. Langerak,

M. Ritgen, M. Kneba, E. Asikanius, K. Humphrey, M. Wenger and M. Hallek (2014). "Obinutuzumab plus chlorambucil in patients with CLL and coexisting conditions." *N Engl J Med* 370(12): 1101–1110.

104. Niederfellner, G., A. Lammens, O. Mundigl, G. J. Georges, W. Schaefer, M. Schwaiger, A. Franke, K. Wiechmann, S. Jenewein, J. W. Slootstra, P. Timmerman, A. Brannstrom, F. Lindstrom, E. Mossner, P. Umana, K. P. Hopfner and C. Klein (2011). "Epitope characterization and crystal structure of GA101 provide insights into the molecular basis for type I/II distinction of CD20 antibodies." *Blood* 118(2): 358–367.

105. Wolchok, J. D., H. Kluger, M. K. Callahan, M. A. Postow, N. A. Rizvi, A. M. Lesokhin, N. H. Segal, C. E. Ariyan, R. A. Gordon, K. Reed, M. M. Burke, A. Caldwell, S. A. Kronenberg, B. U. Agunwamba, X. Zhang, I. Lowy, H. D. Inzunza, W. Feely, C. E. Horak, Q. Hong, A. J. Korman, J. M. Wigginton, A. Gupta and M. Sznol (2013). "Nivolumab plus ipilimumab in advanced melanoma." *N Engl J Med* 369(2): 122–133.

106. Tumeh, P. C., C. L. Harview, J. H. Yearley, I. P. Shintaku, E. J. Taylor, L. Robert, B. Chmielowski, M. Spasic, G. Henry, V. Ciobanu, A. N. West, M. Carmona, C. Kivork, E. Seja, G. Cherry, A. J. Gutierrez, T. R. Grogan, C. Mateus, G. Tomasic, J. A. Glaspy, R. O. Emerson, H. Robins, R. H. Pierce, D. A. Elashoff, C. Robert and A. Ribas (2014). "PD-1 blockade induces responses by inhibiting adaptive immune resistance." *Nature* 515(7528): 568–571.

107. Barker, E., B. M. Mueller, R. Handgretinger, M. Herter, A. L. Yu and R. A. Reisfeld (1991). "Effect of a chimeric anti-ganglioside GD2 antibody on cell-mediated lysis of human neuroblastoma cells." *Cancer Res* 51(1): 144–149.

108. Mueller, B. M., C. A. Romerdahl, S. D. Gillies and R. A. Reisfeld (1990). "Enhancement of antibody-dependent cytotoxicity with a chimeric anti-GD2 antibody." *J Immunol* 144(4): 1382–1386.

109. Yu, A. L., A. L. Gilman, M. F. Ozkaynak, W. B. London, S. G. Kreissman, H. X. Chen, M. Smith, B. Anderson, J. G. Villablanca, K. K. Matthay, H. Shimada, S. A. Grupp, R. Seeger, C. P. Reynolds, A. Buxton, R. A. Reisfeld, S. D. Gillies, S. L. Cohn, J. M. Maris, P. M. Sondel and G. Children's Oncology (2010). "Anti-GD2 antibody with GM-CSF, interleukin-2, and isotretinoin for neuroblastoma." *N Engl J Med* 363(14): 1324–1334.

110. Bargou, R., E. Leo, G. Zugmaier, M. Klinger, M. Goebeler, S. Knop, R. Noppeney, A. Viardot, G. Hess, M. Schuler, H. Einsele, C. Brandl, A. Wolf, P. Kirchinger, P. Klappers, M. Schmidt, G. Riethmuller, C. Reinhardt, P. A. Baeuerle and P. Kufer (2008). "Tumor regression in cancer patients by very low doses of a T cell-engaging antibody." *Science* 321(5891): 974–977.

111. Di Gaetano, N., E. Cittera, R. Nota, A. Vecchi, V. Grieco, E. Scanziani, M. Botto, M. Introna and J. Golay (2003). "Complement activation determines the therapeutic activity of rituximab in vivo." *J Immunol* 171(3): 1581–1587.

112. Bruggemann, M., G. T. Williams, C. I. Bindon, M. R. Clark, M. R. Walker, R. Jefferis, H. Waldmann and M. S. Neuberger (1987). "Comparison of the effector functions of human immunoglobulins using a matched set of chimeric antibodies." *J Exp Med* 166(5): 1351–1361.

113. Taube, J. M., A. Klein, J. R. Brahmer, H. Xu, X. Pan, J. H. Kim, L. Chen, D. M. Pardoll, S. L. Topalian and R. A. Anders (2014). "Association of PD-1, PD-1 ligands, and other features of the tumor immune microenvironment with response to anti-PD-1 therapy." *Clin Cancer Res* 20(19): 5064–5074.

114. Clynes, R. A., T. L. Towers, L. G. Presta and J. V. Ravetch (2000). "Inhibitory Fc receptors modulate in vivo cytotoxicity against tumor targets." *Nat Med* 6(4): 443–446.

115. Bhatt, S., B. M. Ashlock, Y. Natkunam, V. Sujoy, J. R. Chapman, J. C. Ramos, E. A. Mesri and I. S. Lossos (2013). "CD30 targeting with brentuximab vedotin: a novel therapeutic approach to primary effusion lymphoma." *Blood* 122(7): 1233–1242.

116. van de Donk, N. W. and E. Dhimolea (2012). "Brentuximab vedotin." *MAbs* 4(4): 458–465.

117. Park, J. W., K. Hong, D. B. Kirpotin, G. Colbern, R. Shalaby, J. Baselga, Y. Shao, U. B. Nielsen, J. D. Marks, D. Moore, D. Papahadjopoulos and C. C. Benz (2002). "Anti-HER2 immunoliposomes: enhanced efficacy attributable to targeted delivery." *Clin Cancer Res* 8(4): 1172–1181.

118. Park, J. W., D. B. Kirpotin, K. Hong, R. Shalaby, Y. Shao, U. B. Nielsen, J. D. Marks, D. Papahadjopoulos and C. C. Benz (2001). "Tumor targeting using anti-her2 immunoliposomes." *J Control Release* 74(1–3): 95–113.

119. Herter, S., F. Herting, O. Mundigl, I. Waldhauer, T. Weinzierl, T. Fauti, G. Muth, D. Ziegler-Landesberger, E. Van Puijenbroek, S. Lang, M. N. Duong, L. Reslan, C. A. Gerdes, T. Friess, U. Baer, H. Burtscher, M. Weidner, C. Dumontet, P. Umana, G. Niederfellner, M. Bacac and C. Klein (2013). "Preclinical activity of the type II CD20 antibody GA101 (obinutuzumab) compared with rituximab and ofatumumab in vitro and in xenograft models." *Mol Cancer Ther* 12(10): 2031–2042.

120. Shan, D., J. A. Ledbetter and O. W. Press (1998). "Apoptosis of malignant human B cells by ligation of CD20 with monoclonal antibodies." *Blood* 91(5): 1644–1652.

121. Montalvao, F., Z. Garcia, S. Celli, B. Breart, J. Deguine, N. Van Rooijen and P. Bousso (2013). "The mechanism of anti-CD20-mediated B cell depletion revealed by intravital imaging." *J Clin Invest* 123(12): 5098–5103.

122. Bologna, L., E. Gotti, F. Da Roit, T. Intermesoli, A. Rambaldi, M. Introna and J. Golay (2013). "Ofatumumab is more efficient than rituximab in lysing B chronic lymphocytic leukemia cells in whole blood and in combination with chemotherapy." *J Immunol* 190(1): 231–239.

123. Pawluczkowycz, A. W., F. J. Beurskens, P. V. Beum, M. A. Lindorfer, J. G. van de Winkel, P. W. Parren and R. P. Taylor (2009). "Binding of submaximal C1q promotes complement-dependent cytotoxicity (CDC) of B cells opsonized with anti-CD20 mAbs ofatumumab (OFA) or rituximab (RTX): considerably higher levels of CDC are induced by OFA than by RTX." *J Immunol* 183(1): 749–758.

124. Marcus, R. (2005). "Use of 90Y-ibritumomab tiuxetan in non-Hodgkin's lymphoma." *Semin Oncol* 32(1 Suppl 1): S36–S43.

125. Lane, H. A., A. B. Motoyama, I. Beuvink and N. E. Hynes (2001). "Modulation of p27/Cdk2 complex formation through 4D5-mediated inhibition of HER2 receptor signaling." *Ann Oncol* 12(Suppl 1): S21–S22.

126. Park, S., Z. Jiang, E. D. Mortenson, L. Deng, O. Radkevich-Brown, X. Yang, H. Sattar, Y. Wang, N. K. Brown, M. Greene, Y. Liu, J. Tang, S. Wang and Y. X. Fu (2010). "The therapeutic effect of anti-HER2/neu antibody depends on both innate and adaptive immunity." *Cancer Cell* 18(2): 160–170.

127. Adams, C. W., D. E. Allison, K. Flagella, L. Presta, J. Clarke, N. Dybdal, K. McKeever and M. X. Sliwkowski (2006). "Humanization of a recombinant monoclonal antibody to produce a therapeutic HER dimerization inhibitor, pertuzumab." *Cancer Immunol Immunother* 55(6): 717–727.

128. Clark, M., G. Hale and H. Waldmann (1985). "Interaction of rat monoclonal antibodies with human killer cells." *Adv Exp Med Biol* 186: 797–803.

129. Lazar, G. A., W. Dang, S. Karki, O. Vafa, J. S. Peng, L. Hyun, C. Chan, H. S. Chung, A. Eivazi, S. C. Yoder, J. Vielmetter, D. F. Carmichael, R. J. Hayes and B. I. Dahiyat (2006). "Engineered antibody Fc variants with enhanced effector function." *Proc Natl Acad Sci U S A* 103(11): 4005–4010.

130. Shields, R. L., J. Lai, R. Keck, L. Y. O'Connell, K. Hong, Y. G. Meng, S. H. Weikert and L. G. Presta (2002). "Lack of fucose on human IgG1 N-linked oligosaccharide improves binding to human Fcgamma RIII and antibody-dependent cellular toxicity." *J Biol Chem* 277(30): 26733–26740.

131. Mossner, E., P. Brunker, S. Moser, U. Puntener, C. Schmidt, S. Herter, R. Grau, C. Gerdes, A. Nopora, E. van Puijenbroek, C. Ferrara, P. Sondermann, C. Jager, P. Strein, G. Fertig, T. Friess, C. Schull, S. Bauer, J. Dal Porto, C. Del Nagro, K. Dabbagh, M. J. Dyer, S. Poppema, C. Klein and P. Umana (2010). "Increasing the efficacy of CD20 antibody therapy through the engineering of a new type II anti-CD20 antibody with enhanced direct and immune effector cell-mediated B-cell cytotoxicity." *Blood* 115(22): 4393–4402.

132. Galon, J., A. Costes, F. Sanchez-Cabo, A. Kirilovsky, B. Mlecnik, C. Lagorce-Pages, M. Tosolini, M. Camus, A. Berger, P. Wind, F. Zinzindohoue, P. Bruneval, P. H. Cugnenc, Z. Trajanoski, W. H. Fridman and F. Pages (2006). "Type, density, and location of immune cells within human colorectal tumors predict clinical outcome." *Science* 313(5795): 1960–1964.

133. Rabinovich, G. A., D. Gabrilovich and E. M. Sotomayor (2007). "Immunosuppressive strategies that are mediated by tumor cells." *Annu Rev Immunol* 25: 267–296.

134. Staerz, U. D. and M. J. Bevan (1985). "Cytotoxic T lymphocyte-mediated lysis via the Fc receptor of target cells." *Eur J Immunol* 15(12): 1172–1177.

135. Staerz, U. D., O. Kanagawa and M. J. Bevan (1985). "Hybrid antibodies can target sites for attack by T cells." *Nature* 314(6012): 628–631.

136. Martinelli, E., R. De Palma, M. Orditura, F. De Vita and F. Ciardiello (2009). "Anti-epidermal growth factor receptor monoclonal antibodies in cancer therapy." *Clin Exp Immunol* 158(1): 1–9.

137. Friedman, L. M., A. Rinon, B. Schechter, L. Lyass, S. Lavi, S. S. Bacus, M. Sela and Y. Yarden (2005). "Synergistic down-regulation of receptor tyrosine kinases by combinations of mAbs: implications for cancer immunotherapy." *Proc Natl Acad Sci U S A* 102(6): 1915–1920.

138. Pedersen, M. W., H. J. Jacobsen, K. Koefoed, A. Hey, C. Pyke, J. S. Haurum and M. Kragh (2010). "Sym004: a novel synergistic anti-epidermal growth factor receptor antibody mixture with superior anticancer efficacy." *Cancer Res* 70(2): 588–597.

139. Ben-Kasus, T., B. Schechter, S. Lavi, Y. Yarden and M. Sela (2009). "Persistent elimination of ErbB-2/HER2-overexpressing tumors using combinations of monoclonal antibodies: relevance of receptor endocytosis." *Proc Natl Acad Sci U S A* 106(9): 3294–3299.

140. Spangler, J. B., J. R. Neil, S. Abramovitch, Y. Yarden, F. M. White, D. A. Lauffenburger and K. D. Wittrup (2010). "Combination antibody treatment down-regulates epidermal growth factor receptor by inhibiting endosomal recycling." *Proc Natl Acad Sci U S A* 107(30): 13252–13257.

141. Vitale, C., C. Romagnani, M. Falco, M. Ponte, M. Vitale, A. Moretta, A. Bacigalupo, L. Moretta and M. C. Mingari (1999). "Engagement of p75/AIRM1 or CD33 inhibits the proliferation of normal or leukemic myeloid cells." *Proc Natl Acad Sci U S A* 96(26): 15091–15096.

142. Kirpotin, D. B., D. C. Drummond, Y. Shao, M. R. Shalaby, K. Hong, U. B. Nielsen, J. D. Marks, C. C. Benz and J. W. Park (2006). "Antibody targeting of long-circulating lipidic nanoparticles does not increase tumor localization but does increase internalization in animal models." *Cancer Res* 66(13): 6732–6740.

143. Saito, R., M. T. Krauze, J. R. Bringas, C. Noble, T. R. McKnight, P. Jackson, M. F. Wendland, C. Mamot, D. C. Drummond, D. B. Kirpotin, K. Hong, M. S. Berger, J. W. Park and K. S. Bankiewicz (2005). "Gadolinium-loaded liposomes allow for real-time magnetic resonance imaging of convection-enhanced delivery in the primate brain." *Exp Neurol* 196(2): 381–389.

4

Therapeutic Monoclonal Antibodies and Their Targets

Jose A. Figueroa[1], Camilo Pena[2], Leonardo Mirandola[2,3], Adair Reidy[1], J. Drew Payne[2], Nattamol Hosiriluck[2], Natallia Suvorava[2], Rakhshanda Layeequr Rahman[4], Adrienne R. Whitlow[5], Rashmi Verma[2], Everardo Cobos[1], and Maurizio Chiriva-Internati[1,2,3]

[1] *Immuno-oncology Division, Kiromic LLC, Lubbock, TX, USA*
[2] *Department of Internal Medicine, Division of Hematology & Oncology, Texas Tech University Health Sciences Center, Lubbock, TX, USA*
[3] *Laura W. Bush Institute for Women's Health and Center for Women's Health and Gender-Based Medicine, Texas Tech University Health Sciences Center, Amarillo, TX, USA*
[4] *Division of Surgical Oncology, Texas Tech University Medical Center, Amarillo, TX, USA*
[5] *Internal Medicine, Texas Tech University Health Sciences Center, Lubbock, TX, USA*

4.1 Summary

Neoplastic diseases are currently treated by a variety of mAbs that use both natural cytotoxic mechanisms and cytotoxic molecules or radioisotopes in order to produce clinical activity against various cancers. Targets presently used in mAb treatments include CD20 for patients with CLL and NHL; CD30 for HL and ALCL; CTLA-4 and PD-1 for melanoma; HER2 for HER-expressing breast cancer and gastroesaphageal cancer; EGFR-1 for CC and squamous cell carcinoma; VEGF for CC, glioblastoma, NSCLC, ovarian cancer, and metastatic renal cell carcinoma; and VEGFR-2 for GE cancer, gastric cancer, and NSCLC. While no specific cancer-related antigens have yet to be identified, novel tumor targets CTAs such as SP-17 are highly restricted to cancer cells with normal expression limited to testis and placenta.

Since the first generation of monoclonal antibodies in 1975, this novel therapeutic strategy has achieved great success in its development of targeted treatment strategies for a variety of diseases. The success achieved with mAbs during the past four decades represents only the beginning of a spectacular progress in the field. While challenges such as immunogenicity and target expression character still pose problems to current mAb treatments, the field

Biosimilars of Monoclonal Antibodies: A Practical Guide to Manufacturing, Preclinical, and Clinical Development, First Edition. Edited by Cheng Liu and K. John Morrow, Jr.

is quickly expanding its efforts in order to address the diverse needs of patients. With advancing knowledge and understanding of the molecular mechanisms underlying immunity, as well as improvement in bioengineering methods and technological developments used in the design and manufacturing of these remarkable molecules, the future looks brighter than ever. The use of new protein expression platforms and glycoengineering techniques, the design of multispecific binding domains, heterologous fusion constructs, molecular safety switches, and effective linkers, will result in an explosion of more effective and safe products. Therefore, for those of us interested in the development and application of therapeutic mAbs, this journey is just beginning.

4.2 Introduction

In 1975, Köhler and Milstein presented novel findings describing the discovery and generation of monoclonal antibodies (mAbs) [1]. This first fundamental step toward the development of therapeutic antibodies was hampered by the strong immunogenicity and lack of activity observed with first-generation murine antibodies. During the 1980s, the use of molecular cloning and recombination techniques led to the development of mAbs with combined murine and human sequences (chimeric, humanized mAbs) with significant clinical activity, lower immunogenicity, lower toxicity, and enhanced target affinity. In the early 1990s, the development of phage display and the use of transgenic mice allowed for the manufacturing of mAbs with a 100% human sequence [2]. This technological leap opened the door for the design of novel human or humanized mAbs with an improved therapeutic profile. Since the early 2000s, more than 40 mAbs have been approved in Europe and the United States for the treatment of a variety of diseases. Table 4.1 summarizes some of the goals of mAb development.

mAbs are stable, mono- or bispecific molecules capable of binding to a wide array of antigenic targets with high affinity. Using their specificity, mAbs linked to potent cytotoxic drugs and radioisotopes have been developed for the targeted delivery of treatments to cancer cells with nonlethal toxicity to patient [3]. mAbs are produced from a single B-cell clone in abundant quantities using different technologies including immortalized hybridoma cell lines and yeast- or plant-based systems. Depending on the manufacturing process, mAbs are potentially cheaper and easier to produce compared to procedures that rely on procuring industrial amounts of animal or human plasma-derived immunoglobulins. Therapeutic mAbs are currently used for the treatment of various infectious diseases, autoimmune disorders, malignancies, and cardiovascular conditions. They represent an active area of research and development in the biotechnology industry with over 20 new agents currently under clinical

Table 4.1 An overview of the therapeutic goals of mAbs.

1.	To produce a novel therapy for patients with malignancies, infectious diseases, and autoimmune processes
2.	To take advantage of mAb effects resulting from direct action, immune-mediated cell killing mechanisms, payload delivery, and interaction with vasculature and stroma
3.	To support various mAb applications, approaches, and targets for mAb use with clinical trial evidence
4.	To develop comprehensive serological, genomic, proteomic, and bioinformatic databases for the identification of overexpressed antigens and receptors in various cell populations
5.	To exploit mAb targeting capabilities through the use of immunoconjugates composed of an antibody, a linker, and a therapeutic entity for treatments of diseases like cancer
6.	To provide various immunotherapeutic approaches to malignancy management with novel methods for early diagnosis, tumor killing, and disease curing provided that methods involve decreased toxicity, diminished side effects, and fewer complications
7.	To efficiently produce low-cost, effective mAbs

evaluation (ClinicalTrials.gov). To date, the most notable obstacles faced by mAb manufacturers include:

- Discovery and availability of disease-specific targets
- "Antigenic drift" related to mutations and/or downregulation of target antigens
- Antigen heterogeneity
- Unpredictable and/or variable pharmacokinetic (PK) profiles
- Ineffective FcγR binding
- Immune suppression
- *In vivo* accessibility to molecular targets
- Toxicities related to on-target/off-tissue effects [4]

Table 4.2 summarizes some of the mechanisms for unresponsiveness to mAbs.

Currently approved mAbs are members of the immunoglobulin gamma (IgG) family that display simple, linear PKs with small volumes of distribution, low rates of clearance, and longer biological half-lives in healthy subjects [5]. Compared to traditional chemotherapeutic agents, mAbs have several pharmacological advantages including a long circulating half-life and high target specificity that often lead to complex, nonlinear PKs in patients. In general, two pathways are responsible for mAb elimination: a nonspecific IgG elimination pathway and a target-mediated drug disposition pathway [6]. The neonatal Fc receptor (FcRn) plays an important role in the nonspecific elimination pathway, while the target-mediated pathway is influenced by the

Table 4.2 Mechanisms for unresponsiveness to mAb therapy.

mAb target capacity	Mechanism of failure
Antigen or receptor	Heterogeneity: initial or acquired
	Mutations: initial or acquired
	Downregulation of antigen or receptor expression
Kinetics	mAb stability
	mAb half-life
Penetrance	Vascular permeability
	mAb size
	Antibody affinity
Receptor occupancy	Low receptor concentration
	Receptor saturation
	Ineffective blockade
Immune function	FcγR polymorphisms
	Immune escape
	Complement inhibition

mAb's target-binding affinity and the extent of target expression in cells versus soluble target circulation. High-affinity mAb-target binding creates a high ratio of target-bound mAbs relative to the concentration of unbound mAbs in the blood. In many cases, mAb-target binding accelerates mAb-target complex endocytosis and subsequent intracellular mAb catabolism and elimination. Therefore, high-affinity target binding may expedite mAb elimination. As mAb dose increases, the target saturates, leading to a decreased volume of distribution and rate of antibody clearance. Therapeutic effects may also influence PK as they determine the availability of cell-bound versus circulating levels of target and dictate volume of distribution and clearance [7]. A special situation occurs when the target is present in the interstitial fluid space rather than the plasma compartment. In this case, the distribution volume and drug clearance depend on tissue penetration instead of target-binding affinity to plasma components [7].

Besides nonspecific and target-mediated disposition and elimination pathways, other factors than can affect the PK profile of mAbs include:

- Disease burden
- Target protein plasma levels
- Pharmacodynamic profiles
- Patient characteristics (i.e., age, gender, body size) [4, 8]

In this chapter, we will summarize the role that therapeutic mAbs play in three main categories of human diseases: infectious, autoimmune, and neoplastic.

4.3 Monoclonal Antibody Therapies for Infectious Diseases

4.3.1 Respiratory Syncytial Virus (RSV)

The role of mAbs in the treatment of infectious diseases has been widely recognized in recent years. Palivizumab is a 148 kDa humanized mAb (95% human and 5% murine, IgG1k) directed against an epitope in the F protein of RSV. Palivizumab's mechanism of action involves neutralization and inhibition of RSV fusion, and its PK profile is similar to human IgG1 antibodies, with a half-life of approximately 18–20 days. In palivizumab's pivotal clinical study, pediatric patients randomly received five monthly injections of either the mAb or placebo. When compared to placebo, palivizumab treatment reduced the incidence of RSV-related hospitalizations by 55%. Based on these results, palivizumab was approved in 1998 for the prevention of serious lower respiratory tract disease caused by RSV in pediatric patients at high risk of infection [9].

4.3.2 *Bacillus anthracis* (*B. anthracis*, Anthrax)

Actively dividing *B. anthracis* produces protective antigen (PA), lethal factor (LF), and edema factor (EF). PA facilitates binding of LF and EF to the anthrax toxin receptor (ATR) (also known as capillary morphogenesis protein 2, CMG2) on mammalian cell surfaces—resulting in internalization of LF and EF into cells [10]. *Raxibacumab* is an anti-PA fully human recombinant IgG1λ mAb that binds PA and prevents LF and EF from entering cells [11]. Raxibacumab's activity in inhalational anthrax was demonstrated in monkeys and rabbits exposed to aerosolized *B. anthracis* spores and treated with 40 mg/kg of the mAb. Up to 65% of monkeys and 44% of rabbits survived the exposure and all surviving animals developed toxin-neutralizing antibodies [12]. The safety of raxibacumab has been evaluated in 326 healthy human volunteers without major side effects [13]. The US Food and Drug Administration (FDA) has approved raxibacumab for the treatment and prevention of inhalational anthrax when alternative therapies are not available or not appropriate. Raxibacumab is also the first mAb approved under the Animal Efficacy Rule, which accepts efficacy findings derived from adequate and well-controlled animal studies in the absence of human data [14].

4.3.3 Human Immunodeficiency Virus (HIV)

In 1988, plasma infusions from an asymptomatic HIV-positive donor with high p24 antibody titers were administered to a patient with advanced acquired immunodeficiency syndrome (AIDS) resulting in improved clinical symptoms, decreased HIV antigenemia, and increased T-cell immunity in the recipient [15]. Antibodies harvested from the donor were found to have high affinity to gp120 and gp41 glycoproteins of type 1 HIV, making those targets attractive for vaccine and therapeutic mAb development [16]. Unfortunately, HIV has demonstrated resistance toward humoral immunity against gp120 during the HIV epidemics in 1987–1991, 1996–2000, and 2006–2010. Interestingly, neutralizing antibodies against gp41 membrane-proximal external region showed stable neutralizing activity, which suggests that this antigen may represent the next therapeutic target in this disease [17]. Although elusive thus far, clinical development of HIV-specific mAbs, as an alternative to current toxic retroviral therapies, remains a priority.

4.3.4 Ebola Virus Family (EBOV)

The EBOV family-related hemorrhagic fever, an infectious condition with a high fatality rate in humans, has recently become a global health priority. There are five recognizable EBOV that differ by 40–50% in their primary amino acid sequence. Neutralizing Abs bind GP1,2 the only virally encoded surface protein with a crucial role in viral adhesion, entry, and internalization into host cells [18]. The GP1,2 gene can be transcribed into either a small soluble glycoprotein (sGP) or the full-length GP1,2. The sGP protein acts as a decoy to "capture" circulating antibodies and disrupt their ability to attack the virus through sGP engagement. The GP1 subunit is comprised of the GP core unit, a glycan cap, a receptor-binding domain (RBD), and a mucin-like domain encompassing the RBD in the form of a bowl. The GP1 subunit plays a role in receptor-binding and immune evasion. The GP2 subunit contains an internal fusion loop, a membrane-proximal external region, heptad repeats 1 and 2, a transmembrane region, and a cytoplasmic tail. After endosomal cleavage, GP2 unfolds and the internal fusion loop inserts into the endosomal membrane [18]. During recent years, it has been shown that treatment with antibody-based therapies can improve survival in nonhuman primates (NHPs) infected with EBOV when administered early after infection. ZMapp™, a combination of the two Ab cocktails ZMapp and MB-003, protected animals when treatment was initiated within 5 days postinfection [19]. The ZMapp cocktail consists of three mouse-derived mAbs (1H3, 2G4, and 4G7), which target GP1,2. The 1H3 mAb binds to a conformational region shared by EBOV sGP and GP1,2. This is the only mAb in the ZMapp cocktail that binds sGP. The 2G4 mAb recognizes GP1,2 but does not bind to either sGP or GP1 alone, while the 4G7 antibody binds to

Table 4.3 Examples of mAb targets for infectious diseases.

Target	Antibody		Mechanism of action	Infectious disease
F protein	Palivizumab		Neutralizes and inhibits RSV fusion	Respiratory syncytial virus
PA	Raxibacumab		Binds PA to prevent internalization of lethal factor and edema factor	*Bacillus anthracis*
gp41	Antibodies against gp41 membrane-proximal external region		Neutralizing activity	Type 1 HIV
GP1,2 + sGP	1H3	ZMapp cocktail	Binds to a conformational region shared by sGP and GP1,2	Ebola virus
GP1,2	2G4		Binds to GP1,2 but not sGP or GP1 alone	
GP1,2 and GP1	4G7		Binds to GP1 and GP1,2, but not sGP	
GP1 and sGP	13C6	MB-033 cocktail	Binds GP1 and sGP	
GP1,2	13F6		Targets linear epitopes in the mucin-like domain of GP1,2	
GP1,2	6D8			
sGP, GP1, GP1,2	ZMapp		ZMapp + MB-033 cocktails	

GP1 and GP1,2, but not to sGP [20]. The MB-003 cocktail consists of three chimeric antibodies: 13C6, 13F6, and 6D8. Antibody 13C6 binds GP1 and sGP (like the 1H3 antibody), while 13F6 and 6D8 mAbs target linear epitopes in the mucin-like domain of GP1,2. More recently, a combination therapy using two chimeric mAbs (ch133 and ch226) has been developed [21]. All three cocktails contain at least one antibody that binds close to the glycan cap or glycan cap itself (ch226, c13C6, 1H3), suggesting the importance of this region for viral neutralization [20]. The success of some of these EBOV-specific mAbs during the recent EBOV outbreak has fueled global efforts to develop effective therapeutic antibodies against this disease. Unfortunately, and despite these advances, development of a mAb therapy for infectious agents has lagged behind that achieved for other conditions, including autoimmune disorders and cancer. Table 4.3 summarizes the targets of the infectious diseases discussed.

4.4 Monoclonal Antibody Therapies for Autoimmune Diseases

Aberrant physiological immune responses to autoantigens (antigens expressed by host tissues) result in the pathogenesis of autoimmune diseases by serving as targets for antibody binding and aberrant T-cell activation [22]. Activated T cells and aberrant B cells induce tissue injury through a variety of mechanisms resulting in clinical conditions such as rheumatoid arthritis (RA) and systemic lupus erythematosus (SLE). Thus, mAbs directed at antigens expressed on activated T cells and B cells have emerged as viable, less toxic, and more effective alternatives to traditional treatments (i.e., cytotoxic agents and corticosteroids) for these diseases [23].

4.4.1 Multiple Sclerosis (MS)

MS is an autoimmune disease associated with demyelination and axonal loss in the central nervous system (CNS) resulting from the migration of activated lymphocyte across the blood–brain barrier [24]. Natalizumab is a 149 kDa humanized IgG4k mAb that binds to the $\alpha4$ subunit of $\alpha4\beta1$ and $\alpha4\beta7$ integrins expressed on leukocytes and inhibits $\alpha4$-mediated adhesion to vascular cell adhesion molecule 1 (VCAM-1) in CNS blood vessels and osteopontin, a molecule necessary for lymphocyte survival, priming, and activation [25]. Disruption of these molecular interactions prevents transmigration of leukocytes across the endothelium into inflamed parenchymal tissue. Natalizumab treatment results in significant reduction of demyelinated plaques within the CNS and has been approved for patients with relapsing-remitting MS [26]. In this pivotal study, natalizumab monotherapy decreased the risk of progression and relapse rate by 50% and 68%, respectively [26]. Unfortunately, 9% of patients develop anti-natalizumab antibodies with the potential to cause drug resistance and autoimmunity, and natalizumab treatment is associated with the risk of progressive multifocal leukoencephalopathy (PML), an opportunistic brain infection that is caused by the JC virus [27]. PML appears to occur after a complex interaction between host and viral factors, leading to the development of a pathogenic form of JC virus that can infect and destroy oligodendrocytes in the CNS [28]. Natalizumab was voluntarily withdrawn from the market in 2005 after three cases of PML had been identified in clinical trials but was reintroduced in the market in 2006 after an intensive global risk management program was established.

4.4.2 Inflammatory Bowel Disease (IBD)

IBD is defined by an abnormal and exacerbated immune response against mucosal constituents and components of the luminal flora in individuals with a genetic predisposition. Crohn's disease (CD) and ulcerative colitis (UC) are the two major forms of IBD and proinflammatory T cells and their secreted cytokines are the main effectors in induction and perpetuation of the intestinal inflammation [29]. Treatment strategies for IBD aim at suppression and modulation of this aberrant immune response, with the use of mAbs against inflammatory cytokines becoming the mainstay of treatment for these conditions.

IBD can occur through two immune mechanisms, both of which produce chronic, active inflammation: (i) deregulated responses against luminal bacteria or their products or (ii) inappropriate responses to favorable intestinal organisms. CD is characterized by increased production of interleukin-12 (IL-12) by antigen-presenting cells (APCs), tumor necrosis factor (TNF)α, and interferon-γ (IFN-γ), all of which activate phagocytic macrophages [30]. TNFα is first synthesized as a transmembrane protein that is later cleaved to release a soluble TNFα homotrimer. Both the TNFα transmembrane protein and cleaved homotrimer are considered biologically active and are produced by microglia, astrocytes, and oligodendrocytes in response to trauma, pathogens, stroke, protein modification, and other stimuli [31]. IFN-γ is a cytokine produced by APCs that stimulates TNFα gene expression in macrophages, which ultimately facilitates CD40 production [32]. A receptor of the TNF superfamily, CD40, is often aberrantly expressed on APCs of patients with autoimmune inflammatory diseases. CD40 on B cells interacts with CD154 (also known as CD40 ligand or CD40L) on T cells to produce primary and secondary thymus-dependent humoral immune responses and trigger memory B-cell maturation [33]. When CD40 interacts with CD154, expression is not limited to B and T lymphocytes but may be detected on dendritic cells, macrophages, monocytes, and other nonlymphoid cells. Activation of CD154 produces an immune response that includes B-cell growth, differentiation, immunoglobulin class switching, and upregulation of costimulatory molecules like VCAM-1 and CD40. It addition, CD154 triggers the production of cytokines and chemokines such as IL-12 and TNFα [32, 33]. Interaction between T cells expressing CD154 and CD40-expressing monocytes and macrophages induces IL-12 expression by monocytes and macrophages, which amplify the aberrant immune response and increase T-cell CD154 expression. When innate and acquired immune responses are unable to resolve stimuli triggering chronic interactions between CD40, deregulated CD154, and TNFα expression, chronic inflammation ensues and manifests in diseases such as IBD encephalomyelitis, RA, MS, thyroiditis, uveitis, diabetes, and other autoimmune diseases.

Infliximab is a 149 kDa chimeric IgG1k mAb specific for human TNFα, a cytokine involved in the generation of a robust inflammatory response. Infliximab blocks the biological activity of TNFα by binding both the soluble and transmembrane forms of TNFα and inhibiting its interaction with its receptor [34]. Infliximab neutralizes the biological activities of TNFα, which include induction of interleukin-1 (IL-1) and interleukin-6 (IL-6) expression, enhancement of leukocyte migration, and induction of tissue-degrading enzymes produced by synoviocytes and/or chondrocytes, as well as induction of apoptosis of activated T cells and macrophages [34].

The safety and efficacy of single and multiple doses of infliximab were assessed in two randomized, double-blind, placebo-controlled clinical studies in patients with moderate to severe CD with an inadequate response to prior conventional therapies. In the single-dose trial, 81% of patients receiving infliximab (5 mg/kg) achieved a clinical response compared with 16% of those receiving placebo [35]. In the multidose trial (ACCENT I), patients were treated with infliximab (5 mg/kg) and then randomized to maintenance therapy with either placebo or two different doses of infliximab (5 and 10 mg/kg) [36]. Following induction, 57% of patients responded and a greater proportion of those receiving maintenance treatment infliximab remained in clinical remission. Two studies also evaluated the safety and efficacy of infliximab in patients with UC and demonstrated that a greater percentage of patients achieved sustained clinical response, clinical remission, and mucosal healing than in the placebo group [37, 38]. Based on these results, infliximab became the first mAb approved for the treatment of IBD. While anti-TNF agents have been employed successfully in the treatment of IBD and other autoimmune diseases, autoantibodies can develop resulting in additional immune dysfunction. This aggravated autoimmunity can trigger drug resistance and rare drug-induced lupus [39]. Other possible TNF targeting side effects include potential injection site reactions, risk of severe opportunistic infections, and development of malignancies such as lymphoma [40].

Vedolizumab is a 147 kDa humanized IgG1 mAb that binds to the human $\alpha 4\beta 7$ integrin blocking the interaction of $\alpha 4\beta 7$ integrin with mucosal addressin cell adhesion molecule-1 (MAdCAM-1) and inhibiting the migration of memory T cells across the endothelium into inflamed gastrointestinal parenchymal tissue [41]. The $\alpha 4\beta 7$ integrin is expressed on the surface of a discrete subset of memory T cells that preferentially migrate into the gastrointestinal tract. MAdCAM-1 is mainly expressed on gut endothelial cells and plays a critical role in the homing of T-cell to gut lymphoid tissue. The interaction of the $\alpha 4\beta 7$ integrin with MAdCAM-1 has been implicated as an important contributor to the chronic inflammation that is a hallmark of UC and CD [41]. Vedolizumab is different from natalizumab in that it does not antagonize the interaction of $\alpha 4$ integrins with VCAM-1. The safety and effectiveness of vedolizumab for UC and CD were established in two clinical trials involving approximately 900 patients and three studies involving 1500 patients who had not responded adequately to corticosteroids, immunomodulators, or TNF blocker medications [41–44]. Results showed that a greater percentage of participants treated with vedolizumab achieved and maintained a significant clinical and corticosteroid-free response compared to those that received the placebo. Participants in these trials were actively monitored for PML, but no cases were observed.

As mentioned earlier, natalizumab also binds to the α4 subunit of α4β7 and, by inhibiting MAdCAM-1, may display activity in IBD [45].

4.4.3 Inflammatory Arthritis (IA)

IA is a hallmark of many autoimmune disorders and several mAbs have been approved for the treatment of IA associated with a variety of autoimmune disorders including RA, ankylosing spondylitis (AS), psoriatic arthritis (PsA), and SLE. The safety and efficacy of infliximab were assessed in two large randomized trials (ATTRACT and ASPIRE) in patients with active RA despite treatment with methotrexate (MTX) [46, 47]. All doses and schedules of infliximab resulted in improved signs and symptoms of active disease. Similar results were seen in patients with psoriatic arthritis and AS [48, 49].

Adalimumab and golimumab are two recombinant human IgG1 mAbs specific for human TNFα with potent anti-inflammatory activity. Compared to placebo, treatment with either adalimumab or golimumab has resulted in improved signs and symptoms of IA in patients with RA, AS, and psoriatic arthritis [50, 51]. Adalimumab and golimumab gained approval in the United States in 2002 and 2009, respectively.

Other anti-TNFα strategies have been devised that take advantage of the distinct structures present within anti-TNFα mAbs without using the entire molecule. For example, certolizumab pegol, a humanized monoclonal Fab fragment with high binding affinity for TNFα attached to two polyethylene glycol molecules, has been successful in targeting various autoimmune diseases by binding and inactivating both soluble and membrane-bound TNFα [52, 53]. Etanercept, on the other hand, is a humanized IgG1 Fc fragment fused with two humanized p75 TNFα receptors that has been successfully used in RA treatment but demonstrated no significant activity against CD due to its inability to bind membrane-bound TNFα [54, 55].

Tocilizumab is a recombinant humanized antihuman IL-6 receptor mAb of the IgG1k subtype [56]. Tocilizumab binds specifically to both soluble and membrane-bound IL-6 receptors and has been shown to inhibit IL-6-mediated signaling. IL-6 is a pleiotropic proinflammatory cytokine produced by a variety of cell types including T and B cells, monocytes, and fibroblasts and has been implicated in T-cell activation, induction of immunoglobulin secretion, and stimulation of hematopoietic precursor cell proliferation and differentiation. IL-6 is also locally produced by synovial and endothelial cells contributing to the inflammatory process in RA. IL-6 also induces B-cell maturation into immunoglobulin-producing plasma cells, promotes CD4+ T-cell differentiation into Th17 cells in cooperation with TGF-β, and inhibits TGF-β-induced regulatory T (Treg) cell development [57]. IL-6 also promotes the production of vascular endothelial growth factor (VEGF), which causes RA's synovial tissue inflammation due to angiogenesis and increased vascular permeability [57]. The safety and efficacy of tocilizumab have been evaluated in over 4000 RA patients using doses of 4 and 8 mg/kg. Compared to placebo plus disease-modifying drugs, patients treated with 8 mg/kg of tocilizumab had significantly improved signs and symptoms of active IA (average improvement in five studies; 26% vs. 57%) [58, 59]. Tocilizumab has been approved for the treatment of RA and polyarticular and systemic juvenile arthritis and is currently being reviewed for potential use in treating large-vessel vasculitis, polymyalgia rheumatica, amyloid A amyloidosis, neuromyelitis optica (NMO), adult-onset Still's disease, and cytokine release syndrome [60].

4.4.4 Systemic Lupus Erythematosus (SLE)

SLE is the classic systemic autoimmune disease affecting a multiplicity of organs and resulting in significant morbidity. Our current understanding of SLE pathophysiology revolves around the concept of B-cell deregulation resulting in autoantibody formation, immune complex deposition, antigen presentation, cytokine activation, and consequent tissue injury [61]. The difficulty in maintaining clinical remission coupled with an increasing number of patients refractory to standard therapies has stimulated the design of novel mAbs that specifically target critical B-cell functions fundamental to SLE pathogenesis. Belimumab is a human IgG1 mAb specific for the soluble human B-cell-activating factor (BAFF), also referred to as BLyS and TNFSF13B [62]. BAFF is a B-cell survival factor with a key role in B-cell homeostasis. Belimumab blocks the binding of soluble BAFF to its receptors on B cells and inhibits the survival of autoreactive B cells as well as their differentiation into immunoglobulin-producing plasma cells. Deregulated BAFF expression may contribute to amyloidosis (a hematological disorder, associated with plasma cell dyscrasia) or B-cell malignancies via effects on abnormal B-lymphocyte activation, proliferation, survival, and immunoglobulin secretion [62].

Belimumab was evaluated in three randomized, placebo-controlled studies involving 2133 patients with SLE according to the American College of Rheumatology criteria. Patients with severe active lupus nephritis and severe active CNS lupus were excluded. The proportion of SLE patients achieving an SLE Response Index (SRI) response, as defined for the primary endpoint, was significantly higher in the belimumab 10 mg/kg dose group than in the placebo group. The effect on the SRI was not consistently significantly different for the 1 mg/kg group relative to placebo in both trials. Thus, the 1 mg/kg dose is not recommended [63–65].

Other anti-BAFF therapies are currently under phase III evaluation in patients with SLE. As an example, tabalumab is a human IgG4 antibody with high affinity for human BAFF able to neutralize both soluble and membrane-bound BAFF [66]. In patients treated with multiple infusions of tabalumab, a transient increase in circulating B cells has been observed, followed by a significant reduction in their numbers and activity. Future studies would likely compare the efficacy of different anti-B-cell biologics among different subpopulations of SLE patients in order to establish the more effective approach to this dreadful disease.

4.4.5 Psoriasis (Ps)

Ps is a chronic systemic inflammatory disease characterized by inflammation and keratinocyte hyperproliferation thought to be the consequence of T-cell-mediated immune response to unidentified autoantigen [67]. Immune pathways mediated by a variety of cytokines have an important role in Ps. Among these cytokines, TNFα, IL-12, IL-23, and IL-17 appear to play major roles [68]. Ustekinumab is an approved recombinant human IgG1k mAb that binds with high affinity and specificity for the p40 protein subunit used by both the IL-12 and IL-23 cytokines [69]. IL-12 and IL-23 are naturally occurring cytokines involved in inflammatory and immune responses, such as natural killer cell activation and CD4+ T-cell differentiation and activation. *In vitro,* ustekinumab has been shown to disrupt IL-12- and IL-23-mediated signaling and cytokine cascades by disrupting the interaction of these cytokines with a shared cell surface receptor chain, IL-12 β1 [70]. Two multicenter, randomized, placebo-controlled studies enrolled a total of 1996 subjects 18 years of age and older with plaque Ps who had a minimum body surface area involvement of 10%, and Psoriasis Area and Severity Index (PASI) score greater than 12, and who were candidates for phototherapy or systemic therapy [71, 72]. In both studies, patients were treated with either 45 or 90 mg of ustekinumab or placebo and the endpoints were the proportion of subjects who achieved at least a 75% reduction in PASI score (PASI 75) from baseline to week 12. Both studies showed a significant improvement in PASI score in patients treated at both dose levels compared to placebo (3–4% vs. 59–67%). Based on these results, ustekinumab has been approved for adult patients with plaque Ps who are candidates for phototherapy or systemic treatment.

Since Th17 cells secrete the proinflammatory cytokine IL-17A in Ps, this cytokine has become the target for therapeutic mAb development in this disease [73]. For example, ixekizumab is a humanized IgG4 mAb that neutralizes IL-17A undergoing late-stage clinical evaluation in patients with Ps [74].

4.4.6 Cryopyrin-Associated Periodic Syndromes (CAPS)

CAPS are a group of rare, inherited, autoinflammatory diseases with the same genetic basis and overlapping symptomatology [75]. There are three subtypes of CAPS: familial cold autoinflammatory syndrome (FCAS), Muckle–Wells syndrome (MWS), and neonatal-onset multisystem inflammatory disease (NOMID) (also called chronic infantile neurologic cutaneous articular (CINCA) syndrome). CAPS are generally caused by autosomal-dominant mutations of the nucleotide-binding domain, leucine-rich family [NLR], pyrin domain-containing three (NLRP3) gene (formerly known as the CIAS1 gene), and resultant alterations in cryopyrin, the NLRP3-encoded protein. Cryopyrin is critical to the production of the inflammatory cytokine interleukin-1β (IL-1β) and alterations in cryopyrin lead to IL-1β overproduction that creates the inflammatory response and symptoms seen in CAPS. Different from autoimmune diseases, CAPS are characterized by an inflammatory reaction that is unprovoked. Unlike autoimmune diseases, CAPS are not associated with high-titer autoantibodies or antigen-specific T cells [76]. Canakinumab is a recombinant human antihuman IL-1β mAb of the IgG1k subclass. Both heavy chains of canakinumab contain oligosaccharide chains linked to the protein backbone at asparagine 298. Canakinumab binds to human IL1β and blocks its interaction with IL-1 receptors, but it does not bind IL-1α or IL-1 receptor antagonist (IL-1ra) [77].

The efficacy and safety of canakinumab for the treatment of CAPS were demonstrated in a 3-part trial in patients with the MWS phenotype of CAPS. Complete clinical response was observed in 71% of patients 1 week following initiation of treatment and in 97% of patients by week 8 [78]. In a second trial, patients 4–74 years of age with both MWS and FCAS phenotypes of CAPS were treated in an open-label study. The treatment resulted in clinically significant improvement of signs and symptoms in the majority of patients within 1 week [79]. Canakinumab has been approved for the treatment of patients with CAPS and active systemic juvenile idiopathic arthritis.

4.4.7 Antineutrophil Cytoplasmic Antibody (ANCA)-Associated Vasculitis

Wegener's granulomatosis, microscopic polyangiitis, and renal-limited ANCA-associated vasculitis are the main ANCA-associated vasculitis variants [80]. Standard therapeutic strategies for these conditions combine glucocorticoids and cyclophosphamide to induce remission, a method that has dramatically improved survival rates over the past few decades. However, this strategy frequently causes early and late side effects. *Rituximab* is a chimeric murine/human monoclonal IgG1k mAb directed against the CD20 antigen (human B-lymphocyte-restricted differentiation antigen, Bp35), a hydrophobic transmembrane protein expressed on pre-B and mature B lymphocytes that regulates early steps in the activation process for cell cycle initiation and differentiation [81]. The results of two trials (RAVE and RITUXVAS) showed that rituximab was not inferior to daily oral or pulse intravenous cyclophosphamide for the induction of complete remission in patients with ANCA-associated vasculitis [82, 83]. In a recent study, more patients with ANCA-associated vasculitis were shown to have sustained remission at month 28 with rituximab than with azathioprine [84]. Although the combination of cyclophosphamide and glucocorticoids leads to remission in most patients with ANCA-associated vasculitis, maintenance treatment with azathioprine or methotrexate is associated with high relapse rates. Therefore, and given its activity in these conditions, rituximab treatment may play a role in the maintenance phase of such diseases. Rituximab is approved for the treatment of B-cell lymphomas and is considered a standard option for the management of disorders associated with autoantibody production such as refractory autoimmune hemolytic anemia and idiopathic thrombocytopenic purpura [85].

4.4.8 Type 1 Diabetes Mellitus (DM)

T-cell-mediated type 1 DM is characterized by the loss of pancreatic insulin-secreting β-cells due to autoreactive CD4+ and CD8+ T cells and their soluble products such as IFN-γ, TNFα, and others [86]. Such observations have led to clinical trials with immunomodulatory drugs such as cyclosporine, azathioprine, and prednisone, which were shown to cause transient improvement in clinical measures but resulted in significant toxicity. Unfortunately, the toxic effects of such drugs, the concern about the risk associated with immune suppression, and the need for continuous treatment in an otherwise healthy, young population limit the use of these agents. CD3 is a pan-T-cell marker that has been targeted with mAbs to induce profound immunosuppression [87]. Over the past years, the development of anti-CD3 mAbs as a potential treatment for type 1 DM has been revisited with encouraging results. Sprangers et al. reported success in treating patients with otelixizumab (ChAglyCD3), a non-mitogenic, humanized Fc-mutated CD3 mAb [88]. The study concluded that short-term otelixizumab treatment prevents an increase in exogenous insulin needs for a minimum of 18 months, by preserving residual β-cell function [89]. Side effects included headache, fever, gastrointestinal symptoms, arthralgia, and myalgia. Teplizumab [hOKT3ϒL (Ala-Ala)] is a humanized anti-CD3 mAb that has been mutated to greatly reduce Fc receptor and complement binding [90]. This modification decreases FcR binding that could potentially result in toxicity and thus increase binding specificity to CD3. Teplizumab has been shown to protect β-cells for as long as 5 years after treatment [91]. Table 4.4 summarizes the mAb targets and diseases described in this section on autoimmune diseases.

4.5 Therapeutic Monoclonal Antibodies Against Neoplastic Diseases

Since the identification of carcinoembryonic antigen (CEA) in 1965, the search for new cancer-specific antigens has driven the discovery of biomarkers to diagnose and treat malignancies by targeting these molecules in the cancer

Table 4.4 Examples of mAb targets for autoimmune diseases.

Target	Antibody	Mechanism	Autoimmune disease
α4 subunit of α4β1 and α4β7 integrins	Natalizumab	Inhibits α4-mediated adhesion to VCAM-1	MS
α4β7 integrin	Vedolizumab	Blocks α4β7 integrin interaction with MAdCAM-1, inhibits memory T-cell migration into the GI tract	CD, UC
TNFα	Infliximab	Binds soluble and transmembrane TNFα to block biological activity of TNFα to inhibit receptor interaction	IBD, RA, AS, psoriatic arthritis, CD
	Adalimumab, golimumab	Blocks biological activity of TNFα	IA of patients with RA, AS, and psoriatic arthritis
	Certolizumab pegol	Binds soluble and transmembrane TNFα to block biological activity of TNFα	Psoriatic arthritis, AS, axial spondyloarthritis
	Etanercept	Binds soluble TNFα to block biological activity of TNFα	RA
IL-6	Tocilizumab, rituximab	Binds soluble and membrane-bound IL-6 receptors, inhibits IL-6-mediated signaling	IA, RA, polyarticular and systemic juvenile arthritis
BAFF	Belimumab	Blocks binding of soluble BAFF to its B-cell receptors	SLE
p40 protein subunit of IL-12 and IL-23	Ustekinumab	Binds p40 to disrupt IL-12 and IL-23 interaction with IL-12 β1	Ps
IL-17A	Secukinumab	Neutralizes IL-17A	Ps
IL-1β	Canakinumab	Binds IL-1β and blocks interaction with IL-1 receptors	MWS, FCAS, CAPS, active systemic juvenile idiopathic arthritis

Table 4.4 (Continued)

Target	Antibody	Mechanism	Autoimmune disease
CD20	Rituximab, obinutuzumab, ofatumumab	Prevents CD20-regulated pre-B and mature B-lymphocyte cell cycle initiation and differentiation	ANCA-associated vasculitis, B-cell lymphoma, refractory autoimmune hemolytic anemia, idiopathic thrombocytopenic purpura, management of disorders associated with autoantibody production
Omalizumab	Xolair	IgE blocker	Asthma
CD3	Otelixizumab Teplizumab	Induces immunosuppression via CD3 binding	Type 1 DM (both drugs are in phase III clinical trial)

cells. Although specific cancer-related antigens (CRAs) have not been identified, some of these CRAs have proven useful as targets for mAb therapies. Notwithstanding the lack of universal specificity, the majority of available therapeutic mAbs have been designed to target CRAs taking advantage of natural cytotoxic mechanisms such as antibody-dependent cell cytotoxicity (ADCC) or complement-mediated cytotoxicity (CMC), as well as incorporating cytotoxic molecules or radioisotopes to mAbs as "anticancer payloads" [3]. Current or previously FDA-approved antibody targets for neoplastic diseases include CD20, CD30, CD52, PD-1 and PD1 ligands, CTL-4, epidermal growth factor receptor (EGFR), and VEGF family ligands and receptors. Here we review some of the currently approved therapeutic mAbs for selected malignancies.

4.5.1 Hematologic Malignancies

NHL and CLL comprise a heterogeneous group of B-cell neoplasms characterized by the expression of the CD20 in more than 90% of cases. The chimeric anti-CD20 mAb rituximab, either as monotherapy or in combination with standard systemic therapy, has become the standard of care in NHL since the late 1990s [92]. Following the success of rituximab in the treatment of CD20-positive lymphoproliferative disorders, other anti-CD20 mAbs have been developed. Ofatumumab is an IgG1k human mAb that binds specifically to both the small and large extracellular loops of the CD20 molecule [93]. The Fab domain of ofatumumab binds to the CD20 molecule and the Fc domain mediates B-cell lysis via CMC and/or antibody-dependent cell-mediated cytotoxicity (ADCC). The clinical efficacy of ofatumumab was evaluated in 154 patients with relapsed or refractory CLL. In patients with CLL refractory to fludarabine and alemtuzumab, the overall response rate (ORR) was 42% with a median response duration of 6.5 months [94]. Obinutuzumab is a 150 kDa humanized anti-CD20 mAb of the IgG1 subclass, with the same mechanism of action as rituximab and ofatumumab [95]. The efficacy of obinutuzumab was evaluated in a three-arm, open-label, active control, randomized, multicenter trial in 356 patients with previously untreated CD20-positive CLL [96]. Patients were treated with chlorambucil or the combination of chlorambucil and obinutuzumab as first-line therapy. The median progression-free survival (PFS) of patients treated in the combination arm was found to be 23.0 months compared to 11.1 months for those in the chlorambucil alone arm. Both ofatumumab and obinutuzumab have been approved for the treatment of relapsed CLL.

Building on the success of anti-CD20 mAbs in these diseases, two mAbs conjugated to radioisotopes were developed in the 1990s. ^{131}Iodine-tositumomab is a radioiodinated murine IgG2aλ mAb directed against the CD20 antigen [97]. ^{131}Iodine-tositumomab is administered in two discrete steps: the dosimetric and therapeutic steps. Each step is administered 7–14 days apart and consists of a sequential infusion of "naked" tositumomab followed by ^{131}Iodine-tositumomab. Besides inducing B-cell apoptosis through CMC and ADCC, γ and β-emissions from ^{131}Iodine decay results in free radical formation, chromosomal breakage, and cell death. The efficacy of ^{131}Iodine-tositumomab was evaluated in a multicenter, single-arm study in patients with low-grade or transformed NHL whose disease had not responded to or had progressed after rituximab therapy. Overall responses were evident in 63 and 68% of the rituximab-refractory and responsive patients, respectively [98]. These results were confirmed in four single-arm studies enrolling 190 patients with rituximab-naïve, follicular NHL with or without transformation, who had relapsed following or were refractory to chemotherapy. In these studies, the ORR ranged from 47 to 64% and the median durations of response ranged from 12 to 18 months [99].

Ibritumomab tiuxetan is a radioimmunoconjugate formed via a stable thiourea covalent bond between the murine IgG1k anti-CD20 mAb ibritumomab and the linker-chelator tiuxetan, which provides a high-affinity, restricted chelation site for ^{111}Indium or ^{90}Yitrium [100]. Ibritumomab tiuxetan therapeutic regimen is also administered in two steps consisting of a rituximab infusion followed by ^{111}Indium-ibritumomab tiuxetan and 7–9 days later another rituximab infusion followed by ^{90}Yitrium-ibritumomab tiuxetan. The β-emission from ^{90}Yitrium induces cellular damage and death through the formation of free radicals in the target and neighboring cells. The safety and efficacy of ibritumomab tiuxetan were evaluated in two multicenter trials enrolling a total of 197 subjects [101, 102]. The first study demonstrated an ORR of 70% in patients deemed refractory to rituximab. In the second study, treatment of patients with refractory or progressive NHL following chemotherapy with rituximab or ibritumomab tiuxetan resulted in an ORR of 56% and 80%, respectively. In these studies, the more serious complications were related to profound bone marrow suppression, including severe thrombocytopenia. Both radioimmunoconjugates are currently approved in the United States for the treatment of refractory NHL.

HL is a B-cell-derived lymphoid malignancy achieving long-term remission following multiagent chemotherapy and/or radiotherapy in 80% of affected individuals [103]. Unfortunately, available therapies are only palliative for patients with relapses following salvage therapies. CD30 is a cytokine receptor belonging to the TNF superfamily (TNFRSF8) that acts as a regulator of apoptosis and its expression in HL and anaplastic large-cell lymphoma (ALCL) has opened the door for the design of novel targeted therapies [104]. Brentuximab vedotin is a CD30-directed antibody–drug conjugate (ADC) consisting of the CD30-specific chimeric IgG1 mAb cAC10, the microtubule-disrupting agent monomethyl auristatin E (MMAE), and a protease-cleavable linker that covalently attaches MMAE to cAC10 [105]. Anticancer activity of brentuximab vedotin is due to its binding to CD30, followed by internalization of the brentuximab vedotin–CD30 complex and release of MMAE via proteolytic cleavage. Binding of MMAE to tubulin disrupts the microtubule network within the cell and induces cell cycle arrest and apoptotic death. The efficacy of brentuximab vedotin was evaluated in a single-arm, multicenter study of 102 patients with HL who relapsed after autologous stem cell transplant [106]. Patients were treated with 1.8 mg/kg of the ADC every 3 weeks with an ORR of 73% and a response duration of 1.3 to 21-plus months. A similar study was conducted in patients with ALCL who had progressed following standard therapy. The ORR in this study was 86% [107]. Brentuximab vedotin is currently approved for relapsed HL following salvage therapy (including autologous stem cell transplant) and relapsed ALCL after first-line multiagent chemotherapy.

4.5.2 Solid Malignancies

4.5.2.1 Melanoma

One of the hallmarks of malignancy is the maintenance of an immunosuppressive tumor environment through complex interactions between cancer cells and immune checkpoint molecules, such as CD-28, cytotoxic T-lymphocyte-associated antigen 4 (CTLA-4), programmed death receptor-1 (PD-1), and their ligands [108]. The development of mAbs against these immunomodulatory targets represents one of the major therapeutic breakthroughs in the treatment of patients with melanoma and other malignancies. Ipilimumab is a recombinant human IgG1k mAb that binds to CTLA-4, a negative regulator of

T-cell activation, with high affinity [109]. Ipilimumab binding to CTLA-4 blocks the interaction of CTLA-4 with its ligands, CD80/CD86 resulting in enhanced T-cell activation and proliferation. The mechanism of action of ipilimumab in patients with melanoma is indirect, possibly through T-cell-mediated antitumor immune responses. The safety and efficacy of ipilimumab were investigated in a randomized (3 : 1 : 1), double-blind study of 676 patients with previously treated unresectable or metastatic melanoma [110]. Although ORR was low in the ipilimumab-treated groups (5–11%), a significant advantage in overall survival was evident in the ipilimumab-treated population. Unfortunately, ipilimumab treatment was associated with several serious immune adverse effects including colitis, endocrinopathies, dermatitis, hepatitis, and uveitis. Despite these drawbacks, ipilimumab was the first therapeutic intervention associated with survival advantage in this disease. Building on the unprecedented success experienced with ipilimumab, other immune checkpoints have been recently explored as therapeutic targets in melanoma. Two mAbs targeting PD-1 on T cells and its ligands (PDL-1) have recently been approved for the treatment of melanoma patients. Nivolumab is a human IgGk mAb that blocks the interaction between PD-1 and its ligands, PD-L1 and PD-L2. Binding of the PD-1 ligands to the PD-1 receptor found on T cells inhibits T-cell proliferation and cytokine production. Upregulation of PD-1 ligand occurs in some tumors and signaling through this pathway contributes to inhibition of active T-cell immune surveillance of tumors. Thus, the blockade of PD-1 pathway-mediated inhibition of the immune response caused by nivolumab results in enhanced T-cell function and decreased tumor growth [111]. The safety and efficacy of nivolumab were evaluated in a multicenter, open-label trial that randomized (2 : 1) patients with unresectable or metastatic melanoma to nivolumab or chemotherapy. Patients were required to have disease progression on or following ipilimumab treatment and, if BRAF V600 mutation positive, a BRAF inhibitor [112]. The trial excluded patients with autoimmune disease, medical conditions requiring systemic immunosuppression, ocular melanoma, active brain metastasis, or a history of serious or unresolved ipilimumab-related adverse reactions. The ORR was 32% and of 38 responsive patients, 87% had response durations of 2.6 to 10-plus months. As anticipated, immune-related adverse events were seen in the study. Pembrolizumab is a humanized IgGk mAb that also blocks the interaction between PD-1 and its ligands, PD-L1 and PD-L2 [113]. The efficacy of pembrolizumab was investigated in a multicenter, open-label, randomized (1 : 1), dose comparative study in patients with unresectable or metastatic melanoma refractory to two or more doses of ipilimumab and, if BRAF V600 mutation positive, a BRAF or MEK inhibitor [114]. Patients were randomized to receive 2 mg/kg ($n = 89$) or 10 mg/kg ($n = 84$) of pembrolizumab every 3 weeks until unacceptable toxicity or disease progression. Among the 173 patients enrolled, the ORR was 24% in the 2 mg/kg arm, with 86% of those responding for a duration of 1.4 to 8.5-plus months. Based on these remarkable results, both nivolumab and pembrolizumab were recently approved for the treatment of ipilimumab-refractory melanoma.

4.5.2.2 Breast Cancer (BC)

BC is one of the most prevalent neoplasms in the world, accounting for more than 200,000 new cases in the United States. Approximately 15–20% of BC overexpress the EGFR-2 oncoprotein (otherwise known as HER2), a transmembrane receptor protein of 185 kDa involved in cancer cell proliferation, disease aggression, and poor prognosis [115]. Several mAbs have been developed that target HER2 resulting in improved prognosis for patients with early as well as advanced BC. Trastuzumab is a humanized IgG1k mAb that selectively binds with high affinity to the extracellular domain of HER2 and causes cell death predominantly through ADCC [116]. More than 10,000 patients with HER-overexpressing BC have been evaluated in four large randomized trials of standard chemotherapy with or without trastuzumab in the adjuvant setting. These studies demonstrated that adding trastuzumab to standard adjuvant chemotherapy resulted in a 40–50% improvement in DFS [117]. Studies in patients with metastatic HER2-expressing BC treated with the combination of trastuzumab and chemotherapy have also demonstrated a significant improvement in ORR and time to progression (TTP) when compared to chemotherapy alone [118]. Importantly, the extensive experience with trastuzumab therapy has revealed a risk of reversible cardiotoxicity in up to 15% treated [119]. Thus, cardiac monitoring is necessary in all patients treated with this mAb. Based on these results, trastuzumab has become the standard of care in patients with HER-expressing BC and gastroesophageal cancer.

Pertuzumab is a humanized mAb that targets the extracellular dimerization domain (subdomain II) of HER2 and blocks ligand-dependent heterodimerization of HER2 with other HER family members, including EGFR, HER3, and HER4 [120]. By interfering with ligand-initiated signaling, pertuzumab inhibits the mitogen-activated protein (MAP) kinase and phosphoinositide 3-kinase (PI3K) pathways resulting in cell arrest and apoptosis in addition to inducing cell death through ADCC. The role of pertuzumab in the treatment of HER2-expressing BC was evaluated in patients with metastatic disease and in the preoperative setting. In the metastatic study, patients treated with the combination of pertuzumab, trastuzumab, and docetaxel demonstrated an improvement in DFS of approximately 38% [121]. Treatment with the same combination prior to surgery was associated with pathologic complete remissions in 39% of treated patients [122]. Based on these encouraging results, pertuzumab has been approved for the treatment of HER2-expressing BC, in combination with trastuzumab and docetaxel, in these two settings.

Ado-trastuzumab emtansine is a HER2-targeted ADC composed of trastuzumab covalently linked to the maytansine derivative DM1, a potent microtubule inhibitory drug, via a stable thioether linker [123]. Upon binding to subdomain IV of the HER2 receptor, ado-trastuzumab emtansine undergoes receptor-mediated internalization and subsequent lysosomal degradation, resulting in intracellular release of DM1-containing cytotoxic catabolites that cause cellular arrest and apoptosis. The efficacy of ado-trastuzumab emtansine was evaluated in a randomized, multicenter, open trial of 991 patients with HER2-positive, unresectable locally advanced or metastatic BC who had failed prior taxane- and trastuzumab-based therapy [124]. Patients were randomized to treatments consisting of ado-trastuzumab emtansine or lapatinib plus capecitabine. An improved PFS (35%) was observed on the ADC-treated group and ado-trastuzumab emtansine is currently approved for this indication.

4.5.2.3 Colorectal Cancer (CC) and Upper Gastrointestinal Cancers (UGC)

EGFR-1 is a transmembrane type I receptor tyrosine kinase involved in cellular proliferation and growth. EGFR-1 is constitutively expressed in many normal epithelial tissues, including the skin and hair follicle, and its overexpression has been detected in many human cancers, including CC and head and neck malignancies. Thus, EGFR-1 expression plays an important role in the pathophysiology of CC and other malignancies [125]. Cetuximab is a recombinant human/mouse chimeric mAb that binds specifically to the extracellular domain of the EGFR-1 [126]. Cetuximab is composed of the Fv regions of a murine anti-EGFR antibody with human IgG1k light and heavy chain constant regions and binds to the EGFR-1, on both normal and tumor cells, competitively inhibiting binding of EGFR ligands. Cetuximab binding to EGFR-1 blocks receptor phosphorylation and activation in order to inhibit cell growth, decrease production of matrix metalloproteinase and VEGF by cancer cells, and induce apoptosis. The efficacy of cetuximab was evaluated in a multicenter, randomized study of patients refractory to standard chemotherapy (oxaliplatin and/or irinotecan) [127]. Patients were randomized to receive irinotecan plus cetuximab or cetuximab monotherapy and both ORR and TTP were found to favor the combination group (23% vs. 12%, 4.1 vs. 1.5 months).

Panitumumab is a recombinant human IgG2k mAb that binds specifically to the EGFR-1 [126]. Signal transduction through the EGFR-1 results in activation of the wild-type KRAS protein and downstream signaling pathways. However, in cells with KRAS somatic mutations, the mutant KRAS protein is continuously active independently of EGFR-1 regulation [128]. Therefore, panitumumab (and cetuximab) have no significant effect in cancer cells carrying a "driver" KRAS mutation [129]. The safety and efficacy of panitumumab were studied in a multinational, randomized trial of 463 patients with EGFR-expressing metastatic CC who had progressed following treatment with a fluoropyrimidine, oxaliplatin, and irinotecan [130]. Patients were treated with panitumumab versus best supportive care (BSC) with a superior PFS observed in the panitumumab group (13.7 vs. 8.6 weeks). Unexpectedly, panitumumab was found to shorten PFS, decrease survival time, and increase toxicity when given in combination with bevacizumab and chemotherapy [131]. Besides being approved for the treatment of metastatic CC, cetuximab and panitumumab are used for the treatment of locally advanced or metastatic squamous cell carcinoma [132].

Bevacizumab is a recombinant humanized IgG1 mAb that binds and inhibits the biological activity of human VEGF [133]. Bevacizumab binds VEGF and prevents the interaction of VEGF to its receptors on the surface of endothelial cells in this way inhibiting endothelial cell proliferation and angiogenesis. The efficacy of bevacizumab has been extensively evaluated, in combination with chemotherapy, in patients with metastatic CC [134, 135]. The majority of these studies have confirmed a superior DFS (average of 40% improvement) in patients treated with bevacizumab. Bevacizumab is currently approved for the treatment of patients with CC, glioblastoma, non-small-cell lung cancer, ovarian cancer, and metastatic renal cell carcinoma.

Ramucirumab is a recombinant human IgG1 mAb that specifically binds to the VEGF receptor-2 (VEGFR-2) [136]. Ramucirumab binds to VEGFR-2 and blocks binding of its ligands, VEGF-A, VEGF-C, and VEGF-D. As a result, ramucirumab inhibits ligand-stimulated activation of VEGFR-2 inhibiting proliferation and migration of endothelial cells and angiogenesis. A multinational, randomized study of ramucirumab plus BSC versus BSC alone was conducted on 355 patients with locally advanced or metastatic UGC (i.e., gastric and gastroesophageal cancer) who had failed prior therapy [137]. In this study, both DFS and OS favored the ramucirumab-treated group (HR: 0.48 and 0.78, respectively). Ramucirumab has been approved for the treatment of advanced, progressive gastric and GE cancer and more recently NSCLC based on similar results [138]. Table 4.5 summarizes all targets and mAbs discussed involving cancer treatments.

4.5.3 Novel Targets for Antibody-Based Tumor Immunotherapy: Cancer–Testis Antigens (CTAs)

A variety of genes have been described that are expressed in malignancies of different histologies, but not in normal tissues, except germ cells and placenta. The term cancer–testis antigens (CTAs) was originally coined by Old and Chen to encompass this distinct class of molecules [139]. So far, over 140 members of the CTA family have been identified. CTAs can be divided in those that are encoded by genes on the X chromosome (X-CTAs) and those that are not (non-X-CTAs) [140]. A few examples of X-CTA genes include SSX, GAGE/PAGE/XAGE and MAGE-A, MAGE-C, and NY-ESO-1. The non-X-CTA genes are distributed throughout the genome. X-CTA genes are expressed primarily on the spermatogonia, while non-X-CTA are expressed in later

Table 4.5 Examples of mAbs for cancer treatment.

Target	Antibody	Description	Cancer
CD19	Blinatumomab	Murine origin CD19/CD3 bispecific tandem scFv, direct T cells to specifically kill CD19-expressing B cells	ALL
CD20	Rituximab	Chimeric anti-CD20 mAb	NHL, CLL
	Ofatumumab	IgG1k human mAb; binds to both the small and large extracellular loops of CD20	CLL
	Obinutuzumab	Humanized anti-CD20 mAb of the IgG1 subclass	CLL
	^{131}Iodine-tositumomab	Radioiodinated murine IgG2aλ mAb directed against CD20	NHL
	Ibritumomab tiuxetan	Radioimmunoconjugate	NHL
CD30	Brentuximab vedotin	ADC consisting of the CD30-specific chimeric IgG1 mAb cAC10, MMAE, and a protease-cleavable linker that covalently attaches MMAE to cAC10	HL, ALCL
CD52	Alemtuzumab	Humanized IgG1, kill CD52-expressing B cells and T cells	MS
CTLA-4	Ipilimumab	Recombinant human IgG1k mAb	Melanoma
PD-1	Nivolumab	Human IgGk mAb, blocks interaction of PD-1 with PD-L1 and PD-L2	
	Pembrolizumab	Humanized IgGk mAb, blocks interaction of PD-1 with PD-L1 and PD-L2	

HER2	Trastuzumab	Humanized IgG1k mAb, binds to the extracellular domain of HER2 to cause cell death	HER-expressing breast cancer, gastroesophageal cancer
	Pertuzumab	Humanized mAb, targets HER2 subdomain II and blocks ligand-dependent heterodimerization of HER2 with HER family members	HER-expressing breast cancer
	Ado-trastuzumab emtansine	ADC composed of trastuzumab linked to the maytansine derivative DM1	HER-expressing breast cancer
EGFR-1	Cetuximab	Recombinant human/mouse chimeric mAb, binds to the extracellular domain of EGFR-1	CC, squamous cell carcinoma
	Panitumumab	Recombinant human IgG2k mAb, binds to EGFR-1	
VEGF	Bevacizumab	Recombinant humanized IgG1 mAb, binds and inhibits biological activity of human VEGF	CC, glioblastoma, NSCLC, ovarian cancer, metastatic renal cell carcinoma
VEGFR-2	Ramucirumab	Recombinant human IgG1 mAb, binds to the VEGFR-2 to block ligand binding in order to inhibit proliferation, endothelial cell migration, and angiogenesis	GE cancer, gastric cancer, NSCLC
EPCAM	Catumaxomab	Rat/mouse bispecific mAb, kill EPCAM-expressing tumor cells in ascites	Malignant ascites
GD2	Dinutuximab	Recombinant chimeric IgG1 mAb, binds to GD2 and induces ADCC and CDC against GD2-expressing tumor cells	Neuroblastoma

stages of germ-cell differentiation, such as spermatocytes [140]. Although the function of X-CTAs is largely unknown, most members of the non-X-CTA families have known roles in spermatogenesis, fertilization, and possibly flagellar motility [141, 142]. Some X-CTAs have been found to be expressed in the placenta, and certain somatic tissues (i.e., pancreas, liver, and spleen) have been shown to express very low levels of CTA mRNA accounting for less than 1% of that present in testis [143]. It is thought this restricted expression relates to epigenetic events (i.e., DNA methylation and histone posttranslational modifications) that regulate CTA expression in normal and neoplastic cells [143].

The lack of HLA class I expression on the surface of germ cells and the blood–sperm barrier prevent the immune system from recognizing CTAs as nonself molecules [144]. Thus, from an immunological viewpoint, CTAs may be regarded as highly immunogenic tumor-specific targets. In fact, expression of CTAs in tumors varies greatly among different histologies and members of different CTA families are expressed in tumors as diverse as melanomas, non-small-cell lung cancer, breast, bladder, prostate, esophageal, kidney, CC, and hematologic malignancies [142]. The characterization of immunogenic peptides from selected CTA, together with the identification of their B-cell immunodominance, represents the rationale for the development of immunotherapeutic approaches with anti-CTA mAbs. One area of CTA exploration involves the use of a specific group of CTAs, known as fibrous sperm sheath antigens (FSSA), in cancer immunotherapy. It has been shown that FSSA are expressed in different malignancies including multiple myeloma, ovarian cancer, and non-small-cell lung cancer [145–150]. More importantly, it has been demonstrated that specific epitopes within these CTAs are able to generate cytotoxic T cells capable of killing CTA-expressing tumor cells [150, 151]. These observations are the basis for the effort to develop a CTA-specific immunotherapies for solid tumors and hematologic malignancies (ClinicalTrials.gov). Recent preclinical work by Song and coworkers has demonstrated that an anti-Sp17 mAb was capable of targeting ovarian cancer cells in a specific and safe manner [152]. The same group has also reported the efficacy of a novel anti-SP17–doxorubicin immunoconjugate against SP17-expressing ovarian cancer cells, both *in vitro* and *in vivo* using a xenograft model [153].

4.6 Conclusion

The generation of mAbs for the development of targeted treatments involves a variety of factors. Each targeted mAb therapy involves the choice of a disease-specific target capable of generating an immune response. Chosen targets must be absent in healthy tissues or expressed only on noncrucial tissues in order to avoid toxicity and achieve clinical activity against a disease without further harming patients. Furthermore, mAbs must then be generated to specifically bind with high affinity to the chosen targets in order to prevent unpredictable outcomes such as toxicity or autoimmunity.

A variety of techniques are involved in the design and production of mAbs. Designing mAbs involves both molecular cloning and recombination in order to enhance mAb stability and target affinity, achieve desirable immunogenicity, lower toxicity, and attain an overall beneficial clinical outcome for each patient. Currently, mAbs approved for patient use belong to the IgG family and are characterized by simple, linear PK profiles, small distribution volume, low clearance rates, and longer biological half-lives in healthy subjects [5]. Often, the development and patient use of novel targeted treatments are hindered by steep research, production, and market costs. However, there are promising options for lowering mAb production costs through the use of yeast- or plant-based systems.

In this chapter, three main categories of human diseases were discussed including infectious, autoimmune, and neoplastic. Currently, mAb treatments for infectious diseases are not as advanced as those for neoplastic and autoimmune diseases. However, mAbs are currently being used to treat a variety of infectious diseases including RSV using palivizumab (a humanized anti-F protein mAb) and anthrax with raxibacumab (a human recombinant anti-PA mAb). Meanwhile, active research for creation of mAbs against diseases such as HIV and Ebola viruses is ongoing.

Creating mAb treatments for both autoimmune and neoplastic diseases presents a different challenge compared to infectious diseases as autoimmune and neoplastic diseases often arise as a result of native changes in patient tissue genetics and epigenetics as opposed to the introduction of foreign genetic information involved with infectious diseases. In both autoimmune and neoplastic diseases, targets expressed on diseased tissue are often expressed on healthy tissue as part of normal tissue phenotypes. Therefore, selection of proper mAb targets involves the determination of target expression not only in diseased tissues but also healthy tissues. Expression of a chosen target must be low or absent in healthy tissues or limited to nonessential tissues.

With regard to autoimmune diseases, mAbs against antigens expressed on activated T and B cells are most effective. Currently, mAbs used to treat patients suffering from autoimmune diseases include natalizumab (humanized anti-$\alpha 4$ subunit of $\alpha 4\beta 1$ and $\alpha 4\beta 7$ mAb) for MS; vedolizumab (humanized antihuman $\alpha 4\beta 7$ integrin mAb) for UC and CD; infliximab (chimeric antihuman TNFα mAb) for IBD, RA, AS, and psoriatic arthritis; tocilizumab (recombinant humanized antihuman IL-6 mAb) for IA, RA, and polyarticular and systemic juvenile arthritis; and belimumab (human anti-BAFF mAb) for SLE.

References

1. Tansey, E. M.; Catterall, P. P., Monoclonal antibodies: a witness seminar in contemporary medical history. *Medical History* 1994, 38 (3), 322–327.
2. Thie, H.; Meyer, T.; Schirrmann, T.; Hust, M.; Dübel, S., Phage display derived therapeutic antibodies. *Current Pharmaceutical Biotechnology* 2008, 9 (6), 439–446.
3. Wu, A. M.; Senter, P. D., Arming antibodies: prospects and challenges for immunoconjugates. *Nature Biotechnology* 2005, 23 (9), 1137–1146.
4. Mould, D. R.; Green, B., Pharmacokinetics and pharmacodynamics of monoclonal antibodies: concepts and lessons for drug development. *BioDrugs* 2010, 24 (1), 23–39.
5. Wang, W.; Wang, E. Q.; Balthasar, J. P., Monoclonal antibody pharmacokinetics and pharmacodynamics. *Clinical Pharmacology and Therapeutics* 2008, 84 (5), 548–558.
6. Mager, D. E.; Jusko, W. J., General pharmacokinetic model for drugs exhibiting target-mediated drug disposition. *Journal of Pharmacokinetics and Pharmacodynamics* 2001, 28 (6), 507–532.
7. Cao, Y.; Jusko, W. J., Incorporating target-mediated drug disposition in a minimal physiologically-based pharmacokinetic model for monoclonal antibodies. *Journal of Pharmacokinetics and Pharmacodynamics* 2014, 41 (4), 375–387.
8. Dirks, N. L.; Meibohm, B., Population pharmacokinetics of therapeutic monoclonal antibodies. *Clinical Pharmacokinetics* 2010, 49 (10), 633–659.
9. Hemming, V. G.; Prince, G. A.; Groothuis, J. R.; Siber, G. R., Hyperimmune globulins in prevention and treatment of respiratory syncytial virus infections. *Clinical Microbiology Reviews* 1995, 8 (1), 22–33.
10. Collier, R. J.; Young, J. A. T., Anthrax toxin. *Annual Review of Cell and Developmental Biology* 2003, 19 (1), 45–70.
11. Mazumdar, S., Raxibacumab. *mAbs* 2009, 1 (6), 531–538.
12. Migone, T. S.; Subramanian, G. M.; Zhong, J.; Healey, L. M.; Corey, A.; Devalaraja, M.; Lo, L.; Ullrich, S.; Zimmerman, J.; Chen, A.; Lewis, M.; Meister, G.; Gillum, K.; Sanford, D.; Mott, J.; Bolmer, S. D., Raxibacumab for the treatment of inhalational anthrax. *New England Journal of Medicine* 2009, 361 (2), 135–144.
13. Subramanian, G. M.; Cronin, P. W.; Poley, G.; Weinstein, A.; Stouqhton, S. M.; Zhong, J.; Ou, Y.; Zmuda, J. F.; Osborn, B. L.; Freimuth, W. W., A phase 1 study of PAmAb, a fully human monoclonal antibody against *Bacillus anthracis* protective antigen, in healthy volunteers. *Clinical Infectious Diseases* 2005, 41 (1), 12–20.
14. Dolgin, E., Animal rule for drug approval creates a jungle of confusion. *Nature Medicine* 2013, 19 (2), 118–119.

15. Jackson, G. G.; Perkins, J. T.; Rubenis, M.; Paul, D. A.; Knigge, M.; Despotes, J. C.; Spencer, P., Passive immunoneutralization of human immunodeficiency virus in patients with advanced AIDS. *Lancet* 1988, 2 (8612), 647–652.
16. Burton, D. R.; Barbas, C. F., 3rd; Persson, M. A.; Koenig, S.; Chanock, R. M.; Lerner, R. A., A large array of human monoclonal antibodies to type 1 human immunodeficiency virus from combinatorial libraries of asymptomatic seropositive individuals. *Proceedings of the National Academy of Sciences of the United States of America* 1991, 88 (22), 10134–10137.
17. Bouvin-Pley, M.; Morgand, M.; Meyer, L.; Goujard, C.; Moreau, A.; Mouquet, H.; Nussenzweig, M.; Pace, C.; Ho, D.; Bjorkman, P. J.; Baty, D.; Chames, P.; Pancera, M.; Kwong, P. D.; Poignard, P.; Barin, F.; Braibant, M., Drift of the HIV-1 envelope glycoprotein gp120 toward increased neutralization resistance over the course of the epidemic: a comprehensive study using the most potent and broadly neutralizing monoclonal antibodies. *Journal of Virology* 2014, 88 (23), 13910–13917.
18. Bale, S.; Dias, J. M.; Fusco, M. L.; Hashiguchi, T.; Wong, A. C.; Tong, L.; Keuhne, A. I.; Sheng, L.; Woods, V. L., Jr.; Chandran, K.; Dye, J. M.; Saphire, E. O., Structural basis for differential neutralization of ebolaviruses. *Viruses* 2012, 4 (4), 447–470.
19. Qiu, X.; Fernando, L.; Alimonti, J. B.; Aviles, J.; Wong, G.; Audet, J.; Bello, A.; Fausther-Bovendo, H.; Wei, H.; Hiatt, E.; Johnson, A.; Morton, J.; Swope, K.; Bohorov, O.; Bohorova, N.; Goodman, C.; Kim, D.; Pauly, M. H.; Velasco, J.; Whaley, K., Reversion of advanced Ebola virus disease in nonhuman primates with ZMapp. *Nature* 2014, 514 (7520), 47–53.
20. Audet, J.; Wong, G.; Wang, H.; Lu, G.; Gao, G. F.; Kobinger, G.; Qiu, X., Molecular characterization of the monoclonal antibodies composing ZMAb: a protective cocktail against Ebola virus. *Scientific Reports* 2014, 4, 6881.
21. Marzi, A.; Yoshida, R.; Miyamoto, H.; Ishijima, M.; Suzuki, Y.; Higuchi, M.; Matsuyama, Y.; Igarashi, M.; Nakayama, E.; Kuroda, M.; Saijo, M.; Feldmann, F.; Brining, D.; Feldmann, H.; Takada, A., Protective efficacy of neutralizing monoclonal antibodies in a nonhuman primate model of Ebola hemorrhagic fever. *PLoS One* 2012, 7 (4), e36192.
22. Davidson, A.; Diamond, B., Autoimmune diseases. *The New England Journal of Medicine* 2001, 345 (5), 340–350.
23. Steinman, L., Immune therapy for autoimmune diseases. *Science* 2004, 305 (5681), 212–216.
24. Frohman, E. M.; Racke, M. K.; Raine, C. S., Multiple sclerosis—the plaque and its pathogenesis. *The New England Journal of Medicine* 2006, 354 (9), 942–955.
25. Steinman, L., Blocking adhesion molecules as therapy for multiple sclerosis: natalizumab. *Nature Reviews. Drug Discovery* 2005, 4 (6), 510–518.
26. Polman, C. H.; O'Connor, P. W.; Havrdova, E.; Hutchinson, M.; Kappos, L.; Miller, D. H.; Phillips, J. T.; Lublin, F. D.; Giovannoni, G.; Wajgt, A.; Toal, M.;

Lynn, F.; Panzara, M. A.; Sandrock, A. W.; AFFIRM Investigators, A randomized, placebo-controlled trial of natalizumab for relapsing multiple sclerosis. *The New England Journal of Medicine* 2006, 354 (9), 899–910.

27. Berger, J. R.; Koralnik, I. J., Progressive multifocal leukoencephalopathy and natalizumab—unforeseen consequences. *The New England Journal of Medicine* 2005, 353 (4), 414–416.
28. Weissert, R., Progressive multifocal leukoencephalopathy. *Journal of Neuroimmunology* 2011, 231 (1/2), 73–77.
29. Xavier, R. J.; Podolsky, D. K., Unravelling the pathogenesis of inflammatory bowel disease. *Nature* 2007, 448 (7152), 427–434.
30. Zhang, Y.-Z.; Li, Y.-Y., Inflammatory bowel disease: pathogenesis. *World Journal of Gastroenterology: WJG* 2014, 20 (1), 91–99.
31. Olmos, G.; Lladó, J., Tumor necrosis factor alpha: a link between neuroinflammation and excitotoxicity. *Mediators of Inflammation* 2014, 2014, 861231.
32. Nguyen, V. T.; Benveniste, E. N., Critical role of tumor necrosis factor-alpha and NF-kappa B in interferon-gamma -induced CD40 expression in microglia/macrophages. *The Journal of Biological Chemistry* 2002, 277 (16), 13796–13803.
33. Schönbeck, U.; Libby, P., The CD40/CD154 receptor/ligand dyad. *Cellular and Molecular Life Sciences: CMLS* 2001, 58 (1), 4–43.
34. Tracey, D.; Klareskog, L.; Sasso, E. H.; Salfeld, J. G.; Tak, P. P., Tumor necrosis factor antagonist mechanisms of action: a comprehensive review. *Pharmacology & Therapeutics* 2008, 117 (2), 244–279.
35. Rutgeerts, P. J., Review article: efficacy of infliximab in Crohn's disease—induction and maintenance of remission. *Alimentary Pharmacology & Therapeutics* 1999, 13, 9–15.
36. Hanauer, S. B.; Feagan, B. G.; Lichtenstein, G. R.; Mayer, L. F.; Schreiber, S.; Colombel, J. F.; Rachmilewitz, D.; Wolf, D. C.; Olson, A.; Bao, W.; Rutgeerts, P., Maintenance infliximab for Crohn's disease: the ACCENT I randomised trial. *Lancet* 2002, 359 (9317), 1541.
37. Rutgeerts, P.; Sandborn, W. J.; Feagan, B. G.; Reinisch, W.; Olson, A.; Johanns, J.; Travers, S.; Rachmilewitz, D.; Hanauer, S. B.; Lichtenstein, G. R.; de Villiers, W. J. S.; Present, D.; Sands, B. E.; Colombel, J. F., Infliximab for induction and maintenance therapy for ulcerative colitis. *The New England Journal of Medicine* 2005, 353 (23), 2462–2476.
38. Levin, A.; Shibolet, O., Infliximab in ulcerative colitis. *Biologics: Targets and Therapy* 2008, 2 (3), 379–388.
39. Atzeni, F.; Talotta, R.; Salaffi, F.; Cassinotti, A.; Varisco, V.; Battellino, M.; Ardizzone, S.; Pace, F.; Sarzi-Puttini, P., Immunogenicity and autoimmunity during anti-TNF therapy. *Autoimmunity Reviews* 2013, 12 (7), 703–708.
40. Singh, J. A.; Wells, G. A.; Christensen, R.; Tanjong Ghogomu, E.; Maxwell, L.; Macdonald, J. K.; Filippini, G.; Skoetz, N.; Francis, D.; Lopes, L. C.; Guyatt, G. H.;

Schmitt, J.; La Mantia, L.; Weberschock, T.; Roos, J. F.; Siebert, H.; Hershan, S.; Lunn, M. P.; Tugwell, P.; Buchbinder, R., Adverse effects of biologics: a network meta-analysis and Cochrane overview. *The Cochrane Database of Systematic Reviews* 2011 (2), CD008794.

41. Eksteen, B., Targeting of gut specific leucocyte recruitment in IBD by vedolizumab. *Gut* 2015, 64 (1), 8–10.

42. Sandborn, W. J.; Feagan, B. G.; Rutgeerts, P.; Hanauer, S.; Colombel, J.-F.; Sands, B. E.; Lukas, M.; Fedorak, R. N.; Lee, S.; Bressler, B.; Fox, I.; Rosario, M.; Sankoh, S.; Xu, J.; Stephens, K.; Milch, C.; Parikh, A., Vedolizumab as induction and maintenance therapy for Crohn's disease. *The New England Journal of Medicine* 2013, 369 (8), 711–721.

43. Feagan, B. G.; Rutgeerts, P.; Sands, B. E.; Hanauer, S.; Colombel, J.-F.; Sandborn, W. J.; Van Assche, G.; Axler, J.; Kim, H.-J.; Danese, S.; Fox, I.; Milch, C.; Sankoh, S.; Wyant, T.; Xu, J.; Parikh, A., Vedolizumab as induction and maintenance therapy for ulcerative colitis. *The New England Journal of Medicine* 2013, 369 (8), 699–710.

44. Lam, M. C.; Bressler, B., Vedolizumab for ulcerative colitis and Crohn's disease: results and implications of GEMINI studies. *Immunotherapy* 2014, 6 (9), 963–971.

45. Sakuraba, A.; Keyashian, K.; Correia, C.; Melek, J.; Cohen, R. D.; Hanauer, S. B.; Rubin, D. T., Natalizumab in Crohn's disease: results from a US tertiary inflammatory bowel disease center. *Inflammatory Bowel Diseases* 2013, 19 (3), 621–626.

46. Maini, R.; St. Clair, E. W.; Breedveld, F.; Furst, D.; Kalden, J.; Weisman, M.; Smolen, J.; Emery, P.; Harriman, G.; Feldmann, M.; Lipsky, P., Infliximab (chimeric anti-tumor necrosis factor alpha monoclonal antibody) versus placebo in rheumatoid arthritis patients receiving concomitant methotrexate: a randomised phase III trial. *Lancet* 1999, 354 (9194), 1932–1939.

47. Smolen, J. S.; van der Heijde, D. M. F.; St. Clair, E. W.; Emery, P.; Bathon, J. M.; Keystone, E.; Maini, R. N.; Kalden, J. R.; Schiff, M.; Baker, D.; Han, C.; Han, J.; Bala, M., Predictors of joint damage in patients with early rheumatoid arthritis treated with high-dose methotrexate with or without concomitant infliximab: results from the ASPIRE trial. *Arthritis and Rheumatism* 2006, 54 (3), 702–710.

48. Cantini, F.; Niccoli, L.; Nannini, C.; Kaloudi, O.; Cassarà, E., Infliximab in psoriatic arthritis. *The Journal of Rheumatology. Supplement* 2012, 89, 71–73.

49. Baraliakos, X.; Listing, J.; Fritz, C.; Haibel, H.; Alten, R.; Burmester, G.-R.; Krause, A.; Schewe, S.; Schneider, M.; Sörensen, H.; Schmidt, R.; Sieper, J.; Braun, J., Persistent clinical efficacy and safety of infliximab in ankylosing spondylitis after 8 years—early clinical response predicts long-term outcome. *Rheumatology* 2011, 50 (9), 1690–1699.

50. Burmester, G. R.; Panaccione, R.; Gordon, K. B.; McIlraith, M. J.; Lacerda, A. P. M., Adalimumab: long-term safety in 23 458 patients from global clinical trials in rheumatoid arthritis, juvenile idiopathic arthritis, ankylosing spondylitis, psoriatic arthritis, psoriasis and Crohn's disease. *Annals of the Rheumatic Diseases* 2013, 72 (4), 517–524.

51. Kay, J.; Fleischmann, R.; Keystone, E.; Hsia, E. C.; Hsu, B.; Mack, M.; Goldstein, N.; Braun, J.; Kavanaugh, A., Golimumab 3-year safety update: an analysis of pooled data from the long-term extensions of randomised, double-blind, placebo-controlled trials conducted in patients with rheumatoid arthritis, psoriatic arthritis or ankylosing spondylitis. *Annals of the Rheumatic Diseases* 2015, 74 (3), 538–546.

52. Landewé, R.; Braun, J.; Deodhar, A.; Dougados, M.; Maksymowych, W. P.; Mease, P. J.; Reveille, J. D.; Rudwaleit, M.; van der Heijde, D.; Stach, C.; Hoepken, B.; Fichtner, A.; Coteur, G.; de Longueville, M.; Sieper, J., Efficacy of certolizumab pegol on signs and symptoms of axial spondyloarthritis including ankylosing spondylitis: 24-week results of a double-blind randomised placebo-controlled Phase 3 study. *Annals of the Rheumatic Diseases* 2014, 73 (1), 39–47.

53. Mease, P. J.; Fleischmann, R.; Deodhar, A. A.; Wollenhaupt, J.; Khraishi, M.; Kielar, D.; Woltering, F.; Stach, C.; Hoepken, B.; Arledge, T.; van der Heijde, D., Effect of certolizumab pegol on signs and symptoms in patients with psoriatic arthritis: 24-week results of a Phase 3 double-blind randomised placebo-controlled study (RAPID-PsA). *Annals of the Rheumatic Diseases* 2014, 73 (1), 48–55.

54. Smolen, J. S.; Nash, P.; Durez, P.; Hall, S.; Ilivanova, E.; Irazoque-Palazuelos, F.; Miranda, P.; Min-Chan, P.; Pavelka, K.; Pedersen, R.; Szumski, A.; Hammond, C.; Koenig, A. S.; Vlahos, B., Maintenance, reduction, or withdrawal of etanercept after treatment with etanercept and methotrexate in patients with moderate rheumatoid arthritis (PRESERVE): a randomised controlled trial. *Lancet* 2013, 381 (9870), 918–929.

55. Perrier, C.; de Hertogh, G.; Cremer, J.; Vermeire, S.; Rutgeerts, P.; Van Assche, G.; Szymkowski, D. E.; Ceuppens, J. L., Neutralization of membrane TNF, but not soluble TNF, is crucial for the treatment of experimental colitis. *Inflammatory Bowel Diseases* 2013, 19 (2), 246–253.

56. Hashizume, M.; Tan, S.-L.; Takano, J.; Ohsawa, K.; Hasada, I.; Hanasaki, A.; Ito, I.; Mihara, M.; Nishida, K., Tocilizumab, a humanized anti-IL-6R antibody, as an emerging therapeutic option for rheumatoid arthritis: molecular and cellular mechanistic insights. *International Reviews of Immunology* 2015, 34 (3), 265–279.

57. Tanaka, T.; Narazaki, M.; Ogata, A.; Kishimoto, T., A new era for the treatment of inflammatory autoimmune diseases by interleukin-6 blockade strategy. *Seminars in Immunology* 2014, 26 (1), 88–96.

58. Navarro-Millán, I.; Singh, J. A.; Curtis, J. R., Systematic review of tocilizumab for rheumatoid arthritis: a new biologic agent targeting the interleukin-6 receptor. *Clinical Therapeutics* 2012, 34 (4), 788–802.e3.

59. Singh, J. A.; Beg, S.; Lopez-Olivo, M. A., Tocilizumab for rheumatoid arthritis: a Cochrane systematic review. *Journal of Rheumatology* 2011, 38 (1), 10–20.
60. Tanaka, T.; Narazaki, M.; Kishimoto, T., Anti-interleukin-6 receptor antibody, tocilizumab, for the treatment of autoimmune diseases. *FEBS Letters* 2011, 585 (23), 3699–3709.
61. Anolik, J. H., B cell biology: implications for treatment of systemic lupus erythematosus. *Lupus* 2013, 22 (4), 342–349.
62. Stohl, W.; Hilbert, D. M., The discovery and development of belimumab: the anti-BLyS-lupus connection. *Nature Biotechnology* 2012, 30 (1), 69–77.
63. Navarra, S. V.; Guzmán, R. M.; Gallacher, A. E.; Hall, S.; Levy, R. A.; Jimenez, R. E.; Li, E. K. M.; Thomas, M.; Kim, H.-Y.; León, M. G.; Tanasescu, C.; Nasonov, E.; Lan, J.-L.; Pineda, L.; Zhong, Z. J.; Freimuth, W.; Petri, M. A., Efficacy and safety of belimumab in patients with active systemic lupus erythematosus: a randomised, placebo-controlled, phase 3 trial. *Lancet* 2011, 377 (9767), 721–731.
64. Furie, R.; Petri, M.; Zamani, O.; Cervera, R.; Wallace, D. J.; Tegzová, D.; Sanchez-Guerrero, J.; Schwarting, A.; Merrill, J. T.; Chatham, W. W.; Stohl, W.; Ginzler, E. M.; Hough, D. R.; Zhong, Z. J.; Freimuth, W.; van Vollenhoven, R. F., A phase III, randomized, placebo-controlled study of belimumab, a monoclonal antibody that inhibits B lymphocyte stimulator, in patients with systemic lupus erythematosus. *Arthritis and Rheumatism* 2011, 63 (12), 3918–3930.
65. Petri, M. A.; van Vollenhoven, R. F.; Buyon, J.; Levy, R. A.; Navarra, S. V.; Cervera, R.; Zhong, Z. J.; Freimuth, W. W., Baseline predictors of systemic lupus erythematosus flares: data from the combined placebo groups in the phase III belimumab trials. *Arthritis and Rheumatism* 2013, 65 (8), 2143–2153.
66. Manetta, J.; Bina, H.; Ryan, P.; Fox, N.; Witcher, D. R.; Kikly, K., Generation and characterization of tabalumab, a human monoclonal antibody that neutralizes both soluble and membrane-bound B-cell activating factor. *Journal of Inflammation Research* 2014, 7, 121–131.
67. Cai, Y.; Fleming, C.; Yan, J., New insights of T cells in the pathogenesis of psoriasis. *Cellular & Molecular Immunology* 2012, 9 (4), 302–309.
68. Coimbra, S.; Figueiredo, A.; Castro, E.; Rocha-Pereira, P.; Santos-Silva, A., The roles of cells and cytokines in the pathogenesis of psoriasis. *International Journal of Dermatology* 2012, 51 (4), 389–398.
69. Zaghi, D.; Krueger, G. G.; Callis Duffin, K., Ustekinumab: a review in the treatment of plaque psoriasis and psoriatic arthritis. *Journal of Drugs in Dermatology: JDD* 2012, 11 (2), 160–167.
70. Benson, J. M.; Sachs, C. W.; Treacy, G.; Zhou, H.; Pendley, C. E.; Brodmerkel, C. M.; Shankar, G.; Mascelli, M. A., Therapeutic targeting of the IL-12/23 pathways: generation and characterization of ustekinumab. *Nature Biotechnology* 2011, 29 (7), 615–624.

71. McInnes, I. B.; Kavanaugh, A.; Gottlieb, A. B.; Puig, L.; Rahman, P.; Ritchlin, C.; Brodmerkel, C.; Li, S.; Wang, Y.; Mendelsohn, A. M.; Doyle, M. K., Efficacy and safety of ustekinumab in patients with active psoriatic arthritis: 1 year results of the phase 3, multicentre, double-blind, placebo-controlled PSUMMIT 1 trial. *Lancet* 2013, 382 (9894), 780–789.
72. Kimball, A. B.; Papp, K. A.; Wasfi, Y.; Chan, D.; Bissonnette, R.; Sofen, H.; Yeilding, N.; Li, S.; Szapary, P.; Gordon, K. B., Long-term efficacy of ustekinumab in patients with moderate-to-severe psoriasis treated for up to 5 years in the PHOENIX 1 study. *Journal of the European Academy of Dermatology and Venereology: JEADV* 2013, 27 (12), 1535–1545.
73. Mitra, A.; Raychaudhuri, S. K.; Raychaudhuri, S. P., IL-17 and IL-17R: an auspicious therapeutic target for psoriatic disease. *Actas Dermo-Sifiliográficas* 2014, 105 (Suppl 1), 21–33.
74. Reich, K., Anti-interleukin-17 monoclonal antibody ixekizumab in psoriasis. *The New England Journal of Medicine* 2012, 367 (3), 274; author reply 275.
75. Posch, C.; Kaulfersch, W.; Rappersberger, K., Cryopyrin-associated periodic syndrome. *Pediatric Dermatology* 2014, 31 (2), 228–231.
76. Doria, A.; Zen, M.; Bettio, S.; Gatto, M.; Bassi, N.; Nalotto, L.; Ghirardello, A.; Iaccarino, L.; Punzi, L., Autoinflammation and autoimmunity: bridging the divide. *Autoimmunity Reviews* 2012, 12 (1), 22–30.
77. Chakraborty, A.; Tannenbaum, S.; Rordorf, C.; Lowe, P. J.; Floch, D.; Gram, H.; Roy, S., Pharmacokinetic and pharmacodynamic properties of canakinumab, a human anti-interleukin-1β monoclonal antibody. *Clinical Pharmacokinetics* 2013, 52 (2), e1–e18.
78. Koné-Paut, I.; Piram, M., Targeting interleukin-1β in CAPS (cryopyrin-associated periodic) syndromes: what did we learn? *Autoimmunity Reviews* 2012, 12 (1), 77–80.
79. Marsaud, C.; Marie, I.; Koné-Paut, I., Longterm followup of quality of life in patients with cryopyrin-associated periodic syndrome treated with canakinumab, an anti-interleukin 1β monoclonal antibody. *The Journal of Rheumatology* 2014, 41 (8), 1721–1722.
80. Lyons, P. A.; Rayner, T. F.; Trivedi, S.; Holle, J. U.; Watts, R. A.; Jayne, D. R.; Baslund, B.; Brenchley, P.; Bruchfeld, A.; Chaudhry, A. N.; Cohen Tervaert, J. W.; Deloukas, P.; Feighery, C.; Gross, W. L.; Guillevin, L.; Gunnarsson, I.; Harper, L.; Hrusková, Z.; Little, M. A.; Martorana, D.; Neumann, T.; Ohlsson, S.; Padmanabhan, S.; Pusey, C. D.; Salama, A. D.; Sanders, J. S.; Savage, C. O.; Segelmark, M.; Stegeman, C. A.; Tesa, V.; Vaglio, A.; Wieczorek, S.; Wilde, B.; Zwerina, J.; Rees, A. J.; Clayton, D. G.; Smith, K. G., Genetically distinct subsets within ANCA-associated vasculitis. *The New England Journal of Medicine* 2012, 367 (3), 214–223.
81. Boross, P.; Leusen, J. H. W., Mechanisms of action of CD20 antibodies. *American Journal of Cancer Research* 2012, 2 (6), 676–690.

82. Guillevin, L., Rituximab for ANCA-associated vasculitides. *Clinical and Experimental Rheumatology* 2014, 32 (3 Suppl 82), S118–S121.
83. Geetha, D.; Specks, U.; Stone, J. H.; Merkel, P. A.; Seo, P.; Spiera, R.; Langford, C. A.; Hoffman, G. S.; Kallenberg, C. G. M.; St Clair, E. W.; Fessler, B. J.; Ding, L.; Tchao, N. K.; Ikle, D.; Jepson, B.; Brunetta, P.; Fervenza, F. C., Rituximab versus cyclophosphamide for ANCA-associated vasculitis with renal involvement. *Journal of the American Society of Nephrology: JASN* 2015, 26 (4), 976–985.
84. Hebert, L. A.; Alvarado, A.; Rovin, B., Rituximab or azathioprine maintenance in ANCA-associated vasculitis. *The New England Journal of Medicine* 2015, 372 (4), 385–387.
85. Danés, I.; Agustí, A.; Vallano, A.; Martínez, J.; Alerany, C.; Ferrer, A.; López, A.; Cortés-Hernández, J.; Bosch, J., Available evidence and outcome of off-label use of rituximab in clinical practice. *European Journal of Clinical Pharmacology* 2013, 69 (9), 1689–1699.
86. Maire, V.; Baldeyron, C.; Richardson, M.; Tesson, B.; Vincent-Salomon, A.; Gravier, E.; Marty-Prouvost, B.; De Koning, L.; Rigaill, G.; Dumont, A.; Gentien, D.; Barillot, E.; Roman-Roman, S.; Depil, S.; Cruzalegui, F.; Pierré, A.; Tucker, G. C.; Dubois, T., TTK/hMPS1 is an attractive therapeutic target for triple-negative breast cancer. *PLoS One* 2013, 8 (5), e63712.
87. Daifotis, A. G.; Koenig, S.; Chatenoud, L.; Herold, K. C., Anti-CD3 clinical trials in type 1 diabetes mellitus. *Clinical Immunology* 2013, 149 (3), 268–278.
88. Sprangers, B.; Van der Schueren, B.; Gillard, P.; Mathieu, C., Otelixizumab in the treatment of type 1 diabetes mellitus. *Immunotherapy* 2011, 3 (11), 1303–1316.
89. Keymeulen, B.; Vandemeulebroucke, E.; Ziegler, A. G.; Mathieu, C.; Kaufman, L.; Hale, G.; Gorus, F.; Goldman, M.; Walter, M.; Candon, S.; Schandene, L.; Crenier, L.; De Block, C.; Seigneurin, J.-M.; De Pauw, P.; Pierard, D.; Weets, I.; Rebello, P.; Bird, P.; Berrie, E.; Frewin, M.; Waldmann, H.; Bach, J.-F.; Pipeleers, D.; Chatenoud, L., Insulin needs after CD3-antibody therapy in new-onset type 1 diabetes. *The New England Journal of Medicine* 2005, 352 (25), 2598–2608.
90. Herold, K. C.; Gitelman, S. E.; Ehlers, M. R.; Gottlieb, P. A.; Greenbaum, C. J.; Hagopian, W.; Boyle, K. D.; Keyes-Elstein, L.; Aggarwal, S.; Phippard, D.; Sayre, P. H.; McNamara, J.; Bluestone, J. A., Teplizumab (anti-CD3 mAb) treatment preserves C-peptide responses in patients with new-onset type 1 diabetes in a randomized controlled trial: metabolic and immunologic features at baseline identify a subgroup of responders. *Diabetes* 2013, 62 (11), 3766–3774.
91. Herold, K. C.; Gitelman, S.; Greenbaum, C.; Puck, J.; Hagopian, W.; Gottlieb, P.; Sayre, P.; Bianchine, P.; Wong, E.; Seyfert-Margolis, V.; Bourcier, K.; Bluestone, J. A., Treatment of patients with new onset Type 1 diabetes with a single course of anti-CD3 mAb Teplizumab preserves insulin production for up to 5 years. *Clinical Immunology* 2009, 132 (2), 166–173.

92. Fisher, R. I.; Gaynor, E. R.; Dahlberg, S.; Oken, M. M.; Grogan, T. M.; Mize, E. M.; Glick, J. H.; Coltman, C. A., Jr.; Miller, T. P., Comparison of a standard regimen (CHOP) with three intensive chemotherapy regimens for advanced non-Hodgkin's lymphoma. *The New England Journal of Medicine* 1993, 328 (14), 1002–1006.
93. Du, J.; Yang, H.; Guo, Y.; Ding, J., Structure of the Fab fragment of therapeutic antibody Ofatumumab provides insights into the recognition mechanism with CD20. *Molecular Immunology* 2009, 46 (11–12), 2419–2423.
94. Wierda, W. G.; Kipps, T. J.; Mayer, J.; Stilgenbauer, S.; Williams, C. D.; Hellmann, A.; Robak, T.; Furman, R. R.; Hillmen, P.; Trneny, M.; Dyer, M. J. S.; Padmanabhan, S.; Piotrowska, M.; Kozak, T.; Chan, G.; Davis, R.; Losic, N.; Wilms, J.; Russell, C. A.; Osterborg, A., Ofatumumab as single-agent CD20 immunotherapy in fludarabine-refractory chronic lymphocytic leukemia. *Journal of Clinical Oncology: Official Journal of the American Society of Clinical Oncology* 2010, 28 (10), 1749–1755.
95. Abou-Nassar, K.; Brown, J. R., Novel agents for the treatment of chronic lymphocytic leukemia. *Clinical Advances in Hematology & Oncology: H&O* 2010, 8 (12), 886–895.
96. Rai, K. R.; Peterson, B. L.; Appelbaum, F. R.; Kolitz, J.; Elias, L.; Shepherd, L.; Hines, J.; Threatte, G. A.; Larson, R. A.; Cheson, B. D.; Schiffer, C. A., Fludarabine compared with chlorambucil as primary therapy for chronic lymphocytic leukemia. *The New England Journal of Medicine* 2000, 343 (24), 1750–1757.
97. Lewington, V., Development of 131I-tositumomab. *Seminars in Oncology* 2005, 32 (1 Suppl 1), S50–S56.
98. Kaminski, M. S.; Zelenetz, A. D.; Press, O. W.; Saleh, M.; Leonard, J.; Fehrenbacher, L.; Lister, T. A.; Stagg, R. J.; Tidmarsh, G. F.; Kroll, S.; Wahl, R. L.; Knox, S. J.; Vose, J. M., Pivotal study of iodine I 131 tositumomab for chemotherapy-refractory low-grade or transformed low-grade B-cell non-Hodgkin's lymphomas. *Journal of Clinical Oncology: Official Journal of the American Society of Clinical Oncology* 2001, 19 (19), 3918–3928.
99. Davies, A. J., A review of tositumomab and I(131) tositumomab radioimmunotherapy for the treatment of follicular lymphoma. *Expert Opinion on Biological Therapy* 2005, 5 (4), 577–588.
100. Marcus, R., Use of 90Y-ibritumomab tiuxetan in non-Hodgkin's lymphoma. *Seminars in Oncology* 2005, 32 (1 Suppl 1), S36–S43.
101. Witzig, T. E.; Gordon, L. I.; Cabanillas, F.; Czuczman, M. S.; Emmanouilides, C.; Joyce, R.; Pohlman, B. L.; Bartlett, N. L.; Wiseman, G. A.; Padre, N.; Grillo-López, A. J.; Multani, P.; White, C. A., Randomized controlled trial of yttrium-90-labeled ibritumomab tiuxetan radioimmunotherapy versus rituximab immunotherapy for patients with relapsed or refractory

low-grade, follicular, or transformed B-cell non-Hodgkin's lymphoma. *Journal of Clinical Oncology: Official Journal of the American Society of Clinical Oncology* 2002, 20 (10), 2453–2463.

102. Witzig, T. E.; Flinn, I. W.; Gordon, L. I.; Emmanouilides, C.; Czuczman, M. S.; Saleh, M. N.; Cripe, L.; Wiseman, G.; Olejnik, T.; Multani, P. S.; White, C. A., Treatment with ibritumomab tiuxetan radioimmunotherapy in patients with rituximab-refractory follicular non-Hodgkin's lymphoma. *Journal of Clinical Oncology: Official Journal of the American Society of Clinical Oncology* 2002, 20 (15), 3262–3269.

103. Skoetz, N.; Trelle, S.; Rancea, M.; Haverkamp, H.; Diehl, V.; Engert, A.; Borchmann, P., Effect of initial treatment strategy on survival of patients with advanced-stage Hodgkin's lymphoma: a systematic review and network meta-analysis. *Lancet Oncology* 2013, 14 (10), 943–952.

104. Pizzolo, G.; Vinante, F.; Chilosi, M.; Romagnani, S.; Del Prete, G., CD30 antigen and cellular biology of Reed-Sternberg cells. *Blood* 1994, 84 (11), 3983–3984.

105. Senter, P. D.; Sievers, E. L., The discovery and development of brentuximab vedotin for use in relapsed Hodgkin lymphoma and systemic anaplastic large cell lymphoma. *Nature Biotechnology* 2012, 30 (7), 631–637.

106. Younes, A.; Gopal, A. K.; Smith, S. E.; Ansell, S. M.; Rosenblatt, J. D.; Savage, K. J.; Ramchandren, R.; Bartlett, N. L.; Cheson, B. D.; de Vos, S.; Forero-Torres, A.; Moskowitz, C. H.; Connors, J. M.; Engert, A.; Larsen, E. K.; Kennedy, D. A.; Sievers, E. L.; Chen, R., Results of a pivotal phase II study of brentuximab vedotin for patients with relapsed or refractory Hodgkin's lymphoma. *Journal of Clinical Oncology: Official Journal of the American Society of Clinical Oncology* 2012, 30 (18), 2183–2189.

107. Pro, B.; Advani, R.; Brice, P.; Bartlett, N. L.; Rosenblatt, J. D.; Illidge, T.; Matous, J.; Ramchandren, R.; Fanale, M.; Connors, J. M.; Yang, Y.; Sievers, E. L.; Kennedy, D. A.; Shustov, A., Brentuximab vedotin (SGN-35) in patients with relapsed or refractory systemic anaplastic large-cell lymphoma: results of a phase II study. *Journal of Clinical Oncology: Official Journal of the American Society of Clinical Oncology* 2012, 30 (18), 2190–2196.

108. Mocellin, S.; Benna, C.; Pilati, P., Coinhibitory molecules in cancer biology and therapy. *Cytokine & Growth Factor Reviews* 2013, 24 (2), 147–161.

109. Lipson, E. J.; Drake, C. G., Ipilimumab: an anti-CTLA-4 antibody for metastatic melanoma. *Clinical Cancer Research: An Official Journal of the American Association for Cancer Research* 2011, 17 (22), 6958–6962.

110. Robert, C.; Thomas, L.; Bondarenko, I.; O'Day, S.; Weber, J.; Garbe, C.; Lebbe, C.; Baurain, J. F.; Testori, A.; Grob, J. J.; Davidson, N.; Richards, J.; Maio, M.; Hauschild, A.; Miller, W. H., Jr.; Gascon, P.; Lotem, M.; Harmankaya, K.; Ibrahim, R.; Francis, S.; Chen, T. T.; Humphrey, R.; Hoos, A.; Wolchok, J. D., Ipilimumab plus dacarbazine for previously untreated metastatic melanoma. *New England Journal of Medicine* 2011, 364 (26), 2517–2526.

111. Webster, R. M., The immune checkpoint inhibitors: where are we now? *Nature Reviews Drug Discovery* 2014, 13 (12), 883–884.

112. Weber, J. S.; Minor, D. R.; D'Angelo, S. P.; Hodi, F. S.; Gutzmer, R.; Neyns, B.; Hoeller, C.; Khushalani, N. I.; Miller, W. H.; Grob, J.-J.; Lao, C.; Linette, G.; Grossmann, K.; Hassel, J. C.; Lorigan, P.; Maio, M.; Sznol, M.; Lambert, A.; Yang, A.; Larkin, J., LBA3_PRA phase 3 randomized, open-label study of nivolumab (Anti-Pd-1; Bms-936558; Ono-4538) versus investigator's choice chemotherapy (ICC) in patients with advanced melanoma after prior anti-CTLA-4 therapy. *Annals of Oncology* 2014, 25 (Suppl 4). 10.1093/annonc/mdu438.34.

113. McDermott, J.; Jimeno, A., Pembrolizumab: PD-1 inhibition as a therapeutic strategy in cancer. *Drugs of Today (Barcelona, Spain: 1998)* 2015, 51 (1), 7–20.

114. Bagcchi, S., Pembrolizumab for treatment of refractory melanoma. *The Lancet Oncology* 2014, 15 (10), e419.

115. Gutierrez, C.; Schiff, R., HER2: biology, detection, and clinical implications. *Archives of Pathology & Laboratory Medicine* 2011, 135 (1), 55–62.

116. Pegram, M. D.; Konecny, G.; Slamon, D. J., The molecular and cellular biology of HER2/neu gene amplification/overexpression and the clinical development of herceptin (trastuzumab) therapy for breast cancer. *Cancer Treatment and Research* 2000, 103, 57–75.

117. Slamon, D.; Eiermann, W.; Robert, N.; Pienkowski, T.; Martin, M.; Press, M.; Mackey, J.; Glaspy, J.; Chan, A.; Pawlicki, M.; Pinter, T.; Valero, V.; Liu, M.-C.; Sauter, G.; von Minckwitz, G.; Visco, F.; Bee, V.; Buyse, M.; Bendahmane, B.; Tabah-Fisch, I.; Lindsay, M.-A.; Riva, A.; Crown, J., Adjuvant trastuzumab in HER2-positive breast cancer. *The New England Journal of Medicine* 2011, 365 (14), 1273–1283.

118. Arteaga, C. L.; Sliwkowski, M. X.; Osborne, C. K.; Perez, E. A.; Puglisi, F.; Gianni, L., Treatment of HER2-positive breast cancer: current status and future perspectives. *Nature Reviews. Clinical Oncology* 2011, 9 (1), 16–32.

119. Fiuza, M.; Magalhães, A., Trastuzumab and Cardiotoxicity. INTECH Open Access Publisher, European Union, Croatia, 2012. Available at http://www.intechopen.com/contact.html (accessed June 15, 2016).

120. Capelan, M.; Pugliano, L.; De Azambuja, E.; Bozovic, I.; Saini, K. S.; Sotiriou, C.; Loi, S.; Piccart-Gebhart, M. J., Pertuzumab: new hope for patients with HER2-positive breast cancer. *Annals of Oncology: Official Journal of the European Society for Medical Oncology/ESMO* 2013, 24 (2), 273–282.

121. Baselga, J.; Cortés, J.; Kim, S.-B.; Im, S.-A.; Hegg, R.; Im, Y.-H.; Roman, L.; Pedrini, J. L.; Pienkowski, T.; Knott, A.; Clark, E.; Benyunes, M. C.; Ross, G.; Swain, S. M., Pertuzumab plus trastuzumab plus docetaxel for metastatic breast cancer. *The New England Journal of Medicine* 2012, 366 (2), 109–119.

122. Amiri-Kordestani, L.; Wedam, S.; Zhang, L.; Tang, S.; Tilley, A.; Ibrahim, A.; Justice, R.; Pazdur, R.; Cortazar, P., First FDA approval of neoadjuvant therapy for breast cancer: pertuzumab for the treatment of patients with HER2-positive breast cancer. *Clinical Cancer Research: An Official Journal of the American Association for Cancer Research* 2014, 20 (21), 5359–5364.

123. Corrigan, P. A.; Cicci, T. A.; Auten, J. J.; Lowe, D. K., Ado-trastuzumab emtansine: a HER2-positive targeted antibody-drug conjugate. *The Annals of Pharmacotherapy* 2014, 48 (11), 1484–1493.

124. Verma, S.; Miles, D.; Gianni, L.; Krop, I. E.; Welslau, M.; Baselga, J.; Pegram, M.; Oh, D.-Y.; Diéras, V.; Guardino, E.; Fang, L.; Lu, M. W.; Olsen, S.; Blackwell, K., Trastuzumab emtansine for HER2-positive advanced breast cancer. *The New England Journal of Medicine* 2012, 367 (19), 1783–1791.

125. West, C. M.; Joseph, L.; Bhana, S., Epidermal growth factor receptor-targeted therapy. *The British Journal of Radiology* 2008, 81 (Special Issue 1), S36–S44.

126. You, B.; Chen, E. X., Anti-EGFR monoclonal antibodies for treatment of colorectal cancers: development of cetuximab and panitumumab. *Journal of Clinical Pharmacology* 2012, 52 (2), 128–155.

127. Cunningham, D.; Humblet, Y.; Siena, S.; Khayat, D.; Bleiberg, H.; Santoro, A.; Bets, D.; Mueser, M.; Harstrick, A.; Verslype, C.; Chau, I.; Van Cutsem, E., Cetuximab monotherapy and cetuximab plus irinotecan in irinotecan-refractory metastatic colorectal cancer. *The New England Journal of Medicine* 2004, 351 (4), 337–345.

128. Normanno, N.; Tejpar, S.; Morgillo, F.; De Luca, A.; Van Cutsem, E.; Ciardiello, F., Implications for KRAS status and EGFR-targeted therapies in metastatic CRC. *Nature Reviews. Clinical Oncology* 2009, 6 (9), 519–527.

129. Di Nicolantonio, F.; Martini, M.; Molinari, F.; Sartore-Bianchi, A.; Arena, S.; Saletti, P.; De Dosso, S.; Mazzucchelli, L.; Frattini, M.; Siena, S.; Bardelli, A., Wild-type BRAF is required for response to panitumumab or cetuximab in metastatic colorectal cancer. *Journal of Clinical Oncology: Official Journal of the American Society of Clinical Oncology* 2008, 26 (35), 5705–5712.

130. Van Cutsem, E.; Peeters, M.; Siena, S.; Humblet, Y.; Hendlisz, A.; Neyns, B.; Canon, J.-L.; Van Laethem, J.-L.; Maurel, J.; Richardson, G.; Wolf, M.; Amado, R. G., Open-label phase III trial of panitumumab plus best supportive care compared with best supportive care alone in patients with chemotherapy-refractory metastatic colorectal cancer. *Journal of Clinical Oncology: Official Journal of the American Society of Clinical Oncology* 2007, 25 (13), 1658–1664.

131. Hecht, J. R.; Mitchell, E.; Chidiac, T.; Scroggin, C.; Hagenstad, C.; Spigel, D.; Marshall, J.; Cohn, A.; McCollum, D.; Stella, P.; Deeter, R.; Shahin, S.; Amado, R. G., A randomized phase IIIB trial of chemotherapy, bevacizumab, and panitumumab compared with chemotherapy and bevacizumab alone for metastatic colorectal cancer. *Journal of Clinical Oncology: Official Journal of the American Society of Clinical Oncology* 2009, 27 (5), 672–680.

132. Haddad, R. I.; Shin, D. M., Recent advances in head and neck cancer. *New England Journal of Medicine* 2008, 359 (11), 1143–1154.

133. Ellis, L. M., Mechanisms of action of bevacizumab as a component of therapy for metastatic colorectal cancer. *Seminars in Oncology* 2006, 33 (5 Suppl 10), S1–S7.

134. Hurwitz, H.; Fehrenbacher, L.; Novotny, W.; Cartwright, T.; Hainsworth, J.; Heim, W.; Berlin, J.; Baron, A.; Griffing, S.; Holmgren, E.; Ferrara, N.; Fyfe, G.; Rogers, B.; Ross, R.; Kabbinavar, F., Bevacizumab plus irinotecan, fluorouracil, and leucovorin for metastatic colorectal cancer. *The New England Journal of Medicine* 2004, 350 (23), 2335–2342.

135. Giantonio, B. J.; Catalano, P. J.; Meropol, N. J.; O'Dwyer, P. J.; Mitchell, E. P.; Alberts, S. R.; Schwartz, M. A.; Benson, A. B., 3rd, Bevacizumab in combination with oxaliplatin, fluorouracil, and leucovorin (FOLFOX4) for previously treated metastatic colorectal cancer: results from the Eastern Cooperative Oncology Group Study E3200. *Journal of Clinical Oncology* 2007, 25 (12), 1539–1544.

136. Spratlin, J. L.; Mulder, K. E.; Mackey, J. R., Ramucirumab (IMC-1121B): a novel attack on angiogenesis. *Future Oncology* 2010, 6 (7), 1085–1094.

137. Fuchs, C. S.; Tomasek, J.; Yong, C. J.; Dumitru, F.; Passalacqua, R.; Goswami, C.; Safran, H.; dos Santos, L. V.; Aprile, G.; Ferry, D. R.; Melichar, B.; Tehfe, M.; Topuzov, E.; Zalcberg, J. R.; Chau, I.; Campbell, W.; Sivanandan, C.; Pikiel, J.; Koshiji, M.; Hsu, Y.; Liepa, A. M.; Gao, L.; Schwartz, J. D.; Tabernero, J.; REGARD Trial Investigators, Ramucirumab monotherapy for previously treated advanced gastric or gastro-oesophageal junction adenocarcinoma (REGARD): an international, randomised, multicentre, placebo-controlled, phase 3 trial. *Lancet* 2014, 383 (9911), 31–39.

138. Garon, E. B.; Ciuleanu, T.-E.; Arrieta, O.; Prabhash, K.; Syrigos, K. N.; Goksel, T.; Park, K.; Gorbunova, V.; Kowalyszyn, R. D.; Pikiel, J.; Czyzewicz, G.; Orlov, S. V.; Lewanski, C. R.; Thomas, M.; Bidoli, P.; Dakhil, S.; Gans, S.; Kim, J.-H.; Grigorescu, A.; Karaseva, N.; Reck, M.; Cappuzzo, F.; Alexandris, E.; Sashegyi, A.; Yurasov, S.; Perol, M., Ramucirumab plus docetaxel versus placebo plus docetaxel for second-line treatment of stage IV non-small-cell lung cancer after disease progression on platinum-based therapy (REVEL): a multicentre, double-blind, randomised phase 3 trial. *Lancet* 2014, 384 (9944), 665–673.

139. Old, L. J.; Chen, Y. T., New paths in human cancer serology. *The Journal of Experimental Medicine* 1998, 187 (8), 1163–1167.

140. Simpson, A. J.; Caballero, O. L.; Jungbluth, A.; Chen, Y. T.; Old, L. J., Cancer/testis antigens, gametogenesis and cancer. *Nature Reviews Cancer* 2005, 5 (8), 615–625.

141. Türeci, O.; Sahin, U.; Zwick, C.; Koslowski, M.; Seitz, G.; Pfreundschuh, M., Identification of a meiosis-specific protein as a member of the class of cancer/testis antigens. *Proceedings of the National Academy of Sciences of the United States of America* 1998, 95 (9), 5211–5216.

142. Whitehurst, A. W., Cause and consequence of cancer/testis antigen activation in cancer. *Annual Review of Pharmacology and Toxicology* 2014, 54, 251–272.

143. Caballero, O. L.; Chen, Y.-T., Cancer/testis (CT) antigens: potential targets for immunotherapy. *Cancer Science* 2009, 100 (11), 2014–2021.

144. Kalejs, M.; Erenpreisa, J., Cancer/testis antigens and gametogenesis: a review and "brain-storming" session. *Cancer Cell International* 2005, 5, 4–11.

145. Lim, S. H.; Wang, Z.; Chiriva-Internati, M.; Xue, Y., Sperm protein 17 is a novel cancer-testis antigen in multiple myeloma. *Blood* 2001, 97 (5), 1508–1510.

146. Grizzi, F.; Franceschini, B.; Hamrick, C.; Frezza, E. E.; Cobos, E.; Chiriva-Internati, M., Usefulness of cancer-testis antigens as biomarkers for the diagnosis and treatment of hepatocellular carcinoma. *Journal of Translational Medicine* 2007, 5, 3–11.

147. Straughn, J. M., Jr.; Shaw, D. R.; Guerrero, A.; Bhoola, S. M.; Racelis, A.; Wang, Z.; Chiriva-Internati, M.; Grizzle, W. E.; Alvarez, R. D.; Lim, S. H.; Strong, T. V., Expression of sperm protein 17 (Sp17) in ovarian cancer. *International Journal of Cancer* 2004, 108 (6), 805–811.

148. Chiriva-Internati, M.; Ferrari, R.; Yuefei, Y.; Hamrick, C.; Gagliano, N.; Grizzi, F.; Frezza, E.; Jenkins, M. R.; Hardwick, F.; D'Cunha, N.; Kast, W. M.; Cobos, E., AKAP-4: a novel cancer testis antigen for multiple myeloma. *British Journal of Haematology* 2008, 140, 465–468.

149. Chiriva-Internati, M.; Yu, Y.; Mirandola, L.; D'Cunha, N.; Hardwicke, F.; Cannon, M. J.; Cobos, E.; Kast, W. M., Identification of AKAP-4 as a new cancer/testis antigen for detection and immunotherapy of prostate cancer. *The Prostate* 2012, 72 (1), 12–23.

150. Mirandola, L.; Figueroa, J. A.; Phan, T. T.; Grizzi, F.; Kim, M.; Rahman, R. L.; Jenkins, M. R.; Cobos, E.; Jumper, C.; Alalawi, R.; Chiriva-Internati, M., Novel antigens in non-small cell lung cancer: SP17, AKAP4, and PTTG1 are potential immunotherapeutic targets. *Oncotarget* 2015, 6 (5), 2812–2826.

151. Chiriva-Internati, M.; Wang, Z.; Salati, E.; Bumm, K.; Barlogie, B.; Lim, S. H., Sperm protein 17 (Sp17) is a suitable target for immunotherapy of multiple myeloma. *Blood* 2002, 100 (3), 961–965.

152. Song, J.-X.; Cao, W.-L.; Li, F.-Q.; Shi, L.-N.; Jia, X., Anti-Sp17 monoclonal antibody with antibody-dependent cell-mediated cytotoxicity and complement-dependent cytotoxicity activities against human ovarian cancer cells. *Medical Oncology (Northwood, London, England)* 2012, 29 (4), 2923–2931.

153. Song, J.-X.; Li, F.-Q.; Cao, W.-L.; Jia, X.; Shi, L.-N.; Lu, J.-F.; Ma, C.-F.; Kong, Q.-Q., Anti-Sp17 monoclonal antibody-doxorubicin conjugates as molecularly targeted chemotherapy for ovarian carcinoma. *Targeted Oncology* 2014, 9 (3), 263–272.

5
Antibody Posttranslational Modifications

Roy Jefferis

The School of Immunity and Infection, University of Birmingham, Birmingham, UK

5.1 Summary

The clinical efficacy of recombinant antibody therapeutics has resulted in their worldwide adoption as drugs of choice, for example, in oncology and inflammation. Antibodies are often referred to as "adapter molecules" as they form a bridge between a specific target and the activation of downstream molecular and cellular effector functions. Immune complexes of each of the four human IgG subclasses may bind and activate multiple host ligands with varying downstream outcomes. Recognition/activation of host ligands is dependent on interactions with the IgG–Fc region and can vary depending on the precise structure of the naturally occurring isoforms, for example, subclasses: glycoforms. Antibody therapeutics must therefore be fully characterized, structurally and functionally, as must a candidate biosimilar. This chapter enumerates and discusses the multiple posttranslational and chemical heterogeneities that may arise within recombinant antibodies and possible consequences for mechanisms of action *in vivo*.

5.2 Introduction

Humans have been somewhat chastened to learn that they possess a similar number of genes (~20,000) to that of the roundworm [1, 2] and only one-fifth that of wheat. However, gene number alone does not account for biologic complexity, and it is estimated that, for humans, alternate RNA splicing, co- and posttranslational modifications (CTM, PTM), and the production of active metabolites generate ~1,000,000 distinct biologically active molecules [3–5]; the functional profile for most remains to be determined. Additional

Biosimilars of Monoclonal Antibodies: A Practical Guide to Manufacturing, Preclinical, and Clinical Development, First Edition. Edited by Cheng Liu and K. John Morrow, Jr.

structural complexity may arise from epigenetic influences [6]. Structural and functional diversity may be illustrated by the ubiquitous and dynamic PTM of phosphorylation and dephosphorylation (phosphoproteomics) [7–9]. It has been estimated that one third of all expressed proteins may be phosphorelated, at a given time, and that up to 50% of all proteins may be phosphorelated during their lifetime [8]. Another ubiquitous PTM that contributes significantly to structural and functional diversity is glycosylation [10, 11]. It has been estimated that ~50% of human genes having an open reading frame encode a -N-X-S/T- amino acid sequence (sequon) that comprises a potential site for N-linked glycosylation of the translated protein (X can be any amino acid other than proline). However, this may overestimate the number of glycoproteins, since proteins may express N-glycosylation sequons that are not occupied or bear several sequons having variable occupation. Proteins may also bear oligosaccharides O-linked to serine, threonine, or tyrosine residues that are attached during transit through the Golgi; to date no sequence or structural feature that correlates with O-linked glycosylation has been identified.

It will be appreciated, therefore, that the structure of a nascent human protein, *in vivo,* cannot be predicted from the gene sequence and must be determined by structural analysis of the isolated, purified molecule. This presents a further challenge since proteins are subject to structural alterations *in vivo,* prior to being subject to multiple isolation, purification, and analytical protocols. This constitutes a biological "Heisenberg uncertainty principle"; the purer the protein, the less certain one can be that the product is representative of natural or "wild type" structure, including PTM variants. In practice, a consensus structure is established both for functional investigations and possible development as a recombinant molecule. When producing recombinant protein as a potential therapeutic, it is mandatory that natural PTMs are present while unnatural ones are not. This is a further challenge since recombinant proteins are produced in "alien" tissues (CHO, NS0, Sp2/0 cells etc.), exposed to culture medium, products of the intact and effete producer cells, and subjected to rigorous downstream purification, formulation, and storage conditions. Production in a prokaryotic system (e.g., *Escherichia coli*) may result in a protein recovered as an inclusion body that has to be solubilized and refolded *in vitro* to yield a product that may lack natural PTMs or bearing unnatural ones.

In addition to the demonstration of clinical efficacy, an innovator company seeking approval for a potential protein therapeutic is required to characterize the drug product, functionally and structurally, employing multiple orthogonal analytical technologies. The parameters established define the drug substance, and/or drug product, and must be maintained throughout the lifetime of the drug. In the process, critical quality attributes (CQAs) are identified that contribute to drug efficacy and are achieved employing quality by design (QbD) parameters unique to the production platform and downstream protocols employed. The parameters defining CQAs and QbD are the undisclosed intellectual property of the innovator company; consequently, it is deemed essentially impossible to produce an identical product within another facility; that is, in principle, it is not possible to develop generic biopharmaceuticals [12, 13]. It is often claimed that "the process defines the product."

Initially, innovator companies generating approved protein therapeutics emphasized their large size, structural complexity, unavoidable heterogeneity, and hence the impossibility of another provider being able to generate an identical "copy." However, as protein therapeutics achieved "blockbuster" status, particularly monoclonal antibodies (mAbs), and the period of patent protection shortened, "Big Pharma" recognized the financial potential of a share in "blockbuster" markets. Consequently, pharmaceutical companies, large and small, initiated programs to generate "copies" of successful protein therapeutics. It was recognized that "copies" cannot be structurally identical to innovator product; however, regulatory authorities demand that they be demonstrated to be "comparable" to the innovator product [12, 13]. Structural comparability is established by applying multiple orthogonal analytical protocols to innovator product, sourced from a pharmacy, and the proposed biosimilar. Inevitably, structural differences will be revealed and must be demonstrated not to compromise functional efficacy. If approved, they are not designated as generics but as "biosimilars" (Europe) or follow-on biologics (United States). The criteria for approval of a biosimilar biologic differ between national regulatory authorities. Biosimilar mAb drugs that have been approved in India (Reditux/Rituxan) and South Korea (Remsima/Remicade) have not been automatically approved by the European Medicines Authority (EMA) or the Federal Drug Administration (FDA). However, the biosimilar candidate for Remicade (Remsima, Celltrion; Inflectra, Hospira) that was developed under EMA guidelines received approval by the EMA in 2013 [14, 15]. Approval was heralded as a "landmark" event and demonstrates that the EMA has confidence in its ability to evaluate the comparability of biosimilar antibody products; the FDA has recently published revised guidelines while seeking industry responses [12].

5.3 Overview of Co- and Posttranslational Modifications

Variations in protein structure from that predicted by open reading frames genes are frequently referred to, collectively, as posttranslational modifications; however, as previously indicated, heterogeneity may be introduced cotranslationally (CTM) and at earlier stages in gene expression. Such CTMs and PTMs may be critical to the function of the molecule systemically, within a given tissue or an environmental niche. Additionally, the CTM/PTM profile may vary with age, sex, health, and disease. Heterogeneity of structure may arise as a result of errors in transcription of DNA, amino acid misincorporation, the aging process *in vivo*, for example, glycation [16], deamidation [17, 18], etc. Qualitative and quantitative analysis has been revolutionized in the past decade by developments in mass spectrometry, and in excess of 300 different structural CTM/PTMs are documented [19, 20]. Analysis of 530,264 sequences in the Swiss-Prot database was shown to yield 87,308 experimentally identified PTMs and 234,938 putative PTMs [4, 5, 19, 20]. The level of functional complexity can be appreciated from the finding that the human genome encodes 518 protein kinases and 200 phosphatases [21, 22]. The second most frequent CTM/PTM is glycosylation. Oligosaccharides may be attached to asparagine residues to generate N-linked glycoproteins or to the hydroxyl groups of serine, threonine, or tyrosine to generate O-linked glycoforms. The N-linked repertoire is provided for by >500 different oligosaccharide structures that may be differentially attached at multiple glycosylation sites to generate >1000 different types of glycan, a consequence of the activities of >250 glycosyltransferases [23, 24]. It is not my purpose to present an all-encompassing review of protein CTM/PTMs; however, more specialized reviews may be found within the *Nature Reviews* series [25]. Recombinant mAbs present a unique challenge because each is constituted of a unique amino acid sequence. An innovator company will have applied clone selection criteria at an early stage to identify candidates and rejected those with vulnerable PTMs in CDRs; other criteria may be applied to optimize solubility, stability, etc. The selected IgG isotype and expression platform will determine the overall identity of the final drug substance and drug product gaining approval. A biosimilar candidate must be demonstrated to be structurally and functionally comparable [12, 13].

5.3.1 Co- and Posttranslational Modifications of Antibody Molecules

The hallmark of a protective humoral immune response is specificity, achieved through structural heterogeneity, that is, a population (polyclonal) of antigen-binding sites (paratope) each specific for a given structure expressed on the surface of the antigen (epitope). For the purposes of this review, I will only consider the IgG response, which itself may be comprised of molecules of each of the four IgG subclasses, IgG1, IgG2, IgG3, and IgG4. While structural information was gained from the study of polyclonal IgG, isolated from human sera, the first full-length sequence was determined for a monoclonal human IgG1 protein (paraprotein, Eu) [26], purified from the blood of a patient with multiple myeloma, a cancer of antibody-secreting plasma cells. In advanced cases, normal IgG synthesis is suppressed, while the paraprotein may reach levels ~50 g/l; such proteins have been shown to be representative of normal IgG molecules although in no case has the antigen been identified.

5.3.2 N- and C-Terminal Residues

Little information can be gleaned from N-terminal amino acid sequencing of polyclonal IgG, due to V-region diversity. Unique N-terminal sequence is obtained for the heavy and light chains of monoclonal IgG paraprotein; however, for some proteins the N-terminal amino acid yield was not quantitative or appeared to be "blocked." It was later appreciated that this results when a gene encodes for the incorporation of N-terminal glutamic acid or glutamine residues with subsequent cyclization and the generation pyroglutamic acid (pGlu), both *in vivo* and *in vitro* [27–31]. The formation of pGlu in antibodies [28] and therapeutic proteins generally is a concern for the biopharmaceutical industry since it introduces charge heterogeneity and may be considered evidence for lack of process control [23]. Importantly, N-terminal pGlu is also implicated in Alzheimer's disease and dementia since it increases the tendency for proteins to form insoluble fibrils; light chains are particularly prone to processing with the formation of fibrils [28, 32]. When producing recombinant mAbs in mammalian cells, the balance between production of heavy and light chains is critical, and an excess of light chain production is optimal, the excess being secreted into the medium [33]. As there is no evidence of a specific benefit attached to the presence of N-terminal pGlu, to either the heavy or light chain, it may be considered to be a CQA and best avoided during clone selection for a potential mAb therapeutic.

The C-terminal residue reported for the heavy chains of serum-derived IgG has always been glycine; however, when the genes for the IgG subclasses were sequenced, it was seen that they encoded for a C-terminal lysine residue. It was later shown that the lysine residue is cleaved *in vivo* by an endogenous carboxypeptidase. Recombinant IgG molecules produced in mammalian cells exhibit mixed populations of molecules with lysine present or absent on each heavy chain. The level of C-terminal lysine can vary between clones of cells, culture conditions, and other parameters of the production platform and introduces charge heterogeneity [28]. Although there is no evidence that the presence or absence of C-terminal lysine is a CQA, it presents an issue for QbD [34]. A concept paper produced by the European Biopharmaceutical Enterprises (EBE), a specialized group of European Federation of Pharmaceutical Industries and Associations (EFPIA) included the statements, "A number of scientific publications suggest that C-terminal lysine truncation has no impact on biological activity, PK/PD, immunogenicity, and safety," and elsewhere in the document, "Lysine truncation does not appear to adversely affect product potency or safety. However, taking a conservative approach potential C-terminal lysine effects on all antibodies cannot be ruled out. Thus, lysine truncation should be characterized, and process consistency should be demonstrated during product development. This is in line with the agency input that emphasized the importance of reporting the ranges of C-terminal lysine heterogeneity observed during the characterization and development phases" [35]. The heavy chain C-terminal lysine content can vary with culture conditions [36].

5.3.3 Cysteine and Disulfide Bond Formation

The gene sequence for a human IgG1 subclass protein encodes for 14 cysteine residues, 28 for a H2L2 protein dimer. The standard structural cartoon for the human IgG1 protein (Eu) exhibits 12 intrachain and 4 interchain disulfide bridges. This pattern of intrachain disulfide bridge formation is maintained for each of the other IgG subclasses except for the number of interheavy chain disulfide bridges [11, 37]. However, heterogeneity of disulfide bridge formation has been reported for normal and monoclonal IgG: paraproteins and recombinant mAbs. Formation of the H2L2 dimer occurs following release of heavy and light chains into the endoplasmic reticulum (ER), with evidence that the presence of light chain is essential to the generation of a folded H2L2 structure [33]. This nascent form explores multiple dynamic structures; with the formation of native and nonnative disulfide bonds that are transiently formed and reduced until a low energy, conformation is achieved [38]; it must be noted that little or no processing of the high mannose oligosaccharide will have occurred at this point; therefore, the final native conformation will not be achieved until oligosaccharide processing is completed.

The predominant pattern of disulfide bridge formation differs between the IgG subclasses. While IgG1 establishes the "standard" pattern with two interheavy chain disulfide bridges, IgG2 has 2, IgG3 up to 11 and IgG4 2. The conformation of the IgG molecule results in close proximity for cysteine residues forming the light–heavy chain bridges and those forming the interheavy chain bridges. This is particularly so for IgG2 and IgG4 molecules that have relatively short hinge regions. Heterogeneity in disulfide bond formation in IgG2 was first reported for recombinant IgG2 but, later, shown to be present in normal serum-derived IgG2 also [39, 40]. The interconversion of these isoforms was shown to be dynamic and promoted by a reducing environment; such an environment can be provided by the presence of thioredoxin/thioredoxin reductase, released into culture media by effete cells; it can be ameliorated by control of dissolved oxygen levels [41–46]. An *in vitro* model revealed that susceptibility to reduction/oxidation differed between IgG subclasses and light chain types with sensitivity being in the order IgG1λ > IgG1k > IgG2λ > IgG2k [41].

A core hinge region sequence of -Cys-Pro-Pro-Cys-, present in IgG1, IgG2, and IgG3, forms a partial helical structure that does not allow for intraheavy chain disulfide bridge formation; however, the homologous sequence in the IgG4 subclass is -Cys-Pro-Ser-Cys-, and this does allow for intraheavy chain disulfide bridge formation. Consequently, natural and recombinant IgG4 antibody populations are a mixture of these isoforms that are generated *in vivo* and *in vitro* [44]. An added heterogeneity results from "Fab-arm exchange" with the generation of monovalent bispecific antibody molecules *in vivo* and *in vitro*. The exchange is facilitated by the presence of an arginine residue at position 409 (R409), in place of the lysine 409 (K409) present in IgG1, IgG2, IgG3, and a polymorphic variant of IgG4 [44–46]. For the R409 isoform lateral noncovalent interactions between the two C_H3 domains are reduced such that, under physiologic conditions and in the absence of hinge region intraheavy chain disulfide bridges, they dissociate to form HL heteromonomers. Reassociation is nonselective with the formation of monovalent bispecific antibodies.

5.3.4 Oxidation of Methionine

Methionine residues exposed on the surface of proteins are susceptible to oxidation. While methionine residues within Fab framework regions have not been reported to be vulnerable to oxidation, exposed methionine residues within CDRs may be [47]. Consequently, early sequencing and selection for potential antibody clones are employed to eliminate this possibility. It has been demonstrated, for antibodies of the IgG1 and IgG2 subclasses, that methionine residues M252 and M428 are prone to oxidation [47–51]. Although these residues are distant in linear sequence, they are proximal at the C_H2C_H3 interface. Minimal M252 oxidation levels of 2–5% are reported for purified IgG antibodies in formulation buffers, while lower levels of oxidation are reported for M428. Oxidation of M252 and M428 increases under conditions of accelerated stability testing and on prolonged storage. The interaction site for the FcRn receptor that regulates IgG catabolism and placental passage is formed at the C_H2C_H3 interface, and M252 and M428 oxidation has been shown to reduce both the affinity of binding to FcRn and half-life *in vivo*, in mice transgenic for human FcRn [48, 49]. However, it was shown that oxidation of both M252 and M428 on both heavy chains was required to impact IgG1 half-life, and this was only observed when M252 oxidation was in excess of ~80% [49]. Both staphylococcal protein A (SpA) and streptococcal protein G (SpG) bind IgG-Fc at the C_H2C_H3 interface, and it has been shown that methionine oxidation impacts binding affinity for both [50, 51]; therefore, M252 and 428 oxidation levels may be a useful QpD CQA/CbD parameter. The impact of M252 and M428 oxidation on FcγR binding has been shown to be minimal, but a subtle decrease in binding to the FcγRIIa 131H allele was reported [48]. Analysis of Herceptin, obtained from a pharmacy, and a potential biosimilar demonstrated that care has to be exercised when resuspending antibody therapeutics since a discrepancy was observed for the level of M252 oxidation between the innovator product (4.39%) and the proposed biosimilar (10.33%); even so a note added in proof commented that the value of 4.39% was greater than that determined by the innovator company; no oxidation of M428 was recorded [52].

5.3.5 Amidation and Deamidation

Deamidation of asparagine residues, with the generation of aspartic acid or isoaspartic acid, and glutamine, with the generation of glutamic acid, is a frequently encountered PTM [17, 18]. Deamidation of asparagine residues is influenced by the degree of exposure to external environment and immediately adjacent amino acid residues, particularly the presence of a glycine residue C-terminal to the asparagine [-N-G-]. Studies of IgG1 and IgG2 proteins, *in vitro* and *in vivo*, have shown that asparagine residues 315 and 384 are susceptible to deamidation with the formation of isoaspartic and aspartic acid residues, respectively [53–55]. The relative susceptibility to deamidation at these sites varied between studies; however, the significance may be ameliorated by the finding that ~23% of normal polyclonal IgG has an aspartic acid 384 residue; thus it may be concluded that healthy humans are constantly exposed to IgG bearing this PTM. These studies did not identify asparagine deamidation within the constant region of the kappa light chains.

In contrast, deamidation within variable regions, particularly within CDRs, of recombinant antibodies has been shown to compromise antibody specificity and/or binding affinity [56–59]. Interestingly, the approved blockbuster antibody therapeutic trastuzumab (Herceptin) has asparagine residues in light chain CDR1 (Asn 30) and heavy chain CDR2 (Asn 55) that were shown to be susceptible to deamidation under *in vitro* conditions. The approved drug substance did not exhibit these PTMs; therefore, their presence or absence could be used as a lot release criterion [59]. The levels of deamidation, both within variable and constant region asparagine residues, for innovator and proposed Herceptin biosimilar were higher than that reported by the innovator [52], underlining the susceptibility to deamidation of these residues and the care that has to be exercised when resuspending this antibody.

In contrast, glutamine residues are relatively resistant to deamidation, and no glutamine residues were reported to be subject to deamidation in the study reported in Ref. [51]. A study of pH dependence for deamidation of glutamine in a mAb showed that six Gln residues were susceptible to deamination; four in variable regions and residues 295 and 418 in the IgG-Fc [60]. The cyclization of N-terminal glutamine to form pyroglutamic acids has been discussed previously.

5.4 Glycosylation

The commonly perceived "hallmark" of an antibody is its specificity for a target pathogen (antigen); however, protection is dependent also on the activation of a cascade of downstream biologic mechanisms, triggered by the antigen/antibody (immune) complexes formed, resulting in the killing and elimination of pathogenic organisms or pathologic targets. These biologic mechanisms are triggered following the binding of immune complexes (ICs) to one or more of multiple IgG-Fc effector molecules, with consequent activation [11]. It is established that N-linked glycosylation of the IgG-Fc, at asparagine 297 (N297; Eu numbering), is essential for optimal activation. Glycoanalysis of normal polyclonal human IgG reveals a considerable heterogeneity of diantennary oligosaccharide structures; similar analysis of monoclonal human IgG proteins, produced by neoplastic plasma cells (multiple myeloma), also reveals heterogeneity, but the pattern is restricted and characteristic for each malignant plasma cell population (i.e., patient). Detailed structural studies have mostly shown that the oligosaccharide(s) is not directly involved in ligand binding but exerts influence through its impacts on the tertiary/quaternary conformation of the IgG-Fc. Given the significant impact of different IgG-Fc glycoforms on the activation of downstream effector mechanisms, an understanding has been sought through characterization of resulting structural changes; intriguingly only, apparently, minor differences have been detected. Before discussing the functional consequences for individual IgG-Fc glycoforms(s) of normal, myeloma, and recombinant IgG, an overview of IgG-Fc structure is appropriate.

5.4.1 Quaternary Structure of IgG-Fc: The Protein Moiety

The first crystal structure of IgG-Fc, resolved at 2.9-Å, was published by Deisenhofer in 1981 [61]. The IgG-Fc was generated by papain cleavage of polyclonal IgG at the Lys_{222}-Thr_{223} peptide bond, within the hinge region, and extending to a C-terminus residue at 446. It was reported that interpretable electron density was not obtained for residues 223–237, which comprise most of the core and lower hinge region, or the C-terminal residues 444–446; it was not known at this time that the C_H3 exon codes for a C-terminal 447 lysine residue that is removed by endogenous serum carboxypeptidase B. Similar structures have been reported for human, rabbit, and mouse IgG-Fc and chicken IgY-Fc fragments and have provided further structural insight due to progressively higher resolution An α-carbon IgG ribbon structure is shown in Figure 5.1. The common structural features reported [11] are:

1) The C_H3 domains are well defined due to noncovalent pairing, involving ~2000 Å^2 of accessible surface area in the (C_H3)2 module.
2) The area of noncovalent contact between the C_H2 and C_H3 domains is ~800 Å^2. This suggests that the C_H2–C_H3 contact contributes to the relative stability observed for the C-terminal proximal region of C_H2 domains, as opposed to the "softness" of the C_H2 domain proximal to the hinge region.
3) The C_H2 domains do not pair, and the hydrophobic surface of each C_H2 domain is "overlaid" by the carbohydrate. Hydrophobic and polar interactions between the oligosaccharide and the C_H2 domain surface occupy ~500 Å^2 and substitute for domain pairing [20, 46].

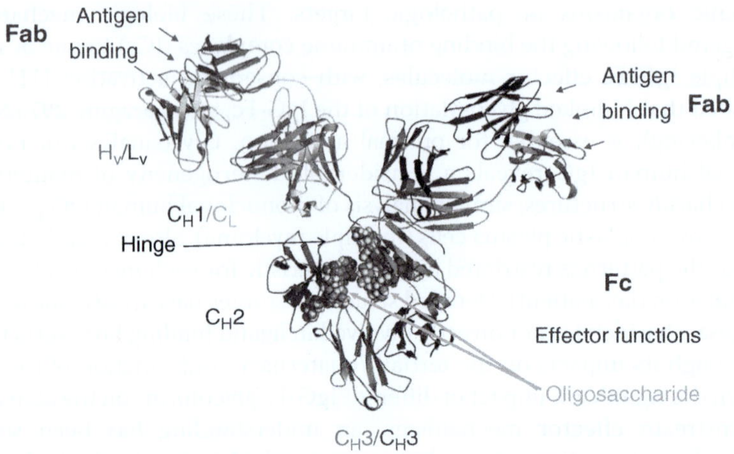

Figure 5.1 Domain structure of IgG. Courtesy of R. Emery and D. Fernandez, Ludger United Kingdom; based on the pdb 1IGY molecule. (*See insert for color representation of the figure.*)

4) One C_H2 domain was more ordered than the other, due to its crystal contact with a neighboring C_H2 domain.
5) A more disordered structure for the hinge proximal region of the C_H2 domain is reflected in higher temperature factors.
6) The intrinsic stability of the immunoglobulin fold is reflected in higher structural resolution for the β-sheets regions than for β-bends.

These interpretations and conclusions have been confirmed in X-ray structures obtained for human IgG-Fc alone or in complex with SpA, SpG, rheumatoid factor (RF) and human sFcγRIIa, sFcγRIIIb, and sFcγRIIIa [11, 62–64]. The observed internal mobility of the lower hinge and hinge proximal regions of the C_H2 domains allows for an equilibrium of high order conformers to be formed that may differentially bind unique ligands, for example, the three homologous Fcγ receptor types. Previous proposals that different ligands may bind through "overlapping nonidentical sites" may suggest too rigid a structure [11, 65] and may be modified to propose that each FcγR binds to a unique IgG-Fc conformer present within an equilibrium of transient protein structures; however, amino acid residue side chains and/or main chain atoms may be involved in common [66, 67]. The binding sites for sFcγRIIa, sFcγRIIIa, and sFcγRIIIb are asymmetric with both heavy chains being engaged such that monomeric IgG is univalent for Fcγ receptors and the C1 component of complement; this obviates continuous activation of inflammatory cascades by circulating endogenous monomeric IgG *in vivo*; IgG antigen/antibody ICs are multivalent and able to cross-link and activate cellular receptors. Residues of the lower hinge region that are disordered in the Fc crystals are ordered in the

Fc-FcγR complexes and directly involved in binding the receptor [67, 68]. By contrast IgG-Fc is functionally divalent for ligands binding at the C_H2–C_H3 interface, for example, the neonatal Fc receptor (FcRn), RF, SpA, and SpG. Due to the symmetry of the IgG-Fc, the two interaction sites are opposed at ~180°, and each is accessible to bind macromolecular ligands to form multimeric complexes.

5.4.2 The IgG-Fc Oligosaccharide Moiety

The IgG of human, rabbit, mouse, etc. have a consensus glycosylation sequon within the IgG-Fc, at Asn_{297}-Ser-Thr (Eu sequence) [26], and it has been demonstrated that the presence of a core diantennary heptasaccharide at Asn_{297} is essential for optimal activation of Fcγ receptors and the C1 component of complement. The oligosaccharide of normal polyclonal IgG-Fc is heterogeneous and essentially comprised of the core heptasaccharide with variable addition of fucose, galactose, bisecting *N*-acetylglucosamine, and sialic acid residues. Recent analyses employing high sensitivity mass spectrometry suggest the possible presence of further minor oligosaccharides, for example, high mannose and hybrid glycoforms [69].

Several systems of nomenclature are currently in use to represent oligosaccharide structures, and carbohydrate chemists, glycobiologists, and mass spectrometry scientists, among others, have developed different systems of nomenclatures [70, 71]. Among antibody "practitioners" a shorthand nomenclature has evolved for oligosaccharides released from normal polyclonal IgG. A core heptasaccharide, highlighted in blue in Figure 5.2, is designated G0 (zero galactose); the core bearing one or two galactose residues is designated G1 and G2, respectively. The core + fucose is designated G0F, the core + fucose + galactose G1F, G2F, etc; when bisecting *N*-acetylglucosamine is present, a B is added, for example, G0B, G0BF, G1BF, etc; sialylation at the galactose residues is designated as G1FS, G2FBS, etc. The approximate composition of neutral oligosaccharides released from normal polyclonal human IgG-Fc is G0, 3%; G1, 3%; G2, 6%; G0F, 23%; G1F, 30%; G2F, 24%; G0BF, 3%; G1BF, 4%; and G2BF, 7% [67–69, 72–75]. It is important to define the glycoform of the intact IgG molecule, for example, [G0/G1F], [G1F/G2BF], etc. since it has been shown that individual IgG molecules may be comprised of symmetrical or asymmetrical heavy chain glycoform pairs [75, 76]. Consequently, enhanced FcγRIIIa-mediated antibody-dependent cellular cytotoxicity (ADCC) may be observed for IgG in which one heavy chain only bears oligosaccharide devoid of fucose [76, 77]; thus the [G0/G0F] glycoform could be as potent in ADCC as the [G0/G0] glycoform.

Minor oligosaccharide structures present in polyclonal IgG-Fc may be significant since each could be the predominant form present of an individual antibody secreted from a single plasma cell; analysis of monoclonal human

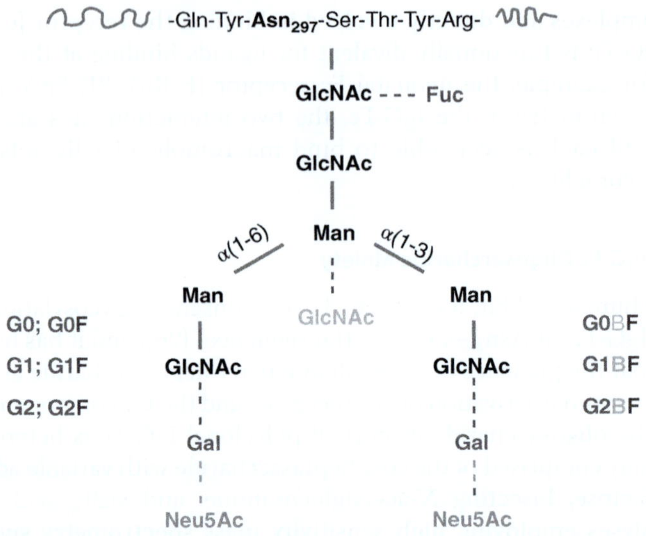

Figure 5.2 Representative IgG-Fc complex diantennary oligosaccharide. The "core" heptasaccharide has the composition: (GlcNAc)2Man3(GlcNAc)2, in blue [69, 72–75]. (*See insert for color representation of the figure.*)

IgG, isolated from the sera of patient having plasma cell cancer (multiple myeloma), has shown that the IgG-Fc glycoform profile of the paraprotein is essentially unique for each protein analyzed [78–80]. Subtle differences in oligosaccharide processing were also observed with a preference for addition of galactose to the α(1–6) arm of IgG1-Fc and the α(1–3) arm of IgG2-Fc; for the IgG3 subclass the arm preference correlated with allotype [78–80]. These data suggest a critical balance between the conformation of the IgG-Fc and the steric requirements of glycosyltransferases that may be sensitive to niche environments within the Golgi apparatus.

The glycoform profile of total polyclonal IgG can vary significantly in health and disease, particularly in autoimmune and inflammatory diseases [11, 81–84]. Methods have been developed that allow the glycoform profile of antigen specific polyclonal IgG autoantibodies to be analyzed, and significant differences in the glycoform profiles of IgG autoantibodies and the bulk IgG have been reported [85, 86]. The oligosaccharide profiles of recombinant IgG proteins produced in mammalian cells are significantly influenced by the cell type, the culture method, and the precise conditions employed; however, the [G0F/G0F] glycoform predominates (see in the following text). Under conditions of stress, such as nutrient depletion, acid pH, etc., deviant glycosylation may be observed, for example, the presence of high mannose forms and/or incomplete occupancy [87].

5.4.3 IgG-Fc Protein/Oligosaccharide Interactions

The Deisenhofer IgG-Fc structure indicated a potential for 72 protein/oligosaccharide interactions including six C_H2 protein/oligosaccharide hydrogen bonds and six hydrogen bonds within each oligosaccharide moiety [11, 61, 88–93], shown in Figure 5.3. These interactions include the sugar residues of the α(1–6) arm, while residues of the α(1–3)-Man-GlcNAc arms are orientated toward the internal space between the C_H2 domains; weak lateral interactions between sugar residues present on opposed heavy chains have been suggested for some structures [61, 88–92, 94].

Structural and functional studies of normal, truncated, and aglycosylated glycoforms of IgG1-Fc generated *in vitro* through X-ray crystallographic analysis, differential scanning microcalorimetry (DSC), and Fcγ receptor binding (isothermal microcalorimetry) [94–96]. These studies established that while the galactose residue on the α(1–6) arm has substantial contacts with the protein structure, it does not impact C_H2 domain stability sequential removal of the terminal GlcNAc, and the two arm mannose residues, generating a (GlcNAc2Man)2 glycoform, resulted in destabilization of the C_H2 domain. DSC has proved to be a valuable tool for probing the contributions of buffers and Fab sequence to stability and solubility of intact IgG and antibody fragments [97–99].

X-ray crystallography of the truncated IgG-Fc glycoforms revealed a progressive increase in temperature factors for the protein moiety of the C_H2 domain, as evidence of increasing structural disorder (destabilization) [94–96]. Minimal (weak) FcγRI and C1 binding and activation were observed for the

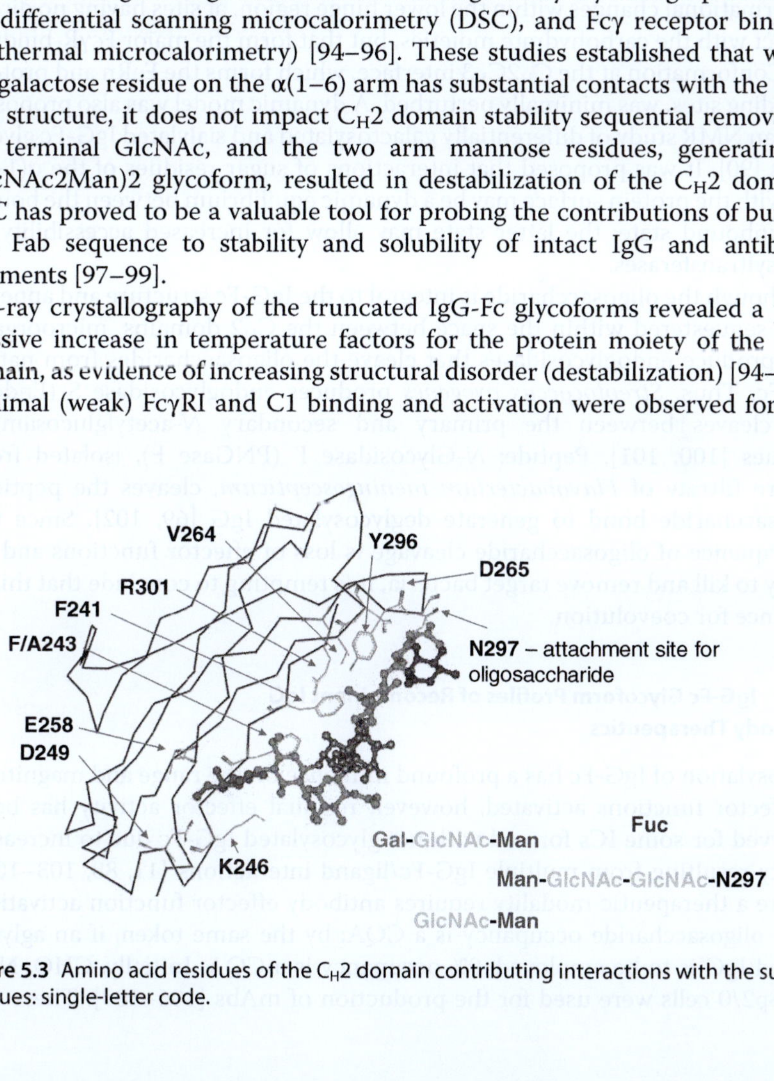

Figure 5.3 Amino acid residues of the C_H2 domain contributing interactions with the sugar residues: single-letter code.

[GlcNAc2Man]2 glycoform that has the potential to form 31 noncovalent contacts with the protein, including at least three hydrogen bonds [92]. Truncation of the sugar residues results in the mutual approach of C_H2 domains with the generation of a "closed" conformation; in contrast to the "open" conformation observed for the fully galactosylated IgG-Fc [96]. The dramatic reduction of FcγR and C1 binding and activation for aglycosylated IgG-Fc contrasts with consistent reports of minimal structural change within the protein structure.

An extensive NMR study of a series of truncated glycoforms showed that trimming of the oligosaccharide was accompanied by concomitant increase in the number of amino acid residues perturbed within the C_H2 domains [89]. Cleavage between the primary and secondary GlcNAc sugar residues induced conformational changes within the lower hinge region, at sites having no direct contact with the carbohydrate moieties, but that form the major FcγR-binding site. Conformation at the C_H2C_H3 interface, which forms the FcRn and protein A binding sites, was minimally perturbed. A dynamic model was also proposed from an NMR study of differentially galactosylated and sialylated IgG-Fc glycoforms [90]. It was proposed that interactions of sugar residues of the $\alpha(1-6)$ arm with the protein surface may be a dynamic equilibrium between the bound and unbound state; the latter state may allow for increased accessibility to glycosyltransferases.

Although the oligosaccharide is integral to the IgG-Fc structure and appears to be sequestered within the space between the C_H2 domains, microorganisms produce endoglycosidases that cleave the oligosaccharides from native IgG-Fc. Thus, *Streptococcus pyogenes* produces endoglycosidase S (EndoS) that cleaves between the primary and secondary *N*-acetylglucosamine residues [100, 101]. Peptide: *N*-Glycosidase F (PNGase F), isolated from culture filtrate of *Flavobacterium meningosepticum*, cleaves the peptide/oligosaccharide bond to generate deglycosylated IgG [69, 102]. Since the consequence of oligosaccharide cleavage is loss of effector functions and an ability to kill and remove target bacteria, it is tempting to conclude that this is evidence for coevolution.

5.4.4 IgG-Fc Glycoform Profiles of Recombinant IgG Antibody Therapeutics

Glycosylation of IgG-Fc has a profound influence on the range and magnitude of effector functions activated; however, residual effector activity has been observed for some ICs formed with nonglycosylated IgG-Fc due to increased avidity resulting from multiple IgG-Fc/ligand interactions [11, 88, 103–106]. Where a therapeutic modality requires antibody effector function activation, 100% oligosaccharide occupancy is a CQA; by the same token, if an aglycosylated IgG is to be employed, 0% occupancy is a CQA. Initially, CHO, NS0, and Sp2/0 cells were used for the production of mAbs [106–109]. These cell lines produce predominantly G0F IgG-Fc glycoforms with relatively low levels of galactosylated and nonfucosylated IgG-Fc, relative to normal polyclonal IgG-Fc; CHO cells may add *N*-acetylneuraminic acid residues but in α(2–3) linkage rather than the α(2–6) linkage present in humans. In addition, these cell lines may add sugars that are not present in normal serum-derived IgG and can be immunogenic in humans. A particular concern is the addition, by NS0 and Sp2/0 cells, of galactose in α(1–3) linkage to galactose-linked β(1–4) to the *N*-acetylglucosamine residues [87–90]. Humans and higher primates do not have a functional gene encoding the transferase that adds galactose in α(1–3) linkage; however, due to continual environmental exposure to the gal α(1–3) gal epitope, for example, in red meats, humans develop IgG antibody specific for this antigen [109–112]. The (gal α(1–3) gal) epitope is widely expressed on hamster cells, and it has recently been reported that some derived CHO cell lines are capable of (gal α(1–3) gal) addition. Similarly, CH0, NS0, and Sp2/0 cells may add an *N*-glycolylneuraminic acid, in α(2–3) linkage, that is not present in humans and may also be immunogenic [106–112]. A significant population of normal human IgG-Fc bears a bisecting *N*-acetylglucosamine residue that is absent from IgG-Fc produced in CHO, NS0, or Sp2/0 cells. The generation of homogeneous IgG-Fc glycoforms *in vitro* has shown that the effector functions activated, qualitatively and quantitatively, differ between IgG subclasses and antibody glycoforms. It has not proved possible to manipulate culture medium conditions to generate predetermined mAb glycoform profiles; however, significant "tweaking" of the profile can be achieved during a production run [113, 114], and cellular engineering has been employed to enhance production of particular human IgG-Fc glycoforms—see in the following text.

Two apparently unnatural glycoforms that are invariably present in recombinant antibody preparations are the heptasaccharide core bearing a fucose residue but lacking an *N*-acetylglucosamine residue on either the α(1–3) or α(1–6) arm. The levels present are very low (1–5%), and while the impact of this glycoform on mAb performance has not been widely reported, it appears not to generate concern. Of particular moment is the level of fucosylation, both fucosylated and nonfucosylated glycoforms being natural, and oligomannose glycoform that are not reported for natural IgG immunoglobulin. These issues will be explored in turn.

5.4.5 Control of the Addition of Fucose and its Impact on Functional Activities

The influence of recombinant protein glycoforms on biologic activity has been comprehensively explored through their production in mutant CHO cells lacking the ability to add one or more sugar residue [23]. The cell line Lec 13 lacks the ability to add fucose to the primary *N*-acetylglucosamine residue, and antibodies of the IgG1 subclass produced in this cell line were shown to exhibit enhanced abilities to kill cancer cells [115] by the mechanism referred to as ADCC. This finding was confirmed and extended to all IgG subclasses when antibodies were produced in a α(1,6)–fucosyltransferase "knockout" CHO cell line [116, 117]. This cell line is available commercially, from the company Biowa, and provides access to the "Potelligent" platform technology [118]. A nonfucosylated anti-CCR4 antibody (Mogamulizumab) produced by this cell line has been approved in Japan for the treatment of patients with relapsed or refractory CCR4-positive adult T-Cell leukemia–lymphoma (ATL) [119] and is in phase III trials in Europe and the United States. A similar improvement in ADCC was reported for IgG1 antibody produced in a "knock-in" CHO cell line transfected with the human β1,4-*N*-acetylglucosaminyltransferase III (GnTIII) gene, resulting in the addition of bisecting *N*-acetylglucosamine residues [77, 120–122]. The early addition of bisecting *N*-acetylglucosamine during passage through the Golgi was shown to inhibit the addition of fucose by endogenous α(1–6)—fucosyltransferase [77]. It is posited, therefore, that the absence of fucose is the main factor determining increased ADCC for these glycoforms. This platform has been consolidated, by Glycart-Roche, and the anti-CD20 antibody obinutuzumab was approved by the FDA in 2013 for previously untreated chronic lymphocytic leukemia (CLL) [123]. An anti-EGFR (imgatuzumab) produced using this platform is in phase II clinical trials. Multiple technologies are being developed to generate homogeneous examples of naturally occurring antibody glycoforms customized to confer downstream biologic activities appropriate for treatment of specific disease indication [124, 125]. These IgG glycoforms may be minor components of the oligosaccharides released from normal polyclonal human IgG-Fc; however, they are normal components and do not present immunogenicity issues [11, 126]. By contrast some glycoform products of nonhuman (mammalian) cell lines may be immunogenic [6, 11].

The foregoing discussion was essentially confined to ADCC mediated by peripheral blood mononuclear leukocytes; however, the impact of fucosylation is reported to differ for polymorphonuclear cells, for example, neutrophils. A study employing two batches of a monoclonal IgG antibody, one of high fucose content and one of low fucose content, reported that higher fucose content resulted in more active neutrophil-mediated ADCC [127, 128]. A contrary result is reported for neutrophil-mediated phagocytosis and apoptosis [117, 129]. It should be noted that results for *in vitro* ADCC studies employing cell lines expressing cellular receptors may vary since the glycoform of the receptor is also a critical parameter and will differ between cell lines employed [130]. I am not aware of publications reporting a positive or negative impact of fucosylation on complement-dependent cytotoxicity (CDC); however, an IgG1/IgG3 hybrid molecule has been shown to exhibit enhanced CDC for both fucosylated and nonfucosylated IgG-Fc glycoforms [131].

The perceived advantage of nonfucosylated antibodies has led other groups and companies to explore alternative routes for the generation of nonfucosylated glycoproteins. A consequence of engineering the CHO cell line to produce homogeneous Man5 glycoforms results in lack of addition of fucose [23, 24, 132, 133]. Inhibitor targeting of enzymes within the Golgi apparatus provided multiple opportunities for producing nonfucosylated molecules, and the action of kifunensine has been employed by several groups for the generation of nonfucosylated high mannose (Man6–Man9) glycoforms [134–137]. Other platforms that have been generated include GlymaxX [136] that engineers mammalian cells to express a bacterial enzyme that inhibits the pathway leading to the addition of fucose. A further platform employs sugar analogues that, added to the culture medium, inhibit incorporation of the natural sugar [137].

5.4.6 Recombinant Glycoproteins Bearing High Mannose (Man5–Man9) Oligosaccharides

I am not aware of reports of the presence of high mannose glycoforms present in normal serum IgG; however, it is very frequently present in recombinant mAbs. There has been a concern that this glycoform may compromise the efficacy of a mAb therapeutic and/or result in more rapid clearance. A series of mutant CHO cell lines were developed that included CHO-lec 3.2.8.1 cells that lacks GnT1 activity, and consequently processing does not proceed beyond the Man5 glycoform; similarly the human embryonic kidney HEK293S cell line also lacks expression of the GnT1 enzyme [23, 134, 138, 139]. The Man5 oligosaccharide is normally an intermediate in glycoprotein processing, and glycoproteins bearing this oligosaccharide are rarely present on natural glycoprotein products. When present on recombinant glycoproteins, it may be regarded as an artifact of the production platform; however, for some recombinant glycoprotein therapeutics, the presence of terminal mannose residues may be beneficial. The GnT1 deficient cell lines have been exploited to produce homogeneous Man5 glycoforms that may target mannose receptor bearing cells; they are also more amenable to crystallization for structural studies. Ideally, control of the glycoform profile might be achieved by manipulation of culture conditions; while multiple parameters impact Golgi-mediated glycoprotein processing, some control of Man5 levels by manipulation of cell culture conditions has been reported [114, 140, 141]. Inhibition of enzymes within the Golgi apparatus provides another avenue for the production of high mannose glycoforms. Thus, kifunensine inhibits mannosidases with the production of Man6–Man9 glycoforms [133, 142]. It has recently been demonstrated that incomplete processing *in vivo*, with consequent generation of truncated mannose oligosaccharides, can result from restricted access for mannose transferases. Thus, while the surface of recombinant HIV GP120 glycoprotein is almost entirely covered by N-linked high mannose oligosaccharide structures, it has been shown that native GP120, expressed on HIV virus isolates, bears a number of truncated oligomannose structures. It appears that the density of the early oligomannose structures limits accessibility to mannose transferases [143].

Glycoproteins bearing exposed mannose residues may bind and/or be taken up via mannose receptor(s) and/or activate multiple biologic pathways *in vivo*; for some glycoprotein therapeutics uptake via mannose receptor is mandatory for functional benefit while for others inappropriate. An interesting example is the approved biologic Cerezyme, used as enzyme replacement therapy for patients with Gaucher's disease. This is a lysosome storage disease resulting from a deficit in the production of the enzyme glucocerebrosidase by macrophages [144]. The presence of exposed mannose residues on the recombinant enzyme is essential for uptake by macrophages via the mannose receptor. The product produced by CHO cells expresses terminal *N*-acetylglucosamine, galactose, and sialic acid sugar residues and is not taken up efficiently. Consequently, the CHO product is treated with sialidase, galactosidase, and *N*-acetylglucosaminidase to generate a product expressing the exposed trimannose core. A study in which glucocerebrosidase products expressing Man2–Man9 oligosaccharides did not show any dramatic differences in uptake; however, the high mannose products were shown to bind to mannose binding lectin (MBL) with possible unwanted lectin pathway activation of the complement cascade [144, 145].

5.5 Glycation

Glycation results from a nonenzymatic condensation reaction between aldehyde groups of reducing sugars and the primary or the secondary amines of proteins, for example, the ε-amino groups of lysine residues and N-terminal amino acid residues. Glycation occurs under physiological conditions, *in vivo*, and is particularly associated with increased glucose levels in patients with diabetes. The relevance to the production of recombinant antibodies is threefold: during production in glucose containing culture media, formulation, and glycation, *in vivo*, of an approved mAb therapeutic. The latter consequence may be ameliorated by the selection of an antibody that does not have a lysine residue within the CDRs [146]. A study of antibody incubated, *in vitro*, with differing concentrations of glucose reported substantial glycation with loss of IgG-Fc SpA and complement binding, while antigen binding was unimpaired [147]. This contrasts with conclusions reached from analysis of IgG isolated the serum of healthy individuals and diabetic patients. The number of glucose molecules condensed on IgG was 1–5 for healthy individual, 5–10 for well-controlled diabetics, and 10–25 for poorly controlled patients [147]. This study further showed that the IgG-Fab region was more prone to glycation than the IgG-Fc region and that glycation could, potentially, compromise antigen-binding activity [148]. The extent of glycation resulting from extended incubation of antibody formulated in sucrose at differing temperature showed the absence of glycation when stored at 4° for 18 months; however, 10 lysine residues were shown to be substituted after storage for 1 month at 37° [149]. The possible impact of glycation on IgG-Fc function was further evaluated for IgG1 and IgG2 antibodies subjected to glycation conditions, resulting in an average of 42–49 glucose adducts per molecule; binding to FcγRIIa, FcRn, and SpA was reported to be unaffected [150].

5.5.1 Cellular Receptors Mediating Destruction and Removal of IgG/Antigen Immune Complexes

Three types or classes of membrane-bound human FcγR have been defined by immunochemical, biochemical, and gene sequencing studies: FcγRI [CD64], FcγRII [CD32], and FcγRIII [CD16], with an additional six subtypes—[FcγRIIA, FcγRIIB1, FcγRIIB2, FcγRIIC, FcγRIIIA, and FcγRIIIB] [151–155]. These receptors are constitutively, but differentially, expressed on a wide range of leukocytes, and expression may be upregulated and/or induced by cytokines generated and released within an inflammatory response. The effector mechanisms activated are diverse and include "killing" and removal (e.g., phagocytosis, respiratory burst, and cytolysis), accessory functions such as the enhancement of antigen presentation by dendritic cells, and the downregulation of growth and differentiation of lymphocytes. It is evident, therefore, that FcγRs play essential roles in the induction, establishment, and resolution of protective immune responses.

Multiple parameters determine the structure and biological activities of ICs formed within a polyclonal antibody response:

i) Valency
ii) Average affinity/avidity of the antibody population
iii) Isotype profile
iv) IgG-Fc glycoforms
v) Valency and/or epitope density of the antigen
vi) Density of cell surface effector ligands, for example, FcγR
vii) Cumulative valency when multiple ligands are engaged, for example, FcγR and complement receptors
viii) Proportions of each antibody isotype within a polyclonal response [11, 88, 94–96, 102–106, 151–158]

The FcγRI receptor is constitutively expressed on mononuclear phagocytes and dendritic cells; however, expression can be upregulated and/or induced by the action of cytokines. FcγRIIa is the most widely expressed FcγR and is found on most hemopoietic cells. Polymorphic variants of FcγIIa are identified by the presence of histidine (FcγRIIa-131H) or arginine (FcγRIIa-131R) at amino acid residue 131. The higher affinity of the FcγRIIa-131H form for IgG2 results in differing cellular responses to engagement by IgG2 ICs [11, 150–154]. Higher phagocytic capacity for *Streptococcus pneumoniae* opsonized with IgG2 antibody was observed for neutrophils of donors homozygous for FcγRIIa-131H than for FcγIIa-131R. The FcγRIIb receptor is expressed on B lymphocytes and monocytes, and ligation of this receptor results in growth and differentiation inhibition [11, 150–154].

Initially the FcγRIIIa receptor was reported to bind and be activated by IgG1 and IgG3 only; however, recognition of a polymorphism in the receptor and the differential influence of IgG glycoforms have radically changed our understanding, with important clinical consequences. The avidity of binding of IgG differs between the FcγRIIIa-158V and FcγRIIIa-158F polymorphic variants [154]. It was demonstrated, *in vitro*, that IgG1 antibody is more efficient at mediating ADCC through homozygous FcγRIIIa-158V bearing cells than homozygous FcγRIIIa-158F or heterozygous FcγRIIIa-158V/FcγRIIIa-158F cells [154–156]. Similar differences in ADCC efficacy might pertain *in vivo* since more favorable responses were reported for patients diagnosed with systemic lupus erythematosus or leukemia and homozygous for FcγRIIIa-158V when exposed to Rituxan than for homozygous FcγRIIIa-158F patients [154–156]. Similarly, FcγRIIIa polymorphisms were shown to influence the response of Crohn's disease patients to infliximab [99] and red blood cell clearance by anti-D antibody [157].

All FcγRs are transmembrane molecules, with the exception of FcγRIIIb that is glycosylphosphatidylinositol (GPI) anchored within the membrane of neutrophils. FcγRI and FcγRIIIa are members of the multichain immune recognition receptor (MIRR) family and are present in the membrane as heterooligomeric complexes comprised of an α and a γ chain: an IgG/antigen complex binds the α chain to initiate signaling through the γ chain; the FcγRIIIa α chain of NK cells is also associated with a signaling ζ chain. FcγRIIa and FcγRIIb molecules are comprised of an α chain only [150–152]. The FcγR α chains show a high degree of sequence homology in their extra cellular domains (70–98%) but differ significantly in their cytoplasmic domains. The cytoplasmic domains of γ chains and the FcγRIIa α chain express the immunoreceptor tyrosine-based activation motif (ITAM) that is involved in the early stages of intracellular signal generation. By contrast the FcγRIIb receptor α chain expresses an immunoreceptor tyrosine-based inhibition motif (ITIM) [11, 62, 63, 150–152]. Cellular activation may be dependent on the balance between the relative levels of expression of these two isoforms and hence the balance of signals generated through the ITAM and ITIM motifs [150–152].

5.5.1.1 FcγR Binding Sites on IgG

The crystal structure of IgG-Fc in complex with soluble recombinant FcγRIIIb, FcγRIIIa, and FcγRIIa reveals direct involvement of the lower hinge and hinge proximal C_H2 domain residues [11, 67, 68, 159–162]. Analysis of the complex formed between IgG-Fc and recombinant glycosylated FcγRIIIa has confirmed the general features of the interaction but has also revealed a critical role for glycosylation of FcγRIIIa [159, 162]—see later. Thus, the IgG-Fc/FcγRIIIa interaction is significantly influenced by the glycoform of each component. The involvement of both heavy chains in the formation of an asymmetric binding site provides a structural explanation for an essential requirement—which the IgG should be univalent for the FcγR; if monomeric IgG were divalent, then it could cross-link cellular receptors and hence constantly activate inflammatory reactions.

5.5.2 FcRn: Catabolism and Transcytosis

5.5.2.1 Transcytosis

The FcRn receptor was first identified from studies of the transport of IgG across the gut of newborn rats and designated the neonatal Fc receptor, hence FcRn; subsequently the human homologue was identified in human placenta and shown to mediate transport of IgG from the mother to the fetus [156]. The interaction site on IgG-Fc is at the C_H2C_H3 interface, and the C_H3 sequence -H-N-H-Y-H- (Eu: 433–436) is of functional significance since titration of these histidine residues accounts for the observed binding of IgG to FcγRn at pH 6.0–6.5 and its release at pH 7.0–7.5 [163–166]. The interaction of IgG-Fc with FcRn appears not to be influenced by the natural IgG-Fc glycoform profile or indeed the presence or absence of oligosaccharides. On the contrary, the N-glycans in FcRn contribute to the steady-state membrane distribution and direction of IgG transport [165–167].

Each of the four human IgG subclasses is transported across the placenta, however, with differing facility. Cord blood levels of IgG1 may be higher than in matched maternal blood, while IgG3 and IgG4 levels are equivalent. The level of IgG2 is ~80% of the concentration in maternal blood [168]. During pregnancy the level of galactosylation of maternal IgG increases, and there is evidence of preferential transport of galactosylated IgG across the placenta [82, 83, 86, 168]. This provides circumstantial evidence to suggest that the affinity of IgG for FcγRn may differ between glycoforms, under conditions operative at the interface between the mother and the placenta.

Protein therapeutics having short half-lives, for example, cytokines, are being generated as fusion proteins with IgG-Fc to extend serum half-lives. A further development opens a new route for administration since it has been shown that FcRn is expressed in the central and upper airways of the lung and that drug-IgG-Fc fusion proteins delivered to these sites can be transported, by transcytosis, to the systemic circulation. This is an exciting development with considerable promise and significance [169–171].

5.5.2.2 Catabolism

The catabolic pathway of human IgG and albumin is mediated by FcRn-expressing cells that take up IgG within pinocytotic vesicles with the formation of vacuoles [11, 62, 63, 172–176]. Subsequent lowering of pH results in saturation binding to FcRn and protection of bound IgG from cleavage by enzymes present in the vacuole; unbound IgG is degraded [172–176]. When the membrane of the vacuole is recycled to the cellular membrane, the IgG/FcRn complex is exposed to pH 7.2 and released into the extra vascular fluid. The catabolic half-life of human IgG1, 2 and 4, is ~21 days while for IgG3 it is ~7 days; the IgG3 data was generated for IgG3 molecules of G3m(5*) and G3m(21) allotypes in which arginine is present at residue 435, in contrast to the histidine residue that is present in IgG1, IgG2, and IgG4 [11, 172–176]. It has been reported that replacement of the IgG3 arginine 435 residue by histidine increases the affinity for FcRn and consequently the half-life [175]. The IgG allotype G3m(15,16), present within Mongoloid populations, has a histidine residue at position 435 and higher binding affinity for FcRn [166, 167]. The catabolic half-life of IgG mediated through FcRn does not appear to be influenced by IgG-Fc glycoform [167–170, 172]; however, it should be emphasized that only glycoforms of IgG bearing neutral oligosaccharides have been evaluated. It may be anticipated that sialylation could influence IgG-FcRn interactions since it would introduce a negative charge in the vicinity of the histidine residues involved in FcRn binding. Protein engineering has been successfully applied to increase the affinity of IgG-Fc binding to FcRn and hence to increase the catabolic half-life [173–175]. Prolongation of the half-life of an IgG therapeutic could translate into reduced frequency of dosing and attendance at clinic, reducing the cost of treatment (CoT). There is evidence that glycoproteins expressing terminal *N*-acetylglucosamine residues may be cleared through the mannose receptor and account for the enhanced clearance of the IgG-Fc/TNF receptor fusion protein therapeutic (Lenercept) having exposed terminal *N*-acetylglucosamine residues [177, 178].

5.5.3 Complement Activation

5.5.3.1 Classical Pathway: C1q/C1 Binding and Activation

The classical pathway of complement activation results in a cascade of enzymatic reactions with cleavage of downstream complement components, with amplification at each step, and the generation of fragments that (i) bind to an immune complex, (ii) recruit leukocytes to augment an inflammatory response, and (iii) form a multimeric "membrane attack complex" that inserts into cellular and bacteria membranes to generate a "pore" that allows the ingress of water and consequent lysis [11, 62, 63, 131, 179, 180]. Leukocytes expressing FcγR may also express receptors for complement fragments bound to an immune complex, thus enhancing opsonization [131, 179, 180].

Glycosylation of IgG-Fc is essential for C1 binding and activation, and ICs incorporating IgG1 and/or IgG3 antibody are highly active, while only IgG2 complexes formed in antigen excess may be active; there has been a consensus that IgG4 does not activate the classical pathway [11, 62, 63, 181–186]. However, ordered IgG4 hexameric complexes have been reported to activate C1 [127]. Activity can be modulated by protein and glycoengineering, and a hybrid IgG1/IgG3 molecule has been shown to exhibit enhanced activity, relative to either IgG1 or IgG3 alone [183]; this antibody format is available commercially as Complegent. Protein engineering studies suggest that the interaction site of human IgG1 for C1 is localized to the hinge proximal region of the C_H2 domain [11, 62, 63, 182, 186]. This proposal is further supported by the demonstration that replacement of the Pro 331 residue of IgG4 by serine converts it to a molecule that can activate C1; proline 331 is localized at the junction of the b6 bend and the fy3 β-strand that is topographically proximal to hinge region [185]. Extensive studies of mutant chimeric human IgG3 proteins have established that the efficiency of C1 activation is not directly determined by the length of the hinge region but that at least one interheavy chain disulfide bridge is required [91].

5.5.4 Role of IgG Glycoforms in Recognition by Cellular FcγRs

Since glycosylation of IgG-Fc is essential to recognition and activation of FcγRs, the C1 component of complement quantitative and/or qualitative functional differences between glycoforms might be anticipated. This has been confirmed in many studies employing recombinant mAbs glycoengineered *in vitro* and/or produced by engineered cell lines to generate antibodies having a predetermined glycoform profile [11, 73, 77, 94–96, 115, 120, 121, 187]. These findings lead one to speculate whether the immune system responds to pathogens by selective production of antibodies having an optimal antibody isotype and glycoform profile. The high sensitivity of current analytical protocols allows determination of the glycoform profile of specific antibody populations, for example, antiplatelet autoantibodies and anticitrullinated peptide antibodies [85, 188]. It remains to be determined whether these differences relate to disease activity.

5.5.5 The Influence of Fucose and Bisecting *N*-Acetylglucosamine on IgG-Fc Activities

Increased ADCC was reported for antibodies lacking the presence of fucose, produced in Lec 13 CHO cells [115], and expressing bisecting *N*-acetylglucosamine, produced in CHO cells transfected with the human β1,4-*N*-acetylglucosaminyltransferase III (GnTIII) gene [77, 120, 121, 187]. It was further demonstrated that the presence of bisecting *N*-acetylglucosamine inhibits the endogenous α(1,6)-fucosyltransferase and the addition of fucose [73]. This glycoform, bearing bisecting GlcNAc with the absence of fucose, is a minor component (<3%) of oligosaccharides released from normal polyclonal human IgG-Fc; however, it may be a predominant glycoform produced by an individual plasma cell clone [75, 80].

5.5.6 The Influence of Galactosylation on IgG-Fc Activities

Galactosylation of IgG-Fc is a major source of glycoform heterogeneity, in health and disease. In health there is a difference in galactosylation with gender and a decline with age [189, 190]. Increased levels of IgG-Fc galactosylation are observed during the course of normal pregnancy, with levels returning to the adult norm following parturition [81, 82]. Hypogalactosylation of IgG-Fc is reported for a number of inflammatory states associated with autoimmune diseases [83–86, 191–193]. The extent of IgG-Fc galactosylation observed for monoclonal human myeloma IgG proteins is highly variable, indicating that the level of IgG-Fc galactosylation is a clonal property [79, 80, 194]. The antibody products of CHO, Sp2/0, and NS0 cell lines used in commercial production of recombinant antibody are generally highly fucosylated but hypogalactosylated, relative to normal polyclonal human IgG [87, 89, 97, 195, 196]. It is therefore necessary to consider the possible impact of differential IgG-Fc galactosylation on functional activity.

Numerous studies have probed the influence of galactose residues on IgG-Fc structure and function. While little evidence for a significant structural impact has been reported [94–96, 197–200], the impact on function has been variable and may differ depending on epitope specificity of the antibody and consequent immune complex architecture. It should be noted that a majority of these studies have been conducted for IgG1 subclass antibodies. The possible consequences/impact for hypogalactosylated mAbs, *in vivo*, has been extrapolated from *in vitro* assays and animal experiments. Removal of terminal galactose residues from Campath-1H was shown to reduce classical complement activation but to be without effect on FcγR-mediated functions [201]. Similarly, the ability of rituximab to kill tumor cells by the classical complement route has been shown to be maximal for the [G2F]2 glycoform, in comparison to the [G0F]2 glycoform [89]. The product that gained licensing approval was comprised of ~25% of the G1F oligosaccharide; therefore, regulatory authorities required that galactosylation of the manufactured product be controlled to within a few percentage points of this value. In the absence of galactose, the terminal sugar residue is *N*-acetylglucosamine that may be accessible to both the mannose receptor and/or MBL.

5.5.7 Sialylation of IgG-Fc Oligosaccharides

The addition of *N*-acetylneuraminic acid residues, sialylation, requires the presence of a terminal galactose residue. A minority of oligosaccharides released from polyclonal IgG-Fc are sialylated, while ~70% bear one or two galactose residues [69, 72–74, 88, 102–106, 201]. The paucity of sialylation is presumed to reflect restricted access of terminal galactose residues, due to the generation of a "closed" IgG-Fc protein conformation, for the α(2,6) sialyltransferase enzyme, rather than being due to any deficit in the sialylation machinery. This conclusion is supported by the finding that when both IgG-Fc and IgG-Fab are glycosylated the latter bears highly galactosylated and sialylated structures, demonstrating that the glycosylation machinery is fully functional [11, 75]. The presence or absence of terminal galactose and/or sialic acid residues does not influence IgG catabolism since it is not catabolized in the liver, via the asialoglycoprotein receptor (ASGPR), but in multiple cell types that express the FcRn receptor. The balance between structure and accessibility is well illustrated for a panel of IgG-Fcs in which individual amino acid residues making contacts with the oligosaccharide were replaced by alanine. In each case hypergalactosylated and highly sialylated glycoforms resulted, suggesting some relaxation of structure allowing access to glycosyl transferases [94–96, 200, 201].

Recent studies suggest that sialylated human IgG-Fc may inhibit activation of inflammatory cascades. Following binding to the lectin receptor SIGN-R1, in the mouse, or DC-SIGN, in humans, expression of the inhibitory receptor FγRIIb on inflammatory cells is upregulated, attenuating autoantibody-initiated inflammation [202, 203]. However, it has been asserted that caution should be exercised, when extrapolating from mouse models to humans, since the tissue distribution of SIGN-1 and DC-SIGN differs and that anti-inflammatory activity could be demonstrated for intact IgG and $F(ab')_2$ fragments [204–206].

5.6 IgG-Fab Glycosylation

It is established that ~30% of polyclonal human IgG molecules bear N-linked oligosaccharides within the IgG-Fab region, in addition to the conserved glycosylation site at Asn 297 in the IgG-Fc [11, 69, 84, 106, 207]. When present, they are attached within the variable regions of the kappa (Vk), lambda (Vλ), or heavy (VH) chains, sometimes both. In the immunoglobulin sequence database, ~20% of expressed IgG variable regions have N-linked glycosylation consensus sequences (Asn-X-Thr/Ser; where X can be any amino acid except proline). Interestingly, these consensus sequences are mostly not germline encoded but result from somatic hypermutation—suggestive of positive selection for improved antigen binding. Analysis of polyclonal human IgG-Fab reveals the presence of diantennary oligosaccharides that are extensively galactosylated and substantially sialylated in contrast to the oligosaccharides released from IgG-Fc [11, 69, 84, 106, 208]. This pattern was maintained for IgG-Fab prepared from hypogalactosylated IgG isolated from the sera of patients with Wegener's granulomatosis or microscopic polyangiitis [84]. Thus, factors within the local environment of IgG producing plasma cells, *in vivo*, influence the efficacy of glycoprocessing of IgG-Fc but not IgG-Fab during passage through the Golgi apparatus. The functional significance for IgG-Fab glycosylation of polyclonal IgG has not been fully determined but data emerging for mAbs suggests that Vk, Vλ, or VH glycosylation can have a neutral, positive, or negative influence on antigen binding [106, 208, 209]. The differences observed for polyclonal IgG-Fc and IgG-Fab glycoforms have been maintained for recombinant antibodies produced in CHO cells and myeloma IgG proteins [75, 94, 95, 126, 197, 210].

The influence of glycosylation on the thermal stability of human IgG1-Fc has been demonstrated in DSMC and X-ray crystallographic studies [94–96]. Since it is generally observed that the oligosaccharide present in glycoproteins contributes to solubility and stability, it is possible that IgG-Fab glycosylation may similarly be beneficial, particularly when formulating IgG therapeutics at concentrations of 100–150 mg/ml [211]. Such high-concentration formulations allow the development of self-administration protocols and can reduce dosing intervals, resulting in reduced CoT. Controlling glycoform fidelity at two sites offers a further challenge to the biopharmaceutical industry.

The licensed antibody therapeutic Erbitux (cetuximab) bears an N-linked oligosaccharide at Asn 88 of the VH region; interestingly there is also a glycosylation consensus sequence at Asn 41 of the VL, but it is not occupied [212, 213]. Analysis of the IgG-Fc and IgG-Fab oligosaccharides of Erbitux, produced from Sp2/0 cells, reveals highly significant differences in composition. While the IgG-Fc oligosaccharides are typical, that is, comprised predominantly of diantennary G0F oligosaccharides, the IgG-Fab oligosaccharides are extremely heterogeneous and include complex diantennary, triantennary, and hybrid oligosaccharides; nonhuman oligosaccharides were also present, for example, galactose in α(1–3) linkage to galactose and *N*-glycolylneuraminic acid residues.

Severe adverse reaction to cetuximab therapy has been reported, and in a study of 76 patients treated with Erbitux, 25 experienced hypersensitivity reactions; this was shown to be due to the presence of IgE anti-gal α(1,3) gal antibodies. Interestingly, environmental factors appeared to influence the development of IgE anti-gal α(1,3) gal responses, and IgE antibodies were detected in pretreatment samples from 17 of the patients [108, 214–216]. The incidence varied significantly between treatment centers and may be linked to differences in predominant infectious agents present in local environments.

Subsequently, it has been demonstrated that many individuals that consume meat (beef, lamb, pork, etc.) have IgG anti-Gal α(1–3) Gal antibodies and a minority IgE anti-Gal α(1–3) Gal antibodies. It is becoming a routine, therefore, to monitor patients for the presence of IgE anti-Gal α(1–3) Gal antibodies prior to exposure to Erbitux [215–218].

A detailed analysis of the glycoforms of a humanized IgG rMAb bearing oligosaccharides at Asn 56 of the VH and Asn 297, also produced in Sp2/0 cells, reveals the expected IgG-Fc glycoform profile of predominantly G0F oligosaccharides; however, eleven oligosaccharides were released from the IgG-Fab, including diantennary and triantennary oligosaccharides bearing gal α(1,3) gal, *N*-glycolylneuraminic acid and *N*-acetyl galactosamine residues [219]. The consistent observation of higher levels of galactosylation and sialylation for IgG-Fab N-linked oligosaccharides, in comparison to IgG-Fc, is thought to reflect increased exposure and/or accessibility. In view of these experiences, it would seem that the perceived virtues of the NS0 and Sp2/0 cells might best be pursued by engineering to inactivate the gal α(1–3) and *N*-glycolylneuraminic acid transferases.

The double challenge to produce of rMAbs having appropriately glycosylated IgG-Fc and IgG-Fab has led to some companies engineering out VH or VL glycosylation motifs when present in candidate rMAbs [220]; however, present reports suggest that CHO cells can glycosylate VH and/or VL motifs in a similar manner to that observed for normal polyclonal IgG [221]. Since oligosaccharides are hydrophilic, the addition of glycans within VH and/or VL regions may impact the physicochemical properties of an antibody molecule and consequently pharmacokinetics [219, 222], solubility [223], aggregation, etc. While VH glycosylation of a human IgG antibody was shown to have the same pharmacokinetics as the VH deglycosylated molecule, in a mouse model [222], introduction of a glycosylation site within bispecific single-chain diabodies resulted in a significant increase in serum half-lives [224]. An antibody with specificity for IL-13 was generated that included a glycosylation sequon (53NSS55) within the heavy chain CDR2 [225]. Initially this site was engineered out by replacing N53 by an aspartic acid residue; however, the product exhibited very limited solubility (~13 mg/ml) and consequent aggregation. By contrast reintroduction of N53, together with engineering within VL, resulted in a VH glycosylated antibody product with a solubility >110 mg/ml [225].

5.7 The Influence of Expression Platform on CTM/PTMs and Unintended Physicochemical Changes

The expression platform can have a major influence on the CTM/PTM profile of a protein biologic; however, as, at present, we are concerned only with biosimilar antibody therapeutics, it is only the CHO and Sp2/0 cell lines that are of interest and/or concern. Major upstream and downstream improvements will have been achieved in the interval between the launch of an approved mAb therapeutic and the opportunity to enter the market with a biosimilar.

Considerable success has been achieved in increasing the productivity of mammalian cell lines, with levels of 5–10 g/l being routine [223, 226]. However, high production levels may overwhelm the CTM/PTM machinery, resulting in poor product quality. It is essential, therefore, to determine parameters that influence product quality early in clone selection [223]. It will be incumbent on the supplicant to demonstrate that any structural deviations do not compromise the product, for therapeutic efficacy, immunogenicity, etc.

5.8 Human Antibody Isotypes Other than IgG

This review has focused on recombinant IgG antibody therapeutics; however, our understanding of the structure and function of the IgA [223, 227, 228], IgM [229–231], and IgE [232] isotypes are progressing, and therapeutic applications are on the horizon or in the pipeline, if not licensed [233–236]. These isotypes offer a further challenge since each expresses multiple N-linked glycosylation sites and the IgA1 isotype O-linked oligosaccharides in the hinge region [237].

5.9 Conclusion

It is salutary to contemplate the finesse of structural changes induced by conservative amino acid replacements and/or the presence or absence of a fucose and/or bisecting *N*-acetylglucosamine sugar residue can have on the functional activity of the human IgG-Fc. It is evident, therefore, that the conformation of the IgG molecule is a CQA that is relatively robust while being amenable to protein engineering. It is essential, therefore, that multiple orthogonal techniques should be applied to determine structural parameters and that both industry and academia will be best served by having access to a reference standard that has been characterized by all relevant techniques [238]. In this review I have attempted to present the current understanding of IgG structure and function; however, it is far from complete. It is interesting to note that disparate ligands may bind to the Fc through common amino acid residues within the hinge proximal region for FcγR and C1q and the C_H2C_H3 interface for FcRn, SpA, SpG, RFs, and IgG-Fc-like receptors encoded within the genomes of some viruses. The presence of sialic acid might further influence Fc/ligand interactions at this interface. A rationalization for the topography of ligand-binding sites may be the functional necessity for circulating IgG to be monovalent for FcγRs and C1q to prevent continuous cellular activation while providing opportunity for divalency at the C_H2C_H3 interface. The influence of the IgG-Fc glycoform on functional activity may be exploited to generate homogeneous glycoforms selected for a predetermined functional profile considered optimal for a given disease indication. It is important to note that each glycoform is represented within normal polyclonal IgG-Fc; therefore, they do not have potential to be immunogenic. Many innovative studies have explored engineering of the protein moiety to selectively enhance biologic activities; however, these are mutant forms of IgG, for example, nonself that may enhance immunogenicity. This may not be an issue when treating patients for cancer, since they may be immune suppressed; however, it is a concern for long-term treatment of chronic diseases.

References

1. Van Nostrand EL., Kim SK. (2011). Seeing elegance in gene regulatory networks of the worm. *Curr Opin Genet Dev.* 21(6):776–786.
2. Kimball J. (2013). Enclopedia of DNA Elements. http://users.rcn.com/jkimball.ma.ultranet/BiologyPages/E/Encode.html. Accessed May 5, 2016.
3. Walsh G., Jefferis R. (2006). Post-translational modifications in the context of therapeutic proteins. *Nat Biotechnol.* 24:1241–1252.
4. Farriol-Mathis N., Garavelli JS., Boeckmann B., et al. (2004). Annotation of post-translational modifications in the Swiss-Prot knowledge base. *Proteomics.* 4(6):1537–1550.
5. Khoury G., Baliban RC., and Floudas CA. (2011). Proteome-wide post-translational modification statistics: frequency analysis and curation of the swiss-prot database. *Sci Rep.* 1:90.
6. Lauc G., Vojta A., Zoldoš V. (2014). Epigenetic regulation of glycosylation is the quantum mechanics of biology. *Biochim Biophys Acta.* 840(1):65–70.
7. Morandell S., Stasyk T., Grosstessner-Hain K., et al. (2006). Phosphoproteomics strategies for the functional analysis of signal transduction. *Proteomics.* 6:4047–4056.
8. Guo M., Huang BX. (2013). Integration of phosphoproteomic, chemical, and biological strategies for the functional analysis of targeted protein phosphorylation. *Proteomics.* 13:424–437.
9. Beck A. (2013). Glycosylation Engineering of Biopharmaceuticals. Methods and Protocols. Methods in Molecular Biology, Vol. 988. Springer: New York.
10. Aricescu AR., Owens RJ. (2013). Expression of recombinant glycoproteins in mammalian cells: towards an integrative approach to structural biology. *Curr Opin Struct Biol.* 23(3):345–356.
11. Jefferis R. (2012). Isotype and glycoform selection for antibody therapeutics. *Arch Biochem Biophys.* 526(2):159–166.

12. FDA (2014). FDA Guidance for Industry: Scientific Considerations in Demonstrating Biosimilarity to a Reference Product. http://www.fda.gov/downloads/Drugs/GuidanceComplianceRegulatoryInformation/Guidances/UCM397017.pdf. Accessed May 5, 2016.
13. EMEA (2009). EMA Guideline on Development, Production, Characterisation and Specifications for Monoclonal Antibodies and Related Products. http://www.ema.europa.eu/docs/en_GB/document_library/Scientific_guideline/2009/09/WC500003074.pdf. Accessed May 5, 2016.
14. Hospira (2013). Press Release. http://phx.corporate-ir.net/phoenix.zhtml?c=175550&p=irol-newsArticle&ID=1853480&highlight. Accessed May 5, 2016.
15. Beck A., Reichert JM. (2013). Approval of the first biosimilar antibodies in Europe: a major landmark for the biopharmaceutical industry. *MAbs.* 5(5):621–623.
16. Andrew M., Goetze Y., Liu D., et al. (2012). Rates and impact of human antibody glycation in vivo. *Glycobiology.* 22(2):221–234.
17. Khawli LA., Goswami S., Hutchinson R., et al. (2010). Charge variants in IgG1: isolation, characterization, in vitro binding properties and pharmacokinetics in rats. *MAbs.* 2(6):613–624.
18. Wang W., Singh S., Zeng DL., et al. (2007). Antibody structure, instability, and formulation. *J Pharm Sci.* 96(1):1–26.
19. Chicooree N., Unwin RD., Griffiths JR. (2015). The application of targeted mass spectrometry-based strategies to the detection and localization of post-translational modifications. *Mass Spectrom Rev.* 34(6):595–626.
20. Lanucara F., Eyers CE. (2013). Top-down mass spectrometry for the analysis of combinatorial post-translational modifications. *Mass Spectrom Rev.* 32:27–42.
21. Manning G., Whyte DB., Martinez R., et al. (2002). The protein kinase complement of the human genome. *Science.* 298(5600):1912–1934.
22. Sacco F., Perfetto L., Castagnoli L., et al. (2012). The human phosphatase interactome: an intricate family portrait. *FEBS Lett.* 586:2732–2739.
23. Stanley P. (2011). Golgi glycosylation. *Cold Spring Harb Perspect Biol.* 3(4):pii: a005199.
24. Ohtsubo K., Marth JD. (2006). Glycosylation in cellular mechanisms of health and disease. *Cell.* 126(5):855–867.
25. Nature. Post-Translational Modifications. http://www.nature.com/nrm/series/ptm/index.html. Accessed May 5, 2016.
26. Edelman GM., Cunningham BA., Gall WE., et al. (2004). The covalent structure of an entire gamma G immunoglobulin molecule. 1969. *J Immunol.* 173(9):5335–5342.
27. Kumar A., Bachhawat AK. (2012). Pyroglutamic acid: throwing light on a lightly studied metabolite. *Curr Sci.* 102:288–297.

28. Liu YD., Goetze AM., Bass RB., et al. (2011). N-terminal glutamate to pyroglutamate conversion in vivo for human IgG2 antibodies. *J Biol Chem.* 286:11211–11217.
29. Yin S., Pastuskovas CV., Khawli LA., et al. (2013). Characterization of therapeutic monoclonal antibodies reveals differences between in vitro and in vivo time-course studies. *Pharm Res.* 30(1):167–178.
30. Perez-Garmendia R., Gevorkian G. (2013). Pyroglutamate-modified amyloid beta peptides: emerging targets for Alzheimer's disease immunotherapy. *Curr Neuropharmacol.* 11(5):491–498.
31. Liu D., Andrew M., Goetze RB., et al. (2011). Flynn N-terminal glutamate to pyroglutamate conversion in vivo for human IgG2. *J Biol Chem.* 286(13):11211–11217.
32. Merlini G., Comenzo RL., Seldin DC., et al. (2014). Immunoglobulin light chain amyloidosis. *Expert Rev Hematol.* 7(1):143–156.
33. Li J., Menzel C., Meier D., et al. (2007). A comparative study of different vector designs for the mammalian expression of recombinant IgG antibodies. *J Immunol Methods.* 318(1–2):113–124.
34. Tang L., Sundaram S., Zhang J., et al. (2013). Conformational characterization of the charge variants of a human IgG1 monoclonal antibody using H/D exchange mass spectrometry. *MAbs.* 5(1):114–125.
35. EBE (2013). EBE: Concept Paper—Considerations in Setting Specifications. http://www.ebe-biopharma.eu/uploads/Modules/Documents/ebe-concept-paper-%E2%80%93-considerations-in-setting-specifications.pdf. Accessed May 5, 2016.
36. Subramanian K., Perez M., Zeng X., et al. (2013). Methods to modulate C-terminal lysine variant distribution. Patent No: WO2013158273 A1.
37. Liu H., May K. (2012). Disulfide bond structures of IgG molecules: structural variations, chemical modifications and possible impacts to stability and biological function. *MAbs.* 4(1):17–23.
38. Zhong X., Wright JF. (2013). Biological insights into therapeutic protein modifications throughout trafficking and their biopharmaceutical applications. *Int J Cell Biol.* 2013:273086.
39. Wypych J., Li M., Guo A., et al. (2008). Human IgG2 antibodies display disulfide-mediated structural isoforms. *J Biol Chem.* 283(23):16194–16205.
40. Dillon TM., Ricci MS., Vezina C., et al. (2008). Structural and functional characterization of disulfide isoforms of the human IgG2 subclass. *J Biol Chem.* 283(23):16206–16215.
41. Koterba KL., Borgschulte T., Laird MW. (2012). Thioredoxin 1 is responsible for antibody disulfide reduction in CHO cell culture. *J Biotechnol.* 157(1):261–267.
42. Hutterer KM., Hong RW., Lull J., et al. (2013). Monoclonal antibody disulfide reduction during manufacturing: untangling process effects from product effects. *MAbs.* 5(4):608–613.

43. Kao Y-H., Laird MW., Schmidt MT., et al. (2013). Prevention of disulfide bond reduction during recombinant production of polypeptides. Patent No: US8574869 B2.

44. Rispens T., Meesters J., den Bleker TH., et al. (2013). Fc-Fc interactions of human IgG4 require dissociation of heavy chains and are formed predominantly by the intra-chain hinge isomer. *Mol Immunol.* 53(1–2):35–42.

45. Davies AM., Rispens T., den Bleker TH., et al. (2013). Crystal structure of the human IgG4 C(H)3 dimer reveals the role of Arg409 in the mechanism of Fab-arm exchange. *Mol Immunol.* 54(1):1–7.

46. Rispens T., Davies AM., Ooijevaar-de Heer P., et al. (2014). Dynamics of inter-heavy chain interactions in human immunoglobulin G (IgG) subclasses studied by kinetic Fab arm exchange. *J Biol Chem.* 289(9):6098–6109.

47. Chumsae C., Liu H. (2008). Comparison of methionine oxidation in thermal stability and chemically stressed samples of a fully human monoclonal antibody. *J Chromatogr B Analyt Technol Biomed Life Sci.* 870(1):55–62.

48. Bertolotti-Ciarlet A., Wang W., Lownes R., et al. (2009). Impact of methionine oxidation on the binding of human IgG1 to Fc Rn and Fc gamma receptors. *Mol Immunol.* 46(8–9):1878–1882.

49. Wang W., Vlasak J., Li Y., et al. (2011). Impact of methionine oxidation in human IgG1 Fc on serum half-life of monoclonal antibodies. *Mol Immunol.* 48(6–7):860–866.

50. Pan H., Chen K., Chu L., et al. (2009). Methionine oxidation in human IgG2 Fc decreases binding affinities to protein A and FcRn. *Protein Sci.* 18(2):424–433.

51. Gaza-Bulseco G., Faldu S., Hurkmans K., et al. (2008). Effect of methionine oxidation of a recombinant monoclonal antibody on the binding affinity to protein A and protein G. *J Chromatogr B Analyt Technol Biomed Life Sci.* 870(1):55–62.

52. Xie H., Chakraborty A., Ahn J., et al. (2010). Rapid comparison of a candidate biosimilar to an innovator monoclonal antibody with advanced liquid chromatography and mass spectrometry technologies. *MAbs.* 2(4):379–394.

53. Chelius D., Rehder DS., Bondarenko PV. (2005). Identification and characterization of deamidation sites in the conserved regions of human immunoglobulin gamma antibodies. *Anal Chem.* 77(18):6004–6011.

54. Liu YD., van Enk JZ., Flynn GC. (2009). Human antibody Fc deamidation in vivo. *Biologicals.* 37(5):313–322.

55. Sinha S., Zhang L., Duan S., et al. (2009). Effect of protein structure on deamidation rate in the Fc fragment of an IgG1 monoclonal antibody. *Protein Sci.* 18(8):1573–1584.

56. Huang L., Lu J., Wroblewski VJ., et al. (2005). In vivo deamidation characterization of monoclonal antibody by LC/MS/MS. *Anal Chem.* 77:1432–1439.

57. Diepold K., Bomans K., Wiedmann M., et al. (2012). Simultaneous assessment of Asp isomerization and Asn deamidation in recombinant antibodies by

LC-MS following incubation at elevated temperatures. *PLoS One.* 7(1):e30295.

58. Vlasak J., Bussat MC., Wang S., et al. (2009). Identification and characterization of asparagine deamidation in the light chain CDR1 of a humanized IgG1 antibody. *Anal Biochem.* 392(2):145–154.
59. Harris RJ., Kabakoff B., Macchi FD., et al. (2001). Identification of multiple sources of charge heterogeneity in a recombinant antibody. *J Chromatogr B.* 752(2001):233–245.
60. Liu H., Gaza-Bulseco G., Chumsae C. (2008). Glutamine deamidation of a recombinant monoclonal antibody. *Rapid Commun Mass Spectrom.* 22:4081–4088.
61. Deisenhofer J. (1981). Crystallographic refinement and atomic models of a human Fc fragment and its complex with fragment B of protein A from Staphylococcus aureus at 2.9- and 2.8-A resolution. *Biochemistry.* 20(9):2361–2370.
62. Nezlin R., Ghetie V. (2004). Interactions of immunoglobulins outside the antigen-combining site. *Adv Immunol.* 82:155–215.
63. Schroeder HW Jr., Cavacini L. (2010). Structure and function of immunoglobulins. *J Allergy Clin Immunol.* 125(2 Suppl 2):S41–S52.
64. Narciso JE., Uy ID., Cabang AB., et al. (2011). Analysis of the antibody structure based on high-resolution crystallographic studies. *N Biotechnol.* 28(5):435–447.
65. Jefferis R., Lund J., Pound J. (1998). IgG-Fc-mediated effector functions: molecular definition of interaction sites for effector ligands and the role of glycosylation. *Immunol Rev.* 163:59–76.
66. Acuner Ozbabacan SE., Engin HB., Gursoy A., et al. (2011). Transient protein-protein interactions. *Protein Eng Des Sel.* 24(9):635–648.
67. Sondermann P., Huber R., Oosthuizen V., et al. (2000). The 3.2-A crystal structure of the human IgG1 Fc fragment-Fc gammaRIII complex. *Nature.* 406(6793):267–273.
68. Radaev S., Motyka S., Fridman WH., et al. (2001). The structure of a human type III Fcgamma receptor in complex with Fc. *J Biol Chem.* 276(19):16469–16477.
69. Anumula KR. (2012). Quantitative glycan profiling of normal human plasma derived immunoglobulin and its fragments Fab and Fc. *J Immunol Methods.* 382(1–2):167–176.
70. Consortium for Functional Glycomics (2012). Symbol and Text Nomenclature for Representation of Glycan Structure. http://glycomics.scripps.edu/CFGnomenclature.pdf. Accessed May 5, 2016.
71. NIBRT (2012). Glycobase 3.2.4. http://glycobase.nibrt.ie/glycobase/about.action. Accessed May 5, 2016.
72. Reusch D., Haberger M., Selman MH., et al. (2013). High-throughput work flow for IgG Fc-glycosylation analysis of biotechnological samples. *Anal Biochem.* 432(2):82–89.

73. Huhn C., Selman MH., Ruhaak LR., et al. (2009). IgG glycosylation analysis. *Proteomics*. 9(4):882–913.
74. Flynn GC., Chen X., Liu YD., et al. (2010). Naturally occurring glycan forms of human immunoglobulins G1 and G2. *Mol Immunol*. 47(11–12):2074–2082.
75. Mimura Y., Ashton PR., Takahashi N., et al. (2007). Contrasting glycosylation profiles between Fab and Fc of a human IgG protein studied by electrospray ionization mass spectrometry. *J Immunol Methods*. 326(1–2):116–126.
76. Masuda K., Kubota T., Kaneko E., et al. (2007). Enhanced binding affinity for FcgammaRIIIa of fucose-negative antibody is sufficient to induce maximal antibody-dependent cellular cytotoxicity. *Mol Immunol*. 44(12):3122–3131.
77. Ferrara C., Brunker P., Suter T., et al. (2006). Modulation of therapeutic antibody effector functions by glycosylation engineering: influence of Golgi enzyme localization domain and co-expression of heterologous beta1, 4-*N*-acetylglucosaminyltransferase III and Golgi alpha-mannosidase II. *Biotechnol Bioeng*. 93(5):851–861.
78. Kubota T., Niwa R., Satoh M., et al. (2009). Engineered therapeutic antibodies with improved effector functions. *Cancer Sci*. 100(9):1566–1572.
79. Jefferis R., Lund J., Mizutani H., et al. (1990). A comparative study of the N-linked oligosaccharide structures of human IgG subclass proteins. *Biochem J*. 268(3):529–537.
80. Farooq M., Takahashi N., Arrol H., et al. (1997). Glycosylation of polyclonal and paraprotein IgG in multiple myeloma. *Glycoconj J*. 14(4):489–492.
81. Kibe T., Fujimoto S., Ishida C., et al. (1996). Glycosylation and placental transport of immunoglobulin G. *J Clin Biochem Nutr*. 21(1):57–63.
82. Alavi A., Arden N., Spector TD., et al. (2000). Immunoglobulin G glycosylation and clinical outcome in rheumatoid arthritis during pregnancy. *J Rheumatol*. 27(6):1379–1385.
83. Ercan A., Cui J., Chatterton DE., et al. (2010). Aberrant IgG galactosylation precedes disease onset, correlates with disease activity, and is prevalent in autoantibodies in rheumatoid arthritis. *Arthritis Rheum*. 62(8):2239–2248.
84. Holland M., Yagi H., Takahashi N., et al. (2006). Differential glycosylation of polyclonal IgG, IgG-Fc and IgG-Fab isolated from the sera of patients with ANCA-associated systemic vasculitis. *Biochim Biophys Acta*. 1760(4):669–677.
85. Scherer HU., van der Woude D., Ioan-Facsinay A., et al. (2010). Glycan profiling of anti-citrullinated protein antibodies isolated from human serum and synovial fluid. *Arthritis Rheum*. 62(6):1620–1629.
86. Bondt A., Selman MH., Deelder AM., et al. (2013). Association between galactosylation of immunoglobulin G and improvement of rheumatoid arthritis during pregnancy is independent of sialylation. *J Proteome Res*. 12(10):4522–4531.
87. Pascoe DE., Arnott D., Papoutsakis ET., et al. (2007). Proteome analysis of antibody-producing CHO cell lines with different metabolic profiles. *Biotechnol Bioeng*. 98(2):391–410.

88. Borrok MJ., Jung ST., Kang TH., et al. (2012). Revisiting the role of glycosylation in the structure of human IgG Fc. *ACS Chem Biol.* 7(9):1596–1602.
89. Yamaguchi Y., Nishimura M., Nagano M., et al. (2006). Glycoform-dependent conformational alteration of the Fc region of human immunoglobulin G1 as revealed by NMR spectroscopy. *Biochim Biophys Acta.* 1760(4):693–700.
90. Barb AW., Prestegard JH. (2011). NMR analysis demonstrates immunoglobulin G N-glycans are accessible and dynamic. *Nat Chem Biol.* 7(3):147–153.
91. Brekke OH., Michaelsen TE., Sandlie I. (1995). The structural requirements for complement activation by IgG: does it hinge on the hinge. *Immunol Today.* 16(2):85–90.
92. Padlan E. (1990). X-ray diffraction studies of antibody constant regions. In Fc Receptors and the Action of Antibodies, Metzger H., Ed. American Society for Microbiology: Washington, DC, pp 12–30.
93. Girardi E., Holdom MD., Davies AM., et al. (2009). The crystal structure of rabbit IgG-Fc. *Biochem J.* 417:77–83.
94. Mimura Y., Church S., Ghirlando R., et al. (2000). The influence of glycosylation on the thermal stability and effector function expression of human IgG1-Fc: properties of a series of truncated glycoforms. *Mol Immunol.* 37(12–13):697–706.
95. Mimura Y., Sondermann P., Ghirlando R., et al. (2001). Role of oligosaccharide residues of IgG1-Fc in Fc gamma RIIb binding. *J Biol Chem.* 276(49):45539–45547.
96. Krapp S., Mimura Y., Jefferis R., et al. (2003). Structural analysis of human IgG-Fc glycoforms reveals a correlation between glycosylation and structural integrity. *J Mol Biol.* 325(5):979–989.
97. Schaefer JV., Pluckthun A. (2012). Engineering aggregation resistance in IgG by two independent mechanisms: lessons from comparison of Pichia pastoris and mammalian cell expression. *J Mol Biol.* 417(4):309–335.
98. Kameoka D., Ueda T., Imoto T. (2011). Effect of the conformational stability of the CH2 domain on the aggregation and peptide cleavage of a humanized IgG. *Appl Biochem Biotechnol.* 164(5):642–654.
99. Zheng K., Bantog C., Bayer R. (2011). The impact of glycosylation on monoclonal antibody conformation and stability. *MAbs.* 3(6):568–576.
100. Allhorn M., Briceno JG., Baudino L., et al. (2010). The IgG-specific endoglycosidase EndoS inhibits both cellular and complement-mediated autoimmune hemolysis. *Blood.* 115(24):5080–5088.
101. Sjogren J., Struwe WB., Cosgrave EF., et al. (2013). EndoS2 is a unique and conserved enzyme of serotype M49 group A Streptococcus that hydrolyses N-linked glycans on IgG and α1-acid glycoprotein. *Biochem J.* 455(1):107–118.
102. Pound JD., Lund J., Jefferis R. (1993). Human Fc gamma RI triggering of the mononuclear phagocyte respiratory burst. *Mol Immunol.* 30(5):469–478.

103. Nesspor TC., Raju TS., Chin CN., et al. (2012). Avidity confers FcγR binding and immune effector function to aglycosylated immunoglobulin G1. *J Mol Recognit.* 25:147–154.

104. Crispin M. (2013). Therapeutic potential of deglycosylated antibodies. *Proc Natl Acad Sci U S A.* 110(25):10059–10060.

105. Hristodorov D., Fischer R., Linden L. (2013). With or without sugar? (A) glycosylation of therapeutic antibodies. *Mol Biotechnol.* 54(3):1056–1068.

106. Jefferis R. (2009). Glycosylation as a strategy to improve antibody-based therapeutics. *Nat Rev Drug Discov.* 8(3):226–234.

107. Li F., Vijayasankaran N., Shen AY., et al. (2010). Cell culture processes for monoclonal antibody production. *MAbs.* 2(5):466–479.

108. Daguet A., Watier H. (2011). 2nd Charles Richet et Jules Héricourt workshop: therapeutic antibodies and anaphylaxis; May 31–June 1, 2011, Tours, France. *MAbs.* 3(5):417–421.

109. Ghaderi D., Zhang M., Hurtado-Ziola N., et al. (2012). Production platforms for biotherapeutic glycoproteins. Occurrence, impact, and challenges of non-human sialylation. *Biotechnol Genet Eng Rev.* 28:147–175.

110. Galili U. (2013). Discovery of the natural anti-Gal antibody and its past and future relevance to medicine. *Xenotransplantation.* 20(3):138–147.

111. Bosques CJ., Collins BE., Meador JW., et al. (2010). Chinese hamster ovary cells can produce galactose-α-1,3-galactose antigens on proteins. *Nat Biotechnol.* 28(11):1153–1156.

112. Maeda E., Kita S., Kinoshita M., et al. (2012). Analysis of nonhuman N-glycans as the minor constituents in recombinant monoclonal antibody pharmaceuticals. *Anal Chem.* 84(5):2373–2379.

113. Yu M., Hu Z., Pacis E., et al. (2011). Understanding the intracellular effect of enhanced nutrient feeding toward high titer antibody production process. *Biotechnol Bioeng.* 108(5):1078–1088.

114. Pacis E., Yu M., Autsen J., et al. (2011). Effects of cell culture conditions on antibody N-linked glycosylation—what affects high mannose 5 glycoform. *Biotechnol Bioeng.* 108(10):2348–2358.

115. Shields RL., Lai J., Keck R., et al. (2002). Lack of fucose on human IgG1 N-linked oligosaccharide improves binding to human Fcgamma RIII and antibody-dependent cellular toxicity. *J Biol Chem.* 277(30):26733–26740.

116. Niwa R., Natsume A., Uehara A., et al. (2005). IgG subclass-independent improvement of antibody-dependent cellular cytotoxicity by fucose removal from Asn297-linked oligosaccharides. *J Immunol Methods.* 306(1–2):151–160.

117. Shibata-Koyama M., Iida S., Misaka H., et al. (2009). Nonfucosylated rituximab potentiates human neutrophil phagocytosis through its high binding for FcgammaRIIIb and MHC class II expression on the phagocytic neutrophils. *Exp Hematol.* 37(3):309–321.

118. Yamane-Ohnuki N., Satoh M. (2009). Production of therapeutic antibodies with controlled fucosylation. *MAbs.* 1(3):230–236.

119. Subramaniam JM., Whiteside G., McKeage K., et al. (2012). Mogamulizumab: first global approval. *Drugs.* 72(9):1293–1298.

120. Umaña P., Jean-Mairet J., Moudry R., et al. (1999). Engineered glycoforms of an anti-neuroblastoma. IgG1 with optimized antibody-dependent cellular cytotoxic activity. *Nat Biotechnol.* 17(2):176–180.

121. Davies J., Jiang L., Pan LZ., et al. (2004). Expression of GTIII in a recombinant anti-CD20 CHO production cell line: expression of antibodies of altered glycoforms leads to an increase in ADCC thro' higher affinity for FcRIII. *Biotechnol Bioeng.* 74(4):288–294.

122. Mössner E., Brünker P., Moser S., et al. (2010). Increasing the efficacy of CD20 antibody therapy through the engineering of a new type II anti-CD20 antibody with enhanced direct and immune effector cell–mediated B-cell cytotoxicity. *Blood.* 115(22):4393–4402.

123. FDA (2013). Gazyva (Obinutuzumab). U.S. Food and Drug. Available at http://www.fda.gov/Drugs/InformationOnDrugs/ApprovedDrugs/ucm373263.htm. Accessed May 5, 2016.

124. Ogorek C., Jordan I., Sandig V., et al. (2012). Fucose-targeted glycoengineering of pharmaceutical cell lines. *Methods Mol Biol.* 907:507–517.

125. Beck A., Reichert JM. (2012). Marketing approval of mogamulizumab: a triumph for glyco-engineering. *MAbs.* 4(4):419–425.

126. Kobata A. (2008). The N-linked sugar chains of human immunoglobulin G: their unique pattern, and their functional roles. *Biochim Biophys Acta.* 1780(3):472–478.

127. Peipp M., Lammerts van Bueren JJ., Schneider-Merck T., et al. (2008). Antibody fucosylation differentially impacts cytotoxicity mediated by NK and PMN effector cells. *Blood.* 112(6):2390–2399.

128. Derer S., Kellner C., Berger S., et al. (2012). Fc engineering: design, expression, and functional characterization of antibody variants with improved effector function. *Methods Mol Biol.* 907:519–536.

129. Nakagawa T., Natsume A., Satoh M., et al. (2010). Non-fucosylated anti-CD20 antibody potentially induces apoptosis in lymphoma cells through enhanced interaction with FcgammaRIIIb on neutrophils. *Leuk Res.* 34(5):666–671.

130. Zeck A., Pohlentz G., Schlothauer T., et al. (2011). Cell type-specific and site directed N-glycosylation pattern of FcγRIIIa. *J Proteome Res.* 10(7):3031–3039.

131. Natsume A., In M., Takamura H., et al. (2008). Engineered antibodies of IgG1/IgG3 mixed isotype with enhanced cytotoxic activities. *Cancer Res.* 68(10):3863–3872.

132. Haryadi R., Zhang P., Chan KF., et al. (2013). CHO-gmt5, a novel CHO glycosylation mutant for producing afucosylated and asialylated recombinant antibodies. *Bioengineered.* 4(2):90–94.

133. Zhou Q., Shankara S., Roy A., et al. (2008). Development of a simple and rapid method for producing non-fucosylated oligomannose containing antibodies with increased effector function. *Biotechnol Bioeng.* 99(3):652–665.

134. Zhang P., Chan KF., Haryadi R., et al. (2013). CHO glycosylation mutants as potential host cells to produce therapeutic proteins with enhanced efficacy. *Adv Biochem Eng Biotechnol.* 131:63–87.

135. Yu C., Crispin M., Sonnen AF., et al. (2011). Use of the α-mannosidase I inhibitor kifunensine allows the crystallization of apo CTLA-4 homodimer produced in long-term cultures of Chinese hamster ovary cells. *Acta Crystallogr Sect F Struct Biol Cryst Commun.* 67(Pt 7):785–789.

136. Gloster TM., Vocadlo DJ. (2012). Developing inhibitors of glycan processing enzymes as tools for enabling glycobiology. *Nat Chem Biol.* 8(8):683–694.

137. Sealover NR., Davis AM., Brooks JK., et al. (2013). Engineering Chinese hamster ovary (CHO) cells for producing recombinant proteins with simple glycoforms by zinc-finger nuclease (ZFN)-mediated gene knockout of mannosyl (alpha-1,3-)-glycoprotein beta-1,2-*N*-acetylglucosaminyltransferase (Mgat1). *J Biotechnol.* 167(1):24–32.

138. Reeves PJ., Callewaert N., Contreras R., et al. (2002). Structure and function in rhodopsin: high-level expression of rhodopsin with restricted and homogeneous N-glycosylation by a tetracycline-inducible *N*-acetylglucosaminyltransferase I-negative HEK293S stable mammalian cell line. *Proc Natl Acad Sci U S A.* 99(21):13419–13424.

139. Hossler P. (2012). Protein glycosylation control in mammalian cell culture: past precedents and contemporary prospects. *Adv Biochem Eng Biotechnol.* 127:187–219.

140. Franze R., Hirashima C., Link T., et al. (2012). Method for the production of a glycosylated immunoglobulin. Patent No: EP2493922 A1. http://www.google.com/patents/EP2493922A1. Accessed May 5, 2016.

141. Zhong X., Cooley C., Seth N., et al. (2012). Engineering novel Lec1 glycosylation mutants in CHO–DUKX cells: molecular insights and effector modulation of *N*-acetylglucosaminyltransferase I. *Biotechnol Bioeng.* 109(7):1723–1734.

142. Doores KJ., Bonomelli C., Harvey DJ., et al. (2010). Envelope glycans of immunodeficiency virions are almost entirely oligomannose antigens. *Proc Natl Acad Sci U S A.* 107(31):13800–13805.

143. Van Patten SM., Hughes H., Huff MR., et al. (2007). Effect of mannose chain length on targeting of glucocerebrosidase for enzyme replacement therapy of Gaucher disease. *Glycobiology.* 17(5):467–478.

144. Lingg N., Zhang P., Song Z., et al. (2012). The sweet tooth of biopharmaceuticals: importance of recombinant protein glycosylation analysis. *Biotechnol J.* 7(12):1462–1472.

145. Yuk IH., Huynh H., Leach K., et al. (2006). Cell culture efforts to reduce glycation in recombinant humanized antibody. *Microb Cell Fact.* 5(Suppl 1):P57.

146. Dolhofer-Bliesener R., Gerbitz KD. (1990). Impairment by glycation of immunoglobulin G Fc fragment function. *Sc and J Clin Lab Invest.* 50(7):739–746.

147. Lapolla A., Fedele D., Garbeglio M., et al. (2000). Matrix-assisted laser desorption/ionization mass spectrometry, enzymatic digestion, and molecular modeling in the study of nonenzymatic glycation of IgG. *J Am Soc Mass Spectrom.* 11:153–159.

148. Gadgil HS., Bondarenko PV., Pipes G., et al. (2007). The LC/MS analysis of glycation of IgG molecules in sucrose containing formulations. *J Pharm Sci.* 96:2607–2621.

149. Gillis C., Gouel-Chéron A., Jönsson F., et al. (2014). Contribution of human FcγRs to disease with evidence from human polymorphisms and transgenic animal studies. *Front Immunol.* 5:254.

150. Lux A., Yu X., Scanlan CN., et al. (2013). Impact of immune complex size and glycosylation on IgG binding to human FcγRs. *J Immunol.* 190(8):4315–4323.

151. Van de Winkel JG. (2010). Fc receptors: role in biology and antibody therapy. *Immunol Lett.* 128(1):4–5.

152. Anthony RM., Wermeling F., Ravetch JV. (2012). Novel roles for the IgG Fc glycan. *Ann N Y Acad Sci.* 1253:170–180.

153. Cartron G., Dacheux L., Salles G., et al. (2002). Therapeutic activity of humanized anti-CD20 monoclonal antibody and polymorphism in IgG Fc receptor FcgammaRIIIa gene. *Blood.* 99(3):754–758.

154. Sanz I., Anolik JH., Looney RJ. (2007). B cell depletion in autoimmune disease. *Front Biosci.* 12:2546–2567.

155. Louis E., El Ghoul Z., Vermeire S., et al. (2004). Association between polymorphism in IgG Fc receptor IIIa coding gene and biological response to infliximab in Crohn's disease. *Aliment Pharmacol Ther.* 19(5):511–519.

156. Miescher S., Spycher MO., Amstutz H., et al. (2004). A single recombinant anti-RhD IgG prevents RhD immunization: association of RhD-positive red blood cell clearance rate with polymorphisms in the FcgammaRIIA and FcgammaIIIA genes. *Blood.* 103(11):4028–4035.

157. Jefferis R. (2011). Aggregation, immune complexes and immunogenicity. *MAbs.* 3(6):503–504.

158. Ferrara C., Grau S., Jäger C., et al. (2011). Unique carbohydrate-carbohydrate interactions are required for high affinity binding between FcgammaRIII and antibodies lacking core fucose. *Proc Natl Acad Sci U S A.* 108(31):12669–12674.

159. Mizushima T., Yagi H., Takemoto E., et al. (2011). Structural basis for improved efficacy of therapeutic antibodies on defucosylation of their Fc glycans. *Genes Cells.* 16(11):1071–1080.

160. Radaev S., Sun P. (2002). Recognition of immunoglobulins by Fcgamma receptors. *Mol Immunol.* 38(14):1073–1083.

161. Ramsland PA., Farrugia W., Bradford TM., et al. (2011). Structural basis for Fc gammaRIIa recognition of human IgG and formation of inflammatory signaling complexes. *J Immunol.* 187(6):3208–3217.

162. Ward ES., Ober RJ. (2009). Chapter 4: Multitasking by exploitation of intracellular transport functions: the many faces of FcRn. *Adv Immunol.* 103:77–115.

163. Tesar DB., Bjoerkman PJ. (2010). An intracellular traffic jam: Fc receptor-mediated transport of immunoglobulin G. *Curr Opin Struct Biol.* 20(2):226–233.

164. Kuo TT., de Muinck EJ., Claypool SM., et al. (2009). N-Glycan moieties in neonatal Fc receptor determine steady-state membrane distribution and directional transport of IgG. *J Biol Chem.* 284(13):8292–8300.

165. Rath T., Kuo TT., Baker K., et al. (2013). The immunologic functions of the neonatal Fc receptor for IgG. *J Clin Immunol.* 33(Suppl 1):S9–S17.

166. Mathiesen L., Nielsen LK., Andersen JT., et al. (2013). Maternofetal transplacental transport of recombinant IgG antibodies lacking effector functions. *Blood.* 122(7):1174–1181.

167. Einarsdottir HK., Selman MH., Kapur R., et al. (2013). Comparison of the Fc glycosylation of fetal and maternal immunoglobulin G. *Glycoconj J.* 30(2):147–157.

168. Lee CH., Woo JH., Cho KK., et al. (2007). Expression and characterization of human growth hormone-Fc fusion proteins for transcytosis induction. *Biotechnol Appl Biochem.* 46:211–217.

169. Vllasaliu D., Alexander C., Garnett M., et al. (2012). Fc-mediated transport of nanoparticles across airway epithelial cell layers. *J Control Release.* 158(3):479–486.

170. Vallee S., Rakhe S., Reidy T., et al. (2012). Pulmonary administration of interferon Beta-1a-fc fusion protein in non-human primates using an immunoglobulin transport pathway. *J Interferon Cytokine Res.* 32(4):178–184.

171. Stapleton NM., Andersen JT., Stemerding AM., et al. (2011). Competition for FcRn-mediated transport gives rise to short half-life of human IgG3 and offers therapeutic potential. *Nat Commun.* 2:599.

172. Oganesyan V., Damschroder MM., Leach W., et al. (2009). Structural characterization of a mutated, ADCC-enhanced human Fc fragment. *Mol Immunol.* 46(8–9):1750–1755.

173. Zalevsky J., Chamberlain AK., Horton HM., et al. (2010). Enhanced antibody half-life improves in vivo activity. *Nat Biotechnol.* 28(2):157–159.

174. Kapur R., Einarsdottir HK., Vidarsson G. (2014). IgG-effector functions: "the good, the bad and the ugly". *Immunol Lett.* 160(2):139–144.

175. Jefferis R., Lefranc MP. Human immunoglobulin allotypes: possible implications for immunogenicity. *MAbs.* 1(4):332–338.

176. Jones AJ., Papac DI., Chin EH., et al. (2007). Selective clearance of glycoforms of a complex glycoprotein pharmaceutical caused by terminal *N*-acetylglucosamine is similar in humans and cynomolgus monkeys. *Glycobiology*. 17(5):529–540.

177. Keck R., Nayak N., Lerner L., et al. (2008). Characterization of a complex glycoprotein whose variable metabolic clearance in humans is dependent on terminal *N*-acetylglucosamine content. *Biologicals*. 36(1):49–60.

178. Zipfel PF. (2009). Complement and immune defense: from innate immunity to human diseases. *Immunol Lett*. 126(1–2):1–7.

179. Degn SE., Thiel S. (2013). Humoral pattern recognition and the complement system. *Scand J Immunol*. 78(2):181–193.

180. Voice JK., Lachmann PJ. (1997). Neutrophil Fc gamma and complement receptors involved in binding soluble IgG immune complexes and in specific granule release induced by soluble IgG immune complexes. *Eur J Immunol*. 27(10):2514–2523.

181. Idusogie EE., Wong PY., Presta LG., et al. (2001). Engineered antibodies with increased activity to recruit complement. *J Immunol*. 166(4):2571–2575.

182. Natsume A., Shimizu-Yokoyama Y., Satoh M., et al. (2009). Engineered anti-CD20 antibodies with enhanced complement-activating capacity mediate potent anti-lymphoma activity. *Cancer Sci*. 100(12):2411–2418.

183. Moore GL., Chen H., Karki S., et al. (2010). Engineered Fc variant antibodies with enhanced ability to recruit complement and mediate effector functions. *MAbs*. 2(2):181–189.

184. Aalberse RC., Stapel SO., Schuurman J., et al. (2009). Immunoglobulin G4: an odd antibody. *Clin Exp Allergy*. 39(4):469–477.

185. Morgan A., Jones ND., Nesbitt AM., et al. (1995). The N-terminal end of the CH2 domain of chimeric human IgG1 anti-HLA-DR is necessary for C1q, Fc gamma RI and Fc gamma RIII binding. *Immunology*. 86(2):319–324.

186. Diebolder CA., Beurskens FJ., de Jong RN., et al. (2014). Complement is activated by IgG hexamers assembled at the cell surface. *Science*. 343(6176):1260–1263.

187. Shinkawa T., Nakamura K., Yamane N., et al. (2003). The absence of fucose but not the presence of galactose or bisecting *N*-acetylglucosamine of human IgG1 complex-type oligosaccharides shows the critical role of enhancing antibody-dependent cellular cytotoxicity. *J Biol Chem*. 278(5):3466–3473.

188. Yamada E., Tsukamoto Y., Sasaki R., et al. (1997). Structural changes of immunoglobulin G oligosaccharides with age in healthy human serum. *Glycoconj J*. 14(3):401–405.

189. Ruhaak LR., Uh HW., Beekman M., et al. (2011). Plasma protein N-glycan profiles are associated with calendar age, familial longevity and health. *J Proteome Res*. 10(4):1667–1674.

190. Holland M., Takada K., Okumoto T., et al. (2002). Hypogalactosylation of serum IgG in patients with ANCA-associated systemic vasculitis. *Clin Exp Immunol.* 129(1):183–190.
191. Wuhrer M., Porcelijn L., Kapur R., et al. (2009). Regulated glycosylation patterns of IgG during alloimmune responses against human platelet antigens. *J Proteome Res.* 8(2):450–456.
192. Collins ES., Galligan MC., Saldova R., et al. (2013). Glycosylation status of serum in inflammatory arthritis in response to anti-TNF treatment. *Rheumatology (Oxford).* 52(9):1572–1582.
193. Farooq M., Takahashi N., Drayson M., et al. (1998). A longitudinal study of glycosylation of a human IgG3 paraprotein in a patient with multiple myeloma. *Adv Exp Med Biol.* 435:95–103.
194. van Berkel PH., Gerritsen J., Perdok G., et al. (2009). N-linked glycosylation is an important parameter for optimal selection of cell lines producing biopharmaceutical human IgG. *Biotechnol Prog.* 25(1):244–251.
195. Wacker C., Berger CN., Girard P., et al. (2011). Glycosylation profiles of therapeutic antibody pharmaceuticals. *Eur J Pharm Biopharm.* 79(3):503–507.
196. Raju TS., Jordan RE. (2012). Galactosylation variations in marketed therapeutic antibodies. *mAbs.* 4(3):385–391.
197. Barb AW., Meng L., Gao Z., et al. (2012). NMR characterization of immunoglobulin G Fc glycan motion on enzymatic sialylation. *Biochemistry.* 51(22):4618–4626.
198. Houde D., Engen JR. (2013). Conformational analysis of recombinant monoclonal antibodies with hydrogen/deuterium exchange mass spectrometry. *Methods Mol Biol.* 988:269–289.
199. Lund J., Takahashi N., Pound JD., et al. (1996). Multiple interactions of IgG with its core oligosaccharide can modulate recognition by complement and human Fc gamma receptor I and influence the synthesis of its oligosaccharide chains. *J Immunol.* 157(11):4963–4969.
200. Boyd PN., Lines AC., Patel AK. (1995). The effect of the removal of sialic acid, galactose and total carbohydrate on the functional activity of Campath-1H. *Mol Immunol.* 32(17–18):1311–1318.
201. Sondermann P., Pincetic A., Maamary J., et al. (2013). General mechanism for modulating immunoglobulin effector function. *Proc Natl Acad Sci U S A.* 110(24):9868–9872.
202. Oaks M., Taylor S., Shaffer J. (2013). Autoantibodies targeting tumor-associated antigens in metastatic cancer: sialylated IgGs as candidate anti-inflammatory antibodies. *Oncoimmunology.* 2(6):e24841.
203. Bayry J., Bansal K., Kazatchkine MD., et al. (2009). DC-SIGN and alpha2,6-sialylated IgG Fc interaction is dispensable for the anti-inflammatory activity of IVIg on human dendritic cells. *Proc Natl Acad Sci U S A.* 106(9):E24.

204. Käsermann F., Boerema DJ., Rüegsegger M., et al. (2012). Analysis and functional consequences of increased Fab-sialylation of intravenous immunoglobulin (IVIG) after lectin fractionation. *PLoS One.* 7(6):e37243.

205. Yu X., Vasiljevic S., Mitchell DA., et al. (2013). Dissecting the molecular mechanism of IVIg therapy: the interaction between serum IgG and DC-SIGN is independent of antibody glycoform or Fc domain. *J Mol Biol.* 425(8):1253–1258.

206. Stanfield RL. (2014). Determination of antibody structures. *Methods Mol Biol.* 1131:395–406.

207. Youings A., Chang SC., Dwek RA., et al. (1996). Site-specific glycosylation of human immunoglobulin G is altered in four rheumatoid arthritis patients. *Biochem J.* 314(Pt 2):621–630.

208. Coloma MJ., Trinh RK., Martinez AR., et al. (1999). Position effects of variable region carbohydrate on the affinity and in vivo behaviour of an anti-(1→6) dextran antibody. *J Immunol.* 162:2162–2170.

209. Jacquemin M. (2010). Variable region heavy chain glycosylation determines the anticoagulant activity of a factor VIII antibody. *Haemophilia.* 16(102):16–19.

210. Spencer S., Bethea D., Raju TS., et al. (2012). Solubility evaluation of murine hybridoma antibodies. *MAbs.* 4(3):319–325.

211. Qian J., Liu T., Yang L., et al. (2007). Structural characterization of N-linked oligosaccharides on monoclonal antibody cetuximab by the combination of orthogonal matrix-assisted laser desorption/ionization hybrid quadrupole-quadrupole time-of-flight tandem mass spectrometry and sequential enzymatic digestion. *Anal Biochem.* 364:8–18.

212. Wiegandt A., Meyer B. (2014). Unambiguous characterization of N-glycans of monoclonal antibody cetuximab by integration of LC-MS/MS and ^{1}H NMR spectroscopy. *Anal Chem.* 86(10):4807–4814.

213. Chung CH., Mirakhur B., Chan E., et al. (2008). Cetuximab-induced anaphylaxis and IgE specific for galactose-alpha-1,3-galactose. *N Engl J Med.* 358(11):1109–1117.

214. Lammerts van Bueren JJ., Rispens T., Verploegen S., et al. (2011). Anti-galactose-α-1,3-galactose IgE from allergic patients does not bind α-galactosylated glycans on intact therapeutic antibody Fc domains. *Nat Biotechnol.* 29(7):574–576.

215. Pointreau Y., Commins SP., Calais G., et al. (2012). Fatal infusion reactions to cetuximab: role of immunoglobulin e-mediated anaphylaxis. *J Clin Oncol.* 30(3):334.

216. Mullins RJ., James H., Platts-Mills TA., et al. (2012). Relationship between red meat allergy and sensitization to gelatin and galactose-α-1,3-galactose. *J Allergy Clin Immunol.* 129(5):1334–1342.

217. Berg EA., Platts-Mills TA., Commins SP. (2014). Drug allergens and food—the cetuximab and galactose-α-1,3-galactose story. *Ann Allergy Asthma Immunol.* 112(2):97–101.

218. Huang W., Giddens J., Fan SQ., et al. (2012). Chemoenzymatic glycoengineering of intact IgG antibodies for gain of functions. *J Am Chem Soc.* 134(29):12308–12318.

219. Carter P., Presta L., Gorman CM., et al. (1992). Humanization of an anti-p185HER2 antibody for human cancer therapy. *Proc Natl Acad Sci U S A.* 89(10):4285–4289.

220. Lim A., Reed-Bogan A., Harmon BJ. (2008). Glycosylation profiling of a therapeutic recombinant monoclonal antibody with two N-linked glycosylation sites using liquid chromatography coupled to a hybrid quadrupole time-of-flight mass spectrometer. *Anal Biochem.* 375(2):163–172.

221. Millward TA., Heitzmann M., Bill K., et al. (2008). Effect of constant and variable domain glycosylation on pharmacokinetics of therapeutic antibodies in mice. *Biologicals.* 36(1):41–47.

222. Wu SJ., Luo J., O'Neil KT., et al. (2010). Structure-based engineering of a monoclonal antibody for improved solubility. *Protein Eng Des Sel.* 23:643–651.

223. Woof JM., Russell MW. (2011). Structure and function relationships in IgA. *Mucosal Immunol.* 4(6):590–597.

224. Jadhav V., Hackl M., Klanert G., et al. (2014). Stable overexpression of miR-17 enhances recombinant protein production of CHO cells. *J Biotechnol.* 175(100):38–44.

225. Stork R., Zettlitz KA., Müller D., et al. (2008). N-glycosylation as novel strategy to improve pharmacokinetic properties of bispecific single-chain diabodies. *J Biol Chem.* 283(12):7804–7812.

226. Ishii Y., Murakami J., Sasaki K., et al. (2014). Efficient folding/assembly in Chinese hamster ovary cells is critical for high quality (low aggregate content) of secreted trastuzumab as well as for high production: stepwise multivariate regression analyses. *J Biosci Bioeng.* 118(2):223–230.

227. Bakema JE., van Egmond M. (2011). The human immunoglobulin A Fc receptor FcαRI: a multifaceted regulator of mucosal immunity. *MAbs.* 3(4):352–361.

228. Lohse S., Derer S., Beyer T., et al. (2011). Recombinant dimeric IgA antibodies against the epidermal growth factor receptor mediate effective tumor cell killing. *J Immunol.* 186(6):3770–3778.

229. Gautam S., Loh KC. (2011). Immunoglobulin-M purification—challenges and perspectives. *Biotechnol Adv.* 29(6):840–849.

230. Tchoudakova A., Hensel F., Murillo A., et al. (2009). High level expression of functional human IgMs in human PER.C6 cells. *MAbs.* 1(2):163–171.

231. Josephs DH., Spicer JF., Karagiannis P., et al. (2014). IgE immunotherapy: a novel concept with promise for the treatment of cancer. *MAbs.* 6(1):54–72.

232. Odani-Kawabata N., Takai-Imamura M., Katsuta O., et al. (2010). ARG098, a novel anti-human Fas antibody, suppresses synovial hyperplasia and prevents cartilage destruction in a severe combined immunodeficient-HuRAg mouse model. *BMC Musculoskelet Disord.* 11:221.

233. Spillner E., Plum M., Blank S., et al. (2012). Recombinant IgE antibody engineering to target EGFR. *Cancer Immunol Immunother.* 61(9):1565–1573.

234. Rudman SM., Josephs DH., Cambrook H., et al. (2011). Harnessing engineered antibodies of the IgE class to combat malignancy: initial assessment of FcεRI-mediated basophil activation by a tumour-specific IgE antibody to evaluate the risk of type I hypersensitivity. *Clin Exp Allergy.* 41(10):1400–1413.

235. Karagiannis SN., Josephs DH., Karagiannis P., et al. (2012). Recombinant IgE antibodies for passive immunotherapy of solid tumours: from concept towards clinical application. *Cancer Immunol Immunother.* 61(9):1547–1564.

236. Arnold JN., Wormald MR., Sim RB., et al. (2007). The impact of glycosylation on the biological function and structure of human immunoglobulins. *Annu Rev Immunol.* 25:21–50.

237. Davis DL., Schiel J., Borisov O., Eds. (2014). Current State of the Art & Emerging Technologies for the Characterisation of Monoclonal Antibodies. American Chemical Society: Washington, DC.

238. Jefferis R. (2014) Monoclonal antibodies: mechanisms of action. In: State-of-the-Art & Emerging Technologies for the Characterisation of Monoclonal Antibodies, Volume 1. Monoclonal Antibody Therapeutics: Structure, Function, and Regulatory Space (Davis DL, Schiel J, Borisov O, Eds.). ACS Symposium Series; American Chemical Society: Washington, DC, 2014.

6

The Pharmacology, Pharmacokinetics, and Pharmacodynamics of Antibodies

Ningning Xu[1], Meimei Liu[2], and Margaret Liu[1]

[1] *Department of Chemical and Biological Engineering, University of Alabama, Tuscaloosa, AL, USA*
[2] *Thermo Fisher Scientific Inc., Grand Island, NY, USA*

6.1 Summary

Pharmacological studies of anticancer monoclonal antibodies (mAb) focus on the interaction between a patient population and the biopharmaceutical agent. Specifically, these investigations describe the interaction between mAbs and abnormal cells (e.g., cancer cells) as opposed to those dynamics involving normal cells. It deals with the uses, effects, and modes of action of mAbs on the body.

Pharmacokinetics (PK) is concerned with the biological processes that a drug undergoes in the intact organism, that is, what the body does to a drug.

Pharmacodynamics (PD) focuses on the therapeutic action of a drug on an organism, that is, pharmacologic response, as well as the duration and the magnitude of the response associated with the concentration of drug at an effectual site of the organism—in sum, what the body does to the drug.

This review covers clinical pharmacology, PK, and PD of anticancer antibodies, providing a set of general principles describing *in vivo* antibody behavior, to facilitate initial drug development leading to future clinical applications.

6.2 Introduction

The US Food and Drug Administration (FDA) has approved a number of therapeutic mAbs for anticancer treatment. Generally speaking, the mAbs provide better clinical efficacy and/or fewer adverse side effects than conventional cancer therapy. Compared to small molecule drugs, mAbs exhibit unique characteristics in their pharmacology, PK, and PD. Specifically, the interaction

Biosimilars of Monoclonal Antibodies: A Practical Guide to Manufacturing, Preclinical, and Clinical Development, First Edition. Edited by Cheng Liu and K. John Morrow, Jr.

between a living organism and a mAb can be revealed through pharmacology studies investigating the molecular mechanisms of how mAbs target specific antigens associated with core pathways in cancer progression to kill cancer cells. Most mAbs are cleared by multiple pathways in human body, including both clearance mediated by Fc receptor and clearance mediated by the pharmacological targets. This biological process of mAb clearance is analyzed by PK. The anticancer mAbs can directly kill cancer cells through ligand–receptor action or indirectly destroy cancer cells through immune system reaction, that is, antibody-dependent cellular cytotoxicity (ADCC) and/or cell-dependent cytotoxicity (CDC). The resulting pharmacologic response, as well as the duration and magnitude of the response associated with the concentration of the pharmaceutical agent, are investigated by PD.

6.3 Pharmacology of Anticancer MAbs

A number of mAbs have been developed, commercialized, and extensively applied to treat various malignancies including breast cancer, colorectal cancer, non-Hodgkin's lymphoma, leukemia, lung cancer, kidney cancer, and neck/head cancer [45]. In the following sections, their clinical pharmacology is discussed related to their different functions in cancer treatment.

The human epidermal growth factor receptor 2 (HER2) is a 185-kDa transmembrane receptor, overexpressed in about 15–23% of breast cancers. Breast cancer is the second leading cause of death in women, and HER2 provides a good target for antibreast cancer mAbs [41]. For example, trastuzumab (Herceptin®) has been developed to suppress the proliferation of breast cancer cells overexpressing HER2. The recombinant humanized IgG1 Kappa of trastuzumab can attach to HER2 and work as a mediator of ADCC. Ado-trastuzumab emtansine (Kadcyla®) is an antibody-drug conjugate constituted of trastuzumab and microtubule inhibitor (DM1). In addition to ADCC, the conjugated antibody may mobilize the receptor-mediated internalization and subsequent lysosomal degradation after binding to the subdomain IV of HER2. The DM1 binds to the tubulin intracellularly and interrupts the microtubule networks of the tumor cells, which further leads to the arrest of cell cycle and induces the apoptotic cell death. Another anti-HER2 mAb, pertuzumab (Perjeta®), can target the subdomain II of HER2 and inhibit the ligand-dependent heterodimerization of HER2 by blocking the ligand-initiated intracellular signals, that is, mitogen-activated protein (MAP) kinase and phosphoinositide 3-kinase (PI3K). The pertuzumab causes apoptosis and mediates ADCC as well.

Colorectal cancer is the third most common cancer and the fourth lead cause of death. Bevacizumab (Avastin®) has been developed to treat first-line metastatic colorectal cancer by ameliorating immune response, controlling the disease process and thereby improving survival [15]. The *in vitro* model of angiogenesis proposes that the interaction of vascular endothelial growth factor (VEGF) with its receptor (e.g., Flt-1 and KDR) could lead to the proliferation of endothelial cells and the formation of new blood vessels. Bevacizumab can target VEGF, preventing the binding of VEGF to its receptor. In addition to VEGF, the epidermal growth factor receptor (EGFR), which is a transmembrane glycoprotein expressed in many normal epithelial tissues and human cancers such as head/neck, colon, and rectum cancers, is another effective target to treat colorectal cancer. The EGFR can regulate the transcription of the genes involved in cellular growth and survival, motility, and proliferation. The cetuximab (Erbitux®) is a chimeric IgG1 that can specifically bind to the EGFR on both normal and tumor cells. The binding of cetuximab blocks the epidermal growth factor (EGF) and other ligands and inhibits the phosphorylation and activation of receptor-associated kinases [22]. Thus, cetuximab inhibits the cell growth of colorectal cancer by inducing apoptosis, decreasing the matrix metalloproteinase, and inhibiting the production of VEGF.

The CD20 antigen is a 35-kD transmembrane protein expressed on the surface of B lymphocytes. It plays a role in the activation of cell cycle initiation and differentiation and also serves as a calcium ion channel. The CD20 antigen is found on the majority of B-cell non-Hodgkin's lymphomas (NHL), but it is not expressed on hematopoietic stem cells, pro-B cells, normal plasma cells, or other normal tissues. The B cells may take part in the autoimmune/inflammatory process and are vital to the pathogenesis of rheumatoid arthritis and associated chronic synovitis. Rituximab (Rituxan®) has been used to treat the CD20-positive NHL and chronic lymphocytic leukemia (CLL) [3].

Ofatumumab (Arzerra®) binds to both the small and large extracellular loops of the CD20 antigen. The anticancer mechanism of rituximab and ofatumumab is similar, that is, the Fab fragment of rituximab binds to the CD20 antigen of B lymphocytes, and the Fc fragment initiates the B-cell lysis *in vitro* by initiating the function of immune effector. In addition, ofatumumab can activate the intracellular death signaling pathways and/or the activation of the complement cascade. Both ^{90}Y-Ibritumomab tiuxetan (Zevalin®) and ^{131}I-Tositumomab (Bexxar®) have been applied to treat NHL through radioisotopes that release ionizing radiation to CD-expressing lymphocytes or their neighboring cells to cause cell death, ADCC, CDC, and CD20-mediated apoptosis.

To treat the relapsed or refractory Hodgkin's lymphoma and systemic anaplastic lymphoma, brentuximab vedotin has been developed. This antibody-drug conjugate consists of an anti-CD30 chimeric IgG1 antibody and a microtubule-disrupting agent MMAE [38]. It is directed against the CD30 antigen of tumor cells and initiates the internalization of ADC-CD30 complex and the release of MMAE by proteolytic cleavage. After binding to tubulin, the MMEA interrupts the microtubule network inside cell, arrests cell cycle, and causes apoptotic death.

The CD33 antigen is a transmembrane receptor expressed on the cell surface of leukemic blasts in the majority of patients with acute myeloid leukemia (AML). Gemtuzumab ozogamicin (Mylotarg®) can directly bind to the CD33 antigen expressed by hematopoietic cells, which results in the formation of an internalization complex [4]. During internalization, the calicheamicin derivative is liberated inside the lysosomes of the myeloid cell, which can bind to the DNA in the minor groove and break the double strand of DNA, thus causing cell death.

Ipilimumab (Yervoy®) was developed to treat melanoma through T-cell-mediated antitumor immune responses. Ipilimumab can bind to CTLA-4, a protein receptor that downregulates the activation of the T cell, and inhibit the interaction of CTLA-4 with its ligands, CD80/CD86. Thus, blocking CTLA-4 by ipilimumab improves the activation and proliferation of the T cell.

6.4 Antibody Pharmacokinetics

Most antibody-based biopharmaceuticals are administered via intravenous infusion, intramuscular injection, or subcutaneous injection. Postadministration, the concentration of antibodies undergoes a biexponential decline in plasma, while its distribution into tissues is slow because of the large molecular size. Antibodies can be eliminated through fluid-phase endocytosis followed by the lysosomal degradation. Here we discuss the fundamental processes that influence the *in vivo* PK including adsorption, distribution, and clearance (metabolism and excretion); the PK models; and the experimental or clinical PK data of anticancer biopharmaceuticals.

6.4.1 PK Process

Absorption: The mAbs are systemically absorbed when convectively passing through lymphatic vessels to reach the blood and diffusing across blood vessels around the injection sites. The lymphatic transport is the major absorption route of large protein (>20 kDa) after subcutaneous administration [41]. The mAbs finally enter into the systemic circulation through the permeable lymphatic capillaries driven by the interstitial fluid flow. Zhao et al. have reported that the lymphatic flow rate significantly affects the peak concentration (T_{max}) of mAbs, the lymphatic transport dominates the delivery pathway of mAbs [47], and the transit and elimination rate influences the drug bioavailability negatively [11]. It usually takes several days for a mAb to be absorbed after intramuscular or subcutaneous administration.

Distribution: The efficiency of antibodies across cell membranes is usually limited by their large molecular size and polarity. Thus, a significant amount of mAb is stuck in vascular and interstitial spaces, and only a small volume of antibody is distributed. The transcellular and paracellular movements are two main mechanisms to transfer antibodies across or between cell membranes [11]. During transcellular transport, the endocytosis of antibodies is achieved through the movements of phagocytosis, fluid-phase pinocytosis that devour antibodies from the extracellular fluid space, or receptor-mediated endocytosis, triggered when the antibodies bind to the Fc receptors of the antigens on cell surface. In paracellular transport, the antibodies are delivered via convection, a predominant pathway that is caused by fluid flow between the capillaries, interstitial space, and lymphatics. The distribution efficiency of a mAb is determined by the architecture and physiology of the solid tumors and the characteristics of the particular therapeutic proteins. For instance, the tardy diffusion rates, the long distance for diffusional transports, and the raised interstitial pressure in the shortly vascularized tumors could reduce the uptake and distribution efficiency of the mAb.

Clearance: Anticancer biopharmaceuticals can be cleared by metabolism and excretion. Metabolism clearance is a dominant route, which involves a nonspecific route mediated by the complex of Fc domain and receptor, and a specific route mediated by the complex of the Fab domain and its pharmacological target. In Fc-mediated clearance, the binding of the Fc domain to receptors can result in diverse immune defense, for example, phagocytosis and ADCC, and play a critical role in the catabolism of antibodies. The therapeutic proteins can be adopted from the extracellular fluid space to form an endosome for further lysosomal catabolism clearance. The half-life of the antibodies can be extended via the neonatal Fc-receptor (FcRn)-mediated process. Therefore, the antibodies have been engineered by altering the affinity to FcRns [33] via alanine scanning to determine the binding sites for FcRn receptors [47], direct mutagenesis of Fc fragments [8, 20, 36], and *in silico* computer-based modeling analysis to design combined mutations in or near FcRn-binding site [46]. For Fab-mediated clearance, the internalization and lysosomal degradation can be induced through Fab binding to pharmacological targets on the cell surfaces or soluble antigens. A dose-independent linear clearance happens at an extremely low endogenous level of targets (TNF-α, IFN-α, VEGF, and IL-5), while a dose-dependent nonlinear elimination (NLE) is observed at high endogenous level of soluble antigens [14].

6.4.2 PK Models

PK models, including compartmental and noncompartmental models, have been developed to estimate or simulate the kinetics profiles of the plasma concentration of a biopharmaceutical molecule versus time. The compartmental model considers human body as multiple compartments without any physiological or anatomical context. This model describes the biopharmaceutical concentration over time and the biopharmaceutical transport between compartments using nonlinear regression analysis. The assumptions of this model are that (i) the drug is evenly distributed throughout the compartment and (ii) the drug transport or elimination rate is proportional to the drug concentration in the original compartment. The transport rate (dA/dt) of a biopharmaceutical from one compartment is correlated to the drug concentration (A) and the distribution rate constant (k) by the equation

$$\frac{dA_1}{dt} = R_1 + k_{21} \cdot A_2 - k_{12} \cdot A_1 - k_{10} \cdot A_1$$

The two-compartment model with linear elimination (LE) is used as the PK model of mAbs after the infusion administration [44].

The physiologically based pharmacokinetic (PBPK) model has been developed to describe the drug transport between blood and organs/tissues/cells (the physiological or anatomical entities) by integrating both extracellular and intracellular events. The differential equation

$$V_E \cdot \frac{dC_E}{dt} = R_E + Q_E \cdot C_C - Q_E \cdot C_E / K_E - CL_E \cdot C_E$$

is used to describe the drug transport rate (dC_E/dt) using drug concentration (C), blood or plasma flow (Q), tissue/blood partition coefficient (K), and clearance (CL) [40].

The noncompartmental method has been developed to describe the drug concentration *versus* time curve ($AUC_{0-\infty}$) and the mean residence time (MRT) based on statistical moment analysis [43]. This model can be used to analyze any type of PK data, but the disadvantage is that the PK profile cannot be predicted when alterations occur in a dosing regimen.

6.4.3 PK Data of Anticancer Biopharmaceuticals

Four PK parameters have been widely used to describe the properties of biopharmaceuticals, including CL (*clearance*, l/min), V (volume of distribution, l), $t_{1/2}$ (half-life, min), and F (bioavailability, %F) [1]. Table 6.1 summarizes the PK data of mAbs for cancer treatment, PK models, and patient number. The first-order two-compartment (2-COM) model has been used to estimate PK parameters of anti-HER2 mAbs (trastuzumab, ado-trastuzumab, and pertuzumab). Multiple PK models have been applied to the anti-VEGF and anti-EGFR mAbs (such as bevacizumab, cetuximab, and panitumumab). These models include LE two-compartment (2-COM) model , NLE 2-COM, and NLE models [5, 10, 16, 25]. For the anti-CD mAbs, the PK models are not defined for most therapeutic proteins except for rituximab, alemtuzumab, and of atumumab that have been investigated using 2-COM models. The LE models have been used to estimate the PK parameters of ipilimumab and denosumab.

Table 6.1 Pharmacokinetic properties of anticancer antibodies.

Antibodies	V (l)	$t_{1/2}$ (day)	CL (l/day)	Model	Patients no.	References
Trastuzumab/Herceptin	V_c: 2.95	28.5	0.225	LE, 2-COM	476	[5]
Ado-trastuzumab emtansine/ Kadcyla	V_c: 3.33, V_p: 0.89; V_c: 3.127	ND; 3.94	0.7; 0.676	LE, 2-COM; LE, 2-COM	273; 671	[19, 25]
Pertuzumab/Perjeta	V_c: 3.11	18.0	0.235	LE, 2-COM	481	[16]
Bevacizumab/Avastin	V_c: 2.39; V_c: 2.83, V_p: 2.49	20; 19.1	0.207; 0.214	LE, 2-COM; LE, 2-COM	491; 680	[24, 26]
Cetuximab/Erbitux	V_c: 2.83, V_p: 2.43	ND	ND	Non-LE, 2-COM	143	[12]
Panitumumab/Vectibix®	V_c: 3.95, V_p: 2.59	ND	0.273	Non-LE + LE, 2-COM	1200	[28]
Rituximab/Rituxan	V_c: 2.98, V_p: 3.64	ND	0.257	Non-LE, 2-COM	102	[32]
^{90}Y-Ibritumomab tiuxetan/Zevalin	ND	30	ND	ND	ND	[31]
Alemtuzumab/MabCampath®	V_c: 11.3, V_p: 4.5	12	1.05	Non-LE, 2-COM	67	[21, 30]
Ofatumumab/Arzerra	V_c: 5.3	5.20–6.83	0.18	LE, 2-COM	477	[7, 23, 37]
Ipilimumab/Yervoy	V_{ss}: 5.05	15.25	0.246	LE	58	[42]
Gemtuzumab ozogamicin/ Mylotarg	ND; V_{ss}: 6.5 ± 5.5	3 ± 1.75; 2.67 ± 1.83	6.36 ± 5.5; 2.88 ± 3.6	ND; ND	59; 29	[6, 13]
Brentuximab vedotin/Adcetris®	V_{ss}: 6–10	4–6	ND	ND	ND	[10]
^{131}I-Tositumomab/Bexxar	ND	2.8	1.64	ND	ND	[10]
Obinutuzumab/Gazyva®	V_{ss}: 3.8	28.4	0.09	ND	ND	[10]
Denosumab/Xgeva®	V_c: 2.62, V_p: 1.37	ND	0.078	LE, 2-COM	1076	[17]

Notes: 1-COMP, one compartment; 2-COMP, two compartments; CL, clearance; D, dose range; LE, linear elimination; ND, not defined; Non-LE, nonlinear elimination; $t_{1/2}$, terminal plasma half-life, time required to halve the plasma concentration of a drug after reaching pseudoequilibrium; V_c, volume of distribution of central compartment; V_p, volume of distribution of peripheral compartment; V_{ss}, volume of distribution at steady state.

The volumes of distribution of most anticancer mAbs are 2–4 l except alemtuzumab with V_c of 11.3 and V_p of 41.4 l, ofatumumab with V_c of 5.3 l, ipilimumab with V_{ss} of 5.05 l, gemtuzumab ozogamicin with V_{ss} of 6.5 l, and brentuximab vedotin with V_{ss} of 6–10 l. The values of the half-life of these mAbs have a wide distribution, 2.8–30 days. The clearance data of most mAbs are <0.5 l/day, but gemtuzumab ozogamicin has a clearance of up to 6.36 l/day. The patient numbers used to collect the PK data of anticancer mAbs range from 29 to 1200.

6.5 Pharmacodynamics

6.5.1 PD Models

The PD model describes the correlation between the concentration of a biopharmaceutical and the magnitude of the pharmacological response. The sigmoid E_{max} model, described by the equation

$$E = E_0 + E_{max} \times (C_e)^\gamma / \left[(C_e)^\gamma + (EC_{50})^\gamma \right]$$

has been developed to describe the concentration effect relationship of drugs, which tend toward a plateau [2, 34, 39]. The more complex growth/kill models, such as

$$\frac{dN}{dt} = \left[k_s - k_r - k \times f_{uT} \times (C_T - C_{min}) \right] \times N,$$

have been developed to describe the number of tumor cells at any time. The parameters of E_{max} (maximum achievable killing rate) and EC_{50} (drug concentration at which the killing rate is 50% of its maximum value E_{max}) are two important parameters in the growth/kills models.

Table 6.2 summarizes the PD models of the anticancer biopharmaceuticals. Specifically, the model of

$$\frac{dZ_0(t)}{dt} = \frac{\lambda_0 \times Z_0(t)}{1 + (\lambda_0/\lambda_1) \times Z_0(t)^\varphi}$$

establishes that the concentration of bevacizumab to inhibit the maximal tumor growth (IC50) is 3.6 μM [34]. The model of "drug effect = slope × $C(t)$" demonstrates that the 3.6 mg/kg of trastuzumab-DM1 for 3 weeks is a well-tolerated dose with minimal dose delays or reductions for thrombocytopenia [2]. The immunohistochemical analysis of cetuximab using

$$\frac{d(R/R_0)}{dt} = k_{out} \times \left[1 - C_{i.p.}/(EC_{50} + C_{i.p.}) \right] - k_{out} \times \left(\frac{R}{R_0} \right)$$

Table 6.2 PD models of anticancer antibodies.

Antibody	Equation	Parameters	Reference
Trastuzumab/ Herceptin	$\mathrm{LVEF} = \dfrac{\mathrm{LVEF}_0 \times \mathrm{EC}_{50}}{\mathrm{EC}_{50} + C_{\mathrm{EF}}}$	LVEF, left ventricular ejection fraction; EC_{50}, half maximal effect concentration; C_{EF}, effect compartment concentration	[39]
Ado-trastuzumab emtansine/Kadcyla	$\text{Drug effect} = \text{slope} \times C(t)$	$C(t)$, plasma concentration at time t	[2]
Bevacizumab/ Avastin	$\dfrac{dZ_0(t)}{dt} = \dfrac{\lambda_0 \times Z_0(t)}{1 + (\lambda_0 / \lambda_1) \times Z_0(t)^{\varphi}}$	Z_0, weight of the tumor cells; λ, growth rate; φ, shape factor	[34]
Cetuximab/Erbitux	$\dfrac{d(R/R_0)}{dt} = k_{\mathrm{out}} \times \left[1 - C_{\mathrm{i.p.}} / (\mathrm{EC}_{50} + C_{\mathrm{i.p.}})\right] - k_{\mathrm{out}} \times \left(\dfrac{R}{R_0}\right)$	R/R_0, percentage relative to baseline; k_{out}, first-order rate constant; $C_{\mathrm{i.p.}}$, plasma concentration; EC_{50}, drug concentration corresponding to 50% of $E_{\max}$	[27]
Rituximab/Rituxan	$\dfrac{dA}{dt} = k_{\mathrm{in}} \times A_0 - k_{\mathrm{out}} \times A - k_{\mathrm{drug}} \times C^{\gamma} / \left[(\mathrm{EC}_{50}) + C^{\gamma}\right]$	A, bioluminescence; k_{in}, zero-order tumor production constant; k_{out}, first-order constant; C, plasma concentration; k_{drug}, zero-order constant; EC_{50}, concentration leading to 50% of k_{drug}	[9]
Alemtuzumab/ MabCampath	$\dfrac{d\mathrm{WBC}}{dt} = k_{\mathrm{in}} - k_{\mathrm{out}} \times \mathrm{WBC} \times \left[1 + E_{\max} \times C / (\mathrm{EC}_{50} + C)\right]$	WBC, white blood cell count; k_{in}, zero-order rate constant of cell formation; k_{out}, first-order rate constant of cell death; $E_{\max}$, maximum effect on cell death; EC_{50}, concentration at half maximal effect	[30]

(*Continued*)

Table 6.2 (Continued)

Antibody	Equation	Parameters	Reference
Brentuximab vedotin/Adcetris	$\frac{dV1}{dt} = k_{\text{gExponenial}} \times (1 - \text{TV}/V_{\text{Max}}) \times V1 / \left[1 + (k_{\text{gExponenial}}/k_{\text{gLinear}} \times \text{TV})^{\psi}\right]^{1/\psi} - k_{\text{killMax}} \times \text{PL}_{\text{Tumor}} / (\text{KC}_{50} + \text{PL}_{\text{Tumor}}) \times V1$ $\frac{dV2}{dt} = k_{\text{killMax}} \times \text{PL}_{\text{Tumor}} \times V1/(\text{KC}_{50} + \text{PL}_{\text{Tumor}}) - V2/T_{\text{au}}$ $\frac{dV3}{dt} = \frac{(V2 - V3)}{T_{\text{au}}}$ $\frac{dV4}{dt} = \frac{(V3 - V4)}{T_{\text{au}}}$ $\text{TV} = V1 + V2 + V3 + V4$	k_g, growth rate of tumor; $V1-V4$, tumor volume; TV, total tumor volume; V_{Max}, maximum possible tumor volume; ψ, switch function between exponential and linear growth rate; k_{kill}, killing rate constant; PL_{Tumor}, tumor payload concentration; KC_{50}, MMAE concentrations at half of maximum killing rate constant; T_{au}, transduction time between two tumor compartments	[35]
Denosumab/Xgeva	$\frac{d\text{NTX}}{dt} = k_{\text{in}} \times \left[1 - I_{\text{MAX}}\text{Cp} / (\text{IC}_{50} + \text{Cp})\right] - k_{\text{out}} \times \text{NTX}$	NTX, N-telopeptide concentration; k_{in}, zero-order production rate of the response; k_{out}, first-order rate of loss of response; I_{MAX}, maximal fractional extent of inhibition; IC_{50}, drug concentration at 50% maximal inhibition; Cp, free drug concentration	[29]

determines that the phospho-MAPK and Ki67 expression were inhibited between 24 and 72 h at 0.25 and 0.04 mg [27]. The PD study of panitumumab demonstrates a strong correlation between the dosage and skin rash and also found the optimal weekly dose of 2.5 mg/kg. The PD analysis using

$$\frac{dA}{dt} = k_{in} \times A_0 - k_{out} \times A - k_{drug} \times C^{\gamma} \Big/ \left[\left(EC_{50} \right) + C^{\gamma} \right]$$

shows that the concentration of rituximab is inversely correlated with tumor burden, and the 20 mg/kg dose is adequate. A stimulatory loss indirect response model,

$$\frac{d\text{WBC}}{dt} = k_{in} - k_{out} \times \text{WBC} \times \left[1 + E_{max} \times C / \left(EC_{50} + C \right) \right]$$

has been developed to describe the reduction of WBC caused by alemtuzumab. The exposure to obinutuzumab is affected by diagnosis, baseline tumor size (BSIZ), body weight, and gender [18]. It has been found that the integration of a cellular model with clinical data has the distinct advantage predict the role of cellular system variables on drug response.

6.6 Conclusions

In this chapter, we have discussed the clinical pharmacology of the FDA-approved anticancer antibodies, processes, PK models and clinical parameters of mAb PK, and the model and clinical data of mAb PD. The application of these quantitative methods could benefit the biosimilar development by assisting the evaluation of the clinical safety and efficiency, optimizing the dosage, and correlating the *in vitro* behavior with the *in vivo* effect of the developed biopharmaceuticals. In addition, the models and clinical data can also be used to guide the clinical trial, improve the therapeutic efficiency, and guide the rational design of protein engineering of the innovative biopharmaceuticals.

References

1. Benet LZ, Zia-Amirhosseini P. Basic principle of pharmacokinetics. *Toxicol Pathol.* 1995, 23(2): 115–123.
2. Bender BC, Schaedeli-Stark F, Koch R, Joshi A, Chu YW, Rugo H, Gupta M. A population pharmacokinetic/pharmacodynamic model of thrombocytopenia characterizing the effect of trastuzumab emtansine (T-DM1) on platelet counts in patients with HER2-positive metastatic breast cancer. *Cancer Chemother Pharmacol.* 2012, 70(4): 591–601.

3. Braendstrup P, Bjerrum OW, Nielsen OJ, Jensen BA, Clausen NT, Hansen PB, Andersen I, Schmidt K, Andersen TM, Peterslund NA, Birgens HS, Plesner T, Pedersen BB, Hasselbalch HC. Rituximab chimeric anti-CD20 monoclonal antibody treatment for adult refractory idiopathic thrombocytopenic purpura. *Am J Hematol.* 2005, 78: 275–280.
4. Bross PF, Beitz J, Chen G, Chen XH, Duffy E, Kieffer L, Roy S, Sridhara R, Rahman A, Williams G, Pazdur R. Approval summary: gemtuzumab ozogamicin in relapsed acute myeloid leukemia. *Clin Cancer Res.* 2001, 7: 1490–1496.
5. Bruno R, Washington CB, Lu JF, Lieberman G, Banken L, Klein P. Population pharmacokinetics of trastuzumab in patients with HER2+ metastatic breast cancer. *Cancer Chemother Pharmacol.* 2005, 56(4): 361–369.
6. Buckwalter M, Dowell JA, Korth-Bradley J, Gorovits B, Mayer PR. Pharmacokinetics of gemtuzumab ozogamicin as a single-agent treatment of pediatric patients with refractory or relapsed acute myeloid leukemia. *J Clin Pharmacol.* 2004, 44(8): 873–880.
7. Coiffier B, Losic N, Ronn BB, Lepretre S, Pedersen LM, Gadeberg O, Frederiksen H, van Oers MH, Wooldridge J, Kloczko J, Holowiecki J, Hellmann A, Walewski J, Robak T, Petersen J. Pharmacokinetics and pharmacokinetic/pharmacodynamic associations of ofatumumab, a human monoclonal CD20 antibody, in patients with relapsed or refractory chronic lymphocytic leukaemia: a phase 1-2 study. *Br J Haematol.* 2010, 150(1): 58–71.
8. Dall'Acqua WF, Woods RM, Ward ES, Palaszynski SR, Patel NK, Brewah YA, Wu H, Kiener PA, Langermann S. Increasing the affinity of a human IgG1 for the neonatal Fc receptor: biological consequences. *J Immunol.* 2002, 169(9): 5171–5180.
9. Dayde D, Ternant D, Ohresser M, Lerondel S, Pesnel S, Watier H, Cartron G. Tumor burden influences exposure and response to rituximab: pharmacokinetic-pharmacodynamic modeling using a syngeneic bioluminescent murine model expressing human CD20. *Blood.* 2009, 113(16): 3765–3772.
10. Dostalek M, Gardner I, Gurbaxani BM, Rose RH, Chetty M. Pharmacokinetics, pharmacodynamics and physiologically-based pharmacokinetic modelling of monoclonal antibodies. *Clin Pharmacokinet.* 2013, 52(2): 83–124.
11. Dirks NL, Meibohm B. Population pharmacokinetics of therapeutic monoclonal antibodies. *Clin Pharmacokinet.* 2010, 49(10): 633–659.
12. Dirks NL, Nolting A, Kovar A, Meibohm B. Population pharmacokinetics of cetuximab in patients with squamous cell carcinoma of the head and neck. *J Clin Pharmacol.* 2008, 48(3): 267–278.
13. Dowell JA, Korth-Bradley J, Liu H, King SP, Berger MS. Pharmacokinetics of gemtuzumab ozogamicin, an antibody-targeted chemotherapy agent for the

treatment of patients with acute myeloid leukemia in first relapse. *J Clin Pharmacol.* 2001, 41(11): 1206–1214.

14. Fan J, de Lannoy IA. Pharmacokinetics. *Biochem Pharmacol.* 2014, 87(1): 93–120.
15. Friedman HS, Prados MD, Wen PY, Mikkelsen Y, Schiff D, Abrey LE, Yung WK, Paleologos N, Nicholas MK, Jensen R, Vredenburgh J, Huang J, Zheng M, Cloughesy T. Bevacizumab alone and in combination with irinotecan in recurrent glioblastoma. *J Clin Oncol.* 2009, 27: 4733–4740.
16. Garg A, Quartino A, Li J, Jin J, Wada DR, Li H, Cortés J, McNally V, Ross R, Visich J, Lum B. Population pharmacokinetic and covariate analysis of pertuzumab, a HER2-targeted monoclinal antibody, and evaluation of a fixed, non-weight-based dose in patients with a variety of solid tumors. *Cancer Chemother Pharmacol.* 2014, 74(4): 819–829.
17. Gibiansky L, Sutjandra L, Doshi S, Zheng J, Sohn W, Peterson MC, Jang GR, Chow AT, Pérez-Ruixo JJ. Population pharmacokinetic analysis of denosumab in patients with bone metastases from solid tumours. *Clin Pharmacokinet.* 2012, 51(4): 247–260.
18. Gibiansky E, Gibiansky L, Carlile DJ, Jamois C, Buchheit V, Frey N. Population pharmacokinetics of obinutuzumab (GA101) in chronic lymphocytic leukemia (CLL) and non-Hodgkin's lymphoma and exposure–response in CLL. *CPT: Pharmacometrics Syst Pharmacol.* 2014, 3(10): 1–11.
19. Gupta M, Lorusso PM, Wang B, Yi JH, Burris HA 3rd, Beeram M, Modi S, Chu YW, Agresta S, Klencke B, Joshi A, Girish S. Clinical implications of pathophysiological and demographic covariates on the population pharmacokinetics of trastuzumab emtansine, a HER2-targeted antibody-drug conjugate, in patients with HER2-positive metastatic breast cancer. *J Clin Pharmacol.* 2002, 52(5): 691–703.
20. Hinton PR, Xiong JM, Johlfs MG, Tang MT, Keller S, Tsurushita N. An engineered human IgG1 antibody with longer serum half-life. *J Immunol.* 2006, 176(1): 346–356.
21. Frampton JE, Wagstaff AJ. Alemtuzumab. *Drugs.* 2003, 63(12): 1229–1243.
22. Kariolis MS, Kapur S, Cochran JR. Beyond antibodies: using biological principles to guide the development of next-generation protein therapeutics. *Curr Opin Biotechnol.* 2013, 24: 1072–1077.
23. Kurrasch R, Brown JC, Chu M, Craigen J, Overend P, Patel B, Wolfe S, Chang DJ. Subcutaneously administered ofatumumab in rheumatoid arthritis: a phase I/II study of safety, tolerability, pharmacokinetics, and pharmacodynamics. *J Rheumatol.* 2013, 40(7): 1089–1096.
24. Li J, Gupta M, Jin D, Xin Y, Visich J, Allison DE. Characterization of the long-term pharmacokinetics of bevacizumab following last dose in patients with resected stage II and III carcinoma of the colon. *Cancer Chemother Pharmacol.* 2013, 71(3): 575–580.

25. Lu D, Girish S, Gao Y, Wang B, Yi JH, Guardino E, Samant M, Cobleigh M, Rimawi M, Conte P, Jin JY. Population pharmacokinetics of trastuzumab emtansine (T-DM1), a HER2-targeted antibody-drug conjugate, in patients with HER2 positive matastatic breast cancer: clinical implications of the effect of covariates. *Cancer Chemother Pharmacol.* 2014, 74(2): 399–410.

26. Lu JF, Bruno R, Eppler S, Novotny W, Lum B, Gaudreault J. Clinical pharmacokinetics of bevacizumab in patients with solid tumors. *Cancer Chemother Pharmacol.* 2008, 62(5): 779–786.

27. Luo FR, Yang Z, Dong H, Camuso A, McGlinchey K, Fager K, Lee FY. Prediction of active drug plasma concentrations achieved in cancer patients by pharmacodynamic biomarkers identified from the geo human colon carcinoma xenograft model. *Clin Cancer Res.* 2005, 11(15): 5558–5565.

28. Ma P, Yang BB, Wang YM, Peterson M, Narayanan A, Sutjandra L, Rodriguez R, Chow A. Population pharmacokinetic analysis of panitumumab in patients with advanced solid tumors. *J Clin Pharmacol.* 2009, 49(10): 1142–1156.

29. Marathe A, Peterson MC, Mager DE. Integrated cellular bone homeostasis model for denosumab pharmacodynamics in multiple myeloma patients. *J Pharmacol Exp Ther.* 2008, 326(2): 555–562.

30. Mould DR, Baumann A, Kuhlmann J, Keating MJ, Weitman S, Hillmen P, Bonate PL. Population pharmacokinetics–pharmacodynamics of alemtuzumab (Campath®) in patients with chronic lymphocytic leukaemia and its link to treatment response. *Br J Clin Pharmacol.* 2007, 64(3): 278–291.

31. Alazraki N, Shumate MJ, Kooby D. A Clinician's Guide to Nuclear Oncology: Practical Molecular Imaging and Radionuclide Therapies. Reston, VA: Society of Nuclear Medicine, Inc. 2007.

32. Ng CM, Bruno R, Combs D, Davies B. Population pharmacokinetics of rituximab (anti-CD20 monoclonal antibody) in rheumatoid arthritis patients during a phase II clinical trial. *J Clin Pharmacol.* 2005, 45(7): 792–801.

33. Richter WF, Jacobsen B. Subcutaneous absorption of biotherapeutics: knowns and unknowns. *Drug Metab Dispos.* 2014, 42(11): 1881–1889.

34. Rocchetti M, Germani M, Del Bene F, Poggesi I, Magni P, Pesenti E, De Nicolao G. Predictive pharmacokinetic–pharmacodynamic modeling of tumor growth after administration of an anti-angiogenic agent, bevacizumab, as single-agent and combination therapy in tumor xenografts. *Cancer Chemother Pharmacol.* 2013, 71(5): 1147–1157.

35. Shah DK, Haddish-Berhane N, Betts A. Bench to bedside translation of antibody drug conjugates using a multiscale mechanistic PK/PD model: a case study with brentuximab-vedotin. *J Pharmacokinet Pharmacodyn.* 2002, 39(6): 643–659.

36. Shields RL, Namenuk AK, Hong K, Meng YG, Rae J, Briggs J, Xie D, Lai J, Stadlen A, Li B, Fox JA, Presta LG. High resolution mapping of the binding site on human IgG1 for Fc gamma RI, Fc gamma RII, Fc gamma RIII, and

FcRn and design of IgG1 variants with improved binding to the Fc gamma R. *J Biol Chem.* 2001, 276(9): 6591–6604.

37. Struemper H, Sale M, Patel BR, Østergaard M, Österborg A, Wierda WG, Hagenbeek A, Coiffier B, Jewell RC. Population pharmacokinetics of ofatumumab in patients with chronic lymphocytic leukemia, follicular lymphoma, and rheumatoid arthritis. *J Clin Pharmacol.* 2014, 54(7): 818–827.
38. Vaklavas C, Forero-Torres A. Safety and efficacy of brentuximab vedotin in patients with Hodgkin lymphoma or systemic anaplastic large cell lymphoma. *Ther Adv Hematol.* 2012, 3: 209–225.
39. Van Hasselt JGC, Boekhout AH, Beijnen JH, Schellens JHM, Huitema ADR. Population pharmacokinetic-pharmacodynamic analysis of trastuzumab-associated cardiotoxicity. *Clin Pharmacol Ther.* 2011, 90(1): 126–132.
40. Vugmeyster Y, Xu X, Theil FP, Khawli LA, Leach MW. Pharmacokinetics and toxicology of therapeutic proteins: advances and challenges. *World J Biol Chem.* 2012, 3(4): 73–92.
41. Wang WJ, Lei YY, Mei JH, Wang CL. Recent progress in HER2 associated breast cancer. *Asian Pac J Cancer Prev.* 2015, 16(7): 2591–2600.
42. Weber J, Hamid O, Amin A, O'Day S, Masson E, Goldberg SM, Williams D, Parker SM, Chasalow SD, Alaparthy S, Wolchok JD. Randomized phase I pharmacokinetic study of ipilimumab with or without one of two different chemotherapy regimens in patients with untreated advanced melanoma. *Cancer Immun.* 2013, 13: 7–15.
43. Yamaoka K, Nakagawa T, Uno T. Statistical moments in pharmacokinetics. *J Pharmacokinet Biopharm.* 1978, 6: 547–558.
44. Zalevsky J, Chamberlain AK, Horton HM, Karki S, Leung IW, Sproule TJ, Lazar GA, Roopenian DC, Desjarlais JR. Enhanced antibody half-life improves in vivo activity. *Nat Biotechnol.* 2010, 28: 157–159.
45. Zhou L, Xu N, Sun Y, Liu XM. Cancer treatment using targeted biopharmaceuticals. *Cancer Lett.* 2014, 352: 145–151.
46. Zhou J, Johnson JE, Ghetie V, Ober RJ, Ward ES. Generation of mutated variants of the human form of the MHC class I-related receptor, FcRn, with increased affinity for mouse immunoglobulin G. *J Mol Biol.* 2003, 332(4): 901–913.
47. Zhao L, Ji P, Li Z, Roy P, Sahajwalla CG. The antibody drug absorption following subcutaneous or intramuscular administration and its mathematical description by coupling physiologically based absorption process with the conventional compartment pharmacokinetic model. *J Clin Pharmacol.* 2013, 53(3): 314–325.

7

Monoclonal Antibodies

Applications in Clinical Oncology

Jeanene ("Gigi") Robison

Cancer Center, Christ Hospital, Cincinnati, OH, USA

7.1 Summary

This chapter will review the developmental history, target biology, mechanisms of action, clinical efficacy, and Food and Drug Administration (FDA)-approved indications for monoclonal antibodies (mAbs). MAbs are a type of targeted biotherapy, which are derived from human antibodies, mouse antibodies, or combination thereof. They bind with cell surface proteins, such as CD20 or HER2. Monoclonal antibodies may target host tissues or proteins that support tumor growth, such as growth factors (e.g., VEGF) or growth factor receptors (e.g., EGFR). They may also identify targets that are relatively unique to the tumor cell population.

7.2 Introduction

This chapter will review the developmental history, target biology, mechanisms of action, clinical efficacy, and Food and Drug Administration (FDA)-approved indications for monoclonal antibodies (mAbs).

MAbs are a type of targeted biotherapy, which are derived from human antibodies, mouse antibodies, or combination thereof (Fig. 7.1). Human mAbs consist of only human antibody sequences. Humanized mAbs contain some murine sequences but have been modified through the insertion of human sequences to make them less antigenic. Chimeric mAbs are combinations of mouse and human antibodies. Murine mAbs are derived from mouse hybridoma cell lines without modification of the original sequences. For this

Biosimilars of Monoclonal Antibodies: A Practical Guide to Manufacturing, Preclinical, and Clinical Development, First Edition. Edited by Cheng Liu and K. John Morrow, Jr.

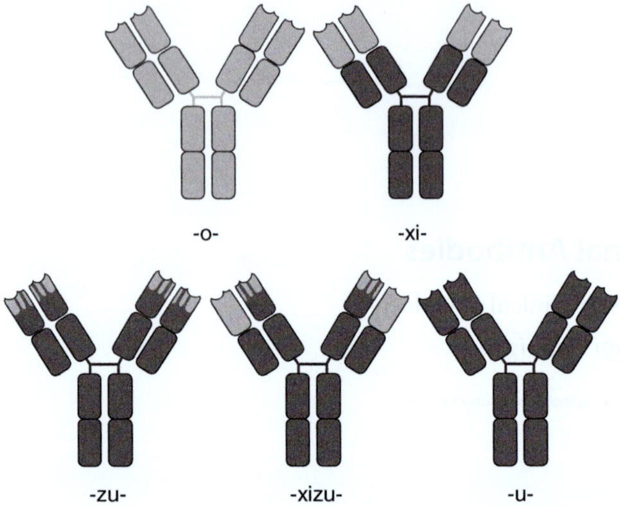

Figure 7.1 Nomenclature of different types of monoclonal antibodies. Comparison of monoclonal antibodies (dark brown, human; light blue, nonhuman or mouse). Top row: murine mAbs (name ends in "-omab"); ibritumomab tiuxetan. Top row: chimeric mAbs (name ends in "-ximab"); example: rituximab. Bottom row: humanized mAbs (name ends in "-zumab"); example: trastuzumab. Bottom row: chimeric/humanized mAbs (name ends in "-xizumab"); non are FDA approved. Bottom row: human mAbs (name ends in "-umab"); example: panitumumab. Note: Public domain image from Wikimedia Commons. Accessed on October 18, 2016, from https://en.wikipedia.org/wiki/Nomenclature_of_monoclonal_antibodies. (*See insert for color representation of the figure.*)

reason, they help invoke an immune response when injected into a human patient and are less effective because they lack the ability to bind to the Fc receptor.

MAbs bind with cell surface proteins, such as CD20 or HER2. They are designed to target host tissues or proteins that support tumor growth, such as growth factors (e.g., VEGF) or growth factor receptors (e.g., EGFR). They may also identify targets that are relatively unique to the tumor cell population.

7.3 Ado-trastuzumab Emtansine (Anti-HER2 Antibody Conjugated with Emtansine, Kadcyla®)

7.3.1 Drug Development History

The US FDA approved ado-trastuzumab emtansine (Kadcyla®, or T-DM1) on February 13, 2013, for the treatment of HER2-positive metastatic breast cancer (MBC) [41].

7.3.2 Target Biology

Ado-trastuzumab emtansine (Kadcyla) targets the HER2 antigen [41].

7.3.3 Mechanism of Action

Ado-trastuzumab emtansine (Kadcyla) is a HER2-targeted antibody-drug conjugate (ADC). The humanized anti-HER2 IgG1 antibody trastuzumab is joined with a small-molecule cytotoxin (e.g., microtubule inhibitor), DM1, using a stable linker. Upon binding to subdomain IV of the HER2 receptor, ado-trastuzumab emtansine undergoes receptor-mediated internalization and subsequent lysosomal degradation. This results in intracellular release of DM1-containing cytotoxic catabolites. Binding of DM1 to tubulin disrupts microtubule networks in the cell, which results in cell cycle arrest and apoptotic cell death. *In vitro* studies have also shown that, similar to trastuzumab, ado-trastuzumab emtansine inhibits HER2 receptor signaling, mediates antibody-dependent cell-mediated cytotoxicity, and inhibits shedding of the HER2 extracellular domain in human breast cancer cells that overexpress HER2 [8, 20, 41].

7.3.4 Clinical Efficacy

7.3.4.1 Metastatic Breast Cancer

The FDA approval of Kadcyla was based on the positive results from the international, randomized, multicenter, open-label phase III EMILIA (TDM4370g/BO21977) study. This trial randomly allocated (1 : 1) patients to receive Kadcyla alone or lapatinib in combination with Xeloda (capecitabine). Study participants ($n = 991$) included patients with HER2-positive, unresectable locally advanced or metastatic breast cancer. Breast tumor samples were required to show HER2 overexpression defined as 3+ IHC or FISH amplification ratio ≥2.0 determined at a central laboratory [41].

In each 21-day cycle, Kadcyla was given intravenously at 3.6 mg/kg (day 1), lapatinib was administered at 1250 mg/day orally once daily, and capecitabine was administered at 1000 mg/m^2 orally twice daily (days 1–14). Patients were treated with Kadcyla or lapatinib plus capecitabine until progression of disease, withdrawal of consent, or unacceptable toxicity. At the time of the primary analysis, median time on study drug was 5.7 months for Kadcyla, 4.9 months for lapatinib, and 4.8 months for capecitabine [41].

The co-primary efficacy endpoints of the study were progression-free survival (PFS) and overall survival (OS). An additional endpoint included PFS (based on investigator tumor response assessments) [41].

Most patients (88%) had received prior systemic treatment in the metastatic setting. Overall, patients received a median of three systemic agents in the metastatic setting. Among patients with hormone receptor-positive tumors,

44.4% received prior adjuvant hormonal therapy, and 44.8% received hormonal therapy for locally advanced/metastatic disease [41].

The randomized trial demonstrated a statistically significant improvement in Independent Review Committee (IRC)-assessed PFS in the Kadcyla-treated group compared with the lapatinib plus capecitabine-treated group and an increase in median PFS of 3.2 months (median PFS of 9.6 months in the Kadcyla-treated group vs. 6.4 months in the lapatinib plus capecitabine group) [41].

At the time of PFS analysis, 223 patients had died. More deaths occurred in the lapatinib plus capecitabine arm (26%) compared with the Kadcyla arm (19%). At the time of the second interim OS analysis, 331 events had occurred. The median duration of survival was 30.9 months in the Kadcyla arm versus 25.1 months in the lapatinib plus capecitabine arm [8, 41].

7.3.5 Approved Indications

Kadcyla®, as a single agent, is indicated for the treatment of patients with HER2-positive MBC who previously received trastuzumab and a taxane, separately or in combination. Patients should have either:

- Received prior therapy for metastatic disease or
- Developed disease recurrence during or within 6 months of completing adjuvant therapy [41]

7.4 Alemtuzumab (Campath®, Campath-1H)

7.4.1 Drug Development History

The antilymphocyte antibody alemtuzumab (Campath-IH) experienced a long road to FDA-approved status. The US FDA approved alemtuzumab in 2001, which was 21 years after the first mAbs for CD52 were generated in Cambridge. This genetically engineered version became licensed as alemtuzumab in the United States and Europe [36, 62].

7.4.2 Target Biology

Alemtuzumab (Campath) is a recombinant DNA-derived humanized mAb directed against the 21–28-kD cell surface glycoprotein, CD52 [21, 36].

7.4.3 Mechanism of Action

Campath binds to CD52, an antigen present on the surface of B and T lymphocytes, a majority of monocytes, macrophages, NK cells, and a subpopulation of granulocytes. A proportion of bone marrow cells, including some CD34+ cells, express variable levels of CD52. The proposed mechanism of action is antibody-dependent cellular-mediated lysis following cell surface binding of Campath to the CD52 antigen on leukemic cells [21, 36].

7.4.4 Clinical Efficacy

7.4.4.1 Previously Untreated B-CLL Patients

Campath was evaluated in an open-label, randomized (1:1) active-controlled study in previously untreated patients ($n = 297$) with B-cell chronic lymphocytic leukemia (B-CLL), Rai stage I–IV, with evidence of progressive disease requiring therapy. Patients received either Campath 30 mg intravenous (IV) three times per week for a maximum of 12 weeks or chlorambucil 40 mg/m^2 orally once every 28 days for a maximum of 12 cycles [36].

Patients randomized to receive Campath experienced longer PFS compared to those randomized to receive chlorambucil (median PFS 14.6 months vs. 11.7 months, respectively). The overall response rates were 83% and 55% ($p < 0.0001$), and the complete response rates (CRR) were 24% and 2% ($p < 0.0001$) for Campath and chlorambucil arms, respectively.

7.4.4.2 Previously Treated B-Cell Patients

Campath was evaluated in three multicenter, open-label, single-arm studies of B-CLL patients ($n = 149$), who were previously treated with alkylating agents, fludarabine, or other chemotherapies. Patients were treated with the recommended dose of Campath, 30 mg intravenously, three times per week for up to 12 weeks. Partial response (PR) rates of 21–31% and CRR of 0–2% were observed [36].

7.4.5 Approved Indications

Alemtuzumab (Campath) is indicated as a single agent for the treatment of B-CLL [36].

7.5 Bevacizumab (Avastin)

7.5.1 Drug Development History

Bevacizumab was the first clinically available angiogenesis inhibitor in the United States. It was initially approved by the FDA in February 2004 for metastatic colorectal cancer (mCRC) in combination with standard chemotherapy (as a first-line treatment) and with 5-fluorouracil (5-FU)-based therapy for second-line mCRC. It has since been approved for use in certain lung cancers, renal cancers, ovarian cancers, and glioblastoma multiforme of the brain. It had been approved for breast cancer, but that approval was withdrawn in 2010 when later studies showed no evidence of effectiveness [28, 29, 31, 34, 51, 53].

7.5.2 Target Biology

Bevacizumab (Avastin) is a recombinant humanized mAb that blocks angiogenesis by inhibiting vascular endothelial growth factor A (VEGF-A) [34].

7.5.3 Mechanism of Action

VEGF-A, while expressed in normal tissues, is also present at physiologically relevant levels in tumors. Vascular endothelial growth factor (VEGF) is a chemical signal that stimulates angiogenesis in a variety of diseases. When the VEGF ligand binds to the VEGF receptor on endothelial cells, this may help drive angiogenesis.

Avastin directly binds to VEGF ligand receptors extracellularly to prevent interaction with VEGF receptors. This may inhibit VEGF's angiogenic activity (e.g., slows the growth of new blood vessels). Early effects of inhibiting VEGF with Avastin may lead to a reduction in tumor size. Sustained VEGF inhibition may be a strategy to inhibit tumor growth and maintain tumor regression [15, 34].

7.5.4 Clinical Efficacy

7.5.4.1 Metastatic Colorectal Cancer (mCRC)

Study 1 In this double-blind, active-controlled study, patients were randomized (1 : 1 : 1) to one of three arms: (1) IV bolus-IFL (irinotecan 125 mg/m^2, 5-FU 500 mg/m^2, and leucovorin (LV) 20 mg/m^2 given once weekly for 4 weeks every 6 weeks) plus placebo, (2) IV bolus-IFL plus Avastin (5 mg/kg every 2 weeks), or (3) 5-FU/LV plus Avastin (5 mg/kg every 2 weeks). Enrollment in Arm 3 was discontinued, as prespecified, when the toxicity of Avastin in combination with the bolus-IFL regimen was deemed unacceptable [34].

The main outcome measure was OS. Based on results from patients ($n = 813$) entered into Arm 1 (IFL + placebo) and Arm 2 (IFL + Avastin) of this study, the addition of Avastin resulted in an improvement in OS (median 15.6 vs. 20.3 months), PFS (median 6.2 vs. 10.6 months), and overall response rate (35 vs. 45%).

In Arm 3 (5FU/LV + Avastin), 110 patients were enrolled. The median OS was 18.3 months, median PFS was 8.8 months, overall response rate was 39%, and median duration of response was 8.5 months [34].

Study 2 Study 2 was a randomized, open-label, active-controlled trial in mCRC patients ($n = 823$) who were previously treated with irinotecan ± 5-FU for initial therapy for metastatic disease or as adjuvant therapy. Patients were randomized (1 : 1 : 1) to one of three arms: (1) IV FOLFOX4 (day 1: oxaliplatin 85 mg/m^2 and LV 200 mg/m^2 concurrently and then 5-FU 400 mg/m^2 bolus followed by 600 mg/m^2 continuously; day 2: LV 200 mg/m^2 and then 5-FU 400 mg/m^2 bolus followed by 600 mg/m^2 continuously; repeated every 2 weeks), (2) FOLFOX4 plus Avastin (10 mg/kg every 2 weeks prior to FOLFOX4 on day 1), or (3) Avastin monotherapy (10 mg/kg every 2 weeks) [34].

The Avastin monotherapy arm was closed to accrual after enrollment of 244 of the planned 290 patients. This followed a planned interim analysis by the data monitoring committee based on evidence of decreased survival in the Avastin monotherapy arm as compared to FOLFOX4 alone [34].

The main outcome measure was OS. The addition of Avastin to FOLFOX4 resulted in significantly longer survival as compared to FOLFOX4 alone (median OS 13.0 vs. 10.8 months), with clinical benefit seen in subgroups defined by age (<65 years, ≥65 years) and gender. PFS and overall response rate based on investigator assessment were higher in the Avastin plus FOLFOX4 arm [34].

Study 3 A single arm study evaluated the activity of Avastin in combination with bolus or infusional 5-FU/LV. Enrolled patients ($N = 339$) had a diagnosis of mCRC with disease progression following both irinotecan- and oxaliplatin-containing chemotherapy regimens. Most patients (73%) received concurrent bolus 5-FU/LV. One objective partial response was verified in the first 100 evaluable patients. The overall response rate was 1% (95% CI 0–5.5%).

Study 4 Study 4 was a prospective, randomized, open-label, multinational, controlled trial in patients ($n = 820$) with histologically confirmed mCRC who had progressed on a first-line Avastin-containing regimen. Patients were randomized (1 : 1) within 3 months after discontinuation of Avastin as first-line therapy to receive fluoropyrimidine/oxaliplatin- or fluoropyrimidine/irinotecan-based chemotherapy with or without Avastin administered at 5 mg/kg every 2 weeks or 7.5 mg/kg every 3 weeks. The choice of second-line therapy was contingent upon first-line chemotherapy treatment. Second-line treatment was administered until progressive disease or unacceptable toxicity [34].

The main outcome measure was OS, which was defined as the time from randomization until death from any cause.

The addition of Avastin to fluoropyrimidine-based chemotherapy resulted in a statistically significant prolongation of OS as compared to chemotherapy alone (median 11.2 vs. 9.8 months) and a statistically significant PFS (median 5.7 vs. 4.0 months). There was no significant difference in overall response rate, a key secondary outcome measure [34].

7.5.4.2 Lack of Efficacy in Adjuvant Treatment of Colon Cancer

Lack of efficacy of Avastin as an adjunct to standard chemotherapy for the adjuvant treatment of colon cancer was determined in two randomized, open-label, multicenter clinical trials. Study results have shown no significant benefit and potential to cause harm in this setting [34].

7.5.4.3 Unresectable Nonsquamous Non-Small Cell Lung Cancer (NSCLC)

The safety and efficacy of Avastin as first-line treatment of patients with locally advanced, metastatic, or recurrent nonsquamous non-small cell lung cancer

(NSCLC) were studied in a single, large, randomized, active-controlled, open-label, multicenter study [34].

7.5.4.4 Advanced Nonsquamous Small Cell Lung Cancer (NSCLC)

In 2006, the FDA approved bevacizumab for use in first-line advanced NSCLC, with a nonsquamous histology, in combination with carboplatin/paclitaxel chemotherapy. The approval was based on a pivotal study, which demonstrated a 2-month improvement in OS in patients treated with bevacizumab. Additional large studies support the use of Avastin in combination with chemotherapy for the treatment of NSCLC [34, 50, 51].

The National Comprehensive Cancer Network recommends bevacizumab as standard first-line treatment in combination with any platinum-based chemotherapy, followed by maintenance bevacizumab until disease progression. Higher doses are usually given with carboplatin-based chemotherapy, whereas the lower dose is usually given with cisplatin-based chemotherapy [28].

7.5.4.5 Glioblastoma

The FDA granted accelerated approval of Avastin for the treatment of recurrent glioblastoma multiforme in May 2009. The efficacy and safety of Avastin when treating glioblastoma were evaluated in two studies. The first study was an open-label, multicenter, randomized, noncomparative study of patients with previously treated glioblastoma. Patients ($n = 85$) received Avastin (10 mg/kg IV) alone or Avastin plus irinotecan every 2 weeks until disease progression or until unacceptable toxicity. All patients received prior radiotherapy (completed at least 8 weeks prior to receiving Avastin) and temozolomide. The efficacy of Avastin was demonstrated using response assessment based on both WHO radiographic criteria and by stable or decreasing corticosteroid use, which occurred in 25.9% of the patients. Median duration of response was 4.2 months. Radiologic assessment was based on MRI (using T1 and T2/FLAIR) [34].

The second study was a single-arm, single-institution trial with 56 patients with glioblastoma. All patients had documented disease progression after receiving temozolomide and radiation therapy (RT). Patients received Avastin 10 mg/kg IV every 2 weeks until disease progression or unacceptable toxicity. The efficacy of Avastin was supported by an overall response rate of 19.6% using the same response criteria as in the first glioblastoma study. Median duration of response was 3.9 months. Effectiveness of bevacizumab in treating glioblastoma is based on improvement in overall response rate and not related to OS or improvement in disease-related symptoms. Thus, it is not recommended in persons who are newly diagnosed with glioblastoma [34].

7.5.4.6 Metastatic Renal Cell Cancer (mRCC)

In 2009, the FDA approved bevacizumab for use in metastatic renal cell cancer. Bevacizumab improves the PFS time but not survival time. Patients with treatment-naïve mRCC were evaluated in a multicenter, randomized, double-blind, international study comparing Avastin plus interferon alfa 2a (IFN-α2a) versus placebo plus IFN-α2a. Patients ($n=649$) who had undergone a nephrectomy were randomized (1:1) to receive either Avastin (10 mg/kg IV infusion every 2 weeks; $n=327$) or placebo (IV every 2 weeks; $n=322$) in combination with IFN-α2a (nine MIU subcutaneously three times weekly for a maximum of 52 weeks). Patients were treated until disease progression or unacceptable toxicity [34].

PFS was statistically significantly prolonged among patients receiving Avastin plus IFN-α2a compared to those receiving IFN-α2a alone (median PFS was 10.2 vs. 5.4 months). Among the 595 patients with measurable disease, overall response rate was also significantly higher (30 vs. 12%). There was no improvement in OS based on the final analysis conducted after 444 deaths [34].

7.5.5 Approved Indications

- mCRC, with intravenous 5-FU-based chemotherapy for first- or second-line treatment
- mCRC, with fluoropyrimidine/irinotecan- or fluoropyrimidine/oxaliplatin-based chemotherapy for second-line treatment in patients who have progressed on a first-line Avastin-containing regimen
- Nonsquamous NSCLC, with carboplatin and paclitaxel for first-line treatment of unresectable, locally advanced, recurrent, or metastatic disease
- Glioblastoma, as a single agent for adult patients with progressive disease following prior therapy
- Metastatic renal cell cancer with interferon alfa
- Cervical cancer, in combination with paclitaxel and cisplatin or paclitaxel and topotecan in persistent, recurrent, or metastatic disease
- Platinum-resistant recurrent epithelial ovarian, fallopian tube, or primary peritoneal cancer, in combination with paclitaxel, pegylated liposomal doxorubicin, or topotecan [34]

7.6 Brentuximab Vedotin (Anti-CD30 Antibody, Adcetris®)

7.6.1 Drug Development History

The US FDA initially approved brentuximab vedotin in 2011 for the treatment of patients with Hodgkin's lymphoma (HL) after failure of autologous stem cell transplant (ASCT) or after failure of at least two prior multiagent chemotherapy regimens in patients who are not ASCT candidates. Subsequently, it was approved for the treatment of patients with systemic anaplastic large cell lymphoma (sALCL) after failure of at least one prior multiagent chemotherapy regimen [32, 61].

7.6.2 Target Biology

Brentuximab vedotin (Adcetris) is a CD30-directed mAb-drug conjugate [32].

7.6.3 Mechanism of Action

Brentuximab vedotin is a monoclonal antibody-drug conjugate (ADC). The antibody is a chimeric IgG1, which is directed against CD30. A small molecule, monomethyl auristatin E (MMAE), is covalently attached to the mAb via a protease-cleavable linker. MMAE is a potent microtubule-disrupting agent.

Nonclinical data suggests that the anticancer activity of Adcetris is due to the binding of the ADC to CD30-expressing cells, followed by internalization of the ADC–CD30 complex and the release of MMAE via proteolytic cleavage. Binding of MMAE to tubulin disrupts the microtubule network within the cell, subsequently inducing cell cycle arrest and apoptotic death of the cells [32, 61].

7.6.4 Clinical Efficacy

7.6.4.1 Hodgkin's Lymphoma

In one open-label, single-arm, multicenter trial, the efficacy of Adcetris was evaluated in patients with HL who relapsed after ASCT. Patients ($N = 102$) were treated with Adcetris 1.8 mg/kg intravenously over 30 min every 3 weeks.

Efficacy evaluations included overall response rate and duration of response. In this trial, 32% of patients achieved a complete response (CR) (median duration of response = 20.5 months), 40% of patients achieved a PR (median duration of response = 3.5 months), and 73% of patients achieved an overall response (median duration of response = 6.7 months) [32].

7.6.4.2 Systemic Anaplastic Large Cell Lymphoma (sALCL)

In one phase II, open-label, single-arm, multicenter trial, the efficacy of Adcetris was evaluated in patients with relapsed sALCL. The clinical trial included sALCL patients that had relapsed after prior therapy. Patients ($N = 58$) were treated with Adcetris 1.8 mg/kg given intravenously over 30 min every 3 weeks.

Efficacy evaluations included overall response rate and duration of response. In this trial, 57% of patients achieved a CR (median duration of response = 13.2 months), 29% of patients achieved a PR (median duration of response = 2.1 months), and 86% of patients achieved an overall response (median duration of response = 12.6 months) [32].

7.6.5 Approved Indications

Adcetris is indicated for:

- The treatment of patients with HL after failure of ASCT or after failure of at least two prior multiagent chemotherapy regimens in patients who are not ASCT candidates [32]

- The treatment of patients with sALCL after failure of at least one prior multiagent chemotherapy regimen [32]

7.7 Cetuximab (Anti-EGFR Antibody, Erbitux®)

7.7.1 Drug Development History

The US FDA approved cetuximab (Erbitux) for the treatment of head and neck cancer and colorectal cancer.

Head and Neck Cancer

March 1, 2006: In combination with RT for the treatment of locally or regionally advanced squamous cell carcinoma of the head and neck (SCCHN) in combination with RT

March 1, 2006: As a single agent for the treatment of patients with recurrent or metastatic SCCHN progressing after platinum-based therapy

November 7, 2011: First-line treatment of recurrent locoregional disease and/or metastatic SCCHN in combination with platinum-based therapy with 5-FU [22, 38].

Colorectal Cancer

February 12, 2004: In combination with irinotecan for the treatment of epidermal growth factor receptor (EGFR)-expressing metastatic colorectal carcinoma in patients who are refractory to irinotecan-based chemotherapy. The effectiveness of Erbitux in combination with irinotecan is based on objective response rate. Currently no data are available that demonstrate an improvement in disease-related symptoms or increased survival.

February 12, 2004: As a single agent for the treatment of EGFR-expressing mCRC after failure of both irinotecan- and oxaliplatin-based regimens or in patients who are intolerant to irinotecan-based regimens.

July 6, 2012: First-line treatment of wild-type KRAS, EGFR-expressing mCRC, in combination with FOLFIRI (irinotecan, 5-FU, and leucovorin), as determined by FDA-approved tests. FDA also approved the therascreen® KRAS RGQ PCR Kit concurrently with this approval [22, 38].

7.7.2 Target Biology

Erbitux (cetuximab) is a recombinant human/mouse chimeric IgG1 mAb that binds specifically to the extracellular domain of the human EGFR [38].

7.7.3 Mechanism of Action

The EGFR (HER1, c-erbB-1) is a transmembrane glycoprotein that is a member of a subfamily of type I receptor tyrosine kinases including EGFR, HER2, HER3, and HER4. The EGFR is constitutively expressed in many normal epithelial tissues, including the skin and hair follicle. Expression of EGFR is also detected in many human cancers including those of the head and neck, colon, and rectum [38].

Cetuximab binds specifically to the EGFR (HER1, c-erbB-1) on both normal and tumor cells and competitively inhibits the binding of epidermal growth factor (EGF) and other ligands, such as transforming growth factor-alpha [38].

In vitro, cetuximab can mediate antibody-dependent cell-mediated cytotoxicity (ADCC) against certain human tumor types. *In vitro* assays and *in vivo* animal studies have shown that binding of cetuximab to the EGFR blocks phosphorylation and activation of receptor-associated kinases. This results in inhibition of cell growth, induction of apoptosis, antitumor effects on cells that express EGFR, and decreased matrix metalloproteinase and VEGF production. No antitumor effects of cetuximab were observed in human tumor xenografts lacking EGFR expression. The addition of cetuximab to RT or irinotecan in human tumor xenograft models in mice resulted in an increase in antitumor effects compared to RT or chemotherapy alone [38].

Signal transduction through the EGFR results in activation of wild-type KRAS protein. However, in cells with activating *KRAS* somatic mutations, the mutant KRAS protein is continuously active and appears independent of EGFR regulation [38].

7.7.4 Clinical Efficacy

7.7.4.1 Squamous Cell Carcinoma of the Head and Neck (SCCHN)

Study 1 Study 1 was a randomized, multicenter, controlled trial of patients with locally or regionally advanced SCCHN ($N = 424$). Patients with stage III/IV SCCHN of the oropharynx, hypopharynx, or larynx with no prior therapy were randomized (1 : 1) to receive either Erbitux plus RT ($N = 211$) or RT alone ($N = 213$). RT was administered for 6–7 weeks as once-daily, twice-daily, or concomitant boost. Erbitux was administered as a 400 mg/m^2 initial dose beginning 1 week prior to initiation of RT, followed by 250 mg/m^2 weekly administered 1 h prior to RT for the duration of RT (6–7 weeks). Median duration of locoregional control was 24.4 months in the Erbitux/RT arm and 14.9 months in the RT-only arm. Median duration of OS was 49.0 months in the Erbitux/RT arm and was 29.3 months in the RT-only arm [38].

Study 2 Study 2 was an open-label, randomized, multicenter, controlled trial of patients with recurrent locoregional disease or metastatic SCCHN conducted outside the United States using EU-approved cetuximab as the clinical trial material ($N = 442$). Patients with no prior therapy for recurrent locoregional disease or metastatic SCCHN were randomized (1 : 1) to Arm 1 of the study (e.g., patients received EU-approved cetuximab plus cisplatin or carboplatin and 5-FU) or to Arm 2 of the study (e.g., patients received cisplatin or carboplatin and 5-FU alone). The treating physician had the discretion to choose either cisplatin or carboplatin. Cisplatin (100 mg/m^2, day 1) or carboplatin (AUC 5, day 1) plus intravenous 5-FU (1000 mg/m^2/day, days 1–4) was administered every 3 weeks (one cycle) for a maximum of six cycles in the absence of disease progression or unacceptable toxicity. Cetuximab was administered at a 400 mg/m^2 initial dose, followed by a 250 mg/m^2 weekly dose in combination with chemotherapy [38].

In Study 2, median duration of OS was 10.1 months for Arm 1 (patients who received the EU-approved cetuximab + platinum-based therapy + 5-FU (N = 222)) and was 7.4 months for Arm 2 (patients who received the platinum-based therapy and 5-FU only (N = 220)). Median duration for PFS was 5.5 months for Arm 1 and 3.3 months for Arm 2 [38].

Study 3 Study 3 was a single-arm, multicenter clinical trial in 103 patients with recurrent or metastatic SCCHN. All patients had documented disease progression within 30 days of a platinum-based chemotherapy regimen. Patients received a 20-mg test dose of Erbitux on day 1, followed by a 400 mg/m^2 initial dose, and 250 mg/m^2 weekly until disease progression or unacceptable toxicity. The objective response rate was 13% (95% CI: 7–21%). Median duration of response was 5.8 months (range 1.2–5.8 months) [38].

7.7.4.2 Erbitux Clinical Trials in EGFR-Expressing Recurrent Metastatic Colorectal Cancer

Study 4 Study 4 was a multicenter, open-label, randomized clinical trial conducted in 572 patients with EGFR-expressing, previously treated, recurrent mCRC. Patients were randomized (1 : 1) to receive either Erbitux plus best supportive care (BSC) or BSC alone. Erbitux was administered as a 400 mg/m^2 initial dose, followed by 250 mg/m^2 weekly until disease progression or unacceptable toxicity. The main outcome measure of the study was OS, and study results demonstrated improvement in OS [38].

Study 5 Study 5 was a multicenter clinical trial conducted in patients with EGFR-expressing recurrent mCRC (N = 329). Patients were randomized (2 : 1) to receive either Erbitux plus irinotecan (218 patients) or Erbitux monotherapy (111 patients). Erbitux was administered as a 400 mg/m^2 initial dose, followed by 250 mg/m^2 weekly until disease progression or unacceptable toxicity. In the Erbitux plus irinotecan arm, the patient received the same dose and schedule for irinotecan as they had previously failed. Acceptable irinotecan schedules were 350 mg/m^2 every 3 weeks, 180 mg/m^2 every 2 weeks, or 125 mg/m^2 weekly times four doses every 6 weeks [38].

Efficacy, based on durable objective responses, was evaluated in all randomized patients and in two prespecified subpopulations: irinotecan refractory patients and irinotecan and oxaliplatin failures. In patients receiving Erbitux plus irinotecan, the objective response rate was 23% (95% confidence interval (CI): 18–29%), median duration of response was 5.7 months, and median time to progression was 4.1 months. In patients receiving Erbitux monotherapy, the objective response rate was 11% (95% CI: 6–18%), median duration of response was 4.2 months, and median time to progression was 1.5 months. Similar response rates were observed in the predefined subsets in both the combination arm and monotherapy arm of the study [38].

7.7.4.3 Lack of Efficacy of Anti-EGFR Monoclonal Antibodies in Patients with mCRC Containing *KRAS* Mutations

Based on a retrospective analysis across seven randomized clinical trials, results suggest that anti-EGFR mAbs are not effective for the treatment of patients with mCRC containing *KRAS* mutations. In these trials, patients received standard of care (i.e., BSC or chemotherapy) and were randomized to receive either an anti-EGFR antibody (cetuximab or panitumumab) or no additional therapy. In all studies, investigational tests were used to detect *KRAS* mutations in codon 12 or 13. The percentage of study populations for which *KRAS* status was assessed ranged from 23 to 92% [38].

7.7.5 Approved Indications

Cetuximab (Erbitux) is indicated for the following.

Head and Neck Cancer

- Locally or regionally advanced SCCHN in combination with RT.
- Recurrent locoregional disease or metastatic SCCHN in combination with platinum-based therapy with 5-FU.
- Recurrent or metastatic SCCHN progressing after platinum-based therapy [38].

Colorectal Cancer

- As a single agent for the treatment of EGFR-expressing mCRC after failure of both irinotecan- and oxaliplatin-based regimens or in patients who are intolerant to irinotecan-based regimens.
- In combination with irinotecan for the treatment of EGFR-expressing metastatic colorectal carcinoma in patients who are refractory to irinotecan-based chemotherapy. Approval is based on objective response rate; no data are available demonstrating an improvement in increased survival [38].

7.8 Denosumab (Anti-RANKL Antibody, Xgeva™; Prolia™)

7.8.1 Drug Development History

The US FDA initially approved denosumab (Xgeva) on November 18, 2010, for the prevention of skeletal-related events (SREs) in patients with bone metastasis from solid tumors. Currently, it is the first and only receptor activator of nuclear factor kappa B ligand (RANKL) inhibitor. The FDA approved denosumab (Xgeva) on June 13, 2013, to treat giant cell tumor of the bone (GCTB). The FDA approved denosumab (Xgeva) on December 8, 2014, for hypercalcemia of malignancy (HCM) refractory to bisphosphonate therapy. Additionally, the FDA approved denosumab for the treatment of postmenopausal women with osteoporosis who are at high risk for fractures. This is marketed under the trade name Prolia [2, 5, 7, 13, 17, 18, 47, 52, 59].

7.8.2 Target Biology

Denosumab targets the RANKL and binds with high affinity and specificity to the RANKL [47].

7.8.3 Mechanism of Action

Denosumab, a fully human IgG_2 mAb, is a RANKL inhibitor. RANKL, which is produced by bone cells, is a protein involved in the formation, function, and survival of osteoclasts. RANKL binds to RANK on osteoclast precursor cells and stimulates them to differentiate into mature osteoclasts. Since osteoclasts are key for bone resorption, the result is bone breakdown and normal turnover of bone tissue [47].

Malignant cells release factors that induce an unregulated release of RANKL from marrow osteoblasts and marrow stromal cells. This increased RANKL production disrupts the natural balance of bone remodeling and leads to an overabundance of osteoclast differentiation. This can lead to bone destruction and devastating bone complications. Also, bone resorption releases growth factors from the bone matrix that may perpetuate tumor activity.

In patients with bone metastasis from solid tumors, denosumab binds to RANKL, prevents activation of the RANK receptor on the osteoclasts, and inhibits osteoclast formation, function, and survival. Denosumab prevents the maturation of osteoclasts, which decreases bone resorption and breaks the cycle of bone destruction.

In patients with GCTB, which is a rare, aggressive benign tumor, RANKL sends signals to GCTB that cause it to grow and lead to bone breakdown. Denosumab works on the RANKL to stop to the activity of the osteoclasts and GCTB and thereby slow down bone breakdown and GCTB growth [18, 47].

7.8.4 Clinical Efficacy

7.8.4.1 Prevention of Skeletal-Related Events in Patients with Bone Metastasis from Solid Tumors

The approval of denosumab is based on results from three international, randomized (1 : 1), double-blind, double-dummy trials in patients with bone metastasis comparing denosumab to zoledronic acid. Trial #1 (20050103) enrolled 1901 patients with castrate-resistant prostate cancer. Trial #2 (20050136) enrolled 2046 patients with breast cancer. Trial #3 (20050244) enrolled 1776 patients with multiple myeloma or solid tumors, other than breast or prostate cancer [7, 13, 18, 47].

In all three studies, patients were randomized to receive either denosumab (120 mg subcutaneously every 4 weeks) or zoledronic acid (4 mg intravenously every 4 weeks, with dose adjusted for reduced renal function). Data from the study results demonstrated that denosumab therapy resulted in a statistically significant delay in the time to first SRE and in the time to first and subsequent SRE compared to zoledronic acid in trial #1 (patients with breast cancer) or trial #2 (castrate-resistant prostate cancer). In trial #3, denosumab was noninferior to zoledronic acid in delaying the time to first SRE and did not demonstrate superiority. No differences in OS were observed in these three groups [7, 13, 18, 47].

7.8.4.2 Giant Cell Tumor of the Bone (GCTB)

Denosumab was approved for treating GCTB based on demonstration of durable objective responses observed in two multicenter, open-label trials. These studies enrolled adult and skeletally mature adolescents with histologically confirmed, measurable GCTB. These tumors were either recurrent, unresectable, or located where planned surgery was likely to result in severe morbidity. All patients received denosumab (120 mg subcutaneously) every 4 weeks and received additional doses on days 8 and 15 of the first month [5, 47].

In this study, 304 patients received denosumab. Radiographic assessments were completed at baseline and following denosumab therapy for 187 (61%) patients. An objective PR was identified in 47 of 187 patients for an overall PR rate of 25% (95% CI). The estimated median time to response was 3 months. In the 47 patients with an objective PR, the median duration of follow-up was 20 months (range 2–44 months), and 51% (24/47) had responses lasting at least 8 months. Three patients experienced disease progression following an objective response [5, 47].

7.8.4.3 Hypercalcemia of Malignancy Refractory to Bisphosphonate Therapy

Patients with advanced cancer may experience HCM due to excessive osteoclast-mediated bone resorption. Patients may not respond to or may relapse after IV bisphosphonate therapy. In this single-arm international study, participants had serum calcium levels corrected for albumin (CSC) >12.5 mg/dl (3.1 mmol/l) despite bisphosphonates given >7 and ≤30 days before screening [47].

Cancer patients ($N = 33$) had solid tumors or hematologic malignancies. By day 10, 21 patients (64%) reached CSC ≤11.5 mg/dl, and 12 patients (33%) reached CSC ≤10.8 mg/dl. During the study, 23 patients (70%) reached CSC ≤11.5 mg/dl, and 21 patients (64%) reached CSC ≤10.8 mg/dl. Estimated median response duration was 104 days. Despite recent IV bisphosphonate treatment in patients with HCM, denosumab lowered patients' serum calcium in 64% within 10 days and induced durable responses [7, 13, 17, 47, 52].

7.8.5 Approved Indications

- Denosumab (Xgeva) is indicated for the prevention of SREs in patients with bone metastases from solid tumors.
- Denosumab (Xgeva) is not indicated for the prevention of SREs in patients with multiple myeloma.
- Denosumab (Xgeva) is used to treat adults and some teenagers with GCTB that cannot be surgically removed (unresectable) or when surgery is likely to result in severe morbidity, such as loss of limbs or joint removal.
- Denosumab (Prolia) is used for the treatment of postmenopausal women with osteoporosis who are at high risk for fractures [47].

7.9 Eculizumab (Anti-C5 Antibody, Soliris®)

7.9.1 Drug Development History

The US FDA initially approved eculizumab (Soliris) on March 16, 2007, for the treatment of patients with paroxysmal nocturnal hemoglobinuria (PNH). Soliris is a first-in-class humanized mAb targeting C5. The FDA granted accelerated approval for the use of eculizumab (Soliris) for the treatment of pediatric and adult patients with atypical hemolytic uremic syndrome (aHUS) on September 23, 2011 [11, 45].

7.9.2 Target Biology

Eculizumab, the active ingredient of Soliris, is a recombinant humanized monoclonal IgG2/4k antibody that specifically binds to the complement protein, C5, with high affinity [45].

7.9.3 Mechanism of Action

Soliris is a complement inhibitor. Eculizumab specifically binds to the complement protein, C5, with high affinity. This inhibits its cleavage to C5a and C5b and prevents the generation of the terminal complement complex C5b-9. By blocking activation of the terminal complement, Soliris inhibits terminal complement-mediated intravascular hemolysis in PNH patients and complement-mediated thrombotic microangiopathy (TMA) in patients with aHUS [45].

The pathophysiologies of PNH and aHUS are both related to the functioning of the complement system. Patients with PNH have a genetic mutation that leads to the generation of populations of abnormal red blood cells (RBCs) that are deficient in terminal complement inhibitors. This deficiency renders PNH

RBCs sensitive to persistent terminal complement-mediated destruction. The result is intravascular hemolysis and anemia, due to the destruction and loss of these PNH RBCs. In aHUS, impairment in the regulation of complement activity leads to uncontrolled terminal complement activation, resulting in platelet activation, endothelial cell damage, and TMA [45].

7.9.4 Clinical Efficacy

7.9.4.1 Paroxysmal Nocturnal Hemoglobinuria (PNH)

Three studies have assessed the safety and efficacy of Soliris in PNH patients with hemolysis. In Study 1, PNH patients were assessed in a randomized, double-blind, placebo-controlled 26-week study. In Study 2, PNH patients were also treated with Soliris in a single-arm 52-week study. In Study 3, PNH patients were evaluated in a long-term extension study [45].

In all studies, PNH patients were given Soliris dosed at 600 mg every 7 ± 2 days for 4 weeks, followed by 900 mg every 7 ± 2 days later and then 900 mg every 14 ± 2 days for the duration of the study. Patients received meningococcal vaccination prior to receipt of Soliris and received Soliris as an intravenous infusion over 25–45 min [45].

7.9.4.2 Atypical Hemolytic Uremic Syndrome (aHUS)

Two prospective single-arm trials enrolling 37 adult and adolescent patients with aHUS and one retrospective trial of 19 pediatric patients and 11 adult patients with aHUS were reviewed [45].

In the first prospective trial of adult and adolescent patients ($N = 17$) with aHUS, patients were resistant to plasma therapy. Treatment with eculizumab resulted in six outcomes:

1) Elimination of the need for dialysis: Four of the five patients, who required dialysis upon entering the study, discontinued dialysis for the remainder of eculizumab therapy.
2) Sustained improvement in the estimated glomerular filtration rate (eGFR): Researchers observed a median improvement in eGFR in 9 of 17 (53%) patients who were enrolled in the study for a median duration of 251 days.
3) Elimination of the need for plasma therapy in most of the patients.
4) Sustained improvement in specific hematologic parameters that correlate with aHUS disease activity: achievement or maintenance of normal platelet counts and LDH levels for at least 4 weeks was achieved in 13 of 17 (76%) patients for a median duration of 37 weeks.
5) Improvement in other laboratory markers of hemolysis.
6) Evidence for suppression of terminal complement activity.

In the second prospective trial of adult and adolescent patients ($N = 20$) with aHUS, patients required chronic plasma therapy. Treatment with eculizumab

(Soliris) allowed cessation of plasma therapy and maintained renal function, as indicated by stable dialysis requirements and eGFR parameters. A total of 18 of 20 (90%) patients maintained hematologic normalization after discontinuing chronic plasma therapy. The median duration of hematologic normalization was 38 weeks [45].

In a retrospective trial, the outcomes of 19 pediatric patients (median age of 6 years) were consistent with the outcomes observed in the prospective studies. Four of eight (50%) pediatric patients, who previously required dialysis, were able to discontinue dialysis after eculizumab therapy. Also, 7 of 19 (37%) pediatric patients exhibited an improvement in eGFR of at least 15 mg/min/1.73 m^2, and 8 of 19 (42%) pediatric patients achieved or maintained normal hematologic parameters for at least 4 weeks. The need for plasma therapy was eliminated for most of the patients in this study [45].

7.9.5 Approved Indications

Indications for Soliris include:

- Treatment of patients with PNH to reduce hemolysis.
- Treatment of patients with aHUS to inhibit complement-mediated TMA.
- Soliris is not indicated for the treatment of patients with Shiga toxin *E. coli*-related hemolytic uremic syndrome (STEC-HUS).

7.10 Ibritumomab Tiuxetan (Anti-CD20 Antibody, Zevalin®)

7.10.1 Drug Development History

The US FDA approved ibritumomab (Zevalin) in February 2002 for the treatment of relapsed or refractory, low-grade, follicular, or transformed B-cell non-Hodgkin's lymphoma (NHL). It is also approved for previously untreated follicular NHL patients who achieve a PR or CR to first-line chemotherapy. Zevalin has been approved as part of a therapeutic regimen including Rituxan® (rituximab). This indication includes patients with Rituxan-refractory follicular NHL. Zevalin was the first radioimmunotherapy drug to receive FDA approval [49, 60].

7.10.2 Target Biology

The complementarity-determining regions of ibritumomab, a mAb, bind specifically to the CD20 antigen (human B-lymphocyte-restricted differentiation antigen, Bp35) on the B lymphocytes with high affinity [49, 60].

7.10.3 Mechanism of Action

Ibritumomab tiuxetan (Zevalin) is a mAb radiotherapy treatment. The drug combines the mAb mouse IgG1 antibody ibritumomab, in conjunction with the chelator tiuxetan, to which a radioactive isotope (yttrium-90) is linked. Ibritumomab, like Rituxan, binds to the CD20 antigen and induces apoptosis in CD20+ B-cell lines *in vitro*. The chelate tiuxetan, which tightly binds Y-90, is covalently linked to ibritumomab. Cytotoxic radiation is delivered directly to malignant NHL cells. The beta emission from Y-90 induces cellular damage by the formation of free radicals in the target and neighboring cells.

In patients with B-cell NHL, the CD20 antigen is expressed on pre- and mature B lymphocytes and on >90% of B-cell NHL. The CD20 antigen is not shed from the cell surface and does not internalize upon antibody binding [49].

7.10.4 Clinical Efficacy

Zevalin received FDA approval based on results from these four studies.

7.10.4.1 Study 1

In Study 1, which supported the full approval of Zevalin, the subjects ($N = 54$) were diagnosed with relapsed follicular lymphoma, and they were refractory to Rituxan treatment. An overall response rate of 74% was achieved with Zevalin treatment, with 15% of subjects experiencing a CR. Median duration of response was 6.4 months, and median time to progression was 6.8 months [49].

7.10.4.2 Study 2

Study 2 was a randomized (1 : 1), open-label, multicenter study, which compared the Zevalin therapeutic regimen with rituximab therapy. The subjects ($N = 130$) were diagnosed with relapsed or refractory low-grade or follicular NHL, and no patient had received rituximab prior to the study. Patients were randomized to receive either the Zevalin therapeutic regimen ($N = 64$) or rituximab ($N = 66$), which was administered as an IV infusion at 375 mg/m^2 weekly times four doses. The overall response rate was significantly higher for patients receiving the Zevalin therapeutic regimen (83 vs. 55%, $p < 0.001$). The CRR was also higher in patients receiving the Zevalin therapeutic regimen (38 vs. 18%). Median duration of response (14.3 vs. 11.5 months), and time to disease progression (12.1 vs. 10.1 months) was not significantly different between study arms [49].

7.10.4.3 Study 3

Study 3 was a single-arm study of lymphoma patients ($N = 30$) who received Y-90 Zevalin. In this sample, 27 patients had relapsed or refractory low-grade follicular NHL and a platelet count of 100,000 to 149,000/mm^3. All patients received Y-90 Zevalin (0.3 mCi/kg (11.1 MBq/kg)). In this study, objective, durable clinical responses were observed (89% overall response rate (95% CI: 70–97%)), and the median duration of response was 11.6 months [49].

7.10.4.4 Study 4

Study 4 was a multicenter, randomized, open-label study conducted in patients ($N=414$) with follicular NHL who had achieved a PR or CR (CRu) after first-line chemotherapy was completed. Randomization was stratified by center and response to first-line therapy (CR or PR). Patients were randomized to receive Zevalin ($n=208$) or no further therapy ($n=206$). Y-90 Zevalin was administered for at least 6 weeks but no more than 12 weeks following the last dose of chemotherapy. PFS was significantly prolonged among Zevalin-treated patients compared to those receiving no further treatment (median PFS 38 vs. 18 months; HR 0.46 (95% CI: 0.35, 0.60) $p<0.0001$ Cox model stratified by response to first-line therapy and initial treatment strategy (immediate vs. watch and wait)). The number of patients who died was too small to permit a reliable comparison on survival [49].

7.10.5 Approved Indications

Ibritumomab tiuxetan (Zevalin) is administered as part of the Zevalin therapeutic regimen when indicated for the treatment of patients with:

- Relapsed or refractory, low-grade, follicular, or transformed B-cell NHL
- Previously untreated follicular NHL who achieve a PR or CR to first line chemotherapy

7.11 Ipilimumab (Anti-CTLA-4 Antibody, Yervoy®)

7.11.1 Drug Development History

The US FDA approved ipilimumab (Yervoy) on March 21, 2011, for the treatment of unresectable or metastatic melanoma [48, 58].

7.11.2 Target Biology

Ipilimumab (Yervoy) is a mAb that binds to cytotoxic T-lymphocyte antigen (CTLA-4), which prevents CTLA-4 from binding with its ligands [48, 58].

7.11.3 Mechanism of Action

CTLA-4 is a negative regulator of T-cell activity. It may play a role in slowing down or turning off the body's immune system and affect the body's ability to kill cancer cells. When CTLA-4 binds with its ligands, CD80/CD86, there are a decrease in activation and proliferation of tumor-infiltrating T-effector cells and a general decrease in T-cell responsiveness [48].

Ipilimumab binds to CTLA-4 and blocks the interaction of CTLA-4 with its ligands. By blocking the binding of CTLA-4 with its ligands, T-cell responsiveness may increase, including the activation and proliferation of tumor-infiltrating T-effector cells and antitumor immune response. Ipilimumab may work by allowing the body's immune system to recognize, target, and attack cells in melanoma tumors [48].

7.11.4 Clinical Efficacy

Yervoy's safety and effectiveness were established in one international study of melanoma patients ($n = 676$). All study participants had stopped responding to other FDA-approved or commonly used treatments for melanoma. Also, participants had metastatic disease or disease that could not be surgically removed [48].

The study was designed to measure OS from the time when this treatment was started until the time of a patient's death. Patients were randomly assigned to one of three groups: (1) Yervoy plus an experimental tumor vaccine (e.g., gp100), (2) Yervoy alone, or (3) vaccine alone. Those who received the combination of Yervoy plus the vaccine or Yervoy alone lived an average of about 10 months, while those who received only the experimental vaccine lived an average of 6.5 months [48].

7.11.5 Approved Indications

- Ipilimumab (Yervoy) is indicated for the treatment of unresectable or metastatic melanoma [48].

7.12 Obinutuzumab (Gazyva®)

7.12.1 Drug Development History

The US FDA approved obinutuzumab (Gazyva) on November 1, 2013, for the treatment of patients with previously untreated chronic lymphocytic leukemia (CLL), in combination with chlorambucil [23, 39, 55].

7.12.2 Target Biology

Gazyva (obinutuzumab) is a humanized anti-CD20 mAb of the IgG1 subclass. It recognizes and binds to a specific epitope of the CD20 molecule, which is found on B cells [23, 39].

7.12.3 Mechanism of Action

Obinutuzumab is a mAb that targets the CD20 antigen expressed on the surface of pre-B and mature B lymphocytes. Upon binding to CD20, obinutuzumab mediates B-cell lysis in three ways: (1) antibody-dependent cell-mediated cytotoxicity (ADCC), which binds to and activates immune effector cells, which trigger the destruction of the mAb-bound cell; (2) direct cell death, which directly activates intracellular death signaling pathways; and/or (3) complement-dependent cytotoxicity (CDC), which triggers activation of the complement cascade. The two mechanisms of action of the immune effector cells include antibody-dependent cellular cytotoxicity (ADCC) and antibody-dependent cellular phagocytosis [14, 39].

Although both Gazyva and Rituxan target the CD20 antigen on B cells, their activity differs. Gazyva is engineered as a mAb with reduced fucose content. In preclinical studies, results demonstrated that decreased fucose content improved the binding and activation of immune effector cells. Obinutuzumab induces greater ADCC activity than rituximab *in vitro* using human cancer cell lines. Obinutuzumab also demonstrated an increased ability to induce direct cell death when compared to rituximab. Obinutuzumab binds to FcγRIII 14 using purified proteins with a higher affinity than rituximab. Obinutuzumab and rituximab bind with similar affinity to overlapping epitopes on CD20 [14, 39].

7.12.4 Clinical Efficacy

7.12.4.1 Chronic Lymphocytic Leukemia

In Study 1, Gazyva was evaluated in a three-arm, open-label, active control, randomized, multicenter trial in patients (N = 781) with previously untreated CD20+ CLL requiring treatment and coexisting medical conditions or reduced renal function as measured by creatinine clearance (CrCl) < 70 ml/min. Patients were randomized into one of three groups: treatment with chlorambucil (Arm 1—control group), treatment with Gazyva in combination with chlorambucil (Arm 2), or treatment with rituximab in combination with chlorambucil (Arm 3) [39].

Study 1 data was evaluated in two stages. In Stage 1, the safety and efficacy of Gazyva were evaluated by comparing data from Arm 1 versus Arm 2 in 356 patients. In Stage 2, the safety and efficacy were evaluated by comparing data from Arm 2 versus Arm 3 in 663 patients [39].

Most patients received Gazyva (1000 mg) on days 1, 8, and 15 of the first cycle, followed by treatment on the first day of five subsequent cycles (total of six cycles, 28 days each). The first dose of Gazyva was divided between day 1 (100 mg) and day 2 (900 mg), which was implemented in 140 patients. Chlorambucil (0.5 mg/kg) was administered orally on day 1 and day 15 of all treatment cycles (1–6) [39].

In the Stage 1 analysis, the median PFS in the Gazyva arm, in combination with chlorambucil, was 27.2 months, and median PFS was 11.2 months in the chlorambucil alone arm (median observation time 22.8 months). This was assessed by independent review and is consistent with investigator-assessed PFS. The median OS was not yet reached with a total of 46 deaths: 22 (9%) in the Gazyva in combination with chlorambucil arm and 24 (20%) in the chlorambucil arm. The hazard ratio for OS was 0.41 (95% CI: 0.23–0.74) [39].

In the Stage 2 analysis, the median PFS was 26.7 months in the Gazyva arm and 14.9 months in the rituximab arm with a median observation time of 18.7 months (HR: 0.42, 95% CI, *p*-value <0.0001). Minimal residual disease (MRD) was evaluated using allele-specific oligonucleotide polymerase chain reaction (ASO-PCR). Among patients who achieved CR and complete response with incomplete marrow recovery (CRi) (94 patients in the Gazyva arm and 34 patients in the rituximab arm), 18 patients (19%) had negative MRD in the bone marrow in the Gazyva arm compared to 2 patients (6%) in the rituximab arm. Out of the patients who achieved CR and CRi, 39 patients (41%) in the Gazyva arm and 4 patients (12%) in the rituximab arm were MRD negative in peripheral blood samples collected at least 3 months after the end of treatment [39].

7.12.5 Approved Indications

Gazyva, in combination with chlorambucil, is indicated for the treatment of patients with previously untreated CLL [39, 55].

7.13 Ofatumumab (Anti-CD20 Antibody, Arzerra®)

7.13.1 Drug Development History

The US FDA approved ofatumumab (Arzerra) on October 26, 2009, for the treatment of patients with CLL refractory to fludarabine and alemtuzumab. On April 17, 2014, the FDA approved ofatumumab (Arzerra) in combination with chlorambucil for the treatment of previously untreated patients with CLL for whom fludarabine-based therapy is considered inappropriate [4, 16, 33, 57].

7.13.2 Target Biology

Ofatumumab binds specifically to both the small and large extracellular loops (epitope) of the CD20 molecule on B cells. This epitope is different from the binding sites targeted by other CD20 antibodies currently available [33].

7.13.3 Mechanism of Action

Arzerra (ofatumumab) is an anti-CD20 cytolytic mAb. The Fab domain of ofatumumab binds to the CD20 molecule on B cells. The Fc domain of ofatumumab mediates immune effector functions, and this results in B-cell lysis *in vitro*. Data suggest that possible mechanisms of cell lysis include CDC and antibody-dependent cell-mediated cytotoxicity [33].

The CD20 molecule is a key target in CLL therapy because it is highly expressed in most B-cell malignancies. The CD20 molecule is expressed on normal B lymphocytes (pre- to mature B lymphocyte) and on B-cell CLL. The CD20 molecule is not shed from the cell surface and is not internalized following antibody binding [33].

7.13.4 Clinical Efficacy

7.13.4.1 Study 1

The first FDA approval of Arzerra was based on a single-arm, multicenter study in patients with relapsed or refractory CLL ($N = 154$). Specifically, the efficacy population included subjects with CLL who were refractory to fludarabine and alemtuzumab. Arzerra was administered by intravenous infusion according to the following schedule: 300 mg (week 0), 2000 mg weekly for seven infusions (weeks 1–7), and 2000 mg every 4 weeks for four infusions (weeks 12–24). The primary endpoint was objective tumor response rate. The overall response rate was 42%, and the median duration of response was 6.5 months. No patients achieved a CR [6, 16, 33].

7.13.4.2 Study 2

The second FDA approval of Arzerra was based on the results of a multicenter, randomized, open-label trial, which compared patients receiving ofatumumab and chlorambucil to patients receiving single-agent chlorambucil. In this trial, 447 patients were enrolled for whom fludarabine-based therapy was considered to be inappropriate by the investigator. Ineligibility reasons included advanced age or presence of comorbidities. In both arms, chlorambucil (10 mg/m^2 orally) was administered on days 1–7 every 28 days. Ofatumumab was administered as an intravenous infusion according to the following schedule: 300 mg (cycle 1, day 1), 1000 mg (cycle 1, day 8), and 1000 mg (day 1 of all subsequent 28 day cycles).

The primary endpoint of the trial was PFS. Median PFS was 22.4 months (95% CI) for patients receiving Arzerra and chlorambucil, as compared to 13.1 months (95% CI) for patients receiving single-agent chlorambucil (hazard ratio 0.57 (95% CI), stratified log-rank p-value <0.001).

Based on the results of this randomized trial, GlaxoSmithKline fulfilled the postmarketing requirement to verify the clinical benefit of ofatumumab; therefore, the FDA approval of ofatumumab was converted from accelerated to regular approval [6, 16, 33].

7.13.5 Approved Indications

Ofatumumab (Arzerra) is approved for treating patients with the following indications:

- For the treatment of patients with CLL refractory to fludarabine and alemtuzumab
- In combination with chlorambucil for the treatment of previously untreated patients with CLL for whom fludarabine-based therapy is considered inappropriate [4, 33, 57]

7.14 Panitumumab (Anti-EGFR Antibody, Vectibix™)

7.14.1 Drug Development History

The US FDA approved panitumumab (Vectibix) on September 27, 2006. Vectibix is the first entirely human mAb approved. Initially, it was approved as monotherapy for treatment of patients with EGFR-expressing mCRC after disease progression on, or following, chemotherapy regimens containing fluoropyrimidine, oxaliplatin, and irinotecan [1, 24, 46].

In 2009, the FDA issued a labeling change to limit use of Vectibix for the treatment of wild-type KRAS (exon 2) mCRC. Vectibix is not indicated for the treatment of patients with *KRAS*-mutant mCRC or for whom *KRAS* mutation status is unknown [46].

On May 23, 2014, Vectibix was approved for use in combination with FOLFOX, an oxaliplatin-based chemotherapy regimen, as first-line treatment in patients with wild-type KRAS (exon 2) mCRC. This approval converts the accelerated monotherapy approval to full approval for Vectibix. The FDA also approved the therascreen KRAS RGQ PCR Kit, which was developed by QIAGEN, as a companion diagnostic test for Vectibix [24, 46].

7.14.2 Target Biology

Panitumumab (Vectibix) binds specifically to EGFR on both normal and tumor cells and competitively inhibits the binding of ligands for EGFR [46].

7.14.3 Mechanism of Action

EGFR is constitutively expressed in normal epithelial tissues, including the skin and hair follicle. EGFR is overexpressed in certain human cancers, including colon and rectum cancers. Interaction of EGFR with its normal ligands (e.g., EGF, transforming growth factor-alpha) leads to phosphorylation and activation of a series of intracellular proteins, which, in turn, regulate transcription of genes involved with cellular growth and survival, motility, and proliferation. Signal transduction through the EGFR results in activation of the wild-type *KRAS* protein. However, in cells with activating *KRAS* somatic mutations, the *KRAS*-mutant protein is continuously active and appears independent of EGFR regulation [46].

Nonclinical studies demonstrate that binding of panitumumab to the EGFR prevents ligand-induced receptor autophosphorylation and activation of receptor-associated kinases. The result is cell growth inhibition, induction of apoptosis, internalization of the EGFR, and decreased proinflammatory cytokine and vascular growth factor production. *In vitro* assays and *in vivo* animal studies demonstrate that panitumumab inhibits the growth and survival of selected human tumor cell lines expressing EGFR [46].

7.14.4 Clinical Efficacy

7.14.4.1 Recurrent or Refractory mCRC

The safety and efficacy of Vectibix were demonstrated in Study 1, an open-label, multinational, randomized controlled trial of 463 patients with EGFR-expressing metastatic carcinoma of the colon or rectum, and in Study 2, an open-label, multicenter, multinational randomized trial of 1010 patients with wild-type KRAS mCRC [46].

Study 1 Patients in Study 1 were required to have progressed on or following treatment with a regimen containing a fluoropyrimidine, oxaliplatin, and irinotecan. Patients were randomized (1:1) to receive panitumumab at a dose of 6 mg/kg given once every 2 weeks plus BSC ($N=231$) or BSC alone ($N=232$) until investigator-determined disease progression. Randomization was stratified based on Eastern Cooperative Oncology Group (ECOG) performance status (PS) (0 and 1 vs. 2) and geographic region (Western Europe, Eastern/Central Europe, or others). When the investigator determined that the disease had progressed, patients in the BSC-alone arm were eligible to receive panitumumab and were followed until disease progression was confirmed by the IRC [46].

Based upon IRC determination of disease progression, a statistically significant prolongation in PFS was observed in patients receiving panitumumab compared to those receiving BSC alone. The mean PFS was 96 days in the panitumumab arm and 60 days in the BSC-alone arm.

The study results were analyzed in the wild-type KRAS subgroup where KRAS status was retrospectively determined using archived paraffin-embedded tumor tissue. KRAS mutation status was determined in 427 patients (92%); of these, 243 (57%) had no detectable KRAS mutations in either codons 12 or 13. The hazard ratio for PFS in patients with wild-type KRAS mCRC was 0.45 (95% CI: 0.34–0.59), favoring the panitumumab arm. The response rate for the panitumumab arm was 17% and for BSC arm was 0%. There were no differences in OS. Patients in the BSC arm (77%) received panitumumab at the time of disease progression [46].

Study 2 Study 2 was an open-label, multicenter, multinational, randomized (1:1) clinical trial, stratified by region (North America, Western Europe, and Australia vs. rest of the world) and ECOG PS (0 and 1 vs. 2) in patients with wild-type KRAS mCRC. Patients ($n=1010$) who received prior treatment with irinotecan, oxaliplatin, and a thymidylate synthase inhibitor were randomized to receive Vectibix 6 mg/kg intravenously over 60 min every 14 days or cetuximab 400 mg/m^2 intravenously over 120 min on day 1 followed by 250 mg/m^2 intravenously over 60 min every 7 days. The major efficacy analysis tested whether the OS of Vectibix was noninferior to cetuximab. Data for investigator-assessed PFS and objective response rate were also collected. The criteria to determine noninferiority was for Vectibix to retain at least 50% of the OS benefit of cetuximab. More patients (62%) had colon cancer than rectal cancer (38%). Most patients (74%) had not received prior bevacizumab. The key efficacy analysis for Study 2 demonstrated that Vectibix was statistically significantly noninferior to cetuximab for OS [46].

7.14.5 Approved Indications

Metastatic Colorectal Cancer

Panitumumab (Vectibix) is indicated for the treatment of patients with wild-type KRAS (exon 2 in codons 12 or 13) mCRC as determined by an FDA-approved test for this use:

- As first-line therapy in combination with FOLFOX
- As monotherapy following disease progression after prior treatment with fluoropyrimidine-, oxaliplatin-, and irinotecan-containing chemotherapy

Limitation of Use

Vectibix is not indicated for the treatment of patients with KRAS-mutant mCRC or for whom KRAS mutation status is unknown [46].

7.15 Pembrolizumab (Keytruda®)

7.15.1 Drug Development History

The US FDA approved pembrolizumab (Keytruda) on September 4, 2014, for the treatment of patients with unresectable or metastatic melanoma and disease progression following ipilimumab (e.g., a type of immunotherapy) and, if BRAF V600 mutation positive, a BRAF inhibitor (e.g., a therapy that blocks activity of BRAF gene mutations). This is the first FDA-approved drug that blocks a cellular pathway known as PD-1 [3, 19, 42, 56].

7.15.2 Target Biology

Pembrolizumab (Keytruda) is a humanized mAb that blocks the interaction between the programmed death receptor-1 (PD-1) and its ligands, PD-L1 and PD-L2 [19, 42].

7.15.3 Mechanism of Action

The PD-1 pathway restricts the body's immune system from attacking melanoma cells. Binding of the PD-1 ligands (e.g., PD-L1 and PD-L2) to the PD-1 receptor inhibits T-cell proliferation and cytokine production. Upregulation of PD-1 ligands occurs in some tumors. Signaling through the PD-1 pathway can contribute to inhibition of active T-cell immune surveillance of tumors [19, 42].

Pembrolizumab (Keytruda) is a humanized mAb that binds to the PD-1 receptor, which is found on T cells. Keytruda blocks the interaction of PD-1 with its ligands, PD-L1 and PD-L2. Keytruda releases the PD-1 pathway-mediated inhibition of the immune response, which includes the antitumor immune response [19, 42].

7.15.4 Clinical Efficacy

Under accelerated approval from the FDA, Keytruda was approved based on tumor response rate and durability of response. The FDA granted Keytruda breakthrough therapy designation as part of this accelerated approval program. Through preliminary clinical evidence, the sponsor demonstrated that the drug may offer a significant improvement over therapies that are currently available on the market. Also, Keytruda received priority review and orphan product designation. Clinical trial results have not yet established an improvement in survival or disease-related symptoms. Continued approval for this indication may be contingent upon verification and description of clinical benefit in the confirmatory trials [19, 42].

The clinical trial evaluated Keytruda's efficacy in an open-label, randomized, comparative dose trial. Clinical trial participants ($N = 173$) were diagnosed with unresectable or metastatic melanoma with progression of disease and were refractory to two or more doses of ipilimumab (3 mg/kg or higher) and, if BRAF V600 mutation positive, a BRAF or MEK inhibitor and disease progression within 24 weeks following the last dose of ipilimumab. All subjects were treated with Keytruda. Subjects were randomized to receive Keytruda 2 mg/kg ($n = 89$) or 10 mg/kg ($n = 84$) every 3 weeks until unacceptable toxicity or disease progression that was symptomatic was rapidly progressive, required urgent intervention, occurred with a decline in PS, or was confirmed at 4–6 weeks with repeat imaging. Tumor status assessment was performed every 12 weeks [19, 42].

The major efficacy outcome measure was confirmed overall response rate, which was assessed according to Response Evaluation Criteria in Solid Tumors (RECIST 1.1). In the Keytruda 2 mg/kg arm, 24% of subjects ($N = 21$) achieved an overall response rate (95% CI: 15, 34) consisting of 1 CR and 20 PR. Among the 21 subjects with an objective response, 3 (14%) had progression of disease at 2.8, 2.9, and 8.2 months after initial response. The remaining 18 subjects (86%) had ongoing responses with durations ranging from 1.4+ to 8.5+ months, which included 8 subjects with ongoing responses of 6 months or longer. One additional subject developed two new asymptomatic lesions at the first tumor assessment concurrent with a 75% decrease in overall tumor burden; Keytruda was continued and this reduction in tumor burden was durable for 5+ months. Subjects with and without BRAF V600 mutation-positive melanoma demonstrated objective responses. In the 10 mg/kg arm of the study, a similar percentage of patients had their tumor shrink [19, 42].

7.15.5 Approved Indications

Pembrolizumab (Keytruda) is indicated for the treatment of patients with unresectable or metastatic melanoma and disease progression following ipilimumab and, if BRAF V600 mutation positive, a BRAF inhibitor [3, 19, 42, 56].

7.16 Pertuzumab (Perjeta®)

7.16.1 Drug Development History

The US FDA initially approved pertuzumab (Perjeta) in 2012 for treatment of MBC. In 2013, Perjeta was approved for neoadjuvant treatment of breast cancer [25, 43].

7.16.2 Target Biology

Pertuzumab (Perjeta) is a recombinant humanized mAb that targets the extracellular dimerization domain (subdomain II) of the human epidermal growth factor receptor 2 protein (HER2) [43].

7.16.3 Mechanism of Action

Pertuzumab targets the extracellular dimerization domain (subdomain II) of the human epidermal growth factor receptor 2 protein and, thereby, blocks ligand-dependent heterodimerization of HER2 with other HER family members, including EGFR, HER3, and HER4. As a result, pertuzumab inhibits ligand-initiated intracellular signaling through two major signal pathways, mitogen-activated protein (MAP) kinase and phosphoinositide 3-kinase (PI3K). Inhibition of these signaling pathways can result in cell growth arrest and apoptosis, respectively. In addition, pertuzumab mediates antibody-dependent cell-mediated cytotoxicity (ADCC) [43].

While pertuzumab alone inhibited the proliferation of human tumor cells, the combination of pertuzumab and trastuzumab augmented antitumor activity in HER2-overexpressing xenograft models.

7.16.4 Clinical Efficacy

7.16.4.1 Metastatic Breast Cancer

The first study was a multicenter, double-blind, placebo-controlled trial that included patients ($N = 808$) with HER2-positive MBC. Patients were randomly allocated 1 : 1 to receive placebo plus trastuzumab and docetaxel or Perjeta plus trastuzumab and docetaxel. Randomization was stratified by prior treatment and geographic region [43].

Perjeta was given intravenously at an initial dose of 840 mg, followed by 420 mg every 3 weeks thereafter. Trastuzumab was given intravenously at an initial dose of 8 mg/kg, followed by 6 mg/kg every 3 weeks thereafter. Patients were treated with Perjeta and trastuzumab until progression of disease, withdrawal of consent, or unacceptable toxicity. Docetaxel was given as an initial dose of 75 mg/m^2 by intravenous infusion every 3 weeks for at least six cycles. The docetaxel dose could be escalated to 100 mg/m^2 at the investigator's discretion if the initial dose was well tolerated. At the time of the primary analysis, the mean number of cycles of study treatment administered was 16.2 in the placebo-treated group and 19.9 in the Perjeta-treated group [43].

The primary endpoint of Study 1 was PFS as assessed by an independent review facility (IRF). Results from this study demonstrated a statistically significant improvement in IRF-assessed PFS in the Perjeta-treated group compared with the placebo-treated group and an increase in median PFS of 6.1 months (median PFS of 18.5 months in the Perjeta-treated group vs. 12.4 months in the placebo-treated group). CR was 80.2% ($n = 275$) in the Perjeta-treated group versus 69.3% ($n = 233$) in the placebo-treated group. PR was 5.5% ($n = 19$) in the Perjeta-treated group versus 4.2% ($n = 14$) in the placebo-treated group [43].

7.16.4.2 Neoadjuvant Treatment of Breast Cancer

The second study was a multicenter, randomized trial that was conducted in patients ($N = 417$) with operable, locally advanced, or inflammatory HER2-positive breast cancer (T2-4d) who were scheduled to receive neoadjuvant therapy. Patients were randomly allocated to receive 1 of 4 neoadjuvant regimens prior to surgery as follows: trastuzumab plus docetaxel, Perjeta plus trastuzumab and docetaxel, Perjeta plus trastuzumab, or Perjeta plus docetaxel. Randomization was stratified by breast cancer type (operable, locally advanced, or inflammatory) and estrogen receptor (ER) or progesterone receptor (PgR) positivity. Perjeta was given intravenously at an initial dose of 840 mg, followed by mg every 3 weeks for four cycles [43].

Trastuzumab was given intravenously at an initial dose of 8 mg/kg, followed by 6 mg/kg every 3 weeks for four cycles. Docetaxel was given as an initial dose of 75 mg/m^2 by intravenous infusion every 3 weeks for four cycles. The docetaxel dose could be escalated to 100 mg/m^2 at the investigator's discretion if the initial dose was well tolerated. Following surgery all patients received three cycles of 5-FU (600 mg/m^2), epirubicin (90 mg/m^2), and cyclophosphamide (600 mg/m^2) (FEC) given intravenously every 3 weeks and trastuzumab administered intravenously every 3 weeks to complete 1 year of therapy. After surgery, patients in the Perjeta plus trastuzumab arm received docetaxel every 3 weeks for four cycles prior to FEC [43].

The primary endpoint of the study was pathological complete response (pCR) rate in the breast (ypT0/is). The FDA-preferred definition of pCR is the absence of invasive cancer in the breast and lymph nodes (ypT0/is ypN0). Statistically significant improvements in pCR rates by both the study and

FDA-preferred definitions were observed in patients receiving Perjeta plus trastuzumab and docetaxel compared to patients receiving trastuzumab plus docetaxel. The pCR rates and magnitude of improvement with Perjeta were lower in the subgroup of patients with hormone receptor-positive tumors compared to patients with hormone receptor-negative tumors [43].

7.16.5 Approved Indications

Pertuzumab (Perjeta) is indicated for:

- Treatment of MBC: Use in combination with trastuzumab and docetaxel for treatment of patients with HER2-positive MBC who have not received prior anti-HER2 therapy or chemotherapy for metastatic disease
- Neoadjuvant treatment of breast cancer: Use in combination with trastuzumab and docetaxel as neoadjuvant treatment of patients with HER2-positive, locally advanced, inflammatory, or early-stage breast cancer (either >2 cm in diameter or node positive) as part of a complete treatment regimen for early breast cancer

Limitations of Use

- The safety of Perjeta as part of a doxorubicin-containing regimen has not been established.
- The safety of Perjeta administered for greater than six cycles for early breast cancer has not been established.

7.17 Ramucirumab (Cyramza®)

7.17.1 Drug Development History

The US FDA approved ramucirumab (Cyramza) in April 2014 for the treatment of patients with advanced or metastatic gastric or gastroesophageal junction (GEJ) adenocarcinoma, with disease progression on or after prior fluoropyrimidine- or platinum-containing chemotherapy [37].

In November 2014, Cyramza was approved, in combination with docetaxel, for the treatment of metastatic NSCLC with disease progression on or after platinum-based chemotherapy. Patients with EGFR or ALK genomic tumor aberrations should have disease progression on FDA-approved therapy for these aberrations prior to receiving Cyramza [37].

In December 2014, Cyramza was approved, in combination with docetaxel, for second-line metastatic NSCLC [37].

7.17.2 Target Biology

Ramucirumab (Cyramza) is a recombinant human IgG1 mAb that specifically binds to VEGF receptor 2. Cyramza has an approximate molecular weight of 147 kDa. Cyramza is produced in genetically engineered mammalian NS0 cells.

7.17.3 Mechanism of Action

Ramucirumab (Cyramza) is a human VEGF receptor 2 antagonist that specifically binds VEGF receptor 2 and blocks binding of VEGFR ligands, VEGF-A, VEGF-C, and VEGF-D. As a result, ramucirumab inhibits ligand-stimulated activation of VEGF receptor 2, thereby inhibiting ligand-induced proliferation, and migration of human endothelial cells. Ramucirumab inhibited angiogenesis in *in vivo* animal models [37].

7.17.4 Clinical Efficacy

7.17.4.1 Gastric Cancer

The first study was a multinational, randomized, double-blind, multicenter study of patients ($N = 355$) receiving Cyramza plus BSC versus placebo plus BSC. Patients were randomized (2:1), including those with locally advanced or metastatic gastric cancer (including adenocarcinoma of the GEJ) who previously received platinum- or fluoropyrimidine-containing chemotherapy. Patients received either an intravenous infusion of Cyramza 8 mg/kg ($n = 238$) or placebo solution ($n = 117$) every 2 weeks. Demographic and baseline characteristics were similar between treatment arms.

The major efficacy outcome measure was OS and the supportive efficacy outcome measure was PFS. OS was 5.2 months in the Cyramza arm ($n = 238$) and 3.8 months in the placebo arm ($n = 117$). PFS was 2.1 months in the Cyramza arm and 1.3 months in the placebo arm.

The second study was a multinational, randomized, double-blind study of patients ($N = 665$) receiving Cyramza plus paclitaxel versus placebo plus paclitaxel. Patients were randomized (1:1), including patients with locally advanced or metastatic gastric cancer (including adenocarcinoma of the GEJ) who previously received platinum- and fluoropyrimidine-containing chemotherapy. Patients were randomized to receive either Cyramza 8 mg/kg ($n = 330$) or placebo ($n = 335$) as an intravenous infusion every 2 weeks (on days 1 and 15) of each 28-day cycle. Patients in both arms received paclitaxel 80 mg/m^2 by intravenous infusion on days 1, 8, and 15 of each 28-day cycle. Demographics and baseline characteristics were similar between treatment arms.

The major efficacy outcome measure was OS and the supportive efficacy outcome measures were PFS and objective response rate. OS was 9.6 months in the Cyramza plus paclitaxel arm and 7.4 months in the placebo arm. PFS was 4.4 months in the Cyramza plus paclitaxel arm and 2.9 months in the placebo plus paclitaxel arm [37].

7.17.4.2 Non-small Cell Lung Cancer

The REVEL study was a multinational, randomized, double-blind, phase III study of Cyramza plus docetaxel versus placebo plus docetaxel. Patients ($N = 1253$) were randomized (1:1), including NSCLC patients with disease progression on or after one platinum-based therapy for locally advanced or metastatic disease. Randomization was stratified by geographic region, gender, prior maintenance therapy, and ECOG PS. Patients were randomized to receive either Cyramza at 10 mg/kg or placebo by intravenous infusion, in combination with docetaxel at 75 mg/m^2 every 21 days. Demographics and baseline characteristics were similar between treatment arms.

The major efficacy outcome measure was OS and the supportive efficacy outcome measures were PFS and objective response rate. OS and PFS were statistically significantly improved in patients randomized to receive Cyramza plus docetaxel compared to patients randomized to receive placebo plus docetaxel. Objective response rate (CR + PR) was 23% for Cyramza plus docetaxel and 14% for placebo plus docetaxel [37].

7.17.5 Approved Indications

Cyramza is a human VEGF receptor 2 antagonist given:

- As a single agent or in combination with paclitaxel for the treatment of advanced gastric or GEJ adenocarcinoma, with disease progression on or after prior fluoropyrimidine- or platinum-containing chemotherapy.
- In combination with docetaxel for the treatment of metastatic NSCLC with disease progression on or after platinum-based chemotherapy. Patients with EGFR or ALK genomic tumor aberrations should have disease progression on FDA-approved therapy for these aberrations prior to receiving Cyramza.
- In combination with docetaxel in second-line metastatic NSCLC.

7.18 Rituximab (Anti-CD20 Antibody, Rituxan)

7.18.1 Drug Development History

The US FDA approved rituximab (Rituxan) initially in 1997 for the treatment of NHL. Rituxan (rituximab) is now approved for CLL, rheumatoid arthritis (RA) in combination with methotrexate (MTX) in adult patients with moderately to severely active RA who have inadequate response to one or more TNF antagonist therapies, and granulomatosis with polyangiitis (GPA) (Wegener's granulomatosis) and microscopic polyangiitis (MPA) in adult patients in combination with glucocorticoids [9, 26, 44, 54].

7.18.2 Target Biology

Rituxan (rituximab) is a genetically engineered chimeric murine/human monoclonal IgG1 kappa antibody directed against the CD20 antigen [44].

7.18.3 Mechanism of Action

Rituximab is a mAb that targets the CD20 antigen expressed on the surface of pre-B and mature B lymphocytes. Upon binding to CD20, rituximab mediates B-cell lysis. Possible mechanisms of cell lysis include CDC and antibody-dependent cell-mediated cytotoxicity (ADCC). The CD20 antibody induced apoptosis in the DHL 4 human B-cell lymphoma cell line.

B cells are believed to play a role in the pathogenesis of RA and associated chronic synovitis. For the RA population, B cells may be acting at multiple sites in the autoimmune/inflammatory process, including through production of rheumatoid factor (RF) and other autoantibodies, antigen presentation, T-cell activation, and/or proinflammatory cytokine production [9, 44].

7.18.4 Clinical Efficacy

7.18.4.1 Relapsed or Refractory, Low-Grade or Follicular, CD20-Positive B-Cell NHL

Three studies were conducted to determine clinical efficacy of Rituxan in relapsed or refractory, low-grade or follicular, CD20-positive B-cell NHL [44].

Study 1 A multicenter, open-label, single-arm study was conducted in patients ($n = 166$) with relapsed or refractory, low-grade or follicular B-cell NHL. These patients received Rituxan (375 mg/m^2), which was administered as a weekly intravenous infusion for four doses. Study results included an overall response rate of 48%, CRR of 6%, and median duration of response of 11.2 months. The median time to onset of response was 50 days. Disease-related signs and symptoms (including B symptoms) resolved in 64% (25/39) of those patients with such symptoms at study entry [44].

Study 2 In a multicenter, single-arm study, patients ($n = 37$) with relapsed or refractory, low-grade NHL received Rituxan (375 mg/m^2) weekly for eight doses. Study results included an overall response rate of 57%, CRR of 14%, and median duration of response of 13.4 months [44].

Study 3 In a multicenter, single-arm study, 60 patients received 375 mg/m^2 of Rituxan weekly for four doses. All patients had relapsed or refractory, low-grade or follicular B-cell NHL and had achieved an objective clinical response to Rituxan administered 3.8–35.6 months (median 14.5 months) prior to retreatment with Rituxan. Of these 60 patients, five received more than one additional course of Rituxan. Study results included an overall response rate of 38%, CRR of 10%, and median duration of response of 15.0 months [44].

7.18.4.2 Previously Untreated, Low-Grade or Follicular, CD20-Positive B-Cell NHL

The safety and effectiveness of Rituxan in previously untreated, low-grade or follicular CD20+ NHL were demonstrated in three randomized controlled trials enrolling 1662 patients.

Study 4 In an open-label, multicenter study, patients with previously untreated follicular NHL ($n = 322$) were randomized (1 : 1) to receive up to eight 3-week cycles of CVP chemotherapy alone (CVP) or in combination with Rituxan (375 mg/m^2) on day 1 of each cycle (R-CVP). The primary outcome measure of the study was PFS. The results for PFS, as determined by a blinded, independent assessment of progression, are 2.4 years for R-CVP ($n = 162$) and 1.4 years for CVP ($n = 160$), with $p < 0.0001$, two-sided stratified log-rank test [44].

Study 5 An open-label, multicenter, randomized (1 : 1) study was conducted in patients with previously untreated follicular NHL ($n = 1018$) who achieved a response (CR or PR) to Rituxan in combination with chemotherapy. Patients were randomized to Rituxan as single-agent maintenance therapy, 375 mg/m^2 every 8 weeks for up to 12 doses or to observation. Rituxan was initiated at 8 weeks following completion of chemotherapy. The main outcome measure of the study was PFS. PFS was longer in patients randomized to Rituxan as single-agent maintenance therapy [44].

Study 6 Patients with previously untreated low-grade B-cell NHL ($n = 322$), who did not progress after six or eight cycles of CVP chemotherapy, were enrolled in an open-label, multicenter, randomized trial. Patients were randomized (1 : 1) to receive Rituxan (375 mg/m^2 IV infusion) once weekly for 4 doses every 6 months for up to 16 doses or no further therapeutic intervention. The main outcome measure of the study was PFS. There was a reduction in the risk of progression, relapse, or death for patients randomized to Rituxan as compared to those who received no additional treatment [44].

7.18.4.3 Diffuse Large B-Cell NHL (DLBCL)

The safety and effectiveness of Rituxan were evaluated in three randomized, active-controlled, open-label, multicenter studies with 1854 patients enrolled. Patients with previously untreated diffuse large B-cell NHL received Rituxan in combination with cyclophosphamide, doxorubicin, vincristine, and prednisone (CHOP) or other anthracycline-based chemotherapy regimens [44].

Study 7 Patients with DLBCL (including primary mediastinal B-cell lymphoma) ($n = 632$), age ≥ 60 years, were randomized in a 1 : 1 ratio to treatment with CHOP or R-CHOP. Patients received six or eight cycles of CHOP, each cycle lasting 21 days. All patients in the R-CHOP arm received four doses of Rituxan 375 mg/m^2 on days −7 and −3 (prior to cycle 1) and 48–72 h prior to cycles 3 and 5. Patients who received eight cycles of CHOP also received a fifth dose of Rituxan prior to cycle 7. The main outcome measure of the study was PFS. Responding patients underwent a second randomization to receive Rituxan or no further therapy. Analysis of results after the second randomization demonstrates that for patients randomized to R-CHOP, additional Rituxan exposure beyond induction was not associated with further improvements in PFS or OS [44].

Study 8 Patients with DLBCL ($n = 399$), age ≥ 60 years, were randomized in a 1 : 1 ratio to receive CHOP or R-CHOP. All patients received up to eight 3-week cycles of CHOP induction; patients in the R-CHOP arm received Rituxan 375 mg/m^2 on day 1 of each cycle. The main outcome measure of the study was event-free survival (EFS), defined as the time from randomization to relapse, progression, change in therapy, or death from any cause. Efficacy results include for R-CHOP: EFS = 2.9 years and for CHOP: EFS = 1.1 years. OS estimates at 5 years were 58% (for R-CHOP) versus 46% (for CHOP) [44].

Study 9 Patients with DLBCL ($n = 823$), aged 18–60 years, were randomized in a 1 : 1 ratio to receive an anthracycline-containing chemotherapy regimen alone or in combination with Rituxan. The main outcome measure of the study was time to treatment failure. Efficacy results are not reliably estimable [44].

7.18.4.4 Ninety-Minute Infusions in Previously Untreated Follicular NHL and DLBCL

Study 10 A total of 363 patients with previously untreated follicular NHL ($n = 113$) or DLBCL ($n = 250$) were evaluated in a prospective, open-label, multicenter, single-arm trial for the safety of 90-min rituximab infusions. Patients with follicular NHL received rituximab 375 mg/m^2 plus CVP chemotherapy. Patients with DLBCL received rituximab 375 mg/m^2 plus CHOP chemotherapy. All patients received premedications prior to Rituxan infusion. Eligible patients received their cycle 2 rituximab infusion as follows: 20% of the total dose given in the first 30 min and the remaining 80% of the total dose given over the next 60 min. During cycle 2, the incidence of Grade 3–4 infusion-related reactions was 1.1% among all patients. During cycles 2–8, the incidence of Grade 3–4 infusion-related reactions was 2.8%. No acute fatal infusion-related reactions were observed [44, 59].

7.18.4.5 Chronic Lymphocytic Leukemia (CLL)

Two randomized (1 : 1), multicenter, open-label studies—which compared fludarabine and cyclophosphamide (FC) (fludarabine 25 mg/m^2/day and cyclophosphamide 250 mg/m^2/day) alone or in combination with Rituxan for up to six cycles in patients with CLL—evaluated the safety and effectiveness of Rituxan. Patients received FC on days 1, 2, and 3 of each cycle, with or without Rituxan. In both studies, 71% of CLL patients received six cycles and 90% received at least three cycles of Rituxan-based therapy. The main outcome measure in both studies was PFS [44].

Study 11 Patients with previously untreated CLL (N = 817) were entered into this trial. The PFS was 39.8 months for patients in the Rituxan–FC arm (n = 408) and was 31.5 months for patients in the FC arm (n = 409). Response rate was 86% for patients in the Rituxan–FC arm and 73% for patients in the FC arm [44].

Study 12 Patients with previously treated CLL (N = 552) were entered into this trial. The PFS was 26.7 months for patients in the Rituxan–FC arm (n = 276) and was 21.7 months for patients in the FC arm (n = 276). Response rate was 54% for patients in the Rituxan–FC arm and 45% for patients in the FC arm [44].

7.18.4.6 Rheumatoid Arthritis (RA)

Reducing the Signs and Symptoms: *Initial and Retreatment Courses*: Two randomized, double-blind placebo-controlled studies evaluated the safety and efficacy of Rituxan in adult patients with moderately to severely active RA who had a prior inadequate response to at least one TNF inhibitor [44].

Study 1: Patients were randomized to receive either one course of Rituxan or two infusions of 1000 mg + MTX or placebo + MTX for 24 weeks. Study results demonstrated a significant difference between the two groups. The American College of Rheumatology (ACR) criteria was used to assess the patient's signs and symptoms. Comparing these two groups at week 24, the proportions of patients achieving ACR 20 were 33% higher, achieving ACR 50 were 21% higher, and achieving ACR 70 were 11% higher in the MTX and Rituxan group versus placebo-controlled group. Improvement was also noted for all components of ACR response following treatment with Rituxan [44].

Radiographic Response: In RA Study 1, structural joint damage was assessed radiographically and expressed as changes in Genant-modified total sharp score (TSS) and its components, the erosion score (ES) and the joint space narrowing (JSN) score. Rituxan + MTX slowed the progression of structural damage compared to placebo + MTX after 1 year [44].

RA Study 3: *Lesser Efficacy of Rituxan 500 mg versus 1000 mg Treatment Courses for Radiographic Outcomes*: This randomized, double-blind, placebo-controlled study evaluated the effect of placebo + MTX compared to a course of Rituxan (two infusions of 500 mg + MTX on day 1 and 15) and a course of Rituxan (two infusions of 1000 mg + MTX on day 1 and 15) in MTX-naïve RA patients with moderately to severely active disease. After 1 year of treatment, the proportion of patients achieving ACR 20/50/70 responses were similar in both Rituxan dose groups and were higher than in the placebo group. Only the Rituxan 1000 mg treatment group demonstrated a statistically significant reduction (67%) in radiographic scores (TSS): a change of 0.36 units compared to 1.08 units for the placebo group [44].

RA Study 4: Physical Function Response: In this randomized, double-blind, placebo-controlled study in adult RA patients with moderately to severely active disease with inadequate response to MTX, patients were randomized to receive an initial course of Rituxan 500 mg, Rituxan 1000 mg, or placebo in addition to background MTX. Physical function was assessed at weeks 24 and 48 using the Health Assessment Questionnaire Disability Index (HAQ-DI). From baseline to week 24, a greater proportion of Rituxan-treated patients had an improvement in HAQ-DI of at least 0.22 (a minimal clinically important difference) and a greater mean HAQ-DI improvement compared to placebo. HAQ-DI results for the Rituxan 500 mg treatment group were similar to the Rituxan 1000 mg treatment group. These improvements were maintained at 48 weeks [44].

7.18.4.7 Granulomatosis with Polyangiitis (GPA) (Wegener's Granulomatosis) and Microscopic Polyangiitis (MPA)

A total of 197 patients with active, severe GPA and MPA were treated in a randomized, double-blind, active-controlled multicenter, noninferiority study, conducted in two phases—a 6-month remission induction phase and a 12-month remission maintenance phase. Patients were diagnosed with GPA (75%) or MPA (24%) [44].

Patients were randomized in a 1 : 1 ratio to receive either Rituxan 375 mg/m^2 once weekly for 4 weeks or oral cyclophosphamide 2 mg/kg daily for 3–6 months in the remission induction phase. Patients also received glucocorticoid therapy and were premedicated with antihistamine and acetaminophen prior to Rituxan infusion. Once remission was achieved or at the end of the 6-month remission induction period, the cyclophosphamide group received azathioprine to maintain remission. The Rituxan group did not receive additional therapy to maintain remission. The main outcome measure for both GPA and MPA patients was achievement of complete remission at 6 months and off glucocorticoid therapy. The study demonstrated noninferiority of Rituxan to cyclophosphamide for complete remission at 6 months [44].

7.18.5 Approved Indications

There are four approved indications for Rituxan (rituximab).

1) **Non-Hodgkin's Lymphoma** (NHL): Rituxan (rituximab) is indicated for the treatment of patients with:
   - Relapsed or refractory, low-grade or follicular, CD20-positive B-cell NHL as a single agent

- Previously untreated follicular, CD20-positive B-cell NHL in combination with first-line chemotherapy and in patients achieving a CR or PR to Rituxan in combination with chemotherapy, as single-agent maintenance therapy
- Nonprogressing (including stable disease), low-grade, CD20-positive B-cell NHL as a single agent after first-line CVP chemotherapy
- Previously untreated diffuse large B-cell CD20-positive NHL in combination with CHOP or other anthracycline-based chemotherapy regimens

2) **Chronic Lymphocytic Leukemia (CLL)**: Rituxan (rituximab) is indicated, in combination with FC, for the treatment of patients with previously untreated and previously treated CD20-positive CLL.
3) **Rheumatoid Arthritis (RA)**: Rituxan (rituximab) in combination with MTX is indicated for the treatment of adult patients with moderately to severely active RA who have had an inadequate response to one or more TNF antagonist therapies.
4) **Granulomatosis with Polyangiitis (GPA) (Wegener's granulomatosis) and MPA**: Rituxan (rituximab), in combination with glucocorticoids, is indicated for the treatment of adult patients with GPA (Wegener's Granulomatosis) and MPA.
5) **Limitations of Use**: Rituxan is not recommended for use in patients with severe, active infections.

7.19 Tositumomab and Iodine I-131 Tositumomab (Anti-CD20 Antibody, Bexxar®)

7.19.1 Drug Development History

In 2003, the FDA approved tositumomab and iodine I-131 tositumomab (Bexxar) for the treatment of patients with CD20-positive, relapsed or refractory, low-grade, follicular, or transformed NHL who have progressed during or after rituximab therapy, including patients with rituximab-refractory NHL [12, 35].

7.19.2 Target Biology

Tositumomab and iodine I-131 tositumomab (Bexxar) is a CD20-directed radiotherapeutic antibody. Bexxar therapeutic regimen is composed of the mAb tositumomab and the radiolabeled mAb, I-131 tositumomab. Tositumomab is a murine IgG2a lambda mAb directed against the CD20 antigen [30, 35].

7.19.3 Mechanism of Action

Tositumomab (a CD-20 antibody) binds specifically to an epitope within the extracellular domain of the CD20 antigen that is expressed on normal B lymphocytes (pre-B lymphocytes to mature B lymphocytes) and on B-cell NHL.

The CD20 antigen is not shed from the cell surface and is not internalized following antibody binding. Since tositumomab is bound with iodine-131, the Bexxar therapeutic regimen induces cell death by emitting ionizing radiation to CD20-expressing lymphocytes on neighboring cells and delivers radioactivity to tumor cells selectively. In addition to cell death mediated by the radioisotope, other possible mechanisms of action include antibody-dependent cellular cytotoxicity, CDC, and CD20-mediated apoptosis [30, 35].

7.19.4 Clinical Efficacy

In five clinical trials, the Bexxar therapeutic regimen was administered to 230 patients with NHL using the recommended dose and schedule. This regimen consists of a dosimetric dose and a therapeutic dose. The two-part dosimetric dose includes (1) tositumomab 450 mg by intravenous infusion and (2) I-131 tositumomab (5 mCi I-131 and 35 mg protein) by intravenous infusion. The two-part therapeutic dose is administered 7–14 days after the dosimetric dose and includes (1) tositumomab 450 mg by intravenous infusion and (2) I-131 tositumomab (35 mg) by intravenous infusion. The iodine-131 dose is calculated based on an assessment of dosimetry and biodistribution obtained following the dosimetric dose and the patient's platelet counts obtained within 28 days prior to dosing [35].

Patients were followed for a median of 39 months; 79% were followed for at least 12 months for survival and selected adverse reactions. Patients had a median of three prior chemotherapy regimens and a median age of 55 years, and 60% were male. Twenty-seven percent (27%) had transformation to a higher-grade histology; 29% had intermediate-grade histology, and 2% had high-grade histology (IWF); 68% had Ann Arbor stage IV disease. Patients enrolled in these studies were not permitted to have prior hematopoietic stem cell transplantation or irradiation to more than 25% of the marrow space [35].

Determination of the effectiveness of the Bexxar therapeutic regimen is based on overall response rates in patients whose disease is refractory to chemotherapy and rituximab. The data from the five trials ($n = 250$) was integrated. The total number of patients who responded to the Bexxar therapeutic regimen was 141 patients (56%), and the median duration of the response was 12.9 months (range 10.6–17.3). The total number of patients who achieved a CR after receiving the Bexxar therapeutic regimen was 75 patients (30%), and the median duration of the CR was 58.4 months (range 28.3 months—longest amount of CR duration not reported). For all patients in these studies, the median time to overall progression or death was 6.4 months (range 5.9–9.4 months). For patients who responded to the Bexxar, the median time to progression or death was 15.0 months (range 12.2–19.2 months) [35].

7.19.5 Approved Indications

Tositumomab and iodine I-131 tositumomab (Bexxar) is indicated for the treatment of patients with CD20-positive, relapsed or refractory, low-grade, follicular, or transformed NHL who have progressed during or after rituximab therapy, including patients with rituximab-refractory NHL [12, 35].

Bexxar therapeutic regimen is only indicated for a single course of treatment and is not indicated for a first-line treatment [35].

7.20 Trastuzumab (Anti-HER2 Antibody, Herceptin®)

7.20.1 Drug Development History

In 1998, the US FDA approved trastuzumab (Herceptin) for treatment of HER2-overexpressing breast cancer. Over the years its indications have expanded and it is now approved for (i) adjuvant treatment of HER2-overexpressing breast cancer, (ii) treatment of metastatic HER2-overexpressing breast cancer, and (iii) treatment of patients with HER2-overexpressing metastatic gastric or GEJ adenocarcinoma, who have not received prior treatment for metastatic cancer, in combination with other chemotherapy [10, 27, 40].

7.20.2 Target Biology

Herceptin (trastuzumab) is a humanized IgG1 kappa mAb that selectively binds with high affinity to the extracellular domain of the human epidermal growth factor receptor 2 protein, HER2 [40].

7.20.3 Mechanism of Action

The HER2 (or c-erbB2) protooncogene encodes a transmembrane receptor protein of 185 kDa, which is structurally related to the EGFR. Herceptin has been shown, in both *in vitro* assays and in animals, to inhibit the proliferation of human tumor cells that over express HER2.

Herceptin is a mediator of antibody-dependent cell-mediated cytotoxicity (ADCC). *In vitro*, Herceptin-mediated ADCC has been shown to be preferentially exerted on HER2-overexpressing cancer cells compared with cancer cells that do not overexpress HER2 [10, 27, 40].

7.20.4 Clinical Efficacy

7.20.4.1 Adjuvant Therapy for Breast Cancer

The safety and efficacy of Herceptin in women, who were receiving adjuvant chemotherapy for HER2-overexpressing breast cancer, were evaluated in four studies.

Table 7.1 Duration of disease-free survival in patients with adjuvant treatment of breast cancer (Studies 1 and 2) [40].

Study arm	0 year	0.5 year	1 year	1.5 years	2 years	2.5 years	3 years	3.5 years
AC→T	1880	1490	1159	926	689	534	375	195
AC→T+H	1872	1529	1240	997	764	575	426	239

Studies 1 and 2 (AC→T versus AC→T+H) Studies 1 and 2 were randomized, open-label clinical trials, with a total of 4063 women at the protocol-specified final OS analysis. For both trials, the study design was a 1:1 randomization of patients to receive four 21-day cycles of AC (doxorubicin (60 mg/m^2) and cyclophosphamide (600 mg/m^2)). This was followed by paclitaxel (AC→T) alone or paclitaxel plus Herceptin (AC→T+H). In Study 1, paclitaxel was administered either weekly (80 mg/m^2) or every 3 weeks (175 mg/m^2), for a total of 12 weeks. In Study 2, paclitaxel (80 mg/m^2) was administered weekly only. Herceptin (loading dose of 4 mg/kg) was administered on the day paclitaxel was initiated, and then patients were given Herceptin (2 mg/kg) weekly for 52 weeks. Herceptin treatment was permanently discontinued in patients who developed congestive heart failure or persistent/recurrent LVEF decline. In the combined efficacy analysis, disease-free survival (DFS) was the primary endpoint and OS was the secondary endpoint [40].

In Study 1 and 2, a total of 3752 patients completed these studies and were included in the joint efficacy analysis of the primary endpoint of DFS. The results for duration of DFS in patients with adjuvant treatment of breast cancer (Studies 1 and 2) are outlined in Table 7.1. There were 92 patient deaths in Studies 1 and 2 in the AC→T arm ($n = 1880$) and 62 patient deaths in the AC→T+H arm ($n = 1872$).

Study 3: Herceptin (Every 3 Weeks) versus Observation Study 3 was a randomized, open-label clinical trial with a total of 3386 women at definitive DFS analysis for 1-year Herceptin treatment versus observation. Study 3 was designed to compare 1 and 2 years of 3-weekly Herceptin treatment versus observation in patients with HER2-positive EBC following surgery, established chemotherapy, and radiotherapy (if applicable). Patients were randomized (1:1:1) upon completion of definitive surgery and at least four cycles of chemotherapy to receive no additional treatment or 1 year of Herceptin treatment or 2 years of Herceptin treatment. Herceptin was administered with an initial dose of 8 mg/kg, followed by subsequent doses of 6 mg/kg once every 3 weeks. The main outcome measure was DFS, defined as in Studies 1 and 2 [40].

Table 7.2 Duration of disease-free survival in patients with adjuvant treatment of breast cancer (Study 4) [40].

Study arm	0 year	1 year	2 years	3 years	4 years
AC→T	1073	971	802	417	103
AC→TH	1074	1023	885	457	126
TCH	1075	1018	877	447	126

After the definitive DFS results comparing observation to 1-year Herceptin treatment were disclosed, a prospectively planned analysis that included comparison of 1 year versus 2 years of Herceptin treatment at a median follow-up duration of 8 years was performed. Based on this analysis, extending Herceptin treatment for a duration of 2 years did not show additional benefit over treatment for 1 year [40].

Study 4 (AC→T versus AC→TH versus TCH) Study 4 was a randomized, open-label clinical trial. Patients were randomized (1 : 1 : 1) to receive drugs in one of three study arms: (1) doxorubicin and cyclophosphamide followed by docetaxel (AC→T), (2) doxorubicin and cyclophosphamide followed by docetaxel plus Herceptin (AC→TH), or (3) docetaxel and carboplatin plus Herceptin (TCH). In both the AC-T and AC-TH arms, doxorubicin 60 mg/m^2 and cyclophosphamide 600 mg/m^2 were administered every 3 weeks for four cycles; docetaxel 100 mg/m^2 was administered every 3 weeks for four cycles. In the TCH arm, docetaxel 75 mg/m^2 and carboplatin (AUC = 6) were administered every 3 weeks for six cycles. Herceptin was administered weekly (initial dose of 4 mg/kg followed by weekly dose of 2 mg/kg) concurrently with either T or TC and then every 3 weeks (6 mg/kg) as monotherapy for a total of 52 weeks [40].

In Study 4, 3222 patients randomized to the three treatment arms. The results for duration of DFS in patients with adjuvant treatment of breast cancer (Study 4) are outlined in Table 7.2. The number of patient deaths in the AC→T arm ($n = 1073$) was 80, in the AC→TH arm ($n = 1074$) was 49, and in the TCH arm ($n = 1075$) was 56.

Study 5: Previously Untreated Metastatic Breast Cancer Study 5 was a multicenter, randomized, open-label clinical trial conducted in 469 women with MBC who had not been previously treated with chemotherapy for metastatic disease. Patients were randomized to receive chemotherapy alone or in combination with Herceptin given intravenously as a 4 mg/kg loading dose followed by weekly doses of Herceptin at 2 mg/kg. Sixty-five percent of patients randomized to receive chemotherapy alone in this study received Herceptin at the time of disease progression as part of a separate extension study [40].

Based upon the determination by an independent response evaluation committee, the patients randomized to Herceptin and chemotherapy experienced a significantly longer median time to disease progression, a higher overall response rate, a longer median duration of response, and a longer median survival as compared with patients randomized to chemotherapy alone. These treatment effects were observed both in patients who received Herceptin plus paclitaxel and in those who received Herceptin plus AC; however, the magnitude of the effects was greater in the paclitaxel subgroup.

Study 6: Previously Treated Metastatic Breast Cancer In Study 6, Herceptin was studied as a single agent in a multicenter, open-label, single-arm clinical trial in patients with HER2-overexpressing MBC who had relapsed following one or two prior chemotherapy regimens for metastatic disease. Herceptin was administered as a loading dose (4 mg/kg) IV and was followed by weekly doses (2 mg/kg) IV [40].

In this study, 222 patients were enrolled. The overall response rate (CR + PR), as determined by an independent response evaluation committee, was 14%, with a 2% CRR and a 12% PR rate. CR were observed only in patients with disease limited to skin and lymph nodes.

Study 7: Metastatic Gastric Cancer The safety and efficacy of Herceptin in combination with cisplatin and a fluoropyrimidine (capecitabine or 5-FU) were studied in patients previously untreated for metastatic gastric or GEJ adenocarcinoma. In this open-label, multicenter trial, 594 patients were randomized 1 : 1 to chemotherapy (cisplatin and a fluoropyrimidine) combined with Herceptin (FC + H) or chemotherapy alone (FC). All patients were either HER2 gene amplified (FISH+) or HER2 overexpressing (IHC 3+). Patients were also required to have adequate cardiac function (e.g., LVEF > 50%) [40].

On the Herceptin-containing arm, Herceptin was administered as an IV infusion at an initial dose of 8 mg/kg followed by 6 mg/kg every 3 weeks until disease progression. On both study arms, cisplatin (80 mg/m^2) was administered on day 1 every 3 weeks for six cycles as a 2 h IV infusion. On both study arms, capecitabine (1000 mg/m^2) was administered orally twice daily (total daily dose of 2000 mg/m^2) for 14 days of each 21-day cycle for six cycles. Alternatively, continuous intravenous infusion (CIV) 5-FU (800 mg/m^2/day) was administered from day 1 through day 5 every 3 weeks for six cycles.

The main outcome measure of Study 7 was OS, analyzed by the unstratified log-rank test. The final OS analysis based on 351 deaths was statistically significant (nominal significance level of 0.0193). An updated OS analysis was conducted at 1 year after the final analysis. The OS in the FCC arm was 11.0 months and in the FCC + Herceptin arm was 13.5 months [40].

7.20.5 Approved Indications

Herceptin has three indications:

1) **Adjuvant Breast Cancer**: Herceptin is indicated for adjuvant treatment of HER2-overexpressing node-positive or node-negative (ER/PR negative or with one high-risk feature) breast cancer:
   - As part of a treatment regimen consisting of doxorubicin, cyclophosphamide, and either paclitaxel or docetaxel
   - With docetaxel and carboplatin
   - As a single agent following multimodality anthracycline-based therapy
2) **Metastatic Breast Cancer**: Herceptin is indicated for the following:
   - In combination with paclitaxel for first-line treatment of HER2-overexpressing MBC
   - As a single agent for treatment of HER2-overexpressing breast cancer in patients who have received one or more chemotherapy regimens for metastatic disease
3) **Metastatic Gastric Cancer**: Herceptin is indicated, in combination with cisplatin and capecitabine or 5-FU, for the treatment of patients with HER2-overexpressing metastatic gastric or GEJ adenocarcinoma, who have not received prior treatment for metastatic disease.

References

1. Amgen (2006) News release. Accessed September 12, 2015 at: http://www.amgen.com/media/media_pr_detail.jsp?year=2006&releaseID=909842
2. Amgen (2015) Xgeva as a treatment for giant cell tumor of bone (GCTB). Accessed July 12, 2015 at: http://www.xgeva.com/giant-cell-tumor-of-bone.html
3. Center Watch (2014) Keytruda (pembrolizumab). Accessed July 13, 2015 at: https://www.centerwatch.com/drug-information/fda-approved-drugs/drug/100034/keytruda-pembrolizumab
4. Center Watch (2015) Ofatumumab (Arzerra). Accessed July 13, 2015 at: https://www.centerwatch.com/drug-information/fda-approved-drugs/drug/1072/arzerra-ofatumumab
5. Chawla, S., Henshaw, R., Seeger, L., Choy, E., Blay, J. Y., Ferrari, S., Kroep, J., Grimer, R., Reichardt, P., Rutkowski, P., Schuetze, S., Skubitz, K., Staddon, A., Thomas, D., Qian, Y., & Jacobs, I. (2013) Safety and efficacy of denosumab for adults and skeletally mature adolescents with giant cell tumour of bone: interim analysis of an open-label, parallel-group, phase 2 study. *The Lancet Oncology*, 14(9), 901–908.

6. Coiffier, B., Lepretre, S., Pedersen, L. M., Gadeberg, O., Fredriksen, H., van Oers, M. H., Wooldridge, J., Kloczko, J., Holowiecki, J., Hellmann, A., Walewski, J., Flensburg, M., Petersen, J., & Robak, T. (2008) Safety and efficacy of ofatumumab, a fully human monoclonal anti-CD20 antibody, in patients with relapsed or refractory B-cell chronic lymphocytic leukemia: a phase 1–2 study. *Blood*, 111(3), 1094–1100.
7. Dietzek, A., Connelly, K., Cotugno, M., Bartel, S., & McDonnell, A. M. (2015) Denosumab in hypercalcemia of malignancy: a case series. *Journal of Oncology Pharmacy Practice*, 21(2), 143–147.
8. Drug Development Technologies (2014) Kadcyla information. Accessed September 15, 2014 at: http://www.drugdevelopment-technology.com/news/newsfda-approves-genentech-metastatic-breast-cancer-drug/
9. Drug Development Technologies (2014) Rituxan information. Accessed September 15, 2014 at: http://www.drugdevelopment-technology.com/projects/rituxan-treatment/
10. Drug Development Technologies (2015) Herceptin information. Accessed February 7, 2015 at: http://www.drugdevelopment-technology.com/search/?q%5B%5D=herceptin
11. FDA (2011) FDA approval of eculizumab (Soliris®). Accessed July 12, 2015 at: http://www.fda.gov/AboutFDA/CentersOffices/OfficeofMedicalProductsandTobacco/CDER/ucm273089.htm
12. Fisher, R. I., Kaminski, M. S., Wahl, R. L., Knox, S. J., Zelenetz, A. D., Vose, J. M., Leonard, J. P., Kroll, S., Goldsmith, S. J., & Coleman, M. (2005) Tositumomab and iodine-131 tositumomab produces durable complete remissions in a subset of heavily pretreated patients with low-grade and transformed non-Hodgkin's lymphoma. *Journal of Clinical Oncology*, 23(30), 7565–7573.
13. Fizazi, K., Carducci, M., Smith, M., Damião, R., Brown, J., Karsh, L., Milecki, P., Shore, N., Rader, M., Wang, H., Jiang, Q., Tadros, S., Dansey, R., & Goessl, C. (2011) Denosumab versus zoledronic acid for treatment of bone metastases in men with castration-resistant prostate cancer: a randomised, double-blind study. *The Lancet*, 377(9768), 813–822.
14. Genentech (2015) Gazyva mechanism of action. Accessed July 15, 2015 at: http://www.gazyva.com/hcp/mechanisms-of-action/overview
15. Genentech (2015) About Avastin: proposed mechanism of action. Accessed October 12, 2015 at: http://www.avastin-hcp.com/about-avastin/proposed-moa
16. Hagenbeek, A., Gadeberg, O., Johnson, P., Pedersen, L. M., Walewski, J., Hellmann, A., Link, B. K., Robak, T., Wojtukiewicz, M., Pfreundschuh, M., Kneba, M., Engert, A., Sonneveld, P., Flensburg, M., Petersen, J., Losic, N., & Radford, J. (2008) First clinical use of ofatumumab, a novel fully human

anti-CD20 monoclonal antibody in relapsed or refractory follicular lymphoma: results of a phase 1/2 trial. *Blood*, 111(12), 5486–5495.

17. Henry, D. H., Costa, L., Goldwasser, F., Hirsh, V., Hungria, V., Prausova, J., Scagliotti, G. V., Sleeboom, H., Spencer, A., Vadhan-Raj, S., von Moos, R., Willenbacher, W., Woll, P. J., Wang, J., Jiang, Q., Jun, S., Dansey, R., & Yeh, H. (2011) Randomized, double-blind study of denosumab versus zoledronic acid in the treatment of bone metastases in patients with advanced cancer (excluding breast and prostate cancer) or multiple myeloma. *Journal of Clinical Oncology*, 29(9), 1125–1132.
18. Hu, M. I., Glezerman, I. G., Leboulleux, S., Insogna, K., Gucalp, R., Misiorowski, W., & Jain, R. K. (2014) Denosumab for treatment of hypercalcemia of malignancy. *The Journal of Clinical Endocrinology & Metabolism*, 99(9), 3144–3152.
19. Merck (2014) Keytruda prescribing information. Accessed February 15, 2015 at: http://www.merck.com/product/usa/pi_circulars/k/keytruda/keytruda_pi.pdf
20. National Institutes of Health (2014) Ado-trastuzumab (Kadcyla) information. Accessed September 14, 2014 at: http://dailymed.nlm.nih.gov/dailymed/lookup.cfm?setid=23f3c1f4-0fc8-4804-a9e3-04cf25dd302e#nlm34090-1
21. National Institutes of Health (2014) Alemtuzumab (Campath®) information. Accessed September 14, 2014 at: http://dailymed.nlm.nih.gov/dailymed/lookup.cfm?setid=4f5f7255-7abc-4328-bd1a-ceaf139ef3e0#nlm34090-1
22. National Cancer Institute (2015) Erbitux® (cetuximab). Accessed July 13, 2015 at: http://www.cancer.gov/about-cancer/treatment/drugs/fda-cetuximab
23. National Institutes of Health (2014) Obinutuzumab (Gazyva) information. Accessed September 14, 2014 at: http://dailymed.nlm.nih.gov/dailymed/lookup.cfm?setid=df12ceb2-5b4b-4ab5-a317-2a36bf2a3cda#nlm34089-3
24. National Institutes of Health (2014) Panitumumab (Vectibix) information. Accessed September 14, 2014 at: http://dailymed.nlm.nih.gov/dailymed/lookup.cfm?setid=e0fa4bca-f245-4d92-ae29-b0c630a315c2
25. National Institutes of Health (2014) Pertuzumab (Perjeta) information. Accessed September 14, 2014 at: http://dailymed.nlm.nih.gov/dailymed/lookup.cfm?setid=17f85d17-ab71-4f5b-9fe3-0b8c822f69ff
26. National Institutes of Health (2014) Rituximab (Rituxan®) information. Accessed September 14, 2014 at: http://dailymed.nlm.nih.gov/dailymed/lookup.cfm?setid=b172773b-3905-4a1c-ad95-bab4b6126563#nlm34067-9
27. National Institutes of Health (2014) Trastuzumab (Herceptin®) information. Accessed September 14, 2014 at: http://dailymed.nlm.nih.gov/dailymed/lookup.cfm?setid=492dbdb2-077e-4064-bff3-372d6af0a7a2#nlm34067-9
28. NCCN (2015) *NCCN Guidelines: Non-Small Cell Lung Cancer*. Version 7.2015.

29. Pazdur, R. (2009) FDA approval for bevacizumab. Drug Information, National Cancer Institute. Accessed May 3, 2015 at: http://www.cancer.gov/cancertopics/druginfo/fda-bevacizumab
30. Polovich, M., Olsen, M., & LeFebvre, K. B. (2014) Chemotherapy and Biotherapy Guidelines and Recommendations for Practice (4th edition). Pittsburgh, PA: Oncology Nursing Society.
31. Pollack, A. (December 16, 2010) FDA rejects use of drug in cases of breast cancer. *New York Times*.
32. Prescribing Information (2011) Adcetris® (Brentuximab vedotin). Accessed October 15, 2015 at: http://www.accessdata.fda.gov/drugsatfda_docs/label/2011/125388s000,125399s000lbl.pdf
33. Prescribing Information (2015) Arzerra (ofatumumab). Accessed July 15, 2015 at: https://www.gsksource.com/pharma/content/dam/GlaxoSmithKline/US/en/Prescribing_Information/Arzerra/pdf/ARZERRA.PDF
34. Prescribing Information (2015) Avastin (bevicizumab). Accessed October 12, 2015 at: http://www.gene.com/download/pdf/avastin_prescribing.pdf
35. Prescribing Information (2013) Bexxar® (tositumomab and iodine I 131 tositumomab). Accessed December 7, 2015 at: https://www.gsksource.com/gskprm/htdocs/documents/BEXXAR.PDF
36. Prescribing Information (2015) Campath (alemtuzumab). Accessed January 15, 2015 at: http://www.campath.com/pdfs/2009-08-Campath%20US%20PI.pdf
37. Prescribing Information (2014) Cyramza (ramucirumab). Accessed October 17, 2015 at: http://pi.lilly.com/us/cyramza-pi.pdf
38. Prescribing Information (2014) Erbitux® (cetuximab). Accessed July 15, 2015 at: http://www.accessdata.fda.gov/drugsatfda_docs/label/2012/125084s229lbl.pdf
39. Prescribing Information (2015) Gazyva (obinutuzumab). Accessed July 15, 2015 at: http://www.gene.com/download/pdf/gazyva_prescribing.pdf
40. Prescribing Information (2014) Herceptin (trastuzumab). Accessed December 3, 2014 at: http://www.gene.com/download/pdf/herceptin_prescribing.pdf
41. Prescribing Information (2015) Kadcyla (ado-trastuzumab). Accessed October 18, 2016 at: https://www.gene.com/download/pdf/kadcyla_prescribing.pdf
42. Prescribing Information (2014) Keytruda (pembrolizumab). Accessed February 15, 2015 at: http://www.merck.com/product/usa/pi_circulars/k/keytruda/keytruda_pi.pdf
43. Prescribing Information (2015) Perjeta® (pertuzumab). Accessed January 30, 2015 at: http://www.gene.com/download/pdf/perjeta_prescribing.pdf
44. Prescribing Information (2014) Rituxan (rituximab). Accessed November 28, 2014 at: http://www.gene.com/download/pdf/rituxan_prescribing.pdf
45. Prescribing Information (2014) Soliris (eculizumab). Accessed July 12, 2015 at: http://soliris.net/sites/default/files/assets/soliris_pi.pdf

46. Prescribing Information (2015) Vectibix™ (panitumumab). Accessed October 17, 2015 at: http://pi.amgen.com/united_states/vectibix/vectibix_pi.pdf
47. Prescribing Information (2015) Xgeva® (denosumab). Accessed October 17, 2015 at: http://pi.amgen.com/united_states/xgeva/xgeva_pi.pdf
48. Prescribing Information (2015) Yervoy (ipilimumab). Accessed October 10, 2015 at: http://packageinserts.bms.com/pi/pi_yervoy.pdf
49. Prescribing Information (2015) Zevalin (Ibritumomab tiuxetan). Accessed July 12, 2015 at: http://www.zevalin.com/downloads/Zevalin_Package_Insert.pdf
50. Sandler, A. B., Johnson, D. H., & Herbst, R. S. (2004) Anti-vascular endothelial growth factor monoclonals in non-small cell lung cancer. *Clinical Cancer Research*, 10(12), 4258s–4262s.
51. Sandler, A., Gray, R., Perry, M. C., Brahmer, J., Schiller, J. H., Dowlati, A., Lilenbaum, R., & Johnson, D. H. (2006) Paclitaxel–carboplatin alone or with bevacizumab for non–small-cell lung cancer. *New England Journal of Medicine*, 355(24), 2542–2550.
52. Stopeck, A. T., Lipton, A., Body, J. J., Steger, G. G., Tonkin, K., de Boer, R. H., Lichinitser, M., Fujiwara, Y., Yardley, D. A., Viniegra, M., Fan, M., Jiang, Q., Dansey, R., Jun, S., & Braun, A. (2010) Denosumab compared with zoledronic acid for the treatment of bone metastases in patients with advanced breast cancer: a randomized, double-blind study. *Journal of Clinical Oncology*, 28(35), 5132–5139.
53. Tanne, J. H. (2011) FDA cancels approval for bevacizumab in advanced breast cancer. *BMJ*, 343, d7684.
54. Tuthill, M., Crook, T., Corbet, T., King, J., & Webb, A. (2009) Rapid infusion of rituximab over 60 min. *European Journal of Haematology*, 82, 322–325.
55. U.S. Food and Drug Administration (2013) FDA approves Gazyva for chronic lymphocytic leukemia. Accessed July 13, 2015 at: http://www.fda.gov/NewsEvents/Newsroom/PressAnnouncements/ucm373209.htm
56. U.S. Food and Drug Administration (2014) FDA approves Keytruda for advanced melanoma. First PD-1 blocking drug to receive agency approval. Accessed July 13, 2015 at: http://www.fda.gov/NewsEvents/Newsroom/PressAnnouncements/ucm412802.htm
57. U.S. Food and Drug Administration (2015) FDA approval of ofatumumab (Arzerra). Accessed July 13, 2015 at: http://www.fda.gov/Drugs/InformationOnDrugs/ApprovedDrugs/ucm393823.htm
58. U.S. Food and Drug Administration (2011) Approval of Yervoy (ipilimumab). Accessed October 10, 2015 at: http://www.fda.gov/NewsEvents/Newsroom/PressAnnouncements/ucm1193237.htm
59. U.S. Food and Drug Association (2015) FDA approval of denosumab for giant cell tumor of bone. Accessed July 12, 2015 at: http://www.fda.gov/Drugs/InformationOnDrugs/ApprovedDrugs/ucm356667.htm

60. U.S. Food and Drug Administration (2015) Zevalin new FDA approval. Accessed July 12, 2015 at: https://www.centerwatch.com/drug-information/fda-approved-drugs/drug/762/zevalin-ibritumomab-tiuxetan

61. van de Donk, N. W. C. J. & Dhimolea, E. (July 1, 2012) Brentuximab vedotin. *MABs*, 4(4), 458–465.

62. Waldmann, H. (2002) A personal history of the CAMPATH-1H antibody. *Medical Oncology*, 19(Supplement), S3–S9.

8

Development of Biosimilar Rituximab and Clinical Experience

Reena Nair

Department of Clinical Hematology, Tata Medical Center, Kolkata, India

8.1 Summary

Biologics find wide application in cancer care for their therapeutic effects, ability to improve outcomes, and as treatment options in secondary care. The complex molecular structure of biologic agents necessitates long periods of development and higher costs when compared to small molecule generics. Biologic agents are increasing economic demand on healthcare systems worldwide. Designed to be highly similar to originator biologic products, biosimilars represent an opportunity to increase access and reduce costs for patients and healthcare systems. A systematic approach for characterization of structural and functional similarity of a biosimilar molecule to the originator molecule provides specifications to guide the development of biosimilars. Identical sequence and similar host cell lines are key to the development of biosimilars. A well-designed development program establishes biosimilarity based on structural, functional, preclinical, and clinical evaluation. Biosimilar endpoints are defined for each tier of characterization, and evaluation of successive tiers addresses additional relevance of biosimilarity. Rituximab was the first monoclonal antibody approved for cancer treatment. Since its launch in 1997, it has been at the forefront of treatment for B-cell non-Hodgkin's lymphoma (NHL). Reditux™ (Dr. Reddy's Laboratories Ltd., Telangana, India) is the first in a series of biosimilar products that has revolutionized the affordability of antibody therapy for patients in India and beyond. Regulatory requirements for biosimilars are evolving. Education of physicians and healthcare providers, patients, and payers about biosimilars may assist informed decision-making and promote acceptance of biosimilars into clinical practice. Scientific bodies should formulate practice guidelines and position statements to establish

Biosimilars of Monoclonal Antibodies: A Practical Guide to Manufacturing, Preclinical, and Clinical Development, First Edition. Edited by Cheng Liu and K. John Morrow, Jr.

biosimilarity in efficacy, safety, comparability, and interchangeability with the reference originator biologic molecule.

8.2 Introduction

The development of monoclonal antibodies (mAbs) has revolutionized the use of therapeutic biological medicines in cancer care [1]. The development of therapeutic antibodies involves two distinct phases: (i) the discovery phase for the identification of target sequences that form the basis for the chimeric, humanized, or human antibody and (ii) the development phase for creation of usable quantities ranging from creating production cell lines, purification, and formulating the created antibodies at commercial scales.

First to be used in oncology, rituximab, is a chimeric IgG1 mouse antihuman CD20 mAbs that initiates immunologic reactions to mediate B-cell lysis. While the exact mechanism of action is still unknown, it is reported that rituximab causes B-cell death by three primary mechanisms, namely, antibody-dependent cell-mediated cytotoxicity (ADCC), complement-mediated cytotoxicity, and apoptosis, leading to a rapid and sustained depletion of B cells [2]. Rituximab is approved for use in adults with CD20-positive non-Hodgkin's lymphoma, chronic lymphocytic leukemia, in combination with methotrexate, as second-line therapy for moderate to severe rheumatoid arthritis, granulomatosis with polyangiitis (Wegener's granulomatosis), and microscopic polyangiitis [3, 4]. Rituximab represents the standard of care in a variety of treatment settings. Rituximab has shown to improve survival when combined with chemotherapy, and various healthcare reimbursement agencies have accorded a positive cost-effectiveness status to the use of rituximab with curative intent in cancer [5].

With the growing global interest in the development of affordable biologics, several regulatory authorities have established guidelines for the development and approval of biosimilars. The European Union (EU) regulatory pathway for the marketing authorization application (MAA) of biosimilars was established in 2005 [6]. The EU approval pathway is governed by the principle "Guideline on Similar Biological Medicinal Products," guidelines on the efficacy, safety, quality, and product-specific class requirements for various biosimilar molecules [7]. The Food and Drug Administration (FDA) mandates proof of biosimilarity based on data derived from analytical, preclinical, and clinical studies. Clinical studies, addressing one or more indications licensed for the reference product, should include an assessment of pharmacokinetics (PK), pharmacodynamics (PD), and immunogenicity. In 2009, the Biologics Price Competition and Innovation Act (BPCI Act, 2009) supported the need for an abbreviated licensure procedure for biologic molecules that are similar to or interchangeable with an FDA-licensed originator biologic product. In February

2012, the FDA issued three guidelines that list the requirements for biosimilar registration, which include product formulation, complexity, and stability in addition to clinical experience with the biosimilar product in comparison with the reference molecule. Both the EMA and FDA support an abridged development program for biosimilars. Additionally, the FDA offers the possibility to adopt full interchangeability for biosimilars [8]. In 2012, the Central Drugs Standard Control Organization (CDSCO), in collaboration with the Department of Biotechnology (DBT), issued the guidelines for similar biologics developed in or imported into India [9]. Approved in 2007, Reditux™, an affordable rituximab, was the first mAbs developed by Dr. Reddy's Laboratories Ltd. in India and is currently licensed for use in 12 countries (Reditux, Prescribing Information. Data on file. Dr Reddy's Laboratories Ltd., Telangana, India). Dr. Reddy's Laboratories Ltd. was the first to obtain approval for the affordable rituximab in Asia, Latin America, and the Middle East [10].

The high costs of these biologic agents pose challenges in cancer care. The introduction of biosimilar molecules provides opportunities to reduce costs and hence improve healthcare access and outcomes [11]. Biosimilarity between the originator molecule and the follow-on biologic agent is established based on a systematic approach that relies on structural, functional, preclinical, and clinical evaluation.

8.3 Reditux Development Overview

In 1988 IDEC Pharmaceutical (now Biogen IDEC) began phase 1 trials with rituximab (IDEC-C2B8) and later commercialized by Genentech in the United States and Roche Pharmaceuticals (Switzerland) globally outside the United States following its approval in 1998 for the treatment of relapsed and refractory low-grade lymphoma treatment. While approved and available globally, many patients, especially in developing countries, were unable to gain access to rituximab due to cost. Nearly 10 years later, in March 2007, the Drug Controller General (India) approved an alternative rituximab, Reditux developed by Dr. Reddy's Laboratories Ltd., India.

Reditux was commercialized at a significant discount to the originator rituximab (MabThera, Hoffman-La Roche, Basel, Switzerland). The development of Reditux involves a development process similar to that of the originator molecule, starting from the desired protein sequence and generation of production cell line that is used to produce the antibody on commercial scales. This use of identical protein sequence and similar host cell lines ensure the basis of similarity assessments between the originator and the biosimilar molecule. Cell banks used for the manufacturing of Reditux have been extensively characterized and tested for identity, purity, and genetic stability in accordance with ICH guidelines [12]. To test for identity, host cells of Chinese hamster origin have been verified by isoenzyme analysis. The testing for the absence of bacteria, fungi, mycoplasma, and viruses is mandated to ensure purity. Tests for genetic stability involve nucleotide sequence of transcripts, recombinant gene content in genome, and transcript quantitation.

As with all biological products, the manufacturing process for Reditux comprises of three broad areas: upstream or cell culture process, downstream or purification process, and aseptic formulation and filling process. The manufacturing complies with the Current Good Manufacturing Practices (CGMP) requirements and has been evaluated for consistency of product quality and removal of process-related impurities [13].

The assessment of similarity of Reditux to the originator rituximab is based on extensive analytical characterization for structural similarity using orthogonal and high-resolution technology. Table 8.1 lists the tests used for comparing the primary structure, conformation and conformational stability, product variants, and purity and safety of products.

Higher-order structural similarity (Fig. 8.1) is critical to ensure the correct binding interfaces and hence similar functionality (Fig. 8.2). Stringent four parametric curve for CDC and ADCC is analyzed to establish comparability of product potency of two products (Fig. 8.2a and b). High-functional similarity was established by similar binding curves at multiple concentrations for the two products (Fig. 8.2c). Finally, product- and process-related impurities are routinely tested as part of final release testing by using techniques like size exclusion chromatography (SEC) and reducing and nonreducing SDS-PAGE.

Table 8.1 Physicochemical characterization tests for Reditux and MabThera comparisons.

Primary structure	• MALDI-TOF • SDS-PAGE • PMF by heavy and light chain by LC-MS/MS • N-terminal sequencing
Conformation and conformational stability	• UV spectral profile • Far-UV CD spectroscopy • SE-HPLC • Fluorescence spectroscopy • Near-UV CD spectroscopy • IEF • Thiol titration • Differential scanning calorimetry
Glycosylation and variants	• Glycan analysis • IEX chromatography, IEF
Purity and safety	• Product related: SE-HPLC, SDS-PAGE • Process related: HCP, HCD, leachable protein A

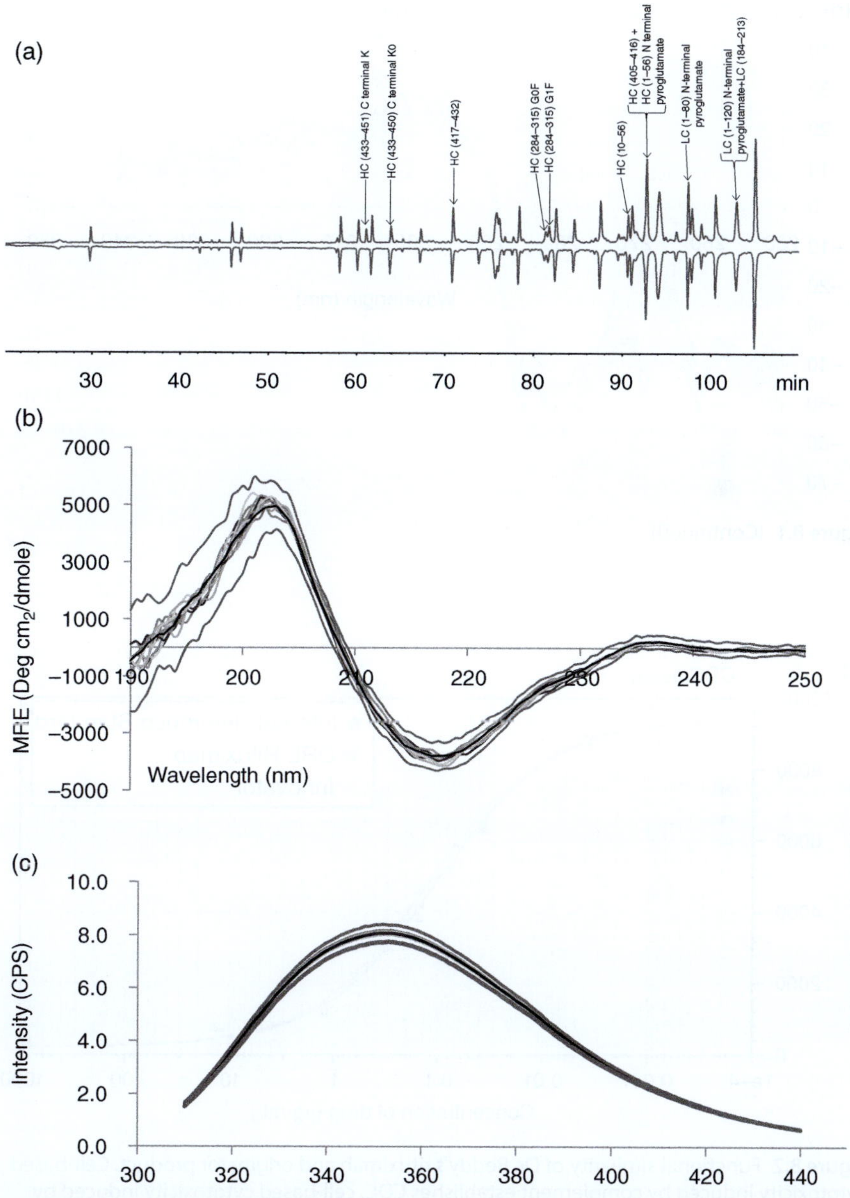

Figure 8.1 Structural similarity of Dr. Reddy's rituximab and originator product. Similar identity in amino acid sequence (a), secondary (b), and tertiary structure (c) for Reditux and originator products. Spectral profiles in red indicate method/batch to batch variability determined from Innovator product sourced in the European Union and the United States. (*See insert for color representation of the figure.*)

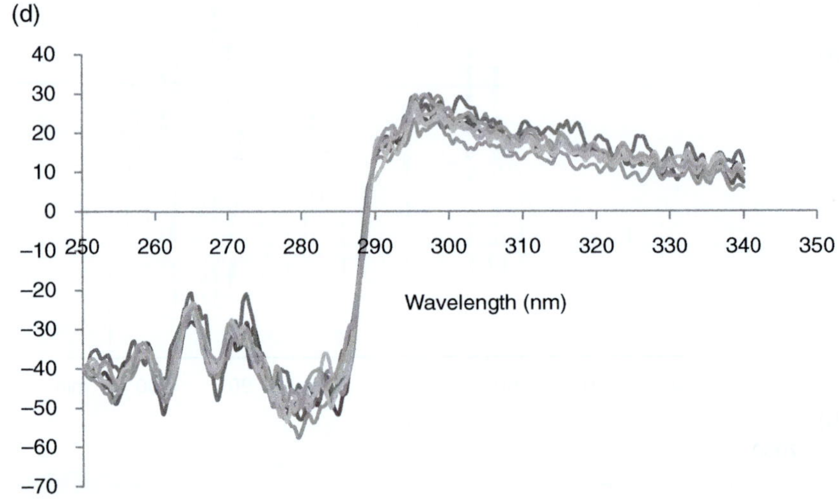

Figure 8.1 (Continued)

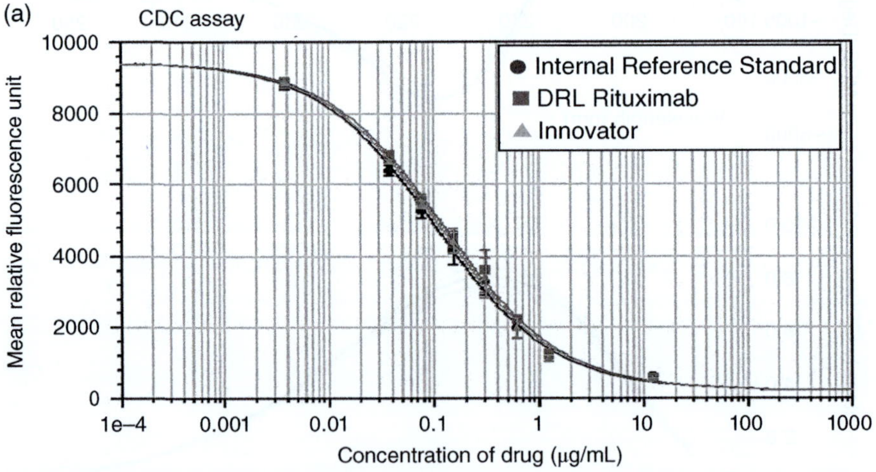

Figure 8.2 Functional similarity of Dr. Reddy's rituximab and originator product. Cell-based cytotoxicity induced by complement establishes CDC, cell-based cytotoxicity induced by PBMCs establishes ADCC, and cell-based assay for detection of annexin V-positive cells establishes apoptosis. Assays used for binding comparisons include cell-based competitive binding FACS assay for CD20 ligand binding; ELISA-based binding assay for C1q, FcγRIIa, and FcγRIIb binding; and binding constant determination by surface plasmon resonance for FcγRI, FcγRIIIa, and FcRn binding. The figure shows CDC (a) and apoptosis (b) assays and the FcγRIIIa binding sensorgrams (c). (*See insert for color representation of the figure.*)

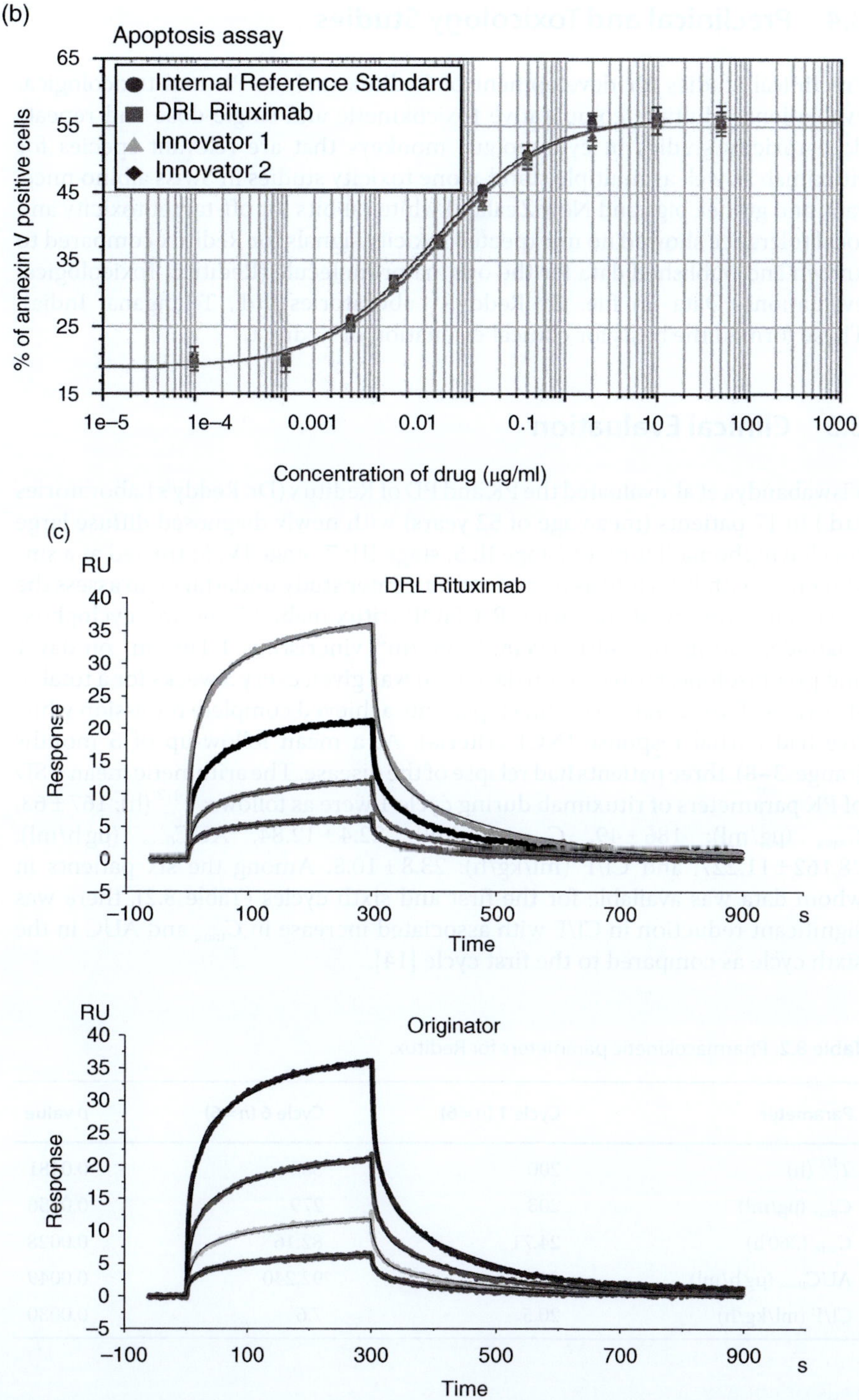

Figure 8.2 (Continued)

8.4 Preclinical and Toxicology Studies

Preclinical studies for development of Reditux include PK and toxicological evaluations. Multiple comparative toxicokinetic and single-dose and repeat-dose toxicity studies in cynomolgus monkeys that are relevant species for rituximab as well as multiple stand-alone toxicity studies in Swiss albino mice, rats, and guinea pigs and New Zealand white rabbits for off-target toxicity and local tolerance showed no unexpected toxicity signals for Reditux compared to known and published data for the originator molecule (Reditux, Toxicological evaluations. Data on file. Dr Reddy's Laboratories Ltd., Telangana, India). These formed the basis for clinical evaluation of Reditux.

8.5 Clinical Evaluation

Viswabandya et al. evaluated the PK and PD of Reditux (Dr. Reddy's Laboratories Ltd.) in 17 patients (mean age of 52 years) with newly diagnosed diffuse large B-cell lymphoma (DLBCL) (stage II: 5, stage III: 7, stage IV: 5) treated at a single center with R-CHOP as part of a multicenter study undertaken to assess the safety and efficacy of this drug. R-CHOP (rituximab, 375 mg/m^2; cyclophosphamide, 750 mg/m^2; adriamycin, 50 mg/m^2; vincristine, 1.4 mg/m^2 on day 1 and prednisolone, 60 mg/m^2 on days 1–5) was given every 3 weeks for a total of six cycles. Twelve out of seventeen patients achieved complete remission while five had partial response (NCI criteria). At a mean follow-up of 5 months (range: 3–8), three patients had relapse of the disease. The arithmetic mean ± SD of PK parameters of rituximab during cycle 1 were as follows: $T^{1/2}$ (h): 167 ± 63, C_{max} (μg/ml): 186 ± 49, C_{min} (μg/ml): 22.4 ± 12.84, $AUC_{0-\infty}$ (μg h/ml): 28,162 ± 11,227, and Cl/F (ml/kg/h): 23.8 ± 10.8. Among the six patients in whom data was available for the first and sixth cycles (Table 8.2), there was significant reduction in Cl/F with associated increase in C_{max} and AUC in the sixth cycle as compared to the first cycle [14].

Table 8.2 Pharmacokinetic parameters for Reditux.

Parameter	Cycle 1 (n = 6)	Cycle 6 (n = 6)	p value
$T^{1/2}$ (h)	200	386	0.0481
C_{max} (μg/ml)	203	279	0.0556
C_{min} (360 h)	24.71	82.16	0.0028
$AUC_{0-\infty}$ (μg h/ml)	31,167	92,240	0.0049
Cl/F (ml/kg/h)	20.5	7.6	0.0030

Table 8.3 Pharmacokinetic parameters of Reditux and MabThera.

PK parameter (mean ± SD)	Reditux (n = 21)	MabThera (published literature)
AUC_{0-t} (μg h/ml)	54,236 ± 47,555	—
C_{max} (μg/ml)	555.74 ± 141.46	408
Vd (l)	1.3 ± 0.64	2.7
CL (l/day)	0.15 ± 0.16	0.14
Half-life (days)	10.9 ± 8.6	22
MRT last (days)	2.78 ± 3.08	—

CL, clearance; MRT, mean residence time; Vd, volume of distribution.

In a prospective, single-center study (November 2011 and June 2013) in patients with DLBCL (n = 21; median age 50 years) who received six cycles of R-CHOP, PK parameters for Reditux were comparable with those of MabThera (Table 8.3). Age and gender were not found to influence the PK profiles of the two products. The median B-cell count on day 3 and day 21 of first cycle was 1.75 ± 0.27 and 5.56 ± 1.24 cells/μl, respectively. At a median follow of 19 months, 70.3% of patients from the Reditux group were progression free. There was no untoward immediate or late toxicity attributable to Reditux on follow-up [15].

Efficacy and safety of Reditux has been evaluated in an open-label, single-arm preregistration study in 76 patients with CD20-positive DLBCL (n = 76) who received CHOP chemotherapy. The efficacy of Reditux in DLBCL for ORR, complete remission (CR), and relapse was reported to be comparable to previous published literature for reference rituximab. No unexpected adverse events were reported. Immunogenicity was assessed by using ELISA for antidrug antibody. The study reported no binding or neutralizing antibodies [16].

A retrospective study (January 2004 and June 2010) compared the efficacy, safety, and toxicity of MabThera with a biosimilar (Reditux) in patients with DLBCL (n = 223) receiving the standard 3-week CHOP with rituximab (R-CHOP) therapies (101 received MabThera, 72 received Reditux). Reditux was found to be as efficacious as MabThera in terms of response rates, progression-free survival (PFS), and overall survival (OS) with comparable toxicity. CR rates were similar with MabThera and Reditux (75 and 82%, respectively; p = 0.294). The PFS rate at 5 years were 72% in MabThera and 81% in Reditux (p = 0.382) (Fig. 8.3). The OS at 5 years were comparable in the two groups (66% in MabThera and 76% in Reditux; p = 0.264) (Fig. 8.4). There were no differences in the infusion reaction rates, grades 3 and 4 neutropenia, and oral mucositis between the two brands. Analysis according to revised IPI showed better outcomes in terms of PFS (5-year PFS in very good risk groups was 86%,

(a)

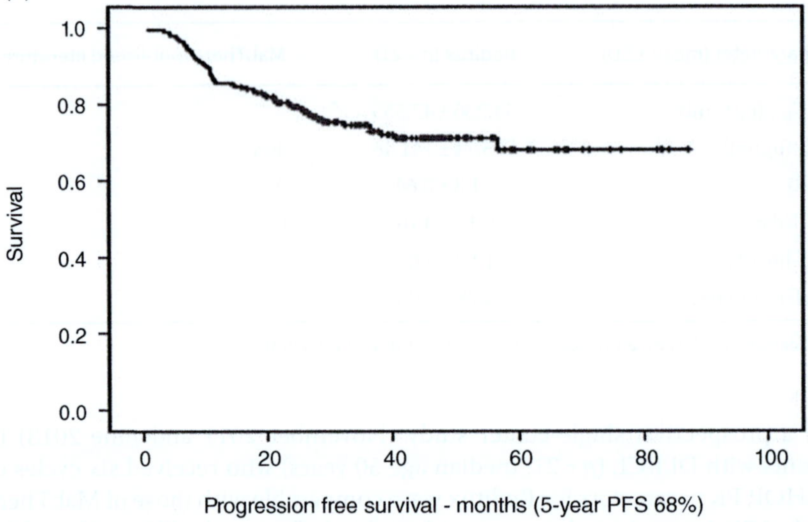

(b)

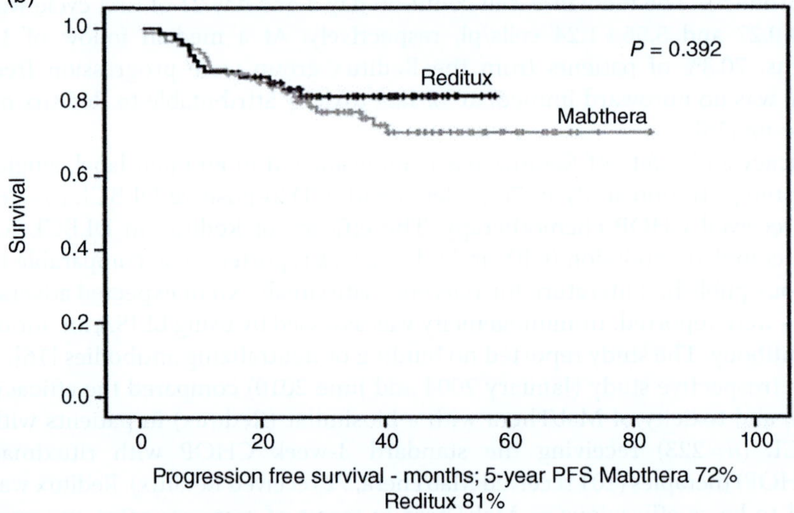

Figure 8.3 Five-year progression-free survival (months) for patients receiving Reditux or MabThera. (a) The 5-year progression-free survival (months) for all patients treated with cyclophosphamide, doxorubicin, vincristine, and prednisolone with rituximab. (b) Five-year progression-free survival (months) for patients receiving MabThera alone or Reditux alone in all cycles of cyclophosphamide, doxorubicin, vincristine, and prednisolone with rituximab.

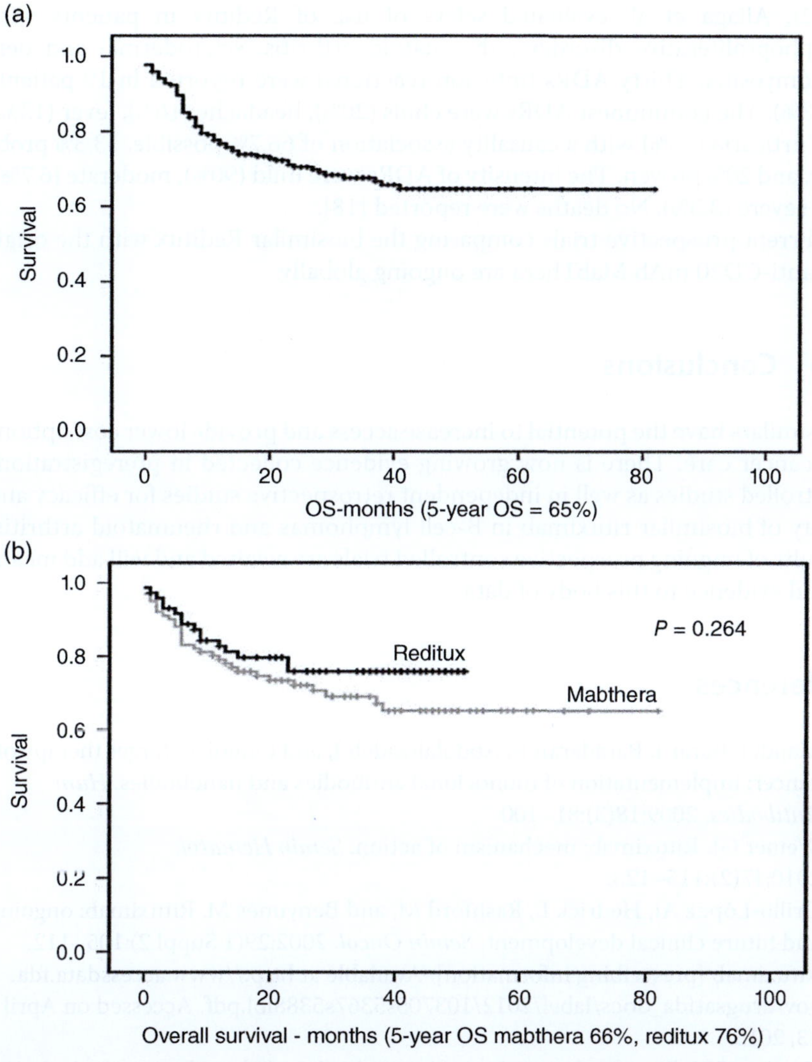

Figure 8.4 Five-year overall survival (months) for patients receiving Reditux or MabThera. (a) The 5-year overall survival (months) of all patients treated with cyclophosphamide, doxorubicin, vincristine, and prednisolone with rituximab. (b) The 5-year overall survival (months) of all patients treated with MabThera alone or Reditux alone in all cycles of cyclophosphamide, doxorubicin, vincristine, and prednisolone with rituximab.

in good risk group 69%, and in poor risk group 49%) and OS (5-year OS in very good risk group 80%, in good risk group 61%, and in poor risk group 37%) [17].

The safety evaluation of Reditux is reported in an independent prospective study at the hospital of social security in Peru (August through November,

2012). Aliaga et al. evaluated safety of use of Reditux in patients with lymphoproliferative disorders, rheumatoid arthritis, scleroderma, and dermatomyositis. Thirty ADRs (infusion reactions) were reported in 19 patients (14.3%). The commonest ADRs were chills (20%), headache (16%), fever (13%), and urticaria (10%) with a causality association of 66.7% possible, 13.3% probable, and 20% proven. The intensity of ADRs were mild (90%), moderate (6.7%), and severe (3.3%). No deaths were reported [18].

Current prospective trials comparing the biosimilar Reditux with the original anti-CD20 mAb MabThera are ongoing globally.

8.6 Conclusions

Biosimilars have the potential to increase access and provide lower cost options for cancer care. There is now growing evidence collected in preregistration-controlled studies as well in independent retrospective studies for efficacy and safety of biosimilar rituximab in B-cell lymphomas and rheumatoid arthritis. Results of ongoing prospective controlled trials are awaited and will add meaningful evidence to this body of data.

References

1. Majidi J, Barar J, Baradaran B, Abdolalizadeh J, and Omidi Y. Target therapy of cancer: implementation of monoclonal antibodies and nanobodies. *Hum Antibodies*. 2009;18(3):81–100.
2. Weiner GJ. Rituximab: mechanism of action. *Semin Hematol.* 2010;47(2):115–123.
3. Grillo-López AJ, Hedrick E, Rashford M, and Benyunes M. Rituximab: ongoing and future clinical development. *Semin Oncol.* 2002;29(1 Suppl 2):105–112.
4. Rituximab [prescribing information]. Available at http://www.accessdata.fda.gov/drugsatfda_docs/label/2012/103705s5367s5388lbl.pdf. Accessed on April 13, 2015.
5. Keating GM. Rituximab: a review of its use in chronic lymphocytic leukaemia, low-grade or follicular lymphoma and diffuse large B-cell lymphoma. *Drugs.* 2010;70(11):1445–1476.
6. Falk JI and Davidson RA. Coping with the regulatory realities of follow-on biologics and the nuances of US and European regulations. *Drug Inf J.* 2010;44:137–145.
7. McCamish M and Woollett G. Worldwide experience with biosimilar development. *mAbs.* 2011;3:209–217.
8. Calvo B and Zuñiga L. The US approach to biosimilars: the long-awaited FDA approval pathway. *BioDrugs.* 2012;26(6):357–361.

9. Guidelines on Similar Biologics: Regulatory requirements for Marketing Authorizations in India. Available at http://dbtbiosafety.nic.in/Files%5CCDSCO-DBTSimilarBiologicsfinal.pdf. Accessed on April 15, 2015.
10. Qureshi ZP, Magwood JS, Singh S, and Bennett CL. Rituximab and biosimilars—equivalence and reciprocity. *Biosimilars*. 2013;2013(3):19–25.
11. Mulcahy AW, Predmore Z, and Mattke S. The Cost Savings Potential of Biosimilar Drugs in the United States, RAND Corporation, 2014. Available at www.rand.org/t/PE127. Accessed on April 13, 2015.
12. ICH Guidelines. Available at http://www.ich.org/products/guidelines/quality/article/quality-guidelines.html. Accessed on April 13, 2015.
13. U.S. Food and Drug Administration. Current Good Manufacturing Practices (CGMPs). Available at http://www.fda.gov/food/guidanceregulation/cgmp/. Accessed on April 13, 2015.
14. Viswabandya A, et al. Pharmacokinetic and Pharmacodynamic Evaluation of a Biosimilar Rituximab in Newly Diagnosed Diffuse Large B-Cell Lymphoma (DLBCL) Treated with R-CHOP (Rituximab, Cyclophosphamide, Adriamycin, Vincristine, Prednisolone). *Blood* (ASH Annual Meeting Abstracts) 2007;110:Abstract 4491.
15. Menon H, Yadav A, Subramanian PG, Sengar M, Rath S, Kavathiya K, and Gota V. Pharmacokinetic and pharmacodynamic properties of a biosimilar rituximab (Reditux®) are identical to the innovator brand MabThera®—experience from a tertiary cancer centre in western India. *Blood*. 2014;124(21).
16. Ranade AA, Bhatt AA, Joshi S, Pradhan S, Kelkar K, Karve A, Dumbre R, Joshi A, Bhattacharya S, Bondarde S, and Parikh PM. Biosimilar reditux in B-cell lymphoma: data from real-life clinical practice. *Indian J Cancer*. 2013;50(Suppl 1):S171.
17. Roy PS, John S, Karnakal S, Kannan S, Pawaskar P, Gawande J, Bagal B, Khattry N, Sengar M, Menon H, Gujral S, and Nair R. Comparison of efficacy and safety of Rituximab (MabThera) and its biosimilar (Reditux) in DLBCL patients treated with chemo immunotherapy: a retrospective analysis. *Indian J Med Paediatr Oncol*. 2013;34(4):292–298.
18. Aliaga L, Fernandez E, Sanchez M, Saavedra H, and Espinoza C. Pharmacovigilance of anti CD 20 monoclonal antibody biosimilar at the Edgardo Rebagliati Martins Hospital—Peru. International Society of Pharmacovigilance. XII Annual Meeting PISA October 2, 2013. Abstract ID: ISP3815-45.

9

Monoclonal Antibodies for Infectious Diseases

Steven J. Projan

MedImmune LLC, The AstraZeneca Group, Gaithersburg, MD, USA

9.1 Summary

This chapter will review the history of antibody-based therapies for infectious diseases and describe multiple ongoing clinical therapeutic antibody programs for treating infections from *Staphylococcus aureus* (MRSA), *Pseudomonas aeruginosa, Clostridium difficile* and influenza. With development of improved manufacturing process, therapeutic antibodies are considered to be not only safe for human use to treat infectious diseases but also poses little risk of drug–drug interactions with other therapeutic agents and thus suitable for combination with other broad-spectrum antibiotics. Therapeutic antibodies with immunomodulatory functions, for example, antibodies that block PD-1/PD-L1 pathway, have the potential to treat chronic infections of both bacteria (e.g., *Mycobacterium tuberculosis*) and viruses (e.g., HBV and HIV). The high price remains a challenge for wide use of antibody-based treatment of infectious diseases, which demands a pharmacoeconomic advantage over existing antibiotics as standard care.

9.2 Into the Future: Prophylaxis and Precision Medicine

The advent of the "age of antibiotics" in the 1940s had a major impact on extending lives, but it also brought with it a false sense of security and led to an overenthusiastic use of antibiotics, sometimes as a substitute for good sanitation and hygiene. We now know that resistance to these important agents can and does arise and this may have dire consequences for patients. Finding

Biosimilars of Monoclonal Antibodies: A Practical Guide to Manufacturing, Preclinical, and Clinical Development, First Edition. Edited by Cheng Liu and K. John Morrow, Jr.

new ways to prevent and treat infections caused by resistant organisms will be a major challenge in the twenty-first century. Identifying "at-risk" patients (e.g., those colonized with *Staphylococcus aureus*) and taking preemptive measures (e.g., decolonization) or perhaps developing a protective vaccine may well obviate the use of the antibiotics that have made the practice of medicine of the late twentieth and early twenty-first century possible. A new term of art has been coined "precision medicine," and we now have (available although not routinely employed) the ability to identify an infecting pathogen within 2 h (if not less). This should allow for the use of narrow-spectrum agents that, in and of themselves, cannot obviate either toxicity or the development of resistance but may well prevent cross-resistance to current antibiotics as well as ameliorate, if not prevent, horizontal gene transfer. In addition to the continued hunt for effective vaccines, currently there are multiple monoclonal antibodies (mAbs) in clinical development to prevent and/or treat infections caused by *Pseudomonas aeruginosa*, *S. aureus*, *Clostridium difficile*, and others. To date these mAbs, mainly of human origin or design, have had an enviable safety record (as may well be predicted), and these may represent the future of antibacterial treatments.

9.3 Immune Therapy: A Noble Undertaking that Went to the Dogs

Emil von Behring pioneered the use of sera, derived from inoculated animals, as a treatment for infections (diphtheria being the first bacterial infection to be treated). For this groundbreaking work, von Behring won the first Nobel Prize in Physiology or Medicine in 1901, and then 7 years later his former associate Paul Ehrlich (sharing the prize with Élie Metchnikoff) also won the award for his contributions to immunotherapeutics for infection [1]. The importance and utility of serum therapy were dramatically demonstrated in 1925 when Nome, Alaska, suffered an outbreak of diphtheria and a "Great Race of Mercy" was undertaken to deliver, by a relay of dog sled teams, antidiphtheria antiserum from the rail head at Nenana near Fairbanks 674 miles westward to Nome. In just over 5 days, the final leg of the relay, with musher Gunnar Kaasen and lead dog Balto, delivered the antiserum to Dr. Curtis Welch without a single broken ampoule (through some of worst weather Alaska had seen for two decades). While there were reportedly only seven deaths, it is not clear how many of the Inuit (native Alaskans) actually perished in the epidemic. Nonetheless, it was clear that the antiserum played a major role in bringing the epidemic under control and underscored the importance of drug delivery (then and now).

9.4 What's Taking So Long?

In 1995, Arturo Casadevall and Matthew Scharff called for novel antibody-based therapies for infectious diseases in what many, even then, referred as a "back to the future" approach [2]. Dr. Casadevall himself has been directly involved in the discovery and development of such agents and continues to be one of its chief proponents. It was a mere 13 years from the publication of Fleming's discovery of a *Penicillium* mold that could lyse *S. aureus* to the first successful therapeutic use of penicillin; more importantly this began a period of natural product research that brought us many new broad-spectrum antibacterial agents from the aminoglycosides (kanamycin, streptomycin) to the tetracyclines to the macrolides (erythromycin) to the rifamycins all in the 15 years since Anne Sheafe Miller was cured of her group A streptococcus infection by penicillin treatment in April of 1942. In the 20 years since Casadevall and Scharff proposed the use of novel immunotherapeutics for infectious disease, a single mAb has been used clinically in the infectious disease space (palivizumab for the prevention of respiratory syncytial virus (RSV) in high-risk infants). However today's picture is far brighter, perhaps necessitated by the continued and inexorable increase of microbial drug resistance as well as improved methodologies in both target identification and antibody discovery, not to mention the continued futility in the discovery and development of the next generation of (broad-spectrum) antibiotics [3]. There are several novel agents currently in clinical development and many more in preclinical development. And we are starting to see a hint of efficacy for those agents in addition to published positive results in animal models of infection.

9.5 *Staphylococcus aureus*: Still Public Enemy Number One?

Recent data published by the CDC and the Veterans Administration suggests that there has been a precipitous decline in rates of hospital infections due to methicillin-resistant staphylococcus (e.g., ~50% drop between 2006 and 2011). However this drop in the United States is not consistent with the fact that linezolid sales have continued to increase both in the United States and worldwide and, even more dramatically, another MRSA drug, daptomycin, has seen sales in the United States go from \$189 M in 2006 to \$908 M in 2013. It is hard to square a 50% drop in infection rates with a 480% increase in sales for a drug whose primary utility is for MRSA infection. There are several ways to interpret this apparent paradox, but a likely one is the application of aggressive prophylaxis in high-risk patients or aggressive empiric therapy where an infection is suspected but as yet undiagnosed. In either event this would represent a significant change in how antistaphylococcal drugs are being used and suggest an important role in the use of specific agents aimed at preventing fulminant staphylococcal infections.

9.5.1 mAbs for Staph

There are currently at least two mAbs in clinical development that target *S. aureus* alpha toxin AR-301 (Aridis, San Jose, CA), and MEDI4893 (MedImmune/AstraZeneca) has been shown to neutralize alpha toxin (formerly known as alpha hemolysin), a virulence factor to have multiple activities that is key in establishing lung and skin and skin found structure infections [4].

SAR279356 (F598) (Sanofi/Alopexx, Concord, MA) is a mAb directed toward a surface carbohydrate alternatively called poly-*N*-acetylglucosamine (PNAG) or polysaccharide intercellular adhesin (PIA). Since PNAG is also found on the surface of coagulase-negative staphylococci (e.g., *Staphylococcus epidermidis*), this mAb may have broader utility as *S. epidermidis* is an important pathogen in high-risk neonates. It is currently listed as in phase 2.

In addition Arsanis, a Vienna-based biotech company, is reportedly planning on beginning clinical trials with a triple mAb combination, likely including an alpha toxin mAb [5].

9.6 *Pseudomonas aeruginosa*: The Bacterial Cockroach

Pseudomonas aeruginosa is not usually associated with infections in otherwise healthy individuals rather than those with compromised immune systems and other underlying diseases, such as cystic fibrosis. Trauma and burns patients, patients in intensive care units, especially those receiving mechanical ventilation, are prime targets for this ubiquitous bacterium. Because *P. aeruginosa* is so adaptable to its environment, it is intrinsically resistant to many antibiotics, and, probably owing to its large (6.3 Mb) genome, it often becomes resistant to virtually any novel agent in a matter of days to weeks. It is an understatement to say that novel approaches to the prevention and treatment of pseudomonal infections are needed.

KB001-A (KaloBios) is a pegylated Fab targeting PcrV, a protein that is best described as the tip of the *P. aeruginosa* type 3 secretion system [6]. As a Fab, KB001-A can block the secretion of toxins through the "injectisome," thereby reducing the virulence of the bacterium, but due to lack of Fc and therefore effector function, KB001-A would not lead to a more rapid clearance of the bacterium and is therefore more suitable as a prophylactic rather than a therapeutic. Currently KB001-A is in a holding pattern as it may require modification of its dosage upward.

Medi3902 (MedImmune/AstraZeneca) that also targets PcrV is a bispecific mAb, which simultaneously targets Psl, a mannose-rich surface polysaccharide [7]. This mAb has been shown to retain effector function as well as act synergistically with antibiotics (even vs. drug-resistant strains). Medi3902 has been entered into clinical development.

9.7 Immune Evasion and Degree of Difficulty

Just because serum therapy has worked for several diverse pathogens does not necessarily mean that this technology is broadly applicable to all bacterial or viral pathogens. It is noteworthy that in Casadevall and Scharff's scholarly paper, none of the serum therapeutics listed were for either staphylococcal or pseudomonal infections. Moreover, most of the bacterial and all of the viral infections for which serum therapies were developed are diseases that today are preventable by vaccination, whereas repeated attempts to develop vaccines to prevent staphylococcal or pseudomonal infections have been somewhere between bitter disappointments and abject failures. Of the 300–400 virulence factors encoded in the *S. aureus* genome, the large majority of them are likely to have a role in immune evasion. Work by Schneewind and colleagues at the University of Chicago has been targeting perhaps the keystone of staphylococcal immune evasion, protein A, with positive preclinical results [8].

9.8 *Clostridium difficile*: You Can't Win for Losing

A disease that arises as a consequence of a therapy is called "iatrogenic." *Clostridium difficile*-associated diarrhea is such a disease. It is most often associated with elderly hospitalized patients who are receiving antibacterial treatments. It is likely that perturbations in the patient's gut microflora allow for the germination of *C. difficile* spores, and this bacterium then expresses two related toxins that give rise to pseudomembranous colitis, a potentially life-threatening condition.

A project began by MassBio and Medarex and subsequently licensed by Merck has resulted in a pair of mAbs that neutralize the A and B toxins of *C. difficile* (TcdA, TcdB) and have been shown to prevent relapse [9]. This project is currently in phase 3 testing.

9.9 If Two Is Enough, Is Six Too Many? mAb Combos

In addition to the *C. difficile* combination currently in development, other combination approaches are being studied preclinically. Symphogen lists Sym009, their polyclonal mAb combo, as "directed against an undisclosed infectious disease target. The project is funded by Genentech." And the Arsanis staph combo is discussed previously.

9.10 Prophylaxis or Therapy? When You Come to a Fork in the Road, Take It

A common question over how these agents should be used is treatment versus prophylaxis. Data is emerging that either approach is viable but pricing would be quite different depending on the use (treatment being more expensive). But the proof of the pudding is in the eating, and the current thinking is that prophylaxis would be a "lower bar" for the demonstration of clinical utility but would require larger clinical trials than for therapeutic use.

9.11 Influenza and Plan "B"

There are currently several mAbs in development for the treatment (and possible prevention) of influenza (both A and B). In many instance these fully human mAbs were discovered from either postconvalescent patients or vaccinated subjects by harvesting B cells and identifying potent antibodies [10, 11]. Most interestingly they appear to have fairly broad coverage across serotypes, maintaining a high degree of potency. The open question is whether these mAbs will be superior to the current marketed neuraminidase inhibitors (superiority in animal models of influenza infection not withstanding).

9.12 Safety: Human Enough for You?

The irony of the replacement of serum therapy by antibiotics was that the antibiotics were perceived to be safer owing to the experience of the early days of serum therapy when "serum sickness" [12] was not uncommon. Although its origins are not completely clarified, it was most likely due to an immune reaction to foreign (mainly equine) protein. Most probably antiantibody reactivity did not play a role, as the structural similarity between human and equine antibodies is quite high. Indeed, today several chronically administered mAbs are chimeras, and even with the development of antidrug antibodies rarely does their application cause serious adverse events.

Therefore the foreign proteins that led to serum sickness were probably not the antibodies in the serum.

Even during the time of Paul Ehrlich in the early twentieth century, safer and safer preparations were being manufactured, and by the mid-1930s serum sickness was rare. Today's ultrapure preparations of mAbs, which are "fully human" in nature, are likely to be safer still. Unlike small-molecule drugs, there is little likelihood of drug–drug interactions with other therapeutics agents, and, as such, they would also be suitable for "adjunctive therapy" in combination with other drugs (even broad-spectrum antibiotics).

9.13 Another Precinct Is Heard from Immunomodulatory Agents for the Treatment of Chronic Infections

A conundrum of protocols for the treatment of chronic infections is why doesn't our immune system rise up and kill the infected cells rather than allowing for these "zombie cells" to harbor bacteria (e.g., *Mycobacterium tuberculosis*) or viruses (e.g., HBV and HIV) where the pathogen can hide out or worse replicate? Starnbach has demonstrated that the obligate intracellular bacterial pathogen *Chlamydia trachomatis* unregulates the host cell's PD-1 pathway [13]. PD-1 ("programmed death") is a receptor for an immune checkpoint inhibitor (a fail-safe system to prevent autoimmune disease) and as such effectively blocks an immune counterattack upon the affected cells. In oncology immune modulation is emerging as a very productive area of research where the potential for actual cures is emerging. It is possible, if not likely, that these agents may well be active versus the chronic infections discussed previously.

9.14 Are We There Yet? Easy to Use, Fast Turnaround, Point-of-Care Diagnostics

As bacterial-specific antibodies and other similarly targeted antibacterial agents enter clinical practice, they are going to necessitate, in parallel, the development of companion diagnostics that are as easy to use as single-serve coffee makers and with a comparable unit cost per diagnosis and a sampling to result time of under 2h. A platform using polymerase chain reaction (PCR) coupled with molecular beacons appears to be capable of delivering with ease and on time with an admirable degree of flexibility while the question of cost remains an open one. Other technologies such as MALDI-TOF and mass spectroscopy are promising, reducing them to practical practice but requiring a significant amount of additional research and development.

9.15 Yeah but Aren't These (Biologic) Drugs Going to Be Expensive?

It is increasingly recognized that broad-spectrum antibiotics, by virtue of that broad spectrum, facilitated the aggressive, empiric use of these drugs. But there is another important fact, also becoming realized: antibiotics are grossly underpriced; even those that are still protected by patents and/or data exclusivity most often cost less than \$1000 for a course of therapy, which in most instances results in a complete cure of the infection. While oncology drugs that extend patients' lives for 2 months can cost upward of \$100,000. (Ironically, oncology patients often require antibacterial therapy by virtue of the immunosuppressive nature of many oncology therapies.) That being stated, in today's resource-constrained (which can be read as "cash-strapped") environment, virtually no drug will gain wide acceptance (much less routine use) if it does not have a pharmacoeconomic advantage over standard of care.

References

1. Baumler E. Paul Ehrlich Scientist for Life. Holmes & Meier, New York 1984.
2. Casadevall A, Scharff MD. Serum therapy revisited: animal models of infection and development of passive antibody therapy. *Antimicrob. Agents Chemother.* 1994 38:1695.
3. Payne D, Gwynn MN, Holmes J, Pompliano DL. Drugs for bad bugs: confronting the challenges of antibacterial discovery. *Nat. Rev. Drug Discov.* 2007 6:29–40.
4. Inoshima I, Inoshima N, Wilke G, Powers M, Wang Y, Bubeck Wardenburg J. A *Staphylococcus aureus* pore-forming toxin subverts the activity of ADAM10 to cause lethal infection. *Nat. Med.* 2011 17:1310–1314.
5. Stulik L, Malafa S, Hudcova J, Rouha H, Henics BZ, Craven DE, Sonnevend AM, Nagy E. α-Hemolysin activity of methicillin-susceptible *Staphylococcus aureus* predicts ventilator-associated pneumonia. *Am. J. Respir. Crit. Care Med.* 2014 190:1139.
6. Francois B, Luyt CE, Dugard A, Wolff M, Diehl JL, Jaber S, Forel JM, Garot D, Kipnis E, Mebazaa A, Misset B, Andremont A, Ploy MC, Jacobs A, Yarranton G, Pearce T, Fagon JY, Chastre J. Safety and pharmacokinetics of an anti-PcrV PEGylated monoclonal antibody fragment in mechanically ventilated patients colonized with *Pseudomonas aeruginosa*: a randomized, double-blind, placebo-controlled trial. *Crit. Care Med.* 2012 40:2320–2326.
7. DiGiandomenico A, Keller AE, Gao C, Rainey GJ, Warrener P, Camara MM, Bonnell J, Fleming R, Bezabeh B, Dimasi N, Sellman BR, Hilliard J, Guenther CM, Datta V, Zhao W, Gao C, Yu X-C, Suzich J, Stover CK. A multifunctional bispecific antibody protects against *Pseudomonas aeruginosa*. *Sci. Transl. Med.* 2014 6:262.

8. Kim HK, Emolo C, DeDent AC, Falugi F, Missiakas DM, Schneewind O. Protein A-specific monoclonal antibodies and prevention of *Staphylococcus aureus* disease in mice. *Infect. Immun.* 2012 80:3460.
9. Lowy I, Molrine DC, Leav BA, Blair BM, Baxter R, Gerding DN, Nichol G, Thomas WD Jr, Leney M, Sloan S, Hay CA, Ambrosino DM. Treatment with monoclonal antibodies against *Clostridium difficile* toxins. *N. Engl. J. Med.* 2010 362:197.
10. Hu W, Chen A, Miao Y, Xia S, Ling Z, Xu K, Wang T, Xu Y, Cui J, Wu H, Hu G, Tian L, Wang L, Shu Y, Ma X, Xu B, Zhang J, Lin X, Bian C, Sun B. Fully human broadly neutralizing monoclonal antibodies against influenza A viruses generated from the memory B cells of a 2009 pandemic H1N1 influenza vaccine recipient. *Virology* 2013 435:320.
11. Pappas L, Foglierini L, Piccoli L, Kallewaard NL, Turrini F, Silacci C, Fernandez-Rodriguez B, Agatic G, Giacchetto-Sasselli I, Pellicciotta G, Sallusto F, Zhu Q, Vicenzi E, Corti D, Lanzavecchia A. Rapid development of broadly influenza neutralizing antibodies through redundant mutations. *Nature* 2014 516:418.
12. Jackson R. Serum sickness. *J. Cutan. Med. Surg.* 2000 4:223.
13. Fankhauser SC, Starnbach MN. PD-L1 limits the mucosal CD8+ T cell response to *Chlamydia trachomatis*. *J. Immunol.* 2014 192:1079.

10

Monoclonal Antibodies for Musculoskeletal, CNS, and Other Diseases

Junming Yie[1,*] *and Tao Wu*[2]

[1] *Amgen, Inc., Thousand Oaks, CA, USA*
[2] *Boehringer Ingelheim Pharmaceuticals, Inc., Ridgefield, CT, USA*
* *Current affiliation: Merck Research Laboratories, Kenilworth, NJ, USA*

10.1 Summary

This chapter covers seven approved monoclonal antibodies (mAbs) for the treatment of various diseases. Natalizumab is a humanized mAb aimed at autoimmune disease such as multiple sclerosis (MS) and Crohn's disease (CD). Eculizumab is a humanized mAb targeting rare diseases associated with the dysfunction of the complement system, part of innate immune system, such as paroxysmal nocturnal hemoglobinuria (PNH) and atypical hemolytic uremic syndrome (aHUS). Ranibizumab is a humanized mAb fragment (Fab) for various macular diseases such as neovascular (wet) age-related macular degeneration (wet AMD), macular edema following retinal vein occlusion (RVO), and diabetic macular edema (DME). Denosumab is a fully human mAb for the therapy of various bone diseases such as osteoporosis, treatment-induced bone loss, skeletal-related events in patients with bone metastases from solid tumors, and giant cell tumor of bone. Daclizumab, basiliximab, and muromonab-CD3 are used as immunosuppressive agents for solid organ transplantations. They were approved at different time frames, reflecting the evolution of the mAb development. Among those, muromonab is a murine mAb, and daclizumab is a human/murine chimeric mAb, whereas basiliximab is a humanized mAb.

Biosimilars of Monoclonal Antibodies: A Practical Guide to Manufacturing, Preclinical, and Clinical Development, First Edition. Edited by Cheng Liu and K. John Morrow, Jr.

10.2 Natalizumab (Tysabri®)

Natalizumab (trade name Tysabri) is a humanized mAb reactive against the cell adhesion molecule α4 subunit of α4β1 (also known as very late antigen 4, VLA-4, or CD49d CD29) and α4β7 integrin [1, 2]. It is used in the treatment of MS and CD, two autoimmune diseases. Natalizumab functions by reducing the ability of inflammatory immune cells to attach to and pass through the cell layers lining the intestines and blood–brain barrier (BBB) [3, 4]. Natalizumab is manufactured by Biogen Idec, Cambridge, MA, and it is comarketed by Biogen Idec and Elan Pharmaceuticals, Inc., South San Francisco, CA.

Natalizumab was approved in 2004 by the FDA for treatment of MS and initially marketed as the trade name Antegren; it was withdrawn voluntarily from the market by its manufacturer in 2005 subsequent to three cases of the rare but deadly neurological condition progressive multifocal leukoencephalopathy (PML) when coadministered with interferon beta-1a, another immunosuppressive drug often used in treating multiple sclerosis. It was reintroduced to the US market with a trade name Tysabri in 2006 for the treatment of MS with a warning label claiming the risk of PML and stating that patients receiving any other immunosuppressive therapy or with underlying immunosuppressive conditions should be precluded from drug usage. Later on, more cases of PML were attributed to natalizumab, but the FDA did not suspend the drug from the market arguing that its clinical benefits outweigh the risks involved. In 2008, natalizumab was also approved by FDA for the treatment of CD [5].

The healthy immune system provides critical defense against the invasion of pathogenic microbes. This beneficial property of the immune system is the cause of a number of autoimmune diseases such as multiple sclerosis, rheumatoid arthritis, inflammatory bowel disease (mainly CD), type I diabetes, lupus, and so on, in which immune system goes uncontrolled and attacks self-tissues in a particular location. For example, in multiple sclerosis, inflammatory immune cells attack the myelin of the spinal cord or central nervous system (CNS), and these cells enter the brain via the blood vessels, crossing BBB. In CD, the immune system attacks the intestinal mucosa. The properties of these diseases demonstrate that the immune cells are able to home to a specific location from the blood, stick to the surface of blood vessels (tethering and rolling), and penetrate the extracellular matrix (ECM) surrounding the vessel (activation, arrest, and diapedesis) in order to gain access to the particular tissue to cause damage (Fig. 10.1).

The increased understanding of the immune cells migration to a specific site on the molecular level allowed immunologists to find strategies to block the migration of lymphocytes. Among these molecules, integrins are of special interest. Integrins are a family of heterodimeric cell surface receptors, composed of noncovalently linked α and β subunits. In mammals, 18 α subunits

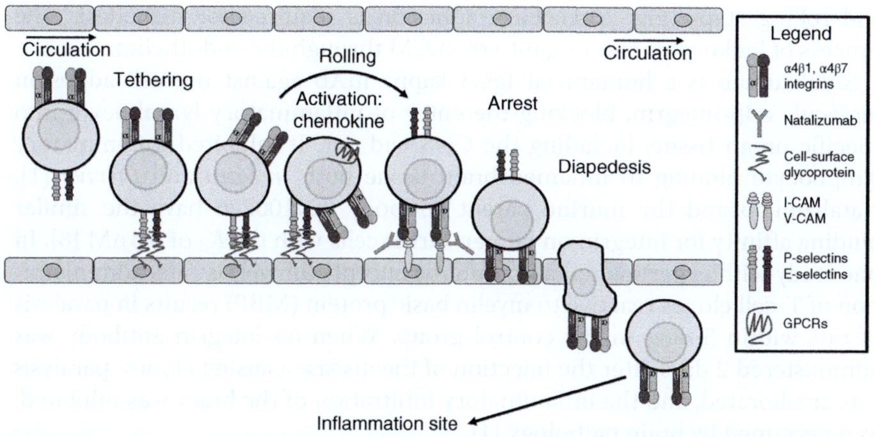

Figure 10.1 Mechanism of immune-cell mediated inflammation and its inhibition by α4 antagonist natalizumab. Lymphocytes flowing in the blood vessels are induced via chemokines to slow down, roll along the blood vessel walls, and finally arrest their movement and attach to the endothelium through a complex sequence of events. The local release of chemokines at the site of inflammation causes upregulation of α4β1 and α4β7 on the circulating immune cells and adhesion molecules such as V-CAM and/or MAdCAM, on the surface of proximal endothelial cells [6]. Increased adhesion molecule density facilitates contacts between circulating immune competent leukocytes and vascular endothelium. As blood flows, binding between P-selectin and E-selectin molecules and cell surface glycoprotein or α4 integrins and endothelial cell adhesion molecules provide the initial contacts needed to capture circulating leukocytes (tethering and rolling). The decreased velocity allows chemokines on the surface of vascular endothelium to bind G-protein coupled receptor (GPCRs) on the leukocyte surface, transducing signals that further activate integrins (activation). The high avidity of activated α4 integrins to their respective ligands results in firm adhesion to the endothelium (arrest). Firm adhesion is required for the process of diapedesis, extravasation of leukocytes through the endothelium into the extracellular matrix (ECM), to occur. When α4 integrins are blocked by natalizumab, lymphocyte adhesion to the endothelial wall is disrupted as their interactions with I-CAM and V-CAM are impaired. (*See insert for color representation of the figure.*)

and eight β subunits have been identified to date [7]. The distinct heterodimer combinations are critical for cell-to-cell and cell-to-ligand specificities relevant to the host cell and the environment in which they function. Among these, α4β1 and α4β7 are expressed on the surfaces of lymphocytes and leukocytes, and they function in leukocyte recruitment from the peripheral circulation to the sites of inflammation with tissue. The local release of chemokines at sites of inflammation causes upregulation of α4β1 and α4β7 on the circulating immune cells and adhesion molecules such as vascular cell adhesion molecule (V-CAM) and mucosal addressin cell adhesion molecule (MAdCAM). The high avidity of activated α4 integrins to their respective ligands (V-CAM and MAdCAM) results in firm adhesion to the endothelium, after initial tethering and rolling steps (Fig. 10.1). Firm adhesion is required for diapedesis, the process of leukocytes entering into the ECM through the endothelium.

Natalizumab is a humanized IgG4 kappa mAb against the cell adhesion molecule α4β-integrin, blocking the entry of inflammatory lymphocyte into specific organ tissue, including the CNS and gut. It inhibited inflammatory lymphocyte binding to inflamed brain tissue both *in vitro* and *in vivo* [1]. Natalizumab and the murine parent antibody AN100226 have the similar binding affinity for integrin on human Jurkat cells with the K_d of 0.3 nM [8]. In one study with experimental autoimmune encephalomyelitis rats, administration of T-cell clones reactive to myelin basic protein (MBP) results in paralysis of rats within 5 days in the control group. When α4-integrin antibody was administered 2 days after the injection of the disease-causing clones, paralysis was ameliorated, and the inflammatory infiltration of the brain was inhibited, as determined by brain pathology [1].

Approval of natalizumab in treating MS was primarily based on 1-year results from the AFFIRM and SENTINEL trials. More than 2000 patients with relapsing–remitting and secondary progressive forms of MS received natalizumab in clinical trials. The Expanded Disability Status Scale (EDSS) is a widely used tool based on functional grading systems that evaluate the degree of neurologic impairment in MS. In the clinical studies, neurological evaluations were performed every 12 weeks and at times of suspected relapse. Magnetic resonance imaging evaluations for T1-weighted gadolinium (Gd)-enhancing lesions and T2-hyperintense lesions were performed annually. The AFFIRM trial included 942 patients aged 18–50 years with relapsing–remitting MS. Patients were randomized in a 2:1 ratio to receive Tysabri 300 mg intravenous infusion ($n = 627$) or placebo ($n = 315$) every 4 weeks for up to 28 months (30 infusions). The difference in the annualized relapse rate (ARR) of 0.25 in the natalizumab group compared with 0.74 in the placebo group was highly significant ($p < 0.001$), representing a 66% reduction in the natalizumab-treated group [9]. In the SENTINEL trial, 1171 patients, who were on the conventional interferon beta-1a therapy but who had experienced at least one relapse while on therapy, were enrolled. The same natalizumab dosing regime was used. Patients receiving combination natalizumab plus interferon beta-1a therapy had an ARR of 36% annually compared with 78% in patients receiving interferon beta-1a alone ($p < 0.001$), representing a 54% reduction in the combination group [10].

The safety and efficacy of natalizumab in treating CD were evaluated in three randomized, double-blind, placebo-controlled clinical trials [11]. A cohort of 1414 adult patients with moderately to severely active CD were studied. The Crohn's Disease Activity Index (CDAI) and the Powell-Tuck score are used to assess clinical response and/or remission. The ENACT-I trial evaluated response and remission in 896 patients with CD randomized in a 4:1 ratio to receive 300 mg of intravenous natalizumab or placebo given every 4 weeks for three doses in patients previously being treated with antitumor necrosis factor therapy. Clinical results were assessed at week 10. Although the efficacy data are not available for the entire study population, *post hoc* analysis of a subset of patients with elevated C-reactive protein (CRP), indicative of active inflammation, natalizumab induced significantly greater clinical response compared with the placebo. It thus led to another trial (ENCORE) [12], which only patients with elevated CRP were enrolled. A cohort of 509 patients were randomized in a 1:1 ratio to receive three monthly infusion of either 300 mg of intravenous natalizumab or placebo. At 6 months, 61% of natalizumab-treated patients continued to meet criteria for clinical response, compared with 29% in placebo group ($p < 0.001$). Maintenance therapy was evaluated in the third clinical trial (ENACT-II). Three hundred and thirty-one patients from the ENACT-I trial that had a clinical response to natalizumab at both weeks 10 and 12 were rerandomized in a 1:1 ratio to treatment with continuing natalizumab treatment or placebo. Continuing natalizumab in the second trial resulted in higher rates of sustained response (61 vs. 28%, $p < 0.001$) and remission (44 vs. 26%, $p = 0.003$) through week 36 than did switching to placebo.

The superior clinical efficacies of natalizumab in MS prompted FDA's approval in 2004 under accelerated review. Shortly after its launch, natalizumab was withdrawn from the market by its manufacturer after it was linked to three cases of rare neurological condition, PML, when coadministered with interferon beta-1a. It was reinstated under a strict monitoring program by regulatory authorities after reviewing its risk/benefit. Since its reintroduction to the market, more cases of PML have been reported. In the meantime, more knowledge has been accumulated to understand natalizumab's benefit and risk. Even though natalizumab is very effective in reducing relapses and halting progression of disease, it also made brains of subjects more susceptible to infection. PML is a potentially fatal CNS opportunistic infection caused by reactivation of a clinically latent John Cunningham (JC) virus. Since JC virus was identified as the causative agent of PML, a biomarker, antibodies to JC virus, has emerged to enable clinicians to know who is at risk [13]. Therefore, JC virus risk stratification in the decision process for patient selection could decrease the risk of PML. Given this new discovery, natalizumab was even proposed as first-line therapy to MS patients that are JC virus negative [14]. It thus made natalizumab a remarkable comeback story.

10.3 Eculizumab (Soliris®)

Eculizumab (trade name Soliris) is a humanized mAb that binds to the terminal complement protein C5 [15]. Eculizumab was approved by FDA in March 2007 as a first-in-class medicine for the treatment of patients with PNH, a rare, progressive, and potentially life-threatening disease characterized by excessive destruction of red blood cells and excessive blood clotting [16]. It is the first therapy approved specifically for this rare blood disorder. In September 2011, eculizumab was also approved by the FDA for the treatment of all pediatric and adult patients with aHUS. aHUS is an ultrarare, life-threatening genetic disease that progressively damages vital organs, leading to stroke, heart attack, kidney failure, and death [17]. Eculizumab was also approved for PNH and aHUS by European Medicines Agency (EMA) following its FDA approvals. Besides current applications, eculizumab is also being investigated as a potential treatment for other severe, ultrarare disorders. Eculizumab was developed and manufactured by Alexion Pharmaceuticals, Inc., Cheshire, CT.

As a humanized IgG2/4 kappa antibody, eculizumab was engineered from a murine antibody derived from hybridoma fusion to reduce immunogenicity and eliminate effector functions. A panel of murine hybridoma specific targeting human C5 was screened for inhibition of complement-mediated hemolysis and C5a generation. A hybridoma clone m5G1.1 that inhibits 100% complement-mediated C5a release and hemolysis at 10 μg/ml was selected and was observed to have a binding affinity of approximately 30 pM to human C5 [18]. The humanization was performed by CDR grafting in which the constant region of IgG was a combination of human IgG2 (CH1 and the hinge region) and human IgG4 (CH2 and CH3) with human kappa light chain to avoid binding to Fc receptors or activate complement cascade. However, comparable binding affinity of parental murine antibody to human C5 at 120 pM (K_d) has been maintained [15]. The half-life of eculizumab is 11 days in humans.

Complement activation is a complex cascade that culminates in the production of the membrane attack complex C5b-9, resulting in cell lysis [19]. There are various pathways initiating this cascade that converge at the terminal complement protein C5 (Fig. 10.2). Cleavage of C5 produces the potent inflammatory mediator C5a and C5b, which initiates that formation of the membrane attack complex (Fig. 10.2). Eculizumab is a recombinant, fully humanized hybrid IgG2/IgG4 mAb from mouse hybridoma. It binds specifically to complement protein C5 with high affinity, preventing its cleavage into C5a and C5b [20]. It thus inhibits complement-mediated intravascular hemolysis. The drug has been engineered to minimize immunogenicity and Fc-mediated functions, including recruitment of inflammatory cells and complement activation.

PNH is a rare form of hemolytic anemia that results from the clonal expansion of hematopoietic stem cells that have an acquired genetic mutation in the phosphatidylinositol glycan complementation class A (PIGA) gene [21]. This mutation blocks the synthesis of a critical cell surface anchor protein, glycolipid glycosylphosphatidylinositol (GPI). The intracellular hemolysis was originated due to the lack of two GPI-anchored proteins, CD55 and CD59, which protect cells from attack by the alternative pathway of complement (Fig. 10.2). Due to lack of GPI-anchored proteins, the PNH red blood cells are vulnerable to hemolysis by terminal complement complex. Primary symptoms include intravascular hemolytic anemia, thrombosis in vessels, and bone marrow failure

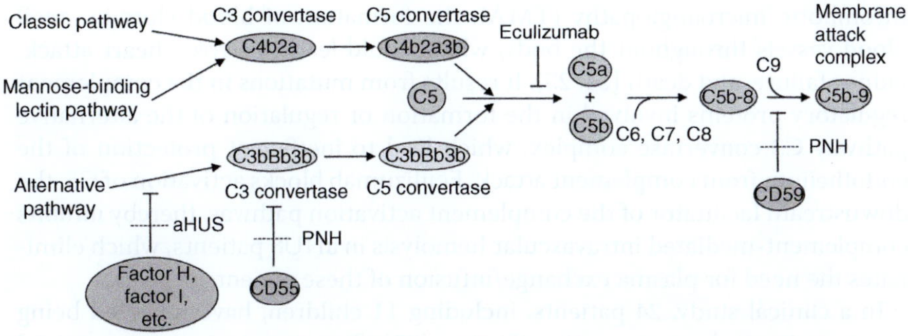

Figure 10.2 Mechanism of complement activation cascades and molecular action of eculizumab. The complement cascade can be activated via the classical, mannose-binding lectin and alternative pathways. All pathways of complement activation converge at the cleavage of the terminal complement protein C5. The C5 convertase cleaves C5 to C5a and C5b. C5b binds C6, C7, and C8 to form C5b-8 complex, which binds to several molecules of C9 to form the mature membrane attack complex (MAC) C5b–9. In PNH patients, the deficiencies of GPI-anchored proteins CD55 and CD59, which are the consequences of mutant PIGA, lost protection of cells from attack by the alternative pathway of complement, thus resulting in intravascular hemolysis. Eculizumab binds to C5 and prevented its entry into the C5 convertase. This effect prevents C5 cleavage and formation of membrane attack complex, thus rescuing PNH patients from complement-mediated intravascular hemolysis. In the case of aHUS, mutations are in the complement regulatory proteins (factor H, factor I, etc.) involved in the formation or regulation of the alternative C3 convertase complex, thus resulting in inefficient protection of the endothelium from complement attack.

[22]. Eculizumab blocks activation of C5 and thereby inhibits complement-mediated intravascular hemolysis in PNH patients.

The efficacy and safety of eculizumab were evaluated in a 26-week, international, double-blind, randomized, placebo-controlled trial (TRIMPH trial) involving 87 patients with PNH [16, 23]. The trial demonstrated that eculizumab stabilized hemoglobin level (49% of the treated group compared with 0% in the placebo group, $p < 0.001$) and reduced transfusion requirements in PNH (a median of 0 units of packed red blood cells in the treated group vs. 10 units in the placebo group, $p < 0.001$). The eculizumab was administered intravenously at 600 mg weekly for 4 weeks, followed by 900 mg the week after, and then 900 mg every 2 weeks for a total of 6 months. A second single-arm extended trail involving 97 patients with PNH studied eculizumab over 52-week period with the same dosing regimen [23]. Eculizumab decreased the intravascular hemolysis, as measured by serum lactate dehydrogenase (LDH) levels throughout the treatment period, and resulted in a reduced need for red blood cell transfusion and decreased fatigue.

aHUS is an ultrarare kidney disease that is recognized as complement-mediated condition characterized by overactivation of the alternative complement pathway. It affects both children and adults and is characterized by systemic thrombotic microangiopathy (TMA), the formation of blood clots in small blood vessels throughout the body, which could lead to stroke, heart attack, kidney failure, and death [24, 25]. It results from mutations in the complement regulatory proteins involved in the formation or regulation of the alternative pathway C3 convertase complex, which lead to inefficient protection of the endothelium from complement attack. Eculizumab blocks activation of C5, the downstream facilitator of the complement activation pathway, thereby inhibits complement-mediated intravascular hemolysis in aHUS patients, which eliminates the need for plasma exchange/infusion of these patients.

In a clinical study, 24 patients, including 11 children, have reported being given eculizumab as a curative therapy [25]. Twenty-two patients received plasma therapy prior to eculizumab, and the other two were given eculizumab as first-line treatment. Remarkably, eculizumab consistently achieved complete remission (TMA response) in the 21 patients with active aHUS, either a first-line therapy or a rescue therapy. These patients showed a prompt and full recovery from aHUS-related hematological disturbances and severe extrarenal symptoms.

Since eculizumab is a complement C5 inhibitor, patients receiving treatment are at higher risk of meningococcal infection. For that reason, patients must receive meningococcal vaccination prior to eculizumab therapy [26].

10.4 Ranibizumab (Lucentis®)

Ranibizumab (trade name Lucentis) is a humanized mAb fragment (Fab) derived from the same parent murine antibody as bevacizumab (Avastin), which binds to and inhibit vascular endothelial growth factor A (VEGF-A), a protein that is believed to play a critical role in angiogenesis, the growth of new blood vessels, and the hyper permeability of the vessels. Ranibizumab (0.5 mg monthly intravitreal injection) was approved by the US FDA in June 2006 for the treatment of neovascular (wet) AMD, a condition in which abnormal blood vessels grow and leak fluid into the macula. Wet AMD is the leading cause of severe vision loss. In June 2010, ranibizumab (0.5 mg monthly intravitreal injection) was approved by FDA for the treatment of macular edema following RVO, another common cause of vision loss due to retinal vascular disease. In August 2012, ranibizumab (0.3 mg monthly intravitreal injection) was also approved for the treatment of DME, a sight-threatening eye disease that occurs in patients with diabetes. Ranibizumab was administered monthly via intravitreal injection. Ranibizumab was developed and manufactured by Genentech/Roche, and is marketed in the United States by Genentech/Roche, and elsewhere by Novartis, under the brand name Lucentis.

Both ranibizumab and bevacizumab are derived from the same parent murine antibody muMAb A4.6.1, which was generated by hybridoma from mice immunized with the predominant 165 residue form of human VEGF-A [27]. Compared with the molecular weight of bevacizumab, ranibizumab was reduced from 150 to 48 kDa. In addition, six mutations were introduced and then selected for affinity maturation, resulting in a variant known as Y0317, which can bind all the VEGF-A isoforms with 100-fold higher affinity than its parental molecule [28]. Because severe side effects are associated with vascular endothelial growth factor (VEGF) inhibitions when delivered systemically, and intravitreal drug delivery resulted in nondetectable serum levels of ranibizumab, local intravitreal injection is much better drug delivery option than systemic drug delivery. Because of its small size, ranibizumab is better suited to intravitreal drug administration. In rhesus monkeys, an intravitreal Fab is capable of penetrating the full thickness of the retina, resulting in high retinal drug concentrations and reaching the retinal pigment epithelium within 1 h and remaining there for up to 7 days [29]. Ranibizumab is produced in an *Escherichia coli* expression system.

AMD is the leading cause of severe vision loss in the elderly in the developed world. There are two forms of AMD: neovascular (wet) AMD and nonneovascular (dry) AMD. The wet AMD involves abnormal blood vessels growth and fluid leakage into the macula, the central part of the retina, and it affects an estimated 1.6 million US adults over the age of 50. In contrast, the dry AMD is associated with atrophic cell death in the macula and does not involve leakage. Even though wet AMD affects only about 10% of patients with AMD, it is responsible for over 80% cases among patients with severe vision loss [30]. Angiogenesis is a key aspect of the wet AMD. There is substantial evidence demonstrating that VEGF-A is a major mediator of angiogenesis and vascular leakage in wet AMD. VEGF-A is the original member of a gene family that now includes PIGF, VEGF-B, VEGF-C, VEGF-D, and the virus-encoded VEGF-E. Several isoforms of VEGF-A could be generated through alternative RNA splicing. Ranibizumab binds to the receptor binding site of all isoforms of VEGF-A. The binding of the ranibizumab to VEGF-A prevents the interaction of VEGF-A with its receptors (VEGFR1 and VEGFR2) on the surface of endothelial cells and reduces endothelial cell proliferation, thus reducing new blood vessel formation and vascular leakage (Fig. 10.3).

The safety and efficacy of ranibizumab in treating wet AMD was demonstrated in three phase III clinical trials (MARINA, ANCHOR, and PIER) [32]. A total of 1323 patients (ranibizumab 879 and control 444) were enrolled in the three studies. The studies measured the number of patients who gained vision, as measured on an eye chart. In the MARINA trial [33], 716 patients with wet AMD received monthly ranibizumab 0.3 or 0.5 mg intravitreal injections or monthly sham injections for 2 years. Patients receiving ranibizumab achieved a vision gain of 15 letters after 2 years in 25 and 33% of eyes treated, respectively, with 0.3 and 0.5 mg versus 5% in the control group ($p < 0.001$ for both doses). In the ANCHOR trial [34], 423 patients received either 0.3/0.5 mg

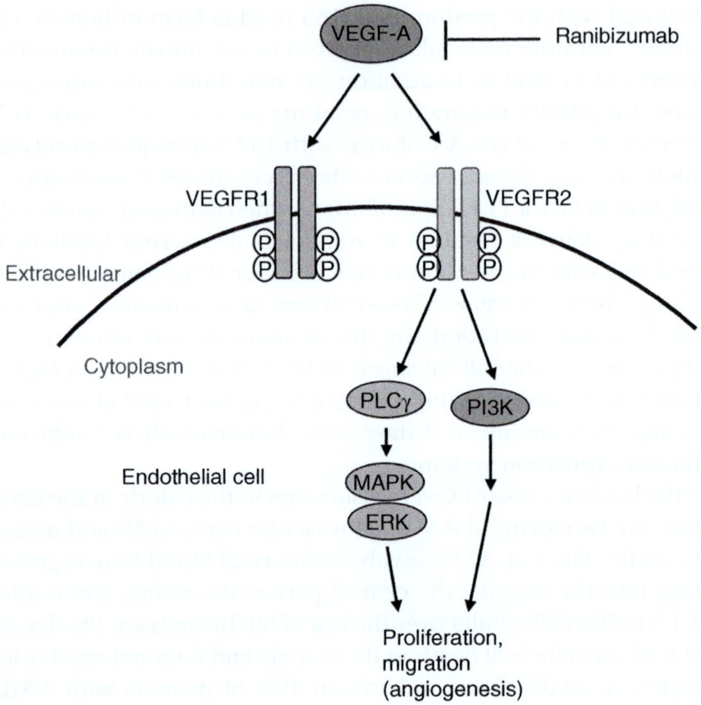

Figure 10.3 Mechanism of VEGF signaling in angiogenesis and its inhibition by ranibizumab. VEGFR1 and VEGFR2 are two major VEGFRs on the endothelial cells involving in angiogenesis, and they are members of the receptor tyrosine kinase family. VEGF-A binds to VEGFR1 and VEGFR2 and induces receptor phosphorylation and subsequent activation of signals downstream. VEGFR1 is required for the recruitment of hematopoietic precursors and migration of monocytes and macrophages, whereas VEGFR2 is essential for the function of vascular endothelial cells. Phosphorylation of VEGFR2 induces the MAPK/ERK signaling pathway, which contributes to cell proliferation and migration. Alternatively, VEGF-A could also activate phosphatidylinositol-3-kinase (PI3K) pathway, which also contributes to cell migration. Downstream signaling of VEGFR2 is very complex; therefore only simplified illustrations are described here. Detail signaling pathways of VEGF receptor signaling in the control of vascular function could be found in Olsson's review [31]. Ranibizumab binds to the receptor binding site of all isoforms of VEGF-A. The binding of the ranibizumab to VEGF-A prevents the interaction of VEGF-A with its receptors on the surface of endothelial cells, thus inhibiting angiogenesis.

of ranibizumab or photodynamic treatment (PDT) with verteporfin, one of the most widely described treatment modalities before ranibizumab. At 1-year follow-up, vision gain of 15 letters was found in 36 and 40% of eyes treated, respectively, with 0.3 and 0.5 mg ranibizumab versus 6% in the verteporfin group ($p < 0.001$ for each comparison). In addition, 94 and 96% of eyes treated, respectively, with 0.3 and 0.5 mg ranibizumab lost less than 15 letters versus 64% in the verteporfin group ($p < 0.001$ for each comparison), thus establishing the long-term superiority of ranibizumab over verteporfin. Reducing the amount of injection without inducing visual loss seems to be challenging. As demonstrated in the PIER study [35], a trial in patients receiving quarterly injections after a 3-month loading of ranibizumab seemed to be less effective than monthly dosing when compared to the MARINA or ANCHOR studies.

RVO affects more than one million people in the United States and is the second most common cause of vision loss due to retinal vascular disease, which can develop either over a long period of time or occur suddenly. It affects working-age adults and the elderly, and its incidence increases with age. Macular edema following RVO occurs when the normal blood flow through a retinal vein was blocked, causing swelling and hemorrhages in the retina, which may result in vision loss. Occlusion can occur in the central retinal vein (CRVO) or branches of the retinal veins (BRVO), and VEGF plays an important role in the pathogenesis of macular edema in both CRVO and BRVO, as they both lead to fluid leakage from capillaries, caused by the secretion of VEGF. Ranibizumab has been studied in patients with macular edema following RVO in two phase III clinical trials (BRAVO and CRUISE). In both studies, patients received monthly intravitreal injections of 0.3 mg ranibizumab or 0.5 mg ranibizumab or sham for 6 months (treatment period), after which they were observed for another 6 months (observational period). Both studies showed significant vision improvement during the 6-month study in patients with macular edema following BRVO (BRAVO) [36] and macular edema following CRVO (CRUISE) [37] in the treatment groups.

DME is the leading cause of vision loss among the working-age adults of most developed countries. It affects more than half million people in the United States, in which more than half of them are unaware of their condition. DME occurs in patients with diabetes and is one major complication (diabetic retinopathy) of diabetes. The exact mechanism by which diabetes leads to DME is not known, but it is known that DME occurs when abnormal blood vessels leak blood and fluid into the macula causing swelling (edema), which results in blurred vision and even vision loss, and research has identified VEGF expression as a key player in its development. Ranibizumab has been studied in patients with DME in two phase III clinical trials (RIDE and RISE), two identically designed, parallel, and double-blinded 3-year trials. Total 759 patients with DME were enrolled to receive monthly intravitreal injections of 0.3 mg ranibizumab or 0.5 mg ranibizumab or sham for 24 months. After 24 months, all patients received monthly ranibizumab either at 0.3 or 0.5 mg. Results showed that between 37 and 51% of those treated with monthly ranibizumab 0.3 mg had vision gain of more than 15 letters compared with 29–22% for the sham group [38]. No additional benefit was observed with the higher monthly ranibizumab dose of 0.5 mg. When combined with focal or grid laser treatments, the amount of edema was reduced, as were the frequency of ranibizumab injections needed [39].

Overall, ranibizumab has no significant toxicity effect observed in the clinic, which is different from toxicity profile of bevacizumab, essentially due to local delivery. The most common adverse event therapy is intraocular inflammation. Since ranibizumab is derived from bevacizumab, and ranibizumab cost significantly more than bevacizumab, bevacizumab has been commonly prescribed in treating macular diseases as off-label usage. Clinical trials have been conducted to compare these two drugs in AMD treatment. Several trials demonstrated that both drugs had equivalent efficacies on visual acuity when administered according to the same schedule; however, serious adverse events were more common with the bevacizumab treatment [40]. It is understandable that both ranibizumab and bevacizumab have similar efficacy since they have the same variable domains, and it is also possible that bevacizumab, as a humanized full-length IgG1 antibody, could have more mechanism-related adverse events due to much longer half-life of bevacizumab and Fc-mediated effects.

Recently, aflibercept (trade name Eylea) became available on the market for treating similar eye diseases as ranibizumab. It represents an alternative therapy based on the same VEGF pathway. Aflibercept was approved in the United States for the treatment of wet AMD in November 2011, for the treatment of macular edema following CRVO in September 2012, and for DME in July 2014. Aflibercept, also called VEGF Trap-Eye, is a human Fc fusion protein that incorporates portions of extracellular domains of the human VEGFR1 and VEGFR2. In contrast to ranibizumab, aflibercept not only blocks VEGF-A but also blocks placental growth factor (PLGF), another growth factor involved in angiogenesis [41]. Thus, the approval of aflibercept offers a much needed new treatment option for patients with macular degeneration.

10.5 Denosumab (Prolia® and Xgeva®)

Denosumab (trade name Prolia and Xgeva) is a fully human IgG2 kappa mAb for the treatment of osteoporosis, treatment-induced bone loss, skeletal-related events in patients with bone metastases from solid tumors, and giant cell tumor of bone [42–45]. Denosumab inhibits receptor activator of nuclear factor kappa-B (NF-κB) ligand (RANKL), a protein that acts as the primary signal for bone removal. In June 2010, FDA approved denosumab for the treatment of osteoporosis in postmenopausal women under the trade name Prolia [42] and later approved Prolia for the treatment of bone loss in patients with prostate or breast cancer undergoing hormone ablation therapy (September 2011) and men with osteoporosis at high risk for fracture (September 2012). In November 2010, FDA approved denosumab for the prevention of skeleton-related events

(SREs) in patients with bone metastases from solid tumors under the trade name as Xgeva [43] and later approved Xgeva for the treatment of giant cell tumor of bone (June 2013). Prolia was administered subcutaneously at 60 mg every 6 months, whereas Xgeva was administered subcutaneously at 120 mg every 4 weeks. Denosumab is developed, manufactured, and marketed by Amgen, Inc. under the trade names of Prolia and Xgeva.

From a large integrative effort to identify novel genes and proteins by extensive sequencing of cDNA libraries from interesting tissue source in early 1990s, a novel secreted glycoprotein, osteoprotegerin (OPG), was identified as a new member of the tumor necrosis factor receptor (TNFR) superfamily by Amgen scientists in 1997 [46]. OPG exists only as a secreted protein, in contrast with most members of TNFR family members, which are membrane proteins. Studies demonstrated that OPG blocked the maturation of osteoclasts as bone-resorbing cells and led to increased bone density [46]. At the same time, receptor activator of NF-kB (RANK) and RANK ligand (RANKL) was identified by Immunex. [47] RANK is a new member of the TNFR superfamily, which is a membrane protein. RANKL was also known as OPGL by Amgen scientists [48], ODF [49], and TRANCE [50]. RANKL is a new TNF family member that interacts with OPG [48]. Recombinant RANKL (OPGL) could stimulate the development of osteoclasts as bone-resorbing cells, and OPG is a natural inhibitor of RANKL [48]. In many bone loss pathological processes, RANKL causes body's bone destruction and overwhelms the body's natural defenses against such destruction (Fig. 10.4).

Since OPG is the natural inhibitor of RANKL, it was first developed as a therapeutic agent by Amgen, but the effort was dropped due to poor pharmacokinetic (PK) and PD properties of the OPG molecule [44]. CHO-produced OPG-Fc demonstrated a good PK/PD profile, but the program also was discontinued due to an observed immune response to endogenous OPG in a clinical model [44]. The antibody development strategy against RANKL became the new focus after the company secured all intellectual property as a part of their acquisition of Immunex in 2001. In addition, the development of the humanized XenoMouse (Abgenix) [51] enabled the generation of fully human antibodies in transgenic mouse. Denosumab was developed through the collaboration between Amgen and Abgenix, and the fully humanized IgG2 mAb was generated by immunization with the RANKL antigen [44]. Abgenix was later acquired by Amgen in 2005.

Denosumab binds to human RANKL with a high affinity, with dissociation equilibrium constant (K_d) of 3 pM. It binds to human and primate RANKL but not to mouse or rat RANKL [52], thus cynomolgus monkeys were used to demonstrate activity *in vivo*. Subcutaneous injection of either 25 or 50 mg/kg/month of denosumab in ovariectomized cynomolgus monkey over a 16-month treatment period markedly reduced osteoclast numbers and other parameters of bone turnover and also significantly increased bone mass and density [53].

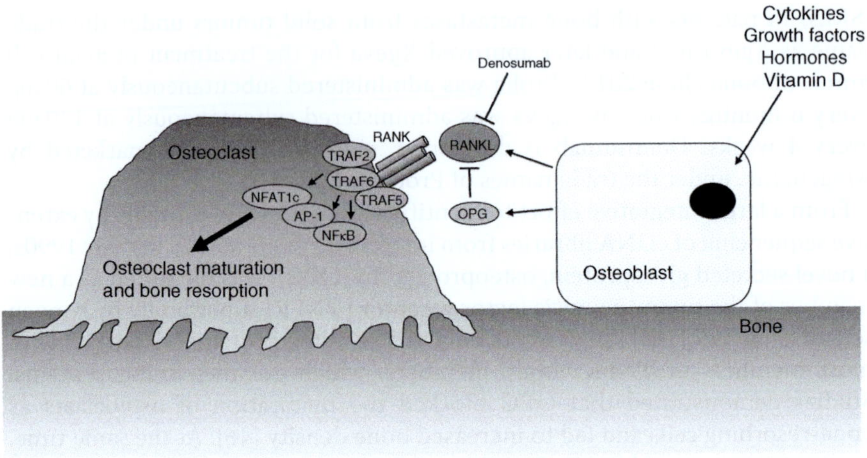

Figure 10.4 Mechanism of RANKL signaling in osteoclast activation and its inhibition by denosumab. Stromal/osteoblast cells can secrete RANKL. RANKL is type II transmembrane protein, which has a small N-terminal intracellular domain, a transmembrane domain, and a C-terminal extracellular domain consisting of a stalk and a receptor binding region to form trimmers [44]. RANKL exists in both membrane-bound and soluble forms. Membrane-bound RANKL exists on marrow stromal/osteoblastic cells, which modulates the production of OPG and RANKL to control the osteoclast activity. Soluble RANKL are generated from cleavage of the extracellular tail region or alternative splicing of membrane-bound RANKL transcripts. The trimer form of soluble RANKL binds to three receptors (RANKs) on the membrane of osteoclast precursor, thus inducing receptor clustering and signaling. Activated RANK recruits TNFR-associated factors (TRAFs) to receptor and activates intracellular signaling downstream, thus leading to osteoclast maturation and bone resorption. OPG is a novel TNFR family member that exists only as a secreted protein, and it is a natural inhibitor of RANKL. Denosumab binds to RANKL with high affinity (K_d of 3 pM) and inhibits its interaction with RANK.

Both denosumab and OPG bind to RANKL and inhibit its activity, but denosumab is specific to human RANKL. In contrast, OPG also binds to TRAIL RANKL [52], another member of TNF family. Denosumab has a half-life of 25.4 days in humans.

The safety and efficacy of denosumab in treating postmenopausal osteoporosis was established in a pivotal 3-year, randomized, double-blind, placebo-controlled, multinational study (FREEDOM trial) [42, 54]. A total of 7868 women between 60 and 90 years old with a bone mineral density (BMD) *T*-score between −2.5 and −4.0 for the lumbar spine or total hip (< −2.5 is the World Health Organization's definition of osteoporosis) were enrolled. Denosumab (subcutaneously administered at 60 mg every 6 months) significantly reduced the risk of new vertebral fractures at 3 years by 68% relative to the placebo ($p < 0.0001$) [54], which is the primary endpoint. In addition, denosumab also significantly reduced the risk of nonvertebral fractures by 20%

and hip fractures by 40%. Relative gain in BMD, a common measured surrogate for bone strength, for denosumab compared with placebo were 9% at the lumbar spine and 6% at the total hip after 3 years [54]. Compared to alendronate, the most widely prescribed antiresorptive agent, denosumab was superior, as demonstrated by the increase in BMD at the total hip at 12 month ($p < 0.0001$) in two head-to-head phase III studies [55, 56].

Cancer metastases to bone result from activation of nonmalignant resident cells such as osteoclasts, stromal cells, and vascular cells in the bone microenvironment by tumor cells. Pathologically increased bone remodeling driven by increased osteoclast activity is a hallmark of bone metastases, which resulted in significant skeletal morbidity such as fractures and spinal cord compression [44]. Therefore, inhibiting RANKL is a potential approach to prevent SREs in patients with bone metastases.

The safety and efficacy of denosumab in treating bone metastases was established in three international, randomized, double-blind, active-controlled phase III studies [43]. The efficacy of subcutaneously administered denosumab (120 mg once every 4 weeks) in preventing SREs was compared with intravenously administered zoledronic acid (4 mg once every 4 weeks). Patients with breast cancer, prostate cancer, and multiple myeloma were studied in three different trials. Denosumab was superior to zoledronic acid with regard to the first on-study SRE in the breast cancer study (hazard ratio (HR) = 0.82; 95% confidence interval (CI): 0.71–0.95; $p = 0.01$) [57] and the prostate cancer study (HR = 0.82; 95% CI: 0.71–0.95; $p = 0.008$) [58]. In the third trial with patients with solid tumor (excluding breast and prostate cancer) and multiple myeloma, denosumab was noninferior to zoledronic acid (HR = 0.84; 95% CI: 0.71–0.98; $p < 0.001$ for noninferiority; $p = 0.06$ for superiority) [59].

Overall, denosumab is safe and well tolerated. Calcium and vitamin D supplements are recommended to patients during the course of treatment. The most common side effect is increased susceptibility to infections [45]. Denosumab could increase the risk of osteonecrosis of the jaw (ONJ) following extraction of teeth or oral surgical procedures, which is similar to bisphosphonates such as zoledronate, alendronate, and risedronate.

10.6 Antibody Therapies for Solid Organ Transplantation (Muromonab-CD3 (Orthoclone OKT3), Basiliximab (Simulect®), and Daclizumab (Zenapax®))

10.6.1 Daclizumab (Zenapax®)

Daclizumab (trade name Zenapax®) is a humanized IgG1 kappa mAb that binds to the interleukin-2 (IL-2) receptor α-chain (CD25). It was used as an add-on therapy to standard immunosuppressive regimen to prevent rejection in solid organ transplantation. Daclizumab was approved in 1997 by the FDA for use in renal allograft transplantation as the first humanized therapeutic mAb. Recently daclizumab has been evaluated for the treatment of multiple sclerosis. Daclizumab was developed by Protein Design Laboratories (PDL), and it was manufactured and marketed by Hoffmann-La Roche, under the brand name Zenapax.

Daclizumab was the first humanized therapeutic antibody, which has 90% human and 10% murine antibody sequences with a binding affinity of 3 nM to IL-2Rα [60]. The original hybridoma clone was generated from the conventional fusion of spleen cells isolated from BALB/c mouse immunized with human T cells derived from peripheral blood T cells from a patient with mycosis fungoides [61]. The murine mAb specifically binds to the 55-kDa subunit (CD25) of the IL-2 receptor and is reactive with activated human T cells or allogeneic cells. To reduce immunogenicity, humanization was carried out by combining the CDRs of the mouse mAb with human framework and constant regions. The variable regions of the heavy and light chains were constructed completely synthetically [61].

Daclizumab was given at 1 mg/kg via intravenous injection within 24 h before transplantation with repeat doses administered every 2 weeks for a total five doses, which will provide IL-2 receptor coverage for 120 days. Due to its long half-life (~20 days in adults) [62], an abbreviated two-dose regimen has also been used in some organ transplantations [63].

Preventing allograft rejection is critical for the preservation of transplanted organs. Allograft rejection depends on the coordinated activation of alloreactive T cells and antigen-presenting cells (APCs) (monocyte/macrophages, dendritic cells, and B cells). Whereas acute rejection is a T-cell-dependent process, several effector mechanisms play roles in the destruction of the allograft. IL-2 has been known for decades as the "T-cell growth factor" and since its discovery has been used for *in vitro* expansion of T cells [64]. IL-2 induced T-cell expansion requires expression of the "high-affinity" IL-2 receptor (IL-2R), which consists of three chains, two signaling proteins γ-chain (CD132) and β-chain (CD122) and the nonsignaling α-chain (CD25) [65]. The γ-chain is used by many cytokines (IL-2, IL-4, IL-7, IL-15, and IL-21), and the β-chain is shared by two closely related cytokines (IL-2 and IL-15), whereas the α-chain (CD25) seems to be selective to IL-2, thus making it a suitable therapeutic target. T-cell activation is initiated when APCs bind and activate the surface T-cell receptor (TCR). Upon activation by the TCR, T cells start expressing high levels of IL-2R as well as large quantities of IL-2. This autocrine IL-2 signaling loop mediates clonal expansion of activated T cells and promotes development of their effector functions. Daclizumab binds to CD25 via the so-called "Tac" epitope and blocks the interaction between CD25 and IL-2 (Fig. 10.5), thus making it the ideal therapy for prevention of rejection of allograft transplants since CD25 is also upregulated on effector T cells.

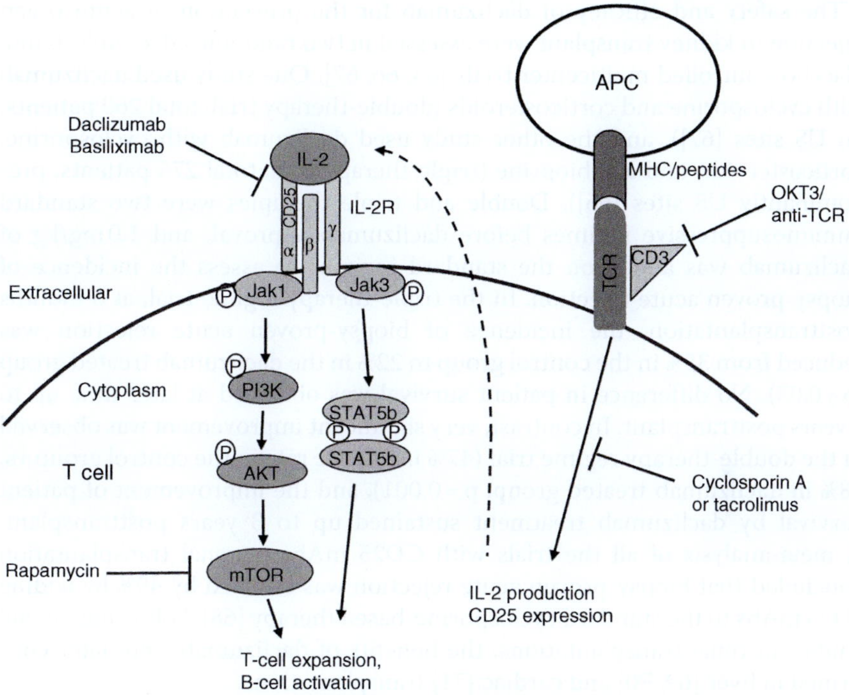

Figure 10.5 Mechanism of T-cell activations and sites of action of available monoclonal antibodies in allograft transplantation. Antigen-presenting cells (APC) presents antigen through MHC, which bind to T-cell receptor (TCR). Activation of TCR triggers various downstream, including IL-2 production and CD25 expression. Secreted IL-2 functions as autocrine and binds to "high-affinity" IL-2R, which includes all three subunits. The IL-2 induced IL-2R stimulation leads to the activation of Janus tyrosine kinases (JAKs) and the signal transducer and activator of transcription (STAT) molecules. Phosphorylated STAT molecules undergo tetramerization and translocate to the nucleus, where they can bind to promoter expression of key genes. PI3Ks are also involved and have key downstream signals, including activation of mTOR, a target of rapamycin. Sites of action of available monoclonal antibodies in allograft transplantation are shown here. Basiliximab and daclizumab bind with high affinity to the interleukin-2 receptor (CD25) and prevent the formation of the IL-2 binding site interrupting the cascade of cellular events such as T-cell activation and expansion and cytokine release to promote B-cell activation and effector functions. OKT3 is an immunoglobulin that targets the membrane glycoprotein CD3 epsilon chain on the surface of circulating human T cells, which is part of the T-cell receptor complex. Thus, OKT3 blocks both the generation and function of cytotoxic T cells clearing them from the circulation. Also shown here are actions of a few small molecule drugs. Rapamycin inhibits cell cycle progression through its interaction with mTOR. Cyclosporin A and tacrolimus inhibit calcineurin, thereby inhibiting NFAT and IL-2 synthesis.

The safety and efficacy of daclizumab for the prevention of acute organ rejection in kidney transplant were assessed in two randomized, double-blind, placebo-controlled multicenter trials [63, 66, 67]. One study used daclizumab with cyclosporine and corticosteroids (double-therapy trial, total 260 patients, no US sites [67]), and the other study used daclizumab with cyclosporine, corticosteroids, and azathioprine (triple-therapy trial, total 275 patients, predominantly US sites [66]). Double and triple therapies were two standard immunosuppressive regimes before daclizumab approval, and 1.0 mg/kg of daclizumab was added on the standard regimes to assess the incidence of biopsy-proven acute rejection. In the triple-therapy regime trial, at 6 months posttransplantation, the incidence of biopsy-proven acute rejection was reduced from 35% in the control group to 22% in the daclizumab treated group ($p < 0.03$). No difference in patient survival was observed at later time up to 3 years posttransplant. In contrast, very significant improvement was observed in the double-therapy regime trial (47% incidence rate in the control group vs. 28% in daclizumab treated group, $p = 0.001$), and the improvement of patient survival by daclizumab treatment sustained up to 3 years posttransplant. A meta-analysis of all the trials with CD25 mAbs in renal transplantation concluded that biopsy-proven acute rejection was reduced by 49% by adding cD25 mAbs to the standard cyclosporine-based therapy [68]. Following pivotal studies in renal transplantations, the benefits of daclizumab were later confirmed in liver [69, 70] and cardiac [71] transplantations.

Since daclizumab was considered to be a therapy that selectively inhibits T-cell activation, it seems to be an ideal therapy for T-cell-mediated autoimmune diseases. Daclizumab was subsequently tested in inflammatory uveitis and MS (codeveloped by PDL and Biogen Idec). The recent studies for MS used a new preparation of daclizumab called daclizumab high-yield process (DAC HYP), which shares identical amino acids as the original daclizumab (Zenapax®), but a different production process that resulted in changed glycosylation pattern of the antibody. The glycosylation change affected the binding of daclizumab to Fc receptors, which may promote or inhibit effector functions such as CDC and ADCC. In addition, DAC HYP is being developed for monthly subcutaneous administration in contrast to intravenous injection of original daclizumab (Zenapax®). DAC HYP was jointly developed by Biogen Idec and AbbVie.

On June 16, 2014, Biogen Idec and AbbVie announced positive top-line results from a phase III clinical trial investigating DAC HYP in MS (DECIDE study; Biogen Idec press release) [72]. This 2–3-year study was designed to evaluate the superiority of monthly subcutaneous DAC HYP when compared to intramuscular IFN β-1α, as a potential treatment for relapsing–remitting multiple sclerosis (RRMS), the most common form of MS. Results showed that DAC HYP was superior on the study's primary endpoint, demonstrating a statistically significant 45% reduction in ARR compared to IFN β-1α ($p < 0.0001$).

10.6.2 Basiliximab (Simulect®)

Basiliximab (trade name Simulect) is a mouse–human chimeric IgG1 kappa mAb that binds to the interleukin-2 (IL-2) receptor α-chain (CD25) but to a different epitope from that of daclizumab. Basiliximab was engineered from a murine hybridoma cell line (RFT5, IgG2a) as a chimeric mAb to keep high avidity to CD25 and reduce immunogenicity in humans [73], which has around 75% human and 25% murine protein (the variable regions of the heavy and light chains are of murine origin, and the constant regions are of human origin), whereas daclizumab is a humanized mAb with <90% human and 10% murine. Similar to daclizumab, basiliximab is also used as add-on therapy to the standard immunosuppressive regime to prevent rejection in solid organ transplantation. For adults, basiliximab was given at 20 mg by intravenous injection within 2 h prior to transplantation surgery and the second dose at 20 mg given 4 days after transplantation. In adults, basiliximab has a half-life of 7 days [74]. Basiliximab was first approved by FDA in 1998. Basiliximab was manufactured and marketed by Novartis, under the brand name Simulect.

The safety and efficacy of basiliximab for the prevention of acute organ rejection in kidney transplant were assessed in four randomized, double-blind, placebo-controlled multicenter trials [74]. The first two phase III trials, one in the United States [75] and the other in Europe and Canada [76], studied basiliximab with cyclosporine and corticosteroids (double-therapy trials, total 722 patients). In both studies, basiliximab significantly reduced the incidence of acute rejection at 6 months (US study: placebo 55% vs. basiliximab 38%, $p < 0.01$ and Europe/Canada study: placebo 57% vs. basiliximab 42%, $p < 0.01$). These two trials were followed by two more trials, in which basiliximab was compared to placebo in patients treated with standard triple immunosuppression therapy including azathioprine (total 340 patients) [77] or mycophenolate mofetil (MMF; total 123 patients) [78]. Basiliximab significantly reduced the incidence of acute rejection at 6 months in the third trial (placebo 35% vs. basiliximab 21%, $p < 0.01$), but not in the fourth trial (placebo 27% vs. basiliximab 15%, not significant), probably due to small sample size.

Since basiliximab is a chimeric mouse–human mAb, immunogenicity was a potential concern. Of the renal transplantation patients treated with basiliximab and tested for anti-idiotype antibodies, 4 out of 339 developed an anti-idiotype antibody response, with no deleterious clinical effect detected [74]. Not a single case was found in which the presence of anti-idiotype antibody accelerated basiliximab clearance or decreased the period of IL-2 receptor saturation.

Both basiliximab and daclizumab are efficacious and safe therapies in organ allograft transplantation. Clinical trials were conducted to compare these two IL-2 receptor antagonists. Two prospective, randomized, open label, single-center studies of patients undergoing deceased donor renal transplant using triple immunosuppression therapy, including cyclosporine, mycophenolate mofetil, and methylprednisolone, were performed. No difference was observed between IL-2 receptor antagonists (basiliximab vs. five-dose daclizumab) in the incidence of acute rejection episodes, the renal graft function, the safety, and the patient and graft survivals at 1 year posttransplant [79]. A two-dose protocol of daclizumab was also found to be comparable to basiliximab in another study [80]. A perspective study of use basiliximab and daclizumab in renal transplantation found no major short-term or long-term side effects, and both drugs were well tolerated [81]. In summary, the data available suggest that the differences between basiliximab and daclizumab are very small, if present at all. With both reagents, a significant reduction in the number of acute rejection episodes following solid organ transplantation has been observed without an increase in adverse effects. They have been widely accepted in the transplant community and will be continuously studied for the purpose of reducing or eliminating the more toxic elements of the standard immunosuppression therapy.

10.6.3 Muromonab-CD3 (Orthoclone OKT3®)

Muromonab-CD3 (trade name Orthoclone OKT3) is an IgG2a murine mAb against membrane glycoprotein CD3 epsilon chain, as a part of the TCR complex on the surface of T lymphocytes [82]. It is involved in antigen recognition and cell stimulation. Similar to daclizumab and basiliximab, muromonab is also used as an immunosuppressant agent to reduce corticosteroid-resistant acute rejection in solid organ transplantation including renal, liver, and cardiac procedures. In contrast to daclizumab and basiliximab, which are humanized mAb and chimeric murine–human mAb, respectively, muromonab is a completely murine IgG2a mAb derived from hybridoma technology. Muromonab, approved by the FDA in 1986, is the first ever mAb approved for clinical use in humans [83]. Muromonab was marketed by Janssen–Cilag, under the brand name Orthoclone OKT3. However, manufacturer of muromonab voluntarily withdrew it from US market due to limited clinical utilization, probably due to potentially higher chance of immunogenicity from human antimouse IgG antibody (HAMA) response and lower general efficacy and safety profile than competitive therapies including daclizumab and basiliximab [83, 84].

Allograft rejection depends on the coordinated activation of alloreactive T cells and APCs. The first step of T-cell activation is through the binding of MHC and antigen peptide of APCs to the TCR of the T cells, thus triggering the activation of T cells (Fig. 10.5). Muromonab targets the CD3 receptor on the surface of circulating human T cells, which is part of the TCR complex, initially leading to activation, but subsequently inducing blockage and apoptosis of the T cells. Thus, OKT3 blocks both the generation and function of cytotoxic T cells clearing them from the circulation.

The safety and efficacy of muromonab was studied in several large multicenter trials [83]. In one trial, 173 patients from allograft renal transplantation who were unresponsive to conventional treatments were treated with muromonab, and rejection reversal rate was 70%. In a randomized, prospective trial with 215 renal transplant patients, muromonab was compared with standard triple therapy (cyclosporine, prednisone, and azathioprine). The muromonab treatment group had a greater time to first rejection in the OKT3 group (46 vs. 8 days, $p = 0.001$), fewer rejection episodes (0.82 vs. 1.14/patient, $p = 0.014$), and fewer patients with more than one rejection episode (17 vs. 32%, $p = 0.025$). In another trial involving 313 centers, patients with cadaver renal transplants were studied by comparing prophylactic muromonab and sequential cyclosporine administration with cyclosporine administration without muromonab. Patients receiving muromonab prophylaxis had a significantly higher success rate (75%) at 3 years than did patients who did not receive muromonab prophylaxis (71%, $p < 0.0001$).

Several clinical studies have been conducted to compare muromonab with IL-2R antagonist (daclizumab or basiliximab). In a nonrandomized, prospective study of 233 renal transplant recipients treated with daclizumab that were retrospective compared with a control group of 225 renal transplant recipients treated with muromonab, acute rejection rates (<6 months) were lower in the daclizumab group as compared with the muromonab group (2.1 vs. 7.1%, $p = 0.011$). No difference was observed at 12 months [85]. In another prospective study daclizumab was administered in 40 heart transplant patients. Results were retrospective compared with a historical group of 40 subjects who received muromonab. Mortality was low in both groups, and no differences in rejection profile or time to the first significant rejection event was detected. The only difference is 6-month prevalence for complications (33 vs. 55%, $p = 0.04$) [86]. A comparative study comprising 52 patients who had undergone cardiac transplantation examined the difference in outcomes in patients receiving daclizumab versus muromonab. The results suggest that daclizumab is associated with a decreased incidence of grade 3A or greater biopsy-confirmed acute rejection (45.5 vs. 73.3%, $p < 0.0.32$) and to a reduced necessity to modify the immunosuppressive therapy during follow-up (10 vs. 51.7%, $p < 0.003$) due to serious side effects or repetitive acute rejection. There were no significant differences in mortality [87]. There are three studies conducted to compare muromonab CD3 and basiliximab. With a total of 368 patients, studies reported that basiliximab was associated with lower incidences of acute rejection than muromonab CD3 and fewer adverse events [88–90].

Muromonab was the first therapeutic antibody, and it is a powerful immunosuppressive murine antibody that inhibits CD3. For this reason, chimeric and fully human therapeutic mAbs have been developed for better safety and efficacy, including otelixizumab, teplizumab, and visilizumab. They are being studied for the treatment of autoimmune diseases such as type I diabetes, CD, and ulcerative colitis. Certain side effects associated with muromonab such as T-cell activation were due to FcR-binding capability of muromonab. These newer anti-CD3 antibodies are humanized FcR nonbinding mAbs [91]. Reduced side effects have been observed in both preclinical and clinical studies. Otelixizumab and teplizumab have been tested in patients with type I diabetes. They both have shown ability to reduce the loss of insulin production in clinical trials; however, they did not restore normal glucose control. The combination therapies with these anti-CD3 mAbs might be a better approach for the treatment of autoimmune diseases [92].

10.7 Conclusion

The mAbs covered in this chapter span more than two decades of investigation and application. For the convenience to reader, the fact sheet of the mAbs mentioned in this chapter was provided in the Table 10.1, and the CDR sequences were provided in the Table 10.2. Muromonab-CD3 was approved by the FDA in 1986, whereas denosumab gained approval in 2010. Significant advancements have been made over the last three decades. In this chapter, we chronicle the technological improvements that have been made from murine antibody (muromonab-CD3), through human/murine chimera (basiliximab) and humanized antibody (daclizumab, natalizumab, eculizumab, and ranibizumab), to fully human antibody (denosumab) (Fig. 10.6). The radical enhancement of antibody structure and function significantly decreases the risk of immunogenicity. In the future, we strongly believe that continued progress in antibody engineering will significantly increase antibody stability, lower immunogenicity, drive higher production yield, and result in better PK properties with longer half-life. For example, modulating the PH-dependent affinity between the Fc and the neonatal Fc receptor (FcRn) has been shown to extend the PK profile of therapeutic antibodies because of the protective effect of FcRn against lysosomal degradation of immunoglobulin. Antibodies are normally prevented from access to the CNS due to the BBB. We believe that receptors such as the low-density lipoprotein receptor-related protein (LRP), the transferrin receptor (TfR), and the insulin receptor (IR) may help to mediate active transportation to allow biologics such as antibodies to cross the BBB. Moreover, the BBB FcRn may mediate the efflux of immunoglobulin from brain to the blood [93]. All seven therapeutic antibodies in this chapter are competitive orthosteric antagonist. Agonist antibodies and noncompetitive allosteric antibodies with different mechanism of action (MOA) are being developed and will in time move to the marketplace.

It is instructive to compare MOA of orthosteric and allosteric antibodies. In the latter case the allosteric antibody binds to a nonactive (nonligand- or -substrate binding) site, which is physically distinct from active site on a receptor or an enzyme. This could result in a conformational change and a resulting

Table 10.1 Fact sheet of the monoclonal antibodies mentioned in this chapter.

Generic Name	Brand Name	Major Indication	Antibody Type	Manufacturing cell line	Molecular Target	First US (EU) approval year	Originator	Marketing Companies
Natalizumab	Tysabri®	Multiple sclerosis Crohn's disease	Humanized IgG4κ	NS0	cell adhesion molecule α4-integrin	2004 (2006)	Biogen Idec	Biogen Idec and Elan
Eculizumab	Soliris®	Paroxysmal nocturnal hemoglobinuria (PNH) atypical hemolytic uremic syndrome (aHUS)	Humanized IgG2/4κ	NS0	complement protein C5	2007 (2007)	Alexion	Alexion
Ranibizumab	Lucentis®	"wet" age-related macular degeneration (AMD) Macular edema	Humanized IgG1κ fragment (Fab)	*E.coli*	VEGF-A	2006 (2007)	Genentech	Genentech/Roche (US) and Novartis (elsewhere)
Denosumab	Prolia® Xgeva®	Osteoporosis (Prolia) Bone metastases (Xgeva)	Fully human IgG2κ	CHO	RANKL (RANK ligand)	2010 (2010)	Amgen	Amgen
Daclizumab	Zenapax®	Immunosuppression in patients with organ transplants	Humanized IgG1κ	NS0	α chain (CD25) of the IL-2 receptor	1997 (1999)	PDL Biopharm	Hoffmann-La Roche
Basiliximab	Simulect®	Immunosuppression in patients with organ transplants	Human/murine chimeric IgG1κ	Sp2/0	α chain (CD25) of the IL-2 receptor	1998 (1998)	Novartis	Novartis
Muromonab-CD3	Orthoclone OKT3®	Immunosuppression in patients with organ transplants	Murine IgG2aκ	Hybridoma	CD3 receptor	1986 (1986)		Janssen-Cilag

Note: Data sources are from EMA public assessment reports (http://www.ema.europa.eu), US FDA (drugs@fda, http://www.accessdata.fda.gov/scripts/cder/drugsatfda/), the international ImMunoGeneTics information system® (www.imgt.org/mAb-DB/index), and data mentioned in the main text.

Table 10.2 Variable domain sequences of listed antibody drugs (CDRs are in Gray).

Natalizumab	HCVR	vklqqsgaelvkpgasvklfctasgfnikdtymhwvkqrpqqglewigridpasgdtkydpkfqvkatitadtssntawlqlssltsedtavyycadgmwvstgyaldfwgqgttvtvss
	LCVR	sivmtqtpkfllvsagdrvtitckasqsvtndvawyqqkpgqspklliyyasnrytgvpdrftgsgygtdftftistvqaedlavyfcqqdysspytfgggtklei
Eculizumab	HCVR	qvqlqqsgaelmkpgasvkmsckatgyifsnywiqwikqrpghglewigeilpgsgsteytenfkdkaaftadtssntaymqlssltsedsavyycaryffgsspnwyfdvwgagttvtvss
	LCVR	diqmtqspaslsasvgetvtitcgaseniygalnwyqrkqgkspqlliygatnladgmssrfsgsgsgrqyylkisslhpddvatyycqnvlntpltfgagtklelk
Ranibizumab	HCVR	evqlvesggglvqpggslrlscaasgydfthygmnwvrqapgkglewvgwintytgeptyaadfkrrftfsldtskstaylqmnslraedtavyycakypyyygtshwyfdvwgqgtlvtvss
	LCVR	diqltqspsslsasvgdrvtitcsasqdisnylnwyqqkpgkapkvliyftsslhsgvpsrfsgsgsgtdftltisslqpedfatyycqqystvpwtfgqgtkveik
Denosumab	HCVR	evqllesggglvqpggslrlscaasgftfssyamswvrqapgkglewvsgitgsggstyyadsvkgrftisrdnskntlylqmnslraedtavyycakdpgttvimswfdpwgqgtlvtvss
	LCVR	eivltqspgtlslspgeratlscrasqsvrgrylawyqqkpgqaprlliygassratgipdrfsgsgsgtdftltisrlepedfavfycqqygssprtfgqgtkveik
Daclizumab	HCVR	qvqlvqsgaevkkpgssvkvsckasgytftsyrmhwvrqapgqglewigyinpstgyteynqkfkdkatitadestntaymelsslrsedtavyycargggvfdywgqgtlvtvss
	LCVR	diqmtqspstlsasvgdrvtitcsasssisymhwyqqkpgkapklliyttsnlasgvparfsgsgsgteftltisslqpddfatyychqrstypltfgqgtkvevk
Basiliximab	HCVR	qlqqsgtvlarpgasvkmsckasgysftrywmhwikqrpgqglewigaiypgnsdtsynqkfegkakltavtsastaymelsslthedsavyycsrdygyyfdfwgqgttltvss
	LCVR	qivstqspaimsaspgekvtmtcsasssrsymqwyqqkpgtspkrwiydtsklasgvparfsgsgsgtsysltissmeaedaatyychqrssytfgggtkleik
Muromonab	HCVR	qvqlqqsgaelarpgasvkmsckasgytftrytmhwvkqrpgqglewigyinpsrgytnynqkfkdkatlttdkssstaymqlssltsedsavyycaryyddhycldywgqgttltvss
	LCVR	qivltqspaimsaspgekvtmtcsasssvsymnwyqqksgtspkrwiydtsklasgvpahfrgsgsgtsysltisgmeaedaatyycqqwssnpftfgsgtklein

Note:
Natalizumab: Chothia CDRs, source Patent US20120022236
Eculizumab: Chothia CDRs, source Patent US6355245
Ranibizumab: Chothia CDRs, source NCBI
Denosumab: Chothia CDRs, source Patent US 20120087923
Daclizumab: Chothia CDRs, source Patent US7361740
Basiliximab: Chothia CDRs, source NCBI
Muromonab: Drug Bank Mouse CDRs, source: http://www.drugbank.ca/drugs/DB00075

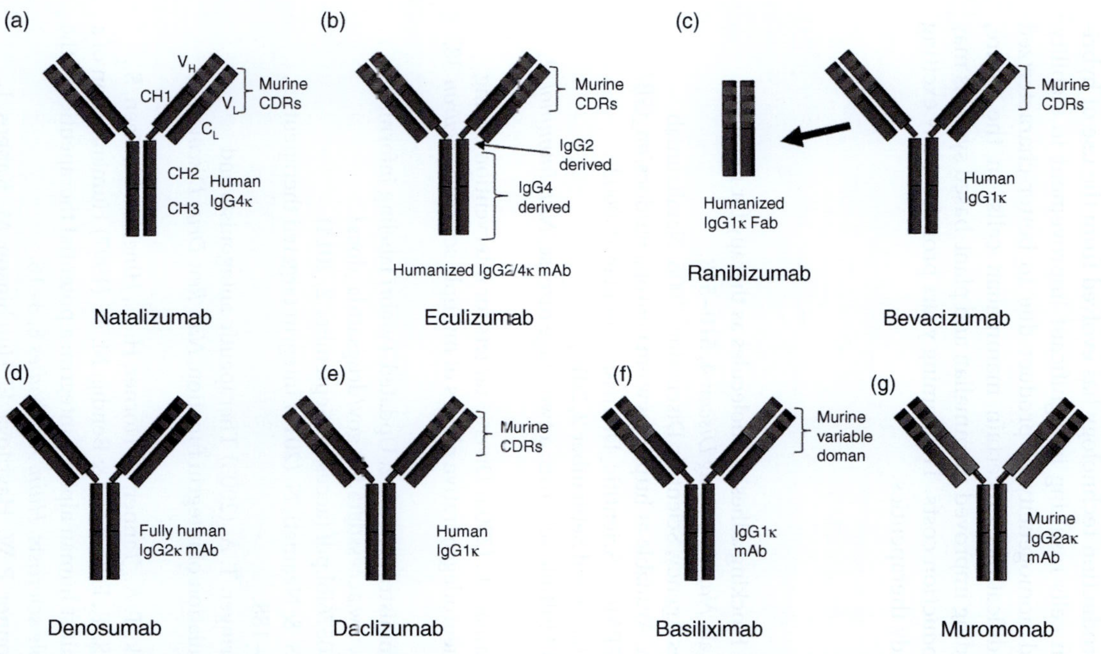

Figure 10.6 The schematic structures of studied antibodies in this chapter. Monoclonal antibodies covered in this chapter illustrate the evolution of monoclonal antibody therapeutics. Muromonab (g) is a murine antibody. Basiliximab (f) is a human/murine chimera with whole variable domain from murine origin. Natalizumab (a), eculizumab (b), bevacizumab (c), and daclizumab (e) are humanized monoclonal antibodies with only murine CDRs. Ranibizumab (c) is a humanized antibody fragment (Fab) derived from bevacizumab. Denosumab (d) is a fully human monoclonal antibody. (*See insert for color representation of the figure.*)

modulation of the receptor or enzyme's function with or without competing with the ligand or substrate. An allosteric antibody could potentially provide more desirable safety profile, avoid ligand deposition and possible adverse effect, and be independent from ligand concentration for dosing.

Finally, upstream production technology has evolved from the use of hybridomas to mammalian cells resulting in significant improvement in stability, production yield, and homogeneity of product due to better-characterized posttranslational modifications in certain mammalian cells. In the future, other platforms including improved mammalian and plant-based systems may bring much lower production costs. The coming years promise many exciting possibilities in antibody therapeutics.

References

1. Steinman, L. (2005) Blocking adhesion molecules as therapy for multiple sclerosis: natalizumab, *Nat. Rev. Drug Discov.* 4, 510–518.
2. European Medicines Agency, Scientific Discussion. 2006. Natalizumab scientific discussion. Available at http://www.ema.europa.eu/docs/en_GB/document_library/EPAR_-_Scientific_Discussion/human/000603/WC500044690.pdf (accessed September 2, 2014).
3. Steinman, L. (2001) Multiple sclerosis: a two-stage disease, *Nat. Immunol.* 2, 762–764.
4. Zamvil, S. S. & Steinman, L. (2003) Diverse targets for intervention during inflammatory and neurodegenerative phases of multiple sclerosis, *Neuron* 38, 685–688.
5. Food and Drug Administration. 2012. Updated Tysabri labeling information. Available at http://www.accessdata.fda.gov/drugsatfda_docs/label/2012/125104s0576lbl.pdf (accessed September 2, 2014).
6. Millard, M., Odde, S. & Neamati, N. (2011) Integrin targeted therapeutics, *Theranostics* 1, 154–188.
7. Shimaoka, M. & Springer, T. A. (2003) Therapeutic antagonists and conformational regulation of integrin function, *Nat. Rev. Drug Discov.* 2, 703–716.
8. Leger, O. J., Yednock, T. A., Tanner, L., Horner, H. C., Hines, D. K., Keen, S., Saldanha, J., Jones, S. T., Fritz, L. C. & Bendig, M. M. (1997) Humanization of a mouse antibody against human alpha-4 integrin: a potential therapeutic for the treatment of multiple sclerosis, *Hum. Antibodies* 8, 3–16.
9. Polman, C. H., O'Connor, P. W., Havrdova, E., Hutchinson, M., Kappos, L., Miller, D. H., Phillips, J. T., Lublin, F. D., Giovannoni, G., Wajgt, A., Toal, M., Lynn, F., Panzara, M. A. & Sandrock, A. W. (2006) A randomized, placebo-controlled trial of natalizumab for relapsing multiple sclerosis, *N. Engl. J. Med.* 354, 899–910.

10. Rudick, R. A., Stuart, W. H., Calabresi, P. A., Confavreux, C., Galetta, S. L., Radue, E. W., Lublin, F. D., Weinstock-Guttman, B., Wynn, D. R., Lynn, F., Panzara, M. A. & Sandrock, A. W. (2006) Natalizumab plus interferon beta-1a for relapsing multiple sclerosis, *N. Engl. J. Med.* 354, 911–923.
11. Sandborn, W. J., Colombel, J. F., Enns, R., Feagan, B. G., Hanauer, S. B., Lawrance, I. C., Panaccione, R., Sanders, M., Schreiber, S., Targan, S., van Deventer, S., Goldblum, R., Despain, D., Hogge, G. S. & Rutgeerts, P. (2005) Natalizumab induction and maintenance therapy for Crohn's disease, *N. Engl. J. Med.* 353, 1912–1925.
12. Targan, S. R., Feagan, B. G., Fedorak, R. N., Lashner, B. A., Panaccione, R., Present, D. H., Spehlmann, M. E., Rutgeerts, P. J., Tulassay, Z., Volfova, M., Wolf, D. C., Hernandez, C., Bornstein, J. & Sandborn, W. J. (2007) Natalizumab for the treatment of active Crohn's disease: results of the ENCORE Trial, *Gastroenterology* 132, 1672–1683.
13. Rudick, R., Polman, C., Clifford, D., Miller, D. & Steinman, L. (2013) Natalizumab: bench to bedside and beyond, *JAMA Neurol.* 70, 172–182.
14. Nicholas, J. A., Racke, M. K., Imitola, J. & Boster, A. L. (2014) First-line natalizumab in multiple sclerosis: rationale, patient selection, benefits and risks, *Ther. Adv. Chronic Dis.* 5, 62–68.
15. Rother, R. P., Rollins, S. A., Mojcik, C. F., Brodsky, R. A. & Bell, L. (2007) Discovery and development of the complement inhibitor eculizumab for the treatment of paroxysmal nocturnal hemoglobinuria, *Nat. Biotechnol.* 25, 1256–1264.
16. Hillmen, P., Young, N. S., Schubert, J., Brodsky, R. A., Socie, G., Muus, P., Roth, A., Szer, J., Elebute, M. O., Nakamura, R., Browne, P., Risitano, A. M., Hill, A., Schrezenmeier, H., Fu, C. L., Maciejewski, J., Rollins, S. A., Mojcik, C. F., Rother, R. P. & Luzzatto, L. (2006) The complement inhibitor eculizumab in paroxysmal nocturnal hemoglobinuria, *N. Engl. J. Med.* 355, 1233–1243.
17. Caprioli, J., Noris, M., Brioschi, S., Pianetti, G., Castelletti, F., Bettinaglio, P., Mele, C., Bresin, E., Cassis, L., Gamba, S., Porrati, F., Bucchioni, S., Monteferrante, G., Fang, C. J., Liszewski, M. K., Kavanagh, D., Atkinson, J. P. & Remuzzi, G. (2006) Genetics of HUS: the impact of MCP, CFH, and IF mutations on clinical presentation, response to treatment, and outcome, *Blood* 108, 1267–1279.
18. European Medicines Agency, Scientific Discussion. 2007. Eculizumab scientific discussion. Available at http://www.ema.europa.eu/docs/en_GB/document_library/EPAR_-_Scientific_Discussion/human/000791/WC500054212.pdf (accessed September 2, 2014).
19. Rosse, W. F., Hillmen, P. & Schreiber, A. D. (2004) Immune-mediated hemolytic anemia, *Hematol. Am. Soc. Hematol. Educ. Program*, 2004(1), 48–62.
20. Thomas, T. C., Rollins, S. A., Rother, R. P., Giannoni, M. A., Hartman, S. L., Elliott, E. A., Nye, S. H., Matis, L. A., Squinto, S. P. & Evans, M. J. (1996)

Inhibition of complement activity by humanized anti-C5 antibody and single-chain Fv, *Mol. Immunol.* 33, 1389–1401.

21. Takeda, J., Miyata, T., Kawagoe, K., Iida, Y., Endo, Y., Fujita, T., Takahashi, M., Kitani, T. & Kinoshita, T. (1993) Deficiency of the GPI anchor caused by a somatic mutation of the PIG-A gene in paroxysmal nocturnal hemoglobinuria, *Cell* 73, 703–711.
22. Johnson, R. J. & Hillmen, P. (2002) Paroxysmal nocturnal haemoglobinuria: nature's gene therapy?, *Mol. Pathol.* 55, 145–152.
23. Food and Drug Administration. 2011. Soliris labeling information. Available at http://www.accessdata.fda.gov/drugsatfda_docs/label/2011/125166s172lbl.pdf (accessed September 2, 2014).
24. Noris, M. & Remuzzi, G. (2009) Atypical hemolytic-uremic syndrome, *N. Engl. J. Med.* 361, 1676–1687.
25. Zuber, J., Fakhouri, F., Roumenina, L. T., Loirat, C. & Fremeaux-Bacchi, V. (2012) Use of eculizumab for atypical haemolytic uraemic syndrome and C3 glomerulopathies, *Nat. Rev. Nephrol.* 8, 643–657.
26. Zareba, K. M. (2007) Eculizumab: a novel therapy for paroxysmal nocturnal hemoglobinuria, *Drugs Today (Barc.).* 43, 539–546.
27. European Medicines Agency, Scientific Discussion. 2007. Ranibizumab scientific discussion. Available at http://www.ema.europa.eu/docs/en_GB/document_library/EPAR_-_Scientific_Discussion/human/000715/WC500043550.pdf (accessed September 2, 2014).
28. Chen, Y., Wiesmann, C., Fuh, G., Li, B., Christinger, H. W., McKay, P., de Vos, A. M. & Lowman, H. B. (1999) Selection and analysis of an optimized anti-VEGF antibody: crystal structure of an affinity-matured Fab in complex with antigen, *J. Mol. Biol.* 293, 865–881.
29. Mordenti, J., Cuthbertson, R. A., Ferrara, N., Thomsen, K., Berleau, L., Licko, V., Allen, P. C., Valverde, C. R., Meng, Y. G., Fei, D. T., Fourre, K. M. & Ryan, A. M. (1999) Comparisons of the intraocular tissue distribution, pharmacokinetics, and safety of 125I-labeled full-length and Fab antibodies in rhesus monkeys following intravitreal administration, *Toxicol. Pathol.* 27, 536–544.
30. Fine, S. L., Berger, J. W., Maguire, M. G. & Ho, A. C. (2000) Age-related macular degeneration, *N. Engl. J. Med.* 342, 483–492.
31. Olsson, A. K., Dimberg, A., Kreuger, J. & Claesson-Welsh, L. (2006) VEGF receptor signalling—in control of vascular function, *Nat. Rev. Mol. Cell Biol.* 7, 359–371.
32. Food and Drug Administration. 2006. Lucentis labeling information. Available at http://www.accessdata.fda.gov/drugsatfda_docs/label/2006/125156lbl.pdf (accessed September 2, 2014).
33. Rosenfeld, P. J., Brown, D. M., Heier, J. S., Boyer, D. S., Kaiser, P. K., Chung, C. Y. & Kim, R. Y. (2006) Ranibizumab for neovascular age-related macular degeneration, *N. Engl. J. Med.* 355, 1419–1431.

34. Brown, D. M., Kaiser, P. K., Michels, M., Soubrane, G., Heier, J. S., Kim, R. Y., Sy, J. P. & Schneider, S. (2006) Ranibizumab versus verteporfin for neovascular age-related macular degeneration, *N. Engl. J. Med.* 355, 1432–1444.

35. Regillo, C. D., Brown, D. M., Abraham, P., Yue, H., Ianchulev, T., Schneider, S. & Shams, N. (2008) Randomized, double-masked, sham-controlled trial of ranibizumab for neovascular age-related macular degeneration: PIER Study year 1, *Am. J. Ophthalmol.* 145, 239–248.

36. Campochiaro, P. A., Heier, J. S., Feiner, L., Gray, S., Saroj, N., Rundle, A. C., Murahashi, W. Y. & Rubio, R. G. (2010) Ranibizumab for macular edema following branch retinal vein occlusion: six-month primary end point results of a phase III study, *Ophthalmology* 117, 1102–1112.e1.

37. Brown, D. M., Campochiaro, P. A., Singh, R. P., Li, Z., Gray, S., Saroj, N., Rundle, A. C., Rubio, R. G. & Murahashi, W. Y. (2010) Ranibizumab for macular edema following central retinal vein occlusion: six-month primary end point results of a phase III study, *Ophthalmology* 117, 1124–1133.e1.

38. Nguyen, Q. D., Brown, D. M., Marcus, D. M., Boyer, D. S., Patel, S., Feiner, L., Gibson, A., Sy, J., Rundle, A. C., Hopkins, J. J., Rubio, R. G. & Ehrlich, J. S. (2012) Ranibizumab for diabetic macular edema: results from 2 phase III randomized trials: RISE and RIDE, *Ophthalmology* 119, 789–801.

39. Nguyen, Q. D., Shah, S. M., Khwaja, A. A., Channa, R., Hatef, E., Do, D. V., Boyer, D., Heier, J. S., Abraham, P., Thach, A. B., Lit, E. S., Foster, B. S., Kruger, E., Dugel, P., Chang, T., Das, A., Ciulla, T. A., Pollack, J. S., Lim, J. I., Eliott, D. & Campochiaro, P. A. (2010) Two-year outcomes of the ranibizumab for edema of the mAcula in diabetes (READ-2) study, *Ophthalmology* 117, 2146–2151.

40. Schmucker, C., Ehlken, C., Agostini, H. T., Antes, G., Ruecker, G., Lelgemann, M. & Loke, Y. K. (2012) A safety review and meta-analyses of bevacizumab and ranibizumab: off-label versus goldstandard, *PLoS ONE* 7, e42701.

41. Nguyen, Q. D., Shah, S. M., Hafiz, G., Quinlan, E., Sung, J., Chu, K., Cedarbaum, J. M. & Campochiaro, P. A. (2006) A phase I trial of an IV-administered vascular endothelial growth factor trap for treatment in patients with choroidal neovascularization due to age-related macular degeneration, *Ophthalmology* 113, 1522 e1–1522 e14.

42. Food and Drug Administration. 2011. Updated Prolia labeling information. Available at http://www.accessdata.fda.gov/drugsatfda_docs/label/2011/125320s5s6lbl.pdf (accessed September 2, 2014).

43. Food and Drug Administration. 2013. Updated Xgeva labeling information. Available at http://www.accessdata.fda.gov/drugsatfda_docs/label/2013/125320s094lbl.pdf (accessed September 2, 2014).

44. Lacey, D. L., Boyle, W. J., Simonet, W. S., Kostenuik, P. J., Dougall, W. C., Sullivan, J. K., San Martin, J. & Dansey, R. (2012) Bench to bedside: elucidation of the OPG–RANK–RANKL pathway and the development of denosumab, *Nat. Rev. Drug Discov.* 11, 401–419.

45. European Medicines Agency, denosumab CHMP assessment report. 2010. Available at http://www.ema.europa.eu/docs/en_GB/document_library/EPAR_-_Public_assessment_report/human/001120/WC500093529.pdf (accessed September 2, 2014).
46. Simonet, W. S., Lacey, D. L., Dunstan, C. R., Kelley, M., Chang, M. S., Luthy, R., Nguyen, H. Q., Wooden, S., Bennett, L., Boone, T., Shimamoto, G., DeRose, M., Elliott, R., Colombero, A., Tan, H. L., Trail, G., Sullivan, J., Davy, E., Bucay, N., Renshaw-Gegg, L., Hughes, T. M., Hill, D., Pattison, W., Campbell, P., Sander, S., Van, G., Tarpley, J., Derby, P., Lee, R. & Boyle, W. J. (1997) Osteoprotegerin: a novel secreted protein involved in the regulation of bone density, *Cell* 89, 309–319.
47. Anderson, D. M., Maraskovsky, E., Billingsley, W. L., Dougall, W. C., Tometsko, M. E., Roux, E. R., Teepe, M. C., DuBose, R. F., Cosman, D. & Galibert, L. (1997) A homologue of the TNF receptor and its ligand enhance T-cell growth and dendritic-cell function, *Nature* 390, 175–179.
48. Lacey, D. L., Timms, E., Tan, H. L., Kelley, M. J., Dunstan, C. R., Burgess, T., Elliott, R., Colombero, A., Elliott, G., Scully, S., Hsu, H., Sullivan, J., Hawkins, N., Davy, E., Capparelli, C., Eli, A., Qian, Y. X., Kaufman, S., Sarosi, I., Shalhoub, V., Senaldi, G., Guo, J., Delaney, J. & Boyle, W. J. (1998) Osteoprotegerin ligand is a cytokine that regulates osteoclast differentiation and activation, *Cell* 93, 165–176.
49. Yasuda, H., Shima, N., Nakagawa, N., Yamaguchi, K., Kinosaki, M., Mochizuki, S., Tomoyasu, A., Yano, K., Goto, M., Murakami, A., Tsuda, E., Morinaga, T., Higashio, K., Udagawa, N., Takahashi, N. & Suda, T. (1998) Osteoclast differentiation factor is a ligand for osteoprotegerin/osteoclastogenesis-inhibitory factor and is identical to TRANCE/RANKL, *Proc. Natl. Acad. Sci. U. S. A.* 95, 3597–3602.
50. Wong, B. R., Rho, J., Arron, J., Robinson, E., Orlinick, J., Chao, M., Kalachikov, S., Cayani, E., Bartlett, F. S., 3rd, Frankel, W. N., Lee, S. Y. & Choi, Y. (1997) TRANCE is a novel ligand of the tumor necrosis factor receptor family that activates c-Jun N-terminal kinase in T cells, *J. Biol. Chem.* 272, 25190–25194.
51. Green, L. L. (1999) Antibody engineering via genetic engineering of the mouse: XenoMouse strains are a vehicle for the facile generation of therapeutic human monoclonal antibodies, *J. Immunol. Methods* 231, 11–23.
52. Kostenuik, P. J., Nguyen, H. Q., McCabe, J., Warmington, K. S., Kurahara, C., Sun, N., Chen, C., Li, L., Cattley, R. C., Van, G., Scully, S., Elliott, R., Grisanti, M., Morony, S., Tan, H. L., Asuncion, F., Li, X., Ominsky, M. S., Stolina, M., Dwyer, D., Dougall, W. C., Hawkins, N., Boyle, W. J., Simonet, W. S. & Sullivan, J. K. (2009) Denosumab, a fully human monoclonal antibody to RANKL, inhibits bone resorption and increases BMD in knock-in mice that express chimeric (murine/human) RANKL, *J. Bone Miner. Res.* 24, 182–195.
53. Kostenuik, P. J., Smith, S. Y., Jolette, J., Schroeder, J., Pyrah, I. & Ominsky, M. S. (2011) Decreased bone remodeling and porosity are associated with

improved bone strength in ovariectomized cynomolgus monkeys treated with denosumab, a fully human RANKL antibody, *Bone* 49, 151–161.

54. Cummings, S. R., San Martin, J., McClung, M. R., Siris, E. S., Eastell, R., Reid, I. R., Delmas, P., Zoog, H. B., Austin, M., Wang, A., Kutilek, S., Adami, S., Zanchetta, J., Libanati, C., Siddhanti, S. & Christiansen, C. (2009) Denosumab for prevention of fractures in postmenopausal women with osteoporosis, *N. Engl. J. Med.* 361, 756–765.

55. Brown, J. P., Prince, R. L., Deal, C., Recker, R. R., Kiel, D. P., de Gregorio, L. H., Hadji, P., Hofbauer, L. C., Alvaro-Gracia, J. M., Wang, H., Austin, M., Wagman, R. B., Newmark, R., Libanati, C., San Martin, J. & Bone, H. G. (2009) Comparison of the effect of denosumab and alendronate on BMD and biochemical markers of bone turnover in postmenopausal women with low bone mass: a randomized, blinded, phase 3 trial, *J. Bone Miner. Res.* 24, 153–161.

56. Kendler, D. L., Roux, C., Benhamou, C. L., Brown, J. P., Lillestol, M., Siddhanti, S., Man, H. S., San Martin, J. & Bone, H. G. (2010) Effects of denosumab on bone mineral density and bone turnover in postmenopausal women transitioning from alendronate therapy, *J. Bone Miner. Res.* 25, 72–81.

57. Stopeck, A. T., Lipton, A., Body, J. J., Steger, G. G., Tonkin, K., de Boer, R. H., Lichinitser, M., Fujiwara, Y., Yardley, D. A., Viniegra, M., Fan, M., Jiang, Q., Dansey, R., Jun, S. & Braun, A. (2010) Denosumab compared with zoledronic acid for the treatment of bone metastases in patients with advanced breast cancer: a randomized, double-blind study, *J. Clin. Oncol.* 28, 5132–5139.

58. Fizazi, K., Carducci, M., Smith, M., Damiao, R., Brown, J., Karsh, L., Milecki, P., Shore, N., Rader, M., Wang, H., Jiang, Q., Tadros, S., Dansey, R. & Goessl, C. (2011) Denosumab versus zoledronic acid for treatment of bone metastases in men with castration-resistant prostate cancer: a randomised, double-blind study, *Lancet* 377, 813–822.

59. Henry, D. H., Costa, L., Goldwasser, F., Hirsh, V., Hungria, V., Prausova, J., Scagliotti, G. V., Sleeboom, H., Spencer, A., Vadhan-Raj, S., von Moos, R., Willenbacher, W., Woll, P. J., Wang, J., Jiang, Q., Jun, S., Dansey, R. & Yeh, H. (2011) Randomized, double-blind study of denosumab versus zoledronic acid in the treatment of bone metastases in patients with advanced cancer (excluding breast and prostate cancer) or multiple myeloma, *J. Clin. Oncol.* 29, 1125–1132.

60. Queen, C., Schneider, W. P., Selick, H. E., Payne, P. W., Landolfi, N. F., Duncan, J. F., Avdalovic, N. M., Levitt, M., Junghans, R. P. & Waldmann, T. A. (1989) A humanized antibody that binds to the interleukin 2 receptor, *Proc. Natl. Acad. Sci. U. S. A.* 86, 10029–10033.

61. European Medicines Agency, Zenapax scientific discussion. 2005. Available at http://www.ema.europa.eu/docs/en_GB/document_library/EPAR_-_Scientific_Discussion/human/000198/WC500057572.pdf (accessed September 2, 2014).

62. Vincenti, F., Pace, D., Birnbaum, J. & Lantz, M. (2003) Pharmacokinetic and pharmacodynamic studies of one or two doses of daclizumab in renal transplantation, *Am. J. Transplant.* 3, 50–52.
63. Food and Drug Administration. 2002. Zenapax labeling information. Available at http://www.fda.gov/downloads/drugs/developmentapprovalprocess/howdrugsaredevelopedandapproved/approvalapplications/therapeuticbiologicapplications/ucm113486.pdf (accessed September 2, 2014).
64. Ruscetti, F. W., Morgan, D. A. & Gallo, R. C. (1977) Functional and morphologic characterization of human T cells continuously grown in vitro, *J. Immunol.* 119, 131–138.
65. Rickert, M., Wang, X., Boulanger, M. J., Goriatcheva, N. & Garcia, K. C. (2005) The structure of interleukin-2 complexed with its alpha receptor, *Science* 308, 1477–1480.
66. Vincenti, F., Kirkman, R., Light, S., Bumgardner, G., Pescovitz, M., Halloran, P., Neylan, J., Wilkinson, A., Ekberg, H., Gaston, R., Backman, L. & Burdick, J. (1998) Interleukin-2-receptor blockade with daclizumab to prevent acute rejection in renal transplantation. Daclizumab Triple Therapy Study Group, *N. Engl. J. Med.* 338, 161–165.
67. Nashan, B., Light, S., Hardie, I. R., Lin, A. & Johnson, J. R. (1999) Reduction of acute renal allograft rejection by daclizumab. Daclizumab Double Therapy Study Group, *Transplantation* 67, 110–115.
68. Adu, D., Cockwell, P., Ives, N. J., Shaw, J. & Wheatley, K. (2003) Interleukin-2 receptor monoclonal antibodies in renal transplantation: meta-analysis of randomised trials, *BMJ* 326, 789.
69. Niemeyer, G., Koch, M., Light, S., Kuse, E. R. & Nashan, B. (2002) Long-term safety, tolerability and efficacy of daclizumab (Zenapax) in a two-dose regimen in liver transplant recipients, *Am. J. Transplant.* 2, 454–460.
70. Sellers, M. T., McGuire, B. M., Haustein, S. V., Bynon, J. S., Hunt, S. L. & Eckhoff, D. E. (2004) Two-dose daclizumab induction therapy in 209 liver transplants: a single-center analysis, *Transplantation* 78, 1212–1217.
71. Joyal, D., Cantarovich, M., Cecere, R. & Giannetti, N. (2004) Early experience with two-dose daclizumab in the prevention of acute rejection in cardiac transplantation, *Clin. Transplant.* 18, 493–496.
72. Biogen Idec press release on June 16, 2014. Available at http://www.biogenidec.com/press_release_details.aspx?ID=14712&M=News&PID=61997&NewsID=2358 (accessed on September 2, 2014).
73. European Medicines Agency, Simulect scientific discussion. 2005. Available at http://www.ema.europa.eu/docs/en_GB/document_library/EPAR_-_Scientific_Discussion/human/000207/WC500053538.pdf (accessed September 2, 2014).
74. Food and Drug Administration. 1998. Simulect labeling information. Available at http://www.fda.gov/downloads/drugs/

developmentapprovalprocess/howdrugsaredevelopedandapproved/approvalapplications/therapeuticbiologicapplications/ucm113359.pdf (accessed September 2, 2014).

75. Kahan, B. D., Rajagopalan, P. R. & Hall, M. (1999) Reduction of the occurrence of acute cellular rejection among renal allograft recipients treated with basiliximab, a chimeric anti-interleukin-2-receptor monoclonal antibody. United States Simulect Renal Study Group, *Transplantation* 67, 276–284.
76. Nashan, B., Moore, R., Amlot, P., Schmidt, A. G., Abeywickrama, K. & Soulillou, J. P. (1997) Randomised trial of basiliximab versus placebo for control of acute cellular rejection in renal allograft recipients. CHIB 201 International Study Group, *Lancet* 350, 1193–1198.
77. Ponticelli, C., Yussim, A., Cambi, V., Legendre, C., Rizzo, G., Salvadori, M., Kahn, D., Kashi, H., Salmela, K., Fricke, L., Heemann, U., Garcia-Martinez, J., Lechler, R., Prestele, H. & Girault, D. (2001) A randomized, double-blind trial of basiliximab immunoprophylaxis plus triple therapy in kidney transplant recipients, *Transplantation* 72, 1261–1267.
78. Lawen, J. G., Davies, E. A., Mourad, G., Oppenheimer, F., Molina, M. G., Rostaing, L., Wilkinson, A. H., Mulloy, L. L., Bourbigot, B. J., Prestele, H., Korn, A. & Girault, D. (2003) Randomized double-blind study of immunoprophylaxis with basiliximab, a chimeric anti-interleukin-2 receptor monoclonal antibody, in combination with mycophenolate mofetil-containing triple therapy in renal transplantation, *Transplantation* 75, 37–43.
79. Grego, K., Arnol, M., Bren, A. F., Kmetec, A., Tomazic, J. & Kandus, A. (2007) Basiliximab versus daclizumab combined with triple immunosuppression in deceased donor renal graft recipients, *Transplant. Proc.* 39, 3093–3097.
80. Pham, K., Kraft, K., Thielke, J., Oberholzer, J., Sankary, H., Testa, G. & Benedetti, E. (2005) Limited-dose Daclizumab versus Basiliximab: a comparison of cost and efficacy in preventing acute rejection, *Transplant. Proc.* 37, 899–902.
81. Nair, M. P., Nampoory, M. R., Johny, K. V., Costandi, J. N., Abdulhalim, M., El-Reshaid, W., Al-Muzairai, I., Ninan, V. T., Samhan, M. & Al-Mousawi, M. (2001) Induction immunosuppression with interleukin-2 receptor antibodies (basiliximab and daclizumab) in renal transplant recipients, *Transplant. Proc.* 33, 2767–2769.
82. Overington, J. P., Al-Lazikani, B. & Hopkins, A. L. (2006) How many drug targets are there? *Nat. Rev. Drug Discov.* 5, 993–996.
83. Smith, S. L. (1996) Ten years of Orthoclone OKT3 (muromonab-CD3): a review, *J. Transpl. Coord.* 6, 109–19; quiz 120-1.
84. Sgro, C. (1995) Side-effects of a monoclonal antibody, muromonab CD3/orthoclone OKT3: bibliographic review, *Toxicology* 105, 23–29.
85. Ciancio, G., Burke, G. W., Suzart, K., Roth, D., Kupin, W., Rosen, A., Olson, L., Esquenazi, V. & Miller, J. (2002) Daclizumab induction, tacrolimus,

mycophenolate mofetil and steroids as an immunosuppression regimen for primary kidney transplant recipients, *Transplantation* 73, 1100–1106.

86. Chin, C., Pittson, S., Luikart, H., Bernstein, D., Robbins, R., Reitz, B., Oyer, P. & Valantine, H. (2005) Induction therapy for pediatric and adult heart transplantation: comparison between OKT3 and daclizumab, *Transplantation* 80, 477–481.

87. Campos, A., Lage, E., Hinojosa, R., Ordonez, A., Cisneros, J. M., Cabezon, S., Gomez, S., Aguilera, A., Arana, E. & Cayuela, A. (2005) Comparative study of muromonab-CD3 (OKT3) versus daclizumab (Zenapax) in cardiac transplantation at our center, *Transplant. Proc.* 37, 1548–1549.

88. Chowdhury, S., Kode, R. K., Ranganna, K., Damask, A. M., Lam, A., Fyfe, B., Stabler, S., Kumar, A. M., Tomeny, M. B., Kumar, M. S. & Pankewycz, O. (2001) Induction strategy using basiliximab combined with mycophenolate MMF and immediate low-dose cyclosporin is steroid sparing and more effective than OKT3, *Transplant. Proc.* 33, 1057–1058.

89. Kumar, A. M., Fa, K., Vankawala, R., Vora, M., Kode, R. K., Pankewycz, O. G., Lattavi, M. R., Fyfe, B., Damask, A. M., Ferry, E., Stabler, S., Tomeny, M. B., Phillips, K., Lingaraju, R. & Kumar, M. S. (2001) Simulect, calcineurin inhibitor, mycophenolate mofetil, and prednisone is more effective than OKT3, calcineurin inhibitor, hycophendate mofetil, and prednisone in African American kidney recipients in reducing acute rejections and prolonging graft survival, *Transplant. Proc.* 33, 3195–3196.

90. Kode, R., Fa, K., Chowdhury, S., Ranganna, K., Fyfe, B., Stabler, S., Damask, A., Laftavi, M. R., Kumar, A. M. & Pankewycz, O. (2003) Basiliximab plus low-dose cyclosporin vs. OKT3 for induction immunosuppression following renal transplantation, *Clin. Transplant.* 17, 369–376.

91. Chatenoud, L. & Bluestone, J. A. (2007) CD3-specific antibodies: a portal to the treatment of autoimmunity, *Nat. Rev. Immunol.* 7, 622–632.

92. Kaufman, A. & Herold, K. C. (2009) Anti-CD3 mAbs for treatment of type 1 diabetes, *Diabetes Metab. Res. Rev.* 25, 302–306.

93. Xiao, G. & Gan, L. S. (2013) Receptor-mediated endocytosis and brain delivery of therapeutic biologics, *Int. J. Cell Biol.* 2013, 703545.

11

Manufacture of Recombinant Therapeutic Proteins Using Chinese Hamster Ovary Cells in Large-Scale Bioreactors

History, Methods, and Perspectives

Florian M. Wurm[1,2] *and Maria de Jesus*[2]

[1] *Faculty of Life Sciences, Ecole Polytechnique Fédérale de Lausanne (EPFL), Lausanne, Switzerland*
[2] *ExcellGene SA, Monthey, Switzerland*

11.1 Summary

Immortalized mammalian cells in suspension culture have been used in large-scale bioreactors for the production of recombinant protein therapeutics since the mid-1980s. All recombinant antibody products with greater than $1 billion (US) sales per year on the market today are produced in Chinese hamster ovary (CHO) cells. We wish to discuss briefly the technical status and the potential for further developments in the manufacture of therapeutic proteins, focusing on these remarkable cells and their unique history. Due to the sheer size of the CHO-based industry today, it would be foolish to claim to cover all the diversity in technique, processes, vectors, reactor systems, and protocols; however we will present a broad outline of the field in this chapter. Clearly CHO cells have achieved "superstar" status since more than 50% of all protein pharmaceuticals on the market are produced now in these cells [1–3]. Eight of the ten top selling biologicals in the market today are derived from CHO [4]. When considering products made in animal cells only, CHO cells are probably used in more than 90% of cases. Other immortalized cell lines, such as NS0, BHK, or HEK-293 cells, are used for a few individual products, but they play an insignificant role. All processes here considered involve "stable" cell lines, that is, cells that were engineered by inserting the desired gene(s) of interest into the genome (chromosomes) of the host cells.

The largest vessels in use for CHO cells have a working volume of about 20,000 l. The value of the 10 top selling products, the majority of them antibodies and antibody-like molecules derived from CHO cells, had reached more than

Biosimilars of Monoclonal Antibodies: A Practical Guide to Manufacturing, Preclinical, and Clinical Development, First Edition. Edited by Cheng Liu and K. John Morrow, Jr.

$50 billion (US) already in 2009 [5], and the value of new antibodies and biosimilars is estimated to rise 7–15% per year [6].

This chapter addresses aspects of the development of the upstream part of protein manufacturing, from clone to harvest of cell culture fluid. For a consideration of downstream processing and its relationship to biosimilar antibody products, refer to the chapter in this volume by Langer.

In the last three decades, protein yields from recombinant CHO cells in bioreactors have increased 10–100-fold [7], as the result of improvements in media, bioprocess design, and cell culture process control. The contribution of expression vector constructs and host cell engineering is, contrary to what numerous publications and even recent reviews are claiming, minor [1, 2, 8]. The reason for this possibly surprising statement is the following: the academic world has been very active in solving perceived problems in CHO technology by addressing them mostly through cell engineering. The industry however was bound to conservative approaches typical in pharmaceutical manufacturing and by the benefit of "sticking to things that worked." Significant investments (cell banking, testing, established operational handlings) into a relatively solid approach with proven track records from a nonmodified cell host could not easily be abandoned just because a cell cycle gene or an antiapoptotic gene might eventually provide some (minor) benefit. In fact, the historic (proprietary) body of knowledge concerning a given cell host provided ample opportunity to reap the benefits from modifications of media and process, with resultant dramatic improvements.

In general, the same vector components are used today as 20 years ago. Codon optimization of gene sequences for the desired protein of interest has improved the overall expression of some proteins, but did not result in "breakthrough" yield increases. The identification and selection of suitable clonal populations from transfected cells are now facilitated by widely accessible and cost-efficient equipment that allows high-throughput screening, including the use of flow cytometry and sorting. Some DNA elements in plasmids and novel vector/gene transfer approaches have pushed primary expression higher. Currently marketed protein therapeutics are exclusively produced in large-volume steam-sterilizable, stainless steel bioreactors, and the majority of production processes therein can be characterized as "fed-batch" cultures, with processing times of 10–20 days. Over the last 10 years, a strong "disposable" trend has emerged, particularly with innovative products that are entering the clinic but also with "biosimilar" products under development. "Single-use bioreactors" (SUB), presterilized plastic bags, mounted into or onto containers or platforms, serve as a physical barrier between the nonsterile environment and cell culture process liquids. While stirring with marine and pitched blade impellers is still the main method for mixing both sterilizable stainless steel and SUB, other impeller-free approaches have emerged as well that are now applied in SUB. The largest stirred SUB have working volumes of up to 2000 l; the largest prototype disposable bioreactor, being a nonstirred "OrbShake" SUB (Kühner AG, Birsfelden, Switzerland) possesses a working volume of 2500 l.

11.2 Introduction

Stably transfected mammalian cell populations, the so-called stable cell lines, first generated by nonviral gene delivery approaches in the early 1980s [9–11] were derived from immortalized cells, such as cells derived from human tumors, for example, HeLa cells [12], or cells that had been "transformed" by poorly understood genetic modification of cells with virus DNA. HEK-293 cells are typical in this respect, having been transfected with sheared adenovirus DNA [13]. During the late 1950s, researchers working with adherent primary cells derived from the ovary of a Chinese hamster (CHO) observed morphological changes and eventually the capacity of these cells to be cultivated permanently. A "spontaneous" transformation had occurred in these cells (Fig. 11.1).

Because of the ease of handling and rapid growth under standard culture conditions, CHO cell populations were passed on to many laboratories and are now widely distributed within the scientific community. They became key to studies on gene expression and function. Subpopulations of CHO cells were among the very first to be used for stable gene transfer. An attempt to reconstruct the history of industrially relevant CHO cell lines, as established from published records by the authors of this chapter, is presented in Figure 11.2.

The availability of certain auxotrophic CHO mutant lines facilitated the generation and identification of recombinant CHO cells. The term "cell line" always indicates a cell type that is "immortal," other than human primary cells that undergo a senescence phase and die, usually after a maximal cell population doubling number of about 50, the well-known Hayflick barrier or limit [14].

A hallmark of immortalization (or transformation) of animal cells in culture (or in cancer) is frequently a major modification of their karyotype, affecting structures and numbers of numerous chromosomes. Immortalized cell lines thus differ from primary cells and cell strains (purified cells of a specific phenotype that maintain diploidy) with respect to their rearranged and "disturbed" chromosomal status. Tumor cells and immortalized cells in culture have rearranged chromosomes, losses and gains of chromosomes, or chromosome fragments and usually maintain only a reduced number of chromosomes that appear to match their diploid origin. Frequently, the degree of homozygosity increases, so that losses of entire chromosomes or fragments appear to be compensated by a duplication of the remaining segments of the genome, thus reducing its allelic diversity.

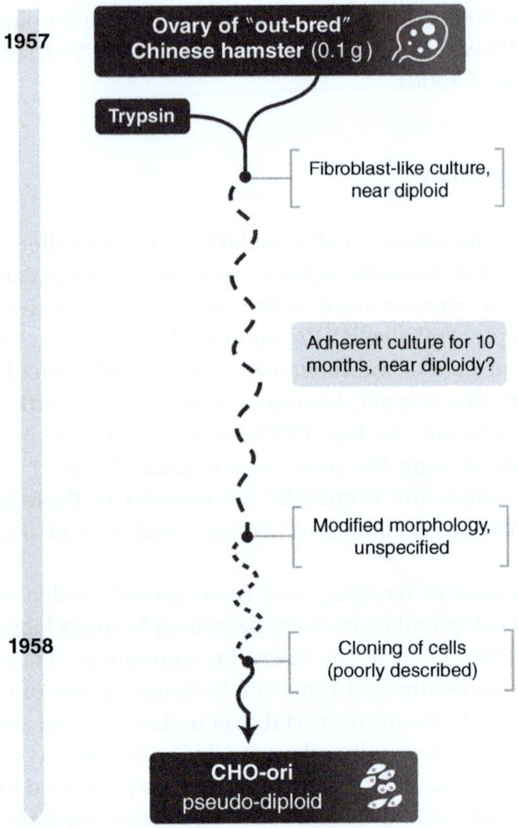

Figure 11.1 Derivation of spontaneously immortalized CHO cells from an ovary explant of a female, outbred Chinese hamster as referred to in the literature. Filled dots indicate events eluded to by traceable description or reference; wavy lines indicate nonreported details on subcultivation (passaging), culture media, and culture vessels. Graphics by C.P. Wurm©, Hamburg, Germany 2014.

Certain lines of CHO cells, in culture now for more than five decades, became popular for research into recombinant protein synthesis after two critical historic developments: (i) the molecular cloning of genes into plasmid DNAs [15] and (ii) the chemically mediated delivery of plasmid DNA into cultured mammalian cells [16–18]. The first CHO-derived recombinant cell lines included those that produced interferons and tissue-type plasminogen activator (tPA) [11, 19, 20]. In 1987, a tPA containing thrombolytic product became the first FDA-approved recombinant therapeutic protein made from animal cells cultivated in bioreactors (Activase®-tPA, Genentech, Inc., South San Francisco, CA, the United States). The market-saturating production of tPA required the cultivation of suspension culture-adapted CHO cells in large-scale

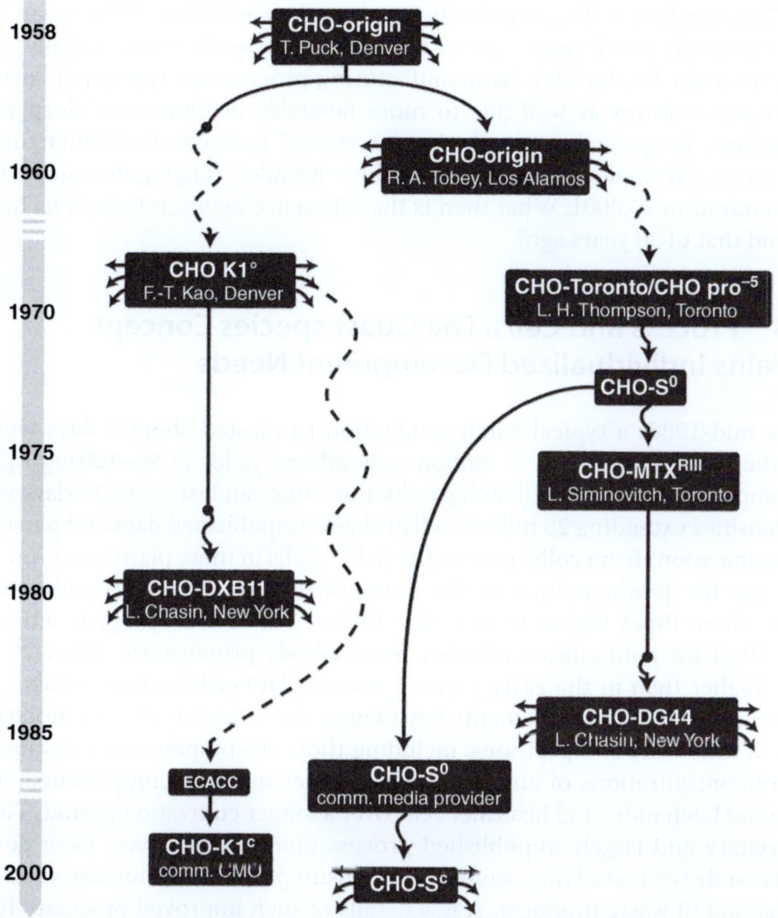

Figure 11.2 History of CHO cell lines as obtained from the literature. Uninterrupted lines indicate known transfers of cell cultures to other labs. Wavy lines indicate uncertain and poorly recorded history, lacking details of culture conditions and number of subcultivations (passaging). The smaller arrows at boxed names of cells indicate probable, but unrecorded, transfer of cell cultures to other laboratories. Historic times indicated in the figure are approximate and deducted from various literature sources. Graphics by C.P. Wurm©, Hamburg 2014.

stirred-tank bioreactors, a technology derived and modified from large-scale manufacturing with bacteria or other robust single-celled organisms [21]. The second protein made in mammalian cell and approved for marketing by the company Amgen, Thousand Oaks, CA, the United States, erythropoietin (Epogen®), did not need deep tank technology because the required quantity for this product was 1000 times smaller, 100 μg/dose, than that for tPA. This product, selling eventually for more than $2 billion (US) a year, and thus financially much more successful than tPA, was therefore initially produced in roller bottles [22]. Its manufacturing process was converted later to suspension cultures as well due to more favorable economics of deep tank technology. In general, technologies for stirred tank-based manufacturing have not much changed during the last three decades, ranging in scale from a few hundred to 20,000 l. What then is the difference between today's technology and that of 30 years ago?

11.3 Process and Cells: The Quasi-species Concept Explains Individualized Development Needs

In the mid-1980s a typical batch production run lasted about 7 days with a maximum cell density of 2–3 million cells/ml and yields of 50–100 mg/l [23]. By comparison, today's fed-batch production runs can last up to 21 days with cell densities exceeding 25 million cells/ml (our unpublished data and personal communications from colleagues in the field). Cells in these platforms typically have specific productivities in the range of 30–60 pg/cell/day—only about two to three times higher than in the 1980s—but volumetric productivities of 1–10 g/l for antibodies and other nonantibody proteins are about 20–50 times higher than in the early years of use of CHO cells in bioreactors [24]. Surprisingly, these improvements have come about mainly due to modifications in the media compositions, including those of complex feeds (optimization of concentrations of amino acids and other medium components). The result has been more and healthier cells over a longer cultivation period. These proprietary and largely unpublished process modifications have been developed mainly while studying carefully the consumption of key nutrients and the production of waste products. A few details of such improved processes have entered the public domain but composition and formulations of such media are largely proprietary. The impact of host cell engineering, touted over decades as responsible for dramatic improvements in yields, or increased robustness, has been overstated. Multigram per liter processes have been developed in our labs without any host cell engineering and are providing satisfactory clinical products (personal information). In our opinion, most processes developed for clinical trials today are still using non-engineered host cell lines.

Each clonal cell line must be individually optimized for its suitability for large-scale manufacturing, irrespectively of any genetic engineering previously performed on the host cell line. At a conference in 2012, the tremendous diversity of more than 50 clonally derived cell lines of CHO cells for one product was presented by Stettler et al. [25]. Growth rates, maximal density, viability trends, glucose consumption rates, lactate production rates, overall productivity, and other metabolic parameters of all clonal cell lines were studied in a high-throughput fed-batch screening format. This and other similar studies were done with hundreds of cell lines (our unpublished data and personal information from colleagues in the field), and the resulting data provide an expanded knowledge base for the CHO research community. Distinctive phenotypes of cell lines can be evaluated as a source of critical information relevant and related to the genetics of the individual CHO cells and their derived clonal populations. In this context, two recent publications on lactate metabolism in clonal CHO cell lines are highly informative with regard to the diversity and apparent mutability of key metabolic pathways in cells [26, 27].

A wealth of observations with CHO cells over decades support the hypothesis that both clonal and nonclonal cell lines are diverse populations of related but distinct quasi-species. This term was used first by Eigen and Schuster for organismal populations that undergo rapid evolution due to high mutation rates [28–31]. The importance of individualized media and feed regimes, as well as the profound impact of reactors, process control regimens, and physicochemical factors with suspension cells in general, is now relatively well understood within the small community of process engineers and scientists handling these cells. The body of knowledge, based on extensive experiences with clonal cell lines derived from one of the commercial CHO quasi-species families, becomes over time a key component of any company's proprietary repertoire and will then be considered a tightly maintained trade secret.

Recently, the genomic sequence of one ancestral CHO genome (K1) was published [32], and the authors acknowledged that this genome may in fact be far from being representative of the clonal cell lines cultivated in today's large-scale bioreactors. In the accompanying commentary [33], the complex and convoluted history of CHO cells (which includes mutagenesis, adaptations to different growth conditions, and cloning), impossible to trace back in detail, was highlighted. In fact, a careful analysis of a few published CHO karyotypes demonstrates their highly variable nature, illustrating the complex genetics/genomics of both recombinant and nonrecombinant CHO cell lines. The high plasticity of the CHO genome with respect to chromosomal restructuring was recently reemphasized in a commentary paper by Omasa and Lee [34]. Thus, the design of "the" ideal mammalian host cell lines for manufacturing is probably a dream, since a definable CHO genome and its corresponding physiology and phenotype do not exist.

However, a body of knowledge on phenotypic characters (e.g., dependence on certain media components, such as proline—as mentioned earlier) and metabolic responses of clonal cell populations derived from a "defined" host cell line can be established. When a nonrecombinant host cell line has been frozen and banked properly, and is then cultivated and handled from a thawed vial through the transfection and cloning procedures in a similar manner toward the establishment of the various manufacturing processes, certain phenotypic response ranges can be expected. With such an approach, a relatively narrowly defined medium and bioreactor "landscape" and "climate" are provided to all derived cell populations and clonal cell lines, thus restricting the potential for further evolutionary divergence to a smaller "environmental window." Protocols exploring the environmental plasticity of CHO cells have contributed to the field of CHO process sciences and CHO manufacturing sciences. Such insights are the foundation for further improved media and feed compositions as well as alternative process interventions to be applied when the next protein project is "on" for development. While this did not free commercial entities from continued investment into cell line and process development for each individual protein product, it provided however a foundation of process options that greatly improved the robustness and longevity of cells in bioreactors. In our own work—dealing with 100 s of different proteins and processes over the last 30 years—we estimate that for any clonal cell line about 200–500 different process conditions (while using design of experiment approaches) can be evaluated/studied and fine-tuned in order to identify a profile of key interacting parameters that work well. Clearly, the number of optimized states is very large, yet pragmatic decisions and timeline constraints will eventually limit choices. Remarkably, very small modifications/adjustments in many of these process conditions can have dramatic effects and result in more than "incremental" improvements.

With respect to cellular productivity for antibodies and other "well-behaved" proteins, it appears that a biological limit in specific productivity of about 100 pg/cell-day can be achieved with CHO cells in bioreactors. However, these very high specific productivities are not required for high yielding processes. The industry consensus is that fast(er) growing cells with specific productivities in the range of 20–40 pg/cell-day are optimal and that these can deliver volumetric yields—with proper cell culture development work—in the range of 1–8 g/l. For the future, extended batch culture yields can go beyond the 10 g/l range, possibly into the 20–30 g/l range. This is achievable mainly by increasing cell densities in bioreactors. Some groups have achieved cell densities of 25×10^6 cell/ml and higher in fed-batch processes with undisclosed process steps and media formulations (personal communications). Under such conditions, the biomass (as cells) represents about 4–5% of the total suspension culture volume. This compares still poorly with bacterial systems that can attain biomasses in reactors of 30–40%. Pushing the mammalian system further, via enriched media, may result in barriers of chemical constraints: physicochemical limits in solubility, opposing pH preferences, and adverse chemistries of individual components in media may be difficult to address. High-throughput, high-density, small-scale cell cultivation systems will be necessary in order to identify even better media compositions. Ultimately these should allow for perhaps one or two additional cell population doublings but also for the identification of media feeds that assure the maintenance of a healthy and productive high-density cell population prior to the eventual decline of viability in a fed-batch process.

Do we need more product and volume of cell cultures? Considering the increasing competitiveness in the market and a dramatic increase in new and diverse products addressing the same diseases, individual product quantities for market supply will be smaller than in the past. "Biobetters" are continuing to enter the market, possibly reducing the probability for new "blockbuster" protein drugs. Products for niche indications will become more attractive, also for larger pharmaceutical companies that struggle today to fill their pipelines.

Downstream processing technologies and capacities have not developed at the same rate as the upstream systems [35]. Frequently, product recovery teams struggle to process the high product loads derived from cell culture today and are therefore forced to break the harvest fluid of a reactor into smaller volumes, thus processing a single batch of upstream volume in multiple downstream processing runs.

We doubt therefore that even higher volumetric yields will be an important objective of future cell line and process development. However, time to clinic and market are always drivers for success. Cell line and process development for a first-generation manufacturing approach can be accomplished within 12 months using an experienced team and with key infrastructures, documentations, and quality systems in place. The "deadline" for this is the delivery of toxicology study materials, derived from a process that is ready for cGMP manufacturing. Shortening this is nearly impossible, since master cell banking and characterization of the bank alone is taking 3–4 months, to which a 3-month stability study needs to be added.

The need for higher economic efficiency will surely further increase the use of disposable container and bioreactor systems for large-scale cell culture, possibly up to the final production scale, and will also open the door for nonclonal cell-based protein production, for example, transient gene expression (TGE) systems or stable cell pool derived manufacturing.

11.4 Choices for Manufacturing: Host Cells for Production and Suitable Selection Systems

The CHO cells used in protein manufacturing originated in 1957 (Fig. 11.1) as a "spontaneously" immortalized cell from a primary culture of ovarian cells from a Chinese hamster (*Cricetulus griseus*) [36]. A proline-dependent clonal strain was derived from the original cell line and referred to as CHO-K1. All industrial cell lines in use today appear to be proline dependent. An adherent K1 cell line was mutagenized to generate CHO-DXB11 (also referred to as CHO-DUKX or CHO-**DUK-XB11**), a cell line lacking DHFR activity [37]. These cells have a deletion of one *dhfr* allele and a missense mutation in the other. Subsequently, the CHO-pro3$^-$ strain, another proline-requiring derivative of the original CHO cell line, was mutagenized to yield CHO-**DG44**, a cell line with deletions of both *dhfr* alleles [31, 38]. These two DHFR-minus strains require hypoxanthine and thymidine (as well as glycine) for growth. Frequently, the nonselective medium for these cells is referred to as GHT medium (glycine, hypoxanthine, thymidine), since these components must be added to standard cell culture media in order to allow growth of these cells.

Although not initially designed for recombinant protein manufacture, DHFR-minus CHO cells were used for a number of pioneering experiments demonstrating stable gene transfer and repair of DHFR activity with an exogenous *dhfr* gene when growing cells in medium lacking hypoxanthine and thymidine. This selection scheme, modified slightly when using other selection markers, remains one of the standard methods to establish stably transfected CHO cell lines for the production of recombinant therapeutic proteins. The process begins with the molecular cloning of the gene of interest (GOI) and the *dhfr* gene into a single or into two separate mammalian expression vectors. The plasmid DNAs carrying the two genes are then delivered into cells by transfection, and the cells are grown under selective conditions in GHT-minus medium. Surviving cells will have one or more copies of the exogenous *dhfr* gene, usually along with the GOI, integrated in its genome [9–11]. The number of integrated plasmids can vary widely from one recombinant cell to another. But when using the "standard" plasmid-based technology without any facilitating elements (to be discussed later), almost always only one integration site per cell is found even when multiple plasmids are transfected [39].

Growth rates and the levels of recombinant protein production of each cell line vary widely, due to the diversity of geno- and phenotypes of cells, as discussed before, but most likely also due to the type and site of integration in the genome. To obtain a few stably transfected cell lines with the desired phenotypic characteristics, it is usually necessary to evaluate several hundred clonal cell lines. Presently available transfection technologies result in 1–5% of exposed cells to be transformed—thus genetically not more than 1000–5000 individually different clonal cells are being generated (since these transfections occur usually in very small volumes and thus only one to two million cells are transfected).

The *dhfr* gene is not the only selection marker available for generating recombinant CHO cell lines. The glutamine synthetase (*gs*) gene, initially considered for the selection of murine NS0-derived cell lines with low or no endogenous GS activity, is also used for the selection of stably transfected CHO cells even though they have a higher endogenous GS activity than NS0 cells [40, 41]. This system is mainly used with CHO-K1 cells, but it can be applied to any of the available host cells, since a deletion of the endogenous glutamine synthetase is not required. After transfection with GOI and *gs* genes, recombinant cells are selected in medium lacking the amino acid glutamine. Both DHFR and GS systems allow enhancing the expression of the GOI by exposing the cells to a drug that blocks the enzymatic activity of the selection marker. DHFR and GS are inhibited by methotrexate (MTX) and methionine sulfoximine (MSX), respectively [20, 40, 42]. This is not a direct effect, but it is based on the selection of clonally arising cells that have enhanced expression of the selective marker gene through the process of gene amplification [9, 33]. For CHO-derived cell lines that express an exogenous *dhfr* gene, a majority of the cells die after 2–3 weeks of exposure to increasing concentrations of MTX. The rare survivors have a higher integrated plasmid copy number than the original cell line as the result of amplification of the *dhfr* gene and the neighboring DNA, including the GOI [10, 20, 42]. Similar observations have been made following the exposure of recombinant NS0 cell lines to MSX [41]. The cultivation procedures that result eventually in gene amplification take many weeks and contribute to significant and unpredictable genomic rearrangements of the host cells. For this reason, gene amplification is less frequently used today and can be replaced by other methods that achieve the same outcome—high productivity with less effort.

We have used CHO-K1 cells, CHO-S cells, and CHO-DG44 cells to select for recombinant clonal and nonclonal cell populations with puromycin. This is a rapidly acting agent, allowing surviving recombinant cell populations to be established in about 2 weeks. Of course, each of these host systems has their own unique media demand, and it takes some time to optimize and assure robust growth in serum-free, animal protein-free media. Some groups use neomycin and its corresponding resistance-providing gene, but we find the stringency of this system unsatisfactory.

In the 1980s and 1990s, most transfections were done in adherent culture, using media containing fetal bovine serum. The industry has now embraced serum-free culture and prefers animal protein- or animal component-free media. Better yet, the so-called chemically defined media (CDM) are now commercially available. As indicated in the preceding text, most commercial media formulations are not released on the grounds that they represent proprietary information. In some cases, companies will disclose whether a given molecule is or is not present in their medium. In our labs, we execute transfections under suspension culture conditions, and thus the first step is the generation of nonclonal recombinant cell populations ("pools") before "clones" are established.

11.5 Methods for Rapid Generation of High-Producing Cell Lines

The major problem with "standard" methods of cell line generation and selection is specific productivities of recovered cell lines that may have a wide range. This necessitates the screening of hundreds if not thousands of individual cell clones to obtain a few acceptable representatives with the desired phenotype combination of high productivity and high growth rate. One way to reduce the number of irrelevant cell lines from the pool of recombinant cells is to increase the stringency of selection. With DHFR-minus CHO cells, for example, the GHT-minus medium may be supplemented with 30–100 nM MTX to increase the probability of selecting cell lines with high *dhfr* activity. Since the GOI is integrated at the same site as the *dhfr* gene, the number of high-producing cell lines can be increased while leaving low-producing lines aside. High-producing cell lines with capacities for 1–3 g/l production [43] have been identified in this way (De Jesus et al., unpublished data). This approach avoids the time-consuming gene amplification approach utilized in the past. As indicated in the preceding text, induced gene amplification is to a large degree unpredictable, and amplified DNA segments in the chromosomes of CHO cells may not be stable in the absence of MTX [34, 39, 44]. Also, high-throughput methods to screen candidate cell lines are being developed to reduce the time necessary for the recovery of high producers [45]. Most of these methods are based on fluorescence-activated cell sorting (FACS). For example, the GFP gene can be co-expressed with the GOI and the cells sorted for GFP-specific fluorescence [46, 47]. The GFP gene may be used as the sole selection marker or in combination with one of the selection markers described earlier. Alternatively, the GOI may be co-expressed with a gene encoding a cell surface protein. The recombinant cells expressing the latter are then stained with a fluorescently labeled antibody specific for this protein and then sorted by FACS [48]. Lastly, recombinant cells selected for the presence of the *dhfr* gene have been incubated with fluorescent MTX that binds to DHFR [49]. The level of MTX-specific fluorescence is expected to correlate with the level of the recombinant protein of interest. It is not known, however, if any of the methods described in the peer-reviewed literature are being actively pursued by companies and entered the manufacturing phase of approved products in the market.

Clonal cell line recovery has been automated with such instruments as the ClonePix System (Genetix Ltd., New Milton, UK) and the CellCelector™ (AVISO GmbH, Aachen, Germany). For these instruments, the putative recombinant cells are suspended in semisolid medium. Under these conditions, the secreted recombinant protein remains near the cell and can be stained with a fluorescent-labeled antibody. These automated systems can detect and transfer cells to another cultivation container for further analysis.

It is also possible to increase the average specific productivity of recombinant cell lines by increasing the amount of plasmid DNA delivered to cells. It was shown that calcium phosphate (CaPi)-mediated transfection of CHO-DG44 cells results in both a higher plasmid copy number and a higher average specific productivity compared to polyethylenimine (PEI)-mediated transfection [50]. Furthermore, the specific productivity of recombinant cell lines generated by microinjection of either BHK-21 or CHO-DG44 cells depended on the amount of plasmid DNA injected per cell [50].

11.6 Silencing: Stability of Expression, Facilitators for High-Level Productivity

Once clonal cell lines are established, they need to be characterized for the "stability" of recombinant protein production. This is necessary because expression of the integrated GOI is not always maintained at a constant level over time. Many clonal cell lines lose or reduce the synthesis of mRNA driven from constitutive promoters. The reasons for this are not fully understood and, in addition, the extent of this phenomenon is variable. Generally, this is referred to as gene silencing (the reduction or elimination of gene-specific transcription). It is assumed that gene silencing occurs through the influence of the DNA composition in the vicinity of the integrated GOI sequences. Thus, candidate cell lines for large-scale productions must be cultivated in the so-called stability studies for several months, prior to their actual use in manufacturing, to exclude stability problems. A major determinant of epigenetic gene silencing is thought to be the structure (sequence composition and modification of histones and other proteins) of the chromatin at the site of integration of the recombinant gene. In general, heterochromatin is condensed and transcriptionally inactive, whereas euchromatin is relaxed and transcriptionally active [51]. The two chromatin states are associated with specific histone modifications including acetylation, methylation, and phosphorylation that function to control chromatin condensation and transcriptional activity [52]. For "standard" plasmid-based transfections without special elements that may stabilize expression, we have observed two types of gene silencing effects in recombinant CHO-DG44 cell lines [50]. Rapid gene silencing occurs in about half the cell lines within days after release of the cells from selective pressure. This type of gene silencing does not appear to be correlated with the level of GOI expression. For about one-third of the cell lines, a slow and gradual reduction in GOI expression occurs within 6 months after removal of the selective pressure (our own data unpublished). Finally, about 15–25% of cell lines derived from standard transfections have a stable level of protein productivity in the absence of selection. For cells treated with MTX, to amplify the copy number of the integrated GOI, the stability of the recombinant protein production in the absence of MTX appeared to be due mainly to gene silencing rather than to loss of transgene copy number [53].

The choice of the promoter/enhancer elements used to drive GOI expression may also influence the extent of gene silencing. It has been documented by numerous workers that DNA elements such as scaffold/matrix attachment regions (S/MARs), insulators, antirepressor elements, and ubiquitous chromatin opening elements (UCOEs) support more stable protein production in recombinant cell lines and increase the percentage of clonal cell lines with high expression levels [54–57]. These DNA elements (facilitators of stable expression) are small enough to be cloned into the expression vector employed for

GOI delivery. Mechanisms associated with their observed function are however not entirely clear. They may ameliorate the effects of gene silencing directly through inhibition of heterochromatin formation or they may affect plasmid DNA integration itself. For example, they may influence the integrated plasmid copy number or the site of integration.

Also, GOI DNA can be incorporated into CHO genomes under use of a lentivirus vector system [58, 59]. This approach significantly reduced the times to generate highly productive cell lines, and moreover, the majority of clonal cell lines generated showed good production stability in the absence of selective culture conditions.

An insect-derived mobile element, the PiggyBac transposon [60], has been shown to transpose in mammalian cells. While it has been used as a tool to generate knockout mutations in transgenic animals [61] and other applications in genome editing [62], this system has now been developed for the generation of highly productive cell lines [63–65]. One advantage of this system is the delivery of expression cassettes into the receiving mammalian genome without any remaining bacterial DNA, since the transposase cuts the inserts out of the plasmid and delivers it, framed with two short repeats, into the receiving genome. We remain reserved toward the newly emerged technologies for target sequence insertions of GOI (such as CRISPR–Cas technology [66]) into CHO cells as adding a significant improvement over the just discussed or older technologies, mostly because of the undefinable genome structures of CHO cells and their rapid evolution.

11.7 High-Throughput Bioprocess Development

Since the 1980s, small-scale process development studies for recombinant protein production have mainly been carried out in spinner flasks lacking monitoring technologies for measuring pH and O_2 levels in real time. However, they are not easily adapted to high-throughput applications. The minimal culture volume for these containers is about 50 ml and increasing the number of such containers requires a substantial footprint, particularly with frequently preferred spinner flasks with working volumes of 100–500 ml. Small volumes of 10 or 20 ml cannot be cultivated in these types of vessels, and all of them have very low volumetric mass transfer coefficients (k_La) of less than 1 h^{-1}. This restricts—due to oxygen limitation—the maximally achievable cell densities to 2–3 × 10^6 cells/ml [67]. In controlled bioreactors with sufficient oxygen supply, high-performance media allow densities of 10 × 10^6 cells/ml or higher. Therefore, cell cultivation in spinner flasks is a poor predictor of performance in large-scale stirred-tank bioreactors.

In an attempt to address the urgent needs for simpler, high-density cultivation systems, we developed orbitally shaken 50 ml "OrbShake tubes." Such tubes, derived from simple centrifuge containers, were first marketed by TPP, Trasadingen, Switzerland, under the name "TubeSpins" and are now sold by providers under a variety of names. These 5–30 ml working volume "bioreactors," when used in an appropriate manner in suitable shaker–incubators, are very efficient and serve as an excellent scale-down system for mammalian cell cultivation in suspension [68]. They have been widely implemented in cell line and process development activities of the industry. Cell densities $>1 \times 10^7$ cells/ml have been achieved in volumes of 5–20 ml with CHO cells [69]. At agitation speeds appropriate for mammalian cell cultivation, k_La values of 10–20 h^{-1} have been observed [70, 71]. Due to the high mass (gas) transfer in these passively ventilated 50 ml containers, oxygen limitation has never been observed at cell densities up to 3×10^7 cells/ml, when shaken at high rpms and at the correct displacement radius for orbital shaking. So far, no other system has been introduced that allows sufficient oxygen supply for such high-density CHO cultures with air. This is quite remarkable since a 0.5 l fully instrumented stirred bioreactor requires addition of pure oxygen into the sparged gas flow from a density of about 1×10^7 cells/ml. Since the OrbShake tubes are used in incubator shakers that provide CO_2 (e.g., Kühner Shaker ISF1-X), a narrow pH range is observed between pH 6.5 and 7.0. The low cost and ease of operation allow 100s, even 1000, of small-scale cultures to be run simultaneously [69]. OrbShake tubes substantially reduce the time necessary for medium optimizations and for the development of feeding strategies for fed-batch cultures [68]. Other scale-down cell culture systems have also been described. Examples include the SimCell microfluidics technology (Invitrogen, the United States) and microbioreactors in 24- and 96-deep well plates [72–74]. All these systems have however a much higher cost basis for the individual container and/or frequently require complex installations and accessory equipment. Also, these smaller reactor systems have the disadvantage of limiting sampling opportunities.

Several small-scale, disposable, and nondisposable stirred-tank bioreactor systems that allow parallel culture runs have been introduced over the last few years. The working volume ranges of specific reactor types are usually small, but well defined, including 10–15 ml, 150–250 ml, 300–500 ml, or 1–2 l. All are fully controlled, that is, the platform allows the regulation of pH, oxygen, and temperature, but not necessarily independently from each other (block control). The multiplicity of the reactors starts with sets of 4 and reaches 48 for some of them (TAP Biosystems/Sartorius, Göttingen, Germany; Infors, Birsfelden, Switzerland; Applikon, Delft, the Netherlands; DASGIP/Eppendorf, Hamburg, Germany). All these systems are highly automated, requiring advanced computer and robotic control, and all are expensive to operate. For the smallest system (with the highest multiplicity) choices for either headspace aeration or sparging are given, but not both. Also, due to excessive foaming, antifoam addition is a must in all these systems.

High-throughput technologies with reliable performance both in terms of quality and quantity in large-scale operation are now de rigueur, since speed to market is a brutal driver. This is particularly true for the emerging market on biosimilars, where an additional criterion has to be fulfilled: matching or at least coming very close in quality attributes to the originator product. The process defines the product; this mandate may be thought of as a statement of highest importance. Modulating process conditions, such as temperature shifts, osmolarity, special feeds, and pH kinetics, has dramatic ramifications for the quality of the product, particularly on secondary modifications.

11.8 Disposable Bioreactors

Today, very large-scale mammalian cell culture is almost exclusively performed in stainless steel, stirred-tank bioreactors. However, in recent years there has been a trend toward the adoption of disposable equipment. In fact, the first cGMP facilities have been installed that are "entirely" disposable for the upstream processing part of manufacturing. These are typically restricted to working volumes in reactors of up to 1500 l. In other facilities, mixing of disposable and nondisposable systems occurs, where the disposable systems are restricted to cell culture volume expansions, but not for the final production phase. The pioneer and most successful of the disposable bioreactors is the wave-type bioreactor that was introduced more than 10 years ago [75]. The cells are cultivated in disposable plastic bags of nominal volumes of up to 500 l mounted on a rocking table. The reported k_La values in wave bioreactors are less than $4\,h^{-1}$, resulting in possible oxygen limitation at densities $>5\times10^6$ cells/ml when air is used as an oxygen providing gas [75]. Modifications of the movement of a horizontally mounted bag on a platform that combines rocking (angle movement) with a horizontal forward and backward movement can improve the gas transfer rates in a CELL-tainer bioreactor [76]. Disposable stirred-tank bioreactors at volumetric scales of up to 2000 l are now available from a number of suppliers (GE–Hyclone, Sartorius Stedim, GE–Xcellerex, Thermo Fisher).

Disposable and ventilated "Erlenmeyer"-type flasks have been used in the past for both microbial and cell culture operations by applying orbital shaking as the mixing principle. Scale-up and multiplicity in these flasks are limited since the working volume is only 10–30% of the nominal volume while using a relatively large "footprint." In our laboratories and in the laboratory of Dr Lui [77], both cylindrical and square-shaped vessels with working volumes in the range of 100 ml to 30 l have been used for suspension cultivation of mammalian cells by orbital shaking [69, 70, 78–81]. Also, we explored disposable bags of 300 and 3500 l mounted within custom-made cylindrical containers on orbital shakers constructed for this purpose [78]. Since 2012, Kühner Shaker AG offers a disposable OrbShake bioreactor with a cultivation volume of up to 250 l. The application of this reactor in our lab and that of others have so far shown high reliability and excellent performance. A prototype disposable OrbShake bioreactor with a working volume of 2500 l has been installed in our labs for a short time now. To our knowledge, this is the largest disposable bioreactor so far built and our limited data show excellent growth and productivity [82].

In contrast to stirred tanks, surface aeration is the operational mode for shaken bioreactors, highlighting the importance of the surface (area) on oxygen transfer. The flow structure and dynamics at or near the liquid–air interface, as well as vertical liquid flows, are dominant factors determining the rate of mass (gas) transfer into and out of the cell culture medium. The wave types resulting from these flows have been studied extensively and show reproducible and consistent forms across all scales so far studied [83]. For cylindrical shaken vessels of up to 100 l working volume, shaking under cell culture-compatible shaking speeds, k_La values of 5–20 h^{-1} were obtained. With the 3500 l disposable OrbShake bioreactor, operated at a working volume of 2000 l and cell culture-compatible shaking speeds, k_La values of 3–12 h^{-1} were seen [82]. In any case, these gas transfer rates allow oxygenation with air only, even at the largest scale tested so far, thus freeing large-scale operations from the exigencies associated with supplying oxygen to the system.

While very large-scale disposable technology is immature and requires improvements in container stability, monitoring and control systems, and provisioning of robust and consistent bag designs, we have noted dramatic improvements in practicability and economics of these devices. It is known that limiting oxygen transfer rates in bioreactors (due to low k_La values) can be overcome by aeration with oxygen-enriched air or pure oxygen, since such gas mixtures or pure oxygen have a higher driving force into the liquid.

11.9 Nonclonal Expression Technologies for Fast Production and Assessment of Expression Potential and Quality

All production with CHO cells for pharmaceutical use is required to be based on single cells, expanded as clonal populations to larger biomasses that serve the manufacturing processes. This approach was necessary and desired since high-level expressing and stable cells were usually rare and cloning and screening were essential to isolate a high expressing line from a heterogeneous population of poor performers.

Today, nonclonal expression technologies, TGE, and "stable pool technology" can deliver large quantities of recombinant proteins, due to more efficient gene delivery and more suitable vector constructs that protect against silencing or mediate integration into more active sites of the CHO genome. TGE for "large scale" (1–100 l) is a relatively new technology considered for recombinant therapeutic protein production [84, 85]. TGE is defined as the production of a recombinant protein over a short period (1–14 days) following highly efficient DNA transfer into suspension cultures. The recombinant gene(s) is expressed from a nonviral expression vector and transfected into cells with phosphate (CaPi), PEI, or lipid-based reagents. In contrast to stable gene expression from recombinant cell lines, genetic selection is not applied at any time in the process.

The technology has been developed mainly with CHO and HEK-293 cells since they are easily transfected, grow in single-cell suspension, and have been used for the production of therapeutic proteins that have gained regulatory approval [86–89]. TGE in suspension culture can be performed in stirred-tank bioreactors or in agitated containers including shake flasks, wave-type bioreactors, and plastic or glass bottles [90]. TGE delivers proteins fast, and with both CHO and HEK-293 cells, volumetric productivities of 1–2 g/l have been achieved in a bioprocess lasting up to 14 days [91–93]. Thus, significant quantities of recombinant protein can be obtained within a few days of transfection. While 1–10 l scale TGE is now feasible, larger scale operations have not yet been reported at the gram per liter range. We reported 100 l scale TGE more than 10 years ago, using calcium phosphate as a transfection reagent, however resulting in only 5 mg/l yields [90, 94, 95]. A drawback of "large-scale" TGE is the very significant amount of vector DNA needed for this technology—1–10 mg/l of transfected cell culture.

So far, no TGE-derived therapeutic protein has gained regulatory approval. This must be accomplished before large-scale TGE becomes a standard method of therapeutic recombinant protein manufacturing.

Another nonclonal manufacturing approach is "stable pool technology." We and others have shown, using more efficient techniques in transfection and better vector systems that favor high, mostly stable expression right after transfection, that pools of stable cell lines can produce several hundreds of milligram per liter for 1–2 months at least [63–65, 96]. These stable pools benefit from an overall improved average expression of recombinant cells, based on improved vectors. Expanded stable pools of cells can be used to inoculate 1000 l scale bioreactors within 6–7 weeks after transfection. This can provide kg quantity production very rapidly. It remains to be seen whether this technique will go beyond the manufacturing of preclinical material. Toxicology materials have been produced this way by several large pharmaceutical companies, as has been stated in conferences. The use of both TGE and stable pool technology on the basis of CHO cells is an excellent predictor of quality range for a given protein before embarking upon the lengthy and costly stable clonal cell line development. This is useful not only for innovator products but also for the rapidly evolving field of biosimilars (of antibodies and other molecules) where the structural demands on the family of molecules generated by a process are expected to be very close to the originator product. Since different process options can be studied early, even with the approaches of transient and stable pools, the impact on product quality can be seen rather early, and not only when a fully developed process from a clonal cell line is emerging.

11.10 Conclusions

We have reviewed here very briefly and superficially a wealth of progress gained in the last three decades. The typical volumetric yields from bioprocesses in large-scale stirred tanks have increased 20–50-fold during this period. With (multi)gram per liter yields in optimized fed-batch processes, the production of tens of kilograms and hundreds of kilograms in large-scale bioreactors has become a routine. Improvements in the generation and identification of high-producing clonally derived cell lines, in the composition of optimized media, and in protocols for their application will result in further yield increases while at the same time shortening production run times. With these trends the overall annual capacity of a given manufacturing facility can be increased dramatically. It remains to be seen if the required bioreactor volumes for individual products may actually decrease, due to the improved yields.

We predict that the time required for establishment of such high yielding processes—when experienced teams conduct this work—will be reduced to a few months (4–6), offering more flexibility and choices. Overall, the improved speed for development, the higher yields, the use of disposable reactors, and the application of "generic" processes will have a positive economic impact on upstream protein manufacturing. Whether similar economic improvements can be made in downstream processing is doubtful in our mind. So far, progress in this area over the last 30 years has been slow and only incremental. Gottschalk, an outspoken expert in the field of downstream processing, has voiced repeatedly disappointment on progress in recovery of complex protein products [97].

We believe that "large-scale" TGE will become an acceptable cGMP manufacturing technology for certain products, expanding the possibility to make high value proteins in mammalian cells rapidly and with high yield. Also, the other nonclonal approach, the "stable pool technology," may eventually become acceptable and required in cases of extreme emergency. The recent widespread outbreaks of influenza H1N5, SARS, or EBOLA have made us wary of future public health challenges, because rapid delivery of product could prevent catastrophic spread of disease. The barrier for these nonclonal approaches is not technological or quality, but rather a legal and regulatory failure to address these issues with openness in mind. Clonal or not, CHO-based cell cultures in bioreactors can deliver the highest quality and highest yield for almost any recombinant protein intended for clinical use while carefully taking into consideration their metabolic and genetic complexity.

Conflict of Interest

Florian M. Wurm is Honorary (retired) Professor of Biotechnology at the Swiss Federal Institute of Technology in Lausanne (EPFL) and founder, CEO, and shareholder of ExcellGene SA and a service provider and CMO using CHO cells in Monthey, Switzerland.

Maria De Jesus is cofounder, shareholder, and chief operating officer of ExcellGene SA.

References

1. Jayapal, KP, Wlaschin, KF, Hu, WS, Yap, MGS. (2007). Recombinant protein therapeutics from CHO cells—20 years and counting. *Chemical Engineering Progress*, 103, 40–47 (CHO Consortium: SBE Special Edition).
2. Zhu, J. (2012). Mammalian cell protein expression for biopharmaceutical production. *Biotechology Advances*, 30, 1158–1170.
3. Casteleijn, M, Richardson, D. (2014). Engineering cells and proteins—creating pharmaceuticals. *European Pharmaceutical Review*, 19, 4, 12–19.
4. Hugget, B. (2013). Public biotech 2012—the numbers. *Nature Biotechnology*, 31, 697–703.
5. Walsh, G. (2010). Biopharmaceutical benchmarks 2010. *Nature Biotechnology*, 28, 9, 917–924.
6. Hiller, A. (2009). Fast growth foreseen for protein therapeutics. *Genetics*, 29, 153–155.
7. Kelly, B. (2009). Industrialization of the mAb production technology—the bioprocess industry at a crossroads. *mAbs*, 1, 5, 443–452.
8. Datta, P, Linhardt, RJ, Sharfstein, ST. (2013). An 'omics approach towards CHO cell engineering. *Biotechnology and Bioengineering*, 110, 5, 1255–1271.
9. Ringold, G, Dieckmann, B, Lee, F. (1981). Co-expression and amplification of dihydrofolate reductase cDNA and the *Escherichia coli* XGPRT gene in Chinese hamster ovary cells. *Journal of Molecular and Applied Genetics*, 1, 165–175.
10. Kaufman, RJ, Sharp, PA. (1982). Amplification and expression of sequences cotransfected with a modular dihydrofolate reductase complementary DNA gene. *Journal of Molecular Biology*, 159, 601–621.
11. Scahill, SJ, Devos, R, Van der Heyden, J, Fiers, W. (1983). Expression and characterization of the product of a human immune interferon cDNA gene in

Chinese hamster ovary cells. *Proceedings of the National Academy of Sciences of the United States of America*, 80, 4654–4658.

12. Gey, GO, Coffman, WD, Kubicek, MT. (1952). Tissue culture studies of the proliferative capacity of cervical carcinoma and normal epithelium. *Cancer Research*, 12, 264–265.
13. Graham, FL, Smiley, J, Russell, WC, Nairn, R. (1977). Characteristics of a human cell line transformed by DNA from human adenovirus type 5. *Journal of General Virology*, 36, 1, 59–74.
14. Hayflick, L, Moorhead, PS. (1961). The serial cultivation of human diploid cell strains. *Experimental Cell Research*, 25, 585–621.
15. Cohen, SN, Chang, AC, Boyer, HW, Helling, RB. (1973). Construction of biologically functional bacterial plasmids *in vitro*. *Proceedings of the National Academy of Sciences of the United States of America*, 70, 3240–3244.
16. Graham, FL, van der Eb, AJ. (1973). A new technique for the assay of infectivity of human adenovirus 5 DNA. *Virology*, 52, 456–467.
17. Svoboda, J, Hlozánek, I, Mach, O, Michlová, A, Rïman, J, Urbánková, M. (1973). Transfection of chicken fibroblasts with single exposure to DNA from virogenic mammalian cells. *Journal of General Virology*, 21, 47–55.
18. Al-Moslih, MI, Dubes, GR. (1973). The kinetics of DEAE-dextran-induced cell sensitization to transfection. *Journal of General Virology*, 18, 189–193.
19. McCormick, F, Trahey, M, Innis, M, Dieckmann, B, Ringold, G. (1984). Inducible expression of amplified human beta interferon genes in CHO cells. *Molecular and Cellular Biology*, 4, 166–172.
20. Kaufman, RJ, Wasley, LC, Spiliotes, AJ, Gossels, SD, Latt, SA, Larsen, GR, Kay, RM. (1985). Coamplification and coexpression of human tissue type plasminogene activator and murine dihydrofolate reductase sequences in Chinese hamster ovary cells. *Molecular and Cellular Biology*, 5, 1750–1759.
21. Birch, J, Arathoon, R. (1990). Suspension culture of mammalian cells. *Bioprocess Technology*, 10, 251–270.
22. Corning Cell Culture selection guide. Available at http://www.sigmaaldrich.com/content/dam/sigmaaldrich/docs/Sigma/General_Information/corning_cc_guide.pdf (accessed April 30, 2016).
23. Wurm, FM. (2004). Production of recombinant protein therapeutics in cultivated mammalian cells. *Nature Biotechnology*, 22, 1393–1398.
24. De Jesus, M, Wurm, FM. (2011). Manufacturing recombinant proteins in kg-ton quantities using animal cells in bioreactors. *European Journal of Pharmaceutics and Biopharmaceutics*, 78, 2, 184–188.
25. Jordan, M, Perilleux, A, Marsaut, M, Stettler, M. (2012). High throughput fed-batch screening for improved cell line selection. Conference Presentation: IBC's 8th Annual Cell Line Development and Engineering, San Francisco, CA, June 7, 2012.

26. Zagari, F, Jordan, M, Stettler, M, Broly, H, Wurm, FM. (2013). Lactate metabolism shift in CHO cell culture: the role of mitochondrial oxidative activity. *New Biotechnology*, 30, 238–245.
27. Zagari, F, Stettler, M, Baldi, L, Broly, H, Wurm, FM, Jordan, M. (2013). High expression of the aspartate–glutamate carrier Aralar1 favors lactate consumption in CHO cell culture. *Pharmaceutical Bioprocessing*, 1, 1, 19–27.
28. Eigen, M, Schuster, P. (1977). The hypercycle. A principle of natural self-organization. Part A: emergence of the hypercycle. *Naturwissenschaften*, 64, 541–565.
29. Eigen, M, Schuster, P. (1978). The hypercycle. A principle of natural self-organisation. Part B: the abstract hypercycle. *Naturwissenschaften*, 65, 7–41.
30. Eigen, M, Schuster, P. (1978). The hypercycle. A principle of natural self-organisation. Part C: the realistic hypercycle. *Naturwissenschaften*, 65, 341–369.
31. Wurm, FM. (2013). CHO Quasispecies—implications for manufacturing processes. *Processes*, 1, 3, 296–311.
32. Xu, X, Nagarajan, H, Lewis, NE, Pan, S, Cai, Z, Liu, X, Chen, W, Xie, M, Wang, W, Hammond, S, Andersen, MR, Neff, N, Passarelli, B, Koh, W, Fan, HC, Wang, J, Gui, Y, Lee, KH, Betenbaugh, MJ, Quake, SR, Famili, I, Palsson, BO, Wang, J. (2011). The genomic sequence of the Chinese hamster ovary (CHO)-K1 cell line. *Nature Biotechnology*, 29, 735–741.
33. Wurm, FM, Hacker, DL. (2011). First CHO genome. *Nature Biotechnology*, 29, 9, 718–720.
34. Omasa, T, Lee, K. (2014). Chromosome rearrangements and gene amplification. In: Hauser, H, Wagner, R. Animal Cell Biotechnology in Biologics Production, DeGruyter, Berlin, pp. 127–143.
35. Gottschalk, U. (2011). The future of downstream processing. *Biopharm International*, 24, 9, 1–6 (August 2, Biopharm International Supplements).
36. Puck, TT. (1985). Development of the Chinese hamster ovary (CHO) cell for use in somatic cell genetics. In: Gottesman, MM. Molecular Cell Genetics, John Wiley & Sons, Inc., New York, pp. 37–64.
37. Urlaub, G, Chasin, LA. (1980). Isolation of Chinese hamster cell mutants deficient in dihydrofolate reductase activity. *Proceeding of the National Academy of Sciences of the United States of America*, 77, 4216–4220.
38. Urlaub, G, Käs, E, Carothers, AM, Chasin, LA. (1983). Deletion of the diploid dihydrofolate reductase locus from cultured mammalian cells. *Cell*, 33, 405–412.
39. Wurm, FM. (1990). Integration, amplification and stability of plasmid sequences in CHO cell culture. *Biologicals*, 18, 159–164.
40. Bebbington, CR, Renner, G, Thomson, S, King, D, Abrams, D, Yarranton, GT. (1992). High-level expression of a recombinant antibody from myeloma cells

using a glutamine synthetase gene as an amplifiable selectable marker. *Biotechnology*, 10, 169–175.

41. Barnes, LM, Bentley, CM, Dickson, AJ. (2000). Advances in animal cell recombinant protein production: GS-NS0 expression system. *Cytotechnology*, 32, 109–123.
42. Nunberg, JH, Kaufman, RJ, Schimke, RT, Urlaub, G, Chasin, LA. (1978). Amplified dihydrofolate reductase genes are localized to a homogeneously staining region of a single chromosome in a methotrexate-resistant Chinese hamster ovary cell line. *Proceedings of the National Academy of Sciences of the United States of America*, 75, 5553–5556.
43. Miescher, S, Zahn-Zabal, M, De Jesus, M, Moudry, R, Fisch, I, Vogel, M, Kobr, M, Imboden, MA, Kragten, E, Bichler, J, Mermod, N, Stadler, BM, Amstutz, H, Wurm, F. (2000). CHO expression of a novel human recombinant IgG1 anti-RhD antibody isolated by phage display. *British Journal of Haematology*, 111, 157–166.
44. Pallavicini, MG, DeTeresa, PS, Rosette, C, Gray, JW, Wurm, FM. (1990). Effects of methotrexate on transfected DNA stability in mammalian cells. *Journal of Molecular Biology*, 10, 1, 401–404.
45. Browne, SM, Al-Rubeai, M. (2007). Selection methods for high-producing mammalian cell lines. *Trends in Biotechnology*, 25, 425–432.
46. Meng, YG, Liang, J, Wong, WL, Chisholm, V. (2000). Green fluorescent protein as a second selectable marker for selection of high producing clones from transfected CHO cells. *Gene*, 242, 201–207.
47. Mancia, F, Patel, SD, Rajala, MW, Scherer, PE, Nemes, A, Schieren, I, Hendrickson, WA, Shapiro, L. (2004). Optimization of protein production in mammalian cells with a co-expressed fluorescent marker. *Structure*, 12, 1355–1360.
48. Brezinsky, SC, Chiang, GG, Szilvasi, A, Mohan, S, Shapiro, RI, MacLean, A, Sisk, W, Thill, G. (2003). A simple method for enriching populations of transfected CHO cells for cells of higher specific productivity. *Journal of Immunological Methods*, 277, 141–155.
49. Yoshikawa, T, Makanishi, F, Ogura, Y, Oi, D, Omasa, T, Katakura, Y, Kishimota, M, Suga, KI. (2001). Flow cytometry: an improved method for the selection of highly productive gene-amplified CHO cells using flow cytometry. *Biotechnology and Bioengineering*, 74, 435–442.
50. Chenuet, S, Martinet, D, Besuchet-Schmutz, N, Wicht, M, Jaccard, N, Bon, AC, Derouazi, M, Hacker, DL, Beckmann, JS, Wurm, FM. (2008). Calcium phosphate transfection generates mammalian recombinant cell lines with higher specific productivity than polyfection. *Biotechnology and Bioengineering*, 101, 937–945.
51. Lamond, AI, Earnshaw, WC. (1998). Structure and function in the nucleus. *Science*, 280, 547–553.

52. Jenuwein, T, Allis, CD. (2001). Translating the histone code. *Science*, 293, 1074–1080.
53. Chusainow, J, Yang, YS, Yeo, JHM, Toh, PC, Asvadi P, Wong NSC, Yap, MGS. (2009). A study of monoclonal antibody-producing CHO cell lines: what makes a stable high producer? *Biotechnology and Bioengineering*, 102, 1182–1196.
54. Stief, A, Winter, DM, Strätling, WH, Sippel, AE. (1989). A nuclear DNA attachment element mediates elevated and position-independent gene activity. *Nature*, 341, 343–345.
55. Zahn-Zabal, M, Kobr, M, Girod, PA, Imhof, M, Chatellard, P, De Jesus, M, Wurm, FM, Mermod, N. (2001). Development of stable cell lines for production or regulated expression using matrix attachment regions. *Journal of Biotechnology*, 87, 29–42.
56. Kwaks, TH, Barnett, P, Hemrika, W, Siersma, T, Sewalt, RG, Satijn, DP, Brons, JF, van Blokland, R, Kwakman, P, Kruckeberg, AL, Kelder, A, Otte, AP. (2003). Identification of anti-repressor elements that confer high and stable protein production in mammalian cells. *Nature Biotechnology*, 21, 553–558.
57. Williams, S, Mustoe, T, Mulcahy, T, Griffiths, M, Simpson, D, Antoniou, M, Irvine, A, Mountain, A, Crombie, R. (2005). CpG-island fragments from the HNRPA2B1/CBX3 genomic locus reduce silencing and enhance transgene expression from the hCMV promoter/enhancer in mammalian cells. *BMC Biotechnology*, 5, 17, doi:10.1186/1472-6750-5-17.
58. Gaillet, B, Gilbert, R, Brossau, S, Pilotte, A, Malenfant, F, Mullick, A, Garnier, A, Massie B. (2010). High-level recombinant protein production using lentiviral vectors and the cumate gene-switch. *Biotechnology and Bioengineering*, 106, 2, 203–215.
59. Oberbek, A, Matasci, M, Hacker, DL, Wurm FM. (2011). Generation of stable, high-producing CHO cell lines by lentiviral vector-mediated gene transfer in serum-free suspension culture. *Biotechnology and Bioengineering*, 108, 3, 600–609.
60. Fraser, MJ, Cary, L, Boonvisudhi, K, Wang, HG. (1995). Assay for movement of Lepidopteran transposon IFP2 in insect cells using a baculovirus genome as a target DNA. *Virology*, 211, 2, 397–407.
61. Izsvak, Z, Fröhlich, J, Grabundzija, I, Shirley, JR, Powell, HM, Chapman, KM, Ivics, Z, Hamra, FK. (2010). Generating knockout rats by transposon mutagenesis in spermatogonial stem cells. *Nature Methods*, 7, 443–445.
62. Li, X, Burnight, ER, Cooney, AL, Malani, N, Brady, T, Sander, JD, Staber, J, Wheelan, SJ, Joung, JK, McCray, PB, Bushman, FD, Sinn, PL, Craig, NL. (2013). PiggyBac transposase tools for genome editing. *Proceedings of the National Academy of Sciences of the United States of America*, 110, 25, E2279–E2287.
63. Matasci, M, Baldi, L, Hacker, DL, Wurm, FM. (2011). The PiggyBac transposon enhances the frequency of CHO stable cell line generation and

yields recombinant lines with superior productivity and stability. *Biotechnology and Bioengineering*, 108, 9, 2141–2150.

64. Li, Z, Michael, IP, Zhou, D, Nagy, A, Rini, JM. (2013). Simple piggyBac transposon-based mammalian cell expression system for inducible protein production. *Proceedings of the National Academy of Sciences of the United States of America*, 110, 13, 5004–5009.

65. Balasubramaniam, S, Matasci, M, Kadlecova, Z, Baldi, L, Hacker, DL, Wurm, FM. (2015) Rapid recombinant protein production from transposon-mediated stable CHO cell pools. *Journal of Biotechnology*, 200, 61–69.

66. Sanders, JD, Joung, JK. (2014) CRISPR Cas systems for editing, regulating and targeting genomes. *Nature Biotechnology*, 32, 347–355.

67. Nienow, AW. (2006). Reactor engineering in large-scale animal cell culture. *Cytotechnology*, 50, 9–33.

68. De Jesus, MJ, Girard, P, Bourgeois, M, Baumgartner, G, Jacko, B, Amstutz, H, Wurm, FM. (2004). TubeSpin satellites: a fast track approach for process development with animal cells using shaking technology. *Biochemical Engineering Journal*, 17, 217–223.

69. Stettler, M, De Jesus, M, Ouertatani-Sakouhi, H, Engelhardt, EM, Muller, N, Chenuet, S, Bertschinger, M, Baldi, L, Hacker, DL, Jordan, M, Wurm, FM. (2007). 1000 non-instrumented bioreactors in a week. In: Smith R. Cell Technology for Cell Products, Springer, The Netherlands, Chapter VI, pp. 489–495.

70. Zhang, XW, Stettler, M, Reif, O, Kocourek, A, De Jesus, M, Hacker, DL, Wurm, FM. (2008). Shaken helical track bioreactors: providing oxygen to high-density cultures of mammalian cells at volumes up to 1000 L by surface aeration with air. *Nature Biotechnology*, 25, 68–75.

71. Xie, K, Zhang, XW, Li H, Wang, YT, Lei, Y, Rong, J, Qian, CW, Xie, QL, Wang, YF, Hong, A, Xiong, S. (2011). On-line monitoring of oxygen in Tubespin, a novel, small-scale disposable bioreactor. *Cytotechnology*, 63, 345–350.

72. Deshpande, RR, Heinzle, E. (2004). On-line oxygen uptake rate and culture viability measurement of animal cell culture using microplates with integrated oxygen sensors. *Biotechnology Letters*, 26, 763–767.

73. Duetz, WA, Witholt, B. (2004). Oxygen transfer by orbital shaking of square vessels and deepwell microtiter plates of various dimensions. *Biochemical Engineering Journal*, 17, 181–185.

74. Chen, A, Chitta, R, Chang, D, Ashraf, A. (2009). Twenty-four well plate miniature bioreactor system as a scale-down model for cell culture process development. *Biotechnology and Bioengineering*, 102, 148–160.

75. Singh, V. (1999). Disposable bioreactor for cell culture using wave-induced agitation. *Cytotechnology*, 30, 149–158.

76. Oosterhuis, NMG, Van Den Berg, HJ. (2011). How multipurpose is a disposable bioreactor? *International BioPharm*, 24, 3, 51–56.

77. Liu, C, Hong, L. (2001). Development of a shaking bioreactor system for animal cell cultures. *Biochemichal Engineering Journal*, 7, 121–125.
78. Stettler, M, Zhang, XW, Hacker, DL, De Jesus, MJ, Wurm, FM. (2007). Novel orbital shake bioreactors for transient production of CHO derived IgGs. *Biotechnology Progress*, 23, 1340–1346.
79. Muller, N, Girard, P, Hacker, DL, Jordan, M, Wurm, FM. (2005). Orbital shaker technology for the cultivation of mammalian cells in suspension. *Biotechnology and Bioengineering*, 89, 400–406.
80. Tissot, S, Farhat, M, Hacker, DL, Anderlei, T, Kühner, M, Comninellis, C, Wurm, FM. (2010). Determination of a scale-up factor from mixing time studies in orbitally shaken bioreactors. *Biochemical Engineering Journal*, 52, 2–3, 181–189.
81. Anderlei, T, Cesana, C, Bürki, C, De Jesus, M, Kühner, M, Wurm, FM, Lohser, R. (2009). Shaken bioreactors provide culture alternative. *Genetic Engineering and Biotechnology News*, 29, 19.
82. Monteil, D. (2014). The engineering of a scalable disposable orbitally shaken bioreactor system for mammalian cell cultivation and therapeutic protein production. PhD thesis. Ecole Polytechnique Federale de Lausanne, Lausanne, Switzerland.
83. Reclari, M Dreyer, M Tissot, S, Obreschkow, D, Wurm FM, Farhat M. (2014). Surface wave dynamics in orbital shaken cylindrical containers. *Physics of Fluids*, 26, 052104.
84. Pham, PL, Kamen, A, Durocher, Y. (2006). Large scale transfection of mammalian cells for the production of recombinant protein. *Molecular Biotechnology*, 34, 225–237.
85. Baldi, L, Hacker, DL, Adam, M, Wurm, FM. (2007). Recombinant protein production by large-scale transient gene expression in mammalian cells: state of the art and future perspectives. *Biotechnology Letters*, 29, 677–684.
86. Meissner, P, Pick, H, Kulangara, A, Chatellard, P, Friedrich, K, Wurm, FM. (2001). Transient gene expression: recombinant protein production with suspension-adapted HEK293–EBNA cells. *Biotechnology and Bioengineering*, 75, 197–203.
87. Durocher, Y, Perret, S, Kamen, A. (2002). High-level and high-throughput recombinant protein production by transient transfection of suspension-growing human 293-EBNA1 cells. *Nucleic Acids Research*, 30, 2, E9.
88. Derouazi, M, Girard, P, Van Tilborgh, F, Iglesias, K, Muller, N, Bertschinger, M, Wurm, FM. (2004). Serum-free large-scale transient transfection of CHO cells. *Biotechnology and Bioengineering*, 87, 537–545.
89. Tait, AS, Brown, CJ, Galbraith, DJ, Hines, MJ, Hoare, M, Birch, JR, James, DC. (2004). Transient production of recombinant proteins by Chinese hamster ovary cells using polyethylenimine/DNA complexes in combination with microtubule disrupting anti-mitotic agents. *Biotechnology and Bioengineering*, 88, 707–721.

90. Geisse, S, Jordan, M, Wurm, FM. (2005). Large-scale transient expression of therapeutic proteins in mammalian cells. *Methods in Molecular Biology*, 308, 87–98.

91. Backliwal, G, Hildinger, M, Chenuet, S, Wulhfard, S, De Jesus, M, Wurm, FM. (2008). Rational vector design and multi-pathway modulation of HEK 293E cells yield recombinant antibody titers exceeding 1 g/l by transient transfection under serum-free conditions. *Nucleic Acids Research*, 36, 15, e96.

92. Rajendra Y, Kiseljak D, Baldi L, Hacker DL, Wurm FM. (2011). A simple high-yielding process for transient gene expression in CHO cells. *Journal of Biotechnology*, 153, 22–26.

93. Daramola, O, Stevenson, J, Dean, G, Hatton, D, Pettman, G, Holmes, W, Field, R. (2014). A high-yielding CHO transient system: coexpression of genes encoding EBNA-1 and GS enhances transient protein expression. *Biotechnology Progress*, 30, 132–141.

94. Girard, P, Derouazi, M, Baumgartner, G, Bourgeois, M, Jordan, M, Wurm, FM. (2002). 100-liter transient transfection. *Cytotechnology*, 38, 15–21.

95. Muller, N, Derouazi, M, Van Tilborgh, F, Wulhfard, S, Hacker, DL, Jordan, M, Wurm, FM. (2007). Scalable transient gene expression in Chinese hamster ovary cells in instrumented and non-instrumented cultivation systems. *Biotechnology Letters*, 29, 703–711.

96. Ye, J, Alvin, K, Latif, H, Hsu, A, Parikh, V, Whitmer, T, Tellers, M, de la Cruz Edmonds, MC, Ly, J, Salmon, P, Markusen, JF. (2010). Rapid protein production using CHO stable transfection pools. *Biotechnology Progress*, 26, 5, 1431–1437.

97. Gottschalk, U. (2011). The future of downstream processing. *Pharmaceutical Technology*, 35, 36–41.

12

Process Development

Samuel D. Stimple and David W. Wood

Department of Chemical and Biomolecular Engineering, The Ohio State University, Columbus, OH, USA

12.1 Summary

This chapter summarizes several advances in the downstream processing of antibodies at industrial scale, with an emphasis on alternatives to conventional protein A affinity batch processing. At the molecular level, these alternatives include the use of newly developed affinity ligands, as well as reemergence of conventional chromatographic processes. In addition, several novel process configurations are described and evaluated, including the use of disposable technologies and the rapidly increasing focus on continuous processing. The general trend of these advances is toward a highly flexible manufacturing environment, which will provide rapid process development as well as production on demand of antibody therapeutics in multiproduct facilities. Each of these developments is described and examined, with the relative strengths, weaknesses, and potential impacts evaluated.

12.2 Introduction

This chapter will discuss several critical advances in unit operations for the production of monoclonal antibodies (MAbs) reported within the last 20 years, with a focus on novel quasi-continuous downstream purification schemes/platforms being explored in academic and industrial settings. The primary drivers for these advances are based on five central ideals: (i) decreased production costs, (ii) equivalent or (ideally) improved product quality/uniformity,

Biosimilars of Monoclonal Antibodies: A Practical Guide to Manufacturing, Preclinical, and Clinical Development, First Edition. Edited by Cheng Liu and K. John Morrow, Jr.

(iii) increased flexibility of processing equipment and facilities, (iv) elimination of unnecessary process/hold steps, and (v) reduction of plant facilities footprint.

Conventional downstream technologies that operate as batch processes will be examined briefly, with a focus on key limitations in satisfying these ideals. Further, recent paradigm-shifting advances, such as single-use technologies and continuous cell culture methods that have made these improvements possible, will also be highlighted.

12.3 Protein A and Protein G Batch Affinity Chromatography

Protein A is an immunoglobulin-binding cell wall-associated staphylococcal protein that has high affinity for the Fc domain of human and other mammalian IgGs [1]. This useful characteristic led to its development into one of the first protein purification platforms that relied on a native protein–protein affinity interaction [2–4]. Protein A, and derivatives thereof, has become widely used as a binding platform for the capture of essentially any commercially relevant antibody of the IgG_1, IgG_2, or IgG_4 subclasses in the context of MAb purification [4]. An analogous affinity platform for MAb purification was developed using another immunoglobulin-binding bacterial protein: streptococcal protein G [5]. A schematic of a typical batch protein A/G purification is shown in Figure 12.1.

Clarified cell culture fluid containing the MAb of interest is loaded onto a protein A column at relatively neutral conditions (pH 6–8) [6]. For a purification performed in batch mode, the mobile-phase flow rate is adjusted such that its column residence time is generally less than 10 min, where longer residence times allow the use of smaller equipment and provide better resin utilization at the price of longer processing times. Protein A binds to the Fc domain of the MAb and immobilizes it onto the solid support while contaminants (host cell protein (HCPs), DNA, viruses, etc.) flow through the column to waste. Upon loading all of the clarified cell culture fluid, the column is washed, and the substantially purified product is then eluted under acidic conditions (pH 2.5–4). A low pH elution not only releases the MAb from the column but also effectively inactivates residual enveloped viruses [7]. The eluted MAb product is then further purified in downstream polishing steps, often consisting of subsequent cation- and anion-exchange chromatography steps. The universal effectiveness of protein A and protein G affinity methods has made them ubiquitous in both research and manufacturing of therapeutic MAbs throughout the world.

An important issue associated with protein A and protein G platforms is the high cost of the resins, which has necessitated their recycling for up to several hundred manufacturing cycles. For this reason, the stability of the ligand

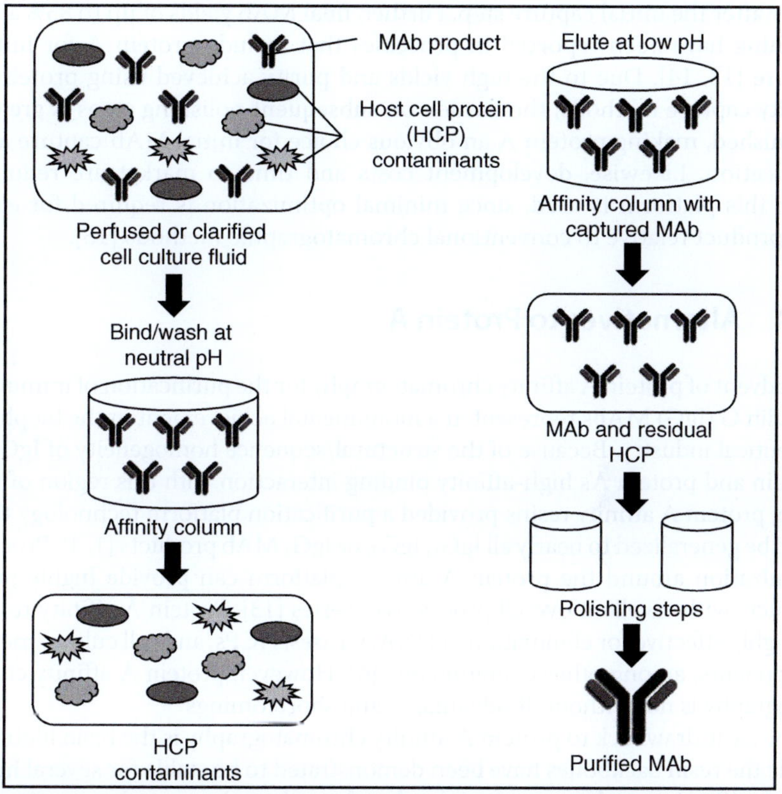

Figure 12.1 Schematic of a protein A/G affinity chromatography purification. Clarified cell culture fluid is loaded onto a column containing protein A resin at a neutral pH (6–8). Protein A binds and captures the MAb, while HCP and other contaminants flow through the column. The column is washed, and relatively pure MAb is eluted from the column under low pH (2.5–4) conditions. Residual contaminants are removed in downstream polishing steps to yield a product of the required purity.

during the binding, washing, elution, and regeneration steps of these cycles is a major concern. Many industrial regeneration steps are performed under highly alkaline conditions (0.1–1 M NaOH), which are generally detrimental to the solid supports used for bioseparations. More importantly, the protein A and protein G capture ligands are sensitive to these conditions as well [8–10], which has led to the development of protein A and G variants with improved alkali resistance and elution behaviors [10–12]. Protein A variants have been particularly marked by enhanced stability as compared to protein G under manufacturing conditions, making protein A the most prominently used affinity capture platform in industry. Protein A affinity chromatography is highly selective and effective at capturing MAbs and typically results in ≥95% product purity after the initial capture step. Further, final MAb yields of up to 85% after polishing have been reported in processes that include protein A for initial capture [13, 14]. Due to the high yields and purity achieved using protein A affinity capture methods, the burden on subsequent polishing steps is greatly diminished, making protein A an obvious choice for initial MAb capture and purification. Likewise, development costs and time to market are reduced when this platform is used, since minimal optimization is required for each new product relative to conventional chromatographic methods [15].

12.4 Alternatives to Protein A

The advent of protein A affinity chromatography for the purification of immunoglobulin G (IgG) MAbs represented a monumental achievement in the biopharmaceutical industry. Because of the structural/sequence homogeneity of IgG Fc domain and protein A's high-affinity binding interaction with this region of the MAb, protein A affinity resins provided a purification platform technology that could be generalized to nearly all IgG_1, IgG_2, or IgG_4 MAb products [1, 4]. Process optimization around the protein A affinity platform can provide highly pure products with excellent overall process recoveries [13]. Protein A affinity resins are highly effective for elimination of DNA, viruses, HCPs, and cell culture media components, among other contaminants [6]. However, protein A affinity chromatography is not without disadvantages and shortcomings.

One clear drawback to protein A affinity chromatography is the resin lifetime. While the resin backbones have been demonstrated to be stable for several hundred purification cycles, protein A leeches from the column under a variety of relevant conditions. In the absence of MAb, protein A leeches from the column under alkaline conditions, requiring the use of expensive chaotropes for column regeneration [16]. More importantly, protein A is also prone to proteolysis and leeches at pH 4 when MAb is present, causing product contamination due to co-elution [17]. This not only decreases the usable life of the protein A resin but also leads to product contamination that must be addressed in downstream polishing steps. Additionally, protein A affinity resins are more than an order of magnitude more costly to purchase than other chromatography resins. Thus, the high cost associated with the development and operation of the protein A affinity platform remains an ongoing concern [16]. As a result, significant effort has been invested in developing alternatives to conventional protein A-based methods. Several of these alternatives will be discussed in the following section.

12.4.1 Small Molecule- and Peptide-Based Affinity Chromatography

While protein A represents the MAb purification platform of choice in the biopharmaceutical industry, alternative affinity platforms based on small-molecule ligands have also been developed and demonstrated to function well with model MAbs. One notable example relies on the multivalent interaction of conserved nucleotide binding sites (NBSs) of the target antibody with the inexpensive small molecule, indole-3-butyric acid (IBA), covalently linked to a resin [18]. The antibody NBS is a pocket on the Fab variable domains of all antibody isotypes that are highly conserved across various species, and each is capable of binding one IBA molecule [19]. Although the monovalent affinity of antibody to IBA is on the order of micromolar, which is not generally sufficient for an effective affinity scheme, each Ig contains multiple NBSs capable of binding IBA. The multivalent nature of the interaction effectively increases the affinity of the antibody for the ligand and provides a good basis for separation from HCPs. In a preliminary study with Rituximab® and Trastuzumab® mixed with mouse ascites fluid, a prototype resin was demonstrated to provide >98% purity with >95% recovery of antibody loaded onto the column, where elution of the antibody was achieved under mild conditions using a sodium chloride gradient at pH 7.0 [18]. These conditions are far less problematic than the acidic pH used for protein A affinity elution, which can cause significant levels of product aggregation [20]. Further, performance of the IBA column remained unchanged after 100 column operations, suggesting that the IBA-linked resin is robust and stable. Since this process was developed recently, however, it has not been validated at large scale and has only been demonstrated with two model MAbs. Nonetheless, because the NBS is conserved among species and antibody isotypes, IBA affinity methods potentially offer a broadly generalizable platform for MAb purification.

There has also been significant effort invested in producing protein A mimetic (PAM) ligands that are less expensive and less prone to degradation under process conditions. Several PAM molecules have been engineered that function well as ligands for affinity purification of MAbs. Some of the notable molecules that have been developed are PAM peptide and artificial protein A (ApA) [21, 22]. PAM is a synthetic tetramer of tripeptides that contains a tetradentate lysine core. It can be easily immobilized on a solid support, and, unlike protein A, it has been demonstrated to be unaffected by denaturants, detergents, and other sanitation reagents [23]. Additionally, the peptide is nonimmunogenic and nontoxic and binds antibody isotypes not recognized by protein A. Thus, PAM offers many advantages compared to protein A, although its affinity for IgG is significantly lower than that of protein A [23].

ApA is a nonpeptidic/nonproteinaceous ligand that mimics the Phe132–Tyr133 dipeptide that has been shown to be responsible for binding of IgGs by protein A [22]. Crucially, the performance of ApA resins has been shown to remain unchanged after 8 days of treatment with 1 M NaOH [23]. Like PAM, however, ApA has lower affinity for IgGs than does protein A, although ApA resins have been used to purify IgG from ascitic fluid and human plasma with purities of 90–99% and yields of 60–70% [23, 24]. Despite these promising demonstrations, however, these and other such mimetic ligands are far less specific for IgGs and generally achieve substantially lower purification factors for IgGs than protein A [25]. Thus, the lower costs and improved resin durability come at the cost of reduced performance; at the present time this has effectively prevented the adoption of these methods as generally accepted alternate platforms for protein A.

12.4.2 Processes Based on Conventional Chromatography

Protein A affinity capture for the purification of MAbs is a highly utilized and very successful technology. Because greater than 99.5% of feed stream impurities are typically removed in a single step, protein A capture has become a fairly standard unit operation in MAb purification [13, 14]. All of these advantages come at a steep price, however, since protein A resins typically cost more than 30 times as much as other nonaffinity (e.g., ion exchange) resins [13]. Because traditional batch purification schemes make use of significantly oversized columns, protein A unit operations can account for a sizable fraction (often >35%) of the total recovery raw material costs for purifications at industrial scale [13]. Additionally, protein A that leeches from the column causes an immunogenic response in humans if not adequately removed in downstream polishing steps [26].

A generally accepted industrial purification strategy for MAb production typically involves the use of a three-column process: protein A affinity capture followed by cation- and anion-exchange chromatography [27, 28]. Due to the high associated costs of protein A, researchers have sought to develop an analogous three-step purification train with equivalent yields and product purity that do not rely on protein A affinity capture. In one key study, eight modes of chromatography were arranged in all possible combinations of three column steps, and the performance of each resulting process was compared to that of the control (protein A affinity, cation exchange, and anion exchange). Three purification schemes were identified that offer equivalent purity and similar or improved overall product yield for a model humanized IgG_1 MAb [13]. These schemes include (i) cation-exchange chromatography followed by hydroxyapatite and anion-exchange chromatography, (ii) cation-exchange chromatography followed by anion-exchange and hydrophobic interaction chromatography, and (iii) cation-exchange chromatography followed by anion-exchange and hydroxyapatite chromatography. Remarkably, the purities and yields for each of these processes were comparable to those achieved using protein A as a capture step, suggesting that there may be viable alternatives to the widely accepted protein A platform methods. An important consideration, however, is that each step of these alternate processes would likely have to be developed, optimized, and validated for each new MAb product. This required investment of time and resources would likely make this option impractical, particularly since the ability to reuse protein A resin several hundred times significantly decreases its cost.

12.5 Disposables and Continuous Downstream Processing

The first licensed MAb was produced in 1986, and over the past 30 years, production technologies and capacities for therapeutic MAbs have improved enormously [29]. Some of the first murine MAbs, such as OKT3, were produced in ascites tumors in mice [30]. However, recombinant cloning and expression technologies have allowed chimeric, humanized, or fully human MAb production from recombinant CHO cells in large-scale bioreactors. Because early MAb processes using recombinant CHO cell culture were marked by low antibody titers, many companies producing MAbs were required to invest in massive production trains containing multiple bioreactors of 10,000l or larger in order to meet market demands [31, 32]. These large-scale plants relied on batch purification processes, which became the norm for MAb production; thus, process development, procedural validation, and process optimization were routinely oriented toward batch production at this scale. As cell culture methodologies matured, bioreactor antibody titers have improved by approximately 100-fold [33]. These technology improvements have produced a situation where the existing production infrastructure greatly exceeds market demand, and downstream purification processes have become the greatest source of bottlenecks in MAb production.

Over the last 20 years, MAbs have become a staple of the biopharmaceutical industry. MAbs represent a substantial fraction of biologics in development pipelines, with 40 novel MAbs currently involved in phase 3 studies as of October 2014 [34]. Importantly, all protein therapeutics in development or on the market are highly regulated and must meet or exceed strict quality requirements. The processes used to manufacture these products necessarily satisfy these requirements, and consequently, these established protocols carry a large amount of momentum due to their credibility with regulatory agencies. As a result, companies have remained hesitant to abandon the large batch production procedures upon which their businesses have been built, which hinders advancement in the field of bioprocess development [35]. Although MAb production is relatively inexpensive when compared to sales prices, cost reduction will become an ever more important process constraint as biosimilars and generic biologics become more prevalent in the market [31]. Furthermore, additional protein therapeutics derived from many different protein classes, with highly variable market sizes, will increasingly demand the adoption of multiproduct manufacturing facilities. These products will also demand the development and utilization of flexible processes and process equipment, as is already becoming increasingly important.

To address the earlier concerns, many companies are attempting to validate and implement continuous and (quasi-)steady-state processing for the production and purification of MAbs. Continuous processing provides many advantages when compared with batch processes for the same level of MAb production. Namely, capital costs are greatly reduced for continuous processes due to the use of smaller equipment, and continuous processes are often marked by increased protein yields and higher process productivity. Importantly, the implementation of continuous bioprocessing has become a much more realizable goal with the coming of age of continuous perfusion cell culture equipment, disposable bag bioreactors, and single-use purification equipment. All of these innovations offer significant gains in process and equipment flexibility to accommodate both the harmonization of practices within a company and production of protein therapeutics at various scales to accommodate different market sizes or changes in demand. Moreover, processes utilizing disposables are more practical and less costly from a labor perspective, considering that they do not need to be cleaned or sterilized during changeover and downtime; it is also noteworthy that the use of disposables circumvents the need for exhaustive validation of such procedures [36]. Disposables are far less prone to failures in sterilization protocol because of their single-use nature and have been demonstrated to have minimal or no detrimental effects on the productivity of cell culture processes, the titers of antibody at harvest, or antibody quality and purity [37]. Studies have determined that the specific productivity of cells cultured in disposable bag bioreactors may even be higher than that of cells grown in stainless steel bioreactors [37].

Until recently, continuous processing efforts have focused almost entirely on upstream applications such as cell culture in continuous perfusion bioreactors. Yet, as has been suggested in the literature, continuous processing can only reach its full potential upon successful integration of continuous upstream and downstream unit operations [31, 38, 39]. A principal obstacle to fully continuous bioprocessing has been the antibody capture/purification step. Traditional capture occurs in batch mode, using oversized protein A columns and long residence times in order to minimize or eliminate breakthrough, which is defined as the point at which the target protein (or a specific percentage of the loaded target protein) washes through the column without binding. In the absence of another downstream column, breakthrough contributes directly to product loss. Because breakthrough occurs long before column saturation (the point where the protein is bound to the column at or very near the static binding capacity), the protein A affinity columns that are used in batch purification processes are generally oversized and never reach full saturation. This contributes to an underutilization of resin, increased buffer requirements, and product dilution. Several capture and/or purification methodologies have been developed to address these limitations and are discussed in the succeeding text. Each of these allows for continuous downstream processing/purification, and each has its own advantages and disadvantages.

12.5.1 Staggered Cycling

Purification of MAbs can be conducted quasi-continuously by simply utilizing multiple columns concurrently or in parallel [40, 41]. Feed is loaded onto a column until it reaches the dynamic binding capacity, at which point breakthrough begins to occur. As the product breakthrough front emerges from the first column, the feed can be redirected to a fresh, second column. Then, while the second column is being loaded, the first column is sequentially washed, eluted, regenerated, and re-equilibrated before being reused for MAb capture. The second column can then be processed in the same fashion, while feed is loaded onto a third column or again onto column 1. The timing of column rotation is dependent on the specifics of the process and the time requirements for each of the steps, and thus the number of columns required for continuous processing of the feed depends entirely upon the relative durations of each of the steps in the purification scheme (i.e., binding, washing, elution, regeneration, and re-equilibration). For example, in a process where each step requires an equivalent amount of time, the number of columns should equal the number of steps in a single-column cycle (in the scenario earlier, five columns would be required) [42].

In the case of protein A affinity chromatography with a high-capacity resin, binding requires significantly more time than the subsequent steps [31]. As a result, in the amount of time required to simply load column 2, multiple subsequent steps may be completed for column 1. This allows the number of columns to be decreased, thus reducing costs. An important limitation of this approach is that the individual columns are only loaded with MAb up to the dynamic binding capacity (initiation of breakthrough) before washing and elution, which is significantly less than the static binding capacity (full saturation of the resin with MAb). Although this approach can lead to a significant decrease in resin requirements relative to a single-column batch process, the limitation of column loading to only its dynamic binding capacity leads to a suboptimal use of resin and associated buffers. This underutilization of separation media is the main shortcoming of staggered cycling purification schemes, and this shortcoming can be compounded by the greater number of regeneration cycles experienced by the resin over the lifetime of a given MAb product.

12.5.2 Multicolumn Periodic Countercurrent (PCC) Chromatography

Multicolumn periodic countercurrent (PCC) chromatography is a relatively simple variation of a staggered cycling purification process that also utilizes multiple purification columns in order to facilitate continuous antibody capture. Whereas the columns in a staggered cycling process are only connected (and loaded with product) in parallel, in PCC the columns undergo countercurrent loading and are connected concurrently. A PCC purification train is depicted schematically in Figure 12.2.

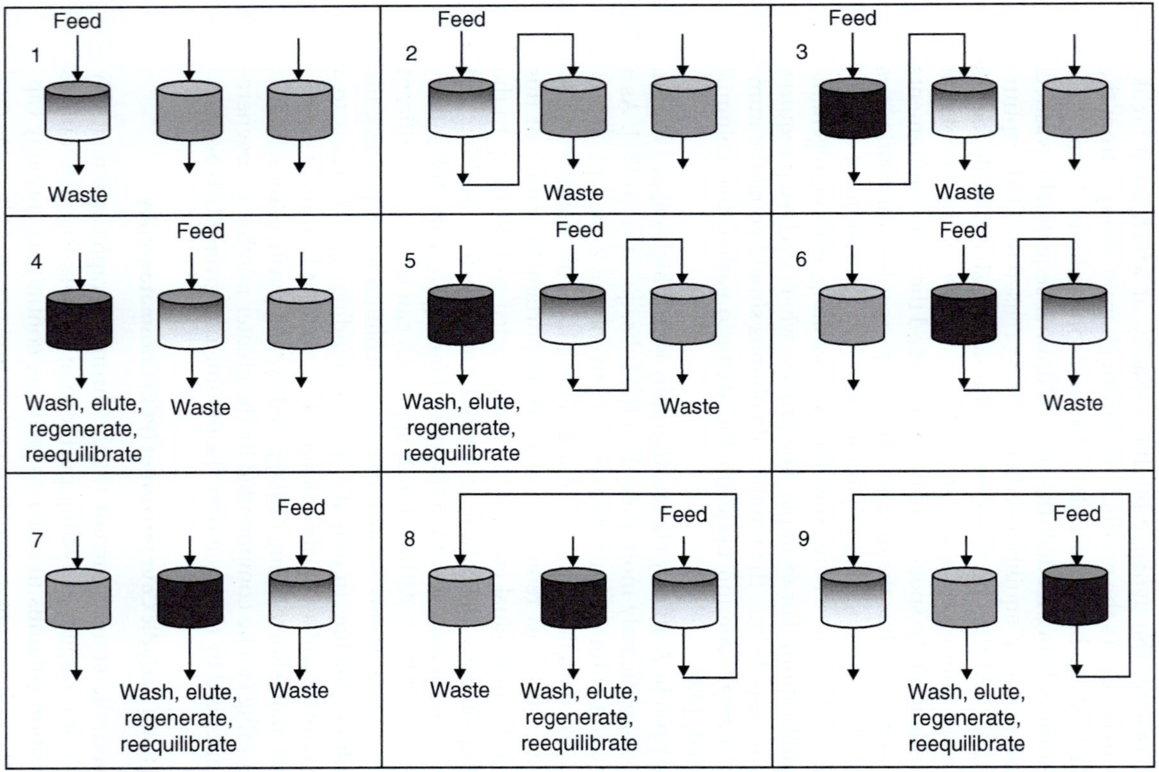

Figure 12.2 Schematic of PCC chromatography. Three (or more) columns are utilized in order to allow for continuous purification of the MAb from the feed. Cells 1 and 2: clarified cell culture fluid is loaded onto the first column with effluent directed to waste until breakthrough occurs. Effluent is redirected to column 2 to avoid product loss. Cells 3 and 4: column 1 is loaded to saturation, feed is loaded onto column 2 with effluent directed to waste, and column 1 is sequentially washed, eluted, regenerated, and re-equilibrated. Cells 5 and 6: at the point of product breakthrough, effluent from column 2 is loaded onto column 3 to capture unbound MAb. Column 2 is then loaded to saturation. Cells 7–9: feed is loaded onto column 3 with effluent directed to waste, and column 2 is processed. At the point of breakthrough from column 3, effluent is redirected to column 1 to capture unbound MAb. Column 3 is loaded to saturation and the process is repeated.

Multiple (at least two) columns are used, and a variety of chromatographic methods may be utilized, including protein A affinity and ion exchange. As noted in the preceding text for staggered cycling, the number of columns required will depend on the relative amount of time required for each of the purification steps. Perfused, clarified cell culture fluid containing the MAb is loaded onto column 1, with the flow-through (containing contaminants) initially directed to waste. As column 1 is loaded to the dynamic binding capacity, breakthrough begins to occur. At this point, the flow-through from column 1 is redirected from waste and loaded onto column 2 in order to capture the breakthrough antibody "missed" by the first column. The additional columns downstream from column 1 minimize product loss and allow column 1 to be loaded with MAb essentially up to its static binding capacity, rather than to the dynamic binding capacity normally used in batch or staggered cycling purification schemes with columns arranged in parallel [31]. In mathematical models and simulations of two-column PCC schemes, up to fourfold enhancements in ligand utilization were calculated compared to a single-column batch purification [43].

Because there are downstream columns to capture breakthrough that is not bound by the first column in PCC, the feed residence time can be reduced in comparison to a batch process [31, 38]. Additionally, in batch purifications the column is generally oversized in order to capture all antibody in the feed in a single step at the column's dynamic binding capacity, whereas columns in PCC trains are able to approach their static binding capacities and are processed many times [31, 38, 44]. This enables PCC trains to utilize substantially smaller columns and buffer volumes for a given production output [45, 46]. In recent proof-of-concept studies, continuous perfusion cell culture was integrated with a PCC protein A affinity purification scheme in order to yield a fully continuous process that operated successfully for more than 30 consecutive days [31, 38]. Despite the conceptual simplicity of this technology, the gains in volumetric productivity and efficiency, with the decreases in process footprint and material consumption, are profound. Not only did these studies report consistently high product quality over the full duration of the experiment, they were also able to quantify the differences between PCC and standard batch purification processes in terms of material consumption and column footprint. Column volume (total volume of the three PCC columns) decreased by over 25-fold for the PCC capture process, while buffer consumption was reduced by approximately 25%, with an accompanying 20–30% increase in resin capacity utilization [31, 38]. Importantly, this process integration resulted in the elimination of a cell culture fluid clarification unit operation and a large hold tank, offering capital investment savings as well [31]. In conjunction with continuous perfusion cell culture, the high productivity of PCC also allows the replacement of a 10,000 l bioreactor with one that is less than 500 l [31]. Other sources have reported reductions of 15–40% in terms of resin and buffer consumption after implementing PCC purification schemes [45, 46]. Therefore, the notable savings in capital investment and raw material consumption are complemented by a better and more efficient use of resin on a per volume basis.

In addition to the economic advantages, continuous processes are also much more suitable for production of labile/unstable proteins that are characterized by short half-lives in the presence of media and/or impurities. By decreasing the durations of hold stages, or by eliminating some of these stages entirely, product loss can be significantly reduced and productivity increased as compared to standard batch purification.

12.5.3 Simulated Moving Bed (SMB) Chromatography

Simulated moving bed (SMB) chromatography is a continuous separation scheme that approximates a countercurrent flow of solids (stationary phase) and fluids (mobile phase) by arranging multiple columns and valves in a specific manner [47]. In practice, complex valve cycling operations are used to simulate the stepwise movements of several columns in a direction opposite to the flow of the mobile phase. The countercurrent movement allows precise tuning of the desired product velocity in the mobile phase, making this method highly effective when used in combination with a number of conventional chromatography resins. This technology has been demonstrated to effectively separate binary and more complex mixtures of proteins, including MAbs contaminated with other proteins [48, 49]. SMB technology offers improved solvent/buffer consumption, higher achievable purity, high overall yield, and overall greater productivity than an analogous single-column batch process or PCC chromatography [42, 50].

A schematic of a four-zone SMB process with four columns, numbered 1 through 4, is depicted in Figure 12.3. A model feed containing a binary mixture of black and light gray (shown in dark gray) is introduced between the separation zones (zones II and III). The mobile phase is pumped clockwise for the example illustrated in Figure 12.3. The mixed feed flows into column 1 in zone III, where the less strongly adsorbed component (A, shown in black) travels quickly through the column, while the more strongly adsorbed product (B, shown in light gray) travels at lower velocity. The columns are then effectively rotated counterclockwise by stepping the input and product streams forward. The timing of the valve steps is calculated to approximate a frequency/velocity that is intermediate to the velocities at which components A and B travel through the column. The calculations are designed such that component A travels clockwise (through the column) at a rate greater than the counterclockwise rotation rate of the columns. Thus, it is carried clockwise with the mobile phase and moves through the entire column, exiting through the raffinate port between zones III and IV. Similarly, product B moves clockwise (through the bed) at a rate slower than the calculated

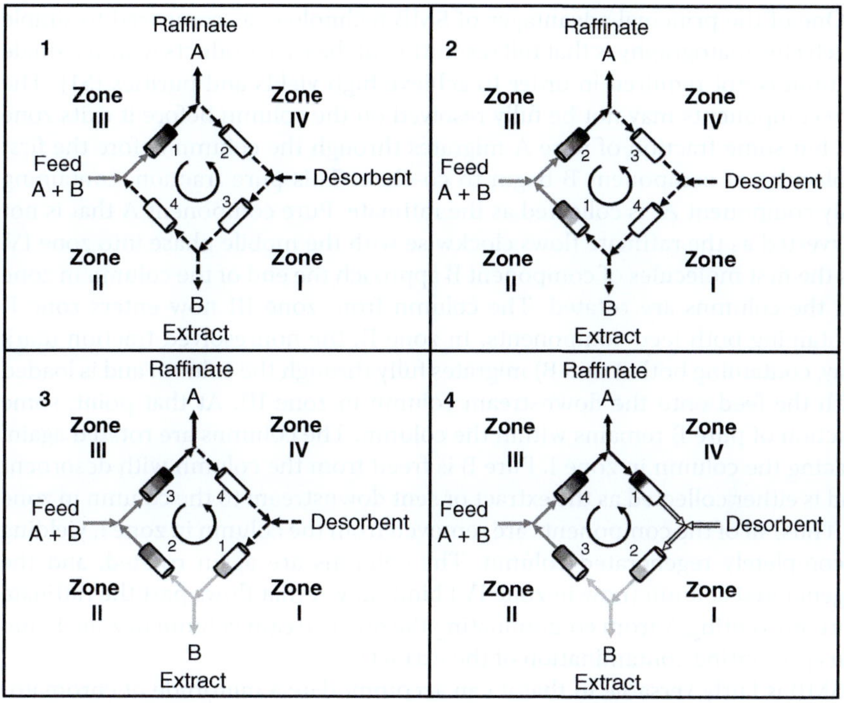

Figure 12.3 Schematic of an SMB purification process. Countercurrent flow of resin and liquid is approximated by periodic rotation of multiple columns (numbered 1–4) in the direction opposite to fluid flow. This figure will describe the activities of one column, starting in zone III, as it makes a full rotation through the system. Cell 1: in zone III, a feed containing A (weakly adsorbed) and B (strongly adsorbed) flows through the column and some pure A elutes before B begins to co-elute. This pure fraction of A is removed as the raffinate. Cell 2: the columns rotate and the unresolved mixture of A and B moves through the column while it remains in zone II. This leaves pure B near the column exit. Cell 3: the columns are rotated, desorbent is added to the column (now in zone I), and this pure fraction of B is eluted from the column. B is collected as the extract. Cell 4: the column, now free of all A and B, rotates into zone IV where it removes all A not collected through the raffinate port from the mobile phase. The columns then rotate and the process is repeated.

counterclockwise rotation of the columns; thus, it travels with the stationary phase in a counterclockwise direction until it has fully migrated through column and is harvested through the extract port between zones I and II. The simulated movement of the columns relative to the mobile phase can be achieved by switching the inlet ports of the columns or by physically rotating the columns. In zone I, desorbent frees component B from the stationary phase and component B is removed through the extract port. Zone IV is used to regenerate the mobile phase by capturing any component A that is not removed through the raffinate port.

One of the principal advantages of SMB technology as compared to simple batch chromatography is that full resolution of the two products within a single column is not required in order to achieve high yields and purities [51]. The two components may not be fully resolved on the column before it exits zone III, but some fraction of pure A migrates through the column before the first molecules of component B begin to co-elute. This pure fraction, containing only component A, is collected as the raffinate. Pure component A that is not harvested as the raffinate flows clockwise with the mobile phase into zone IV. As the first molecules of component B approach the end of the column in zone III, the columns are rotated. The column from zone III now enters zone II containing both feed components. In zone II, the nonresolved fraction (dark gray, containing both A and B) migrates fully through the column and is loaded with the feed onto the downstream column in zone III. At that point, some fraction of pure B remains within the column. The columns are rotated again, placing the column in zone I. Pure B is freed from the column with desorbent and is either collected as the extract or sent downstream to the column in zone II. Thus, all of the components are removed from the column in zone I, yielding a completely regenerated column. The columns are again rotated, and the regenerated column (now in zone IV) binds any A that flows past the raffinate port, preventing A from contaminating the downstream column in zone I (and thus preventing contamination of the extract).

SMB is fairly versatile in that it can accommodate a multitude of chromatographic separation techniques, including protein A affinity chromatography, ion exchange, and size exclusion [42, 48, 52]. This has made it possible for purification trains containing successive SMB trains to separate products in more complex feedstocks containing more than two components [53, 54]. In principle, the individual SMB trains in a tandem arrangement could utilize different chromatographic methods [42]. An eight-column SMB arrangement utilizing protein A affinity chromatography has been demonstrated that achieved ≥90% antibody recovery and more than 99% removal of HCP contaminants from cell culture fluid in a single step [49]. Because protein A binds only MAbs, the feed of clarified cell culture fluid can be considered pseudobinary (i.e., MAbs and non-MAbs), making it an ideal candidate for use in an SMB purification system.

12.5.4 Continuous Annular Chromatography (CAC) and Carrousel Chromatography

Continuous annular chromatography (CAC) is one of the only modes of chromatography that can truly be operated continuously at a steady state (Fig. 12.4). A crosscurrent movement of chromatography resin and the mobile phase, brought about by the rotation of an annular column, allows for product collection at a given angle of rotation, rather than at a specific time [55]. These

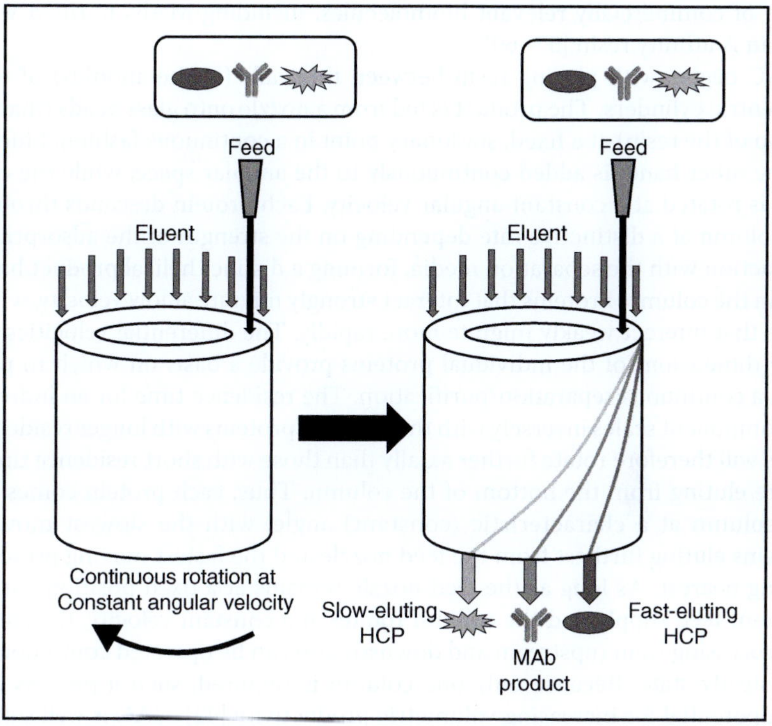

Figure 12.4 Schematic of CAC. Separation media (resin) is loaded into the annulus of two concentric cylinders and covered with glass beads. Clarified cell culture fluid is fed continuously onto the column at a fixed position from a nozzle. Eluent is added at a uniform flow rate to the entire annulus of the column and the column is rotated around its axis at constant angular velocity. Proteins move through the column at different velocities. The faster-migrating proteins elute quickly (the column will have rotated a short distance) and the slower-migrating proteins elute later (the column will have rotated further). Each protein elutes at a constant angle/position and may be collected continuously.

attributes make CAC a very exciting technology that could increase productivity, throughput, and efficiency while simultaneously decreasing costs and offering the potential for highly uniform product quality. This mode of continuous chromatography is also particularly useful because all chromatographic modes, excluding those utilizing a linear gradient elution, have been demonstrated in CAC systems [55]. These advantages come at the cost of introducing a fair degree of process complexity, however, based on the mechanical requirements of the column and valves. This technology was conceptualized more than 65 years ago (in 1949) but is yet to be implemented industrially for MAb purification [42, 56]. However, in recent years there have been demonstrations of its effectiveness and feasibility for purifying a wide range of commercially relevant biomolecules, including MAbs purified with protein A affinity resin [57–60].

CAC begins with placing resin between the walls (in the annulus) of two concentric cylinders. The product is fed from a nozzle onto glass beads (that lie on top of the resin) at a fixed, stationary point in a continuous fashion. Eluent, on the other hand, is added continuously to the annular space, while the column is rotated at a constant angular velocity. Each protein descends through the column at a distinctive rate depending on the strength of the adsorption/interaction with the separation media, forming a distinct helical product band within the column. Proteins that interact strongly migrate at low velocity, while those that interact weakly migrate more rapidly. The differential velocities (in the z-dimension) of the individual proteins provide a basis on which to perform a continuous separation/purification. The residence time for an individual component scales inversely with the velocity; proteins with longer residence times will therefore rotate further axially than those with short residence times before eluting from the bottom of the column. Thus, each protein comes off the column at a characteristic (constant) angle, with the slowest moving proteins eluting furthest from the feed nozzle and the fastest moving proteins eluting nearest. As long as the feed nozzle remains at a fixed position with a constant feed supply and the column rotates at a constant velocity, the entire bioprocessing train (upstream and downstream) can be operated continuously at a steady state. Because only one column is required, such a process has great potential for increasing volumetric productivity [42]. CAC is well suited for purifying large volumes of complex, low titer feedstocks such as those that may be encountered in biotechnology [55].

Approximations of CAC processes have been made using circular arrays of many columns that are rotated in order to continuously process feed [55]. This procedure is also known as carrousel chromatography. Purification occurs via the same mechanism as in CAC, but this arrangement is more practical and feasible [55]. Scale-up of CAC has been successfully demonstrated using relatively straightforward approaches [55, 57, 60]. Many of the process parameters at scale may be estimated simply by studying the chromatogram of a column (of equal length/height) with a diameter equal to the width of the annulus or by maintaining the loading factor [55, 57, 60].

12.5.5 Multicolumn Countercurrent Solvent Gradient Purification (MCSGP)

Multicolumn countercurrent solvent gradient purification (MCSGP) is a modification of SMB technology that allows purification of a desired protein from a complex feedstock using salt gradients to increase yield and purity [42, 51]. This process is used for purifying an intermediate product from strongly and weakly adsorbed impurities and, like SMB, simulates countercurrent flow of solids and liquids in the system with a complex arrangement of multiposition valves [51, 61]. The loading and elution steps occur quasi-continuously, as in PCC chromatography, but the remaining steps are performed concurrently [42, 51, 61]. MCSGP uses a "short-circuited moving column technique" in order to improve resolution of products compared to a traditional SMB process [51]. In an example process containing six columns, feed is loaded onto an individual column (at position 5) that is temporarily removed from the rest of the process, and weakly adsorbed impurities wash through the column. The column is then moved upstream by one space (to position 4) and its effluent is loaded onto the column at position 6, thus "short-circuiting" across column 5. In order to achieve solvent gradient elution, additional pumps are added at the inlet of each column in order to adjust the concentration of modifier. Due to this short-circuiting technique, any weakly adsorbed impurities that are eluted from column 4 are loaded onto column 6, downstream of the intermediately adsorbed product contained in column 5, rather than in front of the intermediately adsorbed product as would happen if effluent from column 4 were loaded onto column 5. This keeps the weakly adsorbed impurities downstream of the product and the strongly adsorbed impurities in column 5. Elution of the product and the strongly adsorbed impurities occurs from columns 3 and 1, respectively, both of which are also short-circuited. Similarly, this ensures that residual product eluted from column 2 is loaded onto a column downstream of the strongly adsorbed impurities contained in column 3, rather than in front of it. Thus, resolution/purity and protein yield are improved for MCSGP compared to batch and SMB purifications [51]. Recent advances have enabled the number of required columns to be decreased from the original 6–3, with concomitant increases in purity and productivity [62].

A study involving the purification of the peptide calcitonin from a multicomponent peptide feedstock demonstrated the promise of MCSGP: yields of nearly 100% were achieved alongside 93% product purity [51]. A mathematical model of MCSGP performance concludes that the volumetric productivity of an MCGSP process is more than tenfold higher than for a single-column batch process [61]. A two-step continuous process has been demonstrated that utilized cation-exchange resin in an MCSGP train followed by polishing with multimodal anion-exchange chromatography in order to purify a model IgG_2 MAb [63]. Yields from the cation-exchange MCSGP capture step were in excess 95% or higher, comparable to those of a batch purification using the much more selective protein A affinity chromatography platform [63]. More importantly, purity was improved for the MCSGP processes; HCP levels in the product were over an order of magnitude less than for protein A batch chromatography. Productivity levels were up to fivefold greater for the MCSGP process as well [63]. While buffer consumption was slightly elevated in comparison to the protein A batch process, achieving equivalent yields to protein A affinity chromatography with higher product purity and less expensive chromatographic resin demonstrates the promise and robustness of MCSGP processes [63]. Most notably, in order to achieve desired purity with protein A affinity chromatography, a three-step process is used; for the cation-exchange MCSGP process, the purity specifications were met using only two. MCSGP therefore offers significant improvements in regard to productivity, process times, and capital investments [63]. Yet another advantage of MCSGP compared to batch protein A affinity chromatography is the ability to discriminate and purify MAb variants in order to isolate high-activity product [64].

12.5.6 Countercurrent Tangential Chromatography (CCTC)

PCC chromatography is highly analogous to traditional batch purification schemes using multiple columns in series that are loaded countercurrently. Despite the ability to operate continuously using a perfusion bioreactor in tandem with PCC, operation at a true steady state is often difficult or unachievable. While the feed from the bioreactor is indeed processed continuously using a multicolumn PCC purification train, product is still collected in batches from multiple columns that may not behave exactly the same. As a result, although PCC gives more consistent product quality over an extended period of time than does batch production, variability in product quality can still present an issue. A truly continuous alternative chromatography method that allows operation at a fully steady state (rather than periodic product elution) is countercurrent tangential chromatography (CCTC). CCTC is a column-free technology that utilizes a slurry of resin that flows throughout the system through a series of mixers and hollow fiber membranes [65]. Because the resin slurry flows throughout the system to distinct compartments where binding, washing, elution, regeneration, and re-equilibration occur, this is a true moving bed technology where buffers move countercurrent to the slurry at each stage.

CCTC has been successfully demonstrated for purification of bovine serum albumin (BSA) from myoglobin using anion-exchange resin, as well as for purification of an IgG_4 MAb from BSA and myoglobin using a protein A affinity resin [65, 66]. In both studies, yield was >94% with purity >97%, demonstrating the utility of CTCC as a purification scheme. Productivity for CCTC processes is much greater than for batch column operation, and the mixers and hollow fiber membranes are inexpensive. As a result, CCTC potentially lends itself to implementation in single-use, disposable chromatography equipment [64]. In addition, because no columns or column packing is required, CCTC appears to offer a fair degree of plant and equipment flexibility [65]. Compared to batch processes, however, product is eluted at lower concentration, and particle attrition is likely very high due to the movement of resin slurry through the equipment [65]. The latter concern may prove to be a significant impediment for this technology for two reasons: (i) particle degradation will shorten the time frame over which a CCTC process may be successfully operated continuously and (ii) replacement of resins may prove to be a considerable cost. While technologies such as PCC chromatography have been operated continuously for longer than a month, CCTC processes reported in the literature generally only operated over the duration of hours, suggesting that the process is perhaps not well suited for continuous operations of extended durations.

12.6 Conclusion

Nonchromatographic MAb purification processes based on commonplace unit operations such as continuous aqueous two-phase extraction and continuous precipitation are currently being thoroughly investigated [67–70]. Although industrial implementation of these processes is less likely in the immediate future, these unit operations have the potential to eliminate some expensive chromatography steps altogether, offering savings in process times, capital investments, and consumables (i.e., resins) [70]. Economic analysis of a theoretical, fully continuous bioprocessing train utilizing antibody precipitation in tandem with perfusion cell culture demonstrated lower costs at all phases of MAb development and production [70]. This chapter focused on alternative chromatographic methods and advances in continuous chromatographic purification schemes because their widespread use, momentum in industry, and the large investments companies have made into chromatography-based equipment and facilities make their use in continuous downstream processes much more realizable in the immediate future than nonvalidated protocols.

The biopharmaceutical industry, like many others as they matured, is moving toward continuous processing technology. Traditional batch cell culture and purification techniques served the industry well in its infancy, and biopharmaceutical companies are currently capable of satisfying market demands with MAbs that cost far less to produce than their selling prices. However, diversification and expansion of product bases require companies to analyze their processes and refine them in the context of greater overall plant/equipment flexibility. Moreover, as patents expire and biosimilars begin to enter the market, cost reduction and consistently high product quality will become concerns of greater significance; investment in additional 10,000 + 1 bioreactors and the large protein A affinity columns that accompany them may no longer suffice when increases in production capacity are desired [31]. All of these obstacles facing the biopharmaceutical industry can be addressed to some degree by a switch from batch to continuous processing.

While more complicated than simple batch protocols, the continuous downstream chromatographic processes relevant to MAb production/purification highlighted in this chapter are characterized by decreased equipment sizes and plant footprints, more efficient use of chromatography/separation media, decreased buffer and solvent requirements (with a few exceptions), high and consistent product quality and purity, and equivalent or improved MAb yields. As implementation of these technologies becomes more commonplace, economic analyses will determine which, if any, of these technologies best serves the biopharmaceutical industry. Although downstream processes continually evolve and new technologies emerge, the continuous purification platforms discussed in this chapter represent a step in the right direction.

References

1. Richman, D. D., Cleveland, P. H., Oxman, M. N., Johnson, K. M. (1982). The binding of staphylococcal protein A by the sera of different animal species. *Journal of Immunology*, 128(5), 2300–2305.
2. Chon, J. H., Zarbis-Papastoitsis, G. (2011). Advances in the production and downstream processing of antibodies. *New Biotechnology*, 28(5), 458–463.
3. Shukla, A. A., Thömmes, J. (2010). Recent advances in large-scale production of monoclonal antibodies and related proteins. *Trends in Biotechnology*, 28(5), 253–261.
4. Hober, S., Nord, K., Linhult, M. (2007). Protein A chromatography for antibody purification. *Journal of Chromatography B*, 848(1), 40–47.
5. Björk, L., Kronvall, G. (1984). Purification and some properties of streptococcal protein G, a novel IgG-binding reagent. *Journal of Immunology*, 133(2), 969–974.
6. Liu, H. F., Ma, J., Winter, C., Bayer, R. (2010). Recovery and purification process development for monoclonal antibody production. *mAbs*, 2(5), 480–499.
7. Brorson, K., Krejci, S., Lee, K., Hamilton, E., Stein, K., Xu, Y. (2003). Bracketed generic inactivation of rodent retroviruses by low pH treatment for monoclonal antibodies and recombinant proteins. *Biotechnology and Bioengineering*, 82(3), 321–329.
8. Asplund, M., Ramberg, M., Johnasson, B.-L. (2000). Development of a cleaning in place protocol and repetitive application of *Escherichia coli* homogenate on STREAMLINE^{TM} Q XL. *Process Biochemistry*, 35(10), 1111–1118.
9. Hale, G., Drumm, A., Harrison, P., Phillips, J. (1994). Repeated cleaning of protein A affinity column with sodium hydroxide. *Journal of Immunological Methods*, 171(1), 15–21.
10. Gülich, S., Linhult, M., Ståhl, S., Hober, S. (2002) Engineering streptococcal protein G for increased alkaline stability. *Protein Engineering*, 15(10), 835–842.

11. Linhult, M., Gülich, S., Gräslund, T., Simon, A., Karlsson, M., Sjöberg, A., Nord, K., Hober, S. (2004). Improving the tolerance of a protein A analogue to repeated alkaline exposures using a bypass mutagenesis approach. *Proteins*, 55(2), 407–416.
12. Xia, H. F., Liang, Z. D., Wang, S. L., Wu, P. Q., Jin, X. H. (2014). Molecular modification of protein A to improve the elution pH and alkali resistance in affinity chromatography. *Applied Biochemistry and Biotechnology*, 172(8), 4002–4012.
13. Follman, D. K., Fahrner, R. L. (2004). Factorial screening of antibody purification processes using 3 chromatography steps without protein A. *Journal of Chromatography A*, 1024(1), 79–85.
14. Ey, P. L., Prowse, S. J., Jenkin, C. R. (1978). Isolation of pure IgG_1, IgG_{2a}, and IgG_{2b} immunoglobulins from mouse serum using protein A–sepharose. *Immunochemistry*, 15(7), 429–436.
15. Nfor, B. K., Verhaert, P. D., van der Wielen, L. A., Hubbuch, J., Ottens, M. (2009). Rational and systematic protein purification process development: the next generation. *Trends in Biotechnology* 27(12), 673–679.
16. Shukla, A. A., Hubbard, B., Tressel, T., Guhan, S., Low, D. (2007). Downstream processing of monoclonal antibodies—application of platform approaches. *Journal of Chromatography B*, 848(1), 28–39.
17. Füglistaller, P. (1989). Comparison of immunoglobulin binding capacities and ligand leakage using eight different protein A affinity chromatography matrices. *Journal of Immunological Methods*, 124(2), 171–177.
18. Alves, N. J., Stimple, S. D., Handlogten, M. W., Ashley, J. D., Kiziltepe, T., Bilgicer, B. (2012). Small-molecule-based affinity chromatography method for antibody purification via nucleotide binding site targeting. *Analytical Chemistry*, 84(18), 7721–7728.
19. Handlogten, M. W., Kiziltepe, T., Moustakas, D. T., Bılgicer, B. (2011). Design of a heterobivalent ligand to inhibit IgE clustering on mast cells. *Chemistry & Biology*, 18(9), 1179–1188.
20. Phillips, J., Drumm, A., Harrison, P., Bird, P., Bhamra, K., Berrie, E., Hale, G. (2001). Manufacture and quality control of CAMPATH-1 antibodies for clinical trials. *Cytotherapy*, 3(3), 233–242.
21. Fassina, G., Verdoliva, A., Palombo, G., Ruvo, M., Cassani, G. (1998). Immunoglobulin specificity of TG 19318: a novel synthetic ligand for antibody purification. *Journal of Molecular Recognition*, 11(1), 128–133.
22. Teng, S. F., Sproule, K., Hussain, A., Lowe, C. R. (1999). A strategy for the generation of biomimetic ligands for affinity chromatography. Combinatorial synthesis and biological evaluation of an IgG binding ligand. *Journal of Molecular Recognition*, 12(1), 67–75.
23. Fassina, G., Ruvo, M., Palombo, G., Verdoliva, A., Marino, M. (2001). Novel ligands for the affinity-chromatographic purification of antibodies. *Journal of Biochemical and Biophysical Methods*, 49(1), 481–490.

24. Teng, S. F., Sproule, K., Husain, A., Lowe, C. R. (2000). Affinity chromatography on immobilized "biomimetic" ligands. Synthesis, immobilization and chromatographic assessment of an immunoglobulin G-binding ligand. *Journal of Chromatography B*, 740(1), 1–15.
25. Ghose, S., Hubbard, B., Cramer, S. M. (2006). Evaluation and comparison of alternatives to protein A chromatography: mimetic and hydrophobic charge induction chromatographic stationary phases. *Journal of Chromatography A*, 1122(1), 144–152.
26. Gagnon, P. In Purification Tools for Monoclonal Antibodies, Validated Biosystems, Tucson, AZ, 1996.
27. Fahrner, R. L., Knudsen, H. L., Basey, C. D. Galan, W., Feuerhelm, D., Vanderlaan, M., Blank, G. S. (2001). Industrial purification of pharmaceutical antibodies: development, operation, and validation of chromatography processes. *Biotechnology and Genetic Engineering Reviews*, 18(1), 301–327.
28. Fahrner, R. L., Whitney, D. H., Vanderlaan, M., Blank, G. S. (1999). Performance comparison of protein A affinity-chromatography sorbents for purifying recombinant monoclonal antibodies. *Biotechnology and Applied Biochemistry*, 30(2), 121–128.
29. Kelley, B. (2009). Industrialization of mAb production technology: the bioprocessing industry at a crossroads. *mAbs*, 1(5), 443–452.
30. Brodeur, B. R., Tsang, P., Larose, Y. (1984). Parameters affecting ascites tumor formation in mice and monoclonal antibody production. *Journal of Immunological Methods*, 71(2), 265–272.
31. Warikoo, V., Godawat, R., Brower, K., Jain, S., Cummings, D., Simons, E., Johnson, T., Walther, J., Yu, M., Wright, B., McLarty, J., Karey, K. P., Hwang, C., Zhou, W., Riske, F., Konstantinov, K. (2012). Integrated continuous production of recombinant therapeutic proteins. *Biotechnology and Bioengineering*, 109(12), 3018–3029.
32. Birch, J. R., Onakunie, Y. (2005). Biopharmaceutical proteins: opportunities and challenges. *Methods in Molecular Biology*, 308, 1–16.
33. Wurm, F. M. (2004). Production of recombinant protein therapeutics in cultivated mammalian cells. *Nature Biotechnology*, 22(11), 1393–1398.
34. Reichert, J. M. (2014). Antibodies to watch in 2015. *mAbs*, 7(1), 1–8.
35. Reay, D., Ramshaw, C., Harvey, A. In Process Intensification: Engineering for Efficiency, Sustainability and Flexibility. Butterworth-Heinemann, Amsterdam, 2008.
36. Brecht, R. (2009). Disposable bioreactors: maturation into pharmaceutical glycoprotein manufacturing. *Advances in Biochemical Engineering/Biotechnology*, 115, 1–31.
37. Diekmann, S., Dürr, C., Herrmann, A., Lindner, I., Jozic, D. Single use bioreactors for the clinical production of monoclonal antibodies—a study to analyze the performance of a CHO cell line and the quality of the produced monoclonal antibody. *BMC Proceedings* 2011, 5(Suppl 8), 103.

38. Godawat, R., Brower, K., Jain, S., Konstantinov, K., Riske, F., Warikoo, V. (2012). Periodic counter-current chromatography—design and operational considerations for integrated and continuous purification of proteins. *Biotechnology Journal*, 7(12), 1496–1508.
39. Vogel, J. H., Nguyen, H., Pritschet, M., Van Wegen, R., Konstantinov, K. (2002). Continuous annular chromatography: general characterization and application for the isolation of recombinant protein drugs. *Biotechnology and Bioengineering*, 80(5), 559–568.
40. Fulton, S., Shahidi, A., Gordon, N., Afeyan, N. (1992). Large-scale processing & high-throughput perfusion chromatography. *Biotechnology*, 10(6), 635–639.
41. Jungbauer, A. (1993). Preparative chromatography of biomolecules. *Journal of Chromatography A*, 639(1), 3–16.
42. Jungbauer, A. (2013). Continuous downstream processing of biopharmaceuticals. *Trends in Biotechnology*, 31(8), 479–492.
43. Arve, B. H., Liapis, A. I. (1988). Biospecific adsorption in fixed and periodic countercurrent beds. *Biotechnology and Bioengineering*, 32(5), 616–627.
44. Heeter, G. A., Liapis, A. I. (1995). Perfusion chromatography: performance of periodic countercurrent column operation and its comparison with fixed-bed operation. *Journal of Chromatography A*, 711(1), 3–21.
45. Mahajan, E., George, A., Wolk, B. (2012). Improving affinity chromatography resin efficiency using semi-continuous chromatography. *Journal of Chromatography A*, 1227, 154–162.
46. Carta, G., Perez-Almodovar, E. X.. (2010). Productivity considerations and design charts for biomolecule capture with periodic countercurrent adsorption systems. *Separation Science and Technology*, 45(2), 149–154.
47. Broughton, D. B., Gerhold, C. G. (1961). Continuous sorption process employing fixed bed of sorbent and moving inlets and outlets. U.S. Patent No. 2,985,589. Washington, DC, USA.
48. Keβler, L. C., Gueorguieva, L., Rinas, U., Seidel-Morgenstern, A. (2007). Step gradients in 3-zone simulated moving bed chromatography: application to the purification of antibodies and bone morphogenetic protein-2. *Journal of Chromatography A*, 1176(1), 69–78.
49. Gottschlich, N., Kasche, V. (1997). Purification of monoclonal antibodies by simulated moving-bed chromatography. *Journal of Chromatography A*, 765(2), 201–206.
50. Song, S.-M., Park, M.-B., Kim, I. H. (2012) Three-zone simulated moving-bed (SMB) for separation of cytosine and guanine. *Korean Journal of Chemical Engineering*, 29(7), 952–958.
51. Aumann, L., Morbidelli, M. (2007). A continuous multicolumn countercurrent solvent gradient purification (MCSGP) process. *Biotechnology and Bioengineering*, 98(5), 1043–1055.
52. Hashimoto, K., Adachi, S., Shirai, Y. (1988). Continuous desalting of proteins with a simulated moving-bed adsorber. *Agricultural and Biological Chemistry*, 52(9), 2161–2167.

53. Mun, S., Xie, Y., Kim, J.-H., Wang, N.-H. L. (2003). Optimal design of a size-exclusion tandem simulated moving bed for insulin purification. *Industrial and Engineering Chemistry*, 42(9), 1977–1993.

54. Chin, C. Y., Wang, N.-H. L. (2004). Simulated moving bed equipment designs. *Separation and Purification Reviews*, 33(2), 77–155.

55. Hilbrig, F., Freitag, R. (2003). Continuous annular chromatography. *Journal of Chromatography B*, 790(1), 1–15.

56. Martin, A. J. P. (1949). Summarizing paper. *Discussions of the Faraday Society*, 7, 332–336.

57. Giovannini, R., Freitag, R. (2001). Isolation of a recombinant antibody from cell culture supernatant: continuous annular versus batch and expanded-bed chromatography. *Biotechnology and Bioengineering*, 73(6), 522–529.

58. Iberer, G., Schwinn, H., Josic, D., Jungbauer, A., Buchacher, A. (2002). Continuous purification of a clotting factor IX concentrate and continuous regeneration by preparative annular chromatography. *Journal of Chromatography A*, 972(1), 115–129.

59. Özdural, A. R., Tanolaç, D., Özdural, N. (2007, April). A novel technology for virus vaccine purification: modeling and operation of continuous annular chromatography unit. In *AIChE Annual Meeting: 2007 Spring Meeting and 3rd Global Congress on Process Safety*, American Institute of Chemical Engineers, New York, p. 18.

60. Giovannini, R., Freitag, R. (2002). Continuous isolation of plasmid DNA by annular chromatography. *Biotechnology and Bioengineering*, 77(4), 445–454.

61. Ströhlein, G., Aumann, L., Mazzotti, M., Morbidelli, M. (2006). A continuous counter-current multi-column chromatographic process incorporating modifier gradients for ternary separations. *Journal of Chromatography A*, 1126(1), 338–346.

62. Aumann, L., Morbidelli, M. (2008). A semicontinuous 3-column countercurrent solvent gradient purification (MCSGP) process. *Biotechnology and Bioengineering*, 99(3), 728–733.

63. Müller-Späth, T., Aumann, L., Ströhlein, G., Kornmann, H., Valax, P., Delegrange, L., Charbaut, E., Baer, G., Lamproye, A., Jöhnck, M., Schulte, M., Morbidelli, M. (2010). Two step capture and purification of IgG_2 using multicolumn countercurrent solvent gradient purification (MCSGP). *Biotechnology and Bioengineering*, 107(6), 974–984.

64. Müller-Späth, T., Krättli, M., Aumann, L., Ströhlein, G., Morbidelli, M. (2010). Increasing the activity of monoclonal antibody therapeutics by continuous chromatography (MCSGP). *Biotechnology and Bioengineering*, 107(4), 652–662.

65. Shinkazh, O., Kanani, D., Barth, M., Long, M., Hussain, D., Zydney, A. L. (2011). Countercurrent tangential chromatography for large-scale protein purification. *Biotechnology and Bioengineering*, 108(3), 582–591.

66. Napadensky, B., Shinkazh, O., Teella, A., Zydney, A. L. (2013). Continuous countercurrent tangential chromatography for monoclonal antibody purification. *Separation Science and Technology*, 48(9), 1289–1297.
67. Rosa, P. A. J., Azevedo, A. M., Sommerfeld, S., Mutter, M., Bäcker, W., Aires-Barros, M. R. (2013). Continuous purification of antibodies from cell culture supernatant with aqueous two-phase systems: from concept to process. *Biotechnology Journal*, 8(3), 352–362.
68. Rosa, P. A. J., Ferreira, I. F., Azevedo, A. M., Aires-Barros, M. R. (2010). Aqueous two-phase systems: a viable platform in the manufacturing of biopharmaceuticals. *Journal of Chromatography A*, 1217(16), 2296–2305.
69. Sim, S.-L., He, T., Tscheliessnig, A., Mueller, M., Tan, R. B. H., Jungbauer, A. (2012). Branched polyethylene glycol for protein precipitation. *Biotechnology and Bioengineering*, 109(3), 736–746.
70. Hammerschmidt, N., Tscheliessnig, A., Sommer, R., Helk, B., Jungbauer, A. (2014). Economics of recombinant antibody production processes at various scales: industry-standard compared to continuous precipitation. *Biotechnology Journal*, 9(6), 766–775.

13

Biosimilars and Biobetters

Impact on Biopharmaceutical Manufacturing and CMOs

Ronald A. Rader

BioPlan Associates, Inc., Rockville, MD, USA

13.1 Summary

Biosimilars are biopharmaceuticals that are rather similar, often nearly identical, to a reference product, while "Biobetters" are biopharmaceuticals that are sufficiently dissimilar from any prior approved product such that they must receive a traditional full approval as a different and innovative product. Biosimilars are subjected to an abbreviated but rigorous approval process involving direct comparisons of the biosimilar to its reference product, including analytical profiles and head-to-head comparative pharmacokinetics and clinical trials.

Biogenerics are follow-on biopharmaceuticals that generally originate in developing countries and do not meet GMP standards applied by the EMA and the FDA. They frequently are promoted as interchangeable with established GMP products, despite lacking related comparative testing.

In contrast, traditional innovative product approvals involve larger placebo-controlled trials. Nearly 20 biosimilars are already marketed in the European Union and some other major market countries having implemented biosimilar approval pathways.

A database tracking the biosimilars/biobetters pipeline reports nearly 700 biosimilars and 500 biobetters in development worldwide, with the great majority targeted to enter the United States and other major markets where patents and other granted exclusivities are expiring [1].

Biosimilars in development include a large number of candidates from established and many new entrants to the biopharmaceutical industry. Many companies view biosimilars (and to a lesser extent, biobetters) as a more affordable, lower risk way to enter the lucrative US and EU markets and gain

Biosimilars of Monoclonal Antibodies: A Practical Guide to Manufacturing, Preclinical, and Clinical Development, First Edition. Edited by Cheng Liu and K. John Morrow, Jr.

credibility as a biopharmaceutical developer. With only comparative testing and trials required for biosimilar approval, this mechanism enables market entry at a cheaper and faster pace, when compared with the pursuit of innovative products receiving traditional full approvals. Risks are also much lower, generally involving much the same active agent that has been on the market for more than two decades.

By the end of the decade, we can anticipate 10–12 entries for every successful reference product. While this may seem a chaotically competitive market, there already are this many or more sources for some major-selling generic drugs.

The United States is projected to be the dominant biosimilars' market. Many new and established companies will be moving aggressively into this market, even if this results in overcrowding and a market so fractured that it becomes impossible to realize desired profit margins.

Biosimilars will raise the bar for product quality expectations for all biopharmaceuticals, with regulators, healthcare professionals, patients, and the public focusing on the details of manufacturing processes and the analytical, bioprocessing, and related quality differences between products.

The biosimilars, biobetters, and biogenerics market will be much like the generic drug market. This includes active pharmaceutical ingredients sources and competing products originating from a range of international companies. However, unlike the generic sector, development and manufacturing will be located in the industrialized countries. As international competition and the available number of biosimilars and other follow-ons increase, buyers will favor the US and EU manufactured products.

13.2 Introduction

Beginning in 2014 and through the rest of this decade, a number of well-established biopharmaceuticals, including highly successful mAbs, are coming off their patent protection. We can anticipate that many follow-on products, which include biosimilars, biobetters, and biogenerics, will be entering the United States and other major markets and will be increasingly competitive with established reference products. Already, a large number of companies are developing biosimilars and biobetters, targeting the United States and other major markets, while biogenerics are mostly developed and marketed in developing countries.

Nearly 20 biosimilars are already marketed in the European Union and some other major market countries having implemented biosimilar or equivalent approval pathways. In the United States the FDA approved its first biosimilar, Zarxio (filgrastim-sndz), in March 2015. However, FDA is still struggling to issue draft regulations 5 years after passage of enabling legislation. Also, in the meantime, several products having received biosimilar approval in the European Union and other countries have received traditional, full approvals in the United States.

"Biobetters" are similar follow-on biopharmaceuticals, with a comparable active agent, but they are sufficiently dissimilar from any prior approved products such that they must receive traditional full approval as a different and innovative product. While biosimilars, defined by regulatory pathways, are relatively new, the biobetters are not an original concept. Companies have long brought out follow-on products designed to be essentially the same, filling the same market niche as established products. This includes classic biologics such as most vaccines and blood/plasma products.

Most biobetter products involve manipulations that alter product performance, such as pegylation of the active agent or a different delivery system such as oral administration of a currently injectable product. Since many biobetters are not really anything new, this article concentrates on the more novel biosimilars.

Biogenerics are those follow-ons that are frequently developed, manufactured, and marketed in developing countries and generally do not meet GMP standards applied by the EMA and the FDA. In most cases, they are not extensively tested beyond minimal trials and are promoted as interchangeable with established GMP products based on sharing common active agent-based generic (INN) names. There are nearly 200 such biogenerics in world commerce, for example, manufactured in China, India, and Brazil, primarily for marketing in these and other lesser-regulated countries.

13.3 The Biosimilar Pipeline

As noted previously, the biosimilars/biobetters pipeline database reports nearly 700 biosimilars in the pipeline. The figure in the following text shows the number of biosimilars in the development pipeline by their expected dates of product launch in the United States. This introduction requires that both relevant patents and regulatory-granted market exclusivities, essentially orphan status and a 12-year new data exclusivity granted to all new full BLAs, be expired at this time. There are large numbers of biosimilars likely to enter the market in the next few years, with another large group, including many mAbs, expected to enter the market late in the decade (Fig. 13.1).

The projected economic impact of biosimilars and biobetters from the US perspective is portrayed in the figure later. This shows the magnitude of cumulative current reference product sales by the probable year of introduction of biosimilars for these products entering the US market. With patents expiring on the majority of currently marketed biopharmaceuticals, including most recombinant proteins and mAbs, these products with current cumulative

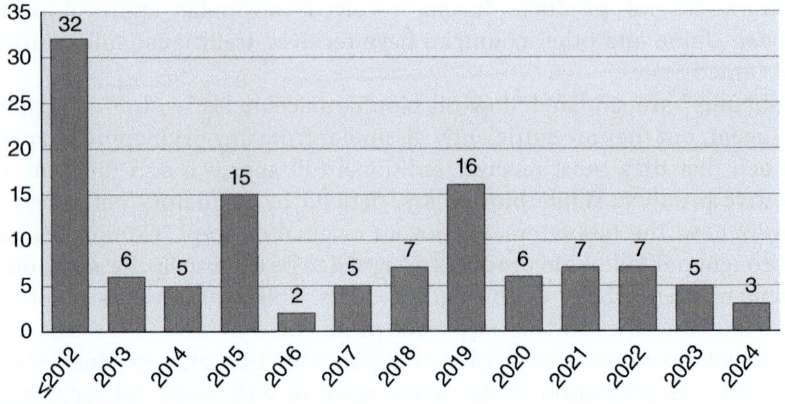

Figure 13.1 US biosimilars launchable dates (by reference products, $n = 133$, not products in the pipeline).

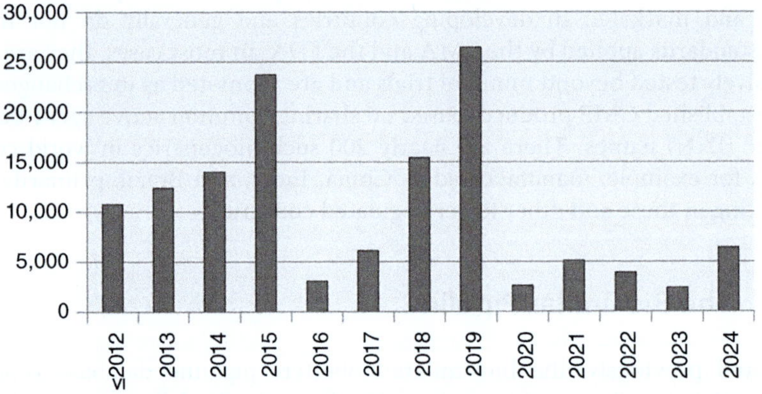

Figure 13.2 Biosimilars launchable dates by sum of current reference products sales ($millions).

annual revenue of about $100 billion will encounter much new competition in the United States and other major markets from biosimilars and biobetters starting in just a few years (Fig. 13.2).

The stages of development of biosimilars and biobetters are shown in the figure later. As with any product pipeline, the great majority of products reported in development are in preclinical phases. Lack of guidelines and uncertainties regarding FDA regulations appear to be the primary reasons many developers are holding back on entering products into trials. Reference product companies are competitively aggressive and have had decades to prepare for the inevitability of patent expiration. There is a variety of legal

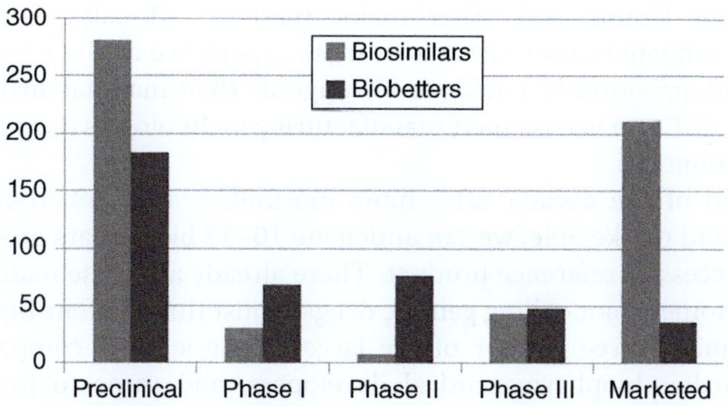

Figure 13.3 Pipelines by phase of development. Note: All but ≥20 of the "marketed" biosimilars are biogenerics in international commerce. (*See insert for color representation of the figure.*)

subterfuges that can delay biosimilars marketing, such as having exclusively licensed third-party patents that effectively cover their products. Such information is not readily obtainable, nor likely to have been publicly reported (Fig. 13.3).

It is important to note that this rather full biosimilars and biobetters pipeline still does not include many future offerings, since some companies have not yet have disclosed their strategic plans. This cautionary admonition applies to the international pharmaceutical, biopharmaceutical, and generic drug companies that possess the financial resources to dominate the marketing of biosimilars and biobetters. Many of these companies are either quietly developing biosimilars or can be expected to license-in or acquire favored products through company buyouts. For the pharmaceutical giants, this constitutes a logical business plan. It rounds out their existing portfolios and avoids the unsavory alternative of having a lesser-extensive portfolio and diminished ability to negotiate bundled discount sales. Many of these portfolios are indications specific, such as treatments for cancer, diabetes, and arthritis.

Biosimilars are new in terms of marketing. As was noted previously, there are nearly 200 biogenerics currently in world commerce, comprising nearly 90% of the "marketed" products portrayed in Figure 13.1, with essentially all but ca. 20 being biogenerics in lesser-regulated international commerce. This excludes nearly 24 biosimilars currently marketed in the European Union and other major markets as genuine biosimilars. Thus, the now approximately 20 genuine biosimilars approved in major markets make up a small portion of the 200+ "marketed" biosimilars portrayed in the figure. Essentially, all these others are biogenerics marketed in lesser-regulated international commerce.

A growing number of these biogenerics will advance to development as biosimilars and will be aimed as biosimilar candidates in the United States, the European Union, and other major markets. Already many foreign biogeneric companies have a decade or more experience manufacturing these products and, presumably, could readily upgrade their manufacturing to US–EU-grade cGMP and license their manufacturing technologies to companies in developed countries.

By the end of the decade, after more biosimilars and biobetters become approvable and marketable, we can anticipate 10–12 biosimilars or even more for every successful reference product. There already are these many or more sources for some major-selling generic drugs. If just three to four Big Pharma-type companies, three to four of the largest generic drug companies, two to three smaller biopharmaceutical developers, and three to four foreign companies each gain US approval for biosimilar versions of a product, this is equivalent to 12 or more biosimilars of this one reference product entering the US market.

And there could be more companies of each type seeking US approval for biosimilars. Many, if not most, biosimilar developers and manufacturers will be new entrants to the biopharmaceutical industry. Thus, the number of biopharmaceutical products and manufacturers will rapidly increase, particularly as US biosimilar approvals come online in the next few years.

Biosimilar developers include a large number of candidates from established and many new entrants to the biopharmaceutical industry. Many companies view biosimilars (and to a lesser extent, biobetters) as a more affordable, lower risk way to enter the lucrative US and EU markets and gain credibility as a biopharmaceutical developer. With only comparative trials required for biosimilar approval, this mechanism enables market entry at a cheaper and faster pace, when compared with the pursuit of innovative products receiving traditional full approvals. Risks are also much lower, involving the same active agent that has been on the market for greater than or equal to two decades.

The United States remains the most promising undeveloped market for biosimilars. With biosimilars moving into a more favorable position and insurers and the federal government likely forcing their use in place of more expensive reference products, the market appears auspicious.

13.4 Developing Countries Will Continue to Prefer Cheaper Biogenerics

Many of the currently marketed products commonly called biosimilars, but actually biogenerics, are already manufactured and marketed in less-regulated international markets. Few of these currently meet high GMP standards, and most lack information regarding product manufacture, quality control, analytical comparisons with reference products, and clinical trials. Most developing countries have long had attempted copies of major reference products in their local markets, whether approved or not. This largely unregulated pharma market will discourage the growth of a genuine biosimilar industry and may stifle the development of an indigenous innovative, R&D-based pharma industry in these regions.

Developing economies may be forced to commit to lower-quality biogenerics, often from government-affiliated or sponsored manufacturers. Prime examples are Brazil and other Latin American economies. For example, a developing country's healthcare system would likely have a limited budget for purchase of epoetin alpha (EPO) for its kidney dialysis patients. With the alternatives being buying less of the reference product, for example, Epogen from Amgen (Thousand Oaks, CA), most governments in developing countries may be forced to commit to a much cheaper biogeneric, even if this has a known higher incidence and severity of adverse events. While this is a far from ideal strategy, it allows many more patients to receive treatment. It also allows developing countries to support local and regional manufacturers, which could include manufacture by the government's own, proxy, captive, or other affiliated companies.

13.5 Biosimilar Candidates in the Pipeline

The table in the following text shows some major-selling reference products and the number of biosimilars targeting these in development worldwide. The great majority of products are in preclinical development. Some reference products, such as G-CSF (Neupogen) and epoetin alpha (Epogen/Procrit), currently have over 50 products reported in development. Such high numbers can saturate a market, with each product on average having too small to compete effectively or be sufficiently profitable. This drawback is especially relevant in highly regulated markets where interchangeability is unlikely to be granted in the near future (Table 13.1).

13.6 Biosimilar Development by Country/Region

Biosimilars in development by country in which the company or its parent based are shown in the figure later. The European Union is the current leader in terms of biosimilars development, closely followed by the United States and India. Asia collectively, including India and China, has more biosimilar products in development than the United States, which is the leading country in terms of biobetter products in development (Fig. 13.4). However, only a portion of biosimilars developed in/by developed country-based companies can be expected to make it to the United States, Europe, or other major highly regulated markets (Table 13.2).

Table 13.1 Some leading reference products and the number of biosimilars targeting these products.

Active agent	Reference product	Biosimilars in development
Tumor necrosis factor mAb	Humira® (AbbVie)	25
Tumor necrosis factor mAb	Remicade® (Janssen/J&J)	14
Erythropoietin; epoetin alpha	Epogen® (Amgen)/Procrit (J&J)	86
Granulocyte colony stimulating factor; filgrastim	Neupogen® (Amgen)	57
Granulocyte colony stimulating factor, pegylated; pegfilgrastim	Neulasta® (Amgen)	20
Tumor necrosis factor, mAb-like fusion protein	Enbrel® (Amgen)	28
CD20 mAb	Rituxan® (Genentech/Roche)	48
Her2 receptor mAb; trastuzumab	Herceptin® (Genentech/Roche)	37
Insulin glargine	Lantus® (Sanofi)	7
Vascular endothelial growth factor mAb; bevacizumab	Avastin® (Genentech/Roche)	22
Insulins	Multiple products	50
Interferon alpha	Multiple products	69
Interferon beta	Multiple products	26
Human growth hormone; somatropin	Nutropin® (Genentech)	34

Note: US trademark and primary US marketing companies are reported. All but one product (labeled) are recombinant proteins or antibodies.

The United States remains the projected dominant biosimilar market. In fact, of all countries, even compared to the European Union, the United States has the largest potential market for biosimilars. Many new and established companies will be moving aggressively into this market, even if this results in overcrowding and a market so fractured that it becomes impossible to realize adequate profit margins. With the US market so unpredictable, and with multi-billion dollar potential openings in the coming years, many companies will be pursuing risky biosimilars development projects.

Much as with generic drugs, it is likely that soon after patent and market, orphan designation, and data exclusivities expired there will be 10–12 or more biosimilars for each current major reference product and also biobetter versions. For example, several months ago, when Singulair went off-patent, 10 generic drugs were approved the same day.

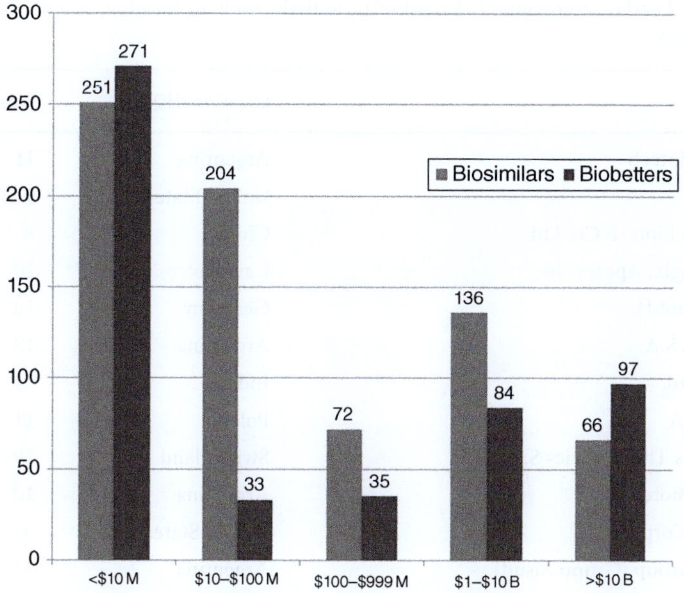

Figure 13.4 Number of companies in involved in biosimilars by size. The previous figure shows the numbers of companies known to be active in biosimilars development by company size. The table in the following shows the number of biosimilars in development by leading companies, in terms of number of biosimilars having been reported in development. (*See insert for color representation of the figure.*)

13.7 Biosimilars Impact on Biopharmaceutical Markets and the Industry

Biosimilars and biobetters will drive an increase in:

a) Biopharmaceuticals being manufactured and marketed worldwide
b) Facilities manufacturing biopharmaceuticals worldwide
c) Companies involved in biopharmaceuticals worldwide

However, while the number of products, manufacturing activity, and companies involved may all increase, the coming waves of biosimilars will result in a contraction and fragmentation of the market. With biosimilars priced below their similar reference products, these cheaper products will take market share from the reference products. So biosimilars and biobetters will not result in any market expansion. Rather existing markets (for similar products) and overall revenues will contract, while the number of biopharmaceutical players will increase. This market contraction is frequently observed with generic

Table 13.2 Leading companies (≥8 biosimilars in their pipeline), location, and number of biosimilars.

Company	Location (HQ)	Biosimilars
Amega Biotech	Argentina	11
Amgen	United States	9
Amoytop Biotech Co. Ltd.	China	8
ApoBiologix, Apotex, Inc.	Canada	10
AXXO GmbH	Germany	12
Bio Sidus S.A.	Argentina	12
Biocon Ltd.	India	14
Bioton S.A.	Poland	11
BioXpress Therapeutics S.A.	Switzerland	19
Cassara Biotech	Argentina	10
Celgene Corp.	United States	9
Chemo Group (Grupo Insud)	Argentina	11
Creative Biomart, Inc.	United States	13
Daiichi Sankyo, Inc.	Japan	10
Green Cross Corp.	Japan	12
Harvest Moon Pharmaceuticals USA, Inc.	United States	28
Hospira	United States	10
Inbiopro Solutions Pvt Ltd.	India	14
InSight Biopharmaceuticals Ltd.	Israel	9
Intas Biopharmaceuticals Ltd.	India	9
LG Life Sciences Ltd.	Korea	10
Mylan Labs	Switzerland	14
Novartis AG/Sandoz AG	Switzerland	10
Pfizer, Inc.	United States	14
Reliance Life Sciences Pvt. Ltd.	India	14
Syd Labs, Inc.	United States	8
Therapeutic Proteins Inc. (TPI)	United States	8
Viropro, Inc.	Canada	10
Zydus Cadila Healthcare Ltd.	India	18

Note: Companies yet to disclose the identity of their candidate products in development are not included. This underreporting likely, in general, counterbalances products having been reported in development (counted previously) but having been more recently had their development abandoned.

drugs, where the generics are discounted much more than the 30% expected with biosimilars, with discounted generics rapidly absorbing 90% market share after the initial launch.

In parallel with market contraction, the number of products and manufacturing facilities will greatly expand. There are currently about 1000 biopharmaceutical manufacturing facilities worldwide, with these reported and ranked by their manufacturing capacities (BioPlan's www.Top1000bio.com) [2]. With biosimilars and biobetters, there will be many new facilities coming online and much expanded use of current in-house and CMO facilities, including many biosimilar manufacturers adopting single-use systems for their upstream manufacturing. These include the largest pharmaceutical companies. Many of these large companies have the capacity and resources to be the leaders in the exploitation of these new opportunities in terms of capturing market share for their products.

13.8 Marketing Biosimilars Will Be a Challenge

Marketing of biosimilars will be challenging in the industrialized countries. Physicians and consumers will need to have confidence in what many will perceive as lower-quality, lesser-tested "generics." Biosimilars and biobetters will have to be marketed much the same as innovative products, since at least for the foreseeable future, no biosimilars can be expected to be approved with generic drug-like ability for interchangeability and substitutability.

Unlike many generic drugs, biosimilar approvals in the United States and the European Union will not include formal interchangeability and substitution with their reference products. Pharmacists will have to fill prescriptions as they are specified by physicians, which explains why the concept of using nonproprietary names is highly controversial. Automatic interchangeability/substitution, as with A–B ratings with generic drugs, enables them to be distributed without the need for marketing. Rather, cheaper generics are simply substituted when prescriptions are specified or filled, with selection made primarily on the basis of price. If the FDA decides to use generic names for biosimilars, this largely removes the need for their marketing, with biosimilars likely often be automatically used to fill prescriptions rather than the reference product.

13.9 Biosimilar Manufacturing Will Be State of the Art

Biosimilars need to be cost competitive and, thus, require cost-effective manufacturing, including using newer, advanced expression systems and disposable upstream and downstream bioprocessing systems. In this context, all biopharmaceutical developers will be increasingly forced to ruthlessly evaluate the cost-effectiveness of their manufacturing systems and to rapidly adopt the most recent improvements in bioprocessing. Thus, many biosimilars will be manufactured at state-of-the-art bioprocessing, while their reference products are still manufactured using the same legacy processing originally adopted decades ago with only minor modification permitted within the restrictions of their FDA approval. Biosimilars can also be expected to disproportionately pioneer many new technologies, such as using novel expression systems or alternatives to protein A and other conventional chromatography processing.

Realizing this, nearly all biosimilar and biobetter developers are not seeking to reverse engineer and duplicate reference products' manufacturing methods developed decades ago. Rather, biosimilar manufacturers are pursuing high yields and efficiencies in product manufacture. For some, this involves having a CMO or in-house facilities plug their product into well-established in-house or licensed-in manufacturing platforms, such as the CMO adapting its current preferred CHO expression system to the manufacture of a biosimilar/biobetter mAb. In many other cases, the in-house or CMO manufacturer will adopt or adapt the latest bioprocessing methods, including novel expressions systems, which may require additional licensing.

13.10 Biosimilars Will Increase Demand for Product Quality and Transparency

Biosimilars will also raise the bar for product quality expectations and related information for all biopharmaceuticals, with regulators, healthcare professionals, patients, and the public focusing on the details of manufacturing processes and the analytical, bioprocessing, and related quality differences between products. Biosimilar approvals are based on similarities and analogies, and related disclosures, product launches, and marketing will increasingly have to discuss similarities and differences between rather similar products.

This will require more public information disclosures by manufacturers, both from the reference product and biosimilar companies, in order to justify their products to the regulatory agencies, physicians, and consumers. Developers will need to prove to skeptical regulators, healthcare professionals, and patients that their new biosimilars are as safe and effective as the decades-old legacy reference. In many cases, new biosimilars will be purer and have fewer detectable contaminants than old reference products.

These conditions create a cycle, in which increased competition drives reference product manufacturers to improve their products and related communications in order to fend off biosimilar developers and to make analytical and clinical trial comparisons with their products more difficult. In this fashion, biosimilars will force the biopharmaceutical industry and regulators to more transparency and making more product information available. The losers, in terms of marketing, could well be those that report the least information about their products.

With biosimilars, the FDA and other regulators will be forced to significantly increase the amount of bioprocessing, analytical, and other product quality-related disclosures required to support publically their regulatory approvals. A citizen petition has been filed with FDA seeking improved nomenclature and substantive public information disclosures regarding all biologics approvals, both full and supplemental [3]. This includes FDA assigning both unique and biosimilar/biogeneric-type names to approved products and FDA disclosing basic requisite information concerning approved product's bioprocessing and quality, including discussion of names prior to approvals, including at public advisory committee meetings.

13.11 CMOs Benefit from Biosimilars

Contract manufacturer organizations are among the major recipients of benefits from the expanding production of biosimilars and biobetters. Many are already experiencing increased demand for bioprocess development, scale-up, and manufacture of preclinical and clinical supplies, with many reporting an overall 15% increase in business due to biosimilar contracts. With a trend among established companies, particularly the largest ones, being to outsource many of their activities, a demand has arisen for increased bioprocessing capacity and in-house expertise among CMOs. Many product developer companies will prefer to devote in-house manufacturing capacity to newer, innovative, and higher-profit products.

Thus, many of the companies engaged in the biosimilar and biobetter industry will be using the services of CMOs to manufacture their products. In other cases, they can also be expected to license-in biosimilars/biobetters from small and foreign developers and then contracted to manufacture these products with a CMO. This analyst projects that 40% or more of biosimilars and biobetters will ultimately be manufactured by CMOs, a much higher portion than with reference products.

Already, many leading biopharmaceutical CMOs are seeing significant increases in business from biosimilar development contracts. For some, the challenge is handling multiple biosimilars from different sponsors at once. CMOs are grappling with this issue, following their established business model of strict separation of protocols for different products. Other issues that must be resolved include sequestration of intellectual property. This may result in redundancy in development, with the CMO incurring additional expense in development and testing in order to avoid overlapping of diverse contractual obligations and risk of IP violations.

Also, the rapid expansion of biosimilar processing technology is leading CMOs to be more competitive in their price quotations for product development. For example, it is rather common for CMOs to quote ca. $3 million as the basic cost for biosimilar development, including selection of a clone expressing a biosimilar protein or antibody with scale-up through cGMP manufacture at ≥100 l scale, generally sufficient for most clinical trials and available for further scale-up for commercial manufacturing. While additional analytical studies, such as mass spectrometry, should be performed, the initial package establishes feasibility. Further development may run to $50–$100 million or even more, including analytical studies, trials, bioprocessing scale-up, regulatory filings, and other preapproval commitments.

An alternative business model is for the CMOs to become biosimilar developers or at least develop and manufacture the initial formulation and transfer it to another entity to build into a finished product. Other biosimilar developers, particularly smaller companies, can be expected to also develop and market biosimilar active pharmaceutical ingredients, with the costs, resources, and time required for biosimilar development, approval, and marketing beyond their capabilities.

There is a reason to believe that the biosimilars (and biobetters and biogenerics) market will be much like the generic drug market. This includes active pharmaceutical ingredients sources and competing products from a range of international companies. However, unlike the generic sector, development and manufacturing will be located in the major market in highly developed countries. As international competition and the available number of biosimilars and other follow-ons increase, buyers will favor the US and EU manufactured products, particularly if these are priced competitively, that is, not priced at too high a premium relative to biogeneric, non-GMP products manufactured in lesser-regulated countries.

13.12 Conclusions

Biosimilars are a very new class of biopharmaceuticals that will be rapidly growing in the coming years. There is a healthy pipeline of biosimilars in development, most of them targeted for the United States and other major markets. Biosimilars will be very competitive, with this having implications including requiring use of highly efficient, if not state-of-the-art, bioprocessing for their commercial manufacture. Biosimilars will bring many new entrants in the biopharmaceutical industry while also increasing business for CMOs.

References

1. Rader, R.A., Biosimilars/Biobetters Pipeline Database, www.biosimilarspipeline.com (accessed 3 May 2016).
2. BioPlan Associates (2015), Top 1000 Global Biopharmaceutical Facilities Index, www.Top1000bio.com (accessed 3 May 2016).
3. Rader, R.A., Citizen Petition: Biologics nomenclature and public information: suitably descriptive names/identifiers and public disclosures are needed. Filed with FDA, 21 June 2013.

14

Cell Line and Cell Culture Development for Biosimilar Antibody-Drug Manufacturing

Jianguo Yang

Abpro Corporation, Woburn, MA, USA

14.1 Summary

Almost all major pharma companies are heavily committed to biosimilar product development resulting in many active players within the industry. Likely, pharma companies with well-integrated drug R&D divisions and robust clinical trial and sales networks will come to dominate the field. Due to a wealth of current biosimilar antibody-drug candidates in various development stages, biopharmaceutical companies are increasingly interested in a platform process for mammalian cell line and cell culture development to meet productivity and quality attributes with maximum efficiency.

Cell line and cell culture processing are the most critical steps for biosimilar antibody development since both can determine productivity, quality, and cost benefits of vital import to the bottom line. For the last several years, average productivity of mammalian cell lines has reached 3–6 g/l. These gains were achieved using fed-batch process, high-productivity cell lines, and optimized media and bioreactor platforms. These innovations are dramatically cutting costs and speeding drug development. Meanwhile, product quality is the key factor for success as many companies compete against one another in a highly competitive landscape.

This chapter will review in detail the current state of mammalian cell line and cell culture process development for biosimilar antibody production.

Biosimilars of Monoclonal Antibodies: A Practical Guide to Manufacturing, Preclinical, and Clinical Development, First Edition. Edited by Cheng Liu and K. John Morrow, Jr.

14.2 Mammalian Cell Line Development

14.2.1 Host Cells

Over 60% of therapeutic proteins are produced in mammalian cell lines although numerous alternatives are available (Fig. 14.1). Chinese hamster ovary (CHO) cells and NS/0 murine myeloma cells are two major mammalian host cells widely used in biopharma industry for antibody-drug production [1–5]. The PER.C6® human cells are also a prime choice for human antibody expression, but because of its much shorter history and more limited market and regulatory experience, it has yet to gain the popularity of the mainstay participants. Choice is determined by the host cell ability to deliver high productivity with acceptable product quality attributes and the historical usage of individual companies, influenced by their biologic drug development experience and patent landscape.

The majority of recombinant antibodies [6] produced today is made in CHO cells, especially in DHFR negative CHO hosts due to the absence of patent restrictions and loyalty/license fees. There are three major CHO hosts (Fig. 14.2), including DUXB11, DG44, and CHOK1 lineages. DUXB11 (with a

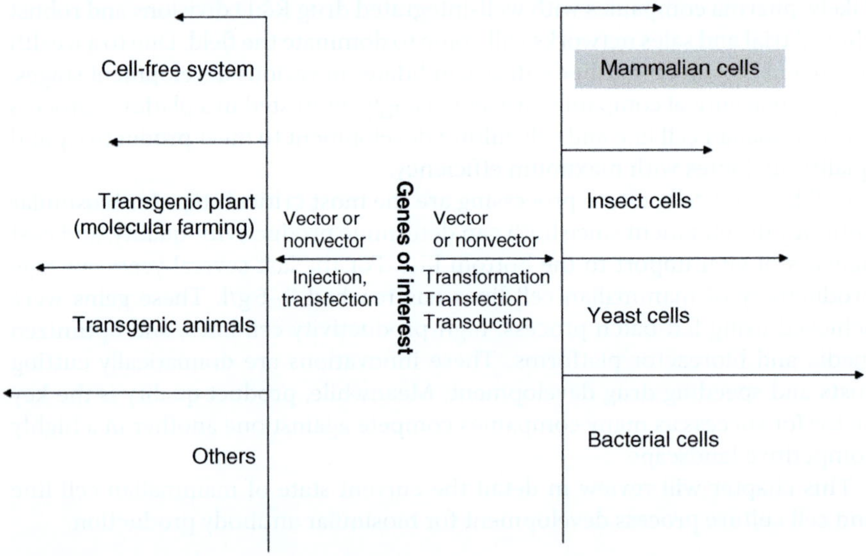

Figure 14.1 Expression systems for biologic products. Mammalian cells are the most popular systems for antibody production, bacterial cells are for antibody fragments or Fab production, yeast cells have heavy glycosylation issue, and insect cells produce proteins with heavy phosphorylation. Plant plan expression platforms for food proteins and animal systems for reagent production. Cell free systems are optimal for protein production toxic to the host cells.

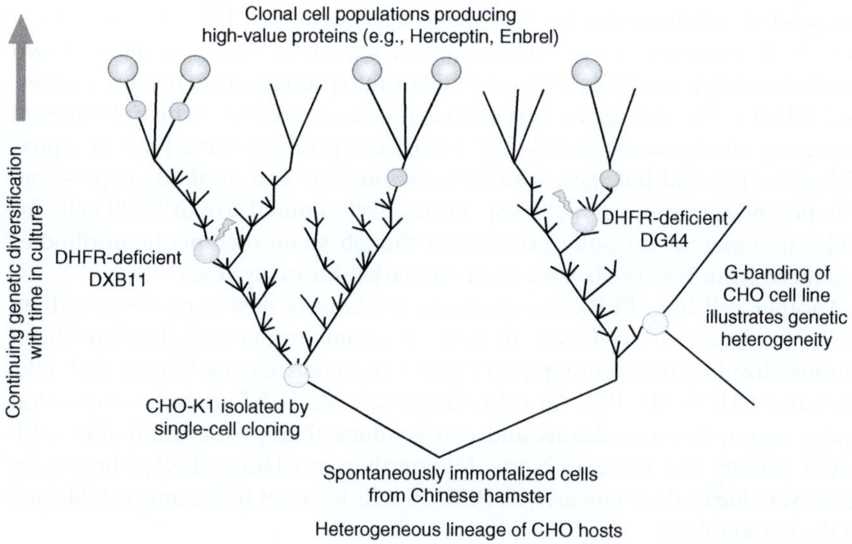

Figure 14.2 Major CHO host cells. CHO-K1 and DXB11 have closer lineage relation, but DG44 is a variant from the CHO prototype. All other cell lines are derived from the original three ancestral parent lines. Courtesy from Dr Florian Wurm, Nature Biotechnology, Vol. 29: 718–720, August 2011.

single DHFR gene deletion) and DG44 (with a double DHFR gene deletion) host cells are from different CHO ancestor lineage but both lack dihydrofolate reductase (DHFR) activity. However, DUXB11 has weak DHFR activity because it carries one copy of the DHFR gene and requires higher methotrexate (MTX) concentrations than the DG44 for effective MTX selection. CHOK1 cells have endogenous DHFR activity and are not satisfactory for DHFR selection. A variant referred to as CHOK1SV with very weak glutamine synthetase (GS) activity has been a very popular host for use in the GS protocol using methionine sulfoximine (MSX) selection pressure.

Murine NS0 nonimmunoglobulin secreting myeloma cells are cholesterol auxotrophs requiring the presence of cholesterol in cell culture medium for growth [7]. Cholesterol-independent NS/0 cells have been established [8–10], but the cholesterol-independent NS/0 cells do not have better growth parameters than their NS/0 ancestor. NS/0 cells lack endogenous antibody and GS enzyme activity making them suitable host cells for GS as a selectable marker for recombinant antibody expression [11]. NS/0 recombinant antibody cell lines have very high antibody productivity. The titer can be over 6 g/l from the fed-batch process, and high antibody titers have been also reported from non-GS NS/0 cell lines [12, 13]. Mouse-derived cell lines, including NS/0, produce *N*-glycolylneuraminic acid (NGNA, a sialic acid that cannot be synthesized by humans) at adequate levels. This sialic acid form was believed to cause a potential immunogenicity reaction in humans. However, it was later shown that humans apparently incorporate NGNA into proteins from dietary sources, and CHO cells also can express low levels of NGNA, which tempered immunogenicity concerns [14–16]. NS/0 cells produce some level of alpha-Gal-alpha(1,3) Gal linkages, which causes humans anti-antibody expression. The potential immunogenicity aspect has likely limited use of NS/0 cells for wider therapeutic antibody production though some therapeutic antibodies produced from NS/0 cells have been on market for many years.

Another cell line, PER.C6, represents a relatively new expression system. PER.C6 cells are derived from human embryonic retina cells that have been immortalized by transfecting the E1 genes from adenovirus 5 DNA [17]. Like NS0 and CHO cells, PER.C6 cells can proliferate indefinitely in suspension under serum-free conditions and can produce therapeutic antibodies with better activity and more authentic folding than in CHO and NS/0 host cells. However, due to their late arrival, PER.C6 cells have yet to become established in the marketplace.

14.2.2 Cell Line Engineering

Engineering of standard cell lines is frequently employed in order to alter the product quality or improve robustness. Controlling glycosylation is very important for antibody quality because glycan forms on antibodies can have substantial impacts on immunogenicity, clearance rate, and bioactivities [18]. Antibodies produced in CHO cells typically have complex biatennary structures with very low or no bisecting *N*-acetylglucosamine (bisecting GlcNAc) and high levels of fucosylation [19]. Overexpression of *N*-acetylglucosaminyltransferase III has been used to increase the fraction of bisecting GlcNAc that resides on antibodies to improve antibody-dependent cellular cytotoxicity (ADCC). RNAi and gene knockout technologies have also been used to reduce fucosylation on antibodies to dramatically increase ADCC activity [20, 21]. In general, glycosylation is hard to control precisely in mammalian cells because it is dependent on multiple factors such as host cell species, recombinant cell line specificity, media, and cell culture process conditions.

It was reported that overexpression of antiapoptotic genes and RNAi technology-mediated knockdown expression of apoptotic genes have been used to improve the culture viability that leads to improved productivity [22]. Due to incompletion of oxidizing glucose to CO_2 and H_2O in mammalian cell cultures, lactate accumulation can cause acidification of culture medium and lead to high osmolarity and low viability because of the alkali added to control the medium pH. A large body of research work [23–25] has been performed to reduce lactate accumulation; however, degree of lactate concentration in cell cultures may be clone and medium dependent.

Another interesting area for cell line engineering is the use of inducible expression systems in mammalian cells. This strategy has the advantage of activating antibody production at the optimal phase of cell growth. Growth arrest in NS0 cultures via the inducible expression of p21 was reported to result in a significant increase in cell-specific productivity but not necessarily an overall improvement in bulk protein titer [26]. Current practice for the selection of high-producing clones involves a time-consuming and laborious screening strategy due to the fact that the random integration of the transgene of interest gives rise to dramatic different expression levels because of variation in the chromosomal surroundings or unpredictable cross-interactions of multiple gene copies [27]. Some strategies have been proposed to overcome the random integration issue. One of the most popular methods is the use of homologous recombination systems for site-specific gene integration such as the Flp/FRT system from yeast (*Saccharomyces cerevisiae*) and the Cre-lox system from the bacteriophage P1 in eukaryotic cells. The Flp/FRT system was used to target gene insertion into a high expression chromosomal locus in order to establish a viral packaging 293 cell line for consistent high-titer virus production [28]. Fukushige and Sauer [29] demonstrated the use of a lox recombination vector to obtain stable transformants with predictable gene expression profiles.

The application of this platform is based on the identification of integration sites that lead to high expression; however, high expression integration sites may be difficult to activate in different host cell lines and different recombinant gene molecules. The cell line specificity of the integration site determines its effective role site for a given recombinant cell line. However, site integration technology combined with high-throughput screening are the conventional technologies for generating higher expression cell lines.

For example, transfection of DHFR deficient CHO cells with a loxP–GFP fusion and DHFR genes and high-throughput screening via FACS was used to identify transcriptionally active sites [30]. Simultaneous transfection with two expression constructs and selection with two different selective markers to generate highly productive stable CHO cell lines appears promising. These methods should result in greater efficiencies and a further reduction in cell line development timelines.

Other optional expression systems such as *Pichia pastoris* and *Escherichia coli* often serve as hosts for mAb secretion [31]. Researchers at Glycofi (a subsidiary of Merck & Co., Inc., Kenilworth, NJ) have successfully demonstrated the feasibility of glycoengineered *P. pastoris* cell lines to produce mAbs with highly specific glycoforms [32]. Various glycoforms of commercially available Rituximab (Rituxan; Genentech, Inc., South San Francisco, CA) were generated, and binding to the Fc receptors and ADCC activity were measured. These research results demonstrated a tenfold increase in binding affinity, as well as enhanced ADCC activity with the glycan-engineered proteins compared with

Rituximab. Controlling the glycan composition and structure of IgGs thus appears to be an effective method for improving the efficacy of therapeutic mAbs that utilize ADCC for biological activity. Coupled with the use of well-established *P. pastoris* as a platform, processes that include high cell density cultures, scalability, cost-effectiveness, and existing large-scale fermentation capacity can allow for high-fidelity production of human glycosylated therapeutic proteins. However, these glycan-engineered proteins have not yet received FDA approval, and possible negative side effects have not been investigated.

E. coli has been often used for production of peptides, small proteins, and antibody Fabs that are utilized when Fc-mediated effector functions are not required or deleterious [32]. Simmons et al. [33] demonstrated that efficient secretion of heavy and light chains in a favorable ratio resulted in the high-level expression and assembly of full-length IgGs in the *E. coli* periplasm. The technology described offers a rapid and potentially inexpensive method for the production of full-length aglycosylated therapeutic antibodies that do not require ADCC functionality. Mazor et al. [34] also showed that it was possible to obtain full-length antibodies from combinatorial libraries expressed in *E. coli*. The full-length secreted heavy and light chains assembled into aglycosylated IgGs that were captured by an Fc-binding protein located on the inner membrane. Flow cytometry was used after permeabilization of the membrane and attachment of the antibody to a fluorescent antigen.

14.2.3 Cell Line Development

As major cell line development steps are articulated in Figure 14.3, production cell line development starts with expression vector construction and transfection. After being transfected with plasmids bearing the recombinant genes, as well as selection marker(s), cells are screened for growth and productivity following growth recovery, serum-free suspension adaptation, amplification (if necessary), and clonal selection. The screening and selection of a highly productive and stable clone from the transfectant population in a limited time frame is a major challenge. For the rapid development of production cell lines, platforms for vector system, transfection, and high-throughput clone selection are vitally important for biosimilar drug development. Mammalian expression vectors normally have two cassettes: one for protein-coding genes (genes of interest (GOI)) and selectable marker gene(s) for expression in mammalian cells and a second cassette for the genes enabling plasmid replication in bacteria. To achieve high levels of protein or antibody expression, strong promoter/enhancers such as the cytomegalovirus (CMV) promoter [35] and elongation factor alpha (EF1α) promoter [36] are used to drive protein or antibody expression. Often an intron sequence in the 5′ untranslated region is included after the promoter/enhancer to increase export of transcribed mRNA to the

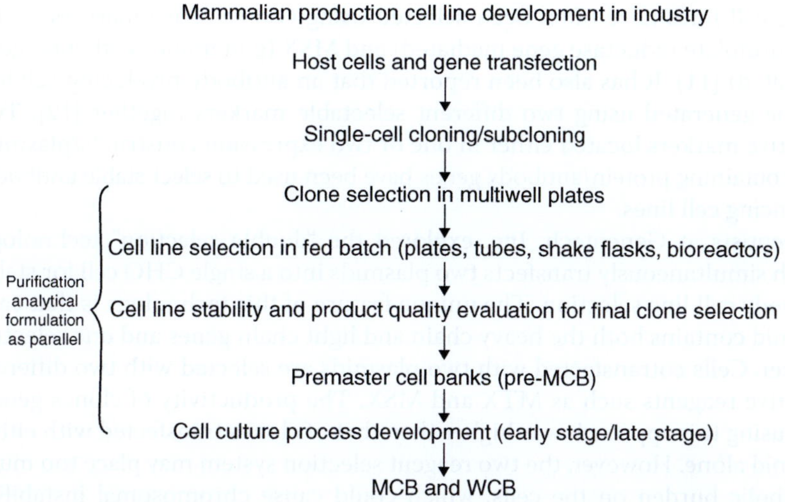

Figure 14.3 Production cell line development steps. From the gene transfection step to the final MCB/WCB point is dependent on the technologies available to each company. The timeline for each step is varied. A 6-month timeline is currently feasible for most companies.

cytoplasm from the nucleus, and one or more 3a polyadenylation signal sequences are included to maximize mRNA levels. Generally used polyadenylation signal sequences are the SV40 late or early polyadenylation signal sequences and the bovine growth hormone polyadenylation sequence. In addition to transcription, translation and secretion are also required for protein production. Normally, a consensus Kozak sequence [37] is created by placing GCC GCC(A/G)CC immediately in front of the first translation initiation codon to enhance translation initiation, while a signal peptide sequence is placed immediately in front of the mature protein peptide to direct protein secretion [11]. The leader sequences can be optimized for different secretion.

Several transfection methods have been developed to introduce vector DNA into chromosomal DNA in mammalian cells, including calcium phosphate, electroporation, cationic lipid-based lipofection, and polymer or dendrimer-based methods [38]. All four methods have been used for stable transfection. However, electroporation is the common choice for industrial cell line development, and lipofection is often used for cell line generation in early research. Transfected cells are then selected, relying on different selectable markers that can be categorized into two groups: metabolic selectable markers and antibiotic selectable markers. DHFR and GS (glutamine synthase) are common metabolic markers for production cell line selection. Antibiotic markers are rarely used for mammalian cell line generation in industry but often in academic labs. NS/0 stable cell lines have GS selection marker, while CHO stable cell lines have often been selected using two selective markers: MTX (dihydrofolate reductase gene mediated) and MSX (glutamine synthetase gene mediated) [11]. It has also been reported that an antibody-producing cell line can be generated using two different selectable markers together [12]. Two selective markers located either in one or two expression constructs/plasmids also containing protein/antibody genes have been used to select stable antibody-producing cell lines.

Scientists at Genentech, Inc. explored the "double selection" technology, which simultaneously transfects two plasmids into a single CHO cell for stable antibody cell line selection. The unique feature of this technology is that each plasmid contains both the heavy chain and light chain genes and one selective marker. Cells cotransfected with two plasmids are selected with two different selective reagents such as MTX and MSX. The productivity of clones generated using this approach was higher than that of clones transfected with either plasmid alone. However, the two reagent selection system may place too much metabolic burden on the cells, which could cause chromosomal instability. A cell line stability study should be conducted to verify this issue before embarking upon cell line production.

After transfection, transfected cells are diluted and cultivated in 96-well plates with production medium and screened for robust cell growth and high productivity under selection pressure. A single medium should be used throughout the process. For production cell lines with adequate productivity for clinical needs, a few thousand clones should be screened. The primary screen is usually a protein/antibody concentration assay with or without cell number normalization, to eliminate non- or low producers. While high producers are scaled up, additional assays are performed to measure cell growth, cell-specific productivity, and volumetric productivity (titer) in order to choose the top candidates (normally 50 clones). These are subsequently evaluated in a fed-batch cell culture scale-down models such as shake flasks or laboratory scale bioreactors. Product quality attributes are further evaluated in bioreactors. At this stage, several clone attributes should be considered and evaluated for features such as product quality, manufacturability, and volumetric productivity.

With increasing demand for protein therapeutic candidates, good platforms are essential to keep the pipeline flowing. For biosimilar antibody manufacturing process development, maintaining acceptable similar quality attributes to innovator antibody drug while reducing time to market, maintaining cost-effectiveness, and providing manufacturing flexibility are key issues in today's competitive biosimilar market, where many companies are often working on same biosimilar drugs and clinical indications.

Since antibody therapies may require large doses over a long period of time, manufacturing capacity becomes a critical issue. In response to the strong demand, many companies have built large-scale manufacturing plants containing multiple 10,000 l or larger cell culture bioreactors. However, in recent years,

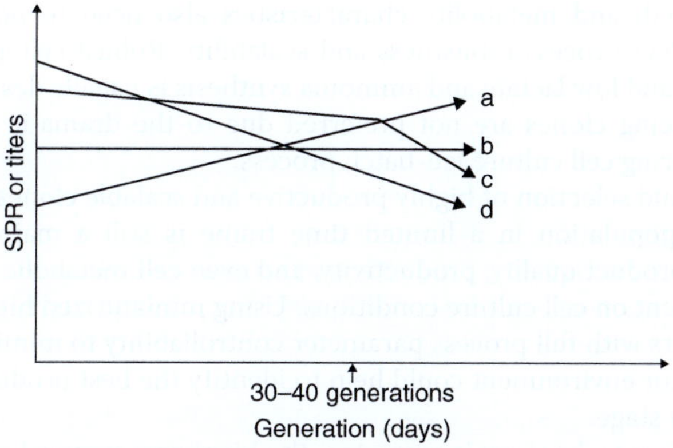

Figure 14.4 Four categories of cell line productivity. The stability falls into four categories: (a) productivity increases over time, (b) productivity stays no change, (c) productivity drops quickly after long period of generation time, and (d) productivity drops gradually over time. a and b are less common, while c and d are more common.

the biopharma industry has dramatically increased cell line productivity to 3–5 g/l or even higher in some cases through application of high-throughput cell line development and enhanced media and bioreactor process improvement.

Selection of the final production clone is generally considered to be one of the most critical decisions in both early and late stage cell culture process development. Since changes in production cell lines during clinical development are considered major process changes, product comparability must be demonstrated if the cell line is changed during late stage development. Changing the cell line after phase III clinical study typically requires additional human clinical studies. It is thus absolutely essential to select the right clone prior to phase III production of drug substance, preferably at the phase I stage. Other features of the platform are critical as the product candidates move forward from small-scale development lab to pilot plant scale and eventually to commercial scale current good manufacturing practices (cGMP) manufacturing [35–37].

Cell line stability is one major issue during production cell line development since volumetric productivity and specific productivity decline over cell culture age for unstable cell lines. In general, there are four types of cell line stability during production cell line development (see Fig. 14.4). Unstable cell lines are not suitable for large-scale production as cell culture process is scaled up through serial culture passages of the seed train and inoculum train and final production stages. In some cases, cell cultures may complete 30 generations or more over a month time span. This could result in a productivity drop of more than 50% when the cell culture reaches harvest stage. In addition to cell line stability, growth and metabolite characteristics also need to be evaluated, which can affect process robustness and scalability. Robust cell growth with high viability and low lactate and ammonia synthesis is usually desirable. High lactate-producing clones are not preferred due to the dramatic increase in osmolality during cell culture fed-batch process.

Screening and selection of highly productive and scalable clones among the transfectant population in a limited time frame is still a major challenge because the product quality, productivity, and even cell metabolic profiles are often dependent on cell culture conditions. Using miniaturized high-throughput bioreactors with full process parameter controllability to mimic the large-scale bioreactor environment could help to identify the best production clone at a very early stage.

As cell culture technology is maturing, the biopharmaceutical industry has applied platform processes to meet material demand and quality requirements within a short period of time. The cell line and cell culture platform often consist of common host cell, expression vector, transfection, and selection methods during cell line generation and standard cell culture media, process control, and scale-up methodologies during process optimization. This approach not only enables fast process development but also provides predictable performances in scale-up, facility fit, and downstream process integration.

Manufacturability and scalability of mammalian cell cultures have historically been considered difficult to work with due to multifactor complexities such as low cell growth and productivity, medium complexity, serum requirement, and shear stress, although the latter has generally been incorrectly overemphasized. After more than two decades of intensive development work in cell line, media, and bioreactor condition optimization, cell-specific productivity of 20–30 pg/cell/day can be routinely achieved for production cell lines [38]; high titers up to <10 g/l and cell densities of over 30 million cells/ml in fed-batch processes have been recently reported by a few companies at major conferences (Fig. 14.5). The enhancement of specific productivity per cell is achieved not only by selection of highly productive clones but also by optimization of medium composition and bioreactor operation conditions. Due to this advance, bioreactors over ten thousand liters will be no longer necessary in future commercialization process. However, the combination of high titers and large capacity has gradually shifted the focus of cell culture process development from pursing even higher titers to controlling product quality and process consistency at all development stages and production scales [39].

14.3 Cell Culture Process Development

Cell culture process development starts with cell line generation and selection, followed by process and media optimization in small-scale systems, including multiwell plates, shake tubes and shaker flasks, and bench-scale bioreactors.

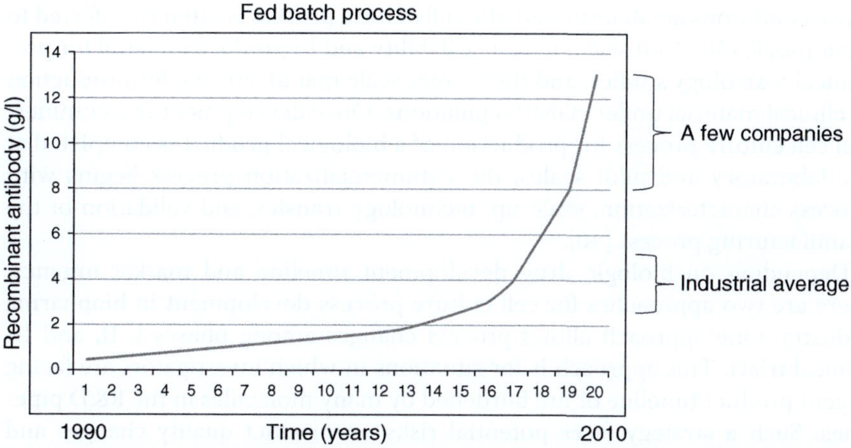

Figure 14.5 Cell line productivity. Numbers can reach over 10 g/l, due to high-productivity cell lines, balanced media, and optimized bioreactor processes. Currently 3–5 g/l is standard in biopharma industry.

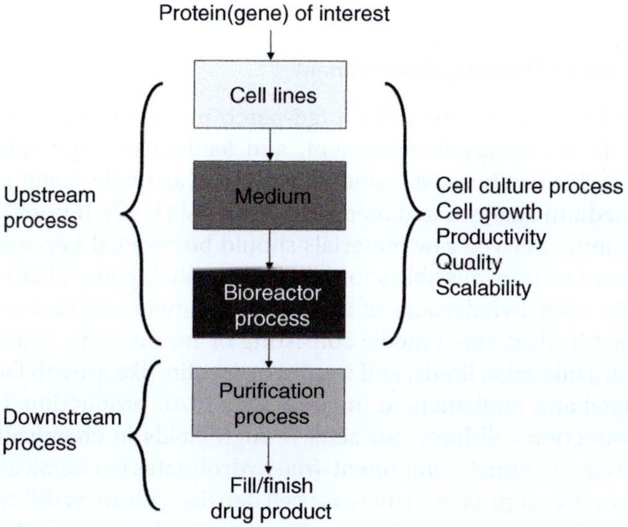

Figure 14.6 Bioprocess steps. Upstream (cell line and cell culture) and downstream (purification and formulation/fill–finish). Analytical processing is ideal for supporting both steps.

There are about five to six cell culture processes (Figure 14.6), but fed batch and perfusion are two of the most pivotal. Under conventional conditions, fed-batch process is optimal for stable antibody molecule drug development, and perfusion cell culture is preferred for unstable protein molecule production.

Once conditions are determined, the cell culture process is often transferred to a pilot scale (50–100 l scale) to test scalability and to produce material for preclinical toxicology studies, and then larger scale manufacturing for production of clinical material under cGMP regulations. Once development of a commercial cell culture process for production of a biological product is completed at the laboratory and pilot scales, the commercialization process begins with process characterization, scale-up, technology transfer, and validation of the manufacturing process [40].

Depending on biologic drug development timeline and market urgency, there are two approaches for cell culture process development in biopharma industry. One approach allows process changes among phases I, II, and III clinical trials. This approach is for situations in which investigators are facing urgent product timeline or are burdened by many molecules in the R&D pipelines. Such a strategy faces potential risks on product quality changes and may involve a need to repeat clinical trial experiments per regulatory agency concerns. A second approach employs standard platform process modifications through phase I to phase III and does not allow for changes at the level of commercial process. This approach is the favored trend in the current biopharma milieu.

14.3.1 Medium and Feeding Development

Typically, medium development for a fed-batch process involves basal batch medium, feed concentrate development, and feeding strategy optimization. Several approaches can be used systematically, such as single-component titration, spent medium analysis, and medium blending [41, 42]. If possible, bovine serum and animal-derived raw materials should be avoided because of safety concerns related to transmissible spongiform encephalopathy (TSE) and other contaminants. After two decades of media development and host cell adaptation, fully chemically defined media consisting of amino acids, vitamins, trace elements, inorganic salts, lipids, and insulin or insulin-like growth factors have been developed and implemented in large-scale mAb production, but not all antibody production cell lines can achieve high yields in chemically defined media. Addition of animal-component-free hydrolysates to chemically defined media is a common approach to increase cell density, culture viability, and productivity, sometimes better product quality in a timely manner. Hydrolysates are protein digests composed of amino acids, small peptides, carbohydrates, vitamins, and minerals that provide nutrient supplements to the media [43, 44]. Nonanimal-derived hydrolysates from soy, wheat, and yeast are used commonly in cell culture media and feeds; however, because of its composition complexity and lot-to-lot variations, hydrolysates can be a significant source of medium variability. To avoid this issue, the amount of hydrolysate usage should be above threshold.

While the most common approach to develop a feed medium is to concentrate the basal medium, sophisticated optimization of feed composition and feeding strategy requires consideration of nutrient consumption, by-product accumulation and the balance between promoting growth versus volumetric productivity [45, 46]. Previous studies indicated that by-products such as lactate and ammonia could be minimized by maintaining low glucose and glutamine concentrations through frequent or continuous feeding or perfusion process [47], but the operational complexity and cost of validating continuous feeding strategies have generally made them less desirable for large-scale manufacturing. However, for some product quality and stability concerns (such as enzymes and growth factors), perfusion or continuous feeding has to be implemented. Scheduled or necessary bolus addition of the feed solution to the production bioreactor is most widely used due to its simplicity and scalability.

Production medium development is labor intensive and time consuming. A combination of high-throughput cell culture scale-down systems with statistical design of experiment (DOE) approaches is commonly applied to shorten development time [48, 49]. Though optimization of cell culture media is often considered to be cell line dependent and best accomplished if based on the metabolism and nutrient consumption of specific cell lines, the aggressive timelines require development of a "platform" approach for optimization of cell culture processes. While the platform process might not provide the greatest volumetric productivity that could be achieved with specific cell lines, it has wide applicability allowing it to be implemented with minimal modifications for cell lines derived within the company. Hence, development of the platform process is accomplished with multiple cell lines that are representative of the various growth, productivity, and metabolic phenotypes commonly observed. If business needs dictate that processes for specific cell lines be optimized further, the platform process could be used as a starting point to enhance the cell culture process to achieve specific process goals such as higher volumetric productivities.

14.3.2 Bioreactor Process and Scale-Up

Bioreactors offer better cell culture growth conditions for most cell lines. Culture operating parameter optimization is required to achieve high expression of product with acceptable product quality profiles. These parameters are physical, chemical, and biological in nature. Physical parameters include temperature, gas flow rate, and agitation speed, while chemical parameters include dissolved oxygen (DO) and carbon dioxide, pH, osmolality, redox potential, and metabolite levels, including substrate, amino acid, and waste by-products. Biological parameters are used for determining the physiological state of the culture and include viable cell concentration, viability, and a variety of intracellular and extracellular measurements such as NADH, LDH levels, mitochondrial activity, and cell cycle analysis. Variations in the microenvironment parameters from optimal levels can have a dramatic impact on culture performance, productivity, and product quality. A typical stirred tank bioreactor is equipped with temperature, pressure, agitation, pH, and DO controls. Due to the complexity of protein products that include isoforms and microheterogeneities, the performance of the cell culture process can have significant effects on product quality and potency, especially with respect to glycosylation, posttranscriptional modifications, and impurity profiles [35–37, 50]. Therefore, the bioreactor operating parameters have to be optimized and characterized thoroughly to improve process performance and understanding of how the process affects product quality. Significant progress has been made in the last two decades in understanding the impact of the cellular environment on culture physiology and its subsequent impact on productivity, product quality, and downstream processes.

The FDA has proposed a process analytical technology (PAT) initiative [51]. The PAT directive is a "system for designing, analyzing, and controlling manufacturing through timely measurements (i.e., during processing) of critical quality and performance attributes of raw and in-process materials and processes, with the goal of ensuring final product quality." It also encourages a better understanding of the manufacturing process by identifying relationships among process parameters, reproducibility, culture performance and product quality attributes. Although originally conceptualized for small molecule manufacturing processes, it is increasingly being extended to biological processes, which are inherently more complex [52].

Process parameters can be measured either online or offline via operator intervention. Typical examples of offline measurements include pH (usually for verification of online pH readings) by BGA, cell counting and viability measurements using automated cell counters or manual count (a hemocytometer), packed cell volume, osmolality, and certain metabolite concentrations. Accurate gas and liquid flow measurements are typically conducted using mass flow meters and magnetic meters. Temperature can be measured accurately using commercially available resistance temperature devices (RTD), although thermocouples can also be used; this parameter is usually shifted during the course of the fed-batch culture to extend culture longevity for higher productivity. One strategy to increase cell culture production period and confine cell growth at peak density range is temperature shift process. Depending cell culture growth and production profile, the temperature shift of the culture is often from 37°C to 28–36°C at near cell peak density or earlier, which can retain cells before the division phase but in G_1 phase longer, and therefore delay cellular apoptosis [53–55].

DO is normally controlled at a specific set point, usually between 20 and 50% of air saturation in order to prevent dissolved oxygen limitation, which might lead to excessive lactate synthesis, and excessively high dissolved oxygen concentrations that could lead to cytotoxicity [56]. The Clark electrode is still the most commonly used. In this type of electrode, oxygen diffuses through a permeable membrane and is reduced at a negatively polarized cathode (platinum) with respect to a reference anode. The cathode and anode are separated by an electrolyte solution, and the reduction of oxygen generates a voltage-dependent current that is directly proportional to the dissolved oxygen concentration. Although cell growth is relatively insensitive to dissolved oxygen in the range 20–100% air saturation [57, 58], the DO level can significantly affect product quality. Kunkel et al. reported that reduction in dissolved oxygen caused decreased glycosylation of antibody N-glycan chains [59].

Dissolved CO_2 is another important process variable factor. This is because it can accumulate to inhibitory levels at values greater than 120–150 mm Hg [60–62] and affect product quality. Accurate online measurement of dissolved CO_2 is still in the development stage. Even though dissolved CO_2 sensors are commercially available, for example, from YSI Inc. (Yellow Springs, OH) or Mettler Toledo (Columbus, OH), dissolved CO_2 is still measured using offline blood gas analyzers such as those from Nova Biomedical (Waltham, MA). The use of *in situ* sensors is not required because it is generally possible to match CO_2 profiles at different scales by the correct choice of bioreactor sparger design, agitation speed, and gas sparge rates. Measurement of oxygen uptake rate (OUR) and CO_2 production rate can also be achieved using online mass spectrometry. Unlike microbial fermentations, this type of analysis is rarely conducted due to the cost of such instrumentation and the lack of sensitivity of measurements at the very low rates of metabolism commonly observed in animal cells.

One of most important process factors is pH, which is difficult to measure and control because even a small deviation (0.1 units) from the optimal pH value can significantly impact culture performance, in particular glucose consumption and lactate production [63, 64]. Cell culture media usually contains sodium bicarbonate as buffering agent, and pH is usually tightly controlled with a combination of CO_2 sparging to reduce pH and base addition to increase it. A pH > 7.0 is usually preferred for the initial cell growth phase, which is accompanied by lactate accumulation. When lactate accumulation exceeds the buffering capacity of the culture medium, pH drifts downward, which could trigger base addition leading to increased osmolality of the cell culture. This could be risky in cell lines that synthesize excessive amounts of lactate since high pH, high lactate, and high osmolality cascade often cause delayed cell growth and accelerated cell death. When cell growth has ceased, lactate is either produced at a much lower rate or consumed. The concomitant upward drift in pH is counteracted by CO_2 sparging. Thus, the pH set point and control range, the so-called dead band, are intimately linked to dissolved CO_2 levels, base consumption for pH control, and therefore osmolality.

Depending on the cell line and medium, a pH control strategy characterized by a reasonable dead band around the set point that would allow the culture pH to drift without triggering the addition of base or CO_2 could be a viable alternative to control either the osmolality or dissolved CO_2. But it is generally not preferred due to potential variability in pH profiles that could be observed in different bioreactor scales. Thus, balancing the impact of pH on cell growth, lactate synthesis, osmolality, and dissolved CO_2 profiles is a key consideration during process optimization. pH probes typically consist of a 0.2–0.5 mm glass membrane fabricated from silicate groups containing sodium that form a hydrated gel layer on the membrane when placed in liquid. An $Ag/AgCl_2$ electrode in KCl electrolyte saturated with $AgCl_2$ is used to measure the change in potential of the outer surface of the membrane caused by changes in the medium pH. Use of pH probes for extended culture duration could lead to drift and reduced sensitivity, and hence online measurements should be verified with periodic offline measurements.

Cell culture media are normally designed to have an osmolality in the range 280–340 mOsm/kg. Culture osmolality is routinely measured using freezing point depression osmometry. Various commercial systems are available, although the analyzers from Advanced Instruments (Norwood, MA) are the most commonly used in cell culture and are capable of measuring osmolality in the range of 0–1500 mOsm/kg of water. The impact of osmolality is cell line and medium specific; deZengotita et al. reported growth inhibition with increasing osmolality and effects on cell-specific productivity [65]. These deleterious effects could be exacerbated when combined with high dissolved CO_2 levels that could occur in high cell density cultures [66]. Therefore it is vital to ensure during process development that the osmolality profile is acceptably low range, especially toward the latter stages of the cell culture process. There are also examples, however, where a high culture osmolality has been used to leverage an increase in specific productivity balanced with a reduction in cell growth to yield an increased volumetric productivity compared to lower osmolality [67, 68], but other examples showing high osmolality has no impact on productivity.

Cell culture metabolites such as glucose, lactate, ammonia, glutamine, and glutamate are commonly measured offline using enzymatic biosensors specific to the measured analyte. These measurements are important to maintain substrate levels above critical levels via feeding strategies and also for reducing by-product formation. Commercially available instruments include the YSI 2700 from YSI, Inc. or Nova Bioprofile 400 from Nova Biomedical and recently CEDEX from Roche Diagnostics Corporation, Indianapolis, IN. These analyzers have biosensors to detect the reaction of a substrate with the enzyme, which produces a product that can be readily measured. For example, glucose can be detected using glucose oxidase in the presence of excess oxygen to produce gluconic acid and hydrogen peroxide. A second reaction between fluorescein and the H_2O_2 produced from the first reaction produces a fluorescent signal that correlates directly to the glucose concentration.

In addition to the quantification of metabolites, the Nova BioProfile 400 can also measure pH, dissolved gases (O_2 and CO_2), NH_4^+ (0.20–25.00 mM), Na^+ (40.0–220.0 mM), K^+ (1.00–25.0 mM), and Ca^{2+} (0.10–10.0 mM). It is important to point out that accurate measurements with these systems some time require verification each other, and proper maintenance and calibration procedures.

Cell concentration and viability are very important indicators of cell growth performance. The cell growth measurements are critical in developing cell culture processes and are used for determining culture physiology in response to operating conditions and calculation of growth rates, specific production rate, and specific consumption/production rates of metabolites. The traditional method for measuring cell concentration and viability is via trypan blue exclusion microscopy. This method depends upon the ability of cells with intact membranes to exclude trypan blue if cells are living; those that stain blue are classified as dead cells. A number of automated analyzers that exploit this method and are coupled to image analysis are commercially available (Cedex, Innovatis, Bielefeld, Germany and Vi-Cell, Beckman Coulter, Fullerton, CA). These analyzers also allow the determination of cell size.

In many cell culture processes, the mean cell size can change by as much as 50% (or over a threefold change cell volume) [69]. In these cases, it is insufficient to measure viable or total cell concentration. Instead, packed cell volume (or viable packed cell volume) is determined using centrifugation-based methods. This may provide better representation of the cell mass since it avoids instrument associated variability, but this method is not suitable for cell count for perfusion culture since cell number is critical for calculating specific perfusion rate, one of important cell culture performance factors in perfusion process.

Guava and other multiparameter flow cytometers using fluorescent dyes such as propidium iodide and annexin V or caspase 5 can be used to quantify live/dead cells and apoptotic cell populations (Guava Technologies, Hayward, CA). These technologies offer faster speed benefit for high-throughput cell culture process in terms of cell counting and cellular assays in 96-well plates, especially DOE for cell culture optimization. Measurements of cell metabolic activity can be quantified by online OUR measurements or offline cell mitochondrial activity using Guava or FACS, which give multilevel of understanding cell culture process, resulting in better process strategy.

Any viable small-scale process has to be scalable. A typical large-scale cell culture process consists of a number of unit operations, including seed train, inoculum train, and production run. Each step has its key process performance indicators such as cell growth, viability and titer, metabolites, and appropriate product quality attributes. The success of process scale-up is usually measured by these key process indicators and product quality attributes that meet acceptable range.

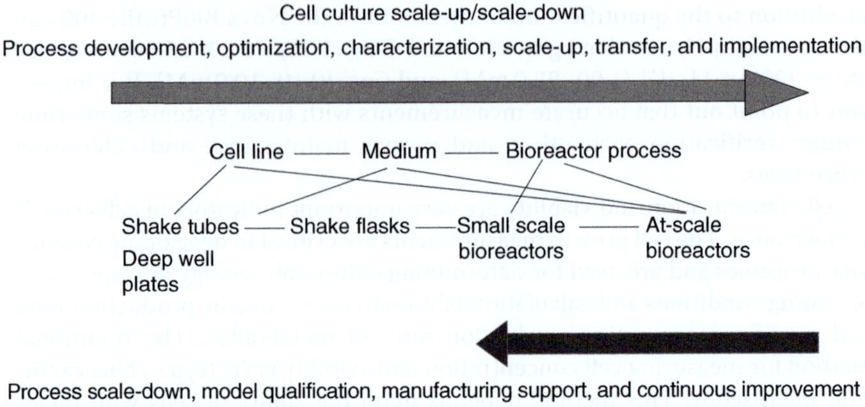

Figure 14.7 Process scale-up. Manufacturing scale process development from multiwell scale to commercial production. Process scale-down is for process characterization, validation, and manufacturing process troubleshooting and commercial process tech support; normally from manufacturing scale to benchtop bioreactor scales.

In general, mammalian cells are difficult to scale up due to their sensitivity to physicochemical conditions. The scale-up of high cell density process to large-scale bioreactors by providing sufficient oxygen supply and CO_2 removal can be very challenging in some cases and requires proper selection of large-scale bioreactor operating parameters [70]. Bioreactor operating parameters can be categorized into volume-dependent parameters, for example, working volume, feed volume, agitation, aeration, and volume-independent parameters, such as pH, dissolved oxygen, and temperature. A general strategy for scale-up is to proportionally scale up the volume-dependent parameters while maintaining the volume-independent parameters at the same set points used in the small-scale process, although some volume-dependent parameters are difficult to scale up/down linearly due to differences in bioreactor geometry, liquid surface to volume ratio, gassing regime, and control capability. Thus nonlinear effects apply but power to volume (P/V) ratio is a critical scale-up factor and often is used to normalized between different scale bioreactors.

Process scale-up generally is manipulated before establishment of the manufacturing process (Fig. 14.7). Agitation and aeration are two important conditions to be scaled up properly to achieve comparable process performance across different scales. Agitation is adjusted to provide sufficient mixing and oxygen mass transfer and is generally scaled using equal specific energy dissipation rate, whereas aeration is modulated for appropriate supply of oxygen and removal of excess CO_2. The CO_2 removal depends on several factors, including concentration of bicarbonate, aeration and agitation rates, bubble size, and the type and positioning of the impeller [61, 66, 70]. In small-scale bioreactors, mixing is generally not a problem even at relatively low agitation speeds, and dissolved CO_2 removal is efficient due to large surface to volume ratios. So the ranges of operating parameters can be wide. In large-scale bioreactors, these parameters need to be selected carefully as agitation-related shear forces, for example, impeller tip speed and aeration-related CO_2 accumulation may increase with scale, and mass transfer can be different.

Though CO_2 and O_2 are among the major factors affecting large-scale bioreactor process, lot-to-lot variations of raw material could cause serious issues for cell culture process scale-up. Relevant factors include consistency of media preparation, media hold stability, and cell line stability during the long duration of the manufacturing process. These issues can often be identified using scale-down models, such as the 2 l bioreactor, shake flasks, or shake tube (tubespin) models. For medium lot variation, a qualified cell culture scale-down model can provide critical information needed to assess and resolve any unexpected performance profile of the large-scale system [40]. Well-established small bioreactor scale-down models can be used for process characterization and validation and manufacturing tech support and troubleshooting.

14.3.3 Cell Culture Process Characterization and Validation

Cell culture process operation needs to be characterized and validated separately because the effectiveness of process parameters on key performance indicators and quality attributes can vary for different cell culture processes and stages. Process characterization and validation usually occurs after the completion of phase III process development, when the manufacturing process is locked down and no further significant changes are expected. Though process characterization and validation work can be performed using qualified scale-down models (Fig. 14.7), full-scale process validation is required to demonstrate consistent process performance and product quality when controlling critical process parameters within predefined acceptable ranges at commercial scale. For validation purposes, three and five consecutive full-scale runs are normally required for biologic license application (BLA) approval in the United States and the European Union, respectively. In any given manufacturing plant, the full-scale process validation often is limited on number of time due to facility capacity.

The purposes of process characterization are to identify critical operational parameters and key performance indicators and to establish acceptable ranges for operational parameters and demonstrate process robustness [40, 71, 72]. Recently, technical information from the characterization studies has also become a regulatory expectation as a prerequisite for manufacturing process validation and for long-term commercial manufacturing support [73, 74]. Scale-down models for the performance of manufacturing scale process are employed since performing a characterization study at the manufacturing scale is not feasible due to limited availability of large-scale bioreactors.

Qualified analytical methods and raw materials should be used throughout the characterization studies for consistency and accuracy. Cell culture process characterization needs support from the purification and analytical groups to monitor process impact and monitor product quality. Sometime, cell culture materials from characterization runs can be used for downstream process robustness studies.

As defined by the FDA, process validation is "establishing documented evidence that a specific process will consistently produce a product meeting its predetermined specifications and quality attributes" [73, 74]. Since product quality can be substantially affected by the manufacturing process, the process needs to be fully characterized and validated to understand the impact of process operating parameters and performance indicators and quality attributes, and the product quality specifications established are approved as part of the BLA.

14.3.4 Cell Culture Tech Transfer and cGMP Manufacturing

After a scalable cell culture process is developed and locked down at pilot scales (50–100 l), the process needs to be transferred to a manufacturing facility for cGMP production of clinical or commercial products. The purpose of technology transfer is to transfer all of the necessary process information, documentation, equipment, materials, and tools for implementing the manufacturing process at a specific facility and obtaining regulatory approval.

The cell culture tech transfer typically requires collaborations among different groups including process development (cell culture, purification, analytical groups, and sometimes including formulation group), manufacturing, and quality [75]. In order to successfully transfer, activities of a cell culture process can be organized into several stages, including process/facility gap analysis, process/facility, equipment modification and qualification, engineering runs at full scale, and then, cGMP commissioning, and cGMP production.

For any new process into an existing facility, a "gap analysis" should be the first step to identify limitations and potential risks to fit the manufacturing process into the facility. The equipment modifications and qualification activities should be completed prior to full-scale production runs. However, when the product moves from early stage (phase I or II clinical studies) to late stage (phase III through commercialization) production at different scale facilities, the process may need to be modified to reflect scale and facility changes. In this case, modifications should be minimized as much as possible to avoid any product quality variation. Product quality comparability has to be assayed before and after such change.

Though regulatory authorities do not require that the pre- and postchange product are identical, high comparability in physicochemical properties and biological activity from process changes is expected. Above all, the process changes must have no impact upon the safety or efficacy of the product [74, 76]. Non-GMP engineering runs are typically conducted to test process performance at full scale to ensure process and product quality consistency. Prior to the initiation of the cGMP production, cGMP commissioning should also be completed to ensure that the facility, raw material, documentation, personnel training, quality control, and production systems are ready.

Currently, new cell culture technologies are being developed to make the processes even more robust and to reduce the cost of operation. These advanced technologies include high-throughput cell culture systems, new online process monitoring and control systems, and disposable technologies.

To deliver a robust and productive process while maintaining aggressive timelines to push biosimilar molecules into the clinic it is essential to adopt experimental approaches to rapidly screen and optimize process parameters [77, 78] assuring that the product quality range that is highly similar to the originator antibody drug. A number of new microbioreactor systems, such as AMBR system (Sartorius—TAP) that enable more sophisticated control of culture conditions, at ten of milliliters of working volumes, are also being evaluated [79–81] for robustness, reliability and scalability. It is conceivable that one or more of these systems could be used more widely for automated screening of a large number of culture conditions.

In recent years, cell culture sensor technologies have undergone great advancement through the use of robust and accurate *in situ* sensors or at-line instruments coupled to automated sampling systems. The BioProfile® FLEX is a good example of an integrated analyzer that combines the functionality of many commonly utilized instruments for pH, dissolved gases, metabolites, cell counting, and osmolality into a single unit. The BioProfile FLEX has been evaluated with fed-batch cultures using multiple cell lines, and the measurements were found to be equivalent to the results using the aforementioned instruments [82]. This advance is especially beneficial due to its ability to reduce expenditures in the conduct of small-scale R&D projects.

Although online optical cell density probes (based on light backscatter) have been used successfully to monitor cell cultures [83], these measurements are generally linear with cell concentration only at high viabilities, and deviate significantly from linearity with decreasing culture viability, which commonly occur in the latter stages of fed-batch cultures. To overcome such limitations, dielectric permittivity and electrical impedance spectroscopy is used to monitor viable cell volume [84]. Spectroscopy, particularly using near- or midinfrared, is an attractive alternative for the measurement of cell culture components, including substrates, waste products, amino acids, cell concentration, and viability. Some studies have been published demonstrating the usefulness of this technique for monitoring cell culture media components [85–88].

Another new development is noninvasive fluorescence sensor technologies, which have been used for online monitoring of cell culture parameters such as optical density, pH, and dissolved oxygen for high-throughput applications [89]. Due to the limitations associated with each of these technologies, they are still in developmental stages. However, these types of sensors have the potential to add to our understanding of bioprocessing and thereby aid in the implementation of PAT.

Due to historical reasons, stainless steel tanks are widely used at laboratory and pilot scales for process development and production of research grade, toxicology, and phase I clinical materials. Stainless steel tanks also dominate large-scale manufacture (> 1,000–25,000 l) of therapeutic antibodies. But, stainless steel bioreactors are costly and inflexible, often, requiring long lead times for installation of the tanks and supporting infrastructure and qualification. There is also a high burden from validation efforts related to sterility and cleaning, as well as maintenance. To overcome some of these challenges using stainless tanks, several single-use bioreactors (SUB) are being currently evaluated [90, 91]. Companies marketing these systems include GE Healthcare, Sartorius, HyClone, Hynetics, ATMI, Xcellerex, Applikon, and Cellution. Disposable technologies offer significant advantages over traditional fixed plant equipment, particularly at pilot scales of operation. They can be introduced rapidly into laboratory and manufacturing facilities since installation, qualification, and personnel training requirements are minimal. In addition, they can increase plant capacity and flexibility by reducing turnaround time, especially in the event of contaminations, decreasing set up times and requiring a smaller footprint due to significantly reduced piping, valve, and instrumentation requirements. Implementing design changes is more convenient and expeditious with disposables, allowing for continuous improvement and integration of new technologies such as online monitoring systems. Overall, these advantages lead to significantly lower capital costs and lower resource requirements, which are key considerations for both large and small companies alike. However, given the fact that many companies have already committed themselves to large capital outlays, stainless steel will be with us for some years to come. No doubt, in the future SUB will be more widely adopted throughout the biosimilar industry.

14.4 Future Trends

Cell line and cell culture processes are critical components of antibody biosimilar drug development. As a result of more than two decades of recombinant antibody cell line R&D, productivity has increased from less than 1 g/l to over 10 g/l. With the resolution of this challenge, commercial entities are free to pursue quality issues that have eluded them for the last two decades.

Future trends for cell line and cell culture development will focused on implementation of high-throughput technologies and multiparameters process development by DOE, understanding cellular metabolic process related to cell line product quality, QbD application in process and feeding strategies, robustness scale-up, cell metabolism, better process understanding, and their impact on product quality and scale-up. Due to high cell line productivity, bioreactor scale for future manufacturing will be at 500–2000 l range. Continuous cell culture process will be a tantalizing option for maintaining quality consistency during cell culture process in biosimilar industry, and manufacturing plant will be small and flexible, with higher efficiency, lower cost, and faster speed.

References

1. Spens E, Hägström L. Defined protein and animal component-free NS0 fed-batch culture. *Biotechnol. Bioeng.* 2007;98:1183–1194.
2. Cruz HJ, Moreira JL, Carrondo MJT. Metabolic shifts by nutrient manipulation in continuous cultures of BHK cells. *Biotechnol. Bioeng.* 1999;66:104–113.
3. Seifert DB, Phillips JA. The production of monoclonal antibody in growth-arrested hybridomas cultivated in suspension and immobilized modes. *Biotechnol. Prog.* 1999;15:655–666.
4. Pau MG, Ophorst C, Koldijk MH, Schouten G, Mehtali M, Uytdehaag F. The human cell line PER.C6 provides a new manufacturing system for the production of influenza vaccines. *Vaccine* 2001;19:2716–2721.
5. Yang J-D, Lu C, Stasny B, Henley J, Guinto W, Gonzalez C, et al. Fed-batch bioreactor process scale-up from 3 L to 2,500 L scale for monoclonal antibody production from cell culture. *Biotechnol. Bioeng.* 2007;98:141–154.
6. Jayapal KP, Wlaschin KF, Hu W-S, Yap MGS. Recombinant protein therapeutics from CHO cells—20 years and counting. *Chem. Eng. Prog.* 2007;103:40–47.
7. Keen MJ, Steward TW. Adaptation of cholesterol-requiring NS0 mouse myeloma cells to high density growth in a fully defined protein-free and cholesterol-free culture medium. *Cytotechnology* 1995;17:203–211.
8. Gorfien S, Paul B, Walowitz J, Keem R, Biddle W, Jayme D. Growth of NS0 cells in protein-free, chemically defined medium. *Biotechnol. Prog.* 2000;16:682–687.
9. Seth G, Ozturk M, Hu WS. Reverting cholesterol auxotroph of NS0 cells by altering epigenetic gene silencing. *Biotechnol. Bioeng.* 2006;93:820–827.
10. Hartman TE, Sar N, Genereux K, Barritt DS, He YM, Burky JE, et al. Derivation and characterization of cholesterol-independent non-GS NS0 cell

lines for production of recombinant antibodies. *Biotechnol. Bioeng.* 2007;96:294–306.

11. Birch JR, Racher JR. Antibody production. *Adv. Drug Deliv. Rev.* 2006;58:671–685.
12. Li F, Kao E, Ryll TA (2004). Generic high cell mass fed-batch process for NS0 cells; Poster Presentation, Cell Culture Engineering IX, Cancun, Mexico, March 7–12, 2004.
13. Burky JE, Wesson MC, Young A, Farnsworth S, Dionne B, Zhu Y, et al. Protein-free fed-batch culture of non-GS NS0 cell lines for production of recombinant antibodies. *Biotechnol. Bioeng.* 2006;96:281–293.
14. Alex Z, Rosa H. Anti-*N*-glycolylneuraminic acid antibodies identified in healthy human serum. *Xenotransplantation* 2002;9:376–381.
15. Byres E, Paton AW, Paton JC, Lofling JC, Smith DF, Wilce MCJ, et al. Incorporation of a non-human glycan mediates human susceptibility to a bacterial toxin. *Nature* 2008;456:648–652.
16. Tangvoranuntakul P, Gagneux P, Diaz S, Bardor M, Varki N, Varki A, et al. Human uptake and incorporation of an immunogenic nonhuman dietary sialic acid. *Proc. Natl. Acad. Sci. U. S. A.* 2003;100:12045–12050.
17. Jones DH, Kroos N, Anema R, Van Montfort B, Vooys A, Van der Kraats S, et al. High-level expression of recombinant IgG in the human cell line PER. C6. *Biotechnol. Prog.* 2003;19:163–168.
18. Davis J, Jiang LY, Pan L-Z, LaBarre MU, Andersen D, Reff M. Expression of GnTIII in recombinant anti-CD20 CHO production cell line: expression of antibodies with altered glycoforms leads to an increase in ADCC through higher affinity for FcγRIII. *Biotechnol. Bioeng.* 2001;74:288–294.
19. Saba JA, Kunkel JP, Jan DCH, Ens WE, Standing KG, Butler M, et al. A study of immunoglobulin G glycosylation in monoclonal and polyclonal species by electrospray and matrix-assisted laser desorption/ionization mass spectrometry. *Anal. Biochem.* 2002;305:16–31.
20. Kanda Y, Imai-Nishiya H, Kuni-Kamochi R, Katsuhiro M, Inoue M, Kitajima-Miyama K, et al. Establishment of a GDP-mannose 4,6-dehydratase (GMD) knockout host cell line: a new strategy for generating completely non-fucosylated recombinant therapeutics. *J. Biotechnol.* 2007;130:300–310.
21. Mori K, Kamochi RK, Ohnuki NY. Engineering Chinese hamster ovary cells to maximize effector function of produced antibodies using FUT8 siRNA. *Biotechnol. Bioeng.* 2004;88:901–908.
22. Sauerwald TM, Figueroa BJ, Hardwick JM, Oyler GA, Betenbaugh MJ. Combining caspase and mitochondrial dysfunction inhibitors of apoptosis to limit cell death in mammalian cell cultures. *Biotechnol. Bioeng.* 2006;94:362–372.
23. Chen K, Liu Q, Xie L-Z, Sharp PA, Wang DI. Engineering of a mammalian cell line for reduction of lactate formation and high monoclonal antibody production. *Biotechnol. Bioeng.* 2001;72:55–61.

24. Kim SH, Lee GM. Downregulation of lactate dehydrogenase-A by siRNAs for reduced lactic acid formation of Chinese hamster ovary cells producing thrombopoietin. *Appl. Microbiol. Biotechnol.* 2007;74:152–159.
25. Jeong D-W, Cho IT, Kim TS, Bae GW, Kim I-H, Kim IY. Effects of lactate dehydrogenase suppression and glycerol-3-phosphate dehydrogenase overexpression on cellular metabolism. *Mol. Cell Biochem.* 2006;284:1–8.
26. Bi J-X, Shuttleworth J, Al-Rubeai M. Uncoupling of cell growth and proliferation results in enhancement of productivity in p21^{cip1}-arrested CHO cells. *Biotechnol. Bioeng.* 2004;85:741–749.
27. Wolf AP. When more is less. *Nat. Genet.* 1998;18:5–6.
28. Coroadinho AS, Schucht R, Gama-Norton L, Wirth D, Hauser H, Carrondo MJ. The use of recombinase mediated cassette exchange in retroviral vector producer cell lines: predictability and efficiency by transgene exchange. *J. Biotechnol.* 2006;124:457–468.
29. Fukushige S, Sauer B. Genomic targeting with a positive-selection lox integration vector allows highly reproducible gene expression in mammalian cells. *Proc. Natl. Acad. Sci. U. S. A.* 1992;89:7905–7909.
30. Kito M, Itami S, Fukano Y, Yamana K, Shibui T. Construction of engineered CHO strains for high-level production of recombinant proteins. *Appl. Microbiol. Biotechnol.* 2002;60:442–448.
31. Andersen DC, Reilly DE. Production technologies for monoclonal antibodies and their fragments. *Curr. Opin. Biotechnol.* 2004;15:456–462.
32. Li H, Sethuraman N, Stadheim TA, Zha D, Prinz B, Ballew N, et al. Optimization of humanized IgGs in glycoengineered *Pichia pastoris*. *Nat. Biotechnol.* 2006;24:210–215.
33. Simmons LC, Reilly D, Klimowski L, Raju TS, Meng G, Sims P, et al. Expression of full-length immunoglobulins in *Escherichia coli*: rapid and efficient production of aglycosylated antibodies. *J. Immunol. Methods* 2002;263:133–147.
34. Mazor Y, Van Blarcom V, Mabry R, Iverson BL, Georgiou G. Isolation of engineered, full length antibodies from libraries expressed in *Escherichia coli*. *Nat. Biotechnol.* 2007;25:563–565.
35. Boshart M, Weber F, Jahn G, Dorsch-Hasler K, Fleckenstein B, Schaffher W. A very strong enhancer is located upstream of an immediate early gene of human cytomegalovirus. *Cell* 1985;41:521.
36. Deer JR, Allison DS. High-level expression of proteins in mammalian cells using transcription regulatory sequences from the Chinese hamster EF-1 alpha gene. *Biotechnol. Prog.* 2004;20:880–889.
37. Kozak M. Point mutations define a sequence flanking the AUG initiator codon that modules translation by eukaryotic ribosomes. *Cell* 1986;44:283–292.
38. Shen AY, Van de Goor J, Zheng L, Reyes AE, Krummen LA. Recombinant DNA technology and cell line development. In: Ozturk SS, Hu W-S, editors.

Cell Culture Technology for Pharmaceutical and Cell-Based Therapies. Boca Raton, FL: CRC Press, Taylor & Francis Group; 2006. pp. 15–40.

39. Butler M. Animal cell cultures: recent achievements and perspectives in the production of biopharmaceuticals. *Appl. Microbiol. Biotechnol.* 2005;68:283–291.
40. Walsh G, Jefferis R. Post-translational modifications in the context of therapeutic proteins. *Nat. Biotechnol.* 2006;24:1241–1252.
41. Gawlitzek M, Estacio M, Furch T, Kiss R. Identification of cell culture conditions to control N-glycosylation site-occupancy of recombinant glycoproteins expressed in CHO cells. *Biotechnol. Bioeng.* 2009;103:1164–1175.
42. Wurm FM. Production of recombinant protein therapeutics in cultivated mammalian cells. *Nat. Biotechnol.* 2004;22:1393–1398.
43. Kelley B. Industrialization of mAb production technology: the bioprocessing industry at a crossroads. *mAbs* 2009;1:443–452.
44. Li F, Hashimura Y, Pendleton R, Harms J, Collins E, Lee B. A systematic approach for scale-down model development and characterization of commercial cell culture processes. *Biotechnol. Prog.* 2006;22:696–703.
45. Fletcher T. Designing culture media for recombinant protein production: a rational approach. *BioProcess Int.* 2005;3:30–36.
46. Vijayasankaran N, Li J, Shawley R, Chen A, Shiratori M, Gawlitzek M, et al. Animal cell culture media. In: Flickinger MC, editor. Encyclopedia of Industrial Biotechnology, Bioprocess, Bioseparation, and Cell Technology. New York: John Wiley & Sons, Inc.; 2010.
47. Heidemann R, Zhang C, Qi H, Larrick Rule J, Rozales C, Park S, et al. The use of peptones as medium additives for the production of a recombinant therapeutic protein in high density perfusion cultures of mammalian cells. *Cytotechnology* 2000;32:157–167.
48. Franek F, Hohenwarter O, Katinger H. Plant protein hydrolysates: preparation of defined peptide fractions promoting growth and production in animal cells cultures. *Biotechnol. Prog.* 2000;16:688–692.
49. Zhou W, Rehm J, Europa A, Hu W-S. Alteration of mammalian cell metabolism by dynamic nutrient feeding. *Cytotechnology* 1997;24:99–108.
50. Zhou W, Rehm J, Hu WS. High viable cell concentration fed-batch cultures of hybridoma cells through on-line nutrient feeding. *Biotechnol. Bioeng.* 1995;46:579–587.
51. Xie LZ, Wang DIC. High cell density and high monoclonal antibody production through medium design and rational control in a bioreactor. *Biotechnol. Bioeng.* 1996;51:725–729.
52. Castro PML, Hayter PM, Ison AP, Bull AT. Application of a statistical design to the optimization of culture medium for recombinant interferon-gamma production by Chinese hamster ovary cells. *Appl. Microbiol. Biotechnol.* 1992;38:84–90.

53. Hammett K, Kuchibhatla J, Hunt C, Holdread S, Brooks J. Developing chemically defined media through DOE: complete optimization with increased protein production in less than 8 months. In: Smith R, editor. Cell Technology for Cell Products. Dordrecht, The Netherlands: Springer; 2007. pp. 683–691.

54. Harcum S. Protein glycosylation. In: Ozturk SS, Hu WS, editors. Cell Culture Technology for Pharmaceutical and Cell-Based Therapies. Boca Raton, FL: CRC Press; 2006. pp. 113–153.

55. Food and Drug Administration (FDA) (2004) Guidance for Industry PAT—A Framework for Innovative Pharmaceutical Development, Manufacturing, and Quality Assurance. U.S. Department of Health and Human Services Food and Drug Administration Center for Drug Evaluation and Research (CDER) Center for Veterinary Medicine (CVM) Office of Regulatory Affairs (ORA), Pharmaceutical CGMPs, September 2004.

56. Gnoth S, Jenzsch M, Simutis R, Lübbert A. Process analytical technology (PAT): batch-to-batch reproducibility of fermentation processes by robust process operational design and control. *J. Biotechnol.* 2007;132:180–186.

57. Furukawa K, Ohsuye K. Enhancement of productivity of recombinant α-amidating enzyme by low temperature culture. *Cytotechnology* 1999;31:85–94.

58. Moore A, Mercer J, Dutina G, Donahue CJ, Bauer KD, Mather JP, et al. Effects of temperature shift on cell cycle, apoptosis and nucleotide pools in CHO cell batch culture. *Cytotechnology* 1997;23:47–54.

59. Fox SR, Tan HK, Tan MC, Wong SC, Yap MG, Wang DI, et al. A detailed understanding of the enhanced hypothermic productivity of interferon-gamma by Chinese-hamster ovary cells. *Biotechnol. Appl. Biochem.* 2005;41:255–264.

60. Fleischaker RJ, Sinskey AJ. Oxygen demand and supply in cell culture. *Appl. Microbiol. Biotechnol.* 1981;12:193–197.

61. Miller WM, Wilke CR, Blanch HW. Effect of dissolved oxygen on hybridoma growth and metabolism in continuous culture. *J. Cell Physiol.* 1987;132:524–530.

62. Ozturk SS, Palsson BO. Effects of dissolved oxygen on hybridoma cell growth, metabolism and antibody production kinetics in continuous cultures. *Biotechnol. Prog.* 1990;6:437–446.

63. Kunkel JP, Jan DCH, Jamieson JC, Bulter M. Dissolved oxygen concentration in serum-free continuous culture affects N-linked glycosylation of a monoclonal antibody. *J. Biotechnol.* 1998;62:55–71.

64. Garnier A, Voyer R, Tom R, Perret S, Jardin B, Kamen A. Dissolved carbon dioxide accumulation in a large scale high cell density production of TGFβ receptor with baculovirus infected Sf-9 cells. *Cytotechnology* 1996;22:53–63.

65. Gray DR, Cgen S, Howarth W, Inlow D, Maiorella L. CO_2 in large-scale and high density CHO cell perfusion culture. *Cytotechnology* 1996;22:65–78.

66. Kimura R, Miller WM. Glycosylation of CHO-derived recombinant tPA produced under elevated pCO_2. *Biotechnol. Prog.* 1997;73:311–317.
67. Wayte J, Boraston R, Bland H, Varley J, Brown M. pH: effects on growth productivity of cell lines producing monoclonal antibodies: control in large scale fermenters. *Cytotechnology* 1997;22:87–94.
68. Schmid G, Blanch HW, Wilke CR. Hybridoma growth, metabolism and product formation in HEPES-buffered media: II: effect of pH. *Biotechnol. Lett.* 1990;12:633–638.
69. deZengotita VM, Schmelzer AE, Miller WM. Characterization of hybridoma cell responses to elevated pCO_2 and osmolality: intracellular pH, cell size, apoptosis and metabolism. *Biotechnol. Bioeng.* 2002;77:369–380.
70. Zhu MM, Goyal A, Rank DL, Gupta SK, Boom TV, Lee SS. Effects of elevated pCO_2 and osmolality on growth of CHO cells and production of antibody-fusion protein B1: a case study. *Biotechnol. Prog.* 2005;21:70–77.
71. Seewöster T, Lehmann J. Cell size distribution as a parameter for the predetermination of exponential growth during repeated batch cultivation of CHO cells. *Biotechnol. Bioeng.* 1997;55:793–797.
72. Xing Z, Kenty BM, Li ZJ, Lee SS. Scale-up analysis for a CHO cell culture process in large-scale bioreactors. *Biotechnol. Bioeng.* 2009;103:733–746.
73. Maiorella B, Inlow D, Howarth W (2001) Method of increasing product expression through solute stress. United States Patent US 6,238,891 B1, May 29, 2001.
74. Chen M, Forman LW (2001) Polypeptide production in animal cell culture. United States Patent US 6,180,401 B1, Jan 30, 2001.
75. Rathore AS, Wang A, Menon M, Riske F, Campbell J, Goodrich E, et al. Optimization, scale-up and validation issues in filtration of biopharmaceuticals—part I. *BioPharm Int.* 2004;17:50–58.
76. Seely JE, Seely R. A rational, step-wise approach to process characterization. *BioPharm Int.* 2003;16:24–34.
77. Rathore AS, Noferi JF, Arling ER, Sofer G, Watler P, O'Leary R. Process validation how much to do and when to do it. *BioPharm Int.* 2002;15:18–28.
78. Food and Drug Administration (2005) FDA Guidance for Industry: Q5E Comparability of Biotechnological/Biological Products Subject to Changes in Their Manufacturing Process, June 2005, http://www.fda.gov/cder/guidance/index.htm (accessed June 16, 2016).
79. Chirino AJ, Mire-Sluis A. Characterizing biological products and assessing comparability following manufacturing changes. *Nat. Biotechnol.* 2004;22:1383–1391.
80. Kensy F, John GT, Hofmann B, Büchs J. Characterization of operation conditions and online monitoring of physiological culture parameters in shaken 24-well microtiter plates. *Bioprocess Biosyst. Eng.* 2005;75:75–81.
81. Micheletti M, Barrett T, Doig SD, Baganz F, Levy MS, Woodley JM, et al. Fluid mixing in shaken bioreactors: implications for scale-up predictions

from microliter-scale microbial and mammalian cell cultures. *Chem. Eng. Sci.* 2006;61:2939–2949.

82. Legmann R, Schreyer HB, Combs RG, McCormick EL, Russo AP, Rodgers ST. A predictive high-throughput scale-down model of monoclonal antibody production in CHO cells. *Biotechnol. Bioeng.* 2009;104:1107–1120.
83. Chen A, Chitta R, Chang D, Amanullah A. Twenty-four well plate miniature bioreactor system as a scale-down model for cell culture process development. *Biotechnol. Bioeng.* 2009;102:148–160.
84. Amanullah A, Otero JM, Mikola MR, Hsu A, Zhang J, Aunins J, et al. (2007) Evaluation of Simcell technology for cell culture process development. Biochemical Engineering XV, Quebec City, Canada, July 15–19, 2007.
85. Derfus GE, Abramzon D, Tung M, Chang D, Kiss R, Amanullah A. Cell culture monitoring via an auto-sampler and an integrated multi-functional off-line analyzer. *Biotechnol. Prog.* 2010;26:284–292.
86. Konstantinov KB, Pambuyan R, Matanguihan R, Yoshida T, Purusich CM, Hu WS. On-line monitoring of hybridoma cell growth using a laser turbidity sensor. *Biotechnol. Bioeng.* 1992;40:1337–1342.
87. Zeiser A, Bedard C, Voyer R, Jardin B, Tom R, Kamen A. On-line monitoring of the progress of infection in Sf-9 insect cell cultures using relative permittivity measurements. *Biotechnol. Bioeng.* 1999;63:122–126.
88. Chung H, Arnold M, Rhiel M, Murhammer D. Simultaneous measurement of glucose and glutamine in aqueous solutions by near infrared spectroscopy. *Appl. Biochem. Biotechnol.* 1995;50:109–125.
89. Riley MR, Okeson CD, Frazier BL. Rapid calibration development for near-infrared spectroscopic monitoring of mammalian cell cultivations. *Biotechnol. Prog.* 1999;15:1133–1141.
90. Riley MR, Crider HM, Nite ME, Garcia RA, Woo J, Wegge RM. Simultaneous measurements of 19 components in animal cell culture media by near infrared spectroscopy. *Biotechnol. Prog.* 2001;17:376–378.
91. Yano T, Harata M. Prediction of the concentration of several constituents in a mouse–mouse hybridoma culture by near infrared spectroscopy. *J. Ferment. Bioeng.* 1994;77:659–662.

15

Product Analysis of Biosimilar Antibodies

Weidong Jiang[1,2]*, Scott Liu*[1]*, and Ziyang Zhong*[1]

[1] *Henlix, Inc., Fremont, CA, USA*
[2] *Shanghai Henlius Biotech. Co. LTD, Shanghai, China*

15.1 Summary

We review the analytical and bioassay methods used to characterize the finished antibody products based on predetermined specifications for biosimilar antibodies. For their identification, it is typical to have the primary sequence confirmed to the reference standard using the LC/MS method. There are many options available on the market for this purpose, including quadrupole time-of-flight (QTOF) and orbitrap. Both Waters and Agilent make excellent TOF instruments, while Thermo–Finnigan is the undisrupted leader in the orbitrap technology sphere. All of these instruments offer techniques to provide both primary sequences and disulfide bonds and information. Their purity and integrity, on the other hand, are demonstrated by conventional chromatography, such as SDS–PAGE and IEF. Similarly, HILIC chromatography is the standard method to characterize the glycan profile, which plays a key role in ADCC. All of these techniques have undergone tremendous advance in the past decade and are thoroughly reviewed in this chapter. This chapter also summarizes methods used to study stability and receptor binding of therapeutic antibodies. The overall purpose of this chapter is to present to the reader the guidelines for biosimilar characterization.

Biosimilars of Monoclonal Antibodies: A Practical Guide to Manufacturing, Preclinical, and Clinical Development, First Edition. Edited by Cheng Liu and K. John Morrow, Jr.

15.2 Introduction

Biosimilar antibodies have generated the most interest out of biopharmaceutical products in recent years due to the expiry of many original drug patents and their importance in many therapeutic areas of cancer and other immunotherapies [1–11]. In September 2013, Celltrion gained European approval for the world's first monoclonal antibody biosimilar, Remsima/Inflectra (infliximab), as part of its collaboration with Hospira, a US-based generic drug company. Many biotech and traditional pharmaceutical companies are increasingly interested in biosimilar antibodies for a simple reason: to gain a share of the multibillion dollar market. However, the path to a successful biosimilar product of an antibody is completely different from the usual approach of pharmaceutical companies to generic drug of small molecules. Many challenges lie ahead for the companies that are trying to enter the field of biosimilars. This chapter will review the analytical methods and bioassays that address some of these challenges in characterizing the final antibody product. These methods will verify whether the product conforms to the predetermined specifications of biosimilar antibodies regarding their **identity, purity, integrity, stability, quantity, activity, efficacy, and safety** and support the claim of the similarity to a referenced product, as outlined in the following chart (Fig. 15.1).

15.3 Identity

Most therapeutic monoclonal antibodies are expressed in Chinese hamster ovary (CHO) cells with two gamma heavy chains (HC) and two kappa light chains (LC); although, recent developments in different antibody formats such as bispecific and domain antibodies are changing the definition of an antibody. A few antibodies are produced in other cell lines such as SP2/0 and PER.C6 cells, which will not be discussed in this chapter. Some companies produce biosimilar versions of drugs using CHO cells instead of the original cell line, resulting in a decreased resemblance of the glycosylation profile to the original drug.

15.3.1 Primary Structure Analysis

To be considered as a biosimilar, the amino acid sequence of a biosimilar antibody must be identical to the originator antibody. Most antibody sequences are disclosed in patent publications and other public literatures. However, some sequences published differ from the originator's sequence, particularly the Fc portion of the antibodies. This is because the Fc regions are usually not covered in the patents. Therefore, all amino acid sequences have to be

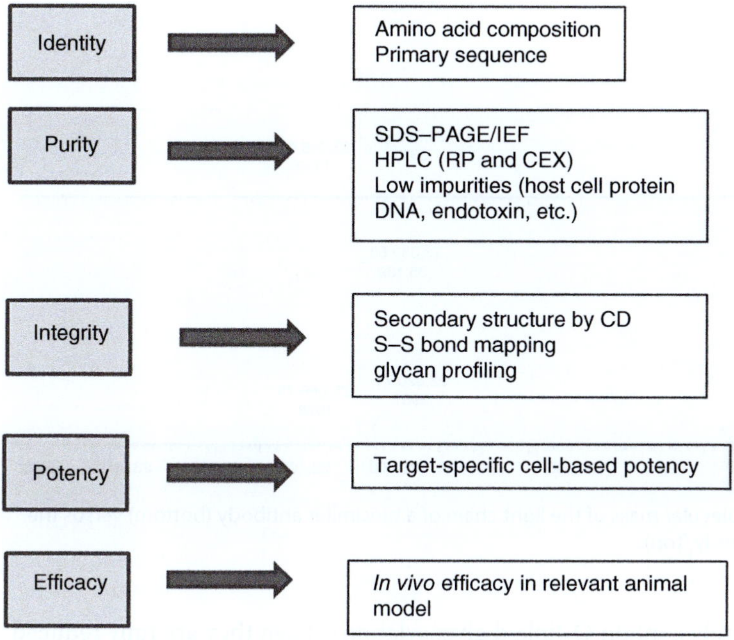

Figure 15.1 Outlines of Product Analysis of Biosimilar Antibodies.

confirmed by sequencing to demonstrate that they are identical to their corresponding original antibody.

After initial characterization of a biosimilar product with routine biochemical methods such as SDS–PAGE gel analysis and size exclusion chromatography (SEC), LC–MS/MS is used to determine molecular masses and a full-length amino acid sequence of a biosimilar antibody [12, 13]. For the molecular determination, an antibody is deglycosylated to remove N-glycans and reduced into its heavy and light chain constituents. Figure 15.2 demonstrates a comparison of the typical LC mass analysis of a biosimilar product to the originator antibody. With a modern high-resolution mass spectrometer, the discrepancy between the observed and expected molecular weights is typically less than 20 ppm. While the intact mass measurement could confirm the overall composition, complete sequence coverages of both HC and LC are still necessary to determine the exact sequence in order to meet the minimum biosimilar antibody standard. It is common to use two to three proteolytic enzymes to digest antibodies before sequencing those peptides with LC–MS/MS. Depending on the need, a combination of trypsin, chymotrypsin, Glu-C, and Asp-N proteases can be used. Before the protease digest, antibodies are deglycosylated with PNGase to remove N-linked glycosylation (monoclonal

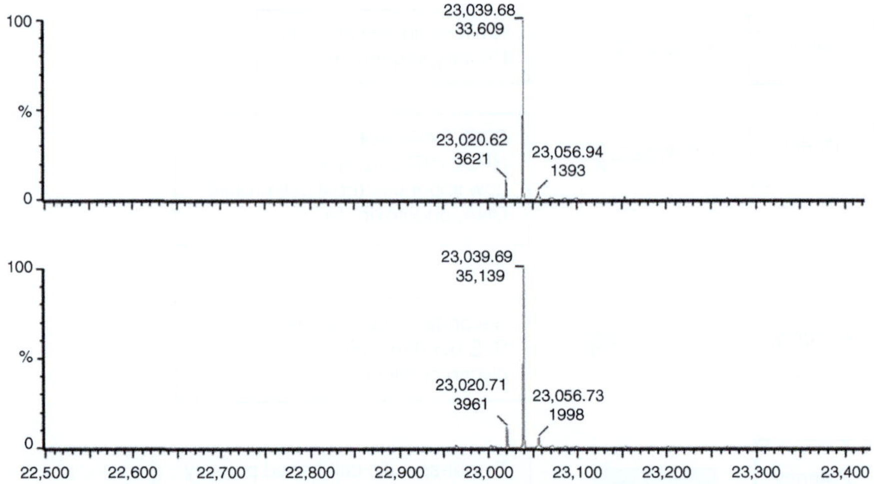

Figure 15.2 Molecular mass of the light chain of a biosimilar antibody (bottom) versus the originator antibody (top).

antibodies rarely contain O-linked glycosylation). Then they are fully reduced by DTT and alkylated with iodoacetamide (IAA) to prevent the reduced cysteines from reforming scrambled disulfide bonds. After treatment, antibodies are fully digested with proteases and analyzed by LC/MS. Figure 15.3 shows a typical comparison of tryptic peptides from a biosimilar product and its reference. The sequence of each peptide can then be precisely confirmed with MS/MS fragmentation spectra as showed in Figure 15.4 for one of the common Fc tryptic peptides. The combination of trypsin and Glu-C will typically achieve >98% coverage. If necessary, employing a third protease such as chymotrypsin will enable us to achieve 100% sequence coverage.

15.3.2 Peptide Mapping: Comparison Studies

Peptide mapping is a common quality control tool to verify the consistency between different lots and to identify single amino acid mutations as well as posttranslation modifications. Following digestion with trypsin, RP-HPLC was performed to obtain the peptide maps.

Antibodies from biosimilar lots and originator lots are denatured in 6M guanidine hydrochloride solution, reduced with DTT and alkylated with IAA. Following buffer exchange with a size exclusion column, digestion was performed with trypsin. The digest was separated with a C18 column. The individual peptide is identified with UV detector based on retention time and further confirmed by mass spectrometer as showed in Figures 15.3 and 15.4. During the initial characterization, the sequences of the peptide fragments and

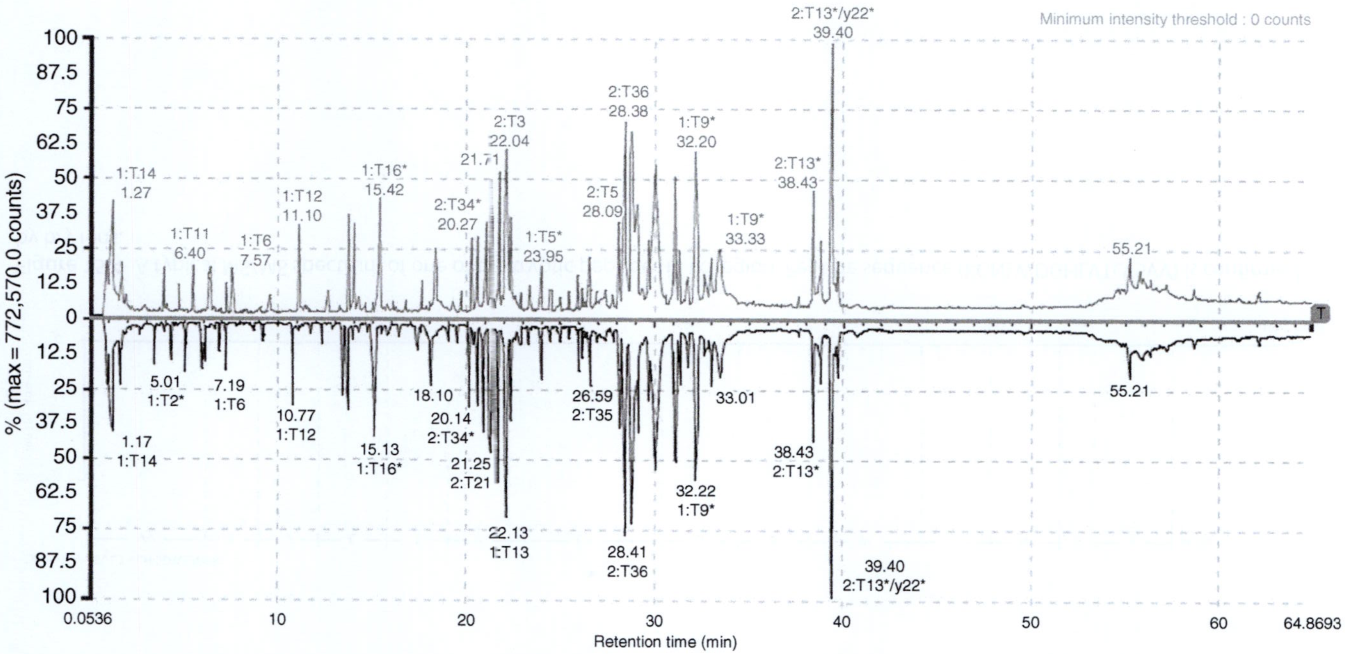

Figure 15.3 Biosimilar and originator products have the same tryptic peptide mapping and mass spectrometry sequences. Top panel: originator drug; bottom panel: biosimilar product. Some antibodies may contain an extra glycan at the Fab region in addition to the Fc site. The extra glycan has to be treated differently from the standard method due to difficulties in cleaving the glycans from the Fab region for analysis (see more details in the glycan analysis section). (*See insert for color representation of the figure.*)

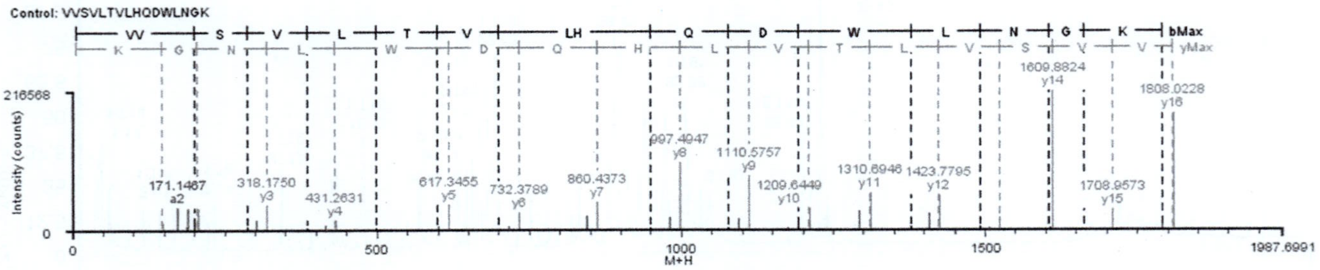

Figure 15.4 A typical MS/MS spectrum of one of the tryptic peptides in Fc region. Peptide sequence (KGNLWDGHLVTLVSVV) is confirmed by b/y ions.

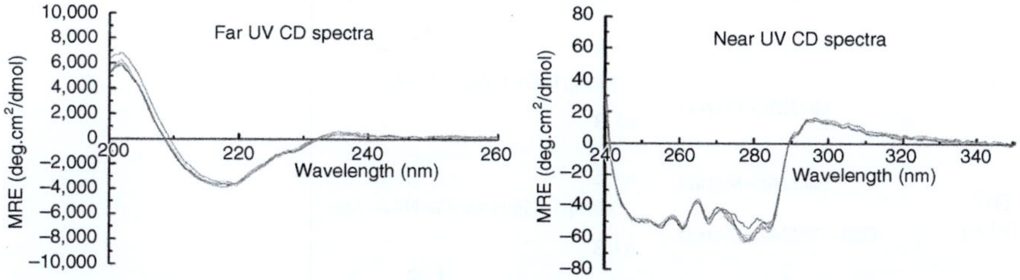

Figure 15.5 Far- and near-UV spectra of three lots of biosimilar antibody and three lots of originator antibody.

the lot-to-lot are identified. Many mass spectrometers are widely used for this purpose, such as the QTOF mass spectrometers from Waters and Agilent and Thermo Fisher's Q Exactive.

15.3.3 Secondary and High-Order Structure Analysis

15.3.3.1 Circular Dichroism Spectropolarimetry

Near-UV circular dichroism (CD) spectrum (>250 nm) provides information on the tertiary structure, while the far-UV CD spectrum (ultraviolet) provides important characteristics of their secondary structure such as alpha-helix, beta-sheet, and beta-turn structure [14–20]. Most antibodies are rich in β-sheet, which is a typical characteristic of immunoglobulins. A comparison of the biosimilar and originator will provide general structural evidence to verify that the biosimilars have indeed retained the overall tertiary structure, as demonstrated in the lot-to-lot comparison of the reference and biosimilar product in Figure 15.5.

15.3.3.2 Posttranslational Modification: Glycan Analysis

Glycosylation is one of the most important posttranslational modifications of all therapeutic antibodies. Differences in glycan structure, such as variation of fucose content in the Fc region, can affect antibody effector functions such as antibody-dependent cell-mediated cytotoxicity (ADCC). Immunogenicity has also been caused by some rare structural moieties of proteins produced from nonhuman cell lines. Some antibodies have an additional glycosylation site in the variable region of Fab fragment. For example, Erbitux has an additional glycosylation site, but it will not be discussed in details in this chapter.

Figure 15.6 summarizes most common glycans from a therapeutic antibody. There are two major classes of glycans with fucose (G0F, G1F, G2F) or without fucose (G0, G1, G2, M5) and with galactose (G1, G1F or G2, G2F) or without galactose (G0 or G0F).

The composition of glycans is determined by normal phase HPLC using a HILIC column. Glycans are first removed from a monoclonal antibody and derivatized with 2-aminobenzoic acid (2-AB) [21–23], which is needed for

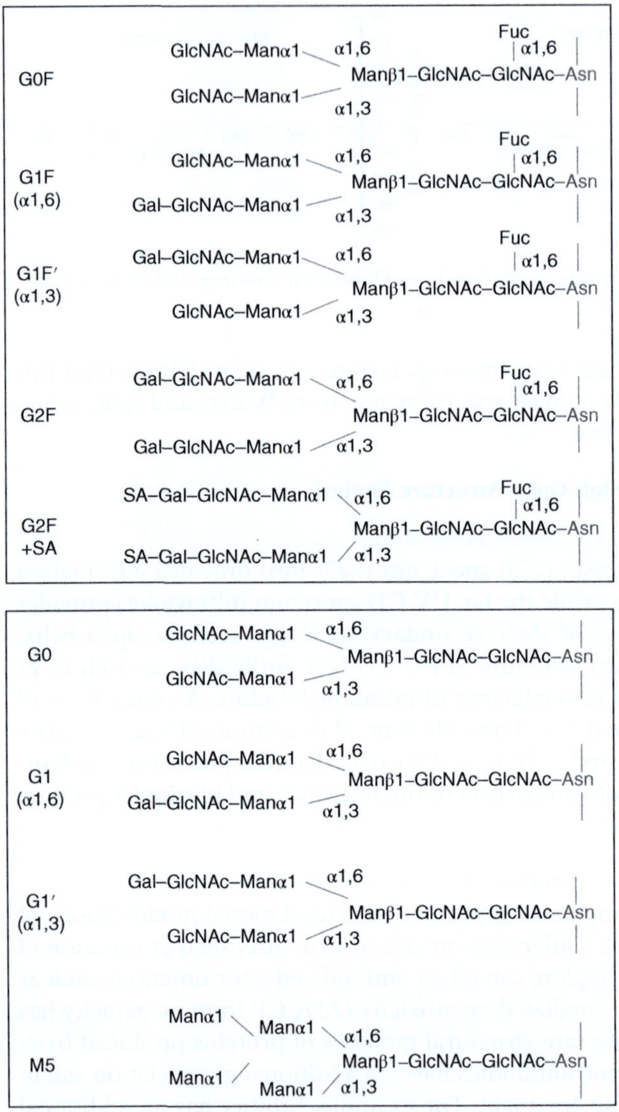

Figure 15.6 Schematic representation of glycan structure in antibodies. G0F, G1F, G1F′, G2F, and G2F with sialic acid, G0, G1, G1′, and M5. Asn is the amino acid residue at the Fc region of monoclonal antibody that is linked to the glycans. Fuc, fucose; Gal, galactose; GlcNAc, *N*-acetylglucosamine; Man, mannose; SA, sialic acid.

sensitive fluorescence detection. The derivatized glycans are separated on a HILIC column, based on relative hydrophilicities, which are determined by the composition and branching of oligosaccharides. The relative retention time on a HILIC column and their intensities are used to determine the glycosylation

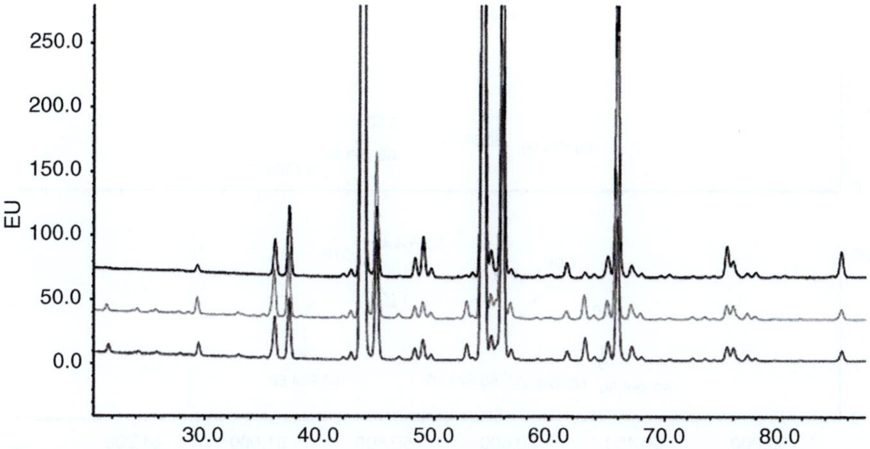

Figure 15.7 HILIC glycan profiles of monoclonal antibodies produced in CHO cells. The black line is from the originator; light gray and gray lines are from two biosimilar lots.

patterns and abundance of each glycan. A sample chromatogram of glycans from a biosimilar product and its originator is given in Figure 15.7. When a HILIC column is connected to a mass spectrometer, the determination of accurate molecular weights and their fragmentation pattern will provide further evidence to demonstrate that biosimilar and originator are glycosylated similarly. The HILIC method, with fluorescence detection, is used during the initial characterization and, more importantly, as a key routine quality control method to ensure the lot-to-lot consistency.

As a second complementary analysis method for the profiling of 2-AB-labeled N-glycans, negative linear mode MALDI–TOF–MS with DHB matrix can be used. For MALDI analysis, samples need to be desalted to remove residual sodium before analysis. Assignments were made based on average masses, measured by MALDI–TOF–MS [24].

The glycosylation site is typically identified in the tryptic mapping [25–27]. When a monoclonal antibody is subjected to the trypsin digestion without prior deglycosylation, the N-linked glycosylation remains attached to the Asn residue. The glycopeptide will be eluted from the reverse phase HPLC column and be detected with the mass spectrometer. The high-energy fragmentation techniques will cause complete dissociation of glycans from the peptide, but the parent molecular mass will contain the mass of peptide plus the mass of respective glycan. While peptide mapping will not characterize the glycosylation patterns, it will provide the exact glycosylation site in the antibody.

Many other analytical techniques can be used to characterize the glycosylation patterns, such as capillary electrophoresis and high-performance anion-exchange chromatography with pulsed amperometric detection [28]. While these methods provide additional information, they will not be discussed in this chapter.

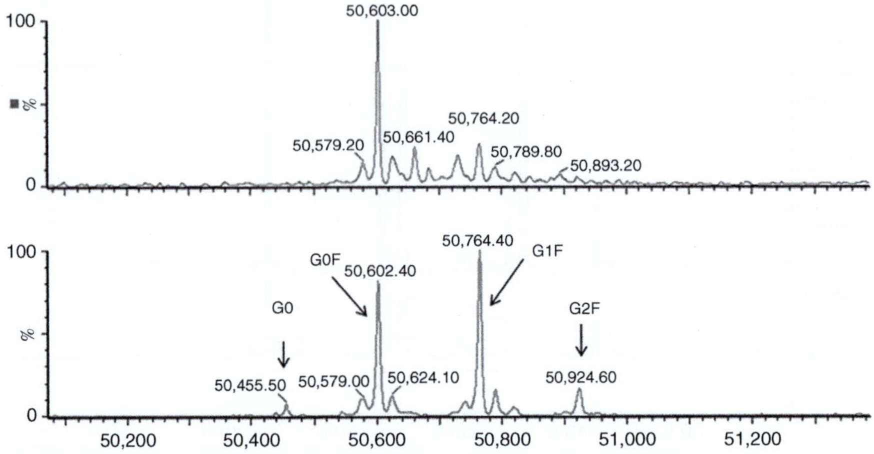

Figure 15.8 Deconvoluted LC–MS mass analysis of heavy chain of an IgG monoclonal antibody either from the originator (bottom panel) or from a biosimilar (low panel).

In addition to using HILIC chromatography for accurate glycan analysis, LC/MS analysis can be a fast alternative to provide nearly quantitative results. After an IgG monoclonal antibody is denatured and fully reduced, the LC and HC are well separated on a C-18 column before entering the mass spectrometer. The glycosylation pattern can be observed in the mass spectrum of the glycosylated HC (Fig. 15.8).

The use of LC/MS as an in-process assay method is particularly useful due to its fast turnaround time. A sample taken from the fermentation vessel can be purified over a protein A column and its glycosylation pattern analyzed on LC/MS, with the entire process lasting only a few hours. In contrast, the glycan analysis using 2-AB labeling method typically takes more than a day.

15.3.3.3 Disulfide Bond Mapping

Disulfide bonds are key components of the tertiary structure of an antibody. The correct disulfide pairings are typically confirmed via peptide mapping techniques under nonreducing condition. Some disulfide bonds can be mapped using a single enzyme digest. Others, however, will require a digest using a combination of two proteases, such as trypsin and chymotrypsin, to generate fragment with single disulfide bond. Due to the fact that any S—S bond link peptides will have a very complex fragmentation pattern, it is impossible to identify S—S bond containing peptides beyond the accurate mass. Fortunately, modern mass spectrometer vendors, such as Waters Corps and Thermo Fisher Scientific, provide tailor-made software to conveniently map the MS/MS fragment to the expected b/y ion pairs [29, 30]. An example of a disulfide-bonded peptide MS/MS spectrum is given in Figure 15.9.

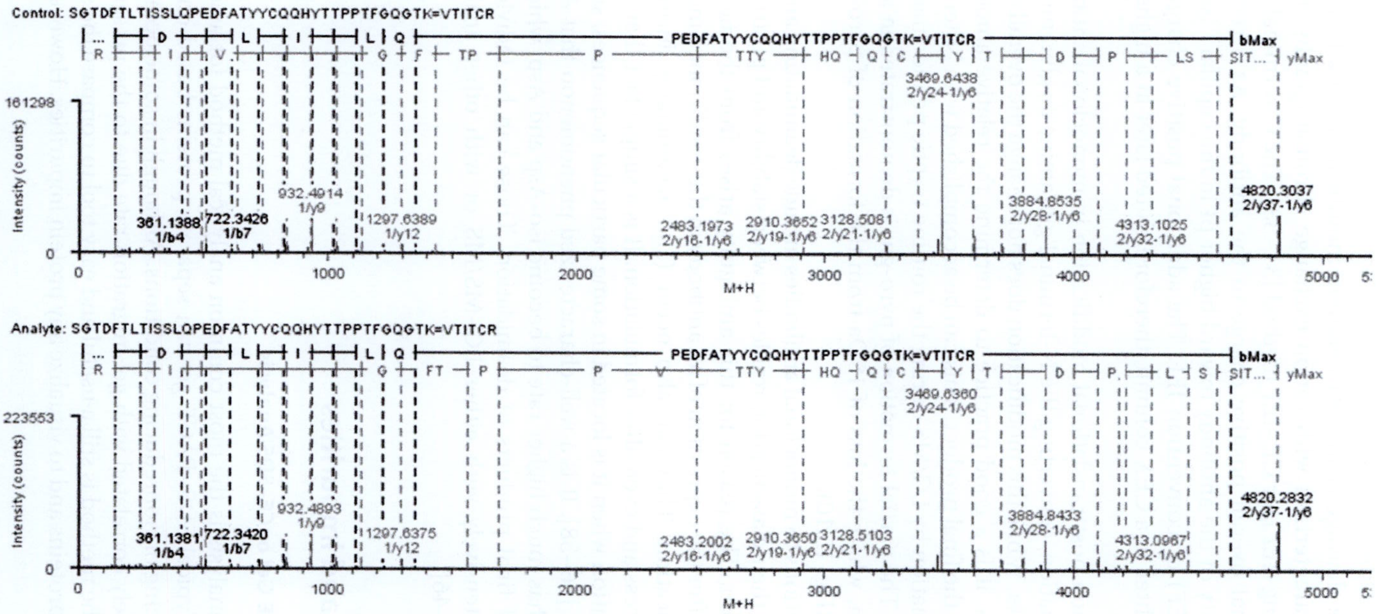

Figure 15.9 MS/MS spectrum of a disulfide peptide VTITCR = SGTDFTLTISSLQPEDFATYYCQQHYTTPPTFGQGTK. Top panel: from originator; bottom panel: from a biosimilar.

15.3.3.4 Other Modifications

C-terminal lysine cleavage is another common modification for all antibodies. It can be characterized with cation exchange chromatography (CEX), isoelectric focusing (IEF), or CE–IEF method [31–33]. The C-terminal lysine adds an additional overall positive charge to the antibody. As a result, it brings the mobility of the antibody toward higher pI in the capillary isoelectric focusing (cIEF) or convention IEF. The additional positive charge also makes it bind tighter to a CEX column, therefore eluted later at a higher salt concentration.

Another common posttranslational modification in monoclonal antibodies is the pyroglutamation, converting the N-terminal glutamine or glutamic acid into pyroglutamate. While this modification does not change the overall activities of antibodies, it is a good practice to determine the relative amount of pyroglutamate in the final product. This can be accomplished with N-terminal sequence confirmation by LC/MS during the routine tryptic peptide mapping analysis [34–36]. The partial formation of pyro-glu is demonstrated in as formation of b4* ion, where the loss of 18 Da from the formation of pyro-glu is clearly observed (Fig. 15.10).

Other modifications to monoclonal antibodies include deamidation and oxidation. However, their classification resides between stability and posttranslational modification. The reason for this ambiguity arises from the fact that these can occur after the expression of an antibody in the conditioned medium before the purification. They can also occur (and accumulate) during the downstream process and even after formulation. It is a simple fact that Asn is prone to deamidation when it is located in some particular sequences, such as SNGQ or ENNY [36–38]. It is a well-characterized phenomenon that Asn in these sequences has much higher rate to become iso-Asp and Asp, which are intermediate and final products of deamidation. Those can be individually characterized extensively with either LC–MS/MS or with other analytical methods [12, 39–46].

15.4 Purity and Impurities

15.4.1 SDS–PAGE Gel or CE–SDS Analysis

SDS–PAGE gel analysis is the most common analytical method for assessing protein purity. Typical SDS–PAGE gel can separate proteins based on their sizes under reducing or nonreducing conditions. Although in some occasions, size is not precisely correlated with gel migration rate due to the presence of rare structures; the method is still a useful and easy tool to compare the size of two comparable proteins and to visualize any protein impurities. However, due

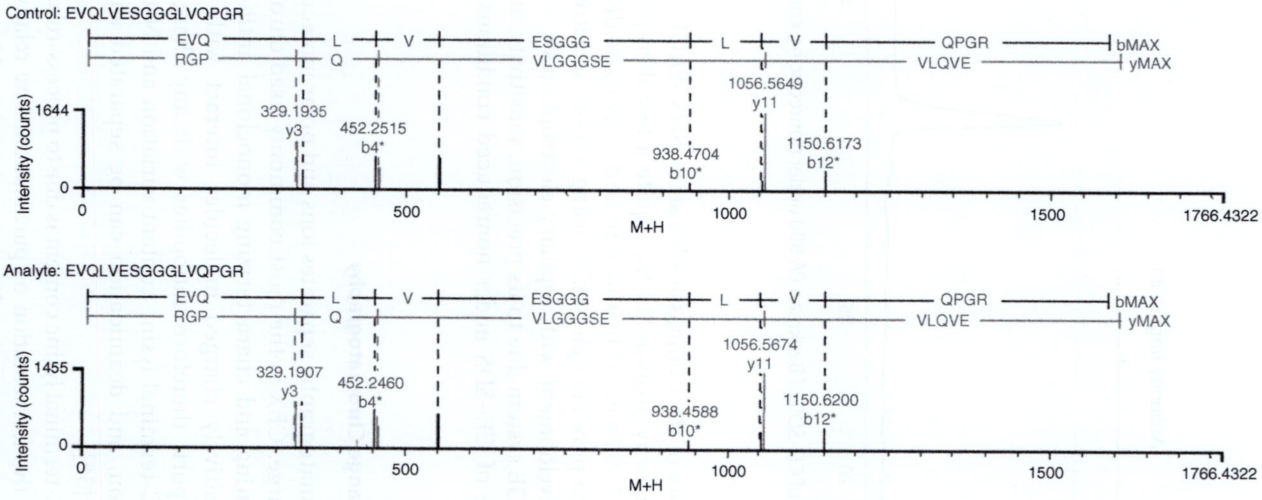

Figure 15.10 Confirmation of N-terminal pyroglutamate from both biosimilar (Analyte) and originator mAb (Control).

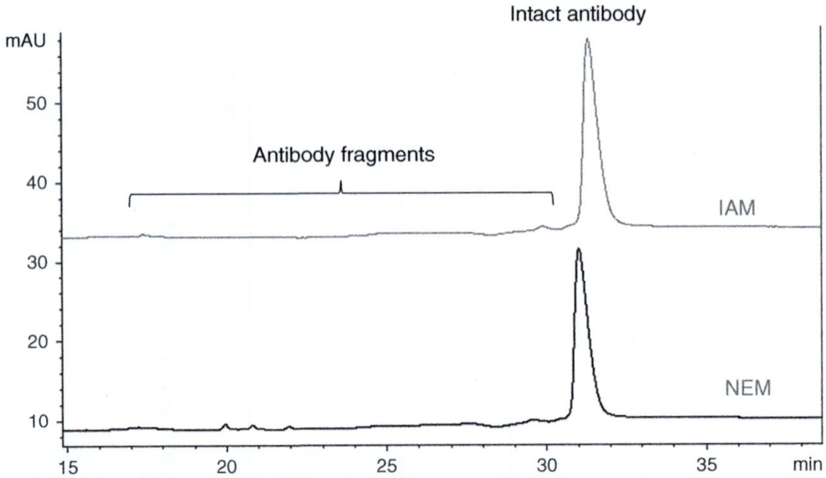

Figure 15.11 Example of CE–SDS of Beckman PA 800 under nonreducing conditions.

to the poor sensitivity of the Coomassie blue stain, SDS–PAGE can only offer limited value in impurity analysis. In fact, quite a few therapeutic proteins were deemed to be ">95% pure" by this standard, only to be discovered that they were borderline pure enough (<85%) using a more sensitive silver stain method. Recent development with capillary electrophoresis will eventually replace the old PAGE system due to its precision, sensitivity, and reproducibility. An example of CE–SDS under nonreduced conditions is shown in Figure 15.11.

15.4.2 Cation Exchange Chromatography

Ion exchange chromatography separates ions and polar molecules based on overall surface charge. CEX is the most commonly used chromatographic method for separating and characterizing monoclonal antibodies due to their high pI. Positively charged molecules interact with a negatively charged solid support; therefore antibodies with any charge variations, such as different C-terminal lysine content variation and N-terminal glutamine modification, and deamidation can be separated and quantified using this method [36].

Variation in the C-terminal lysine content is due to process-related modification, either during the fermentation or purification. The relative amount of C-terminal lysine-containing molecules can be determined by comparing the untreated sample and a sample treated with carboxyl peptidase B (CpB) on

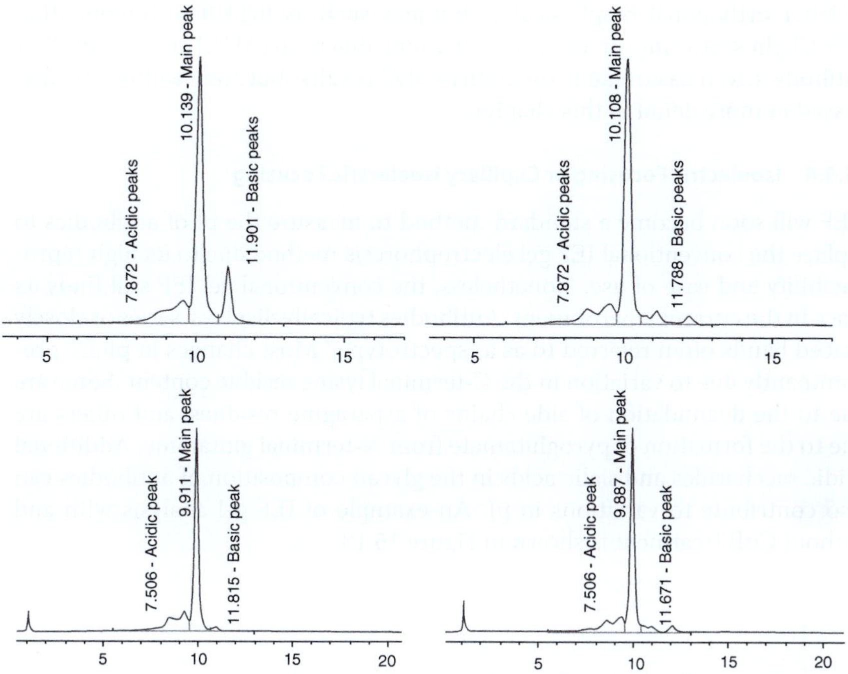

Figure 15.12 The CEX chromatograms of a biosimilar antibody (left panel) and originator antibody sample (right panel) with carboxyl peptidase B (CpB) digestion (lower panel) or without CpB digestion (upper panel).

CEX. Figure 15.12 displays changes in the CEX pattern before and after CpB treatment of a biosimilar and an originator antibody. In this case, the basic peak eluted at 11.9 min represents the lysine-containing portion of the product. Upon the CpB treatment, the basic peak is converted to the main peak, which represents the C-terminal lysine-free portion of the product. Similarities are determined by the relative percentages of acidic, basic peak area and the main peak before and after the treatment.

15.4.3 Size Exclusion Chromatography

SEC is the method of choice for the analysis of the size distribution of the antibody due to its high precision and ease of use. It separates the monomer, which is the main product, from dimers, trimers, and oligomers. The peak area for each represents the relative abundance of each species. Monomers consist of greater than 95% of most antibodies, while other species are the results of aggregations. Because of SEC's unique ability to differentiate the main product from aggregate, one of stability-related end products, it is widely adopted as one of the many QC methods to ensure the integrity of the products.

Other orthogonal biophysical techniques such as field-flow fractionation (FFF), light scattering, or analytical ultracentrifugation (AUC) are also used for antibody size measurement to confirm SEC results, but they will not be discussed in more detail in this chapter.

15.4.4 Isoelectric Focusing or Capillary Isoelectric Focusing

cIEF will soon become a standard method to measure the pI of antibodies to replace the conventional IEF gel electrophoresis method due to its high reproducibility and ease of use. Nonetheless, the conventional gel IEF still finds its place in the current environment. Antibodies typically display a series of closely spaced bands often referred to as a "spectrotype." Most changes in pI are predominantly due to variation in the C-terminal lysine residue content. Some are due to the deamidation of side chains of asparagine residues, and others are due to the formation of pyroglutamate from N-terminal glutamine. Additional acidic saccharides and sialic acids in the glycan composition of antibodies can also contribute to variations in pI. An example of IEF gel analysis with and without CpB treatment is shown in Figure 15.13.

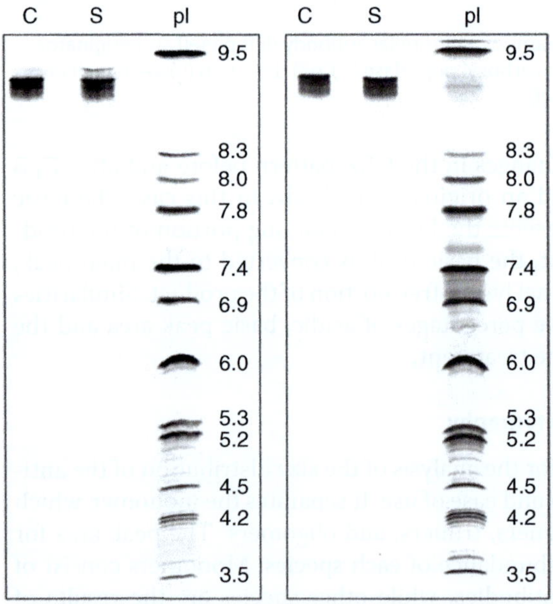

Figure 15.13 IEF gel analysis of antibody samples with CpB treatment (right panel) or without treatment (left panel). C, reference sample; S, biosimilar sample; pI, markers with different pI.

15.4.5 Product-Related Impurities

15.4.5.1 Dimers, Trimers, and Oligomers

Aggregation of therapeutic monoclonal antibodies is well documented. The aggregates can form during cell culture, downstream purification, and storage after formulated into drug products. Many advanced analytical tools have been developed to characterize them. They include elution volume-based SEC, sedimentation, and the dynamic light-scattering techniques [47–51]. Standard SEC chromatography as mentioned previously can detect dimers, tetramers, and small numbers of oligomers. Sedimentation velocity analysis can be used to determine the homogeneity and distribution of long-lived aggregates.

15.4.5.2 C-Terminal Lysine Variants

Standard CEX chromatography can be used to quantify one or two lysine residues on the IgG molecule, which is the cause of charge heterogeneity. Although many publications have shown that the presence or absence of C-terminal lysines in an antibody does not affect functions or serum half-life *in vivo*, it is a good indicator of the lot-to-lot consistency during production. In Figure 15.12, CpB treatment of an antibody can distinguish which peaks in a CEX profile are derived from C-terminal lysine residues.

15.4.5.3 Deamidation and Oxidation Variants

Since it is not easy to determine specific sites of deamidation or oxidation within the whole IgG, it is preferable to characterize the deamidation/oxidation sites as part of peptide mapping using LC–MS/MS. Deamidation or oxidation of amino acids does cause some functional changes to the antibodies; therefore, they should be closely monitored via peptide mapping after initial characterization with mass spectrometry.

15.4.6 Process-Related Impurities

15.4.6.1 Host Cell Proteins

Commercial host cell proteins (HCP) detection kits such as those from Cygnus Technologies (http://www.cygnustechnologies.com) can be used in the early phase of biosimilar antibody development. Different companies offer kits with drastically varying sensitivities to different CHO cell lines. A company even may offer different HCP kits. A cell line and process-specific HCP detection method must be developed before phase III clinical trials. There are many different ways to prepare HCPs from a production cell line, and each may cause different sensitivity in the enzyme-linked immunosorbent assay (ELISA) using antibodies generated from the immunization of animals. HCPs obtained directly from cell culture supernatant or from a similar purification process have been used. It is crucial to validate the detection antibodies for coverage of HCPs in a two-dimensional gel blot. During process development, it is important to track the reduction of HCP before and after each purification step. The results will provide assurance for a robust product that will meet or exceed HCP specifications established by the FDA. Even though the current requirement for maximum HCP concentration in an antibody product is 100 ppm, a lower HCP concentration is universally preferred due to the decreased likelihood of an adverse immunogenic reaction.

15.4.6.2 Host Cell DNA (HCD)

Host cell DNA can be copurified from culture supernatant as contaminants, which may have negative effects on the product. The upper limit for acceptable host cell DNA in the final drug product is 100 pg/therapeutic dose, although the World Health Organization (WHO) has set a residual host cell DNA limit of 10 ng/dose, which is less stringent than the requirement of the US FDA. The most sensitive and commonly used method in the industry is qPCR with standard primers to the host cell type genomic DNA to detect and quantify any contaminating residual DNA [52]. Commercial HCD detection kits are available for common cell lines, such as Applied Biosystems®, DNASEQ™, and Quantitative CHO DNA kit.

15.4.6.3 Other Proteins from Culture Media

If growth factors or proteins, such as recombinant growth factor or insulin, are added to cell growth media, they have to be tested for any residual levels after purification. Commercial ELISA kits are available for recombinant proteins, which are generally highly sensitive and easy to use. For most residual proteins, current maximum limit is 10 ppm, although each protein has to be assessed for their safety and immunogenicity individually. If any serum-derived or animal-derived components are added, additional extensive studies need to be performed to ensure no viral contamination in the final product. Nowadays, the use of animal-derived components has been strongly discouraged by nearly all regulatory agencies, but, nevertheless, it will be wise to avoid adding them during the manufacturing process.

15.4.6.4 Antibiotics or Other Chemicals from Specific Culture Media

Some companies use antibiotics to exert selective pressure during cell growth in the final production stage. Therefore, it is necessary to test residual antibiotics levels in the final product due to the issues of toxicity and allergic reactions in patients. This can be easily achieved by choosing a capable contract research organization (CRO) to perform quantitative analysis of antibiotics in the intermediates and final product.

15.4.6.5 Residual Protein A

Protein A chromatography is a commonly used purification method for almost all antibodies, and therefore leached residual protein A has to be tested in the final product. Conforming to the other standards for protein impurity, protein

A concentration in the final drug product should be below 10 ppm. Commercial protein A detection kits are available from different vendors such as Protein A ELISA Kit from Repligen (Waltham, MA) [53].

15.4.6.6 Bacterial Endotoxins

Bacterial endotoxins, also known as lipopolysaccharides (LPS), are part of the outer membrane of the cell wall of Gram-negative bacteria. Any contamination of such bacteria in the production process may cause increased levels of endotoxins. Therefore, endotoxin levels have to be closely monitored during the downstream process to ensure purity. The most commonly used photometric test is based on a Limulus amoebocyte lysate (LAL), and the colormetric-based method uses a synthetic protease substrate [54, 55]. Commercial endotoxin detection kits are easily available from different vendors. Endotoxin levels in the final drug product should be no greater than 1 endotoxin units (EU)/ml.

15.4.6.7 Sterility Test for Bacterial and Fungal Cell Contamination

Final drug products have to be tested for any bacterial and fungal cell contamination before release. Aliquots will be mixed with different cell growth media for any microbial growth under different conditions. Many contract research companies can provide such professional services.

15.4.6.8 Particulate Matter Test: Visible and Subvisible Particle Test

Visible and subvisible particles can be tested manually or with automated instrument. Those particles can be formed either from aggregated proteins or within drug containers during filling and packaging. Instruments such as FlowCAM can capture real-time images of protein particles, and others such as the Nanosight LM20 can provide size distribution and concentration for particles from 30 nm to 2 μm. Other laser-based instruments can also determine accurate concentrations of particles in the 0.2–2 μm size range to track the progress of aggregation [56, 57].

15.5 Stability

Monoclonal antibodies have to be monitored for their stability under storage conditions. Most antibodies are stored at 2–8°C under refrigerated conditions, while some other antibody products can be stored at room temperature after lyophilization. To avoid any stability issues later on, the antibody is subjected to elevated temperature or highly oxidizing conditions to track the progress of protein degradation. Both analytical and functional assays have to be used simultaneously to monitor stability issues, while LC–MS can detect associated minor structural changes of protein [47].

15.6 Quantity—Concentration Measurement

UV absorption at 280 nm (A280) wavelength is a standard protein concentration measurement. This is a simple and accurate method but require known molecular extinction coefficient constant (ε). Based on the amino acid sequence, ε can be estimated, but it tends to be slightly overestimated due to the additional molecular mass due to glycosylation. Therefore, the exact ε value needs to be experimentally determined. One of the most frequently used methods for ε determination is quantitative amino acid analysis (AAA). For AAA analysis, an antibody solution is prepared, and its absorption at 280 nm is measured. Triplicate aliquots of the solution are submitted for analysis. The complete acidic digest of protein enables the molar quantitation of stable amino acids, such as leucine, isoleucine, phenylalanine, alanine, and so on. The precise molar quantitation of these stable amino acids allows for the accurate calculation of the true extinction coefficient constant of a protein. With the experimentally determined ε value for any given protein, it can be confidently used to determine the concentration of the protein.

Other quick quantitation methods including ELISA and protein A-based HPLC are also commonly used for antibodies. These are useful methods but less precise than UV280 measurement.

15.7 Biological Activity—Functional Bioassays

In 2012, the European Medicines Agency (EMA) published "Guideline on Similar Biological Medicinal Products Containing Monoclonal Antibodies—Non-clinical and Clinical Issues" [58–60]. It requires relevant assays on antigen binding and its Fab-associated functional assay; all binding assays for Fc receptors including FcRn, C1q, and different isoforms of FcγRI, FcγRII, and FcγRIII; and their Fc-associated functional assays including ADCC and complement-dependent cytotoxicity (CDC).

15.7.1 Antigen-Binding Affinity Measurement

There are many affinity measurement methods that are available for antibody–antigen binding such as the standard ELISA assay as well as surface plasmon resonance (SPR), kinetic exclusion assays, and biolayer interferometry (BLI) technology.

15.7.2 Enzyme-Linked Immunosorbent Assay

The ELISA is probably the most commonly used method for testing antibody–antigen binding. It can be used for quantitation of antibodies, comparison of binding affinities, and even test biological functions of antibodies after antigen binding such as inhibiting ligand binding to the receptor.

15.7.3 Surface Plasmon Resonance (SPR)

SPR technology is probably the most commonly used method for measuring antibody–antigen binding kinetics in a label-free environment [61–64]. A variety of instruments have been developed, including the Biacore optical biosensors from Biacore, Inc. (Piscataway, NJ); SPR-2 analytical biosensor system from Sierra Biosensor (Greenville, RI); SensiQ Pioneer system from SensiQ Technologies, Inc. (Oklahoma City, OK); and ProteOn system from BioRad (Hercules, CA). These instruments are capable of determining the on and off rates of antibody–antigen interaction and use these values to derive the final binding constant.

15.7.4 Kinetic Exclusion Assays

Sapidyne Instruments, Inc. (Boise, ID) developed KinExA 3000 and 3200 instrument using Kinetic Exclusion Assay (KinExA®) technology for measuring binding affinity and kinetics [65]. It offers a platform that allows the true binding affinity and kinetic measurement using unmodified molecules in the solution phase. Its core technology uses a solid-phase immobilized molecule to probe for free concentration of one interaction component after allowing sufficient time to reach equilibrium (affinity measurements) or under preequilibrium conditions (kinetics).

15.7.5 BLI Technology

ForteBio, a division of Pall Life Sciences (Menlo Park, CA), has developed a BLI technology that monitors the binding of proteins and other biomolecules to their antigen counterparts [66]. Octet systems enable real-time, label-free analysis for determination of affinity, kinetics, and concentration of 96 samples in parallel.

15.7.6 Biological Functional Assays

15.7.6.1 Ligand–Receptor Binding Competition or Blocking Assay

Some antibodies can compete with ligand–receptor binding. Therefore, its functions can easily be tested *in vitro* using a competition-binding assay. An example of an experimental setup is where an antibody binds to a receptor, hereby blocking ligand binding. This antibody can be premixed with the ligand to investigate its competitive binding with the receptor in an ELISA format. Alternatively, an antibody can bind to the receptor first to see its blocking activity for binding to its ligand.

15.7.6.2 Receptor-Binding Assays

In addition to the antigen-binding functions for the Fab domain, the Fc domain of therapeutic antibodies possesses many biological functions, including many receptor-binding activities affecting serum half-life, CDC, and ADCC activities.

15.7.6.3 Fcγ Receptors: RIIIa, RIIa, RIIb, and RI

FcγR receptors are important for *in vivo* effector functions of IgG antibodies [67–69]. Particularly, FcγRIIIa-binding activity correlates very well with antibody ADCC activity. The FcγRIIIa receptor has two major genotypes: 158V and 158F, where the former is more active in the ADCC. Due to polymorphisms of all FcγR receptors, all receptor-binding activities shall be tested in an ELISA or SPR assay. Almost all of the receptors including FcγRI (CD64), FcγRIIA (CD32), FcγRIIB (CD32), FcγRIIIA (CD16a), and FcγRIIIB (CD16b) can be purchased from commercial vendors. It has been demonstrated that the fucose content of an antibody has an inverse relationship with FcγRIIIa-binding activity [60–72]. Most antibodies have below 10% G0 glycan (without fucose). An increase of G0 from 2 to 8% can increase FcγRIIIa-binding activity by severalfold, which also increases ADCC activity by a similar factor depending on the antibody. However, one cannot oversimplify the importance of FcγR binding. In fact, some antibodies are designed to target antigens that are not located on cell surface, which in those cases the FcγR binding do not provide additional benefits. On the contrary, it may bring unintended consequence if the undesirable binding to cell surface occurs, resulting in cellular toxicities. For the reasons mentioned previously, it is critical to monitor any differences in receptor-binding activities of biosimilar products.

15.7.6.4 Fc Neonatal Receptor (FcRn)

FcRn is important for the recycling of IgG antibodies in serum and is responsible for its serum half-life [73]. The receptor consists of alpha and beta subunits, which binds IgG antibodies at a slightly acidic pH (pH 6.0–6.5) from the intestinal lumen and releases them at a neutral to basic pH (7.0–7.5) to ensure efficient unidirectional transport to the basolateral side to maintain high serum half-life. Structural changes in the Fc region of an IgG antibody or some glycan may influence the FcRn binding at different pH conditions. Therefore, comparison with the originator drug for this human receptor binding may predict its similarity in human pharmacokinetics (PK) in the clinical trials. This is typically tested in a standard ELISA assay.

15.7.6.5 C1q

Complement components of serum are composed of a serial of proteins in the CDC pathway. C1q is the first such protein of the complement cascade to bind to the Fc region of the antibodies and, therefore, is critical for the CDC functions of the antibodies. Its binding affinity correlates with CDC activity, although with some exceptions (authors' unpublished data). These receptors can be purchased commercially, and either ELISA- or SPR-binding assays can be developed for binding assays of receptors.

15.7.7 Whole Cell Function Assays

15.7.7.1 Whole Cell-Binding Assays for Membrane Protein Antigens

Some antigens, such as CD20, cannot be prepared as soluble antigens. In those cases, a whole cell-binding assay is necessary to measure its antigen-binding activity. High antigen-expressing cells either from a primary human cell culture, a tumor cell line, or an antigen-transfected cell line can be used for either whole cell ELISA or FACS analysis. Furthermore, FACS analysis of whole cell-binding antibodies is the most realistic binding assay mimicking *in vivo* conditions.

15.7.7.2 Anti-cell Proliferation Assay

Many cell surface antigens are involved in cell survival and proliferation, typically through the signaling pathway. Antibodies binding to these antigens will prevent binding to their respective ligand and hence resulting in an antiproliferative effect. The inhibitory functions can be tested with different cell lines expressing specific antigens on the cell surface. One example is Erb2, a receptor found on the cell surface of breast carcinoma cells. The binding of trastuzumab prevents Erb2 from binding to EGF, its respective ligand, and blocks signaling. As a result, trastuzumab inhibits the growth of Erb2-positive cells.

15.7.7.3 Neutralization Assay for Blocking Antigen Functions in Cells

Soluble antigens are not associated with cells. Despite the lack of association, most of them have known cellular functions, such as stimulating or inhibiting cell growth. To investigate an antigen's role in the cellular pathway, an antibody neutralization assay can be developed in the whole cell format. For example, anti-TNFα antibody Humira (adalimumab) and Remicade can be tested by blocking TNFα functions in L929 mouse fibroblast cell line growth in a concentration-dependent manner via preincubation of antibody with TNFα prior to addition to the corresponding cells, where L929 cell growth inhibition is dependent on the presence of TNFα.

15.7.7.4 Complement-Dependent Cytotoxicity Assay

The CDC is a critical function of some antibodies *in vivo*. Many therapeutic antibodies have CDC functions against tumor cells when incubated *in vitro* with human or rabbit serum complement. Therefore, an *in vitro* functional assay is needed in evaluating the CDC functions of a biosimilar antibody. It is important to use several cell lines for this test, because different cell lines may exhibit different sensitivities to an antibody due to the different copy numbers of antigen expressed on cell surface. For example, when comparing different lymphoma cell lines such as Ramos, Raji, Daudi, and Wil2-s, they were found to have different sensitivities to rituximab during a CDC assay. As best practice, the most sensitive cell line should be chosen for this assay after careful evaluation.

15.7.7.5 Antibody-Dependent Cell-Mediated Cytotoxicity Assay

The ADCC is part of the adaptive immune response mediated by natural killer (NK) cells, macrophages, neutrophils, and eosinophils, enabled by the Fcγ receptors on the surface of those cells. When a target cell line expressing the specific surface antigen is incubated with a therapeutic antibody that binds to this antigen in the presence of peripheral blood mononuclear cells (PBMCs) containing NK cells, lysis of target cells can be measured by release of intracellular labels such as chromium-51 or Eu, calcein AM, GAPDH, or LDH enzymes. Stable NK cell lines transfected with FcγRIIIa (V158) are also available from the ATCC collection; however, its ADCC function is not as strong as PBMCs.

In addition to CDC and ADCC assays, recent progress with reporter assay may help simplify the experimental setup. Promega has developed a luciferase reporter assay using effector cells that are engineered to express FcγRIIIa (V158) and NFAT-RE luc2 luciferase.

http://www.promega.com/~/media/files/resources/slide%20presentations/adcc_reporter_bioassay_april_2014.pdf

If an antibody does not bind to any membrane proteins, it is not recommended to test CDC and ADCC as stated in EMA guideline published in year 2012.

http://www.ema.europa.eu/docs/en_GB/document_library/Scientific_guideline/2012/06/WC500128686.pdf

Recent review of clinical antitumor efficacy data with FcγRIIIa genotypes showed inconsistent findings. In all three antibody cases, rituximab, cetuximab, and trastuzumab, some data showed that V/V genotype of FcγRIIIa are beneficial, whereas others showed no correlation with FcγRIIIa V/V genotype [74].

15.7.7.6 Apoptosis Assay

Binding of antibodies to cell surface antigens may trigger cell apoptotic signals by causing regulatory proteins to initiate the apoptosis pathway to cause cell death. If an antibody lacks any *in vitro* functions mentioned in the previous sections but demonstrates cell-killing *in vivo*, it is worth testing apoptotic DNA fragmentation of treated cells using the DNA laddering assay [75] and TUNNEL assay [76]. Herceptin is an example of such an antibody, in which it can sensitize taxol-induced apoptosis in HER2-overexpressing tumor cells [77].

15.8 Efficacy and Safety: Animal Studies for Antibody-Drug Efficacy, PK/PD, and Toxicity

Nonclinical studies are critical parts of the IND filing package for a biosimilar product. Comparative studies of the biosimilar and originator mAb are conducted *in vitro* first to assess biological functions as described in the previous section before *in vivo* investigation. Clinical studies are not within this review scope.

Additional *in vivo* animal studies may be implemented if a biosimilar product contains unique attributes that are not detected in or at a different level than the reference product, such as a new posttranslational modification. Any formulation change would also warrant more studies *in vivo*, usually in nonhuman primates.

15.8.1 Efficacy Studies in Mouse Xenograft or Transgenic Mouse Models

After confirming *in vitro* biological functions of biosimilar antibodies, it is essential to confirm their *in vivo* functions in animal models such as mouse xenograft models or transgenic mice. Many tumor cell lines are available for testing *in vivo* on a xenograft mouse model. However, due to species specificity, some antigens cannot be expressed in mouse due to the loss of efficacy of antibodies in the mouse model system. Either a transgenic mouse system has to be established or nonhuman primate (e.g., monkeys) models have to be directly used to mimic human environment. For example, anti-TNFα antibody adalimumab can only be tested in human TNFα transgenic mice for its efficacy in a rheumatoid arthritis model due to its species specificity, while the anti-CD20 antibody rituximab can be tested directly in *Cynomolgus* monkey for its CD20-positive B-cell depletion function *in vivo* as a surrogate for indication of rheumatoid arthritis because rituximab does not recognize mouse CD20 molecules on the B cells.

15.8.2 Pharmacokinetics and Pharmacodynamics Studies in Nonhuman Primates

Antibody PK studies have to be tested in nonhuman primates, usually *Cynomolgus* monkey, due to its species-specific FcRn-binding activity of human Fc sequences. PK studies in mice may not be predictive of antibody PK in humans due to different binding affinities to FcRn at different pH conditions. There is no good correlation between mouse and human FcRn receptor binding for *in vitro* Fc binding (author's unpublished data). Although human FcRn transgenic mice are available, it is still best to use monkey models for such a test before initiation of clinical trials. Most antibodies can also be monitored in monkeys for its pharmacodynamics (PD) to compare their *in vivo* functions. However, it is quite common to observe a lack of specific PD endpoints (PD markers), and, therefore, nonclinical PD evaluations including *in vitro* whole cell assays are critical. A quantitative concentration-dependent assessment covering the therapeutic doses in human is usually required for comparison of PK/PD properties of the biosimilar and reference products.

15.8.3 Acute and Chronic Toxicology Studies in Nonhuman Primates

Acute toxicity study can be done in monkeys in a single or multiple doses, during a 14-day period to closely monitor any adverse effects. A long-term toxicity study with repeated doses in monkeys can last 3–6 months or even longer, depending on the antibody treatment regimen in humans. US FDA and EMA guidelines (2010) do not recommend repeated dose toxicity studies in nonhuman primates if other comparison studies show strong similarity in the data. Nonclinical studies regarding pharmacology and reproductive toxicology are also not recommended for biosimilar mAb.

15.8.4 Immunogenicity

Some posttranslational modifications of antibodies and any product-related or process-related impurities can be immunogenic in humans due to their nonhuman nature. For example, mannose and beta 1, 3-galactose are nonhuman sugar moieties, which may lead to potential immunogenic responses. Differing immunogenic responses should be monitored for biosimilar and originator antibodies, although other immune responses such as those against the nonhuman sequences in the chimeric and humanized antibodies are common to both compounds. In addition, any changes in the formulation and the source of the excipients of an antibody may also alter immunogenicity. Therefore, antidrug antibodies (ADA) shall be monitored initially for all biosimilar antibodies during preclinical tests and clinical trials. However, nonclinical studies aiming at predicting immunogenicity in humans are not a good indicator due to the "foreign" protein nature of human antibodies in other animals. Therefore, ADA measurement in animal studies is mostly used to support adverse effects, PK and PD data, during nonclinical and toxicology studies.

After initial screening of ADA from serum samples, it is critical to validate ADA assays to avoid a false negative probably to "drug interference" effect and to retest all positives to confirm the results were valid.

Further studies may also be required to identify the neutralizing nature of the ADA and determine the ADA isotypes if needed.

References

1. Yang, X., Ambrogelly, A. (2014). Enlarging the repertoire of therapeutic monoclonal antibodies platforms: domesticating half molecule exchange to produce stable IgG4 and IgG1 bispecific antibodies. *Curr. Opin. Biotechnol.*, 30C, 225–229.
2. Lameris, R., de Bruin, R.C., Schneiders, F.L., van Bergen En Henegouwen, P.M.,Verheul, H.M., de Gruijl, T.D., van der Vliet, H.J. (2014). Bispecific

antibody platforms for cancer immunotherapy. *Crit. Rev. Oncol. Hematol.*, 92(3), 153–165.

3. Pegram, M., Hsu, S., Lewis, G., Pietras, R., Beryt, M., Sliwkowski, M., Coombs, D., Baly, D., Kabbinavar, F., Slamon, D. (1999). Inhibitory effects of combinations of HER-2/neu antibody and chemotherapeutic agents used for treatment of human breast cancers. *Oncogene*, 18(13), 2241–2251.
4. Yarden, Y. (2001). Biology of HER2 and its importance in breast cancer. *Oncology*, 61(Suppl 2), 1–13.
5. Sliwkowski, M.X., Lofgren, J.A., Lewis, G.D., Hotaling, T.E., Fendly, B.M., Fox, J.A. (1999). Nonclinical studies addressing the mechanism of action of trastuzumab (Herceptin). *Semin. Oncol.*, 26(4 Suppl 12), 60–70.
6. Choi, B.D., Cai, M., Bigner, D.D., Mehta, A.I., Kuan, C.T., Sampson, J.H. (2011). Bispecific antibodies engage T cells for antitumor immunotherapy. *Expert Opin. Biol. Ther.*, 11(7), 843–853.
7. Robak, T. (2013). Emerging monoclonal antibodies and related agents for the treatment of chronic lymphocytic leukemia. *Future Oncol.*, 9(1), 69–91.
8. Soliman, H. (2013). Immunotherapy strategies in the treatment of breast cancer. *Cancer Control*, 20(1), 17–21.
9. Danylesko, I., Beider, K., Shimoni, A., Nagler, A. (2012). Monoclonal antibody-based immunotherapy for multiple myeloma. *Immunotherapy*, 4(9), 919–938.
10. Weiner, L.M., Murray, J.C., Shuptrine, C.W. (2012). Antibody-based immunotherapy of cancer. *Cell*, 148(6), 1081–1084.
11. Topalian, S.L., Weiner, G.J., Pardoll, D.M. (2011). Cancer immunotherapy comes of age. *J. Clin. Oncol.*, 29(36), 4828–4836.
12. Wang, L., Amphlett, G., Lambert, J.M., Blättler, W., Zhang, W. (2005). Structural characterization of a recombinant monoclonal antibody by electrospray time-of-flight mass spectrometry. *Pharm. Res.*, 22(8), 1338–1349.
13. Chakraborty, A.B., Berger, S.J., Gebler, J.C. (2007). Use of an integrated MS—multiplexed MS/MS data acquisition strategy for high-coverage peptide mapping studies. *Rapid Commun. Mass Spectrom.*, 21(5), 730–744.
14. Roach, C.A., Simpson, J.V., JiJi, R.D. (2012). Evolution of quantitative methods in protein secondary structure determination via deep-ultraviolet resonance Raman spectroscopy. *Analyst*, 137(3), 555–562.
15. Bertucci, C., Pistolozzi, M., De Simone, A. (2011). Structural characterization of recombinant therapeutic proteins by circular dichroism. *Curr. Pharm. Biotechnol.*, 12(10), 1508–1516.
16. Dodero, V.I., Quirolo, Z.B., Sequeira, M.A. (2011). Biomolecular studies by circular dichroism. *Front. Biosci.*, 16, 61–73.
17. Whitmore, L., Wallace, B.A. (2008). Protein secondary structure analyses from circular dichroism spectroscopy: methods and reference databases. *Biopolymers*, 89(5), 392–400.

18. Kelly, S.M., Jess, T.J., Price, N.C. (2005). How to study proteins by circular dichroism. *Biochim. Biophys. Acta*, 1751(2), 119–139.

19. Pelton, J.T., McLean, L.R. (2000). Spectroscopic methods for analysis of protein secondary structure. *Anal. Biochem.*, 277(2), 167–176.

20. Fasman, G.D. (2010) Circular Dichroism and the Conformational Analysis of Biomolecules, Platinum Press, New York, pp. 69.

21. Shang, T.Q., Saati, A., Toler, K.N., Mo, J., Li, H., Matlosz, T., Lin, X., Schenk, J., Ng, C.K., Duffy, T., Porter, T.J., Rouse, J.C. (2014). Development and application of a robust N-glycan profiling method for heightened characterization of monoclonal antibodies and related glycoproteins. *J. Pharm. Sci.*, 103(7), 1967–1978.

22. Ahn, J., Bones, J., Yu, Y.Q., Rudd, P.M., Gilar, M. (2010). Separation of 2-aminobenzamide labeled glycans using hydrophilic interaction chromatography columns packed with 1.7 microm sorbent. *J. Chromatogr. B Analyt. Technol. Biomed. Life Sci.*, 878(3–4), 403–408.

23. Karlsson, G., Swerup, E., Sandberg, H. (2008). Combination of two hydrophilic interaction chromatography methods that facilitates identification of 2-aminobenzamide-labeled oligosaccharides. *J. Chromatogr. Sci.*, 46(1), 68–73.

24. Jensen, P.H., Mysling, S., Højrup, P., Jensen, O.N. (2013). Glycopeptide enrichment for MALDI–TOF mass spectrometry analysis by hydrophilic interaction liquid chromatography solid phase extraction (HILIC SPE). *Methods Mol. Biol.*, 951, 131–144.

25. Shah, B., Jiang, X.G., Chen, L., Zhang, Z. (2014). LC–MS/MS peptide mapping with automated data processing for routine profiling of N-glycans in immunoglobulins. *J. Am. Soc. Mass Spectrom.*, 25(6), 999–1011.

26. Ozohanics, O., Turiák, L., Puerta, A., Vékey, K., Drahos, L. (2012). High-performance liquid chromatography coupled to mass spectrometry methodology for analyzing site-specific N-glycosylation patterns. *J. Chromatogr. A.*, 1259, 200–212.

27. Schiel, J.E., Au, J., He, H.J., Phinney, K.W. (2012). LC–MS/MS biopharmaceutical glycoanalysis: identification of desirable reference material characteristics. *Anal. Bioanal. Chem.*, 403(8), 227922–227989.

28. Hardy, M.R., Townsend, R.R. (1988). Separation of positional isomers of oligosaccharides and glycopeptides by high-performance anion-exchange chromatography with pulsed amperometric detection. *Proc. Natl. Acad. Sci. U. S. A.*, 85, 3289–3293.

29. Li, X., Xu, W., Paporello, B., Richardson, D., Liu, H. (2013). Liquid chromatography and mass spectrometry with post-column partial reduction for the analysis of native and scrambled disulfide bonds. *Anal. Biochem.*, 439(2), 184–186.

30. Zhang, W., Marzilli, L.A., Rouse, J.C., Czupryn, M.J. (2002). Complete disulfide bond assignment of a recombinant immunoglobulin G4 monoclonal antibody. *Anal. Biochem.*, 311(1), 1–9.

31. Wu, S.-L., Jiang, H., Lu, Q., Dai, S., Hancock, W.S., Karger, B.L. (2009). Mass spectrometric determination of disulfide linkages in recombinant therapeutic proteins using on-line LC–MS with electron transfer dissociation (ETD). *Anal. Chem.*, 81(1), 112–122.

32. Vlasak, J., Ionesu, R. (2008). Heterogeneity of monoclonal antibodies revealed by charge-sensitive methods. *Curr. Pharm. Biotechnol.*, 9(6), 468–481.

33. Moorhouse, K.G., Nashabeh, W., Deveney, J. (1997). Validation of an HPLC method for the analysis of the charge heterogeneity of the recombinant monoclonal antibody IDEC-C2B8 after papain digestion. *J. Pharm. Biomed. Anal.*, 16(4), 593–603.

34. Dick, L.W., Jr, Qiu, D., Mahon, D., Adamo, M. (2008). C-terminal lysine variants in fully human monoclonal antibodies: investigation of test methods and possible causes. *Biotechnol. Bioeng.*, 100(6), 1132–1143.

35. Xu, W., Peng, Y., Wang, F., Paporello, B, Richardson, D., Liu, H. (2013). Method to convert N-terminal glutamine to pyroglutamate for characterization of recombinant monoclonal antibodies. *Anal. Biochem.*, 436(1), 10–12.

36. Liu, H., Gaza-Bulseco, G., Faldu, D., Chumsae, C., Sun, J. (2008). Heterogeneity of monoclonal antibodies. *J. Pharm. Sci.*, 97(7), 2426–2447.

37. Pace, A.L., Wong, R.L., Zhang, Y.T., Kao, Y.H., Wang, Y.J. (2013). Asparagine deamidation dependence on buffer type, pH, and temperature. *J. Pharm. Sci.*, 102(6), 1712–1723.

38. Kang, X., Kutzko, J.P., Hayes, M.L., Frey, D.D. (2013). Monoclonal antibody heterogeneity analysis and deamidation monitoring with high-performance cation-exchange chromatofocusing using simple, two component buffer systems. *J. Chromatogr. A*, 1283, 89–97.

39. Sinha, S., Zhang, L., Duan, S., Williams, T.D., Vlasak, J., Ionescu, R., Topp, E.M. (2009). Effect of protein structure on deamidation rate in the Fc fragment of an IgG1 monoclonal antibody. *Protein Sci.*, 18(8), 1573–1584.

40. Huang, H.Z., Nichols, A., Liu, D. (2009). Direct identification and quantification of aspartyl succinimide in an IgG2 mAb by RapiGest assisted digestion. *Anal. Chem.*, 81(4), 1686–1692.

41. Vlasak, J., Bussat, M.C., Wang, S., Wagner-Rousset, E., Schaefer, M., Klinguer-Hamour, C., Kirchmeier, M., Corvaïa, N., Ionescu, R., Beck, A. (2009). Identification and characterization of asparagine deamidation in the light chain CDR1 of a humanized IgG1 antibody. *Anal. Biochem.*, 392(2), 145–154.

42. Liu, H., Gaza-Bulseco, G., Chumsae, C. (2008). Glutamine deamidation of a recombinant monoclonal antibody. *Rapid Commun. Mass Spectrom.*, 22(24), 4081–4088.

43. Huang, L., Lu, J., Wroblewski, V.J., Beals, J.M., Riggin, R.M. (2005). *In vivo* deamidation characterization of monoclonal antibody by LC/MS/MS. *Anal. Chem.*, 77(5), 1432–1439.

44. Chelius, D., Rehder, D.S., Bondarenko, P.V. (2005). Identification and characterization of deamidation sites in the conserved regions of human immunoglobulin gamma antibodies. *Anal. Chem.*, 77(18), 6004–6011.
45. Kroon, D.J., Baldwin-Ferro, A., Lalan, P. (1992). Identification of sites of degradation in a therapeutic monoclonal antibody by peptide mapping. *Pharm. Res.*, 9(11), 1386–1393.
46. Berkowitz, S.A., Engen, J.R., Mazzeo, J.R., Jones, G.B. (2012). Analytical tools for characterizing biopharmaceuticals and the implications for biosimilars. *Nat. Rev. Drug Discov.*, 11(7), 527–540.
47. Lowe, D., Dudgeon, K., Rouet, R., Schofield, P., Jermutus, L., Christ, D. (2011). Aggregation, stability, and formulation of human antibody therapeutics. *Adv. Protein Chem. Struct. Biol.*, 84, 41–61.
48. Drenski, M.F., Brader, M.L., Alston, R.W., Reed, W.F. (2013) Monitoring protein aggregation kinetics with simultaneous multiple sample light scattering. *Anal. Biochem.*, 437(2), 185–197.
49. Demeule, B., Lawrence, M.J., Drake, A.F., Gurny, R., Arvinte, T. (2007). Characterization of protein aggregation: the case of a therapeutic immunoglobulin. *Biochim. Biophys. Acta*, 1774(1), 146–153.
50. Gabrielson, J.P., Brader, M.L., Pekar, A.H., Mathis, K.B., Winter, G., Carpenter, J.F., Randolph, T.W. (2007). Quantitation of aggregate levels in a recombinant humanized monoclonal antibody formulation by size-exclusion chromatography, asymmetrical flow field flow fractionation, and sedimentation velocity. *J. Pharm. Sci.*, 96(2), 268–279.
51. Ye, H. (2006). Simultaneous determination of protein aggregation, degradation, and absolute molecular weight by size exclusion chromatography-multiangle laser light scattering. *Anal. Biochem.*, 356(1), 76–85.
52. Hu, B., Sellers, J., Kupec, J., Ngo, W., Fenton, S., Yang, T.Y., Grebanier, A. (2014). Optimization and validation of DNA extraction and real-time PCR assay for the quantitative measurement of residual host cell DNA in biopharmaceutical products. *J. Pharm. Biomed. Anal.*, 88, 92–95.
53. Rey, G., Wendeler, M.W. (2012). Full automation and validation of a flexible ELISA platform for host cell protein and protein A impurity detection in biopharmaceuticals. *J. Pharm. Biomed. Anal.*, 70, 580–586.
54. Das, A.P., Kumar, P.S., Swain, S. (2014). Recent advances in biosensor based endotoxin detection. *Biosens. Bioelectron.*, 51, 62–75.
55. Pearson, F.C. (1979). The Limulus amebocyte lysate endotoxin assay: current status. *Am. J. Med. Technol.*, 45(8), 704–709.
56. Mickisch, S., Tantipolphan, R., Wiggenhorn, M., Frieß, W., Winter, G., Hawe, A. (2010). Subvisible particles in a monoclonal antibody formulation analyzed by nanoparticle tracking analysis and microflow imaging. 2010 American Association of Pharmaceutical Scientists (AAPS) National Biotechnology Conference, May 16–19, 2010, San Francisco, CA.

57. Jiskoot, W. (2011). Nanoparticle tracking analysis versus dynamic light scattering for measuring sub-micron protein aggregates. Proc FIP Pharmaceutical Sciences 2010 World Congress, November 14–18, 2010, Morial Convention Center, New Orleans, LA, USA.
58. EMA/CHMP/BMWP/403543/2010 Guideline on Similar Biological Medicinal Products Containing Monoclonal Antibodies—Non-clinical and Clinical Issues, 2012.
59. EMA/CHMP/BWP/247713/2012 Guideline on similar biological medicinal products containing biotechnology-derived proteins as active substance: Quality issues, 2012.
60. EMEA/CHMP/BWP/157653/2007 Guideline on development, production, characterisation and specifications for monoclonal antibodies and related substances, 2008.
61. Alfthan, K. (1998). Surface plasmon resonance biosensors as a tool in antibody engineering. *Biosens. Bioelectron.*, 13(6), 653–663.
62. Van Regenmortel, M.H., Altschuh, D., Chatellier, J., Christensen, L., Rauffer-Bruyère, N., Richalet-Secordel, P., Witz, J., Zeder-Lutz, G. (1998). Measurement of antigen–antibody interactions with biosensors. *J. Mol. Recognit.*, 11(1–6), 163–167.
63. Torreri, P., Ceccarini, M., Macioce, P., Petrucci, T.C. (2005). Biomolecular interactions by Surface Plasmon Resonance technology. *Ann. Ist. Super. Sanita*, 41(4), 437–441.
64. Thillaivinayagalingam, P., Gommeaux, J., McLoughlin, M., Collins, D., Newcombe, A.R. (2010). Biopharmaceutical production: applications of surface plasmon resonance biosensors. *J. Chromatogr. B Analyt. Technol. Biomed. Life Sci.*, 878(2), 149–153.
65. Darling, R.J., Brault, P.A. (2004) Kinetic exclusion assay technology: characterization of molecular interactions. *Assay Drug Dev. Technol.*, 2(6), 647–657.
66. Shah, N.B., Duncan, T.M. (2014). Bio-layer interferometry for measuring kinetics of protein–protein interactions and allosteric ligand effects. *J. Vis. Exp.*, 2014 Feb 18 (84), e51383. doi: 10.3791/51383.
67. Gerber, J.S., Mosser, D.M. (2001). Stimulatory and inhibitory signals originating from the macrophage Fcgamma receptors. *Microbes Infect.*, 3(2), 131–139.
68. Cooper, M.A., Fehniger, T.A., Caligiuri, M.A. (2001). The biology of human natural killer-cell subsets. *Trends Immunol.*, 22(11), 633–640.
69. Stevenson, G.T., Anderson, V.A., Leong, W.S. (2002). Engineered antibody for treating lymphoma. *Cancer Res.*, 159, 104–112.
70. Anthony, R.M., Nimmerjahn, F. (2011). The role of differential IgG glycosylation in the interaction of antibodies with FcγRs *in vivo*. *Curr. Opin. Organ Transplant.*, 16(1), 7–14.

71. Zhuang, Y., Xu, W., Shen, Y., Li, J. (2010). Fcγ receptor polymorphisms and clinical efficacy of rituximab in non-Hodgkin lymphoma and chronic lymphocytic leukemia. *Clin. Lymphoma Myeloma Leuk.*, 10(5), 347–352.
72. Nimmerjahn, F., Ravetch, J.V. (2011). FcγRs in health and disease. *Curr. Top. Microbiol. Immunol.*, 350, 105–125.
73. Baker, K., Rath, T., Pyzik, M., Blumberg, R.S. (2014). The role of FcRn in antigen presentation. *Front. Immunol.*, 5, 1–12.
74. Mellor, J.D., Brown, M.P., Irving, H.R., Zalcberg, J.R., Dobrovic, A. (2013). A critical review of the role of Fc gamma receptor polymorphisms in the response to monoclonal antibodies in cancer. *J. Hematol. Oncol.*, 6, 1–10.
75. Wyllie, A.H., Kerr, J.F., Currie, A.R. (1980). Cell death: the significance of apoptosis. *Int. Rev. Cytol.*, 68, 251–306.
76. Gavrieli, Y., Sherman, Y., Ben-Sasson, S.A. (1992). Identification of programmed cell death *in situ* via specific labeling of nuclear DNA fragmentation. *J. Cell Biol.*, 119(3), 493–501.
77. Lee, S., Yang, W., Lan, K.H., Sellappan, S., Klos, K., Hortobagyi, G., Hung, M.C., Yu, D. (2002). Enhanced sensitization to taxol-induced apoptosis by herceptin pretreatment in ErbB2-overexpressing breast cancer cells. *Cancer Res.*, 62(20), 5703–5710.

16

Bioanalytical Development

Rafiq Islam

Senior Director, Celerion, Lincoln, NE, USA; Scientific Adviser, Somru BioScience Inc. Canada, USA

16.1 Summary

This chapter will discuss development of bioanalytical tests of biosimilar antibodies, including glycosylation pattern, ADCC/CDC activity, and FcγRs binding profile. It will explain the test results required to demonstrate similarity to the original product, as stated in the regulation guidelines issued by the FDA and EMEA.

16.2 Introduction

The successful development of a monoclonal antibody (mAb) biosimilar requires the demonstration of biosimilarity in terms of efficacy, safety, and purity to an innovator-approved product. Bioanalytical assays are critical to the demonstration of the biosimilarity. Over the last few years, there have been a number of regulatory guidances posted by the FDA regarding pathways to approval of biosimilars. However none of them details the requirements of bioanalytical method development and validation other than stating the fact that validated bioanalytical methods are required to support the development of biosimilars. There have been increasing numbers of industry white papers published with recommendations on the bioanalysis of biologics [1–3]. While their recommendations are applicable to the bioanalysis of biosimilars, they do not specifically address their bioanalytical challenges and solutions to these issues.

Biosimilars of Monoclonal Antibodies: A Practical Guide to Manufacturing, Preclinical, and Clinical Development, First Edition. Edited by Cheng Liu and K. John Morrow, Jr.

This chapter will focus on specific challenges associated with the bioanalysis of mAb biosimilars in following three key areas:

- Pharmacodynamics (PD) characterization
- Pharmacokinetic (PK) assessment
- Immunogenicity assessment (i.e., antidrug antibody (ADA) detection and characterization)

16.3 Pharmacodynamics Characterization

MAbs are complex molecules with multiple domains that contribute to their mode of action (see Fig. 16.1). The variable domain of the antibody binds with to a specific target and exerts functional effect such as receptor activation, receptor blockade, or neutralization of a soluble target. The constant domain binds to the Fc-gamma receptor (FcγR), to the neonatal Fc receptor (FcRn), or to the complement molecule and triggers signaling pathways associated with antibody-dependent cell-mediated cytotoxicity (ADCC), complement-dependent cytotoxicity (CDC), and complement activation. Minor changes in type and extent of glycosylation in either domains can have a profound impact on the functions of a therapeutic mAb. The complexity of mAbs demands

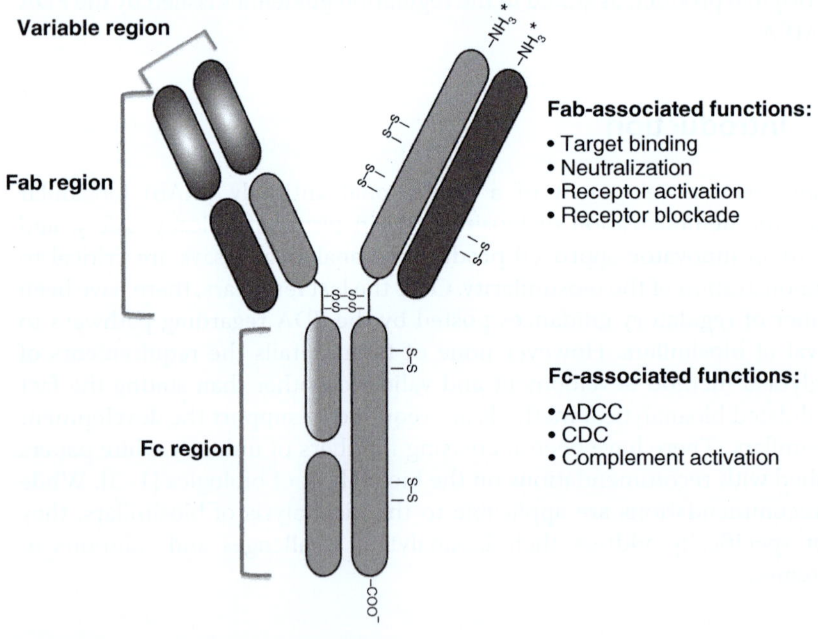

Figure 16.1 Complexity in structure and function of monoclonal antibodies. © Somru Bioscience Inc. (2012–2015).

thorough *in vitro* pharmacodynamics comparability assessment using *in vitro* assays predictive of clinical biosimilarity.

MAbs consist of two light chains (23 kDa each) and two heavy chains (50 kDa each) with a total molecular weight of approximately 150 kDa. The two light chains and two heavy chains are identical and are linked by covalent disulfide bonds and noncovalent interactions. MAbs have two major functional domains: fragment antigen-binding (Fab) and one crystallizable Fc fragment [4].

The Fab fragments contain the antigen binding site and consist of the entire light chain and part of the heavy chain. The Fab fragment is further subdivided into hypervariable (HV) or complementarity-determining regions (CDRs) and framework (FR) regions. There are three HV regions and four FR regions. The antigen binding sites are formed by the interaction of HV regions of the light chain and FR regions of the heavy chain. The combined structures of these regions determine the specificities and affinities of the antibodies for antigen. A slight difference in amino acid sequences may have a significant impact in binding of antigens.

The Fc region also binds to various cell receptors, especially the FcRn. The Fc region also binds to immune molecules such as complement proteins and mediates different biological functions such as the activation of the complement system, which leads to phagocytosis and degranulation of mast cells, basophils, and eosinophils [5].

The Fc-gamma receptors (FcγR) are members of the immunoglobulin superfamily and play a critical role in the function of therapeutic antibodies. There are three types of Fc-gamma receptors (FcγR): Fc-gamma receptor I (FcGRI), CD64; Fc-gamma receptor II (FcGRII), CD32; and Fc-gamma receptor III (FcGRIII), CD16. Both the FcRII and FcRIII can be further divided into two subgroups: FcRIIa/FcRIIb and FcRIIIa/FcRIIIb, respectively. Following the binding of antibody to the target, the subsequent binding of the FcγRs is a crucial step for the initiation and control of cell-mediated effector functions of the immune system. Fc-gamma receptor binding helps evaluate an antibody's ability to recruit specific immune cells, such as natural killer (NK) cells, neutrophils, and macrophages [6]. Therapeutic mAb binding affinity to different FcRn types is critical for understanding the mechanism of antibody action and predicting the therapeutic potential of a particular molecule.

Another key factor in the efficacy of therapeutic antibodies is the serum half-life. This is largely determined by the interaction between the antibody and the FcRn [6]. The serum half-life can have significant impact on antibody's safety and efficacy. The evaluation and understanding of a therapeutic antibody's ability to bind FcRn can play a pivotal role in developing an effective and efficacious mAb biosimilar.

In addition, the *in vivo* functional evaluation of mAb is limited because of the lack of species specificity of antigen and of Fc receptors in animal model. So it is imperative that a thorough assessment of pharmacodynamics parameter be performed using *in vitro* methodologies.

The European Medicines Agency (EMA)'s biosimilar guidance is very detailed and specifically recommends measuring the following parameters for mAbs:

a) Binding to the target antigen
b) Binding to representative isoforms of the relevant three Fc-gamma receptors (FcγRI, FcγRII, and FcγRIII), FcRn and complement (C1q)
c) Fab-associated functions such as activation, blockade, or neutralization of target
d) Fc-associated functions such as ADCC, CDC, and complement activation [7]

On the other hand, the FDA draft guidance is less specific in its recommendations but strongly encourages using a "fingerprint" such as multiple orthogonal assays [8, 9]. The results of these comparability assays can shape the direction of further clinical or nonclinical testing.

Taking all these factors into account, it is recommended that both the binding and functional characteristics be evaluated using orthogonal methods capable of detecting differences in the relationship between the concentration–structure and activity between the biosimilar and the reference antibody. Both the ligand binding methodology and cell-based functional assays should be utilized to compare with each other with respect to parameters mentioned above.

While some of the previous assays may not be considered essential for the therapeutic mode of action, they are critical in the assessment of comparability between innovator and biosimilar. In addition, the more multivariate the fingerprint, the more likely that a match is predictive of clinical biosimilarity. There are two primary groups of technologies that can be applied to generate this multivariate fingerprints: one is ligand binding methodologies and the other the cell-based functional bioassays.

16.3.1 Ligand Binding Methods for Comparison of Pharmacodynamics Parameters

There are several approaches that can be utilized to evaluate the binding properties of a mAb to the target and various receptors (e.g., Fc-gamma receptors, FcRn) and complement (C1q, C3, C4, etc.). This includes the evaluation of binding using immunoassays, label-free technologies such as surface plasmon resonance (SPR), bio-layer interferometry (BLI), and flow cytometry. The use of the latest state-of-the-art technologies is highly recommended for the evaluation of the pharmacodynamics parameters of mAbs.

ELISAs are the most widely used binding assays. While ELISAs are high throughput and are relatively easy to develop and implement, they have some major drawbacks. ELISAs require immobilizing the drug or receptor/target to the solid surface that may mask or alter epitopes. It may also require conjugation of the drug or receptor/target to the reporter molecule or enzyme. The conjugation chemistry may alter the molecular structure of the epitope of interest [10].

Label-free technologies on the other hand can be used to determine the kinetics and affinities of antigen/receptor–antibody interactions in real time. The Biacore, the system that utilizes SPR technology to measure biomolecular interactions, is highly recommended. The interactions occur in flow cells on a functionalized gold biosensor surface. Antigen/receptor is coupled to the sensor surface, and a solution containing the mAb is flowed over the surface of the sensor. When the two molecules interact, a mass change on the sensor surface causes the fluctuations in the refractive index, which is detected in real time. This intermolecular interaction data can be used to measure association rate, dissociation rate, and affinity. While Biacore is a versatile system, it is highly specialized and expensive and requires quite specialized training [11].

The Octet is also a label-free technology, but it utilizes a different optical principle called BLI than SPR to measure biomolecular interactions. In the case of BLI, the tip of a fiber-optic probe is coated with ligand (i.e., receptors, target, etc.) and is immersed in a solution of analyte, that is, mAb. The binding between ligand and analyte in solution produces an increase in optical thickness at the biosensor tip, which results in a wavelength shift, $\Delta\lambda$, which is a direct measure of the change in thickness of the biological layer. This change, measured in nanometers (nm), is reported in real time. This data can be utilized to measure the specificity of biomolecular interactions, on and off rates (kinetics), and binding strength (affinity). In comparison to Biacore, Octet does not employ microfluidics and is easier to use. Since it does not rely on fluctuations in the refractive index of the samples being tested, it is tolerant to samples containing high levels of glycerol and DMSO. While Biacore is the most widely used label-free method, Octet is quickly gaining acceptance due to its ease of use and cost advantage [11].

Other notable biosensors are Corning Epic technology based on the ability of specially patterned surfaces to reflect distinct wavelengths selectively. Roche xCELLigence and Molecular Devices' CellKey System is based on electrical impedance, and TTP LabTech's Resonant Acoustic Profiling (RAP) technology is based on quartz crystal microbalance (QCM).

Flow cytometry is a powerful technology for in-depth analysis of cells and cell surface proteins. It is a laser-based technology that allows simultaneous multiparametric analysis of cells by suspending cells in a stream of fluid and passing them by an electronic detection apparatus, which excites any fluorescent molecules on the cell. Emitted fluorescence is then measured by detectors tuned to specific wavelengths. It is a sensitive, information-rich method for measuring mAb binding to FcRn or target protein in their native state on the cell surface.

Whenever possible, multiple orthogonal technologies should be utilized for the evaluation of pharmacodynamic parameters to demonstrate true differences between the biosimilar and reference products without the variability due to biological variability and statistical noise.

16.3.2 Cell-Based Functional Bioassays

Cell-based bioassays allows for true measurement of the biological activity of a mAb. MAbs operate through multiple mechanisms of actions via Fab-associated functions (e.g., neutralization of a soluble ligand, receptor activation, or blockade) and Fc-associated functions (e.g., ADCC, CDC, complement activation). Understanding the functions and contributions of these two domains is critical in demonstrating similarity between the reference material and a biosimilar product. Cell-based bioassays can play a critical role in understanding the functions and contributions associated with each domain. These assays can also help investigate the impact of the differences observed during structural analysis or binding analysis and their consequences in clinical applications.

While cell-based assays provide true measurement of biological function, they are the most difficult to develop and implement. The challenges involved are identifying cell lines that respond to antibody or its target, high analytical variability, and complex nature of the process. In the absence of cell lines, PBMCs may be used to evaluate Fc or Fab functions. The drawbacks of utilizing PBMCs are poor responsiveness and lack of reproducibility between donors and between lots. An alternative approach can be the screening of a wide range of cell lines and identification of those that will respond to the antibody or its target in both a specific and reproducible manner. A more effective alternative is to generate a reporter cell line that has been genetically modified to contain the antibody target and a reporter gene that will switch on when the mAb binds to its target. It is important to note that the mechanism of action of the bioassay must mimic closely the *in vivo* mechanism of action of the drug. While costly, this technology may provide the most specific and reproducible results [12].

The binding of the antibody (Fab-associated function) to relevant targets should be extensively characterized using cell-based functional assays. In addition, the performance of multiple *in vitro* methods should be utilized to replicate all possible mechanisms of actions (e.g., cell proliferation, cell death, cell signaling, cell migration, receptor binding, receptor activation, ligand binding, cytokine production, etc.) that the biosimilar will mediate *in vivo*.

The confirmation of the absence of certain biological activities can also be helpful in fully characterizing the biosimilar molecule.

Many antibodies are designed to directly bind to a membrane-associated target and in turn may mediate their activity via the immune system (ADCC or CDC) or directly act on cells to inhibit proliferation or induce apoptosis. The evaluation of ADCC and CDC is generally not needed for mAbs directed against nonmembrane-bound targets. However, if a soluble target is also present in a membrane-associated form, the biological activity of the antibody may include some or all of these mechanisms of actions including ADCC or CDC. Where differences occur between the proposed biosimilar and reference product with respect to the additional mechanisms of action, these are likely to be translated into differences in clinical efficacy.

As mentioned previously, the use of state-of-the-art technologies are welcomed by regulatory agencies, who tend to champion the "stepwise" approach to evaluating biosimilarity [8]. The results from sensitive and robust *in vitro* assays can help shape the direction of further testing to establish biosimilarity. As the *in vitro* assays may be more specific and sensitive than studies in animals, based on the results of these assays, additional *in vivo* studies may not be necessary prior to initiation of clinical studies.

16.4 Pharmacokinetic Assessment

Reliable measurement of drug concentrations is critical to understanding the exposure–response relationship (i.e., PK) of mAbs. PK data play a pivotal role in the establishment of biosimilarity.

The most commonly used bioanalytical method for the measurement of mAbs in biological samples is the immunoassay. Unlike chromatographic (LC–MS/MS) methods, which are used for small molecule bioanalysis, immunoassays utilize one (in case of competitive assays) or two (in case of sandwich assay) binding reagents (i.e., antibodies) to measure the drug. The specificity of the method is dependent on the interaction of these binding reagents to the drug (i.e., mAb). The challenges associated with immunoassay methods are summarized in Table 16.1.

A validated bioanalytical method for the determination of both the reference drug and the biosimilar is required by regulatory authorities [7, 8]. There are regulatory guidelines and numerous industry white papers on the method of development and validation of bioanalytical methods to support PK studies. While they are applicable to bioanalytical methods developed to support biosimilar program, they are not comprehensive enough to address nuances specifically involved when a bioanalytical method is used to support biosimilar development.

Table 16.1 Chromatographic assays versus immunoassays.

Component	Chromatographic	Immunoassay
Assay principle	Direct measurement of analyte based on the physical properties of the analyte	Indirect measurement of the analyte based on the binding of antibody to analyte
Analyte	Small molecule	Large molecule
Reagents	In general, reagents are well characterized and commercially available	In general, requires custom reagent production In most cases they are not well characterized
Assay development effort	Low (~1 month)	High (3–6 months)
Standard curve	Linear	Nonlinear
Assay variability	Low	High
Assay range	Broad (>2 logs)	Limited (<2 logs)

Unlike the bioanalytical assays used to support innovator biotherapeutics, where one innovator product is quantified, the assays require measurement of two products: the biosimilar and the reference product. There has been some discussion in the industry regarding the use of one assay to measure both products versus two assays to measure two products. The majority opinion in the bioanalytical community is to use one assay to measure both biosimilar and reference products [13]. In the bioanalytical testing of the biosimilar, the goal is to show that the PK parameters for the biosimilar and the innovator are equivalent. To illustrate equivalency using a PK assay, a common point of comparison is necessary. Using one PK assay will ensure that the comparison is a true representation of the differences in the PK profiles and not the analytical variability between two assays.

If two assays are used for the measurement of biosimilar and reference products separately, it will be nearly impossible to differentiate analytical variability from variability seen in PK profile. During method development every effort should be made to validate one assay capable of measuring both products equivalently. If one assay cannot be validated, it may indicate that the biosimilar product is not truly similar to the reference product. Additional characterization analysis should be performed to understand the differences between the products and may necessitate manufacturing process changes to improve the similarity between biosimilar and reference product.

This chapter will only focus on areas of bioanalytical method development that are unique to supporting biosimilar development. They are:

- Analytical similarity establishment:
  - Protein content and potency verification
  - Reference product and biosimilar comparability assessment
- Robustness and ruggedness
- Impact of ADA on PK measurement

16.4.1 Protein Content and Potency Verification

Despite significant advances in the manufacturing and production of biologics in recent years, the process is still very complex and highly variable. The final product is often comprised of a heterogeneous mixture of similar molecules that are closely related, yet not identical. The manufacturing process for a biosimilar is likely to be different from the originator manufacturing process for proprietary reasons. The differences in the manufacturing processes will most likely result in different protein content per activity unit, and different process- and product-related impurities, microheterogeneities, and excipients [14, 15]. For generic drugs, the FDA requires that "the assayed content of the batch used as test product should not differ more than 5% from that of the batch used as reference product." There are no such criteria for biosimilars. Given the complex nature of manufacturing process, it is anticipated that the assayed content of the biosimilar mAb and the reference mAb will be different. For example, a recent study showed that while Eprex®, an innovator biologic, and Retacrit®, a biosimilar, had identical assigned potency, the protein content was 10% higher for Eprex.

The PK data is generated using immunoassay measuring mass based on binding to assay reagents. It is important that the total protein content of the biosimilar and reference mAb need to be considered when developing bioanalytical assays to support PK studies.

16.4.2 Comparability Assessment: Reference versus Biosimilar Product

The structural differences between the reference product and the biosimilar product may cause them to bind to assay reagents (i.e., capture protein/antibody and detection antibody) differently causing a quantitation bias for biosimilar. The goal of this comparability assessment is to assess analytical variability due to structural differences between the two products.

Comparability assessment includes the evaluation of a set of calibrators and quality control samples prepared in sample matrix and comparing them against each other. An example of assay design, data analysis, and acceptance criteria are included in Table 16.2. An alternative approach may be utilized provided appropriate scientific justification.

Table 16.2 Analytical comparability assessment.

Assay design	In this approach, the two drugs are compared to each other in the same run. Each run should include a set of reference (RF) calibrators, a set of RF Quality control sample (QCs) (five levels including Lower Limit of Quantitation (LLOQ) and Upper Limit of Quantitation (ULOQ)), a set of biosimilar calibrators, and a set of biosimilar QCs (five levels including LLOQ and ULOQ)
Data analysis	The calibrators and QCs for the RF and the biosimilar should first be evaluated separately. Each set of calibrators and their corresponding QCs should meet the predefined acceptance criteria. Once this predefined criteria have been met, the equivalency between the innovator and the biosimilar can then be evaluated
Acceptance criteria	They may be considered equivalent if the following conditions are met: • The biosimilar QCs meet a predefined acceptance criteria (e.g., ±20% bias (25% bias at LLOQ)) when evaluated against the innovator calibration curve • The innovator QCs meet a predefined acceptance criteria (e.g., ±20% bias (25% bias at LLOQ)) when evaluated against the biosimilar calibration curve • The percent difference from mean between the RF QCs and the biosimilar QCs do not exceed ±10% (15% at LLOQ)

QC = quality control sample; LLOQ = Lower Limit of Quantitation; ULOQ = Upper limit of quantitation

If the assay passes analytical comparability, testing further assay development and validation should proceed as per bioanalytical guidances [4–6]. If the assay fails the analytical biocomparability testing, the possible sources of the failure should be investigated.

The most common reason for biocomparability failure is that the assays reagents (e.g., capture protein/antibody, detection antibody, etc.) bind to the reference product and the biosimilar with different affinity. Assay reagents should be carefully chosen so that they have similar immunoreactivity toward both the biosimilar and the reference products. Immunoreactivity of the two products can be evaluated via parallelism test. The parallelism testing procedure is described in Table 16.3.

If two curves are parallel, the only difference between them would be a shift either to the right or to the left of the *X*-axis, and an interpolation of the curves being compared at any *Y*-value would be indicative of the difference (i.e., relative potency) between the two products. This methodology works well only for linear regression analysis, where the slope of the line is constant over the concentration range.

However, dose–response curves generated in immunoassays are not linear. The dose–response is usually four- or five-parameter logistic and does not have a constant slope over the entire tested concentration range. In addition, high variability observed in immunoassays makes it difficult to assess the parallelism of two dose–response curves.

Table 16.3 Parallelism testing.

Steps	Descriptions
1	Perform the assay by using multiple replicates (two or more) of calibrators and quality control samples and at least five dilutions of each test samples
2	Construct a calibration curve by using the desired mathematical model (e.g., four- or five-parameter calibration model with appropriate weighting)
3	Evaluate system suitability results against predetermined acceptance criteria. For example, back-calculated concentration of calibrators are 15% of their nominal values and quality control samples are within 20% of their nominal values
4	Calculate the concentrations against a calibration curve for test samples; calculate within dilution CV for the concentrations that are within the quantitation range. Two-thirds of the dilution must have within dilution CV of 15% or less. If the criteria are not met, samples should be reanalyzed
5	Calculate between dilution %CV for each sample

A detailed discussion of various statistical methods used for the evaluation of parallelism is beyond the scope of this chapter. Briefly, there are two common approaches that are used to establish parallelism on immunoassay data. One is a response comparison method using a constrained model or a free model. The constrained model compares the predicted responses of simultaneously fitted curves that are constrained to be parallel to each other. The free model compares the predicted responses of independently fitted curves with the observed responses. The response comparison method is a direct comparison of the dose–response data.

In contrast, a parameter comparison method is an indirect measurement of the dose–response data. It utilizes one or more parameters of the mathematical function used to independently fit curve models. Some examples are as follows: slope ratio method, EC_{50} (i.e., concentration of analyte that generates 50% of the maximum response) ratio, and model parameter confidence interval method.

The most widely used method in the bioanalytical community for PK and biomarker methods is the recovery estimation method, an extension of the response comparison method in which one curve is used as the calibrator and back-calculated concentrations are measured for the other curve. The in-range measurements of back-calculated concentrations are adjusted for dilutions. The acceptance criterion is less than or equal to 30% coefficient of variation (CV) among neat concentrations (Table 16.4).

While this is a very simple method and among the most frequently applied by bioanalytical scientists, its shortcomings are widely acknowledged. The statistical approach to the evaluation of parallelism should be determined on a

Table 16.4 Parallelism acceptance criteria.

	Dilution adjusted concentration increase or decrease	
%CV	**Yes**	**No**
≤30	Lines are not parallel but within acceptable limits. Determine relative concentration ratio	Yes. Lines parallel. Determine relative concentration
≤30	Lines are not parallel	Greater than 30% CV indicates high variability that requires repeat of the assay or reoptimization of the immunoassay

case-by-case basis based on scientific judgment in consultation with a statistician.

If the reagent passes the parallelism test, it indicates that the interactions of the critical assay reagents with the reference and biosimilar products are similar and that the assay is suitable for measurement of the analyte. It also indicates that there is a concentration difference between the two products. If there is a true difference in concentration of two products, a correction factor may be utilized to normalize the concentration data.

To truly document the concentration differences, multiple lots of reference products should be evaluated. It is not uncommon to see significant variability over multiple lots of reference products. In addition, the World Health Organization (WHO) standard, when available, may be used as a fundamental standard for calibrating both the reference and the biosimilar products.

If the parallelism test fails, it indicates that the reference and biosimilar products do not have the same immune reactivity toward the assay reagents (e.g., capture antibody, detection antibody). Alternative assay reagents that react to the two products in a similar equivalent manner should be evaluated. The selection of appropriate assay reagents requires an in-depth analysis of characterization data. There should be a close collaboration between the bioanalytical scientists and the analytical scientists so that the characterization data may be fully documented. The assay format and the reagents should be selected that will exploit most completely the epitopes that are of closest similarity, if not identical, between the reference and biosimilar product. This will contribute in creating a PK assay that has low analytical variability to enable the detection of potential PK differences between the innovator and the biosimilar.

The lack of parallelism might also indicate that the biosimilar is not adequately similar to the innovator and might affect its efficacy (e.g., if the capture reagent is the same as the target of the innovator). This situation would necessitate an investigation of the manufacturing process to identify the source of the difference between the innovator and the biosimilar.

16.4.3 Robustness and Ruggedness

As stated before, the PK assay is fundamental to uncovering any significant differences between the PK profiles of the innovator and biosimilar drugs. It is crucial that the PK assay that is developed be robust and rugged. Utilizing an assay with high analytical variability could mask the PK variability and make it difficult to interpret the data and uncover true PK variability.

Ensuring the robustness and ruggedness of an assay requires that all variables of the assay (i.e., incubation times, incubation temperature, use of liquid handling systems such as Tecan, Hamilton, etc.) be evaluated objectively. A highly systematic approach such as design of experiments (DOE) could be utilized. The utilization of automation is highly encouraged as it aids in greatly reducing variability in calibrator and QC preparations. Using one preparation of calibrators and QCs for the duration of a study is also highly encouraged.

The in-study validation should utilize both innovator and biosimilar QCs. Both sets of QCs should be run in each assay. Both QCs should be closely monitored for comparability to the results observed in the validation and to detect any possible trends. A significant deviation from the results seen in validation should be investigated and the cause identified. Ideally, the samples should be run in a randomized and blinded fashion (i.e., samples should not be separated based on if a sample has the innovator or the biosimilar).

16.4.4 Impact of ADA on PK

Since ligand binding methods rely on the binding interactions of the analyte in the sample with assay reagents (such as capture and detection antibody), ADAs with overlapping epitopes with assay reagents will interfere with their measurement. The interference may cause over- or underestimation of the quantity of drug present in the samples [13].

The evaluation of ADA impact on PK measurement is challenging due to the highly heterogeneous nature of the ADAs. No single population of positive control can mimic the entire variety of ADA responses that can be expected in a clinical study.

One approach to the evaluation of ADA impact on PK is to perform spike and recovery experiments using validation samples. Various concentrations of surrogate ADAs can be spiked in validation sample, and recovery of PK analyte can be measured. The variability in recovery measurement will provide some insight into potential interferences from ADAs in clinical samples. Alternatively spike and recovery experiments may be performed on actual clinical samples with a high ADA response by spiking various concentration of drug analyte.

As for all aspects of bioanalytical strategy for biosimilar development, a risk-based mitigation approach based on scientific rationale is recommended. For high-risk mAbs such as mouse or chimeric mAbs, it is important to evaluate and implement risk mitigation strategies early in the method development process. For low risk mAbs, such as humanized or fully human mAbs, a retrospective evaluation of ADA impact on PK to confirm the minimal interference may be sufficient.

There are several mitigation strategies that can be employed to minimize the impact of ADA interference on PK. One approach is to select assay reagents that are less prone to ADAs by testing the reagents using surrogate controls. Other assay parameters with high concentration of capture reagents combined with long incubation time will dissociate drug–ADA complex and allow for drug to bind to the capture reagents. Increasing the minimum required dilution (MRD) by itself or combining the previous parameters can also help dissociate the drug–ADA complex, allowing the drug to bind to the capture reagents.

Another approach is to use an acid dissociation step to breakdown the drug–ADA complex and allow for drug to bind to capture reagents. The dissociation buffer and neutralization buffer should be optimized to minimize the loss of epitope binding to capture reagents. Spike and recovery experiments combined with acid dissociation using validation samples containing surrogate ADAs should be conducted to evaluate the accuracy and precision of the acid dissociation approach.

16.5 Immunogenicity Assessment

Despite advances in technologies used in the generation of humanized or fully human mAbs, it is anticipated that they will illicit immune response in human over the course of time and repeated injections. A comparative assessment of unwanted immune responses against the biosimilar and the reference mAb is critical to the establishment of biosimilarity. Regulatory authorities require a validated bioanalytical method for the detection of both the antireference and the antibiosimilar antibodies. There are regulatory guidelines and numerous industry white papers on method development and validation of bioanalytical methods used for the detection of ADAs [2, 16, 17]. While they are applicable to bioanalytical methods developed to support biosimilar program, they do not address the challenges associated with biosimilar product development. This chapter will only focus on issues that are unique to the detection and characterization of ADAs against reference product and biosimilar product in the context of biosimilar product development.

The first issue is whether one assay or two assays should be utilized for the measurement of the reference product and the biosimilar-specific antibodies. There are advantages and disadvantages to both approaches, reflected in the current debate on this issue [18, 19].

Detection of ADA relies on using the drug as a capture reagent to detect a heterogeneous mixture of immunoglobulins that interacts with the drug. In one assay format, only the reference product or biosimilar is used as capture reagent. This may cause false-negative results for the reference product if the capture reagent is the biosimilar product and vice versa. This is a major concern for the one assay strategy as it is anticipated that there will be different impurities and posttranslational modifications between the reference and biosimilar mAb, which can cause differential immune response.

While the two assay strategy resolves this issue by employing one assay for the detection of antireference drug antibodies and another for antibiosimilar antibodies, it is not without limitations. The major limitation of this strategy is the interpretation of data. The use of two assays makes it difficult to interpret data (i.e., incidence of positive response, titer, etc.) due to the analytical variability between the two assays.

In the author's opinion, avoiding the risk of generating false-negative results far outweighs the difficulty in the interpretation of data, and the use of the two assay strategy is the most conservative approach and the most appropriate for clinical risk–benefit evaluation. This is particularly important for mAb due to the fact that it is highly likely that there will be structural difference between biosimilar mAbs and reference mAbs. It should be noted that the ultimate goal here is not to compare the rate of positive responses between the biosimilar versus the innovator per se, rather it is to compare clinically meaningful immunogenicity responses between the biosimilar and the innovator product. If there is a significant difference in data observed solely due to analytical variability, these results will not correlate with clinical parameters. In this scenario, it is possible to have a significant difference in the positive rate of ADAs between the reference and the biosimilar product due to differences in assay sensitivity and still be able to demonstrate biosimilarity based on the effective comparison of clinically meaningful data.

16.5.1 ADA Assay Development Considerations

During assay development all efforts should be made to minimize the variability between the two assays so that it is conducive to data comparison and interpretation. This can be accomplished by utilizing identical assay format, assay conditions, and common reagents.

A positive control should be utilized for the evaluation of assay parameters such as assay precision, sensitivity, drug tolerance, specificity, stability, and titer evaluation). The positive control should be generated in a species closest to the experimental species. The positive control can be generated by immunizing animals either with the biosimilar or the reference product. Alternatively, polyclonal antibodies against the biosimilar and reference product can be generated and pooled together and used for evaluation of both assay parameters.

The assay parameters between the two assays should be compared. If there are significant differences observed, it will necessitate investigation and further redevelopment of the assays. Significant differences in assay parameters (i.e., interaction of drugs, both reference and biosimilar, to the assay reagents) may indicate that the biosimilar is not sufficiently similar to the reference product.

It is well known that the presence of drug products interferes with the detection and measurement of ADA. FDA guidelines specifically recommend that the sponsor should "develop assays capable of sensitively detecting immune responses, even in the presence of circulating drug product (proposed product and reference product)" [20]. Lack of drug tolerance is a critical issue in bridging assays, which is commonly used for the detection of antibodies against mAbs. Methods such as acid capture elution (ACE) and solid-phase extraction and acid dissociation (SPEAD) should be utilized to improve drug tolerance of the assays [21].

16.5.2 ADA Characterization

In addition to the detection of unwanted immune response, the characterization of a positive response is key to understanding the impact of these antibodies and to the facilitation of comparability assessment. The specificity of the positive response should be confirmed by competitive binding experiments with appropriate drugs, biosimilar, or reference product. Confirmatory testing with different competing antibody fragments (e.g., intact biosimilar mAb, intact reference mAb, F(ab')2 domain, and Fc domain) could be used to check specificity of the detected signals.

In addition to specificity the following parameters should also be evaluated: nature of the antibody response (i.e., transient or persistence), titer of the positive immune response, and isotype distribution. The samples that are confirmed as positive in binding assays should also be evaluated in cell-based functional assays to evaluate neutralizing capacity to all relevant functions.

Finally, the incidence and severity of human immune response should be correlated with clinical data to demonstrate that there are no clinically meaningful differences in immune response between a proposed product and the reference product.

16.6 Conclusion

MAbs are large, complex heterogeneous mixture of biomolecules produced by living cells. For biosimilars, it is expected that there will be minor quality differences due to changes in manufacturing processes. This presents a set of unique challenges in developing bioanalytical assays for biosimilars. Despite the advances made in analytical technologies, currently there is no technology that can fully characterize mAbs. As indicated by regulatory authorities, a "stepwise" approach, with an emphasis on nonclinical *in vitro* testing, should be utilized during the development of biosimilar mAbs.

Comparative *in vitro* studies should be undertaken to interpret the relevance of minor quality differences in the physicochemical and biological characterization when comparing a biosimilar mAb to a reference mAb. Sensitive methods, both ligand binding methods and cell-based methods, combined with state-of-the-art technologies, should be utilized for an in-depth analysis of the structure/concentration and activity relationship. The analysis should include binding to target antigen(s), binding to the relevant three Fc-gamma receptors and complement (C1q), and functional analysis of various mAb domains (i.e., Fab-associated functions and Fc-associated functions). The functional analysis should include all mechanisms of action relevant to the mAb products. Since these *in vitro* assays are more likely to be more relevant, sensitive, and specific than *in vivo* animal studies, these assays should be considered the heart of nonclinical testing. If the results of these assays provide strong evidence of biosimilarity, *in vivo* testing may not be necessary.

Comparative clinical studies between the biosimilar and reference product is required by all regulatory authorities worldwide. The focus of the studies is the comparative evaluation of PK parameters, immunogenicity, safety, and efficacy. It is important to remember that differences in detected ADA responses do not provide sufficient evidence of nonbiosimilarity. The ultimate goal of the immunogenicity assessment is to evaluate the clinical impact of these differences. In order to achieve this goal, all adverse events should be correlated to the differences observed in ADA responses using multidimensional analysis. The additional collection of immunogenicity data may be necessary during the postmarketing study to fully understand the clinical impact of the ADA response differences.

References

1. DeSilva B, Smith W, Weiner R et al. (2003). Recommendations for the bioanalytical method validation of ligand-binding assays to support pharmacokinetic assessments of macromolecules. *Pharm. Res.* 20, 1885–1900.
2. Shankar G, Devanarayan V, Amaravadi L et al.(2008). Recommendation for the validation of immunoassays used for the detection of host antibodies against biotechnology products. *J. Pharm. Biomed. Anal.* 48(5), 1267–1281.
3. Viswanathan CT, Bansal S, Booth B et al. (2007). Workshop/conference report-quantitative bioanalytical methods validation and implementation: best practices for chromatographic and ligand binding assays. *AAPS J.* 9(1), E30–E42.
4. Alzari PM, Lascombe M-B, Poljak RJ. (1988). Three-dimensional structure of antibodies. *Annu. Rev. Immunol.* 6, 555–580.
5. Davies DR, Metzger H. (1983). Structural basis of antibody function. *Annu. Rev. Immunol.* 1, 87–117.

6. Burton DR, Woof JM. (1992). Human antibody effector function. *Adv. Immunol.* 51, 1–84.
7. European Medicines Agency. Guideline on similar biological medicinal products containing monoclonal antibodies—non-clinical and clinical issues. London, UK: Committee for Medicinal Products for Human Use, European Medicines Agency; 2012.
8. US Department of Health and Human Services, US FDA, Center for Drug Evaluation and Research, Center for Biologics Evaluation and Research. Scientific considerations in demonstrating biosimilarity to a reference product. U.S. Department of Health and Human Services Food and Drug Administration Center for Drug Evaluation and Research (CDER) Center for Biologics Evaluation and Research (CBER); 2015.
9. US Department of Health and Human Services, US FDA, Center for Drug Evaluation and Research, Center for Biologics Evaluation and Research. Quality considerations in demonstrating biosimilarity to a reference protein product. U.S. Department of Health and Human Services Food and Drug Administration Center for Drug Evaluation and Research (CDER) Center for Biologics Evaluation and Research (CBER); 2015.
10. Mire-Sluis AR, Barrett YC, Devanarayan V et al. (2004). Recommendations for the design and optimization of immunoassays used in the detection of host antibodies against biotechnology products. *J. Immunol. Methods* 289, 1–16.
11. Sadick M. (2014). Studying binding in biosimilars. *Contract Pharma* 16, 88.
12. Parekh BS, Berger E, Sibley S et al. (2012). Development and validation of an antibody-dependent cell-mediated cytotoxicity-reporter gene assay. *mAbs* 4(3), 310–318.
13. Marini JC, Anderson M, Cai XY et al. (2014). Systematic verification of bioanalytical similarity between a biosimilar and a reference biotherapeutic: committee recommendations for the development and validation of a single ligand-binding assay to support pharmacokinetic assessments. *AAPS J.* 16(6), 1149–1158.
14. Berkowitz SA, Engen JR, Mazzeo JR, Jones GB. (2012). Analytical tools for characterizing biopharmaceuticals and the implications for biosimilars. *Nat. Rev. Drug Discov.* 11(7), 527–540.
15. Beck A, Sanglier-Cianférani S, Van Dorsselaer A. (2012). Biosimilar, biobetter, and next generation antibody characterization by mass spectrometry. *Anal. Chem.* 84(11), 4637–4646.
16. Gupta S, Indelicato SR, Jethwa V et al. (2007). Recommendations for the design, optimization, and qualification of cell-based assays used for the detection of neutralizing antibody responses elicited to biological therapeutics. *J. Immunol. Methods* 321, 1–18.
17. Koren E, Smith HW, Shores E et al. (2008). Recommendations on risk-based strategies for detection and characterization of antibodies against biotechnology products. *J. Immunol. Methods* 333, 1–9.

18. Cai XY, Thomas J, Cullen C, Gouty D. (2011). Challenges of developing and validating immunogenicity assays to support comparability studies for biosimilar drug development. *Bioanalysis* 4(17), 2169–2177.
19. Chamberlain P. (2013). Assessing immunogenicity of biosimilar therapeutic monoclonal antibodies: regulatory and bioanalytical considerations. *Bioanalysis* 5(5), 561–574.
20. US Department of Health and Human Services, US FDA, Center for Drug Evaluation and Research, Center for Biologics Evaluation and Research. Guidance for industry: assay development for immunogenicity testing of therapeutic proteins. Silver Spring, MD: FDA; 2009.
21. Butterfield AM, Chain JS, Ackermann BL, Konrad RJ. (2010). Comparison of assay formats for drug-tolerant immunogenicity testing. *Bioanalysis* 2(12), 1961–1969.

17

Preclinical and Clinical Development of Biosimilar Antibodies

João Eurico Fonseca[1,2] *and João Gonçalves*[3]

[1] *Instituto de Medicina Molecular, Faculdade de Medicina da Universidade de Lisboa, Lisboa, Portugal*
[2] *Rheumatology Department, Hospital de Santa Maria, Lisbon Academic Medical Center, Lisboa, Portugal*
[3] *iMed-Research Institute of Medicines, Faculdade de Farmácia da Universidade de Lisboa, Lisboa, Portugal*

17.1 Summary

Biosimilars offer a highly attractive strategy for reducing medical costs and increase accessibility to targeted biologic therapies. However, unlike small molecules, the development and production of candidate original biologics and biosimilars can be hampered by unpredictable variability that should be tackled as far as possible during the preclinical development process. Biological therapies are inherently variable, creating unavoidable differences between even subsequent batches of the same product.

These differing complexities and methods of manufacture create an important difference between biosimilars and conventional generic drugs: while chemical generics can be fully characterized as identical to the originator product, biosimilars cannot. Thus, full guaranty of similarity is only granted after equivalence parallel trials assessing PK, efficacy, safety, and immunogenicity comparing the biosimilar candidate and the originator. A landmark event for the future widespread use of biosimilars was the European Medicines Agency (EMA) approval of infliximab biosimilars. In the EU, biosimilars are licensed through a comparability exercise with the reference product and clinical studies to ensure equivalence of efficacy and safety profiles. Guidelines produced by the EMA detail manufacturing process requirements and the range of protein structure, isoform, aggregate, receptor binding, and biological activity assays necessary to demonstrate biological equivalence. EMA guidelines also outline the required clinical and nonclinical pharmacokinetic (PK), pharmacodynamic (PD), and pharmacotoxicological evaluations necessary to assess safety and efficacy before approval.

Biosimilars of Monoclonal Antibodies: A Practical Guide to Manufacturing, Preclinical, and Clinical Development, First Edition. Edited by Cheng Liu and K. John Morrow, Jr.

EMA guidelines served as a starting point for development of licensing procedures in the United States, where the Food and Drug Administration (FDA) released draft guidance for the regulatory review of biosimilars in early 2012. In addition, the postmarketing surveillance, with special emphasis for the role of national registries, will be crucial for the confidence on the use of these biosimilars and will pave the way for the approval in the near future of other biosimilar candidates of biologics, such as trastuzumab, rituximab, adalimumab, etanercept, bevacizumab, and cetuximab. Therefore, this review will dissect the fundamental issues relevant to the development and confirmation of biosimilarity, emphasizing the relevance of monoclonal antibodies to the success of the biosimilar market.

17.2 Introduction

Biological therapeutics are large, highly complex molecules derived from living cells or organisms. Traditional small molecule pharmaceuticals, by contrast, are usually simple molecules of low molecular weight, synthesized by chemical means. These differing complexities and methods of manufacture create an important difference between biosimilars and conventional generic drugs: while chemical generics can be fully characterized as identical to the originator product, biosimilars cannot. Biological therapies are inherently variable, creating unavoidable differences between even subsequent batches of the same product [1]. An expiring patent does not necessarily provide access to the precise manufacturing conditions used in producing the originator therapy, including, for example, the relevant clone of the cell line and specific growth medium. It therefore cannot be guaranteed that biosimilar products are identical to their reference product on a molecular level. In turn, this difference has important implications for the regulation and licensing of biosimilars. While conventional generic drugs require only a limited comparison and demonstration of identity to the reference product, biosimilars require far more rigorous testing. In general, there must be a thorough comparison of structural and functional characteristics of the biosimilar and originator therapy. Any identified microheterogeneities must then be assessed for their impact on safety and clinical performance.

In the European Union (EU), biosimilars are licensed through a comparability exercise with the reference product and clinical studies to ensure equivalence of efficacy and safety profiles. Guidelines produced by the European Medicines Agency (EMA) detail manufacturing process requirements and the range of protein structure, isoform, aggregate, receptor binding, and biological activity assays necessary to demonstrate biological equivalence [2]. EMA guidelines also outline the required clinical and nonclinical pharmacokinetic (PK), pharmacodynamic (PD), and pharmacotoxicological evaluations necessary to assess safety and efficacy before approval [3, 4]. EMA guidelines served as a starting point for the development of licensing procedures in the United States, where Food and Drug Administration (FDA) released draft guidance for the regulatory review of biosimilars in early 2012 [5, 29].

The initial regulatory experience of EMA involved biologics of relatively small size (insulin, interferon, filgrastim, epoetin, and somatropin) [6]. Among the biosimilars that underwent EMA review from 2006 to 2011, six produced unexpected results that were not foreseen during nonclinical testing. Of these biosimilars, four failed to demonstrate comparable efficacy and/or safety against the originator products and were rejected or withdrawn. For the remaining two manufacturers, one modified the manufacturing process and repeated a clinical trial, and another was denied extension of its label for subcutaneous use, prior to marketing authorization [6].

For monoclonal antibodies (mAbs), the EMA "Guideline on similar biological medicinal products containing monoclonal antibodies: non-clinical and clinical issues" came into effect in 2012 [28]. Stepwise nonclinical *in vitro* and *in vivo* approaches are recommended to evaluate the similarity of the biosimilar to the reference mAb. Subsequently, clinical PK studies need to be performed in order to demonstrate a similar PK profile, prior to conducting clinical efficacy trials. Due to the nature of mAbs, PK is often highly variable even within the same disease, for example, in adjuvant versus metastatic breast cancer, where factors such as comorbidities may alter PK. Therefore, PK studies are a necessary component of the clinical program required to establish similar efficacy to the originator antibody. In addition to PK, PD studies are also important for assessing comparability [28]. For some drugs, such as filgrastim and epoetin, absolute neutrophil counts and hemoglobin concentration/reticulocyte counts are established and validated markers of drug activity. For other drugs, particularly the antineoplastic antibodies, no validated PD markers exist to indicate antitumor activity [7].

17.3 Quality and Preclinical Development of Biosimilar Monoclonal Antibodies

17.3.1 Quality by Design

To better understand the reasoning behind the EMA and FDA approval process for biosimilar products allowing for the extrapolation of indications, one must appreciate the modern biosimilar manufacturing process. The strategy is revolutionary, because it implies that there can be biosimilar developments with no major focus on clinical testing. This new strategy, termed quality by design (QbD), implements a stepwise, knowledge- and science-driven approach of biologic production, thus further guiding nonclinical and clinical development of biosimilars toward the goal of employing suitable test systems, especially those that will give the best results for establishing biosimilarity [9–11]. The manufacturing steps used to create biologicals are complex, and technology has greatly evolved in the last 15 years since the first therapeutic antibodies were introduced. Owing to the complicated nature of the cell-based expression systems, isolation methods, and endogenous "drift" inherent to biological systems, as well as intentional changes introduced in the production processes, all biologics demonstrate some degree of variability over time. Unlike small molecule therapeutics, which are synthesized to obtain homogenous molecules, biologics (reference and biosimilars) are produced by living cells, and expression results in a heterogeneous mixture of similar molecules, with the same primary amino acid sequence and a range of quality attributes (e.g., subtle difference in molecule glycosylation patterns) [12].

17.3.2 Extensive Quality and Preclinical Analysis

As for all biologics, the pharmaceutical quality of a biosimilar must be established including a complete description of the manufacturing process and full characterization of the quality attributes (amino acid sequence, secondary and tertiary structure, posttranslational modifications). In addition, comparability of the range of these attributes for both the biosimilar and reference product must be shown. This interval is determined by obtaining historical data from batches of innovative mAbs present in the market in recent years, allowing the implementation of a range where the quality of the biosimilar is included. A critical feature of the development of biosimilars is that instead of establishing the clinical efficacy and safety of the product per se, it should be demonstrated that the product is *(highly) similar* to the reference product for which substantial clinical evidence already exists [40, 41].

In addition to the comparability of the quality attributes of the reference and biosimilar molecules, the comparability and similarity of the biological activity are determined using assays relevant to the mechanism of action (MOA) of the reference biologic in all indications (Table 17.1) [13].

17.3.3 All Biologics May Be Considered Similar to Themselves

All biological pharmaceutical agents are likely to be modified several times throughout their life cycles [14, 15]. After several changes to the original manufacturing process, the biologics are no longer identical to the original version that underwent clinical testing at the time of marketing authorization [14]. When submitting a modification to a biologic agent, the manufacturer must conduct comparability testing as required by the regulatory authorities to ensure that quality, efficacy, and safety are not adversely affected [13–16]. Over the past decades of biotechnology developments, regulators have accumulated extensive experience in the assessment of such changes. Once approved, the

Table 17.1 Required comparability testing for biosimilar products.

Attribute	How is similarity demonstrated?
Protein structure and manufacturing quality	Extensive laboratory analyses of molecular characteristics (multiple batches)
Pharmacokinetics, pharmacodynamics, and toxicity (animal)	*In vitro* and *in vivo* assays (relevant animal species, only if additional info is needed after *in vitro* experiments)
Pharmacokinetics, pharmacodynamics, and toxicity (human)	Early clinical studies
Clinical efficacy and safety	Pivotal clinical trials
Safety in routine practice	Risk management plan
	Phase IV study
	Routine AE reporting/pharmacovigilance systems

new version is expected to have the same efficacy and safety in all therapeutic indications, without undergoing further clinical testing [14]. For example, with original infliximab, the addition of manufacturing facilities and continual process optimization has required 50 major submissions to the regulatory authorities using the comparability protocol (CP) as a regulatory filing strategy. The CP allows for downgrades to a lesser reporting category with no clinical testing requirement [17]. With each of these process alterations, no label changes were necessary. There were no reported safety or efficacy problems. The comparability that was performed by reference infliximab is the same used by biosimilar infliximab, with the exception that the biosimilar performs its analytics and assesses all known MOA for anti-TNF mAbs. Batch-to-batch variability in the reference product is often minimal, but larger variations can be found after manufacturing changes. Often both pre- and postchange products have already been in use in patients; thus the range of variation between them with respect to product attributes is presumed to be acceptable to regulators and judged as not impacting safety, purity, or potency. All pre- and postchange products are also used interchangeably, and both can be on the market at the same time. Therefore, several switches can and do occur between these products during the course of a treatment regimen [13].

For example, commercial batches of rituximab with expiry dates from September 2007 to October 2011 were recently characterized using glycan mapping, cation-exchange chromatography, and antibody-dependent cellular cytotoxicity (ADCC) *in vitro* bioactivity. In 2008, a sudden change in the quality profile became apparent for batches with expiry dates in 2010 or later, probably due to a manufacturing change. The difference was found in the amount of the

C-terminal lysine and N-terminal glutamine variants when analyzed by cation-exchange chromatography. These changes do not impact the antibody activity strongly. However, another physicochemical difference was detected in the glycan map for afucosylated G0 glycans. The abundance of afucosylated G0 glycan in rituximab was increased in the antibody structure, and the measured ADCC potency also showed an increase [18].

17.3.4 Process-Related Impurities

The raw materials used, the cell lines expressing the biosimilar, and the reference biological may be different for each product, and the purification steps for biosimilar and reference product are unique for each process. Therefore, a potential difference between both products is the different level and type of process-related impurities, such as host cell DNA or host cell proteins. These latter molecules may trigger an immune response in patients through toll-like receptors (TLR) or may cause a hypersensitivity reaction in sensitized individuals (or cause a sensitization). As already stipulated in the International Conference on Harmonization (ICH) S6 guideline on preclinical safety evaluation of biotechnology-derived pharmaceuticals, it is preferable to rely on purification processes to remove impurities and contaminants rather than to establish a clinical testing program for their qualification [16]. The levels of process-related impurities should be kept to a minimum, which is the best strategy to minimize any associated risk.

17.3.5 Product-Related Substances

As for any biological medicinal product, the biosimilar medicinal product is defined by the molecular composition of the active substance, which may introduce its own molecular variants, isoforms, or other product-related substances. When a biosimilar is produced, the final composition of the product is likely to be slightly different from the reference product regarding the relative contribution of all the product-related substances. An important aspect of mAb-related substances to the pharmacology of the medicinal product is the contribution of different glycoforms of a mAb to its function, since parts of the sugar moiety in the Fc region of IgG molecules contribute to the binding to Fc gamma receptors (FcγR). For example, the binding affinity of a mAb for FcγRIIIa is influenced by the presence or absence of fucose, where a higher degree of afucosylation leads to an increased binding affinity for the receptor. Subsequently, the difference in binding may be functionally reflected by an increased potency in an ADCC assay.

Differences in structure, like in the examples earlier, are detected when comparability of quality characteristics is evaluated, including characterization of glycosylation. It is expected that any potential change in pharmacological activity occurring as a consequence of the slight differences in structure will be detected when biosimilar and reference products are compared in binding affinity and functional *in vitro* assays.

17.3.6 Relative Purity or Protein Content

Although there have been significant advances in the manufacture and production of biologics in recent years, the process is still considerably more complex than the chemical synthesis of classic small molecule drugs and often results in impurities [23, 24]. Whereas the assayed content of the batch used as test product should not differ by more than 5% from that of the batch used as reference product in bioequivalence studies, it may not be possible to meet this criterion for biologics [8]. Therefore, the total protein content of drug (both test and reference) needs to be considered when assessing biosimilarity. The importance of this issue is perhaps best illustrated by referencing a recent study performed to compare the pharmacokinetics of Retacrit®, a biosimilar, to that of Eprex®, an innovator biologic [11]. This study highlighted differences in the amount and type of glycoforms between the biosimilar and the reference drug, which is to be anticipated given the different production processes. More notably, however, Eprex comprised more total protein (μg/ml) than Retacrit, which appeared to contribute to the potency (IU/ml) of Eprex being 10% higher than labeled. The EMA accepted that the PK of Eprex and Retacrit were comparable, based on introducing a correction factor to allow for the difference in protein content. Once this was taken into account, the comparison of PK parameters was well within the defined equivalence margins.

17.3.7 Pharmacology Assessment

The challenges that manufacturers may face in establishing similar clinical efficacy and safety parameters of a biosimilar and reference mAbs in the anticancer setting have been recognized. The preferred endpoints for confirming efficacy, such as progression-free, disease-free, and overall survival, may not be feasible for the establishment of biosimilarity, as they may be influenced by external factors, for example, tumor burden, performance status, and previous therapy, unrelated to differences between the biosimilar and reference mAbs [28]. The nonclinical and clinical biosimilar guidelines therefore acknowledge that surrogate endpoints such as overall response rate or change in tumor mass may be more appropriate [39].

Although EMA states that the preferred measures of efficacy in oncology indications are progression-free, disease-free, or overall survival, this recommendation is problematic in clinical practice. The EMA acknowledges that parameters such as performance status, tumor burden or previous treatment may be confounding factors that hamper comparisons of efficacy between biosimilar and originator antibodies using these endpoints. It is important to note that the goal of comparability studies is not patient benefit but merely to show similar safety and efficacy to the originator product. The implicit assumptions are that any difference between the products will be reflected if a sensitive patient population and clinical endpoint is used [19–22].

In the context of immunomodulatory mAbs biosimilars, the MOA of infliximab in rheumatoid arthritis (RA), ankylosing spondylitis (AS), and psoriatic arthritis (PsA) is the inhibition of the inflammatory activity mediated by TNF signaling, via binding to and inhibition of both soluble and membrane TNF. In inflammatory bowel disease (IBD), additional mechanisms have been proposed for the efficacy of mAbs (such as infliximab and adalimumab). These include the ability of these agents to induce reverse signaling through membrane TNF, resulting in reduced cytokine production and increased T-cell apoptosis. The ability of mAbs to induce antibody-dependent cell-mediated cytotoxicity (ADCC) has also been hypothesized to contribute to their efficacy in the therapy of IBD. Besides the binding of the complementarity-determining region (CDR) to its primary target, the Fc portion of the molecule also contains binding sites to different Fc receptors (FcγR and FcRn), which may elicit several effector functions, notably complement activation, complement-dependent cytotoxicity (CDC), and ADCC. These Fc-related binding properties and effector functions can also be evaluated *in vitro*. This knowledge was not present during the initial stages of the commercial life of reference infliximab. Therefore they were not analyzed when the manufacturing changes occurred. Because the biological properties of a biological can be characterized *in vitro*, there is little—if any—further information that would be gained by *in vivo* models. Thus, there was no need to reestablish the pharmacodynamic response in an *in vivo* model. Moreover, using cellular systems, more extensive, precise, and thus more sensitive comparisons can be made using *in vitro* assays, which further strengthens the *in vitro* approach in the evaluation of biosimilar infliximab. These assays are more sensitive than clinical trials to evaluate MOA. Analytical and PK/PD high similarity is the most robust scientific basis for comparing independently sourced biologics. However, it is important to acknowledge that quality attributes of a biological may also affect its PK behavior, which is not covered by the *in vitro* assay. Therefore, it is necessary to establish PK similarity in human volunteers or patients.

17.3.8 Safety Evaluation

Toxicity studies are used for the safety evaluation of pharmaceuticals, including biologicals. Like efficacy, predictable adverse effects of biologicals are considered to be related to exaggerated pharmacology. Once comparable pharmacological activity has been established *in vitro*, there is no need to confirm these properties *in vivo*. Some types of adverse effects are not predicted by animal studies and can be designated as unpredictable adverse effects, as they only occur at a low rate. As an example, infusion reactions, which are associated with the release of cytokines, may be triggered by different causes, including process-related impurities acting through TLRs, which are part of the innate immune system.

When there is a concern related to the presence of process-related impurities, *in vitro* TLR assays may have potential to detect and identify such contaminants. An additional type of adverse effects is due to unexpected toxicity, or "off-target toxicity." Unlike new chemical or protein entities, where it is imperative to perform toxicology studies, for biosimilars of older agents such as infliximab, this is not the case. Examples of unexpected toxicity are scarce, and none has involved biosimilar products but rather biologicals that were still in development. For example, thrombocytopenia occurred after administration of a mAb under development, whereas this did not happen with four other mAbs directed at the same pharmacological target. In this case, mAbs had the same target, but were not biosimilars. Subsequently, it was shown that the functional differences between these antibodies had a structural basis grounded in variances in amino acid sequence. This unexpected toxicity was driven by protein aggregation. Obviously, such structural differences are not within the scope of a biosimilar development, since the amino acid sequence is similar to the reference drug and since structural and functional similarity have already been shown by analytical and *in vitro* methods. Thus, although unexpected toxicity may be encountered in animal studies in rare cases during the development of new biological entities, it has never been shown to occur during the development of a biosimilar. There is no scientific data to indicate realistic safety concerns to extrapolated indications. When analytics demonstrate that the active moieties in two products are "highly similar," then the only "unknowns"—or residual uncertainties—are PK and immunogenicity [13].

17.3.9 Immunogenicity

The elicitation of antidrug antibodies (ADA) against biotherapeutics is an issue concerning drug safety, efficacy, and pharmacokinetics. The immunogenicity of biotherapeutics is, therefore, an important issue. In the case of biosimilars, guidelines that are more product class specific are now drafted, for example, on immunogenicity assessment of mAbs. The clinical outcome of unwanted immune responses differ widely depending on the affinity, class, amount, and persistence of ADA generated, the epitope recognized by the biotherapeutic protein, and the ability of ADA to activate complement. The guidelines of mAbs describe the causes and consequences thoroughly and underline the importance of the systematic evaluation of immunogenicity during clinical trials. We should emphasize that, because of these regulations and guidelines, biosimilars approved in the EU show a close resemblance to their reference product on quality, efficacy, and safety [21].

We know today the quality issues that affect the immunogenicity of a therapeutic antibody. The manufacturing process of a biological medicinal product will be changed several times during its life cycle. The introduction of new production methods, new formulations, and new containers and closures might affect the molecule characteristics and the impurity content of the therapeutic protein during its life cycle. The change will create a new version of the active substance [21, 27]. Therefore, the manufacturer must always compare the new and old versions and demonstrate to the regulatory authorities that the efficacy and safety of the product have not changed. In most cases, this conclusion is achieved by showing comparability using physicochemical and structural analyses, sometimes supplemented by *in vitro* functional assays. The manufacturers and regulators have more than 20 years of experience in assessing the comparability of different versions of a given biological medicinal product [27].

Formation of ADA in patients is an important issue that needs to be evaluated for all biologicals—both originator products and biosimilars—before they enter the market. Binding antibodies may affect the PK, and neutralizing antibodies can lead to loss of efficacy. Knowing that the presence of aggregates may increase the immunogenicity, their levels should be similar (or lower) in a biosimilar compared with the reference product. This can be determined by analytical methods. Most biological therapies elicit an immune response, in most cases with no clinical consequences. However, there are some biologicals for which immune responses have been linked to serious safety issues, notably the pure red cell aplasia (PRCA) caused by cross-reacting neutralizing antibodies against erythropoietin. Even small structural alterations may have an impact on immunogenicity, and analytical or animal data cannot always predict human immune responses. To mitigate this unavoidable risk, extensive nonclinical trial data demonstrating no increase in immunogenicity of the biosimilar compared with the reference product are required before a biosimilar can be licensed. In fact, the risk for detection of new and serious adverse effects after licensing is considered by some to be much lower for a biosimilar than for a biological containing a new or modified active substance [25]. Furthermore, the newer technologies used in manufacturing biosimilars mean that the products are generally of higher purity and quality, and more consistent potency, than their originator reference products [21, 26]. Unfortunately, inadequately produced copies exist and can lead to major issues, as recently exemplified by numerous cases of PRCA in Thailand [27].

17.3.10 Pharmacokinetic Analysis of Biosimilars

Although the route to market for classic generics is well defined and has been successfully applied for many drugs over the years—typically a small number of studies in healthy volunteers are sufficient to prove physiochemical and PK equivalence—the corresponding route to market for biosimilars is relatively new and considerably more complex. There are some key issues that should be considered when performing pharmacokinetic analyses to prove biosimilarity.

Study Design: The most common designs associated with bioequivalence studies are crossover designs; however, as biologics tend to have much longer half-lives than classic drugs, a crossover approach is generally not practical (the washout period that would be required is often prohibitively long) [28–30]. Furthermore, the potential for biologics to illicit an immune response also limits the use of crossover studies; if a patient was to develop an immune response in the first period of a crossover study, the same patient's ability to participate in the second treatment period would be compromised. Therefore, to overcome these issues, it is common to use a parallel group design when conducting biosimilar studies. Only one treatment period is required for each subject, removing the need for a washout period, and the potential problematic effect caused by a patient developing an immune response is limited. However, it is important to note that parallel designs may have other problems: large sample sizes are required to ensure that there is sufficient statistical power to prove biosimilarity, and as treatment differences are estimated between subjects (rather than within subjects), it is important to account for covariates (such as age, weight, and sex) in the statistical assessment.

Acceptance Limits: In classic bioequivalence studies, PK equivalence is demonstrated using bioequivalence limits of (0.80, 1.25) [24]; the test and reference products are considered to be equivalent if the 90% confidence interval of the ratio of geometric least squares means lies entirely within (0.80, 1.25) [8]. There are currently no such limits defined for biosimilars by EMA or FDA. Therefore, it may be possible to justify wider acceptance limits; however, it is still very common for biosimilar studies to apply the same (0.80, 1.25) criteria used in equivalence studies [31, 32].

ADA: Due to their protein nature and complex manufacturing process (often resulting in impurities), biologics have the potential to illicit an immune response. Indeed, nearly all biologics induce ADA; however, the incidence differs widely among products and between individuals [9]. In most instances ADAs have no clinical significance, but high levels can interfere with PK and PD properties of the drug (e.g., increasing clearance and thus reducing the extent of systemic exposure and desired effect of the drug). It is therefore important to have an assay in place to test for the presence of ADAs at suitable intervals during the study so that subjects who elicit an immune response can be identified.

Elimination Characteristics: Whereas bioequivalence studies are typically limited to the area under the concentration time curve (AUC) and maximum concentration (Cmax) as measures of the overall extent of systemic exposure and rates of absorption, respectively, elimination characteristics must also be considered when comparing biologics due to their long half-lives and potential to develop immune responses (which can significantly alter clearance). Indeed, current EMA guidelines state that "the design of comparative PK studies should not necessarily mimic that of the standard 'clinical comparability' design (CHMP/EWP/QWP/1401/98), since similarity in terms of absorption/

bioavailability is not the only parameter of interest. In fact, differences in elimination characteristics between products e.g. clearance and elimination half-life should be explored" [28–30].

In this context it is important to note that to fully characterize elimination kinetics in biosimilarity studies, the sampling schedule guidelines associated with equivalence (or clinical comparability) studies should still be applied [31, 32]. The sampling schedule should also cover the plasma concentration time curve long enough to provide a reliable estimate of the extent of exposure which is achieved if AUC(0–t) covers at least 80% of AUC(0–∞). At least three to four samples should be needed during the terminal log-linear phase in order to reliably estimate the terminal rate constant, which is needed for a reliable estimate of AUC(0–∞) [28–30]. Thus, the last sampling time point may be many weeks after dosing.

17.4 Extrapolation of Indications

17.4.1 Confirming Comparability

As for all biologics, the pharmaceutical quality of a biosimilar must be established by providing a complete description of the manufacturing process and full characterization of the quality attributes (amino acid sequence, secondary and tertiary structure, post-translational modifications). In addition, comparability of the range of these attributes for both the biosimilar and reference product must be shown. A critical feature of the development of biosimilars is that instead of establishing the clinical efficacy and safety of the product per se, it should be demonstrated that the product is highly similar to the reference product for which substantial clinical evidence already exists.

In addition to the comparability of the quality attributes of the reference and biosimilar molecules, the comparability and similarity of the biological activity is determined using assays relevant to the MOAs of the reference biologic in all indications (Table 3) [27]. Together these assays should broadly cover the functional aspects of the mAb, even though some may not be considered essential for the therapeutic mode of action. As the in vitro assays may be more specific and sensitive than studies in animals, these assays can be considered paramount in the nonclinical comparability exercise. If the comparability exercise using the above strategy indicates that the test mAb and the reference mAb cannot be considered biosimilar, it may be more appropriate to consider developing the product as a new drug entity [33].

EMA states in its Guideline on Similar Biological Medicinal Products Containing Monoclonal Antibodies that "Extrapolation of clinical efficacy and safety data to other indications of the reference monoclonal antibody, not specifically studied during the clinical development of the biosimilar monoclonal antibody, is possible based on the overall evidence of comparability provided from the comparability exercise and with adequate justification" [33]. Comparability, as depicted in Figure 17.1 (same dose, same efficacy and safety) is mostly demonstrated by the extensive preclinical assessment, whereas clinical studies are just the final confirmation of these data.

The QbD totality of evidence approach means that all accumulated data, gathered since the reference product was first approved, has been considered. Regarding the most recent example of biosimilar infliximab, current knowledge concerning the MOA in its several clinical indications includes the inactivation of circulating TNF, the induction of membrane-bound TNF reverse signaling [34, 35], and possibly CDC and ADCC [36]. For RA and AS, the synovial effects of infliximab are mainly derived from the prevention of circulating TNF binding to its receptors [12, 33, 40, 41]. However, in IBD, infliximab's binding to membrane-bound TNF may also be essential to achieve maximal efficacy. All of these proposed MOAs were analyzed in the preclinical comparability studies, and the biosimilar was deemed comparable to the reference product in all of them. In fact, EMA's assessment report concludes that in the exhaustive preclinical comparability exercise, only a small potentially relevant difference in the MOA was found between the biosimilar and the originator infliximab molecule [37, 39, 40]. The biosimilar showed a lower amount of afucosylated infliximab, which led to a significantly lower binding

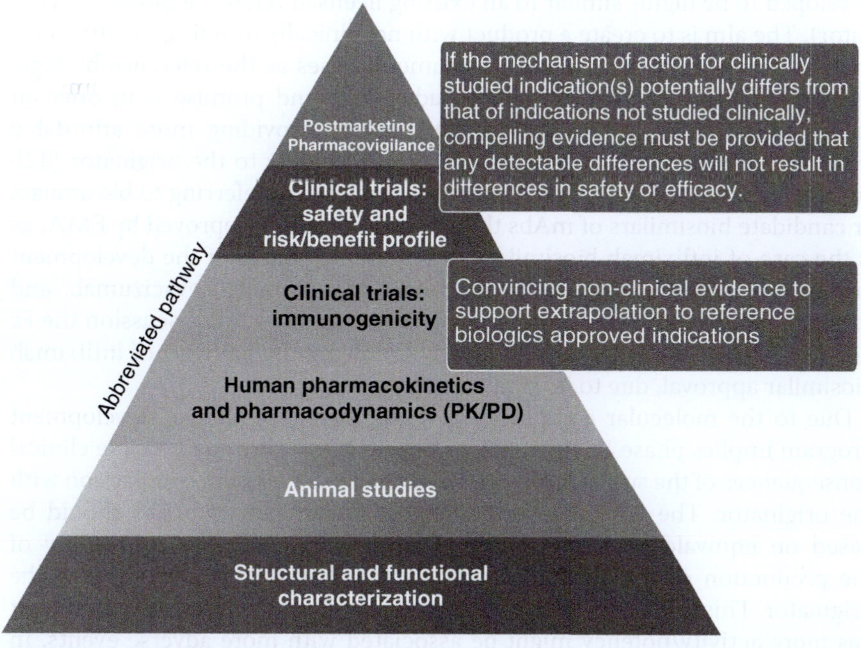

Figure 17.1 Overall evidence of comparability required to support extrapolation to indications not clinically studied [13].

affinity to isolated natural killer (NK) cells' FcγRIIIa and FcγRIIIb receptors, in the high-affinity FcγRIIIa receptor 158 V/V and V/F genotypes. This difference translated into lower ADCC activity, which hypothetically might affect the clinical efficacy of the biosimilar in IBD, while its efficacy in patients with RA would remain unchanged [36]. This small difference, found only in a subgroup of patients and in the most sensitive preclinical assay, was not replicated when conditions closer to real-life conditions were studied. We conclude that it seems extremely unlikely that it will have any clinical impact [37, 38].

These examples provide the information detailing how comparability in all of a biosimilar drug's dimensions has reduced the need for clinical studies after manufacturing changes. These observations can be used as the basis for complete extrapolation between all indications as well as adopted with full interchangeability for each indication. This protocol may be followed irrespective of whether the MOA is known or whether any clinical studies have been conducted on the postchange product.

17.5 Clinical Development of Biosimilars of Monoclonal Antibodies

As already highlighted, the term biosimilar refers to a biologic drug that is developed to be highly similar to an existing licensed reference biologic (originator). The aim is to create a product with no clinically meaningful difference. Biosimilars are intended to treat the same diseases as the reference biologic using the same doses. Their greatest advantage and promise is to offer an increased accessibility to targeted therapies by providing more affordable treatment alternatives with similar efficacy and safety to the originator [42]. For the purpose of this chapter, we have included data referring to biosimilars or candidate biosimilars of mAbs that are either already approved by EMA, as is the case of infliximab biosimilar, or in variable phases of the development process, including trastuzumab, rituximab, adalimumab, bevacizumab, and cetuximab [43]. Although not a mAb, we have added to this discussion the Fc fusion protein etanercept, as it is clearly following the pathway of infliximab biosimilar approval, due to its similar indications for use.

Due to the molecular complexity of biosimilars, the clinical development program implies phase I and III trials in order to capture any possible clinical consequences of the subtle molecular differences that exist in comparison with the originator. The clinical development of biosimilars of mAbs should be based on equivalence trials, as they identify the theoretical possibility of the production of a biosimilar candidate that has higher efficacy than the originator. This is contrary to the concept of similarity, and indeed a drug that has more activity/potency might be associated with more adverse events. In addition, as previously discussed, the design should be parallel, due to the long half-life of mAbs, to allow for adequate monitoring of immunogenicity, which would be hampered by a crossover design. EMA also recommends selection of clinical endpoints and patient populations that would facilitate detection of differences between the products, thus selecting the most sensitive scenario for detecting possible differences. EMA positioning regarding extrapolation has prevented the need for trials in all indications of the originator. Justification for extrapolation depends on clinical experience, on published data, and on the homogeneity of the MOA across indications [42].

17.5.1 Published Trials of Candidate Biosimilars of Monoclonal Antibodies

The biosimilar of infliximab, CT-P13, was the first biosimilar of a mAb to be approved for human use. CT-P13 was compared with infliximab originator in two pivotal trials. The phase I PLANETAS trial [44] included patients with AS. The primary endpoint, PK, and clinical efficacy endpoints were highly similar between CT-P13 and infliximab groups. The safety profile was also comparable. The second trial was the phase III PLANETRA trial [45] in patients with active RA despite methotrexate treatment. CT-P13 and the originator infliximab were shown to be equivalent in terms of ACR 20 response at week 30. The safety profile of CT-P13 was comparable with that of infliximab. The demonstration of pharmacological, efficacy, and safety bioequivalence in the PLANETAS and PLANETRA trials has resulted in the granting to the biosimilar candidate CT-P13 the authorization for use in the EU from September 2013, as branded names Inflectra® and Remsima® [46, 47].

Trastuzumab biosimilars are in late phase of approval; however for the time being only early phase data has been published. FTMB showed bioequivalence with trastuzumab originator in a phase I dose escalation (up to 6 mg/kg) study [24, 48]. PF-05280014 showed PK similarity with trastuzumab in a phase I study (dose 6 mg/kg) [49], after showing comparable results regarding tumor cell growth inhibition in an animal model [43, 51].

Two trials were published assessing etanercept biosimilar candidates. The first one evaluated PK and tolerability of the biosimilar candidate TuNEX. This was a phase I randomized, open-label, single-dose, crossover trial in which healthy male volunteers were administered subcutaneously with 25 mg of TuNEX or original etanercept and, after a one-week washout, they would receive the other drug. PK bioequivalence was demonstrated between TuNEX and the original etanercept, and tolerability outcomes were also similar between the two groups [50]. The second trial evaluated the biosimilar candidate HD203 with a study design very similar to the TuNEX trial and also confirmed PK bioequivalence [51]. The relatively short half-life of etanercept might support the using of a crossover design. However, as already argued, this is not the ideal design for testing biosimilars.

There is available data on preclinical assessment of the candidate biosimilars of rituximab [52, 53], but so far no clinical data was published. No information has been published on adalimumab, bevacizumab, and cetuximab biosimilars.

Despite the relative paucity of published clinical data on biosimilars, there are more than thirty ongoing, or completed, but still not published, clinical trials involving biosimilar candidates of trastuzumab, rituximab, adalimumab, bevacizumab, cetuximab, and etanercept [54].

17.6 Ongoing Trials of Candidate Biosimilars of Monoclonal Antibodies

17.6.1 Ongoing Trials of Biosimilars of Infliximab and Candidate Biosimilars of Infliximab

Even after approval in many countries, infliximab biosimilars are still being evaluated in clinical trials. The NOR-SWITCH study is a phase IV clinical trial, aiming to assess the safety and efficacy of switching from Remicade® to Remsima in patients with RA, spondyloarthritis, PsA, ulcerative colitis, Crohn's disease, and chronic plaque psoriasis. In addition, an observational prospective cohort study to evaluate the safety and efficacy of Remsima in patients with Crohn's disease or ulcerative colitis is also ongoing, reinforced by another observational study aiming at collecting infusion-related reactions in patients treated with any of the trade names of infliximab (Remicade, Remsima and Inflectra). The two extension studies of PLANETAS and PLANETRA have been completed but still not published. Finally the phase III trial evaluating the efficacy, safety, and pharmacokinetics of the biosimilar candidate SB2 in comparison with infliximab originator in RA refractory to methotrexate treatment is ongoing [55].

The infliximab story is paving the way for other biosimilars of mAbs. The successes or drawbacks of the clinical use of Remsima and Inflectra will decisively influence the future of biosimilars. A major contribution to our understanding of the real-life use of infliximab biosimilars will come from national registries, particularly those based on electronic clinical records, such as Reuma.pt, capturing all the biologics used by rheumatologists in one of the first European countries to have introduced biosimilars of infliximab [56].

17.6.2 Ongoing Trials of Candidate Biosimilars of Trastuzumab

As already mentioned trastuzumab biosimilars are in an advance phase of development. Several comparative studies of CT-P6 with trastuzumab originator are registered. An initial phase I/IIb safety, efficacy, and PK equivalence study in patients with metastatic breast cancer was followed by a phase III trial testing the equivalent efficacy and comparable safety in the first-line treatment of metastatic breast cancer of CT-P6 with trastuzumab originator, both in combination with paclitaxel. An additional trial is testing CT-P6 versus trastuzumab originator for equivalence as determined by pathological complete response after neoadjuvant therapy and surgery in HER2-positive early breast cancer. The PF-05280014 phase I trial was already published, and this biosimilar candidate of trastuzumab is now being assessed in 2 phase III trials: a combination with paclitaxel versus trastuzumab originator plus paclitaxel for the first-line treatment of patients with HER2-positive metastatic breast cancer and a combination with taxotere and carboplatin versus trastuzumab originator plus taxotere and carboplatin for the neoadjuvant treatment of patients with operable Her2-positive breast cancer. Other biosimilars of trastuzumab are being developed. A phase III trial comparing the efficacy, safety, PK, and immunogenicity between SB3 and trastuzumab originator in women with newly diagnosed HER2-positive early or locally advanced breast cancer is ongoing. The BCD-022 is being compared with originator trastuzumab in a phase III trial for safety and efficacy in the first-line treatment of HER2-positive metastatic breast cancer. A phase III trial is evaluating the efficacy and safety of ABP 980 compared with trastuzumab in subjects with HER2-positive early breast cancer. Finally, in a different perspective, a phase II trial is comparing the trastuzumab biosimilar candidate, CMAB302, monotherapy to its effect combined with vinorelbine in HER2-overexpressing metastatic breast cancer [57, 58].

In 2014 the patent for exclusivity rights for trastuzumab expired in Europe. It is expected that approval of a biosimilar of trastuzumab will come very soon, becoming the first mAb biosimilar available for the treatment of cancer. So far, data released on CT-P6, only as abstracts and not as full publications, are evaluating the biosimilarity with trastuzumab originator. Equivalent PK and a similar safety profile were demonstrated for CT-P6 versus originator in a phase I/IIb study. Combination with paclitaxel has demonstrated comparability to originator plus paclitaxel in a pooled analysis of data from the phase I/IIb PK study and the phase III efficacy and safety study in women with HER2-positive metastatic breast cancer [58].

Metastatic breast cancer is a highly heterogeneous disease. Indeed, molecular phenotype, disease progression, location of metastasis, prior treatment, and comorbidities are responsible for a great variability in response to treatment [58]. Because early breast cancer represents a more sensitive and homogeneous population, clinical trials of biosimilar trastuzumab should be also done in this setting. Results from ongoing trials in early breast cancer will be very relevant [57].

17.6.3 Ongoing Trials of Candidate Biosimilars of Rituximab

Fourteen current trials are testing candidate biosimilars of rituximab. Three of these trials were performed on patients diagnosed with diffuse large B-cell lymphoma. One is a phase III trial aiming at determining if the response rate obtained with RTXM83 combined with CHOP is lower than the response rate with originator rituximab. TL011 combined with CHOP is also being compared to rituximab originator for PK, pharmacodynamics, and safety. The other one is a phase I trial that evaluates the initial safety, PK, pharmacodynamics, and efficacy of CT-P10 given in combination with dexamethasone, cytosine arabinoside and cisplatin (DHAP) as second-line chemotherapy. Other trials are testing rituximab biosimilars in follicular lymphoma. A phase III trial is comparing Pf-05280586 with originator rituximab for the first-line treatment of patients with CD20-positive, low tumor burden follicular lymphoma. Another ongoing phase III trial in lymphoma patients compares efficacy, safety, and PK of GP2013 with originator rituximab, plus cyclophosphamide, vincristine, and prednisone (CVP) in previously untreated advanced-stage follicular lymphoma. A phase I trial is recruiting patients to evaluate the PK and PD of BI 695500 versus rituximab originator in the induction immunotherapy as a first-line treatment in patients with low tumor burden follicular lymphoma. A phase I trial will evaluate the safety, PK, and antitumor activity of MK-8808, a rituximab biosimilar candidate, in combination with CVP and as a single agent for CD20-positive follicular lymphoma who have had no prior chemotherapy. Finally, a Japanese phase I trial to assess the safety and PK of GP2013 monotherapy administered weekly in patients with CD20-positive low tumor burden indolent B-cell non-Hodgkin's lymphoma is now recruiting patients. In addition, nine other trials are evaluating biosimilar candidates of rituximab in RA. Four phase III trials are in progress in RA patients who had an inadequate response or intolerance to other disease-modifying antirheumatic drugs (DMARDs), including one or more TNF inhibitors. These trials will evaluate the efficacy and safety of the biosimilar candidates BCD-020, GP2013, BI 695500, and Pf-05280586 in comparison with originator rituximab. Extension studies evaluating treatment with BI 695500 or Pf-05280586 versus rituximab originator in RA subjects are already open. Three rituximab biosimilar candidates are in initial phases of development. A phase I/II trial has been terminated, determining the safety and pharmacology of TL011 in comparison with rituximab originator in patients with RA. A phase I trial will access the PK, safety and tolerability of MK-8808 in comparison with rituximab originator in RA patients. Another biosimilar candidate of rituximab, CT-P10, is being tested in a phase I trial for PK equivalence with rituximab originator in RA patients [59, 60].

Rituximab lost market exclusivity in the EU in November 2013 and in the United States in September 2016 [60]. The development of biosimilars of rituximab is in an advanced phase, with unusual feature of addressing two very different diseases, such as RA and lymphoma. Given these properties of the trial, no extrapolation appears to be acceptable.

17.6.4 Ongoing Trials of Candidate Biosimilars of Etanercept

There are seven etanercept candidate biosimilar phase III trials ongoing, six on RA and one on psoriasis. SB4, CHS-0214, HD203, and LBEC0101 are being compared to etanercept originator regarding efficacy and safety in RA patients that failed to respond to methotrexate. LBEC0101 and HD203 phase III trials were preceded by phase I trials for safety and PK versus etanercept originator. ENIA11 is being compared with etanercept originator for efficacy and safety in RA patients in two trials, one in combination with methotrexate and another one in combination with methotrexate or other DMARDs. GP2015 is under study for equivalent efficacy and safety with etanercept originator in patients with moderate to severe chronic plaque-type psoriasis. An additional phase I trial is ongoing, testing the etanercept biosimilar candidate DWP422 [61, 62]. Etanercept will lose market exclusivity in the EU in November 2015 but in the United States only in 2028 [60]. It is likely that one or more of the candidate biosimilars of etanercept will eventually succeed in the EMA approval process, probably sharing the extrapolation opportunities opened by infliximab approval.

17.6.5 Ongoing Trials of Candidate Biosimilars of Adalimumab

Five phase III trials are testing adalimumab biosimilars: three on RA patients, comparing SB5, BI695501, and ABP501 with originator adalimumab in patients refractory to methotrexate, and two on moderate to severe chronic plaque-type psoriasis patients, testing GP2017 and ABP501 versus originator adalimumab. BI695501 phase III trial was preceded by a phase I trial. RA patients exposed to ABP501 are being included in an extension study already opened. In addition, PF-06410293 adalimumab biosimilar candidate is also in phase I studies [63]. Adalimumab will lose market exclusivity in the EU in 2018 and in the United States in 2016. As previously noted for etanercept, it is probable that one or more of the candidate biosimilars of adalimumab will succeed in the EMA approval process, again taking advantage of the extrapolation principles created by infliximab approval.

17.6.6 Ongoing Trials of Candidate Biosimilars of Bevacizumab

BEVZ92 is a candidate biosimilar of bevacizumab that was tested in a phase I trial that evaluated the PK and safety of the association with folinic acid + fluorouracil + oxaliplatin (FOLFOX) or folinic acid + fluorouracil + irinotecan (FOLFIRI) versus bevacizumab originator in association with FOLFOX or FOLFIRI in first-line treatment of metastatic colorectal cancer. BCD-021 in association with paclitaxel + carboplatin is being compared in a phase III trial for efficacy and safety with bevacizumab originator in association with the same drugs in inoperable or advanced nonsquamous non-small cell lung cancer. The same biosimilar candidate is to be tested in a phase III trial comparing efficacy and safety with ranibizumab in patients with neovascular wet age-related macular degeneration [64]. Bevacizumab will lose market exclusivity in the EU in 2018 but in the United States only in 2019, and it is likely that, after trastuzumab and rituximab, it will be the third mAb used in cancer patients to have a biosimilar approved.

17.6.7 Ongoing Trials of Candidate Biosimilars of Cetuximab

It is highly probable that candidate biosimilars of cetuximab will soon be tested in clinical trials in comparison with the originator in indications such as squamous cell carcinoma of the head and neck and metastatic colorectal cancer [65]. Cetuximab lost market exclusivity in the EU in November 2014 and will in the United States in 2016.

17.7 Conclusion

Biosimilars offer a highly attractive strategy for reducing medical costs and increasing accessibility to targeted biologic therapies. Unlike small molecules, biologics are large, and their function is highly dependent on their three-dimensional conformation. Given these properties, candidate biosimilars can be hampered by unpredictable variability that should be resolved as far as possible during the preclinical development process. However, a full guarantee of similarity is only granted after equivalence parallel trials assessing PK, efficacy, safety, and immunogenicity in which the biosimilar candidate is rigorously compared with the originator. An important landmark event that bodes well for the widespread adoption of biosimilars was the EMA approval of infliximab biosimilars. The postmarketing surveillance, with special emphasis for the role of national registries, will be crucial for building the confidence of the public and the medical community for these biosimilars. This will pave the way for the approval in the near future of other biosimilar candidates of biologics, such as trastuzumab, rituximab, adalimumab, etanercept, bevacizumab, and cetuximab.

References

1. Weise M, Bielsky MC, De Smet K, Ehmann F, Ekman N, Narayanan G, et al. Biosimilars-why terminology matters. *Nat Biotechnol.* 2011;29(8):690–3.
2. European Medicines Agency. Guideline on similar biological medicinal products containing biotechnology-derived proteins as active substance: quality issues. 2006a. Available at http://www.ema.europa.eu/docs/en_GB/document_library/Scientific_guideline/2009/09/WC500003953.pdf (accessed on July 5, 2016).

3. European Medicines Agency. Guideline on similar biological medicinal products containing biotechnology-derived proteins as active substance: non-clinical and clinical issues. 2006b. Available at http://www.ema.europa.eu/docs/en_GB/document_library/Scientific_guideline/2009/09/WC500003920.pdf (accessed on July 5, 2016).
4. European Medicines Agency. Guideline on immunogenicity assessment of biotechnology-derived therapeutic proteins. 2007. Available at http://www.ema.europa.eu/docs/en_GB/document_library/Scientific_guideline/2009/09/WC500003947.pdf (accessed on July 5, 2016).
5. US Food and Drug Administration. Scientific considerations in demonstrating biosimilarity to a reference product. Draft guidance. 2012. Available at http://www.fda.gov/downloads/Drugs/GuidanceComplianceRegulatoryInformation/Guidances/UCM291128.pdf (accessed on July 5, 2016).
6. Lee JF, Litten JB, Grampp G. Comparability and biosimilarity: considerations for the healthcare provider. *Curr Med Res Opin.* 2012;28:1053–8.
7. Mellstedt H. Anti-neoplastic biosimilars—the same rules as for cytotoxic generics cannot be applied. *Ann Oncol Off J Eur Soc Med Oncol/ESMO* 2013;24(Suppl 5):v23–8.
8. EMA. Guideline on similar biological medicinal products containing monoclonal antibodies—non-clinical and clinical issues. Available at http://www.ema.europa.eu/ema/pages/includes/document/open_documentjsp?webContentId=WC5001286862012 (accessed on July 5, 2016).
9. Schiestl M, Li J, Abas A, Vallin A, Millband J, Gao K, Joung J, Pluschkell S, Go T, Kang HN. The role of the quality assessment in the determination of overall biosimilarity: a simulated case study exercise. *Biologicals.* 2014;42(2):128–32.
10. Zalai D, Dietzsch C, Herwig C. Risk-based process development of biosimilars as part of the Quality by Design Paradigm. *PDA J Pharm Sci Technol.* 2013;67(6):569–80.
11. Martin-Moe S, Lim FJ, Wong RL, Sreedhara A, Sundaram J, Sane SU. A new roadmap for biopharmaceutical drug product development: integrating development, validation, and quality by design. *J Pharm Sci.* 2011;100:3031–3043.
12. Cook J, Cruañes MT, Gupta M, Riley S, Crison J. quality-by-design: are we there yet? *AAPS PharmSciTech.* 2014;15(1):140–8.
13. McCamish M, Woollett G. The continuum of comparability extends to biosimilarity: how much is enough and what clinical data are necessary? *Clin Pharmacol Ther.* 2013;93:315–317.
14. Weise M, Kurki P, Wolff-Holz E, Bielsky MC, Schneider CK. Biosimilars: the science of extrapolation. *Blood.* October 8, 2014. pii: blood-2014-06-583617.
15. Schneider CK. Biosimilars in rheumatology: the wind of change. *Ann Rheum Dis.* 2013;72(3):315–318.
16. ICH harmonized tripartite guideline. Comparability of biotechnological/biological products subject to changes in their manufacturing process Q5E.

2004. Available at http://www.ich.org/products/guidelines.html (accessed on October 1, 2014).

17. Wojciechowski PW, Smit HI, Myers MM, Voronko PJ, Laverty T, Ramelmeier RA, Siegel RC. Making changes to a biopharmaceutical manufacturing process during development and commercial manufacturing: The REMICADE® Story. In: Shukla AA, Etzel MR, Gadam S (eds.), In Process Scale Bioseparations for the Pharmaceutical Industry, pp. 507–522, 2006. Available at http://www.crcnetbase.com/doi/abs/10.1201/9781420016024.ch18 (accessed on October 1, 2014).
18. Schiestl M, Stangler T, Torella C, Cepeljnik T, Toll H, Grau R. Acceptable changes in quality attributes of glycosylated biopharmaceuticals. *Nat Biotechnol.* 2011 Apr;29(4):310–2. doi: 10.1038/nbt.1839.
19. Aapro M, Cornes P. Biosimilars in oncology: emerging and future benefits. *Eur J Oncol Pharm.* 2012;6 (2):27–9.
20. Höer AdM, C, Häussler B, Haustein R. Saving money in the European healthcare systems with biosimilars. *GaBI J.* 2012;1(3–4): 120–6. doi:10.5639/gabij.2012.0103-4.036
21. Aapro M, Cornes P, Sun D, Abraham I. Comparative cost efficiency across the European G5 countries of originators and a biosimilar erythropoiesis-stimulating agent to manage chemotherapy-induced anemia in patients with cancer. *Ther Adv Med Oncol.* 2012;4(3):95–105.
22. Verpoort K, Mohler TM. A non-interventional study of biosimilar granulocyte colony-stimulating factor as prophylaxis for chemotherapy-induced neutropenia in a community oncology centre. *Ther Adv Med Oncol.* 2012;4(6):289–93.
23. EMA. Guideline on the Investigation of Bioequivalence. European Medicines Agency, London, 2010.
24. Schellekens, H. (2009) Assessing the bioequivalence of biosimilars. *Drug Discov Today.* 14, 9–10.
25. Weise M, Bielsky MC, De Smet K, Ehmann F, Ekman N, Giezen TJ, et al. Biosimilars: what clinicians should know. *Blood.* 2012;120(26):5111–7.
26. Brinks V, Hawe A, Basmeleh AH, Joachin-Rodriguez L, Haselberg R, Somsen GW, et al. Quality of original and biosimilar epoetin products. *Pharm Res.* 2011;28(2):386–93.
27. Praditpornsilpa K, Tiranathanagul K, Kupatawintu P, Jootar S, Intragumtornchai T, Tungsanga K, et al. Biosimilar recombinant human erythropoietin induces the production of neutralizing antibodies. *Kidney Int.* 2011;80(1):88–92.
28. EMA. Guideline on similar biological medicinal products containing biotechnology-derived proteins as active substance: non-clinical and clinical issues. 2006. Available at http://www.ema.europa.eu/docs/en_GB/document_library/Scientific_guideline/2015/01/WC500180219.pdf (accessed on July 6, 2016).

29. FDA. Quality Considerations in Demonstrating Biosimilarity to a Reference Protein Product. 2012. Available at http://www.fda.gov/downloads/drugs/guidancecomplianceregulatoryinformation/guidances/ucm291134.pdf (accessed on July 6, 2016).
30. FDA. Scientific Considerations in Demonstrating Biosimilarity to a Reference Product. Draft Guidance. U.S. Department of Health and Human Services, Food and Drug Administration, CDER, CBER, Rockville, 2012.
31. FDA. Biosimilars: questions and answers regarding implementation of the biologics price competition and innovation act of 2009. Guidance for Industry, U.S. Department of Health and Human Services, Food and Drug Administration, CDER, CBER, Rockville, 2012.
32. EMA. Guideline on the Investigation of Bioequivalence. 2010. Available at http://www.ema.europa.eu/docs/en_GB/document_library/Scientific_guideline/2010/01/WC500070039.pdf (accessed on July 6, 2016).
33. European Medicines Agency. Inflectra Assessment report. EMA/CHMP/589422/2013, June 27, 2013. Available at http://www.ema.europa.eu/docs/en_GB/document_library/EPAR_-_Public_assessment_report/human/002778/WC500151490.pdf (accessed on October 1, 2014).
34. EMA. Guideline on similar biological medicinal products containing monoclonal antibodies – non-clinical and clinical issues. EMA/CHMP/BMWP/403543/2010. Committee for Medicinal Products for Human Use (CHMP), May 2012.
35. Waetzig GH, Rosenstiel P, Nikolaus S, Seegert D, Schreiber S. Differential p38 mitogen-activated protein kinase target phosphorylation in responders and nonresponders to infliximab. *Gastroenterology*. 2003;125(2):633–4.
36. Mitoma H, Horiuchi T, Tsukamoto H, Tamimoto Y, Kimoto Y, Uchino A, To K, Harashima S, Hatta N, Harada M. Mechanisms for cytotoxic effects of anti-tumor necrosis factor agents on transmembrane tumor necrosis factor alpha-expressing cells: comparison among infliximab, etanercept, and adalimumab. *Arthritis Rheum*. 2008;58(5):1248–57.
37. Moroi R, Endo K, Kinouchi Y et al. FCGR3A-158 polymorphism influences the biological response to infliximab in Crohn's disease through affecting the ADCC activity. *Immunogenetics* 2013;65: 265–71.
38. Jung SK, Lee KH, Jeon JW, Lee JW, Kwon BO, Kim YJ, Bae JS, Kim D-I, Lee SY, Chang SJ (2014) Physicochemical characterization of Remsima®. *mAbs*, 6, 5, 1163–1177.
39. Araújo F, Cordeiro I, Teixeira F, Gonçalves J, Fonseca JE. Pharmacology of biosimilar candidate drugs in Rheumatology: a literature review. *Acta Reumatol Port.* 2014;39:19–26.
40. European Medicines Agency. European assessment reports for authorised medicinal products for human use. Available at http://www.ema.europa.eu/ema/index.jsp?curl_pages/medicines/landing/epar_search.jsp&murl_menus/medicines/medicines.jsp&mid_WC0b01ac058001d125 (accessed on October 1, 2014).

41. McKeage K. A review of CT-P13: an infliximab biosimilar. *BioDrugs.* 2014;28:313–21.
42. Bui LA, Taylor C. Developing clinical trials for biosimilars. *Semin Oncol* 2014; 41:S15–S25.
43. Beck A, Reichert JM. Approval of the first biosimilar antibodies in Europe: a major landmark for the biopharmaceutical industry. *mAbs* September–October 2013;5(5):621–3. doi: 10.4161/mabs.25864.
44. Park W, Hrycaj P, Jeka S, et al. A randomised, double blind, multicentre, parallel-group, prospective study comparing the pharmacokinetics, safety and efficacy of CT-P13 and innovator infliximab in patients with ankylosing spondylitis: the PLANETAS study. *Ann Rheum Dis* 2013; 72:1605–12.
45. Yoo D, Hrycaj P, Miranda P, et al. randomised, double-blind, parallel-group study to demonstrate equivalence in efficacy and safety of CT-P13 compared with innovator infliximab when coadministered with methotrexate in patients with active rheumatoid arthritis: the PLANETRA study. *Ann Rheum Dis* 2013; 72:1613–20.
46. European Medicines Agency. EPAR summary for the public – Inflectra. Available at http://www.ema.europa.eu/docs/en_GB/docu- ment_library/EPAR_-_Summary_for_the_public/hu-man/002778/WC500151491.pdf (accessed on October 11, 2013).
47. European Medicines Agency. EPAR summary for the public – Remsima. Available at http://www.ema.europa.eu/docs/en_GB/docu- ment_library/EPAR_-_Summary_for_the_public/hu- man/002576/WC500150872.pdf (accessed on October 11, 2013).
48. Wisman LA, De Cock EP, Reijers JA, Kamerling IM, Van Os SH, de Kam ML, Burggraaf J, Voortman G. A phase I dose-escalation and bioequivalence study of a trastuzumab biosimilar in healthy male volunteers. *Clin Drug Investig.* December 2014;34(12):887–94. doi: 10.1007/s40261-014-0247-5.
49. Yin D, Barker KB, Li R, Meng X, Reich SD, Ricart AD, Rudin D, Taylor CT, Zacharchuk CM, Hansson AG. A randomized phase 1 pharmacokinetic trial comparing the potential biosimilar PF-05280014 with trastuzumab in healthy volunteers (REFLECTIONS B327-01). *Br J Clin Pharmacol.* December 2014;78(6):1281–90. doi: 10.1111/bcp.12464.
50. Gu N, Yi S, Kim T, et al. Comparative Phamarcokinetics and Tolerability of Branded Etanercept (25 mg) and its biosimilar (25 mg): a randomized, open-label, single-dose, two sequence, cross-over study in healthy Korean male volunteers. *Clin Ther* 2011;33:2029–2037.
51. Yi S, Kim S, Park M, et al. Comparative pharmacokinetics of HD203, a biosimilar of etanercept, with marketed etanercept (Enbrel): a double-blind, single-dose, cross-over study in healthy volunteers. *Biodrugs* 2012;26:177–184.

52. Ryan AM, Sokolowski SA, Ng CK, Shirai N, Collinge M, Shen AC, Arrington J, Radi Z, Cummings TR, Ploch SA, Stephenson SA, Tripathi NK, Hurst SI, Finch GL, Leach MW. Comparative nonclinical assessments of the proposed biosimilar PF-05280586 and rituximab (MabThera®). *Toxicol Pathol.* 2014; 42:1069–81.
53. da Silva A, Kronthaler U, Koppenburg V, Fink M, Meyer I, Papandrikopoulou A, Hofmann M, Stangler T, Visser J. Target-directed development and preclinical characterization of the proposed biosimilar rituximab GP2013. *Leuk Lymphoma.* 2014;55:1609–17.
54. www.clinicaltrials.gov (accessed on July 5, 2016).
55. https://www.clinicaltrials.gov/ct2/results?term=infliximab+biosimilar&Search=Search (accessed on July 5, 2016).
56. Canhão H, Faustino A, Martins F, Fonseca JE. Rheumatic Diseases Portuguese Register Board Coordination, Portuguese Society of Rheumatology. Reuma. pt—the rheumatic diseases Portuguese register. *Acta Reumatol Port* 2011; 36:45–56.
57. https://www.clinicaltrials.gov/ct2/results?term=trastuzumab++biosimilar&Search=Search (accessed on July 5, 2016).
58. Cortés J, Curigliano G, Diéras V. Expert perspectives on biosimilar monoclonal antibodies in breast cancer. *Breast Cancer Res Treat.* 2014; 144: 233–9.
59. http://www.clinicaltrials.gov/ct2/results?term=rituximab+biosimilar&Search=Search (accessed on July 5, 2016).
60. Vital EM, Kay J, Emery P. Rituximab biosimilars. *Expert Opin Biol Ther.* 2013; 13: 1049–62.
61. https://www.clinicaltrials.gov/ct2/results?term=etanercept+biosimilar&Search=Search (accessed on July 5, 2016).
62. Azevedo VF, Galli N, Kleinfelder A, D'Ippolito J, Urbano PC. Etanercept biosimilars. *Rheumatol Int.* 2015; 35: 197–209.
63. https://www.clinicaltrials.gov/ct2/results?term=adalimumab+biosimilar&Search=Search (accessed on July 5, 2016).
64. https://www.clinicaltrials.gov/ct2/results?term=BCD-021&Search=Search (accessed on July 5, 2016).
65. Hlavaty T, Letkovsky J. Biosimilars in the therapy of inflammatory bowel diseases. *Eur J Gastroenterol Hepatol.* 2014; 26:581–587.

18

Regulatory Issues

Clarinda Islam

Somru BioScience, Inc., Vice-President, Regulatory Affairs and Project Management, Charlottetown, PE, Canada

18.1 Summary

This chapter will provide an overview of the regulatory process with a focus on biosimilar antibody-based therapeutics. It will explain and explore some of the challenges faced by regulatory authorities globally and at the national level as they seek to evaluate biosimilar products for safety and potency. For example, some questions and concerns may be how similar is similar enough? And what are the implications of the observed differences? This chapter will also address limitations of the regulatory process and the patients' responsibility to report safety concerns and symptoms (Table 18.1).

18.2 Introduction

A pathway for the approval of generic drugs has been well established in many countries for some time. These regulations have well-defined requirements for the approval of the generic versions of small molecule drugs. It has been recognized however that this model is insufficient for the evaluation and approval for the biosimilars of biologically derived therapeutics. Such entities are very sensitive to any changes in the manufacturing process, are not easily characterized, and are more likely to trigger an immune response or other unanticipated side effects. Due to these challenges, a completely different framework is required for the approval of biosimilars. Biosimilars are not exact replicas of the innovator products, but poses the same active portion and uses the same mechanism of action as the innovator. The challenge is to show that the biosimilar is structurally and functionally similar to

Biosimilars of Monoclonal Antibodies: A Practical Guide to Manufacturing, Preclinical, and Clinical Development, First Edition. Edited by Cheng Liu and K. John Morrow, Jr.

Table 18.1 Glossary of terms.

BPCIA	*Biologics Price Competition and Innovation Act* of 2009 (the United States)
EMA	European Medicines Agency
EU	European Union
FDA	Food and Drug Administration
INN	International nonproprietary name
PD	Pharmacodynamic
PHS Act	Public Health Service Act (the United States)
PK	Pharmacokinetic
RP	Reference product
WHO	World Health Organization

the innovator and that any structural differences do not affect the clinical outcomes in any significant way.

This chapter will give a brief synopsis of existing regulatory frameworks for the approval of biosimilars and some of the challenges that face both the industry and the regulatory bodies as additional biosimilars are developed and considered for approval.

18.3 Existing Regulatory Pathways

The European Union (EU) was the first to establish a regulatory pathway for the approval of biosimilars. To date, there have been 22 biosimilars approved in the EU, mostly of biologically derived therapeutics that are structurally similar to naturally occurring hormones or growth factors and are produced through recombinant DNA technology. The European Medicines Agency (EMA) has developed guidelines that establish the requirements for the approval of biosimilars [1]. The process is well defined and comprehensive, with guidance documents and even product class-specific, such as monoclonal antibodies [2], epoetin, and GCSF, guidances that describe the requirements in great detail.

The EMA guidelines describe the requirements for nonclinical studies, clinical studies, PK/PD assessments, clinical efficacy and safety, and postmarketing pharmacovigilance. They represent a mature, comprehensive framework such that the worldwide regulatory community reflects these guidelines in their own documentation covering these issues.

In the United States, the Biologics Price Competition and Innovation Act (BPCIA) signed in 2010 established the pathway for the approval of biosimilars. This pathway is enacted in section 351(k) of the Public Health Service (PHS) Act. The Food and Drug Administration (FDA) has finalized guidances [3, 4] for the industry regarding the quality and safety considerations for biosimilar approval. In addition to having the guidances open for public commentary, the FDA also held a Q&A session to add clarification to the draft guidances [5, 6]. These guidelines reflect the current thinking of the FDA but there are also numerous issues that will need to be considered on a case-by-case basis.

The FDA scientific guidance on assessing biosimilars describes a "stepwise approach" using the "totality of evidence" in demonstrating biosimilarity through structural and functional analysis, nonclinical data, and clinical studies [1]. The first application for a biosimilar in the United States (Zarxio by Sandoz) has recently been approved by the FDA. Zarxio, a biosimilar for filgrastim, has already been approved in 40 other countries, and an extensive amount of patient data has been amassed to prove its safety and efficacy. The approval of this submission will shed light on how the guidances are applied and interpreted by the FDA.

At the international level, other countries including Japan [7], Canada [8], South Korea [9], and Brazil [10] have created their own regulatory pathways for the approval of biosimilars. These frameworks reflect substantial overlap and similarity to the EU and US approaches, with regard to the requirements for comparability data, potency, safety, and efficacy, as well as facing many of the same challenges. The WHO has created a guidance document [11] that can be used as a reference point for those countries that are in the process of drafting their own regulatory pathways for the approval of biosimilars.

The key issues addressed in these guidances are:

- Use of reference product (RP)
- Extrapolation of indication
- Clinical studies/testing
- Interchangeability and substitution
- Naming/labeling
- Pharmacovigilance/postmarketing surveillance
- Intellectual property

18.3.1 Use of Reference Product

Globally, regulatory bodies require that the RP used for comparison should be one that has been approved and/or licensed for marketing in that country. Data generated using an RP not licensed within that country may be considered as part of the application as comparative data, but may not be evaluated solely on this data.

In the United States, bridging data between the US-licensed and non-US-licensed RP is required [3]. In the EU, the consideration of the use of non-EU-licensed RP will be considered on a case-by-case basis, and comparative data between the two is required [1]. For regulations in Japan, Canada, and WHO, the same RP must be used throughout the comparability (i.e., biosimilar application) program. If a nonlicensed RP is considered, its comparability to the licensed product must be demonstrated. Additionally, for Canada, the nonlicensed product may be used only if marketed by the same company and in the same dosage form [8].

18.3.2 Extrapolation of Indication

A biologic must be submitted for approval for each indication (i.e., therapeutic use). Biosimilars, however, may be submitted for approval for extrapolated indications, provided that they meet certain criteria. An extrapolated indication would require less extensive clinical testing and could provide an abbreviated path to the approval process. It is clear that the extrapolation of indications for a biosimilar must be well characterized, the mechanism of action(s) well understood and known, the safety concerns for the different indications or patient populations addressed, that there is no difference in the immunogenicity profile, and that there are no anticipated safety issues. Clinical experience with the RP with respect to these indications would be an important contributing factor to the totality of evidence to be weighed by the regulatory agency in reaching its decision.

Globally, the regulatory authorities have provisions for the extrapolation of indications. The WHO guideline further indicates that extrapolation may be possible if a sensitive clinical test model was used, in addition to the previously stated requirements [12].

18.3.3 Clinical Studies/Testing

All regulatory agencies with established biosimilar approval pathways will require clinical studies that demonstrate the comparability of the biosimilar to the RP in terms of potency, safety, and efficacy. The type and number of clinical studies will be dependent on the application and, for the most part, are determined on a case-by-case basis. Consultation with the regulatory body of the country in which a manufacturer is interested in marketing a biosimilar is necessary for the determination of the clinical studies needed. In general, the clinical studies are required to show that the pharmacokinetic/pharmacodynamic (PK/PD) profile, immunogenicity, and clinical outcomes of the biosimilar are comparable to the RP.

18.3.4 Interchangeability/Substitution

Interchangeability refers to a possible situation in which a patient may switch from the RP to the biosimilar, and from the biosimilar to the RP, during a course of treatment with no adverse effects. Substitution refers to the ability of a clinician or pharmacist to give a patient either the RP or the biosimilar (but not switching between the RP and biosimilar during the course of treatment).

Global regulations regarding biosimilars clearly indicate that biosimilars are not generic forms of the RP. Currently there is no global consensus on a biosimilar being interchangeable or able to be substituted for the RP. Several medical trade associations have issued recommendations against substituting a biosimilar for an RP without notification to the clinician and informing the patient fully on what the substitution entails [13, 14].

In the EU, the substitution of a biosimilar for an RP is determined at the member state level. EMA guidance states that this decision should be made following the opinion of a qualified healthcare professional. Canada explicitly states that the authorization of a biosimilar is not an indication that the biosimilar is equivalent pharmaceutically or therapeutically to the RP. Japan discourages the practice of switching between the RP and the biosimilar during treatment [12]. Substitution and interchangeability are two different issues in the United States. Substitution of a biosimilar for an RP is also determined at the state level for the United States, and no federal guideline exists at the moment.

In the United States, a biosimilar may be submitted as a biosimilar or as an interchangeable biosimilar. The FDA guidances have set a higher bar for an interchangeable biosimilar versus a biosimilar product [4, 6]. For a biosimilar to be considered interchangeable, the safety risk to a patient of alternating or switching to a biosimilar after using an RP should be no greater than if the patient has continued the use of the RP. These are broad requirements and the exact data required to show interchangeability will need additional clarification from the FDA. In a Q&A session regarding the implementation of the BPCIA, the FDA has acknowledged that it "...would be difficult as a scientific matter for a prospective biosimilar applicant to establish interchangeability... FDA is continuing to consider the type of information sufficient to enable FDA to determine that a biological product is interchangeable with the reference product" [6].

18.3.5 Naming/Labeling

Naming a biosimilar in an unambiguous manner will avoid unintended substitutions, minimize pharmacy errors, and aid in the traceability and pharmacovigilance. While there is no standard convention followed internationally, there have been suggestions from industry associations, including having an RP and a biosimilar share a root name but have a distinguishing feature in the name (i.e., suffix or prefix) to indicate biosimilarity with the RP [15]. The recently approved biosimilar Zarxio has a placeholder name of filgrastim-sndz, indicating the manufacturer of the biosimilar. The FDA has indicated that this does not constitute a final ruling on the naming convention for biosimilars.

While the EMA guidance does not specifically address a naming convention, all biosimilars thus far approved have had a unique brand name and/or a distinguishing feature in the nonproprietary name. The FDA has not issued any guidance on naming but it is expected that a guidance document will be forthcoming. Japan has a very detailed naming policy outlined in their guidance and clearly distinguishes not only between the RP and the biosimilar but also among biosimilars (e.g., Sandoz biosimilar for filgrastim is filgrastim BS) [15]. Australia has instituted a unique naming scheme for biosimilars as well [15]. This naming convention would use the nonproprietary name of the RP with a prefix of "sim" and a unique 3-letter code to be issued by the WHO international nonproprietary name (INN) Committee [15]. The WHO has proposed using the INN for the biologic combined with a biological qualifier (e.g., three-letter code or a unique suffix) in order to distinguish between the RP and biosimilar, as well as between all the biosimilars of a particular RP [16].

18.3.6 Pharmacovigilance/Postmarketing

The requirement for pharmacovigilance based on a risk management plan is universal. The EMA guidance requires a pharmacovigilance plan to be submitted and the clinical safety of the biosimilar to be closely monitored [1]. This plan should address any risks identified during the product development process and how those risks will be addressed should they arise after the biosimilar has been approved. The FDA guidance requires a risk evaluation and mitigation strategy (REMS). Additionally, the FDA can require additional postmarketing studies and postmarketing labeling changes should postmarketing surveillance indicate a need [3].

18.3.7 Intellectual Property

As stated previously, biosimilars cannot be considered the generic version of the RP. Even minor differences in manufacturing processes or molecular structure can result in significant changes in potency, purity, or efficacy. There would inevitably be an overlap between the RP and the biosimilar manufacturing processes, and the question of intellectual property around patent protection arises. Manufacturers of RP usually have broad patent protections on their product and can argue that a biosimilar manufacturer has infringed on those IP rights. Patent protection is for a longer period than data exclusivity (i.e., the period of time when no other entity may reference a pioneer manufacturer's data).

Data exclusivity and patent protection for an RP varies globally. The EU grants 8 years of data exclusivity, with an additional 2 years for market exclusivity. Japan grants 6 years of market exclusivity, while in the United States, it is 12 years for biologics (5 years for new pharmaceutical entities and 3 years for new indications).

In the United States, the BPCIA has laid out an extensive guidance for the review of the patents for both the RP (innovator manufacturer/pioneer) and the biosimilar manufacturer (applicant) [17]. This exchange of information was designed to address any IP disputes before a biosimilar enters the market, but does not affect or hinder the FDA's review of the biosimilar for approval. Briefly, the exchange of information is as follows:

- Applicant sends a copy of the accepted application to the pioneer within 20 days of the application being accepted by the FDA.
- The pioneer has 60 days from the receipt of the copy of the application to inform the applicant which patents it may reasonably assert with regard to the pioneer product and which patents it will consider licensing to the applicant.
- The applicant has 60 days from the receipt of the list of patents from the pioneer to respond in detail on its opinion on whether or not the patents are infringed or that the applicant does not intend to market the biosimilar before the patent expiry of the pioneer product.
- The pioneer has 60 days from the receipt of the applicant's detailed statements to reply to the applicant's assertions of enforceability or validity of the pioneer's patents.
- The applicant must inform the pioneer 180 days before the marketing of the biosimilar. The pioneer may seek an injunction at this point.
- The BPCIA requires good faith negotiations between the pioneer and the applicant on resolving the patent disputes. The pioneer may bring an infringement lawsuit within 30 days of the final exchange of the list of patents in dispute between the pioneer and the applicant.

From the steps as outlined in the preceding text, one can see that the exchange of patent information (the "patent dance") can get quite complicated and involves a considerable amount of agreement between parties. The required exchange of information and patent disclosures add a level of detail that is demanding and time sensitive. As the first biosimilar submitted for approval has been approved by the FDA, it remains to be seen what the impact of these disclosure requirements may have to both the pioneer and applicant manufacturers. It can be expected that the resolution of these patent disputes will likely be subject to legal recourse. For example, ahead of its application for Remsima, a biosimilar to Janssen's Remicade, Celltrion has filed a lawsuit against Janssen for trying to "improperly extend its monopoly after its initial patents expired" [18]. Celltrion has not actually given Janssen the information required as stipulated in the BPCIA patent and information disclosure exchange; rather it is anticipating that Janssen will claim patent infringement and bring suit.

Amgen has also brought suit against Sandoz to block the sale of Zarxio, claiming that Sandoz did not follow the FDA's required exchange of information regarding manufacturing processes to determine if there were any patent infringements [19]. Sandoz's argument is that the BPCIA patent disclosure procedures are optional. The most recent ruling of a California federal judge agrees with Sandoz [20]; however an appeal by Amgen is in process and the court has granted an injunction to keep Zarxio off the market until a decision on the appeal is rendered. The appeals process is still ongoing for this case.

In both of these cases, neither Celltrion nor Sandoz has participated in the so-called patent dance as stipulated in the BPCIA, although for different reasons. While the BPCIA has sought to lay out steps in order to resolve patent infringement disputes, it seems it has failed to set out clearly that the procedures are required, as the outcome of any patent infringement conflict does not affect the approval of a biosimilar by the FDA. Both of these cases are still in the process of litigation. The final rulings on both cases will set a precedent on how the intellectual property disclosures will govern future filings for would-be biosimilar manufacturers.

18.4 Challenges

While there have been many biosimilars approved for use around the world under different regulatory agencies and guidances, there remain thorny scientific questions that will have to be resolved through a legal framework. One of the first encountered is the degree of similarity between the RP and the biosimilar—how similar is similar enough? Due to the complexity of the production process, a biological therapeutic can never be identically replicated in the manner of small molecule generic drugs. The regulations address this globally by focusing on the efficacy, potency, and safety, along with the clinical outcomes, of the biosimilar. While there may be some structural differences, the challenge is in showing that they do not cause any critical variance in clinical outcomes. Additionally, if clinical studies do show some differences in patient response or immunogenicity, an evaluation should be made if these differences are of sufficient clinical significance and if they outweigh the clinical benefits.

Regulatory agencies have slightly varying approaches to allowing extrapolation of indications but at this time there is international consensus. Most regulatory agencies have a case-by-case approach to allowing extrapolation of indication. The most important thing is to consider how well the mechanism of action is understood and how the biosimilar applies for the particular indication under consideration. Another equally important consideration is patient safety. Relying heavily on past data and scientific justification would not be sufficient. Safety considerations in different populations and the clinical profile of the RP in the original indication should be carefully monitored.

The policy of extrapolating indications for a biosimilar has been controversial. There are industry and medical groups, however, that are against the extrapolation of indication for biosimilars without some additional clinical testing involving the extrapolated condition or additional patient population. There have been concerns raised by clinicians and medical associations as well, whether it is for another indication for which the RP was approved, and by extension the biosimilar should so be approved, or if the route of administration has changed (e.g., SC vs. IM). Clinicians worry about the safety of the biosimilar if it has not gone through the same level of safety testing for each indication as the RP. A key approach in fully addressing these concerns would be to increase the clinicians' understanding of biosimilars and the extent to which the regulatory agencies have reviewed the totality of evidence in granting the extrapolation of indications. Biologics by their nature cannot be fully replicated in an identical manner. Even RPs are not fully identical once manufacturing changes are made. The key is in ensuring and demonstrating that variances between the RP and the biosimilar do not have any effect on its efficacy.

Once a biosimilar has been deemed equivalent in function and use to the RP, the question then arises as to whether the biosimilar can be substituted or be considered interchangeable with the RP. One of the many challenges in determining interchangeability is the evaluation of an immune response. For a product to be determined not only as biosimilar but also as interchangeable, a study would have to be conducted wherein a patient would be given the RP, switched to the biosimilar, and then switched back to the RP. If an immune response is triggered in the patient, it is difficult to determine which product triggered the immune response. So the fundamental question is the clinical significance of the immune response. If there is an immune response detected, but no correlating adverse effect in patient care outcomes, the significance of the immune response is considerably lessened.

The naming convention of the biosimilar should also be given some due consideration for pharmacovigilance and patient safety. Some pharmaceutical manufacturers have argued for a distinct INN for each biosimilar. Biosimilar manufacturers have advocated for keeping the INN the same as the RP, arguing that a distinct name would indicate that it is pharmacologically different from the RP (hence not a biosimilar) and confusing the clinicians and patients. Ultimately, it is critical that regulatory agencies be able to trace adverse events to a specific biological product. Having a distinct INN from the RP would aid in pharmacovigilance. Sharing the nonproprietary name between manufacturers, as is done with generic drugs, can give a false sense that the biosimilar and RP are structurally similar and might indicate that they can be interchangeable. While some jurisdictions require that a biosimilar be distinctly named from its RP (e.g., Japan, Australia), others do not have the same requirement or do not have any regulatory guidance or a naming convention (e.g., EU, the United States). With its first biosimilar approval of Zarxio, the FDA has not issued a final guidance on naming conventions. In the interim, Zarxio is labeled as filgrastim-sndz, to distinguish it from the RP of Neupogen® (filgrastim).

While there is no global naming convention, a unique biosimilar name, correlating the biosimilar with the RP, similar to the Japanese system would seem to be the most appropriate approach. This gives the stakeholders (i.e., physicians and patients) a clear identification and distinction between the RP and biosimilar and can allow stakeholders to make appropriate decisions, as well as ease in pharmacovigilance.

One of the biggest challenges that face RP and biosimilar manufacturers is the consideration for IP. In the United States, the BPCIA has laid out a fairly specific format of data exchange to resolve possible IP infringement. This "patent dance" requires a timely exchange of sensitive information from both the RP and the biosimilar companies. With the first biosimilar application in the United States (for the filgrastim biosimilar Zarxio), Sandoz has not complied fully with the patent disclosure requirements that have been laid out in the BPCIA (see preceding text). The pending litigations from both the RP and biosimilar companies in this biosimilar application will set the precedent for the future approaches to the IP questions for biosimilars.

18.5 Conclusion

There remain substantial discussions to be had around each of the challenges posed previously, as well as global harmonization on definitions, data requirements, and safety requirements. The first wave of biosimilars has involved biologics derived from endogenous proteins that have been engineered through recombinant procedures (e.g., insulin, erythropoietin). The next wave of biologics will be biosimilars of monoclonal antibodies, which will add a layer of complexity in terms of proving the similarity given the limitations of testing. However the accumulated data over the years regarding biologics involving monoclonal antibodies should greatly contribute to the data-driven comparison, which should mitigate concerns with regard to the approval and use of biosimilars.

References

1. Guideline on Similar Biological Medicinal Products, London, October 30, 2005. CHMP/437/04. Available at http://www.ema.europa.eu/docs/en_GB/document_library/Scientific_guideline/2009/09/WC500003517.pdf (accessed on May 5, 2016).
2. Guideline on Similar Biological Medicinal Products Containing Monoclonal Antibodies—Non-Clinical and Clinical Issues, London, December 1, 2012. EMA/CHMP/BMWP/403543/2010. Available at http://www.ema.europa.eu/docs/en_GB/document_library/Scientific_guideline/2012/06/WC500128686.pdf (accessed on May 5, 2016).
3. Scientific Considerations in Demonstrating Biosimilarity to a Reference Product, Guidance for Industry. FDA, April 2015. Available at http://www.fda.gov/downloads/Drugs/GuidanceComplianceRegulatoryInformation/Guidances/UCM291128.pdf (accessed on May 5, 2016).

4. Quality Considerations in Demonstrating Biosimilarity of a Therapeutic Protein Product to a Reference Product, Guidance for Industry. FDA, April 2015. Available at http://www.fda.gov/downloads/Drugs/GuidanceComplianceRegulatoryInformation/Guidances/UCM291134.pdf (accessed on May 5, 2016).
5. Biosimilars: Questions and Answers Regarding Implementation of the Biologics Price Competition and Innovation Act of 2009, Guidance for Industry. FDA, April 2015. Available at http://www.fda.gov/downloads/Drugs/GuidanceComplianceRegulatoryInformation/Guidances/UCM444661.pdf (accessed on May 5, 2016).
6. Biosimilars: Additional Questions and Answers Regarding Implementation of the Biologics Price Competition and Innovation Act of 2009, Guidance for Industry. Available at http://www.fda.gov/downloads/Drugs/GuidanceComplianceRegulatoryInformation/Guidances/UCM273001.pdf (accessed on May 5, 2016).
7. PMDA Guideline of Quality, Safety and Efficacy of the Biosimilar Product, March 4, 2009. Available at http://www.pmda.go.jp/files/000153851.pdf (accessed on May 5, 2016).
8. Draft Guidance for Sponsors: Information and Submission Requirements for Subsequent Entry Biologics, Minister of Public Works and Government Services, Health Canada, 2008 (SEBs) 2010. Available at http://www.hc-sc.gc.ca/dhp-mps/alt_formats/pdf/brgtherap/applic-demande/guides/seb-pbu/seb-pbu-2010-eng.pdf (accessed on July 25, 2016).
9. Evaluation Guideline for Biosimilars—KFDA. July 2009. Available at http://www.nifds.go.kr/en/research/bio.jsp (accessed on July 25, 2016).
10. Castanheira LG, Barbano DB, Rech N. Current development in regulation of similar biotherapeutic products in Brazil. *Biologicals* 2011; 39(5):308–11. http://www.gabionline.net/Guidelines/Brazilian-guidelines-for-follow-on-biological-products
11. Guidelines on Evaluation of Similar Biotherapeutic Products (SBPs). Available at http://www.who.int/biologicals/areas/biological_therapeutics/BIOTHERAPEUTICS_FOR_WEB_22APRIL2010.pdf (accessed on May 5, 2016).
12. Biosimilars—US and International Update. Elsevier Clinical Decision Support, October 10, 2012. Available at http://www.goldstandard.com/wp-content/uploads/Biosimilars%E2%80%93US-and-International-Update.pdf.
13. American Dermatologists Update Position Statement on Biosimilar Substitution. GaBI Online, November 16, 2012. Available at http://gabionline.net/Biosimilars/General/American-dermatologists-update-position-statement-on-biosimilar-substitution (accessed on May 5, 2016).
14. ABPI Issues Updated Position Paper on Biosimilars. GaBI Online, June 2, 2014. Available at http://gabionline.net/Biosimilars/General/ABPI-issues-updated-position-paper-on-biosimilars (accessed on May 5, 2016).

15. The biosimilar name debate: what's at stake for public health. *GaBI Journal* 2014; 3(1):10–2. Available at http://gabi-journal.net/the-biosimilar-name-debate-whats-at-stake-for-public-health.html (accessed on May 5, 2016).

16. WHO naming of biosimilars. GaBI Online, October 25, 2013. Available at http://gabionline.net/Biosimilars/General/WHO-naming-of-biosimilars (accessed on May 5, 2016).

17. Vatland JA, Siekman MT, Loughran CA. Be prepared for biosimilars. Managing IP, November 2010. Available at http://www.managingip.com/Article/2701899/Search/Be-prepared-for-biosimilars.html?Home=true&OrderType=1&Keywords=janice+vatland&PartialFields=(CATEGORYIDS%3a11398)&Brand=Magazine (accessed on May 5, 2016).

18. Brennan Z. Celltrion Preps for its First Biosimilar Application with Janssen Lawsuit. Biopharma Reporter, April 4, 2014. Available at http://www.biopharma-reporter.com/Markets-Regulations/Celltrion-preps-for-its-first-US-biosimilar-application-with-Janssen-lawsuit (accessed on May 5, 2016).

19. Brennan Z. Amgen Files FDA Petition, Lawsuit Seeking Info on Sandoz Biosimilar Manufacturing. Biopharma Reporter, November 7, 2014. Available at http://www.biopharma-reporter.com/Markets-Regulations/Amgen-files-FDA-petition-lawsuit-seeking-info-on-Sandoz-biosimilar-manufacturing (accessed on May 5, 2016).

20. Williams A. Amgen v. Sandoz—Federal Circuit Oral Argument. Available at http://www.patentdocs.org/2015/06/amgen-v-sandoz-federal-circuit-oral-argument.html (accessed on May 5, 2016).

19

Legal Considerations*

K. Lance Anderson[1], Jennifer R. Moore Meline[2], and Jonathan D. Ball[3]

[1] *Dickinson Wright PLLC, Austin, TX, USA*
[2] *Medtronic, Restorative Therapies Group, Memphis, TN, USA*
[3] *Intellectual Property & Technology Group, Greenberg Traurig, LLP, New York, NY, USA*

19.1 Summary

Follow-on biologics, also known as "biosimilars," are subject to unique statutory and regulatory frameworks. This chapter highlights various legal considerations that may arise as a result of those frameworks during the development of a biosimilar and on the path to its commercialization.

Section 19.2 of this chapter addresses legal issues related to the regulatory scheme for obtaining authorization to market a biosimilar. In order to obtain approval, an entity seeking to market a biosimilar must submit a detailed application, accompanied by supporting data and manufacturing information, to the US Food and Drug Administration ("FDA"). Prior to 2010, applicants seeking approval to market a biological product were required to rely upon their own safety, efficacy, purity, potency, and manufacturing data and information in the biosimilar application, regardless of whether such product was highly similar to a biological product already approved by the FDA. In 2009, in an effort to speed up and reduce the associated costs of market entry for biosimilars, and to increase consumer access and affordability to biologically produced pharmaceuticals, Congress passed the Biologics Price Competition and Innovation Act ("BPCIA") as part of the Patient Protection and Affordable Care Act. The BPCIA substantially altered the regulatory landscape by establishing an abbreviated approval pathway for biosimilars. The change brought the biological product

* The information and comments set out in this Chapter are those of the authors and do not necessarily reflect any official opinion of Medtronic. Neither Medtronic nor any person acting on its behalf may be held responsible for the use which may be made of the information contained in this Chapter.

Biosimilars of Monoclonal Antibodies: A Practical Guide to Manufacturing, Preclinical, and Clinical Development, First Edition. Edited by Cheng Liu and K. John Morrow, Jr.

approval process in line with the abbreviated approval process instituted for generic small molecule drugs under the Drug Price Competition and Patent Term Restoration Act ("the Hatch-Waxman Act") some 25 years earlier, also in an attempt to reduce the delay and costs of those drugs entering the market. One court has commented that the ability to rely on the reference product sponsor (RPS)'s data in an application seeking approval to market a biosimilar is "saving years of research and millions in costs," because the need to generate all of the data required for a first-time reference product is eliminated.[1]

Not surprisingly, the BPCIA shares certain procedures in common with the Hatch-Waxman Act. Under the Hatch-Waxman Act, a company seeking approval to market a generic drug may file an Abbreviated New Drug Application (ANDA), which relies on data relating to the associated branded drug. Likewise, under the BPCIA, an abbreviated "351(k) application" (nicknamed for the statute describing the required application contents) for approval to market a biological product, which is highly similar to a licensed reference biological product, may rely in large part upon data filed by the RPS. ANDAs require only a showing of equivalency with a reference drug. Similarly, 351(k) applications require a showing of biosimilarity or interchangeability with the reference product. In exchange for sharing data, which others may one day rely on for approval of a generic drug, branded drug makers (and first-to-file ANDA submitters) receive the benefit of certain market and data exclusivity periods. RPS as well as those able to establish the interchangeability of their product prior to other follow-on biologics also enjoy market and data exclusivity periods.

However, the BPCIA also differs from the Hatch-Waxman Act in several important respects and contains unique provisions that are designed to take the inherent variability of biosimilars (as compared with small molecule generic drugs) into consideration. For example, as a trade-off for the ability of others to rely upon its data, the BPCIA provides that the RPS has a 4-year exclusivity period in which subsequent applications for biological products related to the reference product may not be submitted to the FDA and another 12-year exclusivity period in which such subsequent applications may not be approved by the FDA. These periods are generally longer than the exclusivity periods afforded by the Hatch-Waxman Act. Another important difference is that, in contrast to ANDA filers, 351(k) applicants do not need to make a certification at the time of application, as to whether any relevant patents exist or are invalid and/or not infringed. Instead, as discussed in more detail in the succeeding text, the biosimilar applicant has the option to engage in a so-called patent dance. This involves an exchange of patent lists, the biosimilar application, and other confidential information between the biosimilar applicant and the RPS, with the goal of resolving patent disputes expediently and, in some cases, without ever involving a court. A more detailed explanation of the BPCIA, informed by recent guidance documents issued by the FDA, is outlined in the succeeding text in Section 19.3. Also, in order to better understand the BPCIA regulatory scheme, various similarities and differences between the relatively new BPCIA and the much older and more familiar Hatch-Waxman Act are highlighted.

Section 19.3 of this chapter describes the interplay between the BPCIA and patent litigation in courts. By taking advantage of the BPCIA's streamlined process to save money and time, biosimilar applicants must show at the very least that their product is "highly similar" to, and in a clinical sense not meaningfully different from, a product that may be claimed in a patent. Therefore, data submitted in the 351(k) application may potentially be offered as evidence of infringement in a patent suit, with the patent holder claiming that, if marketed, the biosimilar (or a method of making it or using it to treat patients) would infringe one or more of its patents. It is therefore useful to have a basic understanding of BPCIA and US patent law, which may be relevant during biosimilar development and commercialization. A summary of some of the most relevant legal considerations is set forth in Section 19.4.

This chapter further explores patent exclusivity that may be available for biosimilars. Potential 351(k) applicants should recognize that FDA exclusivity is not the only type of exclusivity that may be available for newly developed biological products. Patent exclusivity, for example, is distinct from FDA exclusivity. A patent may be obtained during the development of the biological product, as opposed to FDA approval, which is normally sought closer to the completion of product development. Moreover, patent exclusivity generally lasts for up to 20 years from the time a patent application has been filed. Therefore a patent may offer longer protection than the market and data exclusivity given to the RPS or first interchangeable under the BPCIA or a biosimilar that does not have the benefit of any exclusivity period under the BPCIA. For this reason, Section 19.4 of this chapter also briefly describes considerations related to patenting biological products.

19.2 Overview of the Biologics Price Competition and Innovation Act of 2009 ("BPCIA")

19.2.1 Introduction to the BPCIA

The BPCIA of 2009 was signed into law by President Barack H. Obama on March 23, 2010, as Title VII of the Patient Protection and Affordable Care Act.[2] The BPCIA amended section 351 of the Public Health Service ("PHS") Act to create a framework for the FDA to approve follow-on biologics in an abbreviated process.[3]

The BPCIA is modeled after the Drug Price Competition and Patent Term Restoration Act of 1984 (commonly referred to as the "Hatch-Waxman Act"),[4] which established an abbreviated approval pathway for generic drugs. However, the BPCIA is not identical to the Hatch-Waxman statutory scheme. This is in part due to the inherently variable nature of a biosimilar product, whether from its origin, impurities, or method of manufacture. The BPCIA sets forth an abbreviated approval process that is specifically designed to address the inherent product variation of biosimilars as compared with small molecule drugs. Other regulatory authorities, such as the European Medicines Agency (EMA), have developed similar statutory frameworks, which should be consulted for approval in other countries as the specific application and approval requirements of those countries may vary from those of the United States.

In essence, the BPCIA provides an abbreviated approval, or licensure, pathway for biological products shown to be "biosimilar" to, "interchangeable" with, or previously considered by the FDA in connection with an application for an innovator reference biological product.[5] Data respecting the FDA-licensed reference biological product may be relied upon so long as similarities can be shown between the product for which approval is sought and the reference product. The basic framework of the licensure process will be appreciated in light of the reference products to which they pertain. For example, the exclusivities afforded may vary due to the nature of the biological products and the sponsors involved. Certain product exclusivity periods may be obtained, or must be permitted, within the biologics licensure process. As discussed further in the succeeding text, these aspects will remain of importance in the coming years due to known expirations of FDA-approved biological products.

19.2.2 351(k) Application Requirements

The BPCIA amended section 351 of the PHS Act by adding subsection (k), allowing for an application for licensure of a biological product that is biosimilar to a reference product according to an abbreviated pathway. Section 351(k) generally governs the form and substantive content required for biosimilar applications and sets forth applicable exclusivity periods. As a result, an application submitted to obtain FDA approval to market a biosimilar is often referred to as "a 351(k) application."

A 351(k) application must include several types of information generally demonstrating all of the following:

1) The biological product sought to be licensed is "biosimilar"[6] to a "reference product"[7] based upon data derived from analytical studies demonstrating that the biological product is "highly similar" to the reference product notwithstanding minor differences in clinically inactive components, animal studies, and clinical studies sufficient to demonstrate safety, purity, and potency in one or more appropriate conditions in which the biological product is intended to be used.
2) To the extent that the mechanism of action of the reference product is known, the biological product and reference product utilize the same mechanism of action for treating conditions of use prescribed, recommended, or suggested in the proposed labeling.

3) The conditions of use prescribed, recommended, or suggested in the labeling proposed for the biological product have been previously approved for the reference product.
4) The route of administration, the dosage form, and the strength of the biological product are the same as those of the reference product.
5) The facility in which the biological product is manufactured, processed, packed, or held meets standards designed to assure the continued safety, purity, and potency of the biological product.[8]

A higher standard, referred to as "interchangeability," may also be obtained, but additional data beyond biosimilarity must be provided.[9] First, the applicant must also show that the biological product can be expected to produce the same clinical result as the reference product in any given patient. Second, if the proposed product is intended to be administered more than once to an individual, the applicant must show that the risk in terms of safety or diminished efficacy of alternating or switching between the biological product and the reference product is not greater than the risk of using the reference product without such alternation or switch.[10] A product qualifying under the "interchangeability" standard definition could be substituted for the reference product without the intervention of the healthcare provider who originally prescribed the reference product.[11]

In either case, as set forth in the BPCIA, there remains some discretion in determining the nature of these approval requirements. The FDA has significant flexibility to determine the nature of the information submitted in the application and may choose to waive or request additional information as it deems necessary to support the licensure of the biological product.[12]

19.2.3 Exclusivity Periods

The BPCIA provides for certain exclusivity periods that may temporarily bar FDA consideration and approval of subsequent 351(k) applications for a defined period of time. Both a reference product and the first-approved interchangeable product may enjoy such an exclusivity period. For reference products, any period of exclusivity is calculated from the date of first licensure of the reference product. A 351(k) application may not even be submitted until 4 years after the date on which the reference product was first licensed.[13] Furthermore, the FDA cannot approve a 351(k) application until 12 years after the date on which the reference product was first licensed. In addition to the exclusivity periods provided for a reference product, the first-approved interchangeable product is also entitled to exclusivity periods of varying lengths, depending on the nature and orientation of the product to the sponsor and market.

Table 19.1 (later) summarizes the varying exclusivity terms for reference and interchangeable products as provided by the BPCIA.

Table 19.1 Exclusivity periods for various biological products under 351(k) of the PHS Act, as amended by the BPCIA[a].

Subsequent biological products	Exclusivity period for reference product	Exclusivity period for first-approved interchangeable (FIP)
Application submission	**No other 351(k) application may be submitted** until **4 years** after license of reference product	**No determinations of interchangeability** for subsequent biosimilars until the **earlier of**: • **1 year** after the first commercial marketing of FIP • **18 months** after a final court decision in or dismissal of lawsuit instituted under 351(l)(6) • **42 months** after approval of FIP if such litigation is continuing • **18 months** after approval of FIP if there is no 351(l)(6) lawsuit
Application approval	**No approval** of 351(k) applications until **12 years** after reference product license date	—
Application approval	Additional **6 months** for the earlier periods if reference product sponsor conducts pediatric studies meeting pediatric exclusivity of 505(A) of FD&C Act	—
	Reference products having orphan drug exclusivity of 7 **years** under 527(a); term calculated as the **later of** the 12-year period of 7-year orphan drug exclusivity	—

[a] *Id.* at 351(k)(6) (exclusivity for FIP); 351(k)(7) (exclusivity for reference product).

In addition to the time bars outlined in the preceding text in Table 19.1, certain biological products may be eligible for extended exclusivity periods. Products determined to produce health benefits for the pediatric population may be eligible for pediatric exclusivity ("PED"), which allows for an additional 6 months to be tacked on to the application submission and FDA approval exclusivity periods set forth in Table 19.1.[14] In addition, certain biological products indicated for a rare disease or condition and granted 7 years of "orphan drug exclusivity" under section 527(a) of the Food, Drug, and Cosmetic Act may receive an exclusivity period of 6 months.[15]

However, not all biological products are entitled to their own exclusivity periods. No period of exclusivity applies to approvals of (i) applications for supplements for the reference product or (ii) subsequent applications filed by the same sponsor or manufacturer of the reference product for a change resulting from a new indication, route of administration, dosing schedule, dosing form, delivery system, delivery device or strength, or modifications to the structure of the reference product that do not result in a change in safety, purity, or potency.[16]

Determining the date of licensure from which all exclusivity periods will be calculated can be inherently complex. Therefore, the FDA issued a draft "Guidance for Industry: Reference Product Exclusivity for Biological Products Filed under Section 351(a) of the PHS Act" on August 5, 2014 ("Exclusivity Guidance"), to assist with the first date of licensure calculations for reference products under section 351(k)(7)(C).[17] The FDA issues guidance documents including nonbinding recommendations (subject to public comment and further consideration) that are meant to become, when finalized, a representation of the FDA's "current thinking" on the topic introduced.[18] Guidance documents are not binding interpretations of law. Rather, they are intended to provide clarity to assist sponsors that are developing industry products.

The Exclusivity Guidance considers the exclusivity periods for reference products, which are, in effect, a decision on eligibility for reference product exclusivity and the date on which such exclusivity begins to run.[19] As an initial matter, the FDA noted that "Congress did not intend for every biological product licensed under 351(a) of the PHS Act to be eligible for a separate period of reference product exclusivity," and therefore, "the FDA must make a determination regarding the date of first licensure," for "each product licensed under section 351(a)."[20]

Furthermore, because section 351(k)(7)(C) excludes from the date of first licensure the date of approval of supplements and certain subsequent applications filed by the "same sponsor or manufacturer" or "licensor, predecessor in interest, or other related entity," the FDA's examination of business relationships may be critical to the exclusivity period determination. The Exclusivity Guidance confirms that the FDA intends to interpret these terms of the BPCIA in a manner similar to its previous constructions for like terms set forth in connection with other statutes regulating food, drug, and cosmetic approvals.[21] For example, the FDA confirmed that it will interpret the term "predecessor in interest" as it does in the 3-year new drug product exclusivity context, to mean "an entity (e.g., a corporation) that the sponsor has taken over, merged with, or purchased, or from which the sponsor has purchased all rights to the drug [reference product]."[22] The Exclusivity Guidance should be consulted for additional thinking of the FDA with respect to its construction of certain statutory terms relating to the exclusivity period determination.

As mentioned earlier, the BPCIA does not provide for an exclusivity period for every type of biological product. Specifically, the exclusivity periods do not apply to the license for, or approval of, products that supplement, change, or modify a previously approved biologic (section 351(k)(7)(C)).[23] In the Exclusivity Guidance, the FDA suggests that sponsors seeking a determination of first licensure—thereby avoiding exclusivity time bars on license approval—should provide detail on the nature of the structural similarities and differences of the proposed product and currently licensed product, with even more detail for protein products. The FDA will then determine, on a case-by-case basis, whether the modification results in a change in "safety, purity, or potency," which will significantly rely upon data provided by the sponsor. Previous data in the original reference product application (section 351(a)) may be used, as well as evidence showing a change that is beneficial to public health, with citations to a therapeutic advantage or other substantial benefit. The Exclusivity Guidance should be consulted for other types of information that the FDA recommends that sponsors include, upon submission of an application for approval to market a reference biological product or by later amendment or correspondence, to be used in the exclusivity period calculation.

The critical nature of the first licensure determination, and whether or not an exclusivity period should apply, warrants the RPS's compliance with the Exclusivity Guidance, even if it is not technically law. It is important to keep in mind, however, that the Exclusivity Guidance only relates to the calculation of exclusivity periods for reference products. Thus, it remains to be seen whether the FDA will issue guidance and recommendations respecting potentially applicable exclusivity periods under for interchangeable products section 351(k)(6).

19.2.4 Patent Dispute Resolution Process

The BPCIA introduced a unique framework for managing patent issues related to the determination and filing of patent infringement actions that may arise in connection with filing an application for approval to market a follow-on biologic. Unlike the generic drug pathway under the Hatch-Waxman Act, which provides a cause of action for patent infringement based upon the certification by the generic applicant with regard to noninfringement,[24] the 351(k) process provides a procedure for confidential disclosure of information and subsequent good faith negotiation between the 351(k) applicant and the patent holder. This distinguishing feature is significant with respect when patent infringement actions may be brought and will be discussed in more detail in the "Patent Litigation and the BPCIA" section of this chapter. This section sets forth the basic procedure of the BPCIA's patent dispute resolution process as outlined in section 351(l).

19.2.5 Confidentiality

Section 351(l) calls for the biosimilar applicant to provide its 351(k) application and other information to an RPS within 20 days after the application is filed.[25] The purpose of sharing the information is to facilitate a determination as to whether patents owned or controlled by the RPS are possibly being infringed by the biosimilar in question. Confidentiality obligations apply to all of these exchanges.

Under section 351(l)(1)(B), "confidential information" includes the 351(k) application containing information as required by section 351(k)(2), as well as "any other information that the subsection (k) applicant determines, in its sole discretion, to be appropriate."[26] The confidentiality obligations are similar in nature to traditional confidentiality agreements entered into between parties but with significant additional restriction on who is entitled to access the confidential information, including restrictions on employees of the RPS that may access the information.

A 351(k) applicant must provide the confidential information to:

1) Outside counsel—one or more attorneys designated by the RPS who are employees of an entity other than the RPS, provided that such attorneys do not engage, formally or informally, in patent prosecution relevant or related to the reference product
2) In-house counsel—one attorney that represents the RPS who is an employee of the RPS, provided that such attorney does not engage, formally or informally, in patent prosecution relevant or related to the reference product[27]

In the event that there are patents exclusively licensed to the RPS, the confidential information may also be shared with the representative of a patent owner who has retained the right to enforce the patent, so long as the patent owner agrees to be bound by the same confidentiality provisions.[28] These recipients are not allowed to share the confidential information with other persons or entities, including employees of the RPS itself, outside scientific consultants, or other outside counsel retained by the RPS, without prior written consent of the 351(k) applicant, but it should be noted that such written consent "shall not be unreasonably withheld."[29] Access to the biosimilar applicant's confidential information is thus quite limited to specific counsel of the RPS and is meant to keep such information as closely controlled as possible.

There are also limitations on the RPS's use of confidential information. Specifically, the RPS may only use the 351(k) application and other information for "the sole and exclusive purpose of determining, with respect to each patent assigned to or exclusively licensed to a reference product sponsor, whether a claim of patent infringement could reasonably be asserted if a subsection (k) applicant engaged in the manufacture, use, offering for sale, sale, or importation into the United States of the biological product that is the subject of the application under subsection (k)."[30] The BPCIA further provides that the confidential information remains the property of the 351(k) applicant and makes clear that no license or other interest to the confidential information is granted to the RPS or outside counsel by virtue of receiving the information.[31]

In the event that the RPS files a patent infringement action, the confidentiality obligations of the RPS and its legal representatives remain in effect until a court issues a protective order over such information in the litigation, at which point the 351(k) applicant is allowed to redesignate the confidential information under such protective order.[32] If no action is filed, the 351(k) applicant must be provided with certification that the confidential information has been returned or destroyed.[33]

Additional protections for the 351(k) applicant apply under the BPCIA once confidential information has been exchanged. For example, a 351(k) applicant may not be deemed to be admitting the validity, enforceability, or infringement of any patent, or the competency, relevance, or materiality of any information, merely by providing it to the RPS.[34] Further, similar to nondisclosure agreements entered into between the parties, the provisions of section 351(l)(1)(H) include a qualification that the violation of the obligations of confidentiality will cause irreparable harm to the 351(k) applicant and that any court should consider immediate injunctive relief as an "appropriate and necessary" remedy for violation or threatened violation. However, the BPCIA does not expressly authorize any further remedies for breach of confidentiality obligations nor is a civil cause of action specifically authorized under this statute.

19.2.6 Patent Exchange Procedures *(a.k.a. "the Patent Dance")*

The BPCIA sets forth unique procedures for potentially resolving patent disputes between the RPS and the 351(k) applicant, subject to the previously described confidentiality obligations. Because of the back-and-forth nature of these procedures, some people refer to the process as "the patent dance." The procedure involves a series of good faith negotiations between the 351(k) applicant and the RPS designed to illuminate each party's perspective as to whether or not the subject biosimilar may infringe any of the patents owned or controlled by the RPS. The goal of the negotiations is to have the 351(k) applicant and RPS agree on which patents, if any, would be the subject of an infringement action. The negotiations follow an information exchange period that begins no later than 60 days after the RPS receives the 351(k) application and other confidential information.[35]

Each party has the opportunity to identify patents that may be relevant in a lawsuit. The RPS first provides a list of patents for which it believes a claim of patent infringement could reasonably be asserted. The 351(k) must respond with a detailed statement describing, on a claim-by-claim basis, the legal basis of any opinion that the RPS's identified patents are not infringed, invalid, or unenforceable, to which the RPS must ultimately respond.[36] Alternatively, the 351(k) applicant may provide a statement that the biological product will not be marketed until after the expiration of the applicable patents listed along with a response regarding each patent identified by the RPS. The 351(k) applicant also has the option to provide its own list of patents that it believes could be asserted by the RPS, along with a required detailed statement describing.

Another component of the initial information exchange involves the RPS identifying the patents on the initial list that the RPS would be willing to license to the 351(k) applicant.[37] Upon providing the optional response list and required written statement, the 351(k) applicant is also obligated to respond to the patents designated as available for licensing, although the BPCIA provides little additional detail about the nature and content required of such a response. Following the information exchange periods, the parties are required to enter into a negotiation period lasting up to 15 days. The objective of the negotiation is to finalize the patent list for the infringement lawsuit.

Table 19.2 in the succeeding text summarizes the basic framework and timeline for the first round of patent information exchanges following the RPS's receipt of the 351(k) application and information.

Table 19.2 First-round procedure for developing the list and description of patents (see Sections 19.3.1, 19.3.3, and 19.3.4).

Responsible party	Action	Timing
RPS	Provides list of patents (i) believed to be infringed and (ii) willing to license	60 days from receipt of 351(k) application and other confidential information
351(k) applicant	Provides (i) optional list of relevant patents; (ii) statement of legal basis of noninfringement, invalidity, and unenforceability; and (iii) response to license list	60 days from receipt of reference product sponsor list
RPS	Provides statement of legal basis of infringement and response to 351(k) applicant's statement of invalidity and unenforceability	60 days from receipt of 351(k) applicant's list and statement(s)
Both	Negotiate in good faith the patents to be included in infringement matter	15-day period beginning upon receipt of reference product sponsor response

If the parties are able to finalize and mutually agree on a patent list for the infringement action, the RPS must file an infringement action asserting each agreed-upon patent within 30 days of such agreement. However, if the parties are unable to agree, the parties must undertake a second procedure, governed by section 351(l)(5), which requires the parties to simultaneously exchange lists that identify the patents each party believes should be the subject of an infringement action.[38] Notably, the RPS's list is limited to the number of patents identified by the 351(k) applicant, because under the statute, it may not exceed that number.[39] If the 351(k) applicant did not identify any patents, the RPS may only list one patent.

After these patent information exchanges, an infringement suit may be filed according to section 351(l)(6). If the parties have agreed on the patents to be asserted during the first round of negotiations, the RPS must file a patent infringement lawsuit asserting those patents not later than 30 days after such agreement.[40] If there was no agreement, the RPS is required to assert patents identified on each of the RPS and 351(k) applicant's lists not later than 30 days after the simultaneous exchanges of the second round of negotiations. Once filed, the 351(k) applicant has 30 days to provide the FDA with notice of the patent lawsuit and a copy of the complaint, both of which will be published by the FDA in the Federal Register.[41]

19.2.6.1 Newly Issued or Licensed Patents

Under the BPCIA, an RPS may supplement its patent list with additional patents it believes could reasonably be asserted against the 351(k) applicant, provided that the RPS obtained a license to the patent or the patent issued from the US Patent Office *after* the provision of its original list to the 351(k) applicant.[42] The additional patent must be added within 30 days of patent issuance or licensing. The 351(k) applicant then has 30 days to provide a response that addresses, on a claim-by-claim basis, its position concerning noninfringement, invalidity, or unenforceability of the newly listed patent.[43] Though provided in a supplement to which the 351(k) applicant responds, similar to the initial information exchanges, the newly licensed or issued patents do not become part of the parties' negotiations. Rather, such later-identified patents become subject to the provisions regarding notice requirements for commercial marketing, thereby allowing the RPS to seek a preliminary injunction respecting infringement of those patents.[44]

19.2.7 Notice of Commercial Marketing and Preliminary Injunction

Under the BPCIA, an RPS may ask a court to issue a preliminary injunction that will prevent the commercial manufacture or sale of a biosimilar. Section 351(l)(8) requires that the 351(k) applicant provide notice to the RPS no later than 180 days prior to the commercial marketing of the biological product that is the subject of the 351(k) application. Upon receipt of such notice, and prior to the commercial marketing of the biological product, the RPS may seek a preliminary injunction until the court rules on the issue of patent validity, enforcement, and infringement with regard to the patents previously identified and made subject to an infringement action. The patents applicable to such provision are any patents included in the initially exchanged patent lists, provided that such patents are not included in the second-round patent lists. Newly issued or licensed patents may also be subject to the preliminary injunction, if the RPS has timely supplemented its patent list. In the event that the RPS seeks a preliminary injunction, the 351(k) applicant and the RPS are to reasonably cooperate to expedite any necessary discovery in connection with that motion.[45]

19.2.8 Declaratory Judgment Actions

Declaratory actions are a form of nonmonetary remedy allowing for a court to provide a legal determination in a case or controversy.[46] The judgment is considered a statutory remedy and is commonly used in instances where an action is threatened against a party. Sections 505 and 512 of the Federal Food, Drug, and Cosmetic Act, as well as the PHS Act, consider declaratory actions as components for resolving patent infringement actions, pending or threatened. As discussed in more detail in the succeeding text, under the BPCIA, declaratory judgment actions ("DJ actions") may be filed in regard to certain patents in connection with the 351(k) application and patent list and exchange process.

19.3 Patent Litigation and the BPCIA

Both the BPCIA and US patent law have provisions regulating the litigation activities and rights of the 351(k) applicant and the RPS. In particular, there are limits on the circumstances under which a patent lawsuit may be filed. Because the abbreviated approval process for biosimilars is relatively new, the FDA and courts are still interpreting and clarifying the BPCIA's requirements. For example, in 2014, after biosimilar applicant Sandoz refused to turn over its 351(k) application to Amgen, Amgen filed a citizen petition requesting that the FDA require 351(k) applicants to certify that they will "timely comply" with the BPCIA's patent dispute resolution process before accepting any 351(k) application for review.[47] Amgen argued that the BPCIA in essence requires such certification because its patent dispute resolution process is mandatory. Momenta filed a response opposing any certification requirement.[48] In March 2015, the FDA denied Amgen's petition, stating that "neither section 351(k) nor section 351(l) require the FDA to impose a certification requirement as part of the biosimilar review process" and that the BPCIA "generally does not describe any FDA involvement in monitoring or enforcing the information exchange by creating a certification process or otherwise."[49] The FDA determined that it would leave competing views about whether the BPCIA's patent dispute resolution process is mandatory or optional up to the courts to decide.

Courts have recently begun to interpret the BPCIA's provisions with some landmark decisions expected in the near term. The relatively few district court rulings issued to date have involved the issue of whether or not potential or actual 351(k) applicants can forego the BPCIA's patent dispute resolution process altogether in favor of directly resolving patent disputes in court. These include cases such as *Sandoz Inc. v. Amgen Inc.*[50] (biosimilar of Enbrel® (etanercept)), *Celltrion Healthcare Co., Ltd. v. Kennedy Trust for Rheumatology Research,*[51] and *Hospira, Inc. v. Janssen Biotech*[52] (both concerning biosimilar of Remicade® (infliximab)). In each instance, the 351(k) applicant filed a lawsuit asking a court to declare that the patents of the RPS (or the patent holder if the RPS has licensed the patents) were invalid, unenforceable, and not infringed, with the hope of launching after FDA approval without the risk of an injunction or patent damages. More recently in March 2015, in *Amgen Inc. v. Sandoz Inc.*, Amgen sued biosimilar applicant Sandoz because it refused to provide Amgen with its 351(k) application for a biosimilar of Neupogen® (filgrastim).[53] The lower court held that Sandoz was within its rights to do so.

As will be discussed in more detail in the succeeding text, the Federal Circuit (the appellate court with exclusive jurisdiction over appeal in patent cases) recently reviewed the lower court's decision in *Amgen*. The Federal Circuit decided two key questions that inarguably impact the understanding and application of the BPCIA. First, the Federal Circuit determined that the BPCIA's patent exchange provisions are optional, not mandatory. Second, the Federal Circuit held that the BPCIA's 180-day notice of commercial marketing provision is keyed off of the FDA's approval, rather than acceptance of a 351(k) application. These findings have a substantial impact on a 351(k) applicant's liberty to choose between resolving patent disputes in court and through the "patent dance."

There are additional aspects of the BPCIA and patent laws that are relevant for 351(k) applicants to consider. This section will address such questions as:

- Can an applicant be sued for patent infringement for preparing a 351(k) application for submission?
- Can an RPS sue for patent infringement solely based on the submission of 351(k) application to the FDA?
- When can the applicant seek a court's aid to resolve patent disputes rather than participate in the patent resolution process set forth in the BPCIA?
- Can an applicant be sued for manufacturing and testing the biosimilar or reference product when undertaken for the FDA approval process?

Given the rapidly evolving law in this area, it is difficult to provide absolute answers to all of these questions. However, understanding the interplay between patent law and the BPCIA is informative in evaluating the potential risks of, and right to defend against, patent infringement allegations.

19.3.1 Patent Infringement under US Patent Law

A patent bestows its owner with the right to exclude others from exploiting their patented invention during the life of the patent. Most patents in the biologics space claim a product, a method of making a product, or a method of using the product. The acts that traditionally define infringement of such patents are set forth 35 U.S.C. § 271 and generally involve unauthorized making, using, offering for sale, selling, or importing into the United States, a patented invention. If such an act is committed, a patent holder has the right to assert its patents against the accused infringer in a lawsuit filed in federal court.[54]

Patent holders are not the only parties that may institute a lawsuit related to patent infringement allegations. Any party exposed to an "immediate and real" threat of injury, including an accused patent infringer, may request a US federal court to make a declaration of the parties' rights.[55] This type of lawsuit is called a "DJ action." In the context of patent infringement, an "immediate and real" threat may include an overt threat by the patent owner that it intends to sue for patent infringement. In a DJ action, a party accused of patent infringement will ask the court to declare that the patents are (i) invalid, (ii) unenforceable, and (iii) noninfringed. Courts will only entertain a DJ action if they determine that they have jurisdiction to do so. A court has the right to hear a DJ action in a patent case when there is a "substantial controversy, which is sufficiently immediate and real to warrant a declaration that will have the force and effect of a final judgment."[56]

19.3.2 Preapproval Activities and Patent Infringement Lawsuits

Whether or not a 351(k) applicant can be sued for patent infringement based on its 351(k) application (and related activities) appears to depend largely on the stage the applicant is at in the BPCIA process. The BPCIA purposefully keys its patent resolution procedures to the occurrence of certain events on the path to FDA approval. The risk of exposure to patent litigation certainly increases as applicants proceed further into the approval process.

19.3.2.1 Preparation of the 351(k) Application

As described in more detail in section IIB earlier, to establish "biosimilarity," an applicant must demonstrate that its biological product is "*highly similar* to the reference product notwithstanding minor differences in clinically inactive components" and that "there are no clinically meaningful differences between the biological product and the reference product in terms of the safety, purity, and potency of the product."[57] Establishing "interchangeability" requires an even closer identity between the biological product and the reference product, that is, "that the biological product may be *substituted* for the reference product without the intervention of the healthcare provider who prescribed the reference product."[58] If the 351(k) applicant is aware of, or has identified, patents that may potentially cover the reference product, it may perceive that there is a threat of patent litigation merely as a result of compiling information for the 351(k) application or speaking with the FDA about a potential 351(k) application.

Nonetheless, courts have generally held that merely *contemplating or preparing* to submit an application to the FDA does not present a controversy that is "immediate and real" enough for a court to have power to hear the case.[59] The FDA's acceptance of the application and any subsequent approval—acts that may constitute infringement—are considered to be too speculative and far-off when a manufacturer is merely preparing a 351(k) application.

In the 2014 *Celltrion* case, the District Court for the Southern District of New York considered whether it had jurisdiction to consider Celltrion's request for a declaratory judgment that three patents held by Kennedy Trust were invalid. The court declined to hear the case, finding that DJ jurisdiction was lacking, because the threat of Celltrion being sued by Kennedy for marketing Remsima (claimed to be a biosimilar of Remicade) was not sufficiently immediate and real.[60] Celltrion had already made substantial preparations for filing its 351(k) application, including by (i) participating in planning meetings with the FDA, (ii) completing clinical trials, (iii) investing substantial sums of money, (iv) establishing manufacturing facilities, and (v) stockpiling Remsima in preparation for receiving FDA approval.[61] However, the court found that even though Celltrion had exhibited a "true intention" to bring Remsima to market, it was "simply too far from receiving FDA approval ... for the exercise of declaratory judgment jurisdiction to be proper."[62] There were still "numerous steps that would have to occur for there to even be the potential for patent infringement," including FDA acceptance for review of a 351(k) application directed to the same antibody and approval for the same use as the reference product, all of which would have to occur prior to expiration of the patent, for infringement to be possible.[63] The court also determined that Kennedy's suits against other parties for conduct related to Remicade did not create DJ jurisdiction, because Kennedy had "not expressed a clear intent to pursue infringement claims against Celltrion."[64]

Similarly, in the 2014 *Sandoz II* case, the Federal Circuit agreed with the District Court for the Northern District of California that Sandoz's request for declaratory judgment could not be heard because Sandoz had not yet filed its 351(k) application for its follow-on biologic (claimed to be a biosimilar of Neupogen (etanercept)). Sandoz had filed its DJ complaint seeking a declaration that "the manufacture, use, sale, offering for sale, or importation of its etanercept product will not infringe, directly or indirectly, any valid claim of" Amgen's patents, and "that both patents are unenforceable ... and that both patents are invalid," on the same day that it began a phase III trial to compile data for the 351(k) application it was planning to submit to the FDA.[65] Although Sandoz had closely consulted with the FDA in formulating the phase III trial, the court still found that there was no real or immediate controversy to support a lawsuit because "we can hardly proceed by simply assuming that the phase III trial will wholly succeed," and "if the Phase III trial uncovers material problems, Sandoz may, at a minimum, need to delay any FDA application considerably longer."[66]

DJ suits also might be considered to be premature by courts because of additional delays built into the BPCIA framework. For example, even if a 351(k) application is nearly ready to be submitted to the FDA, a biosimilar applicant is precluded from submitting it until 4 years after the FDA licensed the reference product. Likewise, a biosimilar applicant cannot bring the product to market until 8 years after it submits its application, even if already approved.[67]

The earlier cases suggest that it is unlikely that courts will decide patent disputes based solely on the preparation of a 351(k) application. The possibility of infringing activity is considered to be too remote and uncertain because the biosimilar applicant could not legally sell an unapproved product on the open market. For applicants, this means that it may be difficult to get an early resolution of patent liabilities prior to expending the time and cost to prepare a 351(k) application.

19.3.2.2 Submission of 351(k) Application

As discussed in the preceding text, submitting a 351(k) application to the FDA is a prerequisite for obtaining a license to market a biosimilar to the public. Prior to submission and approval, it is therefore unlikely that a 351(k) applicant is engaging in traditionally defined acts of patent infringement (e.g., selling a patented product or treating a patient by a patented method).[68] However, in certain circumstances, the mere act of submitting a 351(k) application is deemed to be an artificial act of patent infringement that can serve as the basis for a lawsuit.

US patent law outlines two situations in which it "shall be an act of infringement ***to submit***" a 351(k) application, provided that its purpose "is to obtain approval under such Act to engage in the commercial manufacture, use, or sale of a drug ... or biological product claimed in a patent or the use of which is claimed in a patent before the expiration of such patent."[69]

First, it is an act of infringement to submit a 351(k) application "with respect to a patent that is identified in the list of patents described in section 351(l)(3)."[70] Notably, this basis for a lawsuit is directly tied to patents identified during the BPCIA's patent resolution process. The applicant has 20 days after the FDA accepts the application to provide it to the RPS, after which the patent list exchanges and negotiations can take up to several months.[71] In practice, there could be a delay of at least a couple of months after a 351(k) application has been submitted before a patent holder could sue based on this provision of the patent law.

Second, "if the applicant for the application fails to provide the application and information required under section 351(l)(2)(A)," that is, a copy of the 351(k) application and other manufacturing information, it is an act of infringement to submit a 351(k) application "for a patent that *could be identified* pursuant to section 351(l)(3)(A)(i)."[72] As already mentioned, the Federal Circuit is set to determine whether the BPCIA's patent resolution regime is mandatory or optional. Proponents of optionality rely in part on this provision of patent law, reasoning that a biosimilar applicant's right to choose whether to turn over its 351(k) application is implicit in a basis for infringement resting upon nonprovision of the application to the RPS.

Both statutory provisions authorizing the RPS to file a lawsuit based on the submission of a 351(k) application depend on the ability to identify relevant patents. This may not be as easy of a task for the RPS as it seems. Relevant patents may not exist or may be expired. To the extent that any nonexpired patents do exist, the RPS may not be the holder or owner of them. Furthermore, the BPCIA's structure does require pre-identification of relevant patents or notice of them to 351(k) applicants. A comparison with the Hatch-Waxman Act for approval of generic drugs is illustrative of this point.

Under US patent law, it is also an artificial act of infringement to submit an ANDA for the purpose of obtaining approval to commercialize "a drug claimed in a patent or the use of which is claimed in a patent."[73] However, the process of identifying patents that potentially cover the reference drug product may be faster and easier than it is for follow-on biologics, because the Hatch-Waxman Act provides for pre-identification of relevant patents in the FDA's Orange Book. Thus at the time an ANDA is filed, at least some patents that potentially cover a drug or its use are readily identifiable. If an ANDA contains a so-called Paragraph IV Certification—that any relevant, nonexpired Orange Book-listed patent "is invalid or will not be infringed by the manufacture, use, or sale of the new drug for which the application is submitted"—the patent holder may be in a position to sue almost immediately after submission of an ANDA.[74]

In contrast, the FDA's List of Licensed Biological Products ("Purple Book") does not list patents.[75] The BPCIA also does not require the biosimilar applicant to ascertain which patents may cover a reference biologic (if any) in its application.[76] In fact, as detailed in the preceding text, the requirement to state whether a particular patent is "invalid, unenforceable, or will not be infringed by the commercial marketing of the biological product," or that the 351(k) applicant "does not intend to begin commercial marketing" until patent expiration, does not arise until the 351(k) applicant receives a list of potentially relevant patents identified by the RPS.[77] As a result of these differences between the Hatch-Waxman Act and BPCIA, the filing of a patent infringement action against a biosimilar applicant based on the submission of its application will not follow as quickly on the heels of the submission as it would against an ANDA applicant.

These, of course, are differences in timing only. The act of submitting a 351(k) application, and receiving subsequent approval, will still likely result in a patent litigation against the biosimilar applicant. It is therefore important to consider additional aspects of the BPCIA and patent law as the approval process continues to move forward.

19.3.2.3 Restrictions on Patent Infringement Lawsuits during the BPCIA Patent Resolution Process

The BPCIA places several limits on when, how, and why the applicant and RPS can engage in litigation. Once the abbreviated approval process is under way, the possibility of resolving disputes in court may be more limited than it would normally be for parties who are not involved in the abbreviated approval process.

Once a 351(k) application has been filed, and so long as the applicant turns over its application and manufacturing information to the RPS, the BPCIA prohibits both parties from filing a DJ action. Specifically, "*neither the reference product sponsor nor the subsection (k) applicant* may, prior to the date" the RPS receives a "notice of commercial marketing" from the biosimilar applicant, file a DJ action for infringement, validity, or enforceability of any patent on the parties' patent lists.[78] DJ actions after application submission have been disfavored because there is a hope that the BPCIA patent exchanges and negotiations will "enable the narrowing of patent disputes and the crystallization of infringement claims"[79] in a "prompt and efficient way ... in order to ensure that approved biosimilars may be sold in the U.S. as soon as they are ready for the market."[80] One court characterized a biosimilar applicant's efforts to bring a DJ action prior to completing the BPCIA patent resolution process as an "attempt to skirt the BPCIA's dispute resolution mechanisms while reaping the benefits of its approval process."[81]

Nonetheless, the BPCIA authorizes the RPS to pursue a DJ action in the event that the applicant fails to do any of the following:

1) Provide the RPS with a copy of the application and related manufacturing information
2) Provide the RPS with a detailed statement as to why it claims the RPS's listed patents are invalid, unenforceable, or not infringed, or else a statement that it does not intend to commercially market its biosimilar prior to patent expiration
3) Provide the RPS with a list of patents the applicant believes should be the subject of a patent infringement action if initial negotiations following the first patent list exchange did not lead to resolution

4) Notify the FDA of any patent infringement action filed and provide a copy of the complaint
5) Provide the RPS with the notice of commercial marketing not later than 180 days before the date of the first commercial marketing of its biosimilar[82]

A mirror provision allowing the applicant to bring a DJ action in the event that the RPS fails to perform during the patent resolution process does not exist. Thus, under the express provisions of the statute, the first time a biosimilar applicant has a clear right to file a DJ action would be after it provides notice of commercial marketing to the RPS.[83] This is markedly different from the Hatch-Waxman Act, which expressly provides for applicants who have submitted an ANDA with a Paragraph IV Certification to file a DJ action against a patent holder who fails to bring suit itself within 45 days of notice.[84]

19.3.2.4 The BPCIA Patent Dance Is Optional

The question of whether the BPCIA's patent dispute resolution process is mandatory or optional was recently decided by the Federal Circuit. A 351(k) applicant may choose not to provide the 351(k) application and other confidential information to the RPS or engage in the series of information exchanges defined in 351(l)(2)–(8). This "opt out" ability allows the 351(k) applicant to provoke the RPS or patent holder to initiate a lawsuit at an earlier date than contemplated by the BPCIA scheme.

Amgen recently petitioned the FDA to require all future 351(k) applicants to certify that they will provide a copy of the 351(k) application and manufacturing information to the RPS within 20 days after the FDA has accepted the application for review.[85] However, the FDA found that "[n]either section 351(k) nor section 351(l) requires FDA to impose a certification requirement as part of the biosimilar review process." The FDA further found that the BPCIA "generally does not describe any FDA involvement in monitoring or enforcing the information exchange by creating a certification process or otherwise."[86] The FDA noted that the BPCIA's lack of an explicit certification requirement stands in contrast to the express patent certification requirements of the Hatch-Waxman Act. Pointing to the *Amgen* case, the FDA ultimately concluded that the "mandatory-or-optional" question was a matter of interpretation for the courts.

The courts have indeed begun to decide this issue—in favor of 351(k) applicants. In the *Amgen* case, Amgen (the RPS for the biological product filgrastrim marketed under the brand name Neupogen) sued Sandoz based on Sandoz's failure to turn over its 351(k) application for its filgrastrim product (to be marketed under the brand name Zarxio) and related manufacturing information. Amgen asked the District Court for the Northern District of California to, among other things, find that Sandoz violated California's Unfair Competition Law and was liable for conversion and patent infringement because it had not turned over its application to Amgen.[87] But the court determined that Sandoz "was within its rights" under section 351(l) to choose not to provide its 351(k) application to Amgen or to participate in the patent exchanges and disclosures of 262(l)(2)–(8). The court explained,

> [S]ubsection (l) lays out a process that could take up to 230 days—just to commence patent litigation. An applicant who values expedience over risk mitigation may believe that the disclosure and negotiation process would introduce needless communications and delay. Such an applicant may have good reason to believe that no unexpired relevant patents relate to its biosimilar, and that it is likely to prevail if challenged with an infringement suit. The applicant may, in such an instance, opt to forego its ability to bring certain types of declaratory actions and receive information about potentially relevant patents from the reference product sponsor, and instead commence litigation immediately ... Sandoz therefore traded in the chance to narrow the scope of potential litigation with Amgen through subsection (l)'s steps, in exchange for the expediency of an immediate lawsuit. The BPCIA's plain language and overall statutory scheme support a reading that renders this decision entirely permissible.[88]

The court further found that "[w]hile engagement in the [BPCIA] process creates a temporary safe harbor from declaratory judgment actions, a party's failure to participate permits the opposing party to commence patent litigation."[89]

The Federal Circuit recently reviewed the lower court's decision and provided its opinion on July 21, 2015.[90] The first impression issue before the Federal Circuit was whether the BPCIA's prescription that the applicant "shall provide" its application to the RPS (262(l)(2)), and certain patent lists and information (e.g., 351(l)(2)(B)) means that the patent resolution process is mandatory, rather than optional. The Federal Circuit acknowledged that the plain language of "shall provide" would suggest that the patent dance is mandatory. However, according to the Federal Circuit, that language cannot be read in isolation from the rest of the statute and particularly sections of the BPCIA such as section 351(l)(9)(C), which contemplate the failure of a 351(k) applicant to disclose the 351(k) application and other confidential information by the statutory deadline and specify that the only remedy for such failure is an infringement action. The Federal Circuit found that "[b]ecause Sandoz took a path expressly contemplated by the BPCIA, it did not violate the BPCIA by not disclosing its [351(k) application] and the manufacturing information by the statutory deadline."[91] Both Amgen and Sandoz are requesting that the Federal Circuit reconsider its findings. Should the Federal Circuit decide to reconsider, this chapter will be updated accordingly.

However, as the law stands now, 351(k) applicants may elect to forego the BPCIA's patent dispute resolution process. But, it will be important to weigh the benefits versus drawbacks of doing so. On one hand, granting access to the application may facilitate resolution of disputes or at least narrow the scope of the disputes. With the opportunity to review data and manufacturing information, there is always the possibility that the RPS or patent holder could determine that there is no infringement. Under the BPCIA scheme, the RPS also provides a list of patents that the RPS "would be prepared to license" to the biosimilar applicant.[92] Moreover, the BPCIA advantageously permits the biosimilar applicant to substantially control the scope of any eventual litigation. The BPCIA calls for the parties to make an initial exchange of patent lists and detailed statements as to their contentions respecting the infringement or validity of any such patents, after which the parties negotiate for 15 days.[93] However, the parties may not agree as to a "final and complete list of which, if any patents" from the parties' lists "shall be the subject of an action for patent infringement."[94] In that instance, the parties must exchange new patent lists.[95] Importantly, the number of patents on the RPS's lists "may not exceed the number of patents" on the biosimilar applicant's patent list, unless the biosimilar applicant declines to list any patents, in which case the RPS "may list 1 patent."[96] In essence, the biosimilar applicant may limit the subject of disputes before a court to one patent, should it so choose. This is in contrast with accused patent infringers who generally do not have the ability to limit the scope of a patent infringement case before it starts.

On the other hand, undertaking the BPCIA patent dispute resolution process can have drawbacks. It may be undesirable to wait for the BPCIA patent resolution process to conclude before litigation can begin. Exchanges and negotiations between the parties can take up to 230 days. Moreover, biosimilar applicants may be concerned about the risk of improper disclosure when sharing the type of sensitive, confidential, and proprietary information contained in a 351(k) application. However, the statute incorporates some safeguards against loss of confidentiality and improper use of the biosimilar manufacturer's information. The number of individuals who may access the information is limited to one or more outside counsel attorneys designated by the RPS and one in-house attorney of the RPS or patent holder, to the extent that it is not the RPS.[97] A so-called prosecution bar specifies that attorneys who receive the biosimilar applicant's information cannot "engage, formally or informally, in patent prosecution relevant or related to the reference product."[98] Moreover, the information may not be disclosed to "any other person or entity, including the RPS employees, outside scientific consultants, or other outside counsel retained by the reverence product sponsor, without the prior written consent of the subsection (k) applicant."[99] Limits on the uses and retention of ownership of confidential information may also help to reduce the risk of improper disclosure.[100]

19.3.3 A Question Remains as to Whether Notice of Commercial Marketing Is Mandatory after Licensure

The Federal Circuit decision in the July 21, 2015, *Amgen v. Sandoz* opinion also made a determination on another landmark issue—whether a 351(k) applicant's notice of commercial marketing can be provided to the RPS *prior to* receipt of FDA approval. Amgen's position was that Sandoz violated the BPCIA "by giving a premature, ineffective, notice of commercial marketing under [section 351(l)(2)(A)] before FDA approval of its biosimilar product."[101] In other words, Amgen was arguing that Sandoz should only have been able to give a notice of commercial marketing *after* the FDA approved its 351(k) application. The Federal Circuit found that "to be effective" under the BPCIA, notice "must be given only after the product is licensed by the FDA." The Federal Circuit explained:

> Requiring that a product be licensed before notice of commercial marketing ensures the existence of a fully crystallized controversy regarding the need for injunctive relief. It provides a defined statutory window during which the court and the parties can fairly assess the parties' rights prior to the launch of the biosimilar product. If a notice of commercial marketing could be given at any time before the FDA licensure, the RPS would be left to guess the scope of the approved license and when commercial marketing would actually begin. Indeed, filing [a 351(k) application] only suggests that a [351(k) applicant] intends to commercially market its product someday in the future.[102]

Ultimately, there is still a question as to whether, according to the Federal Circuit's decision, providing a notice of commercial marketing upon licensure is *mandatory* in all instances.[103] In fact, this question is still being litigated in the courts, including in *Amgen v. Apotex*, Case No. 15-cv-61631 (S.D. Fla.). In that case, Amgen argues that the Federal Circuit decision should be interpreted as requiring the biosimilar applicant to provide a notice of commercial marketing in all instances. Apotex, on the other hand, takes the position that a party who participates in the patent dance is not required to provide a notice of commercial marketing.

The ultimate interpretation of the Federal Circuit's findings will have significant importance for a 351(k) applicant's ability to bring a DJ action earlier on in the application process, because the BPCIA provides that the right of either party to bring a DJ action is triggered when the RPS receives the notice of commercial marketing.[104] A notice of commercial marketing that might be viewed by courts, along with other evidence, as evidencing a dispute is sufficiently "immediate and real" to support DJ jurisdiction, is delayed until licensure.

19.3.4 271(e)(1) "Safe Harbor"

Under the patent law, parties engaged in activities that would normally be infringing are not liable for patent infringement if those activities are conducted "solely for uses reasonably related to the development and submission of information under a Federal law which regulates the manufacture, use, or sale of drugs or veterinary biological products."[105] This exception to infringement is sometimes called the "271(e)(1) safe harbor," and it applies to the FDA approval process for biosimilars. Thus, the manufacture of an infringing monoclonal antibody for purposes of gathering data for a 351(k) application seeking a license from the FDA to market a follow-on biologic would be protected by the safe harbor.

The purpose of the safe harbor is to spur innovation by allowing applicants to perform tasks necessary to obtain approval without the risk of a lawsuit. For the safe harbor to apply, the activities in question must be "solely for uses reasonably related" to seeking FDA approval. To determine whether a given infringing activity is protected by the safe harbor, courts simply ask:

> Would it have been reasonable, objectively, for a party ... to believe that there was a decent prospect that the "use" in question would contribute (relatively directly) to the generation of kinds of information that was likely to be relevant to the process by which the FDA would decide whether to approve the product.[106]

This standard broadly exempts a wide range of activities. Information does not have to be *submitted* to the FDA to be considered "reasonably related" to seeking FDA approval.[107] Moreover, the applicability of the safe harbor is not dependent on the "underlying purposes or attendant consequences of the activity ... as long as the use is reasonably related to the FDA approval."[108] What matters is that the activities are objectively related to compiling data for FDA applications.

The following are some examples of activities that, in certain circumstances, courts have decided do not constitute patent infringement:

- Publicly disseminating or presenting data developed for FDA approval, including to investors, analysts, and journalists[109] and at trade shows[110]
- Stockpiling "commercial quantities" with the FDA's awareness[111]
- Conducting consumer surveys to develop information for the FDA[112]
- Raising investor funds by reporting data compiled for the FDA[113]

Postapproval activities may also be eligible for safe harbor protection under certain circumstances.[114] Testing "carried out to satisfy the FDA's requirements means it falls within the scope of the safe harbor, even though the activity is carried out after approval."[115] On the other hand, activities that relate to "routine submissions" or that are "generated voluntarily by the manufacturer" do not have the benefit of safe harbor protections.[116]

However, biosimilar applicants should recognize that infringing activities related to products that are *not the subject of an application of FDA approval* are not shielded from patent infringement claims. For example, a manufacturer of an optical spray analyzer designed to measure and gather data to include in FDA applications for not-yet-approved aerosol sprays was not entitled to safe harbor protection.[117] Thus infringing activities related to products ancillary to the biosimilar for which a license is sought, for example, equipment to test monoclonal antibodies, would not be protected by the safe harbor rule. Similarly, the safe harbor rule does not protect the manufacture of infringing products in order to obtain foreign regulatory approval.[118] There must be a connection to FDA approval in the United States.

19.4 Patenting Your Biosimilar

A 351(k) applicant who has successfully obtained FDA approval to market a new reference biological product or interchangeable biological product may benefit from the FDA exclusivity periods discussed in the preceding text in Section 19.2. However, that is not the only type of exclusivity available for newly invented products. A patent may also allow for a period of market exclusivity and provide its owner (or an exclusive licensee) with the ability to keep others from making, using, selling, or offering for sale the patented product. To obtain a patent, one must file a patent application including a specification, abstract, drawings, oath of any inventors, and sequence information (where applicable), along with the appropriate filing fee.[119] Specific issues with patent applications may arise due to the nature of biological products, as discussed in the succeeding text.

19.4.1 Confidentiality Considerations

As a general rule, patent applications are "published promptly after the expiration of a period of 18 months from the earliest filing" of the application.[120] But, the specification, drawings, and sequence information that must be provided with the patent application sometimes contain sensitive data and manufacturing information compiled by the inventors, raising concerns that patent exclusivity may come at the cost of disclosing trade secrets and other proprietary information to the world. Patent applicants should note that it is possible to file a nonpublication request with the US Patent Office, subject to several requirements. First, such a request must be submitted *upon filing* of the patent application.[121] The nonpublication request must also be included with the application papers, not filed separately, even if it is on the same date as the filing date of the application.[122]

Second, the request must state, in a conspicuous manner, that the application is not to be published. Third, the requester must certify that the invention disclosed in the application is not and will not be the subject of an application filed in another country, or under a multilateral international agreement, that requires publication of applications 18 months after filing.[123]

19.4.2 The Rigorous Written Description Requirement for Monoclonal Antibodies

Under US patent law, to obtain a valid patent, the patent's specification "shall contain a written description of the invention."[124] In order to determine if the specification satisfies the written description requirement, the patent examiner will look to determine whether the specification describes the claimed invention in sufficient detail that one skilled in the art can reasonably conclude that the inventor had possession of the claimed invention.[125] This long-standing test is applied more rigorously in the context of patents to antibodies. Thus an important consideration is the data that will likely need to be compiled to provide adequate written description for your patent.

Under certain circumstances, claims directed to biomolecules may lack written description, even if they find literal support in the specification. Specifically, claims that recite a broad genus of biomolecules may lack written description if the specification fails to disclose a sufficient number of representative species.[126] In addition, claims that functionally describe a genus of biomolecules may lack written description if the specification does not provide sufficient examples of species that achieve the claimed function.[127]

One case that is helpful in understanding these requirements as they apply to antibody patents is *AbbVie Deutschland GmbH v. Janssen Biotech, Inc.*[128] decided on July 1, 2014. In that case, the Federal Circuit considered whether a jury in the District Court for the District of Massachusetts had properly ruled that AbbVie's US Patent Nos. 6,914,128 ("the '128 patent") and 7,504,485 ("the '485 patent") directed to fully human antibodies that bind to and neutralize the activity of human interleukin 12 ("IL-12") were invalid. AbbVie's patents disclosed the amino acid sequences of nearly 300 antibodies having various ranges of IL-12 binding affinities, all having V_H3 heavy chains and lambda light chains derived from the same lead compound. The antibodies described in the '128 and '485 patents also shared at least 90% amino acid sequence similarity in their variable regions, with more than 200 of the described antibodies differing from Y61 at a single amino acid residue and sharing 99.5% similarity in the variable regions. The claims of the '128 and '485 patents that were at issue defined the claimed antibodies by their function (i.e., IL-12 binding and neutralizing characteristics) rather than by structure. For example, patent claim 29 of the '128 patent recited:

Claim 29. A neutralizing isolated human antibody, or antigen-binding portion thereof that binds to human IL-12 and disassociates from human IL-12 with a k_{off} rate constant of $1x10^{-2}$ s^{-1} or less, as determined by surface plasmon resonance.

AbbVie argued to the court that each of its patent claims met the written description requirement (i.e., that AbbVie had possession of all that was claimed at the time that it filed its patent application) for several reasons. First, AbbVie argued that they were "limited to a small genus of antibodies that are rare and difficult to obtain" and "its patents describe[d] a representative number of antibodies commensurate with the scope of the claims."[129] AbbVie further argued that it disclosed the amino acid sequence of all known species covered by the claims except for one and a k_{off} rate covering the full range of the features claimed in the accused product. Centocor, on the other hand, argued that AbbVie's patent disclosure was limited to a family of closely related, structurally similar antibodies that were all derived from the same lead compound, whereas the claims defined claims covering antibodies having widely varying structures but still meeting the functional limitations of the claims. Consequently, Centocor argued that AbbVie's patents could not be representative of the entire genus claimed. The court agreed with Centocor, based on the following rules for written description:

> When a patent claims a genus using functional language to define a desired result, "the specification must demonstrate that the applicant has made a generic invention that achieves the claimed result and do so by showing that the applicant has invented species sufficient to support a claim to the functionally-defined genus." We have held that "a sufficient description of a genus ... requires the disclosure of either a representative number of species falling with the scope of the genus or structural features common to the members of the genus so that one of skill in the art can 'visualize or recognize' the member of the genus."

Patent applicants should therefore endeavor to include as much functional and structural description about their antibodies as is possible in order to provide adequate written description support for the claims. There are additional concepts of patent that may be relevant, including enablement and what is eligible. A lawyer should be consulted should a patent application be desired.

19.5 Conclusion

Considering the continuing development of the new framework for approval of biosimilars, those in the industry will continue to monitor, if not participate, in the shaping of the ongoing process for the BPCIA. At the time of press, early decisions are emerging, but it is expected that the BPCIA, and the additional clarity provided by the courts, will significantly influence, and hopefully clarify, the remaining confusion with this new process. Those considering how patent rights and related standards for patentability apply, this chapter has provided a foundation for understanding the key elements of patent law and how they will apply to the new standards of the BPCIA and the corresponding patent dance that ensues. For continuing updates of the progress of the BPCIA, the authors welcome further inquiry.

Notes

1 See *Amgen Inc. v. Sandoz Inc.*, 2015 WL 1264756 at *1 (March 19, 2015, N.D. Cal.).
2 Patient Protection and Affordable Care Act, Pub. L. No. 111–148, §§ 7001–7003, 124 Stat. 119, 804–21 (2010) ("Affordable Care Act") (codified as amended at 42 U.S.C. § 262, 35 U.S.C. § 271(e), 28 U.S.C. § 2201(b), and 21 U.S.C. § 355 et seq.).
3 The PHS Act is codified in 42. U.S.C. § 262 et seq.
4 Drug Price Competition and Patent Term Restoration Act of 1984 ("the Hatch-Waxman Act"), Pub. L. 98–417, 98 Stat. 1585 (1984).
5 PHS Act § 351(a) sets forth the application content and other requirements for approval a reference biological product.
6 PHS Act § 351(i)(2).
7 *Id.* at § 351(i)(4).
8 *Id.* at § 351(k)(2)(A)(i).
9 *Id.* at § 351(i)(3) (defining term "interchangeable"); § 351(k)(2)(B) and k(4) (detailing additional data and information required).
10 *Id.* at § 351(k)(4)(A)–(B).
11 *Id.* at § 351(i)(3).
12 *Id.* at § 351(k)(2)(A)(ii)–(iii).
13 *Id.* at § 351(k)(7)(A)–(C).
14 *Id.* at § 351(m)(2) and (3).
15 *Id.* at § 351(m)(2) and (3).
16 *Id.* at § 351(k)(7)(A)–(C).
17 See "FDA Guidance for Industry: Reference Product Exclusivity for Biological Products," August 2014, available at http://www.fda.gov/downloads/drugs/guidancecomplianceregulatoryinformation/guidances/ucm407844.pdf ("Exclusivity Guidance").
18 FDA Guidance documents respecting biological products are generally available at http://www.fda.gov/biologicsbloodVaccines/GuidanceComplianceRegulatoryInformation/guidances/default.htm.
19 See Exclusivity Guidance, p. 1, lines 25–30.
20 See *id.* at p. 4, lines 116–120.
21 See *id.* at p. 4, line 138 to p. 5, line 177.

22 See *id.* at p. 4, lines 145–150 and p. 5, lines 155–165.
23 See *id.* at p. 5, line 179 to p. 7, line 253.
24 See Hatch-Waxman Act § 505(j)(2)(A)(vii)(IV), often referred to as "Paragraph IV."
25 PHS Act § 351(l)(1)(B).
26 *Id.* at § 351(l)(1)(B)(i).
27 *Id.* at § 351(l)(1)(B)(ii).
28 *Id.* at § 351(l)(1)(B)(iii).
29 *Id.* at § 351(l)(1)(C).
30 *Id.* at § 351(l)(1)(D).
31 *Id.* at § 351(l)(1)(E).
32 *Id.* at § 351(l)(1)(F).
33 *Id.*
34 *Id.* at § 351(l)(1)(G).
35 *Id.* at § 351(l)(3)(A).
36 *Id.* at § 351(l)(3)(B).
37 *Id.* at § 351(l)(3)(A)(ii).
38 *Id.* at § 351(l)(5)(B)(i).
39 *Id.* at § 351(l)(5)(B)(ii)(II).
40 *Id.* at § 351(l)(6)(A) and (B).
41 *Id.* at § 351(l)(6)(C).
42 *Id.* at § 351(l)(3)(A).
43 *Id.* at § 351(l)(3)(B).
44 *Id.* at § 351(l)(8).
45 *Id.* at § 351(l)(8)(C).
46 See 28 U.S.C. § 2201.
47 See Letter from Jeffrey P. Kushan of the law firm Sidley Austin LLP to the FDA, dated October 29, 2014, Docket No. FDA-2014-P-171-001, available at http://www.regulations.gov/#!documentDetail;D=FDA-2014-P-1771-0001 ("Amgen Citizen Petition").
48 See Letter from Bruce A Leicher, Senior Vice President and General Counsel of Momenta Pharmaceuticals, Inc., to the FDA, dated November 25, 2014, available at http://www.regulations.gov/#!documentDetail;D=FDA-2014-P-1771-0003 ("Momenta Response").
49 March 25, 2015, FDA Response to Amgen Citizen Petition, available at http://www.regulations.gov/#!documentDetail;D=FDA-2014-P-1771-0004 ("FDA Denial").
50 See 2013 WL 6000069, No. C-13-2904, at *1 (N.D. Cal. November 12, 2013) ("*Sandoz*"), aff'd by, 2014 WL 6845165, No. 2014-1693, at *1 (Fed. Cir. 2014) ("*Sandoz II*").
51 See 2014 WL 6765996, No. 14 Civ. 2256, at *1 (S.D.N.Y. December 1, 2014) ("*Celltrion*").
52 See 2014 WL 6766263, No. 14 Civ. 7049, at *1 (S.D.N.Y. December 1, 2014) ("*Hospira*").
53 See *Amgen Inc. v. Sandoz Inc.*, 2015 WL 1264756 at *1 (March 19, 2015, N.D. Cal.) ("*Amgen*").

54 See 28 U.S.C. §§ 1331, 1338(a), 2201, and 2202.

55 See 28 U.S.C. § 2201 ("[A]ny court of the United States, upon the filing of an appropriate pleading, may declare the rights and other legal relations of any interested party seeking such declaration, whether or not further relief is or could be sought.").

56 See *MedImmune, Inc. v. Genentech, Inc.*, 549 U.S. 118, 127 (2007).

57 PHS Act § 351(i)(2).

58 *Id.* at § 351(i)(3).

59 See, for example, *Sandoz II* at *4 (Fed. Cir. 2014) ("We are aware of no decision in which we have found a case or controversy when the only activity that would create exposure to potential infringement liability was a future activity requiring approval that had not yet been sought.").

60 The court rejected Celltrion's argument that its DJ action was proper despite not having participated in the BPCIA's patent resolution process because Kennedy was the patent owner and not the RPS, stating that "[w]hile it is true that the BPCIA envisions the dispute resolution process to involve the applicant and the RPS, the BPCIA does provide for a level of involvement by the patent owner." See *Celltrion* at *5.

61 *Celltrion* at *3–4.

62 *Id.*

63 *Celltrion* at *3.

64 *Celltrion* at *3–4.

65 *Sandoz II* at *2.

66 *Sandoz II* at *5.

67 PHS Act § 351(k)(7)(A) and (B).

68 See, for example, 35 U.S.C. § 271(a).

69 *Id.* at § 271(e)(2)(C).

70 *Id.* at § 271(e)(2)(C)(i).

71 *Id.* at § 351(l)(2)(A)–(C).

72 *Id.* at § 271(e)(2)(C)(ii).

73 *Id.* at § 271(e)(2)(A).

74 21 U.S.C. § 355(j)(2)(A)(vii)(IV).

75 The FDA's Purple Book is available at http://www.fda.gov/Drugs/DevelopmentApprovalProcess/HowDrugsareDevelopedandApproved/ApprovalApplications/TherapeuticBiologicApplications/Biosimilars/ucm411418.htm. The Purple Book lists the BLA number, proprietary name, active ingredient, date of licensure, and exclusivity expiration date for reference products.

76 PHS Act § 351(k)(2).

77 *Id.* at § 351(l)((3)(B)(ii).

78 *Id.* at § 351(l)(9); see also *Sandoz* at *2 ("[W]ith limited exceptions not applicable here, neither a reference product sponsor, such as Amgen, nor an applicant, such as Sandoz, may file a lawsuit unless and until they have engaged they have engaged in the exchanges of information mandated by the BPCIA at 42 U.S.C. §§ 261(l)(2)–(6).").

79 *Celltrion* at *5.
80 *Celltrion* at *4.
81 *Celltrion* at *5.
82 PHS Act 351(l)(9).
83 *Id.*
84 See 21 U.S.C. § 355(j)(5)(C); see also *Caraco Pharm. Labs., Ltd. v. Forest Labs., Inc.*, 527 F.3d 1278, 1295 (Fed. Cir. 2008) (generic drug maker "has a complete generic drug product that has been submitted to the FDA for approval, and no additional facts are required to determine whether this drug product infringes the claims of [branded drug company's] patent.").
85 See Amgen Citizen Petition and Momenta Response.
86 See FDA Denial.
87 A preliminary injunction essentially restrains a party from continuing with a course of conduct until a case is fully decided.
88 See *Amgen* at *7.
89 See *Amgen* at *1.
90 *Amgen Inc. v. Sandoz Inc.*, 794 F.3d 1347 (Fed. Cir. 2015).
91 *Id.* at 1355–57.
92 PHS Act § 351(l)(3)(A)(ii).
93 *Id.* at § 351(l)(3)–(4).
94 *Id.* at § 351(l)(4).
95 *Id.* at § 351(l)(5).
96 *Id.* at § 351(l)(5)(ii).
97 *Id.* at § 351(l)(A) and (B).
98 *Id.* at § 351(l)(B)(i) and (ii).
99 *Id.* at § 351(l)(C).
100 *Id.* at § 351(l)(D)–(F).
101 *Amgen Inc. v. Sandoz Inc.*, 794 F.3d at 1357.
102 *Id.* at 1358.
103 *Id.* at 1360.
104 *Id.* at § 351(9)(A).
105 35 U.S.C. § 271(e)(1).
106 *Intermedics v. Ventritex*, 775 F. Supp. 1269, 1280 (N.D. Cal. 1991), aff'd by, 1993 WL 87405 (Fed. Cir. 1993).
107 *Merck v. Integra*, 545 U.S. 193, 206–07 (2005).
108 *Abtox v. Exitron*, 122 F.3d 1019, 1030 (Fed. Cir. 1997).
109 *Teletronics v. Ventritex*, 982 F.2d 1520, 1523 (Fed. Cir. 1992).
110 *Teletronics v. Ventritex*, 982 F.2d 1520, 1523 (Fed. Cir. 1992).
111 *NeoRx v. Immunomedics*, 877 F. Supp. 202, 206–07 (D.N.J. 1994).
112 *Chartex v. M.D.*, 1993 WL 306169, at *3 (Fed. Cir. 1993).
113 *Teletronics v. Ventritex*, 982 F.2d 1520, 1523 (Fed. Cir. 1992) ("Congress could not have been unaware of the need of competitors to raise funds for developing and testing competing products, and for preparing to enter the market once controlling patents had expired .... It would appear that manufacturers of

competing medical devices have an even greater need for such fund raising than do generic drug companies.").

114 *Momenta v. Amphastar*, 686 F.3d 1348, 1358 (Fed. Cir. 2012).

115 *Id.*

116 *Id.*

117 *Proveris v. Innovasystems*, 536 F.3d at 1259, 1265–66.

118 *NeoRx v. Immunomedics*, 877 F. Supp. 202, 207–08 (D.N.J. 1994) ("It may be true that most of the vials were used to generate data for the FDA, but in this instance, some vials were clearly used for other purposes.").

119 "*See, e.g.*, 35 USC § 111 *et seq*."

120 Manual of Patent Examination and Procedure ("MPEP") § 1120.

121 *Id.* at § 1120; 1122.

122 *Id.* at § 1122(II).

123 *Id.* at §§ 1120, 1122.

124 35 U.S.C. § 112, first paragraph.

125 MPEP § 2163(l).

126 See, for example, *Regents of the University of California v. Eli Lilly & Co.*, 119 F.3d 1559, 1568–69 (Fed.Cir.1997) (holding that a sufficient description of a genus requires the disclosure of either a representative number of species falling within the scope of the genus or structural features common to the members of the genus); *AbbVie Deutschland GmbH v. Janssen Biotech, Inc.* 759 F.3d 1285, 1300–01 (Fed. Cir. 2014) ("[a]lthough the number of species appears high quantitatively, the described species are all of the similar type and do not qualitatively represent other types of antibodies encompassed by the genus").

127 See, for example, *Ariad Pharm., Inc. v. Eli Lilly and Co.*, 598 F.3d 1336, 1358 (Fed. Cir. 2010) (en banc) (holding that claims reciting methods of reducing NF-kB activity lacked written description because "the specification at best describes decoy molecule structures and hypothesizes with no accompanying description that they could be used to reduce NF-kB activity."); *University of Rochester v. G.D. Searle & Co., Inc.*, 358 F.3d 916, 918–19 (Fed. Cir. 2004) (holding that claims directed to a method of selectively inhibiting the COX-2 enzyme lacked written description because "Rochester did not present any evidence that the ordinarily skilled artisan would be able to identify any compound based on [the specification's] vague functional description"); *Fiers v. Revel*, 984 F.2d 1164, 1170–71 (Fed.Cir.1993) (holding a claim to a genus of DNA molecules not supported by written description of a method for obtaining the molecules).

128 *AbbVie Deutschland GmbH v. Janssen Biotech, Inc.* 759 F.3d 1285, 1300–01 (Fed. Cir. 2014).

129 *Id.* at 1298.

20

ADCC Enhancement Technologies for Next-Generation Therapeutic Antibodies

Cheng Liu and Su Yan

Eureka Therapeutics, Emeryville, CA, USA

20.1 Summary

In summary, ADCC enhancement is a key strategy for improving therapeutic antibody-drug efficacy. Recent clinical studies provided further evidence in support of the technology. It has the potential of lowering effective drug dosage, hereby benefiting patients through lower drug cost. Antibodies with ADCC enhancement are expected to eliminate variations in patient response to antibody treatments caused by genetic polymorphism and improve survival of cancer patients. The commercial value of the technology has been demonstrated by the licensing agreement between Amgen and Kyowa Hakko Kirin on anti-CCR4 antibody, which includes $100 million upfront fee plus $420 million milestone payment and double digit royalties. The FDA approval of obinutuzumab and its superior efficacy compared with the first-generation anti-CD20 antibody, rituximab, in leukemia was a milestone success of ADCC enhancement technology in new mAb drug discovery and development. Meanwhile, ADCC enhancement is also becoming a core technology for developing next-generation therapeutic antibody drugs with favorable clinical outcomes.

20.2 Introduction

Recombinant therapeutic antibodies represent a successful new class of drugs developed over the past two decades. Antibody therapy has a long history tracing back to thousands of years ago, as early as 200 B.C. in China, in the form of vaccination against infectious diseases. In the late 1800s, sera from humans or animals containing antibodies were widely used for prophylaxis and therapy of viral and bacterial diseases [1, 2]. With the advent of molecular biology, it has

Biosimilars of Monoclonal Antibodies: A Practical Guide to Manufacturing, Preclinical, and Clinical Development, First Edition. Edited by Cheng Liu and K. John Morrow, Jr.

become possible to produce recombinant antibodies in mammalian cells [3, 4]. Human chimeric and humanization of antibodies reduced immunogenicity of the drugs, paving the way for broad use in patients and disease applications [5, 6]. Fully human antibodies have become a reality with new technologies such as phage display of antibody libraries and transgenic mouse with human immunoglobulin genes [7].

Therapeutic antibodies have achieved great success in the past decade. As the fastest growing class of innovative drugs, there are more than 350 candidates in different phases of clinical trials. To date, 42 therapeutic antibodies have been approved by the FDA for the treatment of various diseases, including cancer, viral infection, rheumatoid arthritis, and organ graft rejection. It is estimated that about 30% of new drugs in the next decade will be based on antibody products [8, 9]. There is increasing evidence showing that the Fc effector functions, antibody-dependent cell-mediated cytotoxicity (ADCC) activity, and complement-dependent cytotoxicity (CDC) of antibody drugs are major mechanisms of action in antibody therapy [10–13].

20.3 Activation of ADCC Functions

ADCC functions are primarily triggered through direct interaction of the constant region (Fc) of antibodies with Fc receptors (FcγR) on a variety of immune cells [14–16]. In humans, the FcγR family contains three main classes, FcγRI (CD64), FcγRII (CD32, including isoforms FcγRIIa, FcγRIIb, and FcγRIIc), and FcγRIII (CD16, including FcγRIIIa and FcγRIIIb) [17, 18]. Among these FcγRs, FcγRI, FcγRIIa/c, and FcγRIIIa are activating receptors that induce killing events. FcγRIIIa is considered as the key receptor on natural killer cells (NK cells) and other effector cells to mediate ADCC activity. Studies have shown a strong correlation between the clinical efficacy of therapeutic antibodies and polymorphism of FcγRIIIa, in which the patients with the higher affinity genotype (V158) had much better clinical response to therapeutic antibodies than patients with lower affinity type (F158). Pioneering work demonstrated that a better clinical response to rituximab (anti-CD20) was associated with the 158 valine (V) allotype of FcγRIIIa, a key receptor for mediating ADCC activity [10]. Similar results have been obtained in at least four more cohorts of patients with non-Hodgkin's lymphoma treated with rituximab [19]. In solid tumors, it has also been reported that a better response to trastuzumab is associated with the 158V/V genotype [20]. This correlation provides strong evidence linking clinical outcome to ADCC response to antibody treatment. Similar findings have been published that associate the 158V/V genotype of Fc RIIIa with a better response to cetuximab [21].

FcγRIIb functions as an inhibitory receptor and is part of a regulatory mechanism that dampens the activation of naive B cells, mast cells, macrophages, and neutrophils. It was shown *in vivo* that the downregulation of FcγRIIb binding is desirable for optimal antitumor antibody efficacy. Mice deficient in FcγRIIb can induce more efficient antibody-directed tumor killing than wild type [22]. Furthermore, clinical data demonstrated that mantle cell lymphoma (MCL) patients with FcγRIIb^{+ve} lymphoma treated with anti-CD20 antibody-containing regimes have significantly reduced progression-free survival (PFS) compared to FcγRIIb^{-ve} patients [23]. These data indicate that inhibitory receptors have a negative impact on the efficacy of antibody drugs and suggest that antibodies with enhanced affinity to FcγRIIIa and reduced affinity to FcγRIIb might have significantly improved efficacy against tumor targets.

ADCC enhancement has been achieved through a variety of techniques in antibody Fc engineering including mutagenesis and glycol engineering [24–27]. Both methodologies seek to improve Fc binding to low-affinity activating FcγRIIIa and/or to reduce binding to the inhibitory FcγRIIb.

In comparison with the approach of introducing mutations into the Fc region, modification of glycosylation has a lower risk of causing antibody instability or introducing immunogenic structural elements. Because of this, ADCC enhancement through glycoengineering is becoming the preferred technology platform. Figure 20.1 shows an example of tumor cell killing by ADCC-enhanced anti-HER2 antibody by recruiting freshly isolated human PBMC possessing the V158/V158 genotype. This ADCC-enhanced antibody consistently demonstrates higher cell lysis than the unmodified antibody, with its efficacy directly correlated with antibody concentration. More importantly, dramatically improved tumor cell killing is observed with the human PBMC

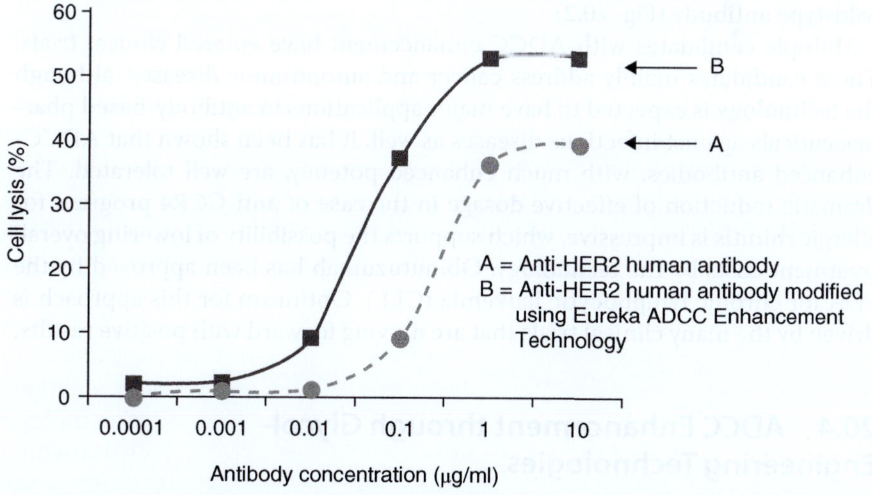

Figure 20.1 Comparison of ADCC-mediated tumor cell killing by enhanced and untreated anti-HER2 antibody preparations. The antibody was treated with freshly recruited, isolated human peripheral blood mononuclear cells possessing the V158/V158 genotype.

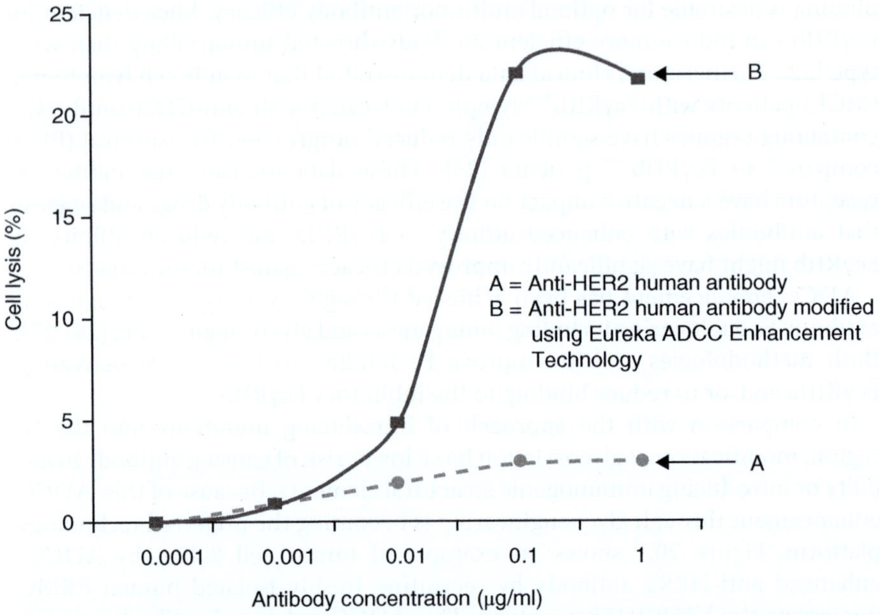

Figure 20.2 Dramatically improved tumor cell killing observed with human peripheral blood mononuclear cells possessing the 158F/158F genotype. Note that minimal cell killing was observed with the wild-type antibody.

possessing the 158F/158F genotype, which showed minimal activity with the wild-type antibody (Fig. 20.2).

Multiple candidates with ADCC enhancement have entered clinical trials. These candidates mainly address cancer and autoimmune diseases, although the technology is expected to have major applications in antibody-based pharmaceuticals against infectious diseases as well. It has been shown that ADCC-enhanced antibodies, with much enhanced potency, are well tolerated. The dramatic reduction of effective dosage in the case of anti-CCR4 program for allergic rhinitis is impressive, which supports the possibility of lowering overall treatment costs by the technology. Obinutuzumab has been approved by the FDA for chronic lymphocytic leukemia (CLL). Optimism for this approach is driven by the many clinical trials that are moving forward with positive results.

20.4 ADCC Enhancement through Glycol-Engineering Technologies

Human IgG1 bears two N-linked biantennary complex-type oligosaccharides covalently attached at the conserved asparagine 297 (Asn297), which is in the CH2 domain of the Fc region. These oligosaccharides are composed of a mannosyl-chitobiose core structure in the presence or absence of a core fucose, a

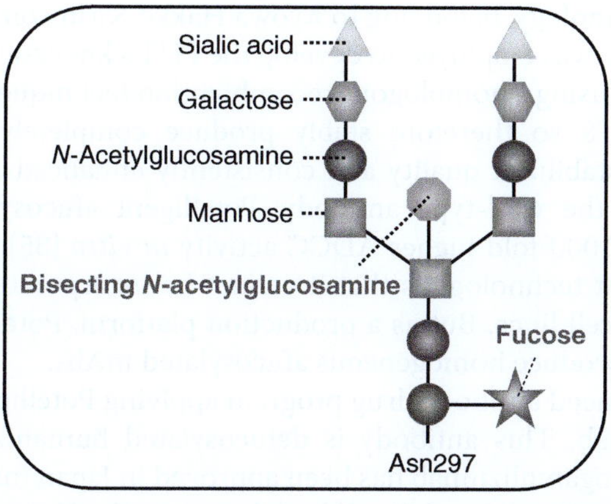

Figure 20.3 Schematic illustrating human IgG1 bearing two N-linked biantennary complex-type oligosaccharides covalently attached at the conserved asparagine 297 (Asn297). This site is located in the CH2 domain of the Fc region. Note composition containing a mannosyl-chitobiose core structure in the presence or absence of a core fucose, a bisecting *N*-acetylglucosamine (GlcNAc), and a terminal galactose and sialic acid.

bisecting *N*-acetylglucosamine (GlcNAc), and a terminal galactose and sialic acid (Fig. 20.3). This structure gives rise to heterogeneity with a mixture of 30 or more N-linked glycoforms [28, 29]. Furthermore, the presence of oligosaccharides and its glycoforms is critical for induced effector functions through their influence on the major interaction of the Fc domain with its Fc receptor counterpart [30]. Crystal structure analysis shows that Fc-linked oligosaccharides influence the conformation of the Fc domain via multiple noncovalent interactions with the CH2 domains. The innermost L-fucose has the most critical role in enhancing ADCC activity, and its elimination markedly increases ADCC activity by improving FcγRIIIa binding, which is independent of the FcγRIIIa functional polymorphism V158F [25, 31]. Antibodies with highly defucosylated oligosaccharides were shown to exhibit approximately 100-fold higher ADCC activity compared to their fully fucosylated counterparts in various antigen/antibody combinations, while possessing enhanced *in vivo* antitumor activity [32, 33].

20.5 Major ADCC Enhancement through Glycol-Engineering Technologies

20.5.1 Kyowa Hakko: Potelligent Technology

In mammals, the fucose on the antibody's N-linked oligosaccharides locates to the innermost GlcNAc residue via an α-1,6 linkage [34]. α-1,6 Fucosyltransferase catalyzes the transfer of fucose from GDP-fucose to the GlcNAc residue in an α-1,6 linkage in the medial Golgi cisternae. FUT8, encoding the only α-1,6-fucosyltransferase in mammals, determines the percentage of fucosylated antibodies in mammalian antibody-producing cells [35].

Potelligent technology, belonging to Kyowa Hakko Kirin company, and later licensed to Lonza, was employed to develop the FUT8 knockout CHO cell line by gene targeting using a homologous recombination technique. These cells do not express FUT8 so therefore stably produce completely defucosylated antibodies with stabilized quality and consistently enhanced ADCC activity. Compared with the wild-type antibody, Potelligent afucosylated antibody exhibited 10- to 1000-fold higher ADCC activity *in vitro* [35]. The disadvantage of Potelligent technology is that it cannot be incorporated into existing mAb-producing cell lines. But as a production platform, Potelligent technology can reliably produce homogeneous afucosylated mAbs.

The most advanced antibody drug program applying Potelligent technology is mogamulizumab. This antibody is defucosylated humanized IgG1 mAb against CCR4. Mogamulizumab has been approved in Japan for relapsed adult T-cell leukemia and peripheral T-cell lymphoma. In the United States, mogamulizumab is in phase III clinical trial for the same indication.

20.5.2 ProBioGen: GlymaxX Technology

The GlymaxX technology for production of afucosylated antibodies was developed by ProBioGen AG, a Germany-based biotech company. This technology is designed to configure the stable integration of a heterologous bacterial enzyme into any antibody producer cell line. This redesigning of the host cell line causes the interference of the intracellular fucose biosynthesis pathway. Antibodies generated from such cell lines are, therefore, defucosylated due to lack of fucose in the cytosol.

In eukaryotic cells, fucose is generated through two routes: (i) from the extracellular space or lysosome through the salvage pathway and (ii) synthesized from mannose through the *de novo* pathway (Fig. 20.4) [36]. Omission of fucose from the culture medium is sufficient to completely block the salvage pathway. To block *de novo* fucose biosynthesis, the GlymaxX approach attacks GDP-4-keto-6-deoxy-D-mannose, which is an unstable intermediate of GDP-fucose in the fucose *de novo* pathway. Using a bacterial enzyme, GDP-6-deoxy-D-lyxo-4-hexulose reductase (RMD), GDP-4-keto-6-deoxy-D-mannose is converted to GDP-D-rhamnose instead of GDP-4-keto-6-deoxy-D-galactose, the precursor of GDP-fucose. The conversion of GDP-4-keto-6-deoxy-D-mannose to GDP-D-rhamnose by RMD proceeds quantitatively because no reverse reaction has been detected [37]. GDP-rhamnose, a 6-deoxyhexose found only in glycoconjugates of bacteria but not in animals [38], is a dead-end product within the context of the vertebrate cytosol. Vertebrate cells lack rhamnosyltransferase and membrane transporters [38], so GDP-D-rhamnose is not able to be incorporated into nascent glycans.

The advantage of GlymaxX technology is that it can be applied to both novel and existing antibody-producing cell lines, as well as to entire production platforms without negatively affecting antibody productivity. On the other hand,

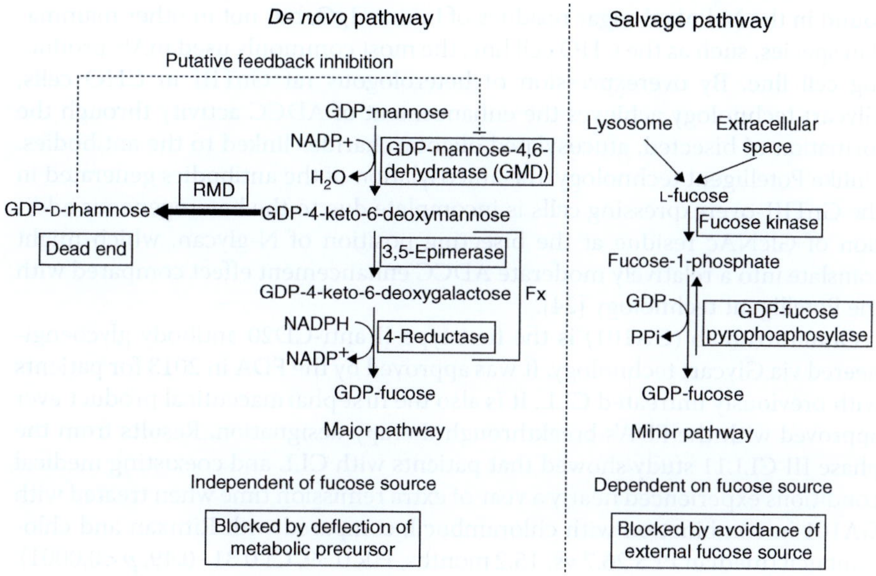

Figure 20.4 Schematic outlining GlycoMab technology. GlycoMab produces defucosylated mAbs in target cell lines through the introduction of a bisecting β-1,4-GlcNAc residue to N-linked oligosaccharides. Subsequently, other central reactions including core fucosylation and conversion of the hybrid to complex glycans are blocked.

when RMD was transfected into an existing antibody-producing cell line, the fucose levels and antibody N-glycan profile varied among different clones [36]. With a wide range of clonal isolates, it is important to carefully examine the N-oligosaccharide profiles including fucose levels to select the most desirable clone.

The GlymaxX technology has been licensed to Boehringer Ingelheim and integrated into the BI-HEX manufacturing platform. Meanwhile, Novartis has also licensed GlymaxX technology for an undisclosed number of clinical and preclinical development programs.

20.5.3 Glycart/Roche: GlycoMab

GlycoMab technology is developed by the Swiss biotech company, Glycart, which was later acquired by Roche. GlycoMab produces defucosylated mAbs in CHO cells through the introduction of a bisecting β-1,4-GlcNAc residue to N-linked oligosaccharides (Fig. 20.4). Once the bisecting GlcNAc is added to an oligosaccharide, other central reactions of the biosynthetic pathway such as core fucosylation and conversion of hybrid to complex glycans are blocked [24].

β-1,4-Acetylglucosaminyltransferase III (GnTIII) is a Golgi-localized enzyme and catalyzes the addition of a GlcNAc residue to a bisecting position of N-linked oligosaccharide chains [39]. The bisecting GlcNAc is commonly found in the N-linked sugar residues of human IgG, but not in other mammalian species, such as the CHO cell line, the most commonly used mAb-producing cell line. By overexpression of heterologous rat GnTIII in CHO cells, Glycart technology achieves the enhancement of ADCC activity through the formation of bisected, afucosylated oligosaccharides linked to the antibodies. Unlike Potelligent technology, the afucosylation of the antibodies generated in the GnTIII-overexpressing cells is incomplete due to the heterogeneous addition of GlcNAc residue at the bisecting position of N-glycan, which might translate into a relatively moderate ADCC enhancement effect compared with the Potelligent technology [24].

Obinutuzumab (GA101) is the first type II anti-CD20 antibody glycoengineered via Glycart technology. It was approved by the FDA in 2013 for patients with previously untreated CLL. It is also the first pharmaceutical product ever approved with the FDA's breakthrough therapy designation. Results from the phase III CLL11 study showed that patients with CLL and coexisting medical conditions experienced nearly a year of extra remission time when treated with GA101 in combination with chlorambucil, compared with Rituxan and chlorambucil (median PFS 26.7 vs. 15.2 months, HR 0.39, CI 0.31–0.49, $p < 0.0001$), meaning patients were in remission for nearly a year longer than the Rituxan control group [40]. As the first FDA-approved ADCC-enhanced mAb drug, obinutuzumab represents the unanimous approval of the promising future of ADCC enhancement technology.

20.5.4 Seattle Genetics: SEA Technology

Using modified sugars that inhibit the incorporation of fucose into the oligosaccharide chains of mAbs, SEA technology, developed by Seattle Genetics, generates defucosylated mAbs with enhanced ADCC activity. The advantages of SEA technology is that the modified sugars can be readily added to standard cell culture media without any genetic manipulation of the antibody-producing cells. This allows substantial saving of time and resources without impacting manufacturing processes or antibody production [41].

An anti-CD40 mAb, SEA-CD40, is now entering phase I clinical study.

20.5.5 Merck: GlycoFi

Several ADCC enhancement technology platforms were developed using yeast strains. Among them, a good example is GlycoFi, acquired by Merck in 2006. GlycoFi's glycol-engineering technology allows the generation of yeast strains capable of replicating the most essential steps of the N-glycosylation pathway found in mammals and the large-scale production of proteins with complex N-glycan modifications, including afucosylated mAbs. More homogeneous glycosylation patterns are observed, as opposed to the large heterogeneity of glycan moieties that is found naturally in other production systems, such as

CHO cell lines. The consistent and homogeneous product is essential given the industrialization of the manufacturing process [42].

20.5.6 Eureka Therapeutics: MAGE Technology

MAGE technology is an N-oligosaccharide-modified CHO cell line generated by chemical-induced mutagenesis of CHO cells. It was developed by Eureka Therapeutics. The mAbs produced by MAGE cells exhibit at least a 10-fold ADCC enhancement compared with their wild-type counterpart antibodies. This mutant cell line bears a unique profile of lectin staining and has comparable cell growth to the parental wild-type CHO cells. The cell line is genetically stable, as the lectin staining profile remained the same through more than 60 passages. The N-linked oligosaccharides synthesized by the wild-type CHO cells consist of three major species, G0, G1, and G2. On the other hand, the N-glycan produced by MAGE cells has a single, homogeneous peak, which is unique among the glycol-engineered mammalian antibody-producing cell lines. Monosaccharide analysis indicates MAGE N-glycan consists of only mannose, glucose, and GlcNAc, but lacks galactose and fucose. Interestingly, mAbs produced by MAGE cells not only have increased binding affinity to the activating receptor FcRIIIa but also decreased binding affinity to the inhibitory receptor FcRIIb. Similar to the behavior toward FcRIIb, the CDC activity of MAGE antibodies is also decreased [43].

20.6 ADCC Enhancement through Fc Mutagenesis

The effector functions of an antibody can be manipulated by engineering the amino acid sequences of the constant regions. In particular, facilitating FcγIIIa binding by mutating antibody sequences is of potential therapeutic value. Indeed, extensive studies have been performed on the effects of amino acid alterations in the Fc that possess improved FcRIIIa binding on the resulting ADCC-enhancing capacity using random or rational designing approaches, such as alanine scanning, computation design, and yeast display [26, 27, 44].

20.7 Major ADCC Enhancement Fc Mutagenesis Technologies

20.7.1 Xencor: XmAb

Combining “directed diversity” and “quality diversity” strategies, the human IgG1 Fc region was computationally optimized for changing FcγR affinity and specificity. When structural information was available (e.g., the Fc/FcγRIII complex), the affinity was directly optimized by designing substitutions that provided more favorable interactions at the Fc/FcγR interface. When structural information was incomplete or lacking (e.g., the Fc/FcγRIIb complex), computational calculations provided a quality set of variants enriched for antibody stability and solubility [27].

Then, with high-throughput screening, a series of Fc variants with optimized FcγR affinity and specificity were generated. Two variants, S239D/I332E and S239D/I332E/A330L, were selected for further analysis because of their dramatically increased binding affinity toward FcγRIIIa, both F158 and V158 alleles. Although both variants also showed significantly increased binding affinity against FcγRIIb, they achieved two to three logs improvement in ADCC potency, reflected by EC50. The ADCC enhancement was not only observed in cell lines expressing elevated levels of the target antigen but also in the low-expressing cell lines. Moreover, the enhanced effector function was found to be mediated by multiple effector cells expressing FcγRIIIa, including NK cells (the major player of ADCC) and phagocytes, such as macrophages, neutrophils, and dendritic cells [27].

Besides enhanced ADCC activity, the S239D/I332E variant retains CDC activity at a level comparable to wild-type Fc. But addition of the A330L mutation ablates CDC activity completely [27].

In many cases, differences between human and mouse FcγRs complicate the use of mouse cancer models for evaluating the human IgG1 ADCC enhancement technology *in vivo*. It was shown that the S239D/I332E and S239D/I332E/A330L variants provide significant improvements in binding to the mouse activating receptor FcγRIII [27].

An anti-CD19 mAb, developed by XmAb technology through a collaboration between Xencor and MorphoSys, is currently in phase II clinical trial against CLL. It has also been licensed to Boehringer Ingelheim, with two programs in clinical trials.

20.7.2 MacroGenics

A comprehensive functional genetic screening of the human IgG1 Fcγ region was undertaken to elucidate the contribution of different amino acid residues to the binding to low-affinity FcγR through yeast surface display. Yeast display was done on a random library of approximately 1×10^7 Fc mutants by using both equilibrium and kinetic screening strategies for binding to a PE-conjugated soluble human FcγRIIIa or FcγRIIa as ligands. To enrich Fc variants with limited or no binding potential to the inhibitory receptor FcγRIIb, the library was negatively selected for binders binding to FcγRIIb and positively selected for binders binding to FcγRIIIa or FcγRIIa.

In total, over 400 amino acid changes were identified from the Fc mutant libraries spanning both the CH2 and CH3 domains. Comparison of off-rates (k_{off}) for selected Fc variants was done to identify an Fc domain optimized for improved binding to the activating receptor FcγRIIIa or FcγRIIa and reduced binding to FcγRIIb. An Fc variant combining five amino acid substitutions, F243L/R292P/Y300L/V305I/P396L, was found to have the strongest binding affinity. It exhibited significantly increased affinity toward FcγRIIIa F158 and V158 alleles and mediated the highest levels of ADCC, showing an approximately 100-fold increased rate of lysis as noted by EC50 and maximum lysis. ADCC enhancement of this variant is effective for both FcγRIIIa F158 and V158 alleles, irrespective of the target antigen expression levels.

The five amino acid substitutions carried by the variant, F243L/R292P/Y300L/V305I/P396L, could only be ascertained to alter FcγR binding through a functional genetic screen. P396 in the CH3 domain is not a soluble residue nor directly interacts with the FcγR, suggesting that these residues affected ligand binding via distant conformational changes. Position F243 directly contacts the carbohydrate moiety and affects sialylation of the oligosaccharide chains, potentially affecting the Fc domain quaternary structure [45, 46]. V305 showed no change in binding to any of the low-affinity receptors and R292 only partially reduced binding to CD32A. Putting five amino acid substitutions together, the variant showed increased binding to the low-affinity activating FcγRIIIa, both 158F and 158V alleles, and reduced binding to FcγRIIb, the low-affinity inhibitory receptor [44].

MacroGenics is actively recruiting patients for clinical studies of its Fc-optimized mAb candidate margetuximab, which targets HER2 (phase II study), and its second Fc-optimized product candidate, MGA271, targeting B7-H3 (phase I study) [47].

20.8 Conclusion

ADCC enhancement is emerging as a key strategy for improving therapeutic antibody efficacy. Recent clinical studies provide further evidence in support of this technology. It has the potential of lowering effective drug dosage, thereby benefiting patients through lower drug costs. Antibodies with ADCC enhancement may eliminate variations in patient response to antibody treatments caused by genetic polymorphism and improve survival of cancer patients.

The commercial value of the technology has been demonstrated by the licensing agreement between Amgen and Kyowa Hakko Kirin for an anti-CCR4 antibody, which includes $100 million upfront fee plus $420 million milestone payment and double digit royalties. The FDA approval of obinutuzumab and its superior efficacy compared with the first-generation anti-CD20 antibody, rituximab, in leukemia was a milestone in the application of ADCC enhancement technology to new mAb drug discovery and development.

ADCC enhancement has been achieved through a variety of techniques in antibody Fc engineering including mutagenesis and glycol engineering. It is now a core technology for developing a next generation of therapeutic antibodies with greatly improved clinical outcomes.

References

1. Casadevall A, Scharff MD. (1995). Return to the past: the case for antibody-based therapies in infectious diseases. *Clin. Infect. Dis.* Jul;21(1):150–161.
2. Zeitlin L, Cone RA, Moench TR, Whaley KJ. (2000). Preventing infectious disease with passive immunization. *Microbes Infect.* May;2(6):701–708.
3. Trill JJ, Shatzman AR, Ganguly S. (1995). Production of monoclonal antibodies in COS and CHO cells. *Curr. Opin. Biotechnol.* Oct;6(5):553–560.
4. Owens RJ, Young RJ. (1994). The genetic engineering of monoclonal antibodies. *J. Immunol. Methods* Feb 10;168(2):149–165.
5. Presta LG. (2006). Engineering of therapeutic antibodies to minimize immunogenicity and optimize function. *Adv. Drug Deliv. Rev.* Aug 7;58(5–6):640–656.
6. Waldmann TA, Morris JC. (2006). Development of antibodies and chimeric molecules for cancer immunotherapy. *Adv. Immunol.* 90:83–131.
7. Weiner LM. (2006). Fully human therapeutic monoclonal antibodies. *J. Immunother.* Jan–Feb;29(1):1–9.
8. Moutel S, Perez F. (2008). Antibodies—Europe. Engineering the next generation of antibodies. *Biotechnol. J.* Mar;3(3):298–300.
9. Reichert JM, Valge-Archer VE. (2007). Development trends for monoclonal antibody cancer therapeutics. *Nat. Rev. Drug Discov.* May;6(5):349–356.
10. Cartron G, Dacheux L, Salles G, Solal-Celigny P, Bardos P, Colombat P, Watier H. (2002). Therapeutic activity of humanized anti-CD20 monoclonal antibody and polymorphism in IgG Fc receptor FcgammaRIIIa gene. *Blood* Feb 1;99(3):754–758.
11. Gennari R, Menard S, Fagnoni F, Ponchio L, Scelsi M, Tagliabue E, Castiglioni F, Villani L, Magalotti C, Gibelli N, Oliviero B, Ballardini B, Da Prada G, Zambelli A, Costa A. (2004). Pilot study of the mechanism of action of preoperative trastuzumab in patients with primary operable breast tumors overexpressing HER2. *Clin. Cancer Res.* 10:5650–5655
12. Weng WK, Levy R. (2003). Two immunoglobulin G fragment C receptor polymorphisms independently predict response to rituximab in patients with follicular lymphoma. *J. Clin. Oncol.* 21:3940–3947.
13. Dall'Ozzo S, Tartas S, Paintaud G, Cartron G, Colombat P, Bardos P, Watier H, Thibault G. (2004). Rituximab-dependent cytotoxicity by natural killer cells: influence of *FCGR3A* polymorphism on the concentration–effect relationship. *Cancer Res.* 64:4664–4669.

14. Taylor RP, Lindorfer MA. (2008). Immunotherapeutic mechanisms of anti-CD20 monoclonal antibodies. *Curr. Opin. Immunol.* 20(4):444–449.
15. Weiner GJ. (2007). Monoclonal antibody mechanisms of action in cancer. *Immunol. Res.* 39(1–3):271–278.
16. Cohen-Solal JF, Cassard L, Fridman WH, Sautes-Fridman C. (2004). Fc gamma receptors. *Immunol. Lett.* 92:199–205.
17. Jefferis R, Lund J. (2002). Interaction sites on human IgG-Fc for FcgammaR: current models. *Immunol. Lett.* 82:57–65.
18. Raghavan M, Bjorkman PJ. (1996). Fc receptors and their interactions with immunoglobulins. *Annu. Rev. Cell Dev. Biol.* 12:181–220.
19. Musolino A, Naldi N, Bortesi B, Pezzuolo D, Capelletti M, Missale G, Laccabue D, Zerbini A, Camisa R, Bisagni G, Neri TM, Ardizzoni A. (2008). Immunoglobulin G fragment C receptor polymorphisms and clinical efficacy of trastuzumab-based therapy in patients with HER-2/neu-positive metastatic breast cancer. *J. Clin. Oncol.* Apr 10;26(11):1789–1796.
20. Mellor JD, Brown MP, Irving HR, Zalcberg JR, Dobrovic A. (2013). A critical review of the role of Fc gamma receptor polymorphisms in the response to monoclonal antibodies in cancer. *J. Hematol. Oncol.* Jan 4;6:1.
21. Bibeau F, Lopez-Crapez E, Di Fiore F, Thezenas S, Ychou M, Blanchard F, Lamy A, Penault-Llorca F, Frébourg T, Michel P, Sabourin JC, Boissière-Michot F. (2009). Impact of Fc{gamma}RIIa-Fc{gamma}RIIIa polymorphisms and KRAS mutations on the clinical outcome of patients with metastatic colorectal cancer treated with cetuximab plus irinotecan. *J. Clin. Oncol.* Mar 1;27(7):1122–1129.
22. Clynes RA, Towers TL, Presta LG, Ravetch JV. (2000). Inhibitory Fc receptors modulate *in vivo* cytotoxicity against tumor targets. *Nat. Med.* 6:443–446.
23. Lim SH, Vaughan AT, Ashton-Key M, Williams EL, Dixon SV, Chan HT, Beers SA, French RR, Cox KL, Davies AJ, Potter KN, Mockridge CI, Oscier DG, Johnson PW, Cragg MS, Glennie MJ. (2011). Fc gamma receptor IIb on target B cells promotes rituximab internalization and reduces clinical efficacy. *Blood* Sep 1;118(9):2530–2540.
24. Umana P, Jean-Mairet J, Moudry R, Amstutz H, Bailey JE. (1999). Engineered glycoforms of an antineuroblastoma IgG1 with optimized antibody-dependent cellular cytotoxic activity. *Nat. Biotechnol.* 17:176–180.
25. Shinkawa T, Nakamura K, Yamane N, Shoji-Hosaka E, Kanda Y, Sakurada M, Uchida K, Anazawa H, Satoh M, Yamasaki M, Hanai N, Shitara K. (2003). The absence of fucose but not the presence of galactose or bisecting *N*-acetylglucosamine of human IgG1 complex-type oligosaccharides shows the critical role of enhancing antibody-dependent cellular cytotoxicity. *J. Biol. Chem.* 278:3466–3473.
26. Shields RL, Namenuk AK, Hong K. (2001). High resolution mapping of the binding site on human IgG1 for FcgRI, FcgRII, FcgRIII, and FcRn and design of

IgG1 variants with improved binding to the FcgR. *J. Biol. Chem.* 276:6591–6604.

27. Lazar GA, Dang W, Karki S. (2006). Engineered antibody Fc variants with enhanced effector function. *Proc. Natl. Acad. Sci. U. S. A.* 103:4005–4010.
28. Kubota T, Niwa R, Satoh M, Akinaga S, Shitara K, Hanai N. (2009). Engineered therapeutic antibodies with improved effector functions. *Cancer Sci.* Sep;100(9):1566–1572.
29. Rademacher TW, Parekh RB, Dwek RA. (1988). Glycobiology. *Annu. Rev. Biochem.* 57:785–838. Review.
30. Wright A, Morrison SL. (1997). Effect of glycosylation on antibody function: implications for genetic engineering. *Trends Biotechnol.* Jan;15(1):26–32. Review.
31. Niwa R, Hatanaka S, Shoji-Hosaka E, Sakurada M, Kobayashi Y, Uehara A, Yokoi H, Nakamura K, Shitara K. (2004). Enhancement of the antibody-dependent cellular cytotoxicity of low-fucose IgG1 Is independent of FcgammaRIIIa functional polymorphism. *Clin. Cancer Res.* Sep 15;10(18 Pt 1):6248–6255.
32. Niwa R, Shoji-Hosaka E, Sakurada M, Shinkawa T, Uchida K, Nakamura K, Matsushima K, Ueda R, Hanai N, Shitara K. (2004). Defucosylated chimeric anti-CC chemokine receptor 4 IgG1 with enhanced antibody-dependent cellular cytotoxicity shows potent therapeutic activity to T-cell leukemia and lymphoma. *Cancer Res.* Mar 15;64(6):2127–2133.
33. Niwa R, Sakurada M, Kobayashi Y, Uehara A, Matsushima K, Ueda R, Nakamura K,Shitara K. (2005). Enhanced natural killer cell binding and activation by low-fucose IgG1 antibody results in potent antibody-dependent cellular cytotoxicity induction at lower antigen density. *Clin. Cancer Res.* Mar 15;11(6):2327–2336.
34. Miyoshi E, Uozumi N, Noda K, Hayashi N, Hori M, Taniguchi N. (1997). Expression of alpha1–6 fucosyltransferase in rat tissues and human cancer cell lines. *Int. J. Cancer* Sep 17;72(6):1117–1121.
35. Yamane-Ohnuki N, Kinoshita S, Inoue-Urakubo M, Kusunoki M, Iida S, Nakano R, Wakitani M, Niwa R, Sakurada M, Uchida K, Shitara K, Satoh M. (2004). Establishment of FUT8 knockout Chinese hamster ovary cells: an ideal host cell line for producing completely defucosylated antibodies with enhanced antibody-dependent cellular cytotoxicity. *Biotechnol. Bioeng.* Sep 5;87(5):614–622.
36. von Horsten HH, Ogorek C, Blanchard V, Demmler C, Giese C, Winkler K, Kaup M, Berger M, Jordan I, Sandig V. (2010). Production of non-fucosylated antibodies by co-expression of heterologous GDP-6-deoxy-D-lyxo-4-hexulose reductase. *Glycobiology* Dec;20(12):1607–1618.
37. Kneidinger B, Graninger M, Adam G, Puchberger M, Kosma P, Zayni S, Messner P. (2001). Identification of two GDP-6-deoxy-D-lyxo-4-hexulose

reductases synthesizing GDP-D-rhamnose in *Aneurinibacillus thermoaerophilus* L420-91T. *J. Biol. Chem.* Feb 23;276(8):5577–5583.

38. Webb NA, Mulichak AM, Lam JS, Rocchetta HL, Garavito RM. (2004). Crystal structure of a tetrameric GDP-D-mannose 4,6-dehydratase from a bacterial GDP-D-rhamnose biosynthetic pathway. *Protein Sci.* Feb;13(2):529–539.

39. Narasimhan S. (1982). Control of glycoprotein synthesis. UDP-GlcNAc:glycopeptide beta 4-*N*-acetylglucosaminyltransferase III, an enzyme in hen oviduct, which adds GlcNAc in beta 1–4 linkage to the beta-linked mannose of the trimannosyl core of *N*-glycosyl oligosaccharides. *J. Biol. Chem.* Sep 10;257(17):10235–10242.

40. Rioufol C, Salles G. (2014). Obinutuzumab for chronic lymphocytic leukemia. *Expert Rev. Hematol.* Oct;7(5):533–543.

41. Okeley NM, Toki BE, Zhang X, Jeffrey SC, Burke PJ, Alley SC, Senter PD. (2013). Metabolic engineering of monoclonal antibody carbohydrates for antibody–drug conjugation. *Bioconjugate Chem.*, 24(10):1650–1655.

42. Beck A, Cochet O, Wurch T. (2010). GlycoFi's technology to control the glycosylation of recombinant therapeutic proteins. *Expert Opin. Drug Discov.* Jan;5(1):95–111.

43. Modified glycoproteins and uses thereof. Cheng Liu, Jingyi Xiang, Su Yan, Pei Wang, Wei Chan Tai. Patent Number 8025879 (US).

44. Stavenhagen JB, Gorlatov S, Tuaillon N, Rankin CT, Li H, Burke S, Huang L, Vijh S, Johnson S, Bonvini E, Koenig S. (2007). Fc optimization of therapeutic antibodies enhances their ability to kill tumor cells *in vitro* and controls tumor expansion *in vivo* via low-affinity activating Fcgamma receptors. *Cancer Res.* Sep 15;67(18):8882–8890. Erratum in: *Cancer Res.* 2008 Sep 15;68(18):7692.

45. Deisenhofer J. (1981). Crystallographic refinement and atomic models of a human Fc fragment and its complex with fragment B of protein A from *Staphylococcus aureus* at 2.9- and 2.8-A resolution. *Biochemistry* Apr 28;20(9):2361–2370.

46. Lund J, Takahashi N, Pound JD, Goodall M, Jefferis R. (1996). Multiple interactions of IgG with its core oligosaccharide can modulate recognition by complement and human Fc gamma receptor I and influence the synthesis of its oligosaccharide chains. *J. Immunol.* Dec 1;157(11):4963–4969.

47. Loo D, Alderson RF, Chen FZ, Huang L, Zhang W, Gorlatov S, Burke S, Ciccarone V, Li H, Yang Y, Son T, Chen Y, Easton AN, Li JC, Rillema JR, Licea M, Fieger C, Liang TW, Mather JP, Koenig S, Stewart SJ, Johnson S, Bonvini E, Moore PA. (2012). Development of an Fc-enhanced anti-B7-H3 monoclonal antibody with potent antitumor activity. *Clin. Cancer Res.* Jul 15;18(14):3834–3845.

21

Antibody Half-Life

Engineering for Optimal Performance

K. John Morrow, Jr.

Newport Biotech Consultants, Newport, KY, USA

21.1 Summary

Biosimilar antibodies include a subcategory of "biobetters," that is, antibodies that actually represent an improvement over the original product. The loss of patent protection and technological improvements is predicted to generate a wave of biosimilar products in the coming years that will force deep inroads into the market share of the original products. The competition brought about through the introduction of biosimilars may cut costs of antibody-based therapeutics by 50% or more.

Because the IgG framework is able to sustain great variability, it lends itself to a range of engineering possibilities. Advances in cloning and engineering technologies accomplished in recent years have opened the door to improving antibody performance, in particular increasing the half-life of therapeutic mAbs. In this chapter, we consider technologies for adjusting the half-life of antibodies to perform optimally according to their therapeutic goals.

Fundamental investigations of the mechanism of half-life extension have shown that the complexing of the Fc region of the antibody with the FcRn protein is critical to the prolongation of serum half-life. This receptor protects the IgG molecule from cellular catabolism by way of a pH-dependent recycling and transcytosis pathway.

Antibody half-life can be extended by increasing the affinity of the antibody for the FcRn molecule. Extensive molecular mapping studies have revealed the mechanism of binding of the two molecules and the best mutational substitutions to the Fc region to optimize affinity. Alternatively, in some cases it may be desirable to interfere with the binding in order to shorten the half-life of the

Biosimilars of Monoclonal Antibodies: A Practical Guide to Manufacturing, Preclinical, and Clinical Development, First Edition. Edited by Cheng Liu and K. John Morrow, Jr.

antibody. This is the strategy that is being applied in therapeutic application to various autoimmune diseases.

The FcRn molecule complexes with immunoglobulin and also with albumin. This property allows for the development of stable albumin fusion molecules with notable therapeutic applications.

As new biosimilars are developed, some preclinical and clinical evaluation will be required in order to establish their safety and efficacy. Since preclinical studies involve the use of mice, the recent controversy concerning the accuracy of the mouse as a model for human disease conditions should be mentioned. While there is a reason for concern, the weight of the evidence indicates that the mouse is an accurate model of the human immune system, with certain caveats.

One of the most promising approaches to protecting the half-life of engineered antibodies is the use of nanotechnological devices. In its most fundamental state, the design concept is a drug-containing core in a soft sheath covered with polyethylene glycol, which attracts water molecules surrounding the nanoparticle with a liquid halo. Currently there are a number of nanotech applications that have received FDA approval, although none involving an antibody therapeutic.

Despite the problems associated with the development of biosimilars and biobetters, the future of the industry appears bright. There is an extremely robust market for biologics, now 27% of pharmaceutical sales in Europe. In the United States alone, there are 907 biologics in development, targeting more than 100 diseases. Extension of the half-life of these molecules could constitute a direct and rapid approach to improving their performance, so it is certain that this will be a major part of an R&D biobetter program.

21.2 Introduction

21.2.1 The Rise of Antibody Therapeutics

FDA-approved monoclonal antibodies (mAbs) comprise a wide range of therapies, targeting cancer, autoimmune dysfunction, infectious disease, neurodegenerative conditions, macular degeneration, osteoporosis, and transplant rejection [1]. The path to approval is arduous, and there are numerous examples of antibodies that fail in trials, even at the late stages of development. Gantenerumab, evaluated for the treatment of Alzheimer's disease, recently suffered such a fate [2]. But for antibodies that reach the marketplace and perform well, the rewards can be staggering. One example is Humira, manufactured by Abbott. This antibody, used in the treatment of chronic inflammatory diseases, was projected to return $11 billion in sales in 2014.

Humira, Enbrel, and Remicade all work through the same mechanism, by blocking tumor necrosis factor-α, and therefore they are referred to as TNF-α inhibitors (or blockers). Rituxan, on the other hand, blocks CD20. Because these agents target specific steps in the process, they do not wipe out the entire immune response as other traditional therapies do.

21.2.2 Biosimilars Come Online

As many therapeutic antibodies reach the end of their patent protection period (e.g., Humira's patent protection will end in 2016), there is great interest in developing biosimilars, as spelled out in many chapters of this report. Analysts predict that at year 3 (2019), biosimilars will achieve 36% US market share in Humira patients and 25% in established Humira patients. Accordingly, Humira revenues are slated to decline from a peak of $16 billion in 2017 to just $6 billion in 2022 [3].

A recent report in Fierce Pharma reinforces this conclusion. Novartis was the first company to gain FDA approval for a biosimilar with the introduction of a biosimilar to Neupogen, which fights infections in cancer patients. A federal judge has denied Amgen's request for a temporary injunction against its release, which may now move ahead. Amgen has threatened that it might bring legal action against Novartis based on the contention that the biosimilar, Zarxio, violates patents of Neupogen.

There have been various estimates of the biosimilar market, ranging from $35 billion this decade to as high as $200 billion in others. The article states once the biosimilar marketplace is up and running, they may be discounted as much as 40–50%, way below the commonly accepted 20% estimates that have been widely circulated; and there are many biosimilar antibodies in various stages of the developmental pipeline [4].

21.2.3 Antibody Half-Life

One of the most compelling features of antibodies as therapeutics is their long half-life in the circulation, as compared to the small molecule-based pharmaceutical agents, whose half-life may be only minutes or at the most hours long [5]. Interest in the understanding and manipulation of antibody half-life centers around the neonatal Fc receptor (FcRn), whose existence was postulated a half century ago as a component of a salvage pathway to rescue IgG from catabolism [6]. FcRn was demonstrated in the rat as the receptor responsible for transporting maternal antibodies to the offspring, but it is now known to take part in a variety of functions. Investigation in a mouse model demonstrated that animals deficient in β_2m, and thus lacking operational FcRn, showed an unusually short serum IgG half-life [7]. This suggested that this system is a component of the regulation and homeostasis of antibodies, both naturally occurring and engineered.

While naturally occurring immunoglobulins and many mAbs have half-lives sufficiently long as to allow them to function effectively as therapeutics, there has been a strong push from the industry to achieve longer and longer half-lives, in order to optimize treatment regimes and meet the demands of patient convenience. This requires manipulation of the amino acid sequence of the binding region. Whereas this is not an insurmountable challenge, it must be carried out in such a manner that the binding of the antibody to its designated antigen will not be compromised. For this reason, computer modeling ("*in silico*") has played a pivotal role in antibody engineering for half-life improvement.

21.2.4 Aims of This Chapter

In this chapter we will review the history of antibody development leading to issues of half-life improvement. We will then consider the role of FcRn molecule in the extension of antibody half-life and mechanisms whereby this performance can be augmented. The chapter concludes with a consideration of technological innovations under way, which may notably improve half-life performance in the coming years.

21.3 The IgG Molecule as a Therapeutic Entity

21.3.1 Historical Considerations

Antibodies have been thought of as potential drug candidates from the inception of modern medicine at the beginning of the twentieth century. Their specificity and ability to protect the host from pathogens suggested that these qualities could be harnessed for the treatment of disease. Moreover, the fact that they are naturally occurring molecules implied that they would raise fewer unacceptable side effects in the patient. Even early in the twentieth century, the long experience with vaccination made it clear that the immune system was a powerful firewall against disease.

But the modern era of antibody therapeutics did not commence until the discovery of mAbs by Köhler and Millstein in the 1970s. Although initially greeted with great optimism, years passed before the many problems associated with this technology were overcome. The difficulties that plagued the early years of therapeutic antibody development resulted in large part from the immunogenicity of mouse monoclonals and their rejection by human hosts. It was not until the 1990s that these issues were resolved through improvements in recombinant DNA manipulation, especially in humanizing mouse mAbs, as well as the development of mice with human immune systems. By the late 1990s a number of antibody candidates had qualified for FDA approval. Even so, the approval of new antibody products has been agonizingly slow, usually at the rate of one or two per year.

21.3.2 Current Status of Immunoglobulins and Protection Against Disease

Immunoglobulins of the IgG class represent a two-edged sword in determining the health status of the individual. On the one hand, they are pivotally responsible for long-term immunity, but on the other, they may behave as pathological effectors of autoimmune disease. Thus, natural selection has evolved two critical behaviors of this system. In the first place, IgGs are designed to be effectively transported to the neonate, to endow the newborn individual with protective immunity. The second structural demand is to ensure the optimum performance of the IgG generated in response to antigenic exposure.

Whereas non-IgG classes have relatively short half-lives of 1–2 days, the IgGs (in man) may have half-lives of 3 weeks or more [8], guaranteeing that they will persist in the circulation long enough to be effective in eliminating pathogens [9]. This homeostatic mechanism is dependent upon the Fc region of the IgG molecule.

21.3.3 MAbs and Disease

MAbs are unique among classes of proteins in that they share a common structural framework, yet each possesses a distinct epitope-binding site. The current design protocols designate therapeutic mAbs for the most part as members of the IgG classes IgG1, IgG2, and IgG4. As such they bind with high affinity to the human FcRn. This binding allows transport and is responsible for the long half-life of the immunoglobulin molecule [10].

Due to their lack of effector functions, IgG4 antibodies represent the preferred IgG subclass for receptor blocking without cell depletion. IgG4 molecules can exchange half-molecules in a dynamic process termed Fab-arm exchange. This phenomenon can occur between therapeutic antibodies and endogenous IgG4 [11].

21.4 FcRn and Antibody Half-Life

21.4.1 FcRn and IgG Stabilization

The extremely long half-life of immunoglobulins is inversely related to their concentration in the serum of the host, a property that suggested that the IgG molecules combined with a saturable receptor that protected them from intracellular degradation. This hypothesis was shown to be correct and led to the discovery of the FcRn protein, which is responsible for protecting the IgG from cellular catabolism via a rigidly pH-dependent recycling and transcytosis pathway [12]. The role of FcRn in IgG stabilization was most convincingly demonstrated in mice lacking a functional transmembrane alpha chain gene, which had reduced levels of IgG and albumin in the plasma compared with wild-type controls [13].

FcRn is assigned the task of transporting IgG across the mucosal epithelium that lines the intestine and the alveolar surfaces as well as moving IgG from the mother to the fetus through the placenta (see Fig. 21.1). It is well known that this feature bolsters the robustness of the fetus through transfer of passive immunity [14]. FcRn is expressed in vascular endothelial cells, epithelial cells, hepatocytes, intestinal macrophages, peripheral blood monocytes, and dendritic cells [15].

However, FcRn expression on the cell surface is limited, and the pH of the extracellular environment is not favorable for an IgG–FcRn interaction; therefore, it is assumed that the IgG molecules enter the cell through nonspecific, fluid-phase pinocytosis. Once inside the cell, the molecule complexes with the FcRn molecule in an acidic microenvironment. The ability to move IgG in either direction across polarized epithelial barriers is a highly conserved property through many species, arguing for its essential function in regulating the immune system. *In vivo* studies have dramatically demonstrated the process of integration of cellular immunity through the critical intervention of FcRn.

Substantial investigation has revealed that the bound complex does not degrade easily and endows the IgG molecules with their extended serum half-life [16]. The understanding of the behavior of the complex has important implications for pharmaceutical design. Internalized into the cell, the complexes

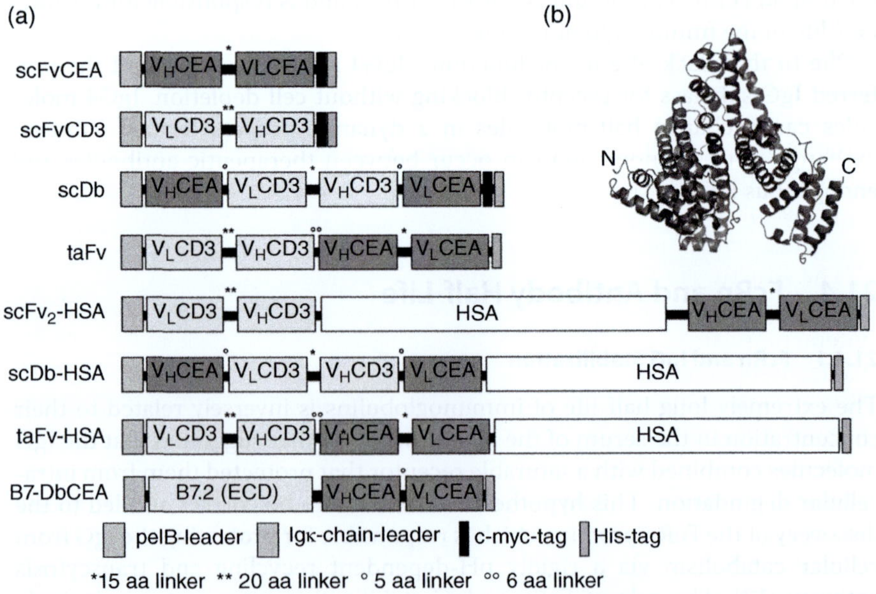

Figure 21.1 Antibody fragments and albumin fusion proteins. From Müller et al. [39].

do not break apart until they are recycled back to the cell surface where they dissociate at neutral or basic pH of the bloodstream, returning them to the circulation [9]. Understanding the nature of the FcRn/IgG interaction at the molecular level is essential for the design of rational therapeutics.

An important distinction between human and murine FcRn is the ability of the latter to bind to various species of IgG, whereas human FcRn is much more limited in its binding ability [17]. Since mice are widely employed as a model of the human immune system, these subtle distinctions need to be taken into account.

Although much of the investigation of the FcRn molecule has been aimed at extending the circulating half-life of antibody therapeutics, in the case of the treatment of autoimmune diseases, research is aimed at decreasing the half-life in the circulation. A long standing approach is the use of human immunoglobulin to saturate the FcRn-binding sites and hasten the degradation of pathogenic antibodies. Approved treatment for various diseases includes Guillain–Barre syndrome, Kawasaki disease, idiopathic thrombocytopenia purpura, and chronic inflammatory demyelinating polyneuropathy [18].

21.4.2 Molecular Structure of FcRn

The FcRn–IgG molecular contact has been studied extensively. The receptor is a heterodimeric protein consisting of 40–46 kDa and 12 kDa subunits. Cloning of the corresponding genes determined that the 12 kDa subunit was β_2-microglobulin, whereas the larger subunit was a heavy chain (HC) related to the major histocompatibility complex class I [19]. This latter molecule is obligatory for the folding transport and functioning of the complex [20]. The molecule contains three soluble domains, a single transmembrane helix, and a cytoplasmic tail. The many functions of FcRn allow it to act as a regulator of mucosal immunity through the process of bidirectional transcytosis of the IgG molecule and the luminal antigens across epithelial boundaries. In addition FcRn drives the active induction of MHC II presentation and MHC I cross-presentation pathways for the generation of antigen-specific T-cell responses [21].

The structure of the FcRn/Fc complex has been studied in detail. The ability of the FcRn to protect the IgG molecule is due to the specific pH-dependent interaction with the Fc portion of the antibody. This is mediated by electrostatic interactions occurring at a pH below 6.5 [22]. The critical sequence consists of residues (H310, H430, I253, and to a lesser extent H433 and Y436) [23] located at the CH2–CH3 domain interface of the Fc region of IgG. Two FcRn molecules bind to a single IgG molecule with a 2:1 stoichiometry. As shown by site-directed mutagenesis, the residues Ile253, His310, and His435 of the IgG molecule are critically involved in the interaction with FcRn [24]. The binding is further augmented by a group of hydrophobic interactions and H bonding between the Fc portion of the antibody and the amino acids complex within the FcRn $\alpha 2$ domains and the β_2m light chain N-terminus [25].

The binding interface is large (18.7 nm^2) and perforce highly complementary. In addition, the shape correlation statistic for the complex is high, indicative of tight binding. There is extensive interaction involving N-linked carbohydrates from FcRn that are in contact with the Fc molecule. The sugars contact four Fc residues, including His433 [22]. The simple but elegant molecular mechanism for pH-dependent binding relies completely on titrating residues on the ligand rather than the receptor so that the histidines on the Fc portion of the complex interact favorably with the negatively charged residues on the FcRn molecule at acidic but not basic pH [19]. Titration of histidines, which have a pK_a in the range of the FcRn/Fc affinity transition, has been proposed to account for the pH dependence.

21.4.3 Molecular Modeling and Modification of FcRn to Improve Performance

Because of the role that the binding of antibody to the FcRn plays in extending its half-life, strategies to increase this affinity would appear to be viable means of optimization and therefore optimizing therapeutic performance.

21.5 Optimizing Antibody Fragments' Half-Life

An important animal model for antibody evaluation is the ferret, which has been used to study the performance of influenza vaccines [26]. However, human mAbs possess an extremely short half-life in the ferret. This observation suggests that the half-life could be extended by using substitutions of amino acids in the binding position. A number of substitutions were found to be effective for this purpose [27]. Lazar et al. engineered a triple mutant (S239D/I332E/A330L) with a higher affinity for FcγRIIIa and a lower affinity for FcγRIIb, resulting in enhanced ADCC [28]. This approach also enabled Richards et al. to generate a different triple mutant (S239D/I332E/G236A) with improved FcγRIIIa affinity and FcγRIIa/FcγRIIb ratio. The construct displays greatly enhanced phagocytosis of target cells by macrophages [29].

It has been noted that when position 252, which ordinarily is occupied by a methionine, is replaced by a tyrosine, the affinity of human mAbs for the Ferret FcRn is increased. A one-residue substitution in the ferret Fc domain, S252Y, was identified that increased binding affinity to the ferret neonatal receptor by 24-fold and extended half-life from 65 ± 27 to 206 ± 28 h or approximately 9 days.

In addition to amino acid substitutions in the Fc region interface, there are a number of other approaches that have been investigated to extend the half-life of antibodies. Table 21.1 shows some of the best documented substitutions in

Table 21.1 Examples of engineered mouse and human Fc regions and their properties.

Engineered Fc	Isotype	Species	Mutations	FcR/C1q binding	Effector function
hIgG1e1-Fc	IgG1	Human	T250Q/M428L	Increased binding to FcRn	Increased half-life
hIgG1e2-Fc	IgG1	Human	1M252Y/S254T/T256E+H433K/N434F	Increased binding to FcRn	Increased half-life
hIgG1e3-Fc	IgG1	Human	E233P/L234V/L235A/?G236+A327G/A330S/P331S	Reduced binding to FcγRI	Reduced ADCC and CDC
hIgG1e4-Fc	IgG1	Human	E333A	Increased binding to FcγRIIIa	Increased ADCC and CDC
hIgG1e5-Fc	IgG1	Human	S239D/A330L/I332E	Increased binding to FcγRIIIa	Increased ADCC
hIgG1e6-Fc	IgG1	Human	P257I/Q311	Increased binding to FcRn	Unchanged half-life
hIgG1e7-Fc	IgG1	Human	K326W/E333S	Increased binding to C1q	Increased CDC
hIgG1e9-Fc	IgG1	Human	S239D/I332E/G236A	Increased FcγRIIa/FcγRIIb ratio	Increased macrophage phagocytosis
hIgG2e1-Fc	IgG1	Human	K322A	Reduced binding to C1q	Reduced CDC
hIgG4e1-Fc	IgG4	Human	S228P	—	Reduced Fab-arm exchange
mIgG2Ae1-Fc	IgG2a	Mouse	L235E+E318A/K320A/K322A	Reduced binding to FcγRI and C1q	Reduced ADCC and CDC

Table 21.2 SPR affinity measures of antibody variants and combinations at steady state on immobilized recombinant human FcRn of IgG-WT and variants produced in YB2/0 cells.

IgG variants	Mutations	ADCC: ratio EC50 WT/ variant	Ratio kDa WT/ variant
WT	None	101.0	1.0
C6A-7B	T256N/A378V/S383N/ N434Y	33.1	3.1
C6A-78A	T256N/A378V/N434Y	26.1	3.9
C6A78B	P228L/T256N/A378V/ N434Y	19.8	5.1
C6A-78-D	P228R/T356N/A378V/ N434Y	25.3	4.0

From Monnet et al. [10].

human and mouse Fc regions [30]. Monnet et al. have developed a number of variants that include multiple mutations that show substantially improved binding to the FcRn protein. Some of the most significant are summarized (see Table 21.2). Other nonmutational platforms include conjugation to polyethylene glycol (PEG), polysialylation, conjugation to the copolymer HPMA, use of dextran, and albumin conjugation (see Section 21.6) [10].

21.5.1 PEG Conjugation

This is a clinically proven method to extend half-life and has been applied to cytokines, hematopoietics, and enzymes [31]. When conjugated to a smaller molecule, PEG increases the overall mass of the complex, endowing it with a larger hydrodynamic volume. This results in a number of auspicious modifications: longer serum half-life, improved bioavailability, protease protection, stability, solubility, reduced immunogenicity, and toxicity [32].

Whereas conjugation extends half-life, it may also reduce immunogenicity. Various antibody fragments, including scFvs and Fabs, have been successfully PEGylated [33]. Pegylation is a covalent coupling of the PEG to the protein through random or site specific conjugation, often through the employment of an amine-reactive linker, which can conjugate to lysine. Site-specific conjugation is clearly preferable, in a position distal to the antigen-binding site. A useful strategy employs an engineered cysteine incorporated into the protein; the thiol coupling approach takes advantage of the greater efficiency, compared to amine conjugation. Cimzia, used to treat rheumatoid arthritis, is an anti-TNF PEGylated Fab fragment that is constructed through thiol coupling [34].

21.5.2 Polysialylation

This is another polymer with hydration properties similar to PEG. One example of the application of this technology was the construction of a PSA-Fab fragment conjugate accomplished through reductive amination, said to be a rather inefficient method [35]. At this time no clinical trial of molecules generated in this fashion is underway.

21.5.3 HPMA

This substance (*N*-(hydroxypropyl)methacrylamide) is the monomer used to make the polymer poly(*N*-[2-hydroxypropyl]methacrylamide). It has the beneficial qualities of high water solubility and hydrophilicity while at the same time being nonimmunogenic and nontoxic. Zhang et al. have evaluated this compound for effective clinical translation. A second-generation molecular variant of the HPMA copolymer demonstrated improved pharmacokinetic profiles and controlled drug release, with markedly improved therapeutic efficacy when compared to the first-generation compound and the free drug. The compound was evaluated in a model system, a human ovarian carcinoma xenograft. Together with an absence of deleterious side effects, these properties hold promise for the future development of anticancer agents based on this strategy [36].

21.6 Albumin Fusions for Half-Life Extension

21.6.1 Albumin Also Complexes with FcRn

A putative albumin receptor that protects albumin from catabolism was proposed many years ago, but it was not until 2003 that the ability of FcRn to bind to albumin was demonstrated [13]. The property of albumin to interact with hydrophobic molecules has been well known for decades, first described by Klotz in 1947 [37]. It was exploited in the development of human growth hormone as a therapeutic entity, coupled noncovalently for administration to children [38].

21.6.2 Large Molecules Require a Carrier to Complex with in Order to Achieve an Extended Half-Life

This places a limitation on the types of modifications that therapeutic molecules can undergo. Although whole mAbs are the only current FDA-approved antibody therapeutics, there are other numerous molecular modifications that are under investigation for their possible application to disease treatment. These include various permutations of bispecific antibodies, including tandem scFv molecules, diabodies, or single-chain diabodies [39].

Bispecific antibodies target two antigens simultaneously, as opposed to naturally produced antibodies, which display two identical antigen-combining regions. The concept was initiated by the work of Coloma and Morrison [40] who reported the fusion of flexible linker peptides to the C-termini of IgG HCs, followed by single-chain variable domains with different binding specificities. Because of this unique structure, they could attach to two different cell types and bring effector cells of the immune system into close proximity to one another [41]. For example, a bispecific antibody with one side reactive against a tumor-specific antibody and the other arm to a trigger molecule on an effector cell could bring a cytotoxic T-lymphocyte cheek to jowl against a tumor cell. This combination could prove to be an effective means of tumor cell destruction.

The recognition that the Fc portion of the antibody molecule could have potential side effects led to experimentation with the bispecific formats mentioned above from which the Fc portion had been omitted. However, lacking the Fc component of the antibody molecule would obviate the binding to the FcRn, which would be expected (and indeed turned out to be the case) [42], resulting in a drastically reduced half-life in circulation. This would mean that in order for these modified antibody bispecific fragments to be therapeutically effective, it would be necessary to use high doses and repeated infusions.

Among a number of strategies that have been evaluated, coupling the smaller antibody fragments to albumin appears to be one of the most effective. Albumin, a monomeric 67 kDa protein produced in the liver, is the most abundant protein in the blood plasma and is responsible for a number of important physiological functions. These include regulating osmotic pressure, transport of metabolites, and binding to numerous molecular species [43].

21.6.3 Albumin–Immunoglobulin Configurations

To exploit these qualities, Müller et al. [39] constructed albumin fusions to bispecific antibodies. The molecules were scFv(2)-HSA, scDb-HSA, and taFv-HSA, all reactive against the tumor antigen carcinoembryonic antigen (CEA) and the T-cell receptor complex molecule CD3. In this configuration they retained full-binding capacity to both antigens when compared with unfused scFv, scDb, and taFv molecules. The tumor antigen-specific retargeting and activation of T cells was observed for scDb, scDb-HSA, taFv-HSA, and to a lesser extent scFv(2)-HSA. The fusion to serum albumin strongly increased the half-life of recombinant bispecific antibodies. Indeed, the scFv-HSA fusion protein showed a 12-fold increase in half-life in comparison with a F(ab')-cys recombinant antibody, when evaluated in a mouse model. Moreover, the single-chain diabody–albumin fusion proteins were the most promising format for elaboration of the cytotoxic properties of these entities *in vitro* and *in vivo*.

The role of albumin in prolonging the half-life of other therapeutic proteins should not be ignored. Recombinant granulocyte colony stimulating factor

(G-CSF) is a widely used biological drug applied in the treatment of neutropenia due to chemotherapy and radiotherapy. However, its evanescence made it unsatisfactory, requiring repeated injections in order to maintain therapeutic titers [44]. It has been determined that albumin binds tightly to the FcRn molecule, a property that allows its protection against lysosomal degradation after being taken up into cells. The domain III of HSA is necessary and sufficient for this binding. Zhao et al. [45] extended the half-life of G-CSF by generating a fusion molecule composed of the domain III albumin sequence fused to G-CSF. The fusion protein was expressed in *Pichia pastoris*, an attractive system for protein production, given that it grows rapidly, and as a eukaryotic yeast, it can accomplish complex posttranslational modifications, including glycosylation, processing, and proper folding.

21.7 Mice as Models for Human Disease

21.7.1 Question Concerning This Model System

Much of our knowledge of human immune disorders comes from the use of mouse models, often taking advantage of transgenic, engineered animals. A recent paper has questioned this reliance [46]. The discussion asserts that the mouse can be a misleading model for human inflammatory diseases. Evidence supporting this argument is based on gene expression studies in mice exposed to trauma, comparing them to a group of human patients who had suffered from trauma, burns, and bacterial endotoxin response. Whereas RNA expression using Affymetrix gene chips was consistent within the human patients, there were major differences between the overall transcriptional response between the human and the mouse arms of the study, driving the conclusion that the mouse models are of no value in understanding the fundamental basis of human inflammatory disorders.

The debate is critical to progress in the field of medical immunology. There have been claims within the research community that the shortcomings of mice as a model of human disease and of inflammatory conditions in particular combined with progress in molecular clinical analysis render the mouse model obsolete. This argument posits that the success of modern genetic analysis makes the human the most satisfactory model of human disease. There is also a political component to this argument. *The New York Times* published an article summarizing these findings for a general audience, concluding that the mouse model has been totally misleading for three major disease categories; sepsis, burns, and trauma, and therefore billions of dollars have been wasted pursuing false leads [47]. This criticism comes at a time when the scientific edifice is under attack, and even the validity of the scientific approach in general is questioned by the public.

21.7.2 Support for Traditional Views

Another research group has argued strongly against the validity of this interpretation and has found results very much to the contrary [48]. The authors reevaluated the data of the previous study, focusing on genes that were significantly changed in both humans and mice. They found that the gene expression patterns in the mouse models showed extremely close correlations with the human data sets. Because the previous analysis included genes that did not show large changes, they argue that the data is skewed toward irrelevant markers. When these are eliminated from the study, the very strong correlation between the mouse and human responses emerges.

Sundberg et al. [49] have provided a balanced assessment of these viewpoints and emphasize that close coordination of analysis of these two systems can provide the most effective and accurate picture of this complex organismal structure.

The authors [49] stress that because of genetic variation, no single mouse strain (as well as no single human) can provide definitive answers to questions concerning complex diseases. Furthermore, mice are young, genetically homogeneous within strains, whereas humans are of various ages, both sexes, and quite heterogeneous genetically. The authors conclude that despite claims to the contrary, the mouse, when carefully selected, can be a vital component of human disease modeling. Given the fact that outbred strains are available and that the mouse genome can be easily manipulated through reverse genetics, mouse models will continue to be invaluable in the assessment of human disease.

Preclinical evaluation of antibody therapeutics is carried out in animal and cell culture models, and the mouse is a critical member of such protocols. As a new generation of biosimilar antibodies undergoes testing, preclinical evaluation will be extensively employed. For this reason it is of utmost importance that reliable animal models be available. It is imperative that the benefits and shortcomings of this approach be carefully taken in account in preclinical evaluation of anticancer agents and therapeutic protocols.

21.8 Half-Life Engineering: Present and Future

The discovery of the FcRn and the elucidation of its functions have opened a number of pathways to drug development that are active areas of investigation. Sockolosky and Szoka [5] have recently reviewed a range of various options that target antibody modification of the binding region to FcRn for the purpose of developing new therapeutic modalities (Table 21.3). Although the most well-trodden paths concentrate on engineering higher affinity IgG Fc domains or developing fusion partners with the albumin molecule, unusual possibilities are generating excitement within the community of immunotherapy researchers.

Table 21.3 Approaches for extending IgG half-life through modification of the FcRn.

Antibody half-life extension strategies			
#	Strategy	Mechanism	Effect
1	Fc-fusion	Hijacking FcRn recycling, increased molecular weight	Half-life extension by reducing catabolism
2	Albumin fusion		
3	Monomeric Fc and CH domains		
4	Fusion to alternative FcRn binding ligands		
5	IgG–Fc or albumin engineering	Increased mAb/albumin affinity for FcRn at pH 6	

Modified from Sockolosky and Szoka [5].

21.8.1 Antibody Fragment Improvements for Half-Life Extension

Antibody fragments hold appeal due to their small size, which makes it easier for them to reach targeted cells; however, the absence of an Fc segment means that they will not interact with the FcRn, and their half-life may be substantially truncated. Ying et al. [50] have generated human soluble monomeric IgG1 Fcs by combining rational protein design with multiple screening strategies. These molecules were stable and of low molecular weight, 27 kDa. They bind to the FcRn, suggesting that the dimeric form of Fc is not necessary for effective interaction, and that binders of small size, long half-lives, and antigen-binding ability may hold promise for development as therapeutic candidates.

21.8.2 Extending Half-Life through Glycosylation

Another strategy for extending the half-life of biosimilar antibodies involves the use of glycosylation-based modifications of the molecule. Ishino et al. [51] constructed a highly soluble Fc monomer that maintained binding affinity for the FcRn in a pH-dependent fashion. They reasoned that a bivalent molecule could work to the detriment of the patient by forming cross-linked complexes in the plasma. In addition the small size of a monomeric antibody could facilitate access to the inner reaches of a tumor. Following this rationale, the authors applied ASN-linked glycosylation in the consensus sequence of Asn-X-Ser/Thr where X is any amino acid except proline. The engineered carbohydrates masked the exposed hydrophobic surface of the C_H3 domain. By improving the avidity of the Fab-monoFc molecule, the authors were able to achieve a threefold longer serum half-life. They found that a tandem repeat of mono-Fc can gain avidity in a similar way to wild-type Fc dimers. They believe that this can occur due to a close proximity of each receptor molecule on the membrane surface. Thus the engineering of glycosylation sites on the CH3 domain stabilizes the monomeric form, allowing for extension of the half-life of these molecules.

21.8.3 Phage Display Technologies

Yet another approach to half-life extension takes advantage of phage display technology to select affibody molecules that can bind to FcRn in a pH-dependent manner [52]. The investigators used biophysical and binding assays to characterize the complexes *in vitro*, and the affibody molecules were found to bind to FcRn more strongly at low pH than at neutral pH. Attachment of the affibody molecules to a recombinant protein, already engineered for increased half-life, resulted in a nearly threefold longer half-life in mice. The authors argue that these engineered modifications will have general use as fusion partners to develop biopharmaceuticals that extend their half-lives *in vivo*.

In a similar approach, Andersen et al. [53] present investigations on phage display-selected nanobodies obtained from a variable domain repertoire library from a llama immunized with recombinant human FcRn. The camelid immune system provides the components for the construction of a VHH library, which was panned on soluble human FcRn. The selected nanobodies bound the receptor at acidic as well as at neutral pH. The nanobody can be attached to horseradish peroxidase and serve as a detection reagent, given it propensity for the receptor. Therefore, the nanobodies can serve as useful detection reagents, although the authors do not discuss the clinical implications of their findings as therapeutics.

21.8.4 FcRn and Albumin

Sand et al. [54] have discussed the properties of the FcRn protein and its unique relationship to the IgG and albumin molecules. Both ligands bind to the receptor in a remarkably similar manner, even though the molecules are very different from one another in form and function. Furthermore, the Fc receptor is widely expressed in a variety of tissues, including kidney, liver, and placenta, no doubt due to the role of FcRn in the maintenance of homeostasis in a variety of systems.

The substantial body of investigation that has advanced our understanding of this system has driven engineering studies of a new class of IgG molecules with unique blocking and transport qualities.

There will be a pressing need in future studies to evaluate the *in vivo* properties and therapeutic potential of these newly generated molecules. Because rodent FcRn binds poorly to human albumin, there is a need to develop new transgenic animal models suitable for studies of human variants and human albumin-based therapeutics.

21.8.5 Organ-Specific Expression of FcRn

There are suggestions that the FcRn molecule may play a role in organ specific functions. There have been observations of FcRn expression in the eye and of a possible role in ocular diseases; however, the association is not well understood at this point. Through the use of RT-qPCR analysis, the expression of Fcgrt mRNA has been documented in the retina, RPE/choroid, and the ciliary body/iris, while FcRn protein expression has been studied in the ciliary body epithelium, macrophages, and endothelial cells in the retina and choroidal vasculature. These observations could have important implication for antibody therapeutic strategies, given the significance of recombinant antibodies in the treatment of ocular diseases [55].

A noteworthy report from Bitonti et al. [56] described an approach to pulmonary delivery using FcRn-mediated transport for delivery of protein therapeutics to the patient through the lung. The authors created a recombinant monomeric erythropoietin–Fc fusion protein and evaluated the transport through the epithelium of the lung in a cynomolgus monkey model. Using immunohistochemical localization, they were able to demonstrate the expression of FcRn molecules in bronchus and alveoli. In further experiments it was demonstrated that the fusion protein could be delivered as aerosols through endotracheal tubes. EpoFc monomers and EpfoFc dimers appeared in the circulation and were biologically active, raising reticulocyte levels in the treated animals.

The delivery of biologics is usually accomplished through injection, so alternative delivery systems are in great demand. In this case the uptake of the EpoFc monomer was of the same order of efficiency as an injection of Epogen in humans, approximately 30–35%. This approach has been the subject of a phase I study that demonstrated dose-dependent concentrations of the fusion protein in the serum and an increase in circulating reticulocytes in the highest dose group. The authors concluded that large therapeutic molecules can be delivered to humans via the lung, with retention of biological activity, using the FcRn-mediated transport pathway [57].

21.8.6 Nanotechnology Applications

Nanomedicine is more or less defined as the design, engineering, and construction of very small devices for use in therapeutic and diagnostic applications. These devices are defined by their size, around 100 nm [58] or 1000 nm, but there is a general agreement that the dimensions are very small, and components within a nanostructure could be down to the 10s of nanometers (or less).

There has been great interest in recent years in the development of nanodevices for transporting drugs through the circulation, especially anticancer agents. The rationale that drives the discovery process is that polymeric particles, metal, and semiconductors have critical structural, magnetic, optical, and electronic properties, which make them suitable drug delivery carriers for targeting [59].

The basic concept in design is a drug-containing core in a soft sheath covered with PEG. The PEG attracts water molecules, surrounding the nanoparticle with a liquid halo. Thus insulated, the particle is much less likely to attract the unwanted attention of the host immune system. In addition, the structure has rounded edges, another mechanism for going under the immune radar. The size of the particle is another intervening factor that extends its half-life in the circulation [60]. A further important advantage of nanoparticle delivery systems is that they could protect an antibody when it is endocytosed into the cell. Naked antibodies are vulnerable at this point, since they may be engulfed and degraded by lysosomal activity. However the nanoparticle can protect the antibody molecule and allow its release into the cell in an intact and functional state [61].

Recently, Godowski et al. [62] have evaluated a carrier system consisting of nanoparticles composed of poly(lactic-co-glycolic acid), which can encapsulate and release antibody molecules without loss of functionality. The authors argue that there are many antibody therapeutic platforms that can be adapted to this approach.

There is a sizable effort to develop functional nanorods; carriers frequently composed of gold and other inert metals that can be functionalized with antibodies. Extra engineering efforts are required to make them stable within the microenvironment of the cancer cell. To prolong their circulation time, gold nanoparticles must be protected from the reticuloendothelial system, a task for which PEG has been put to the test with promising results [63].

Stem cells are a particularly refractory component of many malignancies [64]. They are often in a quiescent G zero state and may be nonresponsive to conventional chemotherapy. In addition, cancerous stem cells frequently share markers in common with normal stem cells, which make the development of targeted therapies even more difficult [65]. Moreover, many stem cells are multidrug resistant and display the efflux transporter phenotype [66]. Nonetheless, there have been a number of studies recently in which nanomedicines were used successfully to target cancer stem cells [67].

One of the most useful features of nanodevices is the fact they can enter the cell with ease, even in cases in which the cells express the multidrug resistance (MDR) phenotype [68]. This phenomenon is the result of mutations in genes controlling drug efflux in the ATP-binding cassette super family. Because it processes so many different drugs, including unrelated compounds, it has been a major stumbling block to successful chemotherapy of tumors [69]. The MDR gene encodes the high molecular glycoprotein, which is amplified in drug-resistant cells, causing their resistance to a host of anticancer agents. The source of the drug-resistant variants appears to be the selection of resistant subclones within heterogeneous tumor cell populations [70]. Given the fact that tumor cells are extremely unstable from a genetic standpoint, it is not surprising that resistant variants are an inevitable cause of failure of chemotherapy. This point has been discussed in detail as a Darwinian model of variation and selection [71].

A substantial number of clinically approved nanomedicines are in the marketplace. These include liposomal doxorubicin (Doxil), albumin-bound paclitaxel (Abraxane), PEG-L-asparaginase (Oncaspar), and others. However, none of these is an antibody nanoconjugate [72].

An important category of nanomedicines is the liposomes, referred to as nanocarriers; they have established their utility as transporters for anticancer antibodies. Li et al. have described the use of a codelivery system for the toxin bufalin and an anti-CD40 antibody [73]. The platform was designed to couple anti-CD40 mAbs to the surface of PEGylated unilamellar liposomes wrapped with bufalin. Bufalin acts by inducing apoptosis through a caspase-dependent pathway. The researchers carried out both *in vitro* and *in vivo* assessments with the B16 mouse model system. In both cases the approach was found to achieve improved therapeutic efficacy, while in the *in vivo* experiments, there was a lowering of systemic toxicity. A major benefit of the technology is that the prolonged release at the tumor site is accompanied by the prevention of the antibody from entering systemic circulation. Combining therapeutic agents can result in a synergistic effect, enhanced through the nanodelivery system.

The elimination of cancer stem cells through the use of nanotechnological delivery systems that protect and extend the half-life of antibodies and other anticancer agents is appealing and represents an active area of research today. However tumor organization is extremely complex: every patient is different, every tumor is different, and tumors evolve in time so that their behavior and response to anticancer treatments may vary dramatically in the course of a protocol. So while these technologies bear a great promise, preclinical and clinical evaluations may take years. It is important for the scientific community and the public to recognize realistic time lines and the current limitations of the biomedical sciences.

21.9 A Bright Future for Biosimilars, Biobetters, and Improved Half-Life Modifications

The science of half-life improvement is tied up with biobetter antibody development, and its future depends on the demand for these products, specifically in the United States, although also in the individual members of the European Union and Asia. Also, smaller markets (Australia, Canada, New Zealand, and South and Central America) come into play, and although their individual contribution is much smaller than the dominant players, their combined weight is significant.

Adoption of biosimilar products varies widely in different countries, with figures ranging from 1 to 99%, depending on the particular product. There are currently at least 452 biobetters in development. Some of the leaders include Kadcyla from Roche, a reengineered Herceptin for treatment of breast cancer, and another version of Herceptin, TrasGEX from Glycotope. There is a widely recognized acceptance within the Big Pharma of the need to build strong pipelines of biosimilars and, especially, biobetters. This is driven by the extremely robust market for biologics, now 27% of pharmaceutical sales in Europe [74]. In the United States alone, there are 907 biologics targeting more than 100 diseases in development. Extension of the half-life of these molecules could constitute a direct and rapid approach to improving their performance, so it is certain that this will be a major part of an R&D biobetter program.

However, the development of biobetters could make biosimilars irrelevant. An example of this rapidly developing approach is Roche' Gazyva, which is heralded as the "son of Rituxan." This anti-CD20 antibody has a glycoengineered Fc region, which allows greater antibody-dependent cell-mediated cytotoxicity, and has been approved for some (but not all) of Rituxan's indications [75]. It has been approved for the treatment of chronic lymphocytic leukemia and has been evaluated for other indications.

Although biobetters are presented as being a more lucrative means of entering the cancer therapeutics market, there are significant risks and drawbacks. The development time may be substantially longer, given the fact that the compound is being reengineered, and it may not function in exactly the same fashion as the original formulation. In addition, the compound will require larger clinical trials to establish its efficacy and safety. Since the biobetter is being measured against the original formulation, it may require a substantial patient population to satisfy the FDAs demands for a substantial improvement over the original [76].

Our understanding of the dynamic role of FcRn has been expanded in recent years from a receptor playing a role in passive immunity through transcytosis of IgG molecules to an active participant in signaling and trafficking, playing a major part in antigen presentation. FcRn plays a fundamental role in MHC II presentation of exogenous antigen in addition to cross presentation on MHC I molecules. These pathways have been shown to guide pathophysiological actions associated with infectious diseases and cancer development along mucosal barriers. Thus, the role that FcRn plays in the transport and preservation of IgG through extension of its half-life promises a wide range of therapeutic applications for treating cancer, autoimmune disease, and infectious conditions (Table 21.4) [21].

Table 21.4 Outlook for therapeutic applications of half-life modifications.

FcRn therapeutic approach	Mode of action	Application
Transporting IgG	Transcytosis by way of respiratory epithelium	Pulmonary delivery of IgG therapeutics and vaccines
	Transcytosis by way of intestinal epithelium	Treatment of enteric pathogens
	Use of Fc-targeted nanoparticles	Oral drug administration
	Transcytosis through placenta	Delivery of therapeutics to the fetus
Extending half-life of IgG	Increase FcRn/Fc binding by engineering FcRn or Fc molecules	Increase bioavailability of IgG and IgG-based therapeutics
Reducing half-life of IgG	Saturating FcRn binding	Amelioration of IgG-mediated diseases
	Engineered antibodies with higher FcRn affinity that enhance IgG degradation (*Abdegs*)	Many examples of autoimmune disorders

Adapted from Rath et al. [21].

References

1. Proetzel G, Roopenian DC. Humanized FcRn mouse model for evaluating pharmacokinetics of human IgG antibodies. *Methods* 2014 Jan 1;65(1):148–153.
2. Strobel G. End of the RoAD for Gantenerumab? Roche Declares Prodromal Alzheimer's Trial Futile. 2014. http://www.alzforum.org/news/research-news/end-road-gantenerumab-roche-declares-prodromal-alzheimers-trial-futile (accessed May 5, 2016).
3. Helfand C. Could a restructuring help protect AbbVie when revenue troubles hit? April 15, 2015. http://www.fiercepharma.com/story/could-restructuring-help-protect-abbvie-when-revenue-troubles-hit/2015-04-15 (accessed May 5, 2016).
4. Palmer E. Judge says Novartis can launch biosimilar of Amgen's Neupogen. *FiercePharma*, March 20, 2015. http://www.fiercepharma.com/?utm_medium=nl&utm_source=internal (accessed May 5, 2016).
5. Sockolosky JT, Szoka FC. The neonatal Fc receptor, FcRn, as a target for drug delivery and therapy. *Adv. Drug Deliv. Rev.* 2015 91:109–124.
6. Brambell FW, Hemmings WA, Morris IG. A theoretical model of gamma globulin catabolism. *Nature* 1964 203:1352–1354.

7. Roopenian DC, Christianson GJ, Sproule TJ, Brown AC, Akilesh S, Jung N, Petkova S, Avanessian L, Choi EY, Shaffer DJ, Eden PA, Anderson CL. The MHC class I-like IgG receptor controls perinatal IgG transport, IgG homeostasis, and fate of IgG-Fc-coupled drugs. *J. Immunol.* 2003 Apr 1;170(7):3528–3533.
8. Waldmann TA, Strober W. Metabolism of immunoglobulins. *Prog. Allergy* 1969 13:1–110.
9. Ghetie V, Ward ES. Multiple roles for the major histocompatibility complex class I-related receptor FcRn. *Annu. Rev. Immunol.* 2000 18:739–766.
10. Monnet C, Jorieux S, Urbain R, Fournier N, Bouayadi K, De Romeuf C, Behrens CK, Fontayne A, Mondon P. Selection of IgG variants with increased FcRn binding using random and directed mutagenesis: impact on effector functions. *Front. Immunol.* 2015 Feb 4;6:39.
11. Schuurman J, Graus YF, Labrijn AF, Ruuls S, Parren PW. Opening the door to innovation. *MAbs* 2014 Jul–Aug;6(4):812–819.
12. Schultze HE, Heremans JF. (1966). Molecular biology of humans proteins: with special reference to plasma proteins. In: Nature and Metabolism of Extracellular Proteins, volume 1, Elsevier, New York.
13. Chaudhury C, Mehnaz S, Robinson JM, Hayton WL, Pearl DK, Roopenian DC, Anderson CL. The major histocompatibility complex-related Fc receptor for IgG (FcRn) binds albumin and prolongs its lifespan. *J. Exp. Med.* 2003 197(3):315–322.
14. Ward ES, Ober RJ. Chapter 4: Multitasking by exploitation of intracellular transport functions the many faces of FcRn. *Adv. Immunol.* 2009 103:77–115.
15. Blumberg RS, Koss T, Story CM, Barisani D, Polischuk J, Lipin A, Pablo L, Green R, Simister NE. A major histocompatibility complex class I-related Fc receptor for IgG on rat hepatocytes. *J. Clin. Invest.* 1995 95(5):2397–2402.
16. Ward ES, Devanaboyina SC, Ober RJ. Targeting FcRn for the modulation of antibody dynamics. *Mol. Immunol.* 2015 67(2):131–141.
17. Ober RJ, Radu CG, Ghetie V, Ward ES. Differences in promiscuity for antibody-FcRn interactions across species: implications for therapeutic antibodies. *Int. Immunol.* 2001 13(12):1551–1559.
18. Łaguna P, Gołębiowska-Staroszczyk S, Trzaska M, Grabarczyk M, Matysiak M. Immunoglobulins and their use in children. *Adv. Clin. Exp. Med.* 2015 24(1):153–159.
19. Simister NE, Mostov KE. An Fc receptor structurally related to MHC class I antigens. *Nature* 1989 337(6203):184–187.
20. Ghetie V, Hubbard JG, Kim JK, Tsen MF, Lee Y, Ward ES. Abnormally short serum half-lives of IgG in beta 2-microglobulin-deficient mice. *Eur. J. Immunol.* 1996 26:690–696.
21. Rath T, Baker K, Pyzik M, Blumberg RS. Regulation of immune responses by the neonatal fc receptor and its therapeutic implications. *Front. Immunol.* 2015 5:664.

22. Martin WL, West AP Jr, Gan L, Bjorkman PJ. Crystal structure at 2.8 A of an FcRn/heterodimeric Fc complex: mechanism of pH-dependent binding. *Mol. Cell* 2001 7(4):867–877.
23. Kuo TT, Aveson VG. Neonatal Fc receptor and IgG-based therapeutics. *mAbs* 2011 3(5):422–430.
24. Raghavan M, Bonagura VR, Morrison SL, Bjorkman PJ. Analysis of the pH dependence of the neonatal Fc receptor/immunoglobulin G interaction using antibody and receptor variants. *Biochemistry* 1995 34(45):14649–14657.
25. Huang X, Zheng F, Zhan CG. Binding structures and energies of the human neonatal Fc receptor with human Fc and its mutants by molecular modeling and dynamics simulations. *Mol. Biosyst.* 2013 9(12):3047–3058.
26. Masher J. The ferret: an animal model to study influenza virus. *Lab Anim.* 2004 33:50–53.
27. Nesspor TC, Scallon B. Chimeric antibodies with extended half-life in ferrets. *Influenza Other Respir. Viruses* 2014 Sep;8(5):596–604.
28. Lazar GA, Dang W, Karki S, Vafa O, Peng JS, Hyun L, Chan C, Chung HS, Eivazi A, Yoder SC, Vielmetter J, Carmichael DF, Hayes RJ, Dahiyat BI. Engineered antibody Fc variants with enhanced effector function. *Proc. Natl. Acad. Sci. U. S. A.* 2006 103(11):4005–4010.
29. Richards JO, Karki S, Lazar GA, Chen H, Dang W, Desjarlais JR. Optimization of antibody binding to FcgammaRIIa enhances macrophage phagocytosis of tumor cells. *Mol. Cancer Ther.* 2008 7(8):2517–2527.
30. Anonymous. Engineered FcRegions Review. http://www.invivogen.com/review-engineered-pfuse-chig (accessed June 17, 2016).
31. Herrington-Symes AP, Farys M, Khalili H, Brocchini S. Antibody fragments: prolonging circulation half-life special-issue antibody research. *Adv. Biosci. Biotech.* 2013 4(5):689–698.
32. Palm T, Esfandiary R, Gandhi R. The effect of PEGylation on the stability of small therapeutic proteins. *Pharm. Dev. Technol.* 2011 16(5):441–448.
33. Germershaus O, Merdan T, Bakowsky U, Behe M, Kissel T. Trastuzumab–polyethylenimine–polyethylene glycol conjugates for targeting Her2-expressing tumors. *Bioconjug. Chem.* 2006 17(5):1190–1199.
34. Choy EH, Hazleman B, Smith M, Moss K, Lisi L, Scott DG, Patel J, Sopwith M, Isenberg DA. Efficacy of a novel PEGylated humanized anti-TNF fragment (CDP870) in patients with rheumatoid arthritis: a phase II double-blinded, randomized, dose-escalating trial. *Rheumatology* 2002 41(10):1133–1137.
35. Constantinou A, Epenetos AA, Hreczuk-Hirst D, Jain S, Deonarain MP. Modulation of antibody pharmacokinetics by chemical polysialylation. *Bioconjug. Chem.* 2008 19(3):643–650.
36. Zhang R, Yang J, Sima M, Zhou Y, Kopeček J. Sequential combination therapy of ovarian cancer with degradable *N*-(2-hydroxypropyl)methacrylamide copolymer paclitaxel and gemcitabine conjugates. *Proc. Natl. Acad. Sci. U. S. A.* 2014 111(33):12181–12186.

37. Klotz IM. The effects of salt and proteins on the spectra of some dyes and indicators. *Chem. Rev.* 1947 41:373–399.

38. Collipp PJ, Snyder RD. Short asthmatic children and human growth hormone. Evaluation of albumin-bound growth hormone. *Clin. Pediatr.* 1968 Nov;7(11):659–664.

39. Müller D, Karle A, Meissburger B, Höfig I, Stork R, Kontermann RE. Improved pharmacokinetics of recombinant bispecific antibody molecules by fusion to human serum albumin. *J. Biol. Chem.* 2007 Apr 27;282(17):12650–12660.

40. Coloma MJ, Morrison SL. Design and production of novel tetravalent bispecific antibodies. *Nat. Biotechnol.* 1997 15:159–163.

41. Müller D, Kontermann RE. Recombinant bispecific antibodies for cellular cancer immunotherapy. *Curr. Opin. Mol. Ther.* 2007 Aug;9(4):319–326.

42. Huhalov A, Chester KA. Engineered single chain antibody fragments for radioimmunotherapy. *Q. J. Nucl. Med. Mol. Imaging* 2004 48(4):279–288.

43. Farrugia A. Albumin usage in clinical medicine: tradition or therapeutic? *Transfus. Med. Rev.* 2010 24(1):53–63.

44. Layton JE, Hockman H, Sheridan WP, Morstyn G. Evidence for a novel *in vivo* control mechanism of granulopoiesis: mature cell-related control of a regulatory growth factor. *Blood* 1989 74(4):1303–1307.

45. Zhao S, Zhang Y, Tian H, Chen X, Cai D, Yao W, Gao X. Extending the serum half-life of G-CSF via fusion with the domain III of human serum albumin. *Biomed. Res. Int.* 2013 2013:10723.

46. Seok J, Warren HS, Cuenca AG, Mindrinos MN, Baker HV, Xu W, Richards DR, McDonald-Smith GP, Gao H, Hennessy L, Finnerty CC, López CM, Honari S, Moore EE, Minei JP, Cuschieri J, Bankey PE, Johnson JL, Sperry J, Nathens AB, Billiar TR, West MA, Jeschke MG, Klein MB, Gamelli RL, Gibran NS, Brownstein BH, Miller-Graziano C, Calvano SE, Mason PH, Cobb JP, Rahme LG, Lowry SF, Maier RV, Moldawer LL, Herndon DN, Davis RW, Xiao W, Tompkins RG; Inflammation and Host Response to Injury, Large Scale Collaborative Research Program. Genomic responses in mouse models poorly mimic human inflammatory diseases. *Proc. Natl. Acad. Sci. U. S. A.* 2013 Feb 26;110(9):3507–3512.

47. Kolata G. Health testing on mice is found misleading in some cases. *New York Times*, February 12, 2013, p. A19. Available at http://www.nytimes.com/2013/02/12/science/testing-of-some-deadly-diseases-on-mice-mislead-report-says.html.

48. Takao K, Miyakawa T. Genomic responses in mouse models greatly mimic human inflammatory diseases. *Proc. Natl. Acad. Sci. U. S. A.* 2015 112(4):1167–1172.

49. Sundberg JP, Roopenian DC, Liu ET, Schofield PN. The Cinderella effect: searching for the best fit between mouse models and human diseases. *J. Invest. Dermatol.* 2013 133(11):2509–2513.

50. Ying T, Feng Y, Wang Y, Chen W, Dimitrov DS. Monomeric IgG1 Fc molecules displaying unique Fc receptor interactions that are exploitable to treat inflammation-mediated diseases. *mAbs* 2014 6(5):1201–1210.

51. Ishino T, Wang M, Mosyak L, Tam A, Duan W, Svenson K, Joyce A, O'Hara DM, Lin L, Somers WS, Kriz R. Engineering a monomeric Fc domain modality by N-glycosylation for the half-life extension of biotherapeutics. *J. Biol. Chem.* 2013 288(23):16529–16537.

52. Seijsing J, Lindborg M, Höidén-Guthenberg I, Bönisch H, Guneriusson E, Frejd FY, Abrahmsén L, Ekblad C, Löfblom J, Uhlén M, Gräslund T. An engineered affibody molecule with pH-dependent binding to FcRn mediates extended circulatory half-life of a fusion protein. *Proc. Natl. Acad. Sci. U. S. A.* 2014 111(48):17110–17115.

53. Andersen JT, Gonzalez-Pajuelo M, Foss S, Landsverk OJ, Pinto D, Szyroki A, de Haard HJ, Saunders M, Vanlandschoot P, Sandlie I. Selection of nanobodies that target human neonatal Fc receptor. *Sci. Rep.* 2013 3:1118.

54. Sand KM, Bern M, Nilsen J, Noordzij HT, Sandlie I, Andersen JT. Unraveling the interaction between FcRn and albumin: opportunities for design of albumin-based therapeutics. *Front. Immunol.* 2015 5:682.

55. Powner MB, McKenzie JA, Christianson GJ, Roopenian DC, Fruttiger M. Expression of neonatal Fc receptor in the eye. *Invest. Ophthalmol. Vis. Sci.* 2014 Mar 19;55(3):1607–1615.

56. Bitonti AJ, Dumont JA, Low SC, Peters RT, Kropp KE, Palombella VJ, Stattel JM, Lu Y, Tan CA, Song JJ, Garcia AM, Simister NE, Spiekermann GM, Lencer WI, Blumberg RS. Pulmonary delivery of an erythropoietin Fc fusion protein in non-human primates through an immunoglobulin transport pathway. *Proc. Natl. Acad. Sci. U. S. A.* 2004 101(26):9763–9768.

57. Dumont JA, Bitonti AJ, Clark D, Evans S, Pickford M, Newman SP. Delivery of an erythropoietin–Fc fusion protein by inhalation in humans through an immunoglobulin transport pathway. *J. Aerosol. Med.* 2005 18(3):294–303.

58. Morrow KJ, Bawa R, Wei C. (2007). Recent advances in basic and clinical nanomedicine. In: Medical Clinics of North America: Nanomedicine, 91(5):805–844, Elsevier Inc., Philadelphia, PA.

59. Sharma A, Jain N, Sareen R. Nanocarriers for diagnosis and targeting of breast cancer. *Biomed. Res. Int.* 2013 2013:960821.

60. Maron DF. Cancer drugs hit their mark. *Sci. Am.* 2015 312(4):44–46.

61. Shih LB, Thorpe SR, Griffiths GL, Diril H, Ong GL, Hansen HJ, Goldenberg DM, Mattes MJ. The processing and fate of antibodies and their radiolabels bound to the surface of tumor cells *in vitro*: a comparison of nine radiolabels. *J. Nucl. Med.* 1994 May;35(5):899–908.

62. Gdowski A, Ranjan A, Mukerjee A, Vishwanatha J. Development of biodegradable nanocarriers loaded with a monoclonal antibody. *Int. J. Mol. Sci.* 2015 16(2):3990–3995.

63. Eghtedari M, Liopo AV, Copland JA, Oraevsky AA, Motamedi M. Engineering of hetero-functional gold nanorods for the *in vivo* molecular targeting of breast cancer cells. *Nano Lett.* 2009 Jan;9(1):287–291.

64. Shackleton M, Quintana E, Fearon ER, Morrison SJ. Heterogeneity in cancer: cancer stem cells versus clonal evolution. *Cell* 2009 138(5):822–829.

65. Welte Y, Adjaye J, Lehrach HR, Regenbrecht CR. Cancer stem cells in solid tumors: elusive or illusive? *Cell Commun. Signal* 2010 May 11;8(1):6.

66. Singh A, Settleman J. EMT, cancer stem cells and drug resistance: an emerging axis of evil in the war on cancer. *Oncogene* 2010 Aug 26;29(34):4741–4751.

67. Zhao Y, Alakhova DY, Kabanov AV. Can nanomedicines kill cancer stem cells? *Adv. Drug Deliv. Rev.* 2013 65(13–14):1763–1783.

68. Markman JL, Rekechenetskiy A, Holler E, Ljubimova JY. Nanomedicine therapeutic approaches to overcome cancer drug resistance. *Adv. Drug Deliv. Rev.* 2013 65(13–14):1866–1879.

69. Xue X, Liang XJ. Overcoming drug efflux-based multidrug-resistant in cancer with nanotechnology. *Chin. J. Cancer* 2012 31:100–109.

70. Gerlinger M, Rowan AJ, Horswell S, Larkin J, Endesfelder D, Gronroos E, Martinez P, Matthews N, Stewart A, Tarpey P, Varela I, Phillimore B, Begum S, McDonald NQ, Butler A, Jones D, Raine K, Latimer C, Santos CR, Nohadani M, Eklund AC, Spencer-Dene B, Clark G, Pickering L, Stamp G, Gore M, Szallasi Z, Downward J, Futreal PA, Swanton C. Intratumor heterogeneity and branched evolution revealed by multiregion sequencing. *N. Engl. J. Med.* 2012 366(10):883–892.

71. Gerlinger M, Swanton C. How Darwinian models inform therapeutic failure initiated by clonal heterogeneity in cancer medicine. *Br. J. Cancer* 2010 103(8):1139–1143.

72. Pillai G. Nanomedicines for cancer therapy: an update of FDA approved and those under various stages of development. 2014. http://www.symbiosisonline.org (accessed May 5, 2016).

73. Li Y, Yuan J, Yang Q, Cao W, Zhou X, Xie Y, Tu H, Zhang Y, Wang S. Immunoliposome co-delivery of bufalin and anti-CD40 antibody adjuvant induces synergetic therapeutic efficacy against melanoma. *Int. J. Nanomed.* 2014 9:5683–5700.

74. Dorey E. How the biologics landscape is evolving. *Pharm. J.* November 17, 2014. http://www.pharmaceutical-journal.com/news-and-analysis/features/forces-driving-the-evolution-of-biologics-into-biosimilars-and-biobetters/20067091.article (accessed May 5, 2016).

75. Barry F. Generation of biobetters could push out biosimilar development, says expert, April 23, 2014. http://www.biopharma-reporter.com/Bio-Developments/Generation-of-biobetters-could-push-out-biosimilar-development-says-expert (accessed May 5, 2016).

76. Ultee ME. Developing biosimilars and biobetters, January 29, 2014. http://www.pda.org/docs/default-source/website-document-library/chapters/presentations/metro/strategies-for-developing-biosimilars-and-biobetters.pdf?sfvrsn=6 (accessed May 5, 2016).

22

Technologies for Antibody-Drug Conjugation

Patrick G. Holder and David Rabuka

Redwood Bioscience, Inc., Emeryville, CA, USA

22.1 Summary

A monoclonal antibody (mAb) binds with high affinity to an antigen. As described in the preceding chapters, therapeutic mAbs leverage this specificity to induce cellular cytotoxicity through a variety of mechanisms. Small-molecule chemical pharmacophores can also induce cytotoxicity. An antibody-drug conjugate (ADC) is prepared by chemically joining therapeutic mAbs with these potent chemical cytotoxins. *In vivo*, a circulating ADC binds its target antigen on the cell surface and is internalized, carrying with it the chemical pharmacophore. This "payload" can then be delivered through a variety of release mechanisms and potentiate cell death. The combination of an mAb that targets a carefully chosen antigen (see Chapter 4) with a highly potent toxin can produce ADCs that significantly increase the therapeutic index of targeted treatments.

A "technology" for antibody-drug conjugation describes any of the modular components that are chemically assembled in order to produce a functional ADC. These include the location on the antibody to which drug is attached, the chemical reaction used for conjugation, the components of the linker joining mAb and cytotoxin, and the chemical functional groups that can trigger release of the cytotoxin. The exact choice of cytotoxin will depend upon the target cell and its sensitivity to a specific mode of action.

Researchers have been designing and building ADCs for over 50 years. As the construction of ADCs has changed, so have the underlying technologies. Pertinent reviews of ADC technologies have been prepared along the way [1–5]; this chapter is intended to survey current knowledge and best practices.

Biosimilars of Monoclonal Antibodies: A Practical Guide to Manufacturing, Preclinical, and Clinical Development, First Edition. Edited by Cheng Liu and K. John Morrow, Jr.

22.2 The Importance of Therapeutic Index

22.2.1 Defining TIs for ADC Research and Development

As with any therapy administered to a patient, treatment with an ADC biotherapeutic requires sufficient efficacy against a clinical target in order to be considered a viable therapeutic approach. However, as with any therapy administered to a patient, the potential failure of an ADC can be dictated by unjustifiable toxicity relative to the observed efficacy [6]. The TI is a summary of the clinical value of a specific treatment, and it is calculated as a ratio of these two critical pharmacological features; in short, it defines the specificity with which the treatment targets the disease, and not the patient. When applied to human subjects, this specificity is a measure of both efficacy and safety.

During design and development of an ADC, calculating and tracking a variety of TIs is an easy way to enable and motivate decision-making about candidate molecules. For technologies used to build an ADC, the stages of research and development involved in candidate selection include *in vitro* efficacy (translational TI) and preclinical efficacy and safety in animals (preclinical TI). These stages lead to clinical evaluation in humans (clinical TI), during which time the ADC technology will not be varied. Following excellent guidelines laid out by Muller and Milton [7], one can calculate TIs for ADC technologies using the following pharmacological and toxicological endpoints.

The translational TI for an ADC correlates the potency (IC_{50}) of a targeted ADC to a nontargeting isotype ADC in an antigen-presenting cell culture. In some cases, it is also prudent to calculate a translational TI from the bystander effect [8] of a targeted ADC *in vitro*. Preclinical evaluation of candidate ADCs is performed in a variety of animals, most commonly immunocompromised mice, rats, cynomolgus monkeys, and dogs. Correlating drug effects across these species—and thus calculating preclinical TIs—is nontrivial, but there are recommendations for comparing exposure [9], efficacy [10, 11], and safety [12–14]. The specific dose response and toxicity profile will vary dramatically depending on target choice, antigen specificity, drug mechanism, nonspecific drug release, and the actual molecular structure of the active drug catabolite [15].

22.2.2 Sources of Efficacy and Toxicity in ADCs

The combination of precise targeting and high potency in a single drug is the hallmark of effective ADCs. Unfortunately, it also potentiates three possible sources of toxicity: the small-molecule drug, antibody, and ADC. This toxic trifecta can occur when the cytotoxic drug molecule prematurely separates from the carrier antibody in circulation. Small-molecule drugs tend to have promiscuous pharmacology and can cause off-target toxicity across a variety of tissue types through nonspecific uptake [6]. In contrast, unconjugated antibodies have very low promiscuity, but they may result in on-target toxic effects in any tissue type that contains the targeted antigen. ADCs may bring about *both* toxicities if the on-target specificity of the antibody delivers the small molecule to an unintended tissue type, causing rapid cell death.

22.2.3 The Design Challenge

In theory, judicious design of ADCs will produce cooperative effects that translate to large TIs. Each component used to build ADCs is modular and can be chosen from a library of technologies. In practice, correlating an ADC technology (a bioconjugation reaction or linker chemistry) with results (translational, preclinical, and clinical studies) is a daunting task. In this chapter, our aim is to provide a survey of technologies for the design of ADCs with high TIs. Our belief is that the method of bond making and bond breaking employed for ADC construction can, in a rational and controlled way, affect ADC stability, potency, specificity, and safety.

22.2.4 Compare and Contrast by TI

Off-target toxicity looms large over the field of ADC development. Often, the toxins used in ADCs failed out of clinical trials specifically because of their toxicity. In the end, the success or failure of an ADC is dictated by the same measure. It cannot be overstated that the primary goal of ADC design must be to ensure the reliable, targeted transport of highly potent molecules—to make it work *and* make it safe.

22.3 ADC Construction: Building from the Protein Out

In 1958, the research team of Jean Bernard at the Hérold Hospital in Paris demonstrated the first covalent attachment of a chemotherapeutic agent to a γ-globulin. The diazotized form of methotrexate was generated and reacted with hamster IgGs. The resulting conjugate was dosed at 20 mg/kg against leukemia 1210 xenografts in hamsters and was found to significantly prolong animal survival versus either IgG, drug, or a noncovalent mixture of IgG and drug [16]. From this auspicious beginning, the technologies for covalent attachment of chemotherapies have expanded to a robust suite of chemistries. While myriad methods for this purpose are now available (and retired—e.g., diazotization), the challenge of building a functional immunoconjugate remains much the same as it was six decades ago; the chemistry of attachment must be compatible with both the carrier antibody and the patient.

Researchers today better understand the many requirements for the reaction conditions tolerated by recombinant antibodies. Modern methods for covalent grafting of small molecules to the surface of proteins are, of course, still performed in aqueous solution containing the IgG substrate. IgGs are sufficiently tolerant of acidity and base, at least for a short time, so that conjugation reactions are performed across a range of pH 4–9. They also tolerate a modest fraction of organic cosolvent (typically <10%) and surfactant, which is often required to solvate the drug of interest.

By way of analogy, during the synthesis of small organic molecules, it is not uncommon for chemists to set reagent concentrations in the range of 100–2000 mM in order to maintain high bimolecular reaction rates. In contrast, proteins, especially large IgGs, are not amenable to such a high concentration in solution, typically reaching a plateau below 1 mM. The reactive pharmacophores also tend to be quite expensive, and they pose a significant safety risk, so minimizing their quantities is vital. Together, these facts predicate that high yields in bioconjugation are most likely if the bond-forming reactions have very large rate constants.

The challenging requirements placed on bioconjugation reaction(s) are supplanted only by the design criteria for ADCs with large TIs. The purified conjugate must (i) be stable during production, purification, and storage; (ii) be stable against the environment experienced during transit through the body, including 37°C, the enzymes, proteins, and small molecules in serum, and environmental changes (pH, redox) that accompany cellular uptake and endosomal recycling; and (iii) successfully deliver the active agent upon arriving at the target tumor cell. To meet these criteria, there has been significant investment in the design of technologies for bioconjugation, linkers, and triggers that allow the active pharmacophore delivered to the cell to be defined with atomic precision.

Considering the litany of reagents and methods for bioconjugation and linker preparation, it's not worthwhile to list the composition of every specific bifunctional reagent or engineered linker synthesized to date. Instead, our survey examines the *modular components* of ADCs (Fig. 22.1). We include both the methods for their preparation and—to what extent it is known—how well they meet the ADC design criteria.

In an effort to review both the successes of ADCs and their potential for improvement, we include both proven and emerging technologies for ADC construction. We take proven technologies to have selective *in vitro* cytotoxicity toward cell lines expressing the target antigen. Figure 22.2 summarizes the *proven* technologies used in ADCs that have entered clinical trials. *Emerging* technologies are unproven, but have significant potential to enhance the TI of ADCs either by increasing their potency or their safety.

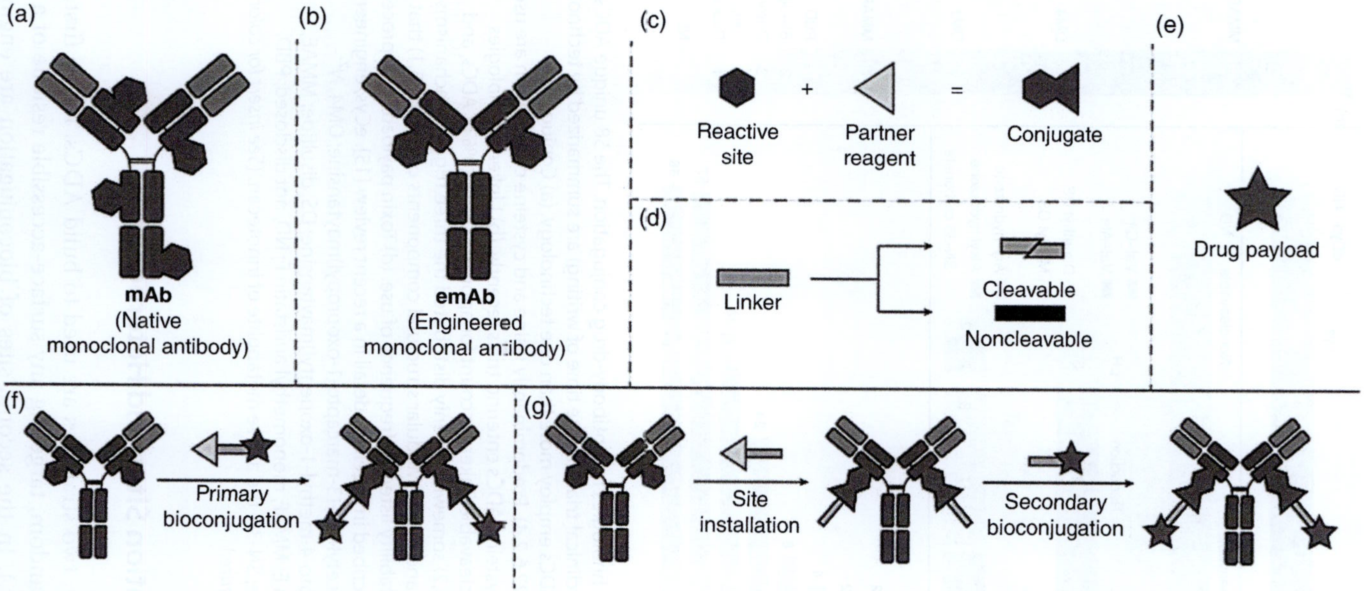

Figure 22.1 Modular components of an antibody-drug conjugate. A monoclonal antibody (mAb) (a) can be engineered (b) to contain reactive sites (c). These sites include amino acids (AAs) and posttranslational modifications (PTMs). Partner reagents are added to form a conjugate with the reactive site (c). A linker (d) is attached to a payload (e), which can include a release trigger. Synthesis of an ADC involves addition of an elaborated drug–linker to an antibody containing a reactive site (f and g). (*See insert for color representation of the figure.*)

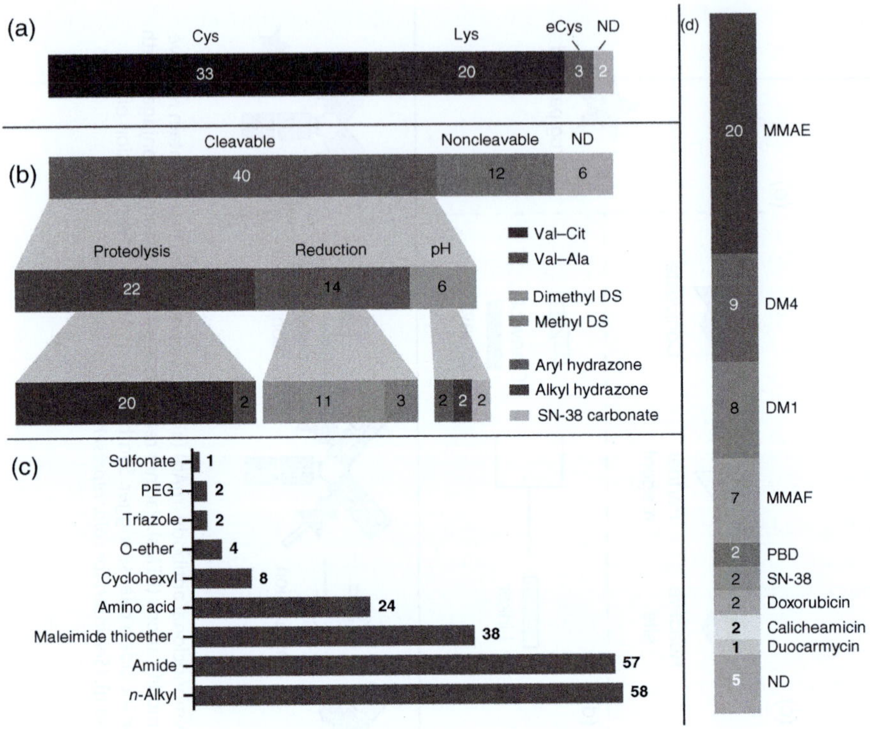

Figure 22.2 Proven technologies for antibody-drug conjugation. The 58 unique ADCs that have advanced to clinical trials (at the time of writing) are summarized by technology used. Note that most ADCs employ more than one technology. (a) Conjugation technologies (Section 22.6.2.4). Predominantly lysine and cysteine conjugation are used, with two engineered cysteine ADCs entering trials recently. (b) Linker technologies (Section 22.7.4.4). Noncleavable linkers account for approximately 20% of ADCs, and cleavable constructs (6.2) somewhat evenly distributed the three trigger mechanisms—proteolysis, reduction, and pH. (c) Modular structural components of linkers (6.1) that impart solubility and stability listed by frequency of use. (d) Toxin payloads. The process of choosing a toxin is described in excellent detail in a recent review [15]. eCys, engineered cysteine; DM1, $N^{2\prime}$-deacetyl-$N^{2\prime}$-(3-mercapto-1-oxopropyl)maytansine; DM4, $N^{2\prime}$-deacetyl-$N^{2\prime}$-(4-mercapto-4-methyl-1-oxopentyl)maytansine; DS, disulfide; MMAE, monomethyl auristatin E; MMAF, monomethyl auristatin F; ND, not disclosed; PBD, pyrrolobenzodiazepine; SN-38, the active metabolite of irinotecan. (*See insert for color representation of the figure.*)

22.4 Conjugation Sites and Heterogeneity

Generally speaking, two strategies are used to build ADCs. In the first, drug bioconjugation is random, targeting any surface-accessible residue of a given type (Lys, Cys, etc.). In the second, sites of bioconjugation are engineered into the mAb structure, and bioorthogonal conjugation chemistries allow researchers to define both the location and the number of modification sites—genetically, chemically, or enzymatically.

22.4.1 Random Bioconjugation

Random protein bioconjugation involves no manipulation of the primary amino acid (AA) sequence before reaction. Each is present in its native position—often multiple times. Addition of an exogenous reagent results in conjugates that vary both in the number of attached drugs and the number of sites of attachment. To describe the heterogeneity of the resulting conjugate, the *d*rug-to-*a*ntibody *r*atio (DAR) was defined to specify the quantity of modifications [17]. The DAR is an average measure, and so it is unable to convey the number of isomeric species produced for a given technology.

22.4.2 Site-Specific Bioconjugation

It is now clear that the specific efficacy and pharmacokinetics of ADC isomers vary greatly. The position and frequency of modification can increase the potency and serum lifetime of conjugates, which has spurred the development of site-specific conjugation methods [18, 19] to produce single ADC isomers. Site-specific ADCs are assembled from mAbs that are engineered to contain bioorthogonal conjugation sites. Like its parent IgG, an engineered monoclonal antibody (emAb) is a symmetric dimer, and so site-specific technologies only generate an even number of locations per antibody (at least with current technologies). As described in the succeeding text, there are numerous methods for site-specific bioconjugation, but it is clear that the site of conjugation influences PK [18–24], efficacy [18, 19, 24–26], conjugate stability [20, 27], and ease of synthesis [28].

22.5 Installation of Conjugation Sites

Monoclonal antibodies from a subclass share a single AA sequence in the constant region. Thankfully, this ensures that the chemical reactivity of AAs and posttranslational modifications (PTMs) is similar among mAbs even though they encode different variable regions. In a random conjugation, changes in the variable region can result in small variations in reactivity. In contrast, in site-specific conjugation researchers engineer reactive sites into a defined location in the constant region, enabling identical chemistry across antibodies independent of the variable region. Since the sequence homology across mAb subclasses is *also* similar, the design of these locations often translates across the constant regions of IgG1–4. The resulting emAbs enable consistent production of homogenous ADCs with defined properties.

22.5.1 Activation of Native Sites of Conjugation

22.5.1.1 Glutamic and Aspartic Acid

Some of the very first covalent bioconjugation reactions on mAbs involved modification of carboxylic acids [29–31]. Water-soluble carbodiimide reagents allowed coupling of amines to Glu, Asp, and the C-terminus to make amides. This method fell out of favor because of cross-reactivity with lysine side chains that resulted in protein cross-linking, precipitation, and poor circulation times *in vivo* [32].

22.5.1.2 Lysine

With a pK_a of approximately 10, the ε-amine of lysine (**1**) is largely (>99%) protonated and positively charged at neutral pH. Proteins display a great many lysine residues on their surface in part to provide aqueous solubility. Trastuzumab, for example, presents 92 lysine residues, of which 70 are known to react with exogenous reagents [33]. Nonetheless, when deprotonated the lysine amine is a powerful nucleophile. Activation of lysine only requires moderate elevation of pH (usually to 8) to generate enough free amine that can react with activated esters, isothiocyanates, aldehydes, and ketones, all with rapid kinetics.

22.5.1.3 Cysteine and Disulfides

In a "perfectly" formed IgG, cysteine residues (**6**) are only present in disulfides (**17**), helping to stabilize the tertiary and quaternary structure of the antibody [34]. Disulfides are generally unreactive, but they can be reduced to generate two equivalents of cysteine. The cysteine thiol can be deprotonated to a thiolate, which reacts rapidly with a variety of electrophiles. Activation of disulfide residues requires the addition of reducing agent and moderate elevation of pH, for example, 6–7.5. The classical IgG reduction–alkylation approach is now over half a century old [35].

The technique of disulfide reduction for drug conjugation began as an experiment measuring the degree to which triaziquone was "taken up" by serum proteins [36]. Researchers at the University of Manitoba found that γ-globulins that were reduced with DTT bound the highest fraction of this benzoquinone alkylating agent and that the conjugate remained cytotoxic to various cancerous cell lines [37, 38].

Partial reduction to reveal cysteine is now relatively common, but a landmark pair of results was reported by researchers at Bristol-Myers Squibb (BMS). In the first [39], they described the maleimidocaproyl (MC) reactive group and demonstrated quantitative modification of the four interchain disulfides of BR96—a chimeric mouse/human antibody that targets the Lewis^y tumor-associated antigen—with eight molecules of doxorubicin. Within a month, they disclosed [40] that this BR96–doxorubicin conjugate has potent, selective *in vivo* cytotoxicity against xenografts of lung, breast, and colon carcinomas.

While development of the BR96 conjugate proceeded rapidly [40, 41], including both phase I [42] and II [43] clinical trials, it took almost 10 years before revisiting the method for activating the conjugation site. This began when the disulfide reduction reaction was optimized with selenol catalysis [44]. Around the same time, Seattle Genetics used maleimide conjugation in order to introduce analogues of the dolastatins onto cAC10 [45], which targets the CD30 antigen, a marker of a subset of lymphomas. In short order, they determined the detrimental effect of loading eight copies of the auristatin onto a single mAb [46]. The conjugates with high DAR were difficult to produce and purify, and they were rapidly cleared from circulation *in vivo.* While they successfully isolated conjugates with 2, 4, or 8 drugs through purification, it became clear that new methods for site-selective incorporation of cysteine residues (Section 22.3.1) were warranted. Site-directed replacement of interchain disulfides with serine residues allowed them to specify the number and location of conjugates [19]. Lastly, recent bioconjugation chemistries are designed to "bridge" disulfides through a single reagent, reforming the structural element lost during classical reduction–alkylation (Section 22.7.1.6).

22.5.1.4 Glycans

The *cis* diol functionality of sugar residues can be cleaved by oxidation with sodium periodate to reveal aldehyde or ketone functional groups, which can be modified with carbonyl-reactive bioconjugation methods (Section 22.7.3.3). This chemistry is used to modify endogenous glycans attached to the C_H2 region of intact mAbs. Hage and coworkers studied the kinetics of the oxidation process. They developed conditions enabling installation of 1, 2, 4, 6, or 8 sites as a function of reaction time, temperature, and $NaIO_4$ stoichiometry [47], and they described a two-stage kinetic model of the oxidation in which exocyclic diols react more rapidly than endocyclic diols [48]. When present, glycans attached on the light chain also become oxidized and can be conjugated [49]. However, antibodies are not insensitive to periodate oxidation. A report from BMS and the University of Bremen disclosed that a variety of antibodies show decreased antigen affinity and disrupted protein structure as a function of increasing oxidant concentration [50].

Antibodies activated by oxidation can be modified with ^{125}I [51], ^{111}In [51], Tc-99m [52], desacetylvinblastine [49, 53–57], doxorubicin [58], and most recently a variant of dolastatin-15 [59], dolastatin-10 [60], and MMAE [60]. In the case of the vinca alkaloids, Lilly performed numerous studies on the desacetylvinblastine conjugates targeting adenocarcinoma. They interrogated conjugation chemistry [53], PK and toxicology [57], preclinical efficacy [54], and phase I clinical safety [55]. In the end, the program was discontinued for two reasons. The first was the immunogenic response to the murine antibodies used at the time. The second was the chemical linkage to the antibody—they postulated that the hydrazone was not sufficiently stable during circulation.

22.5.2 Installation of Engineered Sites of Conjugation

Engineered bioconjugation sites in mAbs are introduced either during production *in vivo* or after the protein is purified. Methods for emAb production *in vivo* include mutagenesis of native AAs, nonnative PTMs, introduction of short peptide tags, and nonnatural AAs (nnAAs).

22.5.2.1 Site-Directed Incorporation of Native Amino Acids: Point Mutations and Tags

Mutagenesis of a single AA is used for the installation of nonnative cysteine residues. This approach enables the use of existing cysteine-reactive bioconjugation chemistries, and it was anticipated that the engineered antibodies (ThioMabs) would not require any reduction because the nonnative residues were not expected to participate in a disulfide bond. However, Junutula et al. [61] were unsuccessful in their initial attempts at direct conjugation of ThioMabs expressed in CHO cells. Subsequent characterization of the ThioMabs revealed that the nonnative cysteines existed as mixed disulfides of glutathione (GSH) and cysteine. To overcome the "blocked" engineered cysteines, standard partial reduction chemistry was used to reduce intermolecular disulfide bonds of the mAb as well as the mixed disulfides at the engineered cysteine sites. Diafiltration was used to eliminate the low-molecular-weight GSH and cysteine components. ThioMab intermolecular disulfides were allowed to reform by either air oxidation or more rapidly in the presence of $CuSO_4$ or dehydroascorbic acid. Following reoxidation, the native cysteines of ThioMab exist as disulfides and are chemically silent from a conjugation perspective, while the engineered cysteines are reduced and available for site-specific conjugation. Overcoming the blocked nonnative cysteines in ThioMab, Genentech pioneered nonnative cysteine residues in the mAb backbone, generating highly homogenous ADCs with a controlled DAR that demonstrates an improved TI in murine xenografts [18].

22.5.2.2 Nonnatural Amino Acids

The prokaryotic and eukaryotic protein synthesis machinery encodes for a small pool of 20 AAs, but recently developed technologies enable a veritable ocean of choices. Synthetic nnAAs containing bioorthogonal functional groups are introduced into the primary sequence of proteins using engineered protein production machinery [62]. The desired AA is attached enzymatically to a tRNA by an engineered tRNA synthetase. Researchers co-opt the amber stop codon to recognize this new tRNA, which delivers the nnAA to the growing polypeptide in the ribosome. To date, both *p*-acetylphenylalanine (pAF, **44**), developed by Ambrx, Inc., San Diego, CA [26, 63], and *p*-azidomethylphenylalanine (pAMF, **50**), from Sutro Biopharma, Inc., South San Francisco, CA are used for the synthesis of site-specific ADCs.

22.5.2.3 Posttranslational Modification *In Vivo*

IgGs natively carry a few PTMs, including acetylation, disulfides, and glycosylation. However, that represents a limited subset of the variety of PTMs that occur on proteins in nature [64, 65]. PTMs are installed by cellular enzymes, often in the ER, using intracellular reagents derived from components of growth medium. Nonnative enzymes can be incorporated into cells that are used for mAb expression. Alternatively nonnatural feed reagents can be added and incorporated into cellular metabolism. In an optimized system, one can affect a synthetic PTM on the mAb during production.

Metabolic oligosaccharide engineering involves selective remodeling of glycoproteins by manipulation of cellular glycan feeds and inhibition of specific glycosyltransferases [66]. Researchers at Seattle Genetics recently disclosed a method that prevents the normal cellular production and incorporation of fucose. The approach includes feeding of nonnatural fucosyl derivatives that are recognized by the fucose salvage pathway in CHO cells. They reported incorporation of a wide variety of nonnatural fucose residues and highlighted thiolfucose (**24**) as a conjugation site for maleimide reagents [67].

Aerobic sulfatases catalyze the hydrolysis of sulfate esters with a formylglycine (FGly, **26**) residue in their active site [68]. FGly is a PTM of cysteine (or serine) in which the AA side chain is converted to an aldehyde [69]. FGly-generating enzyme (FGE) catalyzes FGly installation on nascent polypeptide chains in the early ER [70]. The substrate for FGE is a highly conserved sequence (CXPXR) that can be introduced into any protein sequence, including antibodies [71]. At Redwood Bioscience, we incorporate this machinery into CHO cells. The coexpression of FGE during the production of the protein results in purified emAbs bearing FGly [72]. This "aldehyde tag" can be used to build ADCs in any region and either chain of an antibody [24].

22.5.2.4 Synthetic Posttranslational Modification

Cells have evolved to incorporate PTMs on proteins in order to modulate their activity and function *in vivo*. Chemists can similarly incorporate PTMs with enzymes or synthetic bioconjugation reactions. Activated emAbs are made by conjugation of a bioorthogonal site onto a native site with a heterobifunctional reagent; one half of the reagent modifies the mAb, and the other remains available for "secondary" bioconjugation. In principle, any two functional groups that do not cross-react with each other can be incorporated into a single reagent. In early development, only a few reagents were available, but more recently, increased interest from the academic community and industry has led to an explosion in the combinatorial diversity of heterobifunctional reagents.

Metabolic engineering enables production of nonnatural oligosaccharides *in vivo*, but there is growing use of glycosyltransferases *in vitro* as well [73]. Researchers at Sanofi Genzyme developed methods to enzymatically elaborate native IgG glycans with galactosyltransferase and sialyltransferase to enable subsequent bioconjugation. Sialic acid is particularly susceptible to sodium periodate oxidation as described in the preceding section. They developed conditions in which residues were treated with only 0.5 mM of $NaIO_4$ to afford quantitative yield of reactive ketones (**47**) [74]. In their antibody platform, they closely monitored the effect of oxidant on methionine oxidation in the FcRn binding region, determining that their procedure modestly (10–15%) decreased receptor affinity. They built aminooxyacetyl-terminated (**48**) MMAE and dolastatin-10 payloads that react to form oximes (**49**) with high yield, and they demonstrated antigen-specific regression of SKOV-3 xenografts in SCID mice [60].

Synthetic heterobifunctional cysteine-reactive reagents were developed for ADCs when researchers modified IgGs with anthracycline drugs [75]. Even this early report determined that the type of functional group attached to the drug—maleimide or iodoacetamide, for example—could significantly decrease the toxicity of the payload and hinted at the value of designing the exact molecular structure joining protein and toxin. In fact, installing a prosthetic site for secondary bioconjugation allows introduction of nearly any functional group, a prospect that is encouraging a wave of creative development focused on new synthetic [76–79] and enzymatic [80] bond-forming reactions. The general approach involves choosing a native bioconjugation site, linker components, and a terminating nonnatural functional group that participates in a bioconjugation reaction. The payload is then modified with the orthogonal partner reagent. The nonnatural bioconjugation then forms a covalent connection between the drug and protein. While this may appear at first glance to be a laborious method for building ADCs, the benefit of using extremely fast reactions to make the protein–drug linkage is not to be underestimated. The cost of reagents, reaction time, and yield are worthwhile considerations. The reactions and linker components are discussed in the following section.

22.6 Bioconjugation Reactions

An antibody conjugate is formed by covalent introduction of a partner reagent onto one of the sites detailed in Section 22.1. In this section we describe the technologies for mAb bioconjugation at both native and engineered reaction sites. Theoretically, any bioorthogonal reaction can become an ADC technology; experimentally, only a fraction of known chemistries successfully navigate the gauntlet of *in vitro* assays and stability requirements; practically, fewer still are useful *in vivo* or in the clinic. This winnowing of options encourages further development of new chemistries that meet the stringent design criteria of antibody-drug conjugation.

22.6.1 Conjugation to Native Amino Acids

Native conjugation sites can act either as a primary site of attachment for a drug–linker or as anchor points for the installation of secondary bioorthogonal functional groups. In either case, modification of endogenous protein functional groups is the most common method of bioconjugation (Fig. 22.3).

22.6.1.1 Lysine

Hermanson has prepared a detailed review of lysine (**1**) bioconjugation in general [81]. With respect to ADCs, an early report compared two of the most common methods for preparing lysine conjugates in parallel [31]. The first was

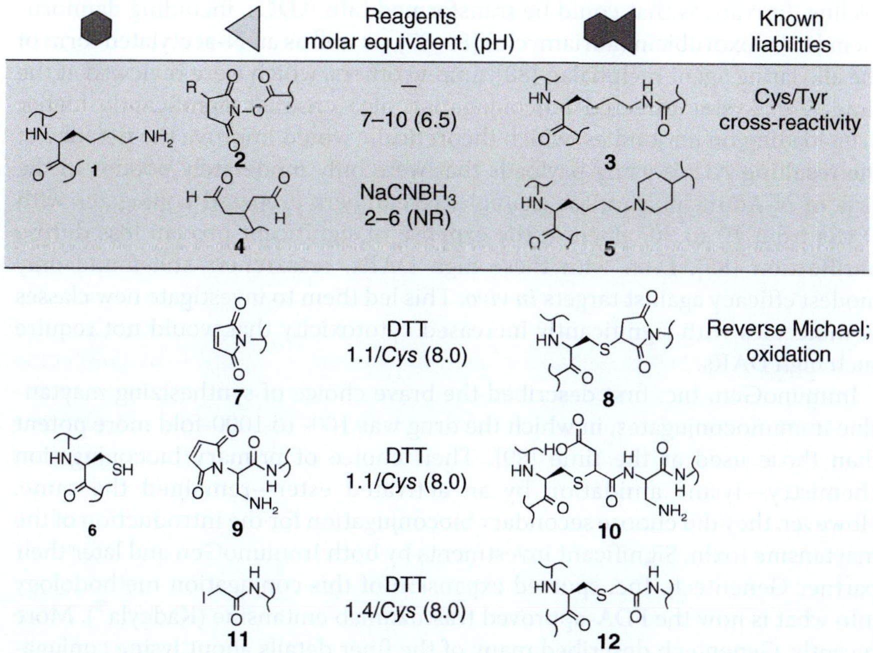

Figure 22.3 Bioconjugation to native AAs. Lysine (1) and cysteine (6) are endogenous sites on mAbs that can react with partner reagents. Common reaction conditions for conjugate formation are shown, along with notable liabilities. NR, not reported.

imine (Schiff base) formation with glutaraldehyde and the amino sugar of daunorubicin. In the second, the amino sugar on the drug was oxidized to produce aldehyde groups, which were then attached to the mAb lysines by reductive amination.

Each of the two methods was synthetically successful, but problems were found. In the imine formation, it was quickly discovered that nonspecific cross-linking and polymerization of proteins are a significant liability [82, 83]. In the subsequent article in the same issue, the oxidized glycan ADCs were tested against murine lymphoid xenografts *in vivo* [84]. The glycan oxidation had a deleterious effect on the potency of the drug.

22.6.1.2 Lysine Amidation

The approach of the community then shifted to modify lysine with an activated ester of *N*-hydroxysuccinimide (NHS, **2**) or its more water-soluble sulfated form [2, 85]. NHS esters do not react at a significant rate with water around neutral pH [81], and their fast reaction with lysine yields an amide bond (**3**) that is stable (Section 22.8) against any physiological amounts of heat, acid, base, reduction, and oxidation.

The ease of lysine-directed bioconjugation amplified the number of anthracycline derivatives that could be transformed into ADCs, including daunorubicin [17], doxorubicin (adriamycin) [86, 87], as well as an N-acetylated form of the alkylating agent melphalan [88], among others, which were reviewed at the time [75]. Lysine-directed bioconjugation also enabled significantly higher drug loading on antibodies, which theoretically would improve the potency of the resulting ADCs using payloads that were only moderately potent. In the case of *N*-Ac melphalan, for example, researchers prepared conjugates with DARs from 10 to 30, albeit at the expense of significant protein loss during purification [88]. Even with these high DARs, researchers still found only modest efficacy against targets *in vivo*. This led them to investigate new classes of molecules with significantly increased cytotoxicity that would not require such high DARs.

ImmunoGen, Inc. first described the brave choice of synthesizing maytansine immunoconjugates, in which the drug was 100- to 1000-fold more potent than those used at the time [89]. Their choice of primary bioconjugation chemistry—lysine amidation by an activated ester—remained the same. However, they did choose secondary bioconjugation for the introduction of the maytansine toxin. Significant investments by both ImmunoGen and later their partner Genentech, Inc. spurred expansion of this conjugation methodology into what is now the FDA-approved trastuzumab emtansine (Kadcyla®). More recently, Genentech described many of the finer details about lysine conjugation, including the statistical distribution of conjugates [33], off-target residues that react with NHS esters [90], and changes that occur to the structural stability of the IgG after modification [91].

22.6.1.3 Lysine Reductive Amination

As mentioned in the preceding text, an alternative to activated esters for lysine bioconjugation is reductive amination. Aldehydes react readily with lysine side chains to form imines, which are in equilibrium with both the hydrated hemiaminal and the free aldehyde. This labile connection is fixed in place as an amine upon reduction with the mild, water-soluble hydride donors $NaBH_4$, $NaCNBH_3$, or $Na(OAc)_3BH$. The newest methods for reductive amination neither cross-link proteins nor do they inactivate pharmacophores. Concortis Biosystems, Inc. markets a reductive amination technology for lysine that leverages a propylene dialdehyde (**4**) that forms an irreversible piperidine-linked conjugate (**5**) [92].

22.6.1.4 Cysteine Alkylation: Maleimide

As with lysine, a salient review of reagents for cysteine (**6**) modification was prepared by Hermanson [81]. In general, technologies for Cys bioconjugation leverage maleimide (**7**) and haloacetyl (**11**) functional groups, but new methods are in development.

Reaction of a cysteine thiol with a maleimide proceeds by the Michael addition, in which nucleophilic attack of sulfur occurs at the β-position of the α,β-unsaturated carbonyl. The hydrogen atom at the β-carbon is then removed by deprotonation, leaving the completed thioether linkage (**8**). Maleimide reaction with Cys is 1000-fold more rapid than Lys at pH 7, enabling selective modification. This rapid reactivity is a practical and efficient choice either for direct bioconjugation to cysteine or secondary bioconjugation to a maleimide-decorated protein [93, 94]. The examples of cysteine–maleimide conjugation are extensive. This reaction has been used for a variety of ADC drugs, including anthracyclines [39], auristatins [45], maytansinoids [89], duocarmycins [95], and pyrrolobenzodiazepines (PBDs) [28]. It is used to link toxins in both clinically approved ADCs, brentuximab vedotin [96] and trastuzumab emtansine [97]. It is not, however, without problems (Section 22.8).

22.6.1.5 Cysteine Alkylation: Iodoacetamide

Haloacetyl reagents (**11**) also enable rapid modification of cysteine at neutral pH to form alkyl thioethers (**12**). Caution is warranted in that haloacetyl reagents can modify other AA side chains, including His, Lys, and Met. Thankfully, reactivity toward thiols is selective when using iodoacetyl functionality. This reaction is having a resurgence of late [21, 28], perhaps as a result of its increased stability relative to maleimide conjugates.

22.6.1.6 Emerging Technologies

Generation of a five-membered thiazolidine ring is efficient at acidic pH by trapping a transient imine with a pendant thiol. This reaction is used to couple an N-terminal cysteine (**13**) with aldehydes. Neri and coworkers incorporated cysteine after the signal peptide sequence in diabodies and mAbs and demonstrated production of proteins bearing the reactive AA [98]. A thiazolidine bridged conjugate (**14**) of a dolastatin-15 variant was synthesized and tested for extracellular cytotoxicity [99].

An alternative method for thiazolidine formation was designed by Rao and coworkers, which involves trapping of 2-cyanobenzothiazole (**15**) with a cysteine residue [100, 101]. The product is a benzothiazole–thiazolidine bridge (**16**), which was used for protein immobilization on surfaces [102, 103]. Importantly, the reaction proceeds with very rapid kinetics, as shown by the preparation of ^{18}F-labeled reagents for PET imaging [104]. The linkage is tolerated *in vivo*, as it is used for live imaging of apoptosis in mice [105].

Three other methods for targeting cysteine were recently reported. First, Kundu and Ball designed rhodium carbenoid reagents for synthesis of thioethers [106]. Second, a collaboration at the University of Strasbourg discovered that 3-arylpropionitrilies are an electrophilic partner with cysteine [107]. These reagents could modify lysozyme and were stable for live cell imaging. Third, PolyTherics, Ltd modified their disulfide bridging reagents to create stable monoconjugates of cysteine [108].

In contrast to the abundance of lysine, tyrosine and tryptophan occur only rarely on the surface of a protein. The nucleophilic reactivity of the phenol and indole is unusual, providing a springboard for selective chemistries. Francis and coworkers developed tyrosine-specific reactions based on π-allyl palladium catalysis [109], the Mannich reaction [110–112], cerium oxidation [113], and the first tryptophan-specific bioconjugation with rhodium carbenoids [114, 115]. Barbas and coworkers pursued tyrosine as well, demonstrating bioconjugation to native trastuzumab tyrosine residues with ene-type reactivity of cyclic diazodicarboxamide [116, 117] reagents.

Lastly, the N-terminus of proteins reacts with pyridoxal-5′-phosphate (PLP) and PLP-like reagents in a transamination to form a single terminal carbonyl. Francis and coworkers discovered this reactivity [118], scanned AA sequence specificity of the reaction, and recently applied the chemistry to the preparation of homogenously modified Fc [119] and trastuzumab constructs [120].

22.6.2 Conjugation to Posttranslational Modifications

Translation of antibody mRNA is directed into the endoplasmic reticulum, where a series of complex PTMs and protein folding result in an active immunoglobulin. Folding of the heavy and light chain subdomains is stabilized by intrachain disulfide formation. Assembly of the intact IgG is stapled through interchain disulfides. Each disulfide formation in the ER relies on enzymes like protein disulfide isomerase, which "proofreads" for reduced or misfolded linkages. N-linked glycans are introduced by glycosyl transferases, which are later elaborated into branched glycoforms. As these chemical functionalities rely on

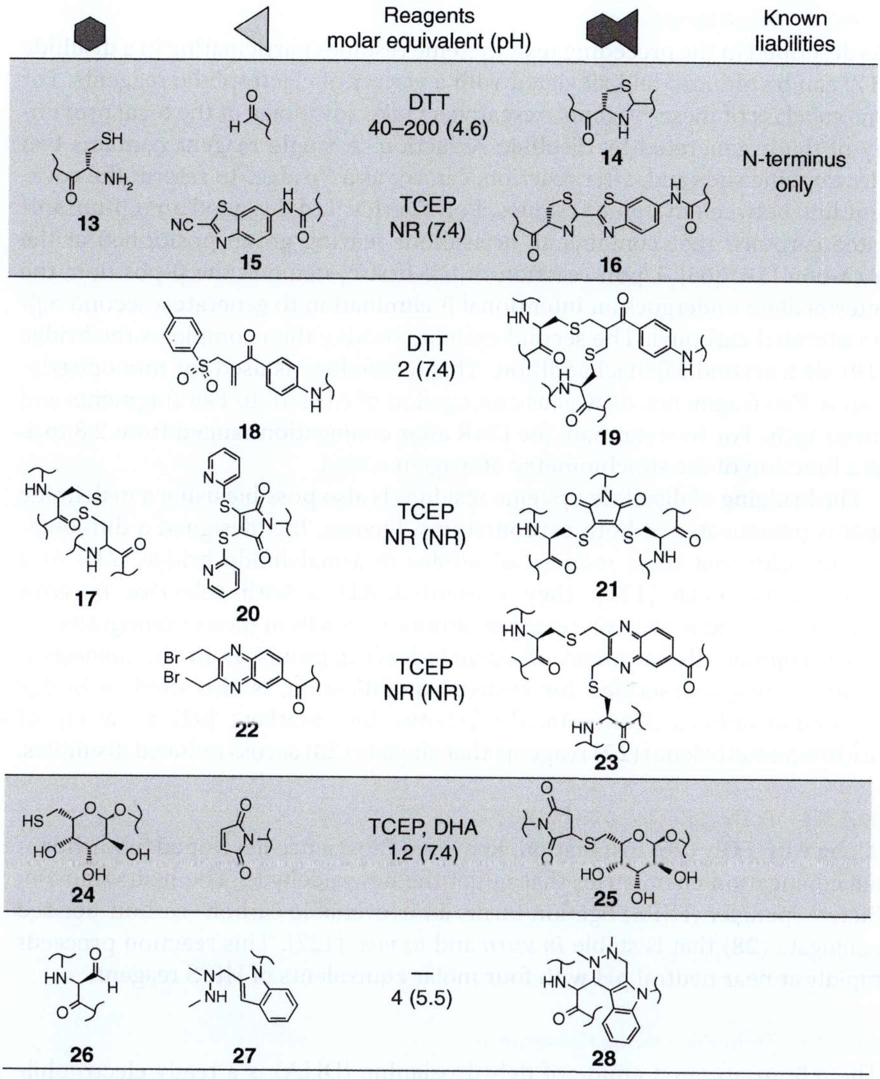

Figure 22.4 Bioconjugation to PTMs. Antibodies can contain native and nonnative PTMs that react with partner reagents. Common reaction conditions for conjugate formation are shown, along with notable liabilities. DHA, dehydroascorbic acid; DTT, D,L-dithiothreitol; TCEP, tris(2-carboxyethyl)phosphine.

enzymatic installation, they are distinct from canonical AAs, enabling their use for orthogonal bioconjugation (Fig. 22.4). Nonnative PTMs can be installed by introducing enzymes into an antibody production cell line, providing new sites for orthogonal bioconjugation.

22.6.2.1 Disulfide Bridge

As described in the preceding text, cysteine residues participating in a disulfide (**17**) can be reduced and alkylated with a variety of electrophilic reagents. For one subclass of these reagents, researchers take advantage of the local proximity of thiols generated by disulfide reduction. A single reagent contains two electrophilic sites and, after reaction, can act as a "bridge" to reform the covalent link between cysteine residues. PolyTherics, Ltd designed an α,β-unsaturated carbonyl that contains an arylsulfone leaving group positioned at the α-carbon (**18**) [65]. Upon reaction of the first cysteine at the β-position, the intermediate undergoes an intentional β-elimination to generate a second α,β-unsaturated carbonyl. The second cysteine residue then completes the bridge (**19**) via a second Michael addition. This technology is used for monopegylation of Fab fragments, as well as conjugation of MMAE to Fab fragments and intact IgGs. For trastuzumab, the DAR after conjugation ranged from 2.8 to 4, as a function of the stoichiometry of reagent added.

The bridging of disulfide cysteine residues is also possible using a maleimide that is preactivated at both carbon atoms. Igenica, Inc. designed a dithiomaleimide (**20**) that traps reduced disulfides in a maleimido bridge (**21**). At a recent conference [121], they presented ADCs with selective *in vitro* cytotoxicity and *in vivo* suppression of tumor growth in mouse xenografts.

Nucleophilic displacement of a halide leaving group, as in the haloacetamide chemistry described for cysteine modification, is also used to bridge reduced disulfides. Concortis Biosystems, Inc. markets [92] a variety of bis(bromomethylene) (**22**) reagents that alkylate (**23**) across reduced disulfides.

22.6.2.2 Formylglycine Alkylation

Along with FGly (**26**) installation, Redwood Bioscience developed bioorthogonal conjugation chemistries that target the new aldehyde. The hydrazino-*iso*-Pictet–Spengler (HIPS) ligation forms an irreversible carbon–carbon-bonded conjugate (**28**) that is stable *in vitro* and *in vivo* [122]. This reaction proceeds rapidly at near neutral pH with four molar equivalents of HIPS reagent.

22.6.2.3 Dehydroalanine (Emerging)

The α,β-unsaturated amide of dehydroalanine (DHA) is a ready electrophile with Michael-type reactivity [123]. The unusual presence of this electrophilic site on a protein enables site-selective bioconjugation. Generation of DHA, however, is challenging to affect under mild conditions. DHA is a naturally occurring functional group used for the cyclization of lantibiotic microbial peptides, and there is an enzyme for its production from serine [124]. Unfortunately, enzymatic DHA formation requires a 24-AA leader peptide to direct the modification. Alternatively, DHA can be produced synthetically from serine, selenocysteine, or cysteine. Early demonstrations used small molecules and short peptides as substrates and required reagents incompatible with protein structure and function (NaOH, H_2O_2). Davis and coworkers have since pioneered selective mild activation of cysteine using a *bis*-alkylation–elimination mechanism, and they demonstrated efficient DHA generation and bioconjugation to subtilisin and a single-domain antibody [123].

22.6.3 Conjugation to Peptide Tags with Enzymes

Enzymes can also catalyze bond-forming reactions on proteins *in vitro*. In parallel to the second-order rate constant for bioconjugation, the Michaelis–Menten parameters (k_{cat}, K_M) describe the efficiency of enzymatic bioconjugation. Substrates in the reaction are consensus sequences of AAs ("tags") that are genetically encoded. The reactive partner varies depending on the native catalytic activity of the enzyme; in principle any bond-forming enzyme that acts on a protein could suffice. We recently reviewed [80] existing technologies for enzymatic bioconjugation in general, and here we will describe those that are used specifically for the synthesis of ADCs (Fig. 22.5).

22.6.3.1 Microbial Transglutaminase

In nature, transglutaminase (TG) cross-links (**31**) the side chains of glutamine (**29**) and lysine by transamidation [125]. The sequence specificity for glutamine and the substrate specificity for the primary amine vary among TGs that evolved in different species. Microbial TG (mTG) has a relatively promiscuous amine preference, but rather a strict sequence requirement. As a result, mTG is site-specific in its reactivity toward proteins, but accepting of a range of primary amine-containing reactive partners (**30**). This duality is put to use by the teams at Rinat-Pfizer [20, 126] and Innate Pharma S.A. [127] for the synthesis of ADCs. The sequence selectivity of mTG is such that it does not modify any natively accessible glutamine residues in the IgG1 class of antibodies. Deglycosylation of N297, or mutation of N297 to remove the glycotransferase recognition sequence (NST), reveals a single glutamine (Q295) that will react with mTG. Alternatively, it is possible to leave the glycans in place and incorporate one or more glutamine tags in the antibody sequence throughout a catalog of locations [125, 128].

22.6.3.2 Phosphopantetheinyl Transferase

During polyketide and non-ribosomal peptide synthesis, incomplete intermediates are covalently attached to carrier proteins by thioester linkages. The sulfur participating in this "anchoring" comes from a phosphopantetheinyl (PP) PTM of the carrier protein. PP transferase (PPTase) catalyzes the installation of this PTM site-specifically at a serine residue. There are a variety of serine-containing tag sequences (**32**) that are substrates for PPTase, and they can be installed into the primary sequence of antibodies. Researchers at Novartis modify these tags with synthetic PP analogues (**33**) carrying cytotoxins

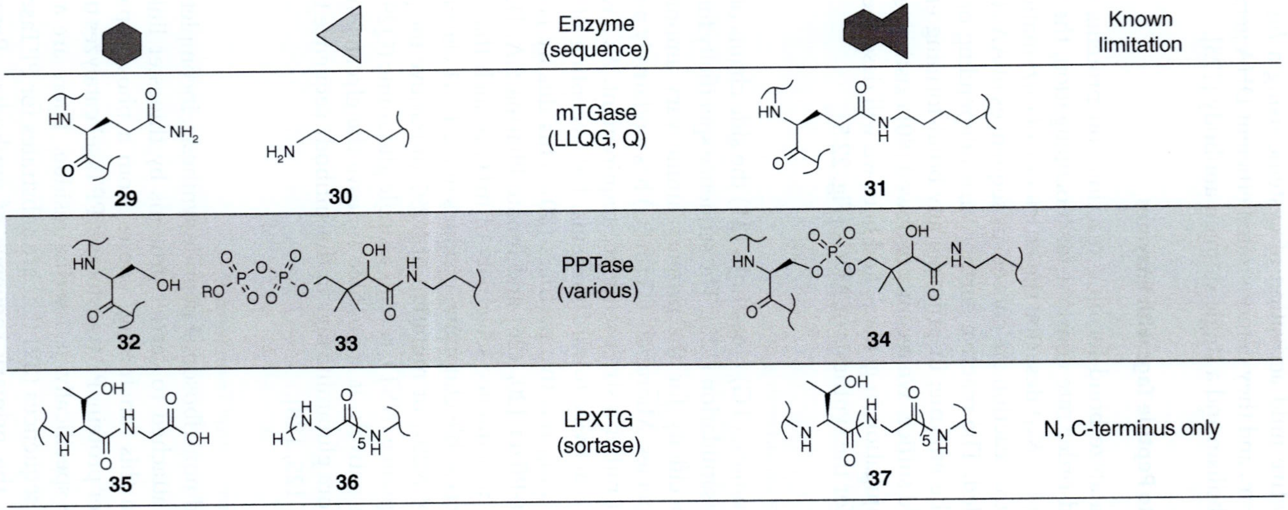

Figure 22.5 Bioconjugation with enzymes on peptide-tagged mAbs. Antibodies can contain AA sequences that are recognized by bond-forming enzymes. Common reaction conditions for conjugate formation are shown. mTGase, microbial transglutaminase; PPTase, phosphopantetheinyl transferase.

[129]. The conjugation efficiency and enzyme requirements are optimized, and *in vitro* and *in vivo* the PP conjugates (**34**) appear stable against degradation during circulation. Across 86 pharmacokinetic timecourse experiments in mice, there was little to no deconjugation of drug from antibody.

22.6.3.3 Emerging Technologies

There is growing interest in bond-forming enzymes that can modify protein sites. Plough and coworkers leverage this approach for N- and C-terminal modification of proteins with sortase. This enzyme natively recognizes a C-terminal LPXTG consensus motif, to which it attaches a polyglycine reagent [130]. It is also possible to encode polyglycine at the N-terminus of a protein to react with a consensus sequence containing reagent [131]. This technology is used for site-specific labeling of HER2-targeting Fab [132] and single-chain antibodies [133].

GSH *S*-transferase (GST) is a relatively promiscuous enzyme that splices the thiol of GSH cysteine residues onto a variety of electrophiles. Pentelute and coworkers developed a method for enzymatic coupling of electron-deficient aryl fluorides to GSH using GST [134, 135]. Installation of GSH into peptide and protein sequences allowed site-selective modification with elaborated aryl substrates.

22.6.4 Conjugation at Nonnatural Substrates

Site-specific installation methods empower researchers with a wide range of orthogonal chemical functionality. The field of bioorthogonal conjugation chemistry is booming, and there is no shortage of choices [76–78, 136]. The difficulty lies in parsing through the available methods for those that might meet the ADC design criteria. This section describes the most developed nonnatural bioconjugation chemistries (Fig. 22.6).

22.6.4.1 Thiol–Disulfide Exchange

Technologies to install disulfides most often rely on reaction with a thiol that is preactivated as a disulfide. Perhaps the most pared down example is *N*-succinimidyl-3-(2-pyridylthio)propionate (SPDP, **41**), which appends a thiopyridine-capped propyl thiol from lysine residues [2]. Incorporation of the desired payload is accomplished by a thiol–disulfide exchange [137] with exogenous thiols (**42**) for secondary bioconjugation to form a new disulfide (**43**). This technology is used for the synthesis of disulfide-bridged immunoconjugates (Section 22.8.1.8) [138–140].

22.6.4.2 Carbonyl Condensation

Reaction of an aldehyde or ketone functional group with an exogenous reagent is rooted early in the development of ADCs. However, most examples include either the formation of an imine or a hydrazone, both of which are more labile

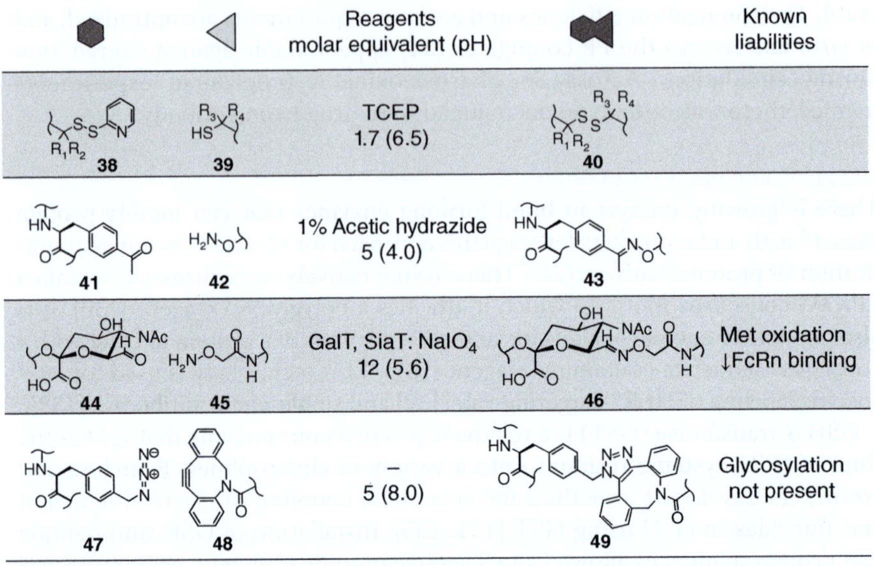

Figure 22.6 Examples of secondary bioconjugation. Nonnatural functional groups are installed with heterobifunctional chemical reagents. Bioconjugation involves a bioorthogonal reaction joining payload and antibody. GalT, galactosyltransferase; SiaT, sialyltransferase.

than desired. In contrast, oxime formation may provide a stable conjugate for use *in vivo*. Aldehyde and ketone derivatives react with O-alkylated hydroxylamines to form oximes. The rate of oxime formation is pH dependent, having a typical maximum at 4–5 [141]. The pH-rate profile follows a bell shape, decreasing at both low and high pH.

Ambrx, Inc. incorporates nnAAs using CHO cell lines engineered with the tRNA/tRNA synthetase machinery. They scanned the sequence space of IgGs with pAF (**44**), an acetophenone variant of phenylalanine. This aryl ketone reacts with alkoxyamines (**45**) to form aryl oximes (**46**) at acidic pH. ADCs prepared in this way are stable against hydrolysis *in vitro* for the lifetime of an antibody in circulation [25, 26]. Similarly, Sanofi Genzyme introduced aminooxy functionality onto auristatin payloads (**48**) that were conjugated to oxidized sialic acids (**47**). The conjugates (**49**) were potent *in vitro* and effective in mouse xenografts [60].

22.6.4.3 Dipolar Cycloaddition

Cycloaddition reactions are perhaps the most popular breeding ground for new bioorthogonal chemistries [142]. Since the report of copper-catalyzed Huisgen cycloaddition [143, 144], and the strain-promoted azide–alkyne cycloaddition (SPAAC) [145], a library of novel approaches to biocompatible cycloaddition have been reported. The primary goal of these new designs has been to enhance the second-order rate constant of reaction. Unfortunately, very little is known about the stability of the resulting products.

That said, a strained cycloaddition technology was reported by Sutro for ADC synthesis. Incorporation of pAMF (**50**) into the primary sequence of antibodies was possible with a cell-free expression system at high titer [146]. In this instance, Sutro chose dibenzocyclooctyne (DBCO, **51**) as the reactive partner. These reagents participate in an SPAAC to form a triazole (**52**). Notably, pAMF and *p*-azidophenylalanine (pAzF) were both tested for reactivity; the SPAAC rate constant for pAMF is sevenfold greater than pAzF. ADCs built with this technology yielded DARs in the range of 1.2–1.9 and showed selective potency *in vitro*.

22.6.4.4 Emerging Technologies

Directed evolution of the phenylalanine tRNA synthetase has enabled a variety of nnAAs that are variants of Phe. These variants can be partners in bioconjugation chemistry. Francis and coworkers developed oxidative coupling technologies that target *p*-aminophenylalanine [147] and *o*-aminotyrosine [148] with aniline-containing reagents. The reactions are mild and biocompatible [148], and perhaps most importantly, they proceed with exceedingly fast rate constants [149]. They have been used to append peptides to intact proteins [150] to enable imaging of fibrin [151], for attachment of aptamers to virus capsids for targeted photodynamic therapy [152], for modification of virus capsids with PEG for PET imaging [153], and for site-selective secondary bioconjugation on Fc domains [148]. There are also technologies that borrow from the rich library of transition metal-catalyzed aryl cross-coupling methods. Installation of iodophenylalanine using nnAA mutagenesis is used for site-selective protein modification with the Mizoroki–Heck [154], Suzuki [155], and Sonogashira [156] reactions.

22.7 Linking Antibodies and Payloads

The connection that spans the space between the protein bioconjugation site and the toxin payload is called a linker. Simply put, linkers must provide a stable junction between antibody and drug, but linkers have a variety of jobs to perform in ADCs. They can increase drug solubility during manufacturing, enable rapid reaction rates during secondary bioconjugation, decrease mAb aggregation after bioconjugation [157], provide multiple toxin attachment sites [157], carry functional groups that can be triggered to release toxin, and even enable the masking of drug from the immune system. To provide some or all of these desired features, and to avoid unwanted properties, linkers must be carefully designed at the molecular level. The backbone materials and synthetic joints used to scaffold linkers must be chosen to meet the ADC design criteria while also enabling desired linker features. Synthetic organic chemists must play a primary role in linker design and construction.

Linkers that include a site of deliberate triggered drug release are deemed "cleavable." In contrast, "noncleavable" linker designs do not include a trigger site, and the active drug is most likely the elaborated drug–linker attached to a single AA as a result of proteolytic digestion of the mAb component in the lysosome. This was confirmed for maytansine–lysine conjugation in trastuzumab emtansine [94] and for an MMAF–cysteine conjugate of brentuximab [158].

Building an elaborated linker is a modular process of assembly. The choice of components will vary depending on the bioconjugation site and the drug solubility and whether or not a trigger is included. In this section we review the methods and components used for linker synthesis. We also describe the functional and stability properties of the product linkages, keeping in mind that catabolic processing of linkers is not well understood.

22.7.1 Linker Backbones and Solubility

When choosing scaffold components to include in the linker backbone, chemists must consider the desired properties of the resulting linkage, which is intended to be permanent. Linkers should provide water solubility, stability against oxidation and reduction, and stability against catabolic degradation *in vivo*. The functional groups (Fig. 22.7) from which linkers can be assembled

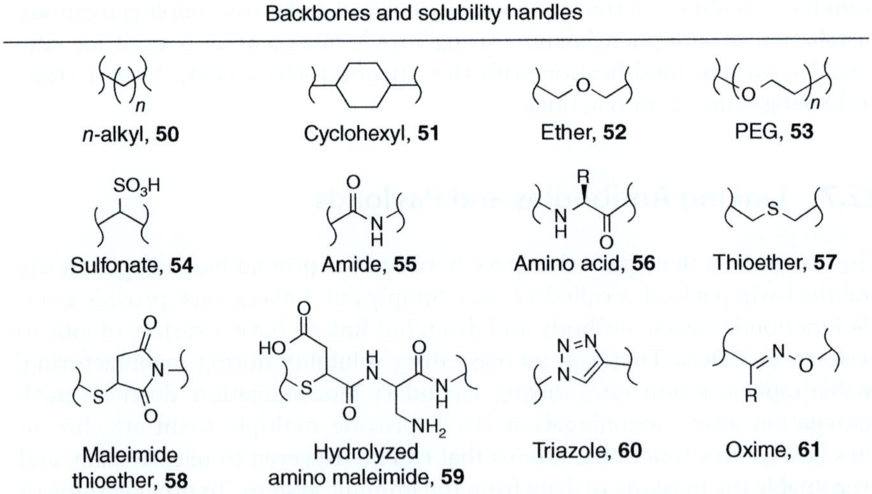

Figure 22.7 Modular components used in linker preparation. Each component must be considered for stability and solubility. Synthetic organic chemistry enables preparation of drug–linker reagents for bioconjugation.

should derive from high-yield bond-forming reactions and easy-to-manipulate chemical intermediates. As will be seen, complex functional groups can sometimes result in unforeseen reactivity or degradation pathways—in general, simpler is better. These factors influence, and in many ways significantly limit, the toolset available to chemists for joining antibody and drug.

22.7.1.1 Alkyl; Cyclohexyl

Linear and branched hydrocarbon chains are stable against most physiological conditions. Hydrocarbons are insensitive to relevant amounts of acid, base, oxidants, and reductants. This simple motif is a good choice for stability against degradation during phase I metabolism [159]. However, the stability of alkyl spacers must be balanced against the hydrophobicity of spacers of increasing length.

Use of a cyclohexyl spacer (**54**) in a heterobifunctional linker (*s*uccinimidyl 4-(*N*-*m*aleimidomethyl)*c*yclohexane-1-*c*arboxylate (SMCC)) was first reported in the conjugation of glucose oxidase to rabbit antibodies [160]. The spacer was chosen primarily because the lifetime of the maleimide was longer for the cyclohexyl group than similar aryl maleimides. This practical fact enabled SMCC to become a reagent of choice for numerous researchers [5] and eventually the commercial production of trastuzumab emtansine.

Linear alkyl maleimides are also stable as reagents. A widely used spacer is the linear C_5H_{10} chain (**53**). Most often this spacer is built from 6-aminocaproic acid, in which the amine becomes a maleimide. The linear analogy to SMCC is the succinimidyl ester of maleimidocaproic acid (succinimidyl 6-(*m*aleimido)*c*aproyl-1-carboxylate, usually abbreviated MC-NHS). Like SMCC, MC-NHS is used to graft a maleimide from Lys residues. As described in Section 22.7, one can also invert the reactivity of the MC spacer so that protein thiols react with the maleimide. This enables attachment of a new functional group by amide bond formation at the caproyl carboxylate. This approach was used by BMS to install hydrazone conjugates of doxorubicin. It is also used by numerous researchers for peptide-linked conjugation, most notably the valine–citrulline (Val–Cit, VC) spacer of brentuximab vedotin (Section 22.8.1.8).

22.7.1.2 *O*-Ether and Poly(Ethylene Glycol)

Alkyl O-linked ethers (**55**) form the backbone of polyethylene glycol (PEG) spacers (**56**), as well as a variety of short bridging linkages. They are stable against heat, acid, and base and are only modestly susceptible to phase I metabolism [161].

PEG chains are an attractive alternative to alkyl spacers because they are highly water soluble. PEG chains are described by the number of repeating ethylene glycol units. At small sizes ($n \sim 2$–10), it is possible to synthesize homogenous reagents with exact lengths. If the chains are longer, their size is defined by an average molecular weight, typically in kDa, and a polydispersity [162].

Small PEG linkers are often used to increase the water solubility of bioconjugation reagents and the hydrophilicity of the product conjugates. There is no consensus on the ideal number of PEG units for this purpose, as conjugates are built with $n = 2$ [24], $n = 3$ [63, 163], $n = 4$ [164–166], and $n = 8$ [60].

Bioconjugates of long PEG chains are used to increase the serum half-life of proteins in general (hormones, cytokines, etc.) [167]. This feature is a result of a number of factors, one of which is that PEG can escape recognition by the immune system, masking the molecule to which it is attached. ADCs are synthesized with long PEG chains as well, and researchers determined that the geometry of attachment affects the utility of PEG conjugation. A branched linker—in which both the toxin and the PEG were attached at the protein surface—was more effective than a linear construct that appended the toxin at the terminus of the PEG chain.

22.7.1.3 Charged Functional Groups

A significant increase in water solubility of hydrophobic molecules is achieved by incorporating charge (positive or negative) into the structure of a molecule. Along with this benefit comes an increased difficulty in synthesis of the reagent. Outside of the incorporation of AAs (see later), there is only one report in which ADCs were prepared with a linker bearing a charged functional group. In an impressive panel of experiments [166], chemists at ImmunoGen installed a propyl sulfonate (**57**) with two different methods. They discovered that these linkers allowed synthesis of ADCs with DARs up to 9 (largely due to less formation of aggregates) without decreasing antigen binding. Their potency increased as a function of this drug loading. They also demonstrated effective toxicity against multiple drug-resistant cancer cell lines. They hypothesized that P-glycoprotein export mechanisms are less effective against highly hydrophilic compounds.

22.7.1.4 Amides, Amino Acids, and Extended Peptides

Perhaps the most common and most versatile junction used to build linkers is the amide (**58**). Synthetically, there are a plethora of methods for forming an amide from an abundance of precursors [168]. Once formed, amides are stable against physiologically relevant amounts of acid, base, reducing and oxidizing agents, and heat [169]. Amides are, however, the substrate for proteases in normal peptide degradation. To avoid undesired cleavage, it is worth a modicum of caution, and usually empirical testing, to determine whether a specific amide is susceptible to proteolysis [170].

Modular construction of linkers from peptide backbones is attractive for a number of reasons. First, solid phase synthesis of peptides is established and straightforward. Second, natural and nonnatural AA side chains (**59**) cover a diverse space of functional groups. Third, peptides can enable triggered release by intracellular proteases.

The lion's share of research on peptide linkers is focused on their intracellular digestion by proteases (Section 22.8.1.8), but a few notable features relate to their structure and solubility. BMS developed [157] a "branched" linker by modifying both the side chain and C-terminal carboxylates of glutamic acid. Increasing drug loading by a factor of two increased the potency of their BR96–doxorubicin conjugates by up to a factor of 10. AA side chains can also increase the solubility of reagents and conjugates. ADCs prepared with lysine [171], citrulline [172], and glutamic acid [24] linkers all enable higher solubility relative to similar hydrophobic groups.

22.7.1.5 S-Ether

Similar to *O*-linked ethers, those formed with sulfur atoms (**60**) are common synthetic products. They are often introduced as a result of nucleophilic displacement by the thiolate anion, which is readily generated by deprotonation at mild pH. Thioethers are generally stable against heat, acid, and reduction. However, the lower ionization potential and increased valency of sulfur—compared to oxygen—make it prone to one- and two-electron oxidation to either the sulfoxide or, subsequently, the sulfone [173–175]. In a biological setting, thioether reactivity is understood from the context of methionine side chains [176]. Met is susceptible to oxidation by reactive oxygen species, specifically hydrogen peroxide. These oxidized forms are much more susceptible to elimination or displacement reactions that can cause undesired linker fragmentation, particularly in the case of the maleimide thioether. For α-CD70 conjugates in mice, the stability of a cysteine acetamide was significantly longer (no loss over 14 days) than its maleimide-linked comparator [177].

22.7.1.6 Maleimide Thioether

An important corollary to the Michael addition (Section 22.5.1) is the β-elimination mechanism, in which the product of a Michael addition can undergo deprotonation at the α-carbon and elimination from the β-position to regenerate the α,β-unsaturated carbonyl. In the case of maleimide thioethers (**61**), this "reverse Michael" is a source of instability [25, 27, 177]. This deconjugation occurs spontaneously at neutral pH *in vitro* and *in vivo*. The latter nonspecifically releases toxin into circulation, the extent of which can be monitored by measuring the presence of maleimide reconjugation onto serum albumin proteins [27, 177, 178]. A rigorous analysis of this liability [27] demonstrates that the reverse Michael can be prevented if the maleimide ring hydrolyzes and that the rate of hydrolysis depends upon the protein environment surrounding the thioether.

In an effort to decrease nonspecific release of drug from maleimide thioethers, a new type of maleimide was designed (**9**) that carries a proximal amine, which is intended to *promote* ring hydrolysis before the reverse Michael reaction can occur [121, 179]. The hydrolyzed maleimide (**62**) cannot undergo the reverse Michael reaction nor can it cyclize to regenerate the maleimide, thereby increasing the stability of the thioether junction. Structurally, the linkage then becomes an S-ether and an amide, both of which are significantly more stable against deconjugation.

Unfortunately, the maleimide thioether is also unstable against oxidation. Researchers at ImmunoGen discovered that physiological conditions (pH 7.4, 37 °C) promote sulfoxide formation on the maleimide thioether sulfur [180]. This sulfoxide is significantly more susceptible to β-elimination, which fragments the linker. The fragmentation is not prevented by the addition of significant excess of a strong reducing agent (DTT), and it is accelerated by the addition of an oxidant (H_2O_2). Considering the oxidizing potential encountered by ADCs during endosomal recycling [181], the possibility that maleimide thioether fragmentation occurs through both the reverse Michael and oxidation is a concern that warrants further exploration.

22.7.1.7 Triazole

Most prominent among emerging bioconjugation technologies are the variety of azide–alkyne cycloadditions (AAC) [76]. The product of this reaction is a triazole (**63**), which is an aromatic five-membered ring. While the reaction chemistries explore a wide range of conditions, very little is specifically known about the stability of triazoles against physiological conditions [182]. Investigators are introducing triazole junctions into ADC linkers, and while preliminary reports suggest they are not a liability, further exploration is warranted.

22.7.1.8 Oxime

Like triazoles, oximes (**64**) typically result from secondary bioconjugation (a notable exception is the nnAA pAF described earlier). As it is the reverse of their formation, the hydrolysis of oximes [183] follows a two-step addition–elimination mechanism. This is the same as for imine and hydrazone hydrolysis. Addition of water into the oxime is dependent upon protonation of the oxime nitrogen, which, as a result of the α-effect of the neighboring oxygen, is significantly depressed. There are very few kinetic studies of oxime hydrolysis, but in one, the pK_a of cyclohexanone oxime was found to be 1.6 in phosphate-buffered solution. This leads to a very small fraction of protonated oxime near neutral pH, and as such the rate of nucleophilic attack of water is almost negligible, despite the high concentration of water.

In the case of benzaldehyde imine hydrolysis, the effect of varying the substitution on the aromatic ring resulted in a range of hydrolysis rates spanning three orders of magnitude [184], which was attributed to the acidity of the protonated imine. Similarly, in the case of pivalaldehyde hydrazone, the rate of hydrolysis depends upon substitution of the terminal N-atom. To the best of our knowledge, for substituted oximes, the change in this pK_a—either as a result of the substitution on the parent carbonyl or the hydroxylamine O-atom—is not known.

22.7.2 Triggered Linker Cleavage

Without a specific mechanism to do so, the release of a toxin from an ADC likely arises from exhaustive proteolytic processing of the antibody. This is known for two examples [94, 158]. For some toxins, this is an acceptable result, in that their potency is not (significantly) modulated by the presence of residual linker and an AA. However, for some toxins, their mechanism of action depends upon release of the unadorned molecule. Prominent examples of this requirement include MMAE [185], duocarmycin [186], and doxorubicin [187]. In addition, endocytosis of ADCs is not a one-way street. Endosomal recycling can eject internalized antibodies into the extracellular space [188]. As such, methods that enable rapid release of the toxin (Fig. 22.8) into the tumor cell before recycling, and without waiting for complete proteolysis, can enhance the therapeutic effect of an ADC.

22.7.2.1 pH

Receptor-mediated endocytosis of an antibody bound to its target antigen is the first step in trafficking an ADC into the cell. The change in pH that accompanies this vesicle formation is significant, dropping from 7.4 in serum to 5.5 in

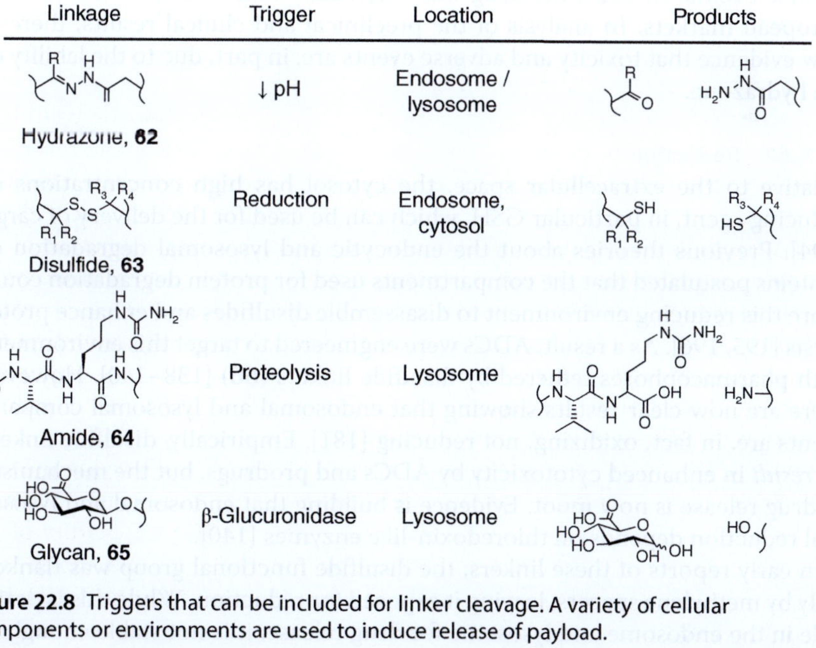

Figure 22.8 Triggers that can be included for linker cleavage. A variety of cellular components or environments are used to induce release of payload.

the endosome [189]. To exploit this cellular trigger, chemists install an acid-labile functional group into linkers. Methods include hydrazones (**65**, from both hydrazines and hydrazides [190]), citraconic acid [191], Kemp's triacid [192], and thiomalic acid [193].

This technology is one of the more troubled stories in the archives of ADC development. Lilly designed and built ADCs with a hydrazone-linked vinca alkaloid (desacetylvinblastine). The drug was appended to chemically modified glycans on monoclonal KS1/4, whose antigen is overexpressed on human ovarian carcinomas. The assembled therapeutic, KS1/4-DAVLB-HY (Lilly LY203725), was selectively potent *in vitro*, induced tumor growth inhibition *in vivo*, and was tested in phase I trials. The program was ended due to dose-limiting toxicity in phase I that was ascribed to systemic instability of the hydrazone linkage [55].

Wyeth Laboratories designed an ADC built from gemtuzumab, a humanized α-CD33 mAb, modified through lysine residues with calicheamicin, a highly potent DNA-targeting cytotoxin. The linker between lysine and the toxin contains both a hydrazone and a hindered disulfide. Gemtuzumab ozogamicin (GO/Mylotarg®) showed antigen-dependent cytotoxicity *in vitro*, with good animal efficacy *in vivo* and acceptable toxicology. GO was granted accelerated approval following abbreviated clinical testing. After being approved for marketing, phase III trials continued, which eventually revealed no benefit to therapy versus the standard of care. In addition, GO resulted in increased adverse events. In 2010, the drug was voluntarily removed from the US and European markets. In analysis of the preclinical and clinical results, there is now evidence that toxicity and adverse events are, in part, due to the lability of the hydrazone.

22.7.2.2 Reduction

Relative to the extracellular space, the cytosol has high concentrations of reducing agent, in particular GSH, which can be used for the delivery of cargo [194]. Previous theories about the endocytic and lysosomal degradation of proteins postulated that the compartments used for protein degradation could share this reducing environment to disassemble disulfides and enhance proteolysis [195, 196]. As a result, ADCs were engineered to target this environment with pharmacophores tethered by disulfide linkers (**66**) [138–140]. However, there are now clear results showing that endosomal and lysosomal compartments are, in fact, oxidizing, not reducing [181]. Empirically, disulfide linkers *do result* in enhanced cytotoxicity by ADCs and prodrugs, but the mechanism of drug release is now moot. Evidence is building that endosomal and lysosomal reduction depends on thioredoxin-like enzymes [140].

In early reports of these linkers, the disulfide functional group was flanked only by methylene spacers, leaving it exposed for reduction. While this is desirable in the endosome and lysosome for drug release, it is not stable enough in serum and other tissues. It is possible to buttress the lability of alkyl disulfides against mild reducing agents by sterically blocking access to the sulfur atoms. Introduction of one or two methyl groups at the disulfide α-carbons decreases the reductive cleavage of disulfides in model systems, *in vitro,* and *in vivo.* Notably, a calicheamicin dimethyl disulfide exhibited lower toxicity and better potency [197].

22.7.2.3 Peptide Hydrolysis (Enzymatic)

The protease expression profile of many cancer types can be significantly different from healthy tissues. Neoplastic cells often overexpress proteases that enhance degradation of the extracellular matrix, promote angiogenesis, prevent caspase activation, and enable increased metabolism. Recognition of this increased expression has led researchers to target lysosomal proteolysis as a trigger to release a toxin from an ADC.

Before attaching them to antibodies, researchers recognized the potential of protease-activated prodrugs. Modifications were made to anthracyclines with short hydrophobic peptides that are substrates of cathepsins. This quickly expanded to albumin-conjugated prodrugs, and stemming from that work, ADCs were developed that incorporated one to four AA linkers. Those linkers with three or four AAs enabled efficient lysosomal proteolysis and release of drug molecules [198]. Faster release of daunorubicin in purified lysosomes correlated with longer peptides, and when tested against a leukemia cell line in mice, the longer peptides led to better survival.

One protease in particular has garnered attention from the community. Cathepsin B (Cat. B) is an abundant lysosomal protease that is overexpressed in many cancer types [199]. It has peptidyl peptidase activity (C-terminal dipeptide hydrolysis) [200] with broad substrate tolerance, but preference for basic residues at P1, either basic or hydrophobic residues at P2, and hydrophobic residues (or functional groups) at P1′ [201, 202]. Development of ADC linkers containing dipeptide substrates of Cat. B led to AA sequences with rapid cleavage kinetics [171]. The utility of these dipeptides for building ADCs was expanded by comparing the relative serum stability across a family of dipeptides in the development of brentuximab vedotin [203]. It was discovered that the Val–Cit (VC, **67**) substrate is rapidly cleaved by Cat. B, but is stable against serum hydrolysis during transit. This motif is now widely used in linker designs.

22.7.2.4 Glycolysis (Enzymatic)

Similar to proteases, the expression profile of glycosidases in cancers is often different from healthy tissues [204]. This effect is observed in the tumor interstitium as neoplasms attempt to remodel their extracellular matrix to enable metastasis. The level of β-glucuronidase, a lysosomal hydrolase, was six times higher in breast cancer tissues [205]. This led to synthesis of doxorubicin prodrugs that respond to β-glucuronidase hydrolysis [204, 206, 207]. Seattle Genetics has since expanded this approach to design ADC linkers that are triggered by β-glucuronidase. Their first application attached amine-containing auristatins (MMAE and MMAF) to the trigger site. Hydrolysis of the sugar (**68**) releases a phenolic residue primed for 1,6-elimination (Section 22.8.2.4) [164]. They succeeded in delivering toxin in an enzyme-dependent manner. This technology was then expanded to a DNA minor groove binder [208], to camptothecin [209], and most recently to phenolic drugs by adding a DMED spacer (Section 22.8.2.4) [210].

22.7.3 Trigger Enhancement by Self-Immolation

The release of drug molecules from trigger sites is not always a straightforward synthetic task. As described in the preceding text, the tools for initiating release depend on quite specific chemical functionality or enzymes. In many cases, compatible groups are not present on a toxin and cannot be introduced without compromising potency.

In these cases, chemists can install a spacer within the linker, allowing them to independently design the trigger and the point of attachment to the toxin. For this approach to succeed, the spacer that is introduced must be compatible with the trigger mechanism, and it must automatically release the unmodified toxin. Spacers that meet these requirements are called "self-immolative." There are two primary technologies that succeed at performing this Rube Goldberg-type cascade, quinone methide elimination and cyclization elimination (Fig. 22.9).

22.7.3.1 (Imino)quinone Methide Elimination

The concept of installing a substituted benzyl group (**69**) as a prodrug spacer was created by Katzenellenbogen and coworkers [211]. Their report provides a complete description of enzymatic activation, the $4n+2$ pi system electronic

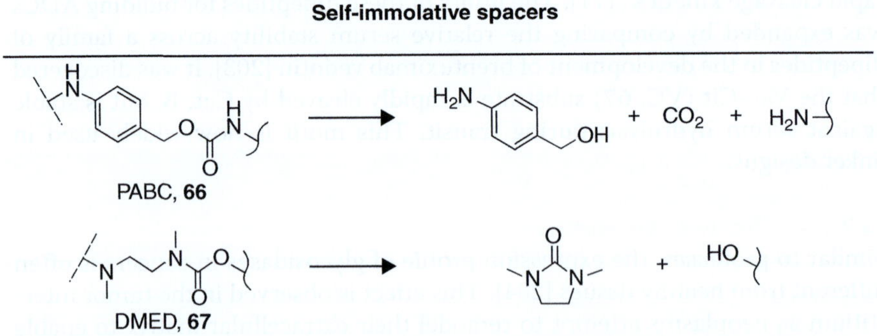

Figure 22.9 Self-immolative spacers. These spacers enable automatic release of unmodified payload after a triggered event. Very few reliable technologies are available for this purpose.

cascade, elimination from the exocyclic benzyl position, and formation of the iminoquinone methide and its probable hydration and even proposed use of "self-immolative" as a descriptor. The method was rapidly adopted with great success, and since this introduction it has been used in PEG prodrugs [212] and antibody-directed enzyme prodrug therapy (ADEPT) [213].

22.7.3.2 Cyclization Elimination

Another self-immolative technology involves triggered release of a functional group that can cyclize upon itself. The prototypical examples of this are carbamates formed from dialkyl ethylenediamines. By varying alkylation patterns in this spacer and the pH of the reaction conditions, the rate of cyclization elimination spanned orders of magnitude [214]. The fastest cyclization was found near neutral pH with *N,N′*-dimethylethylenediamine (DMED, **70**). This spacer has become particularly useful when participating in a carbamate with *p*-aminobenzyl alcohol. Syntarga builds a linear arrangement of these two spacers, which enables two-step self-immolation to release phenolic drugs [210].

22.8 Conclusion

The challenges of ADC design require tenacious, thoughtful, and most of all thorough examination of each ADC component. We have provided a survey of these myriad technologies, with the hope that cataloging and exploring the methods behind this rapidly growing class of treatments can further accelerate their development. As with any dangerous but valuable cargo, care must be taken to craft safe, reliable packaging that arrives at its destination intact. We look forward to your new developments, and we hope for their success.

References

1. Ghose, T.; Blair, A. H. Antibody-linked cytotoxic agents in the treatment of cancer: current status and future prospects. *J. Natl. Cancer Inst.* 1978, 61, 657–676.
2. (a) Ghose, T. I.; Huntley Blair, A.; Kulkarni, P. N. Preparation of antibody-linked cytotoxic agents. *Methods Enzymol.* 1983, 93, 280–333; (b) Langone, J. J.; Vunakis, H. V. (eds) (1983) Immunochemical Techniques Part F: Conventional Antibodies, Fc Receptors, and Cytotoxicity, Academic Press, New York.
3. Pietersz, G. A. The linkage of cytotoxic drugs to monoclonal antibodies for the treatment of cancer. *Bioconjug. Chem.* 1990, 1, 89–95.
4. Chari, R. V. J. Targeted delivery of chemotherapeutics: tumor-activated prodrug therapy. *Adv. Drug Deliv. Rev.* 1998, 31, 89–104.
5. Dubowchik, G. M.; Walker, M. A. Receptor-mediated and enzyme-dependent targeting of cytotoxic anticancer drugs. *Pharmacol. Ther.* 1999, 83, 67–123.

6. Litvak-Greenfeld, D.; Benhar, I. Risks and untoward toxicities of antibody-based immunoconjugates. *Adv. Drug Deliv. Rev.* 2012, 64, 1782–1799.
7. Muller, P. Y.; Milton, M. N. The determination and interpretation of the therapeutic index in drug development. *Nat. Rev. Drug Discov.* 2012, 11, 751–761.
8. Okeley, N. M.; Miyamoto, J. B.; Zhang, X.; Sanderson, R. J.; Benjamin, D. R.; Sievers, E. L.; Senter, P. D.; Alley, S. C. Intracellular activation of SGN-35, a potent anti-CD30 antibody–drug conjugate. *Clin. Cancer Res.* 2010, 16, 888–897.
9. Rocchetti, M.; Simeoni, M.; Pesenti, E.; De Nicolao, G.; Poggesi, I. Predicting the active doses in humans from animal studies: a novel approach in oncology. *Eur. J. Cancer* 2007, 43, 1862–1868.
10. Haddish-Berhane, N.; Shah, D. K.; Ma, D.; Leal, M.; Gerber, H.-P.; Sapra, P.; Barton, H. A.; Betts, A. M. On translation of antibody drug conjugates efficacy from mouse experimental tumors to the clinic: a PK/PD approach. *J. Pharmacokinet. Pharmacodyn.* 2013, 40, 557–571.
11. Jumbe, N. L.; Xin, Y.; Leipold, D. D.; Crocker, L.; Dugger, D.; Mai, E.; Sliwkowski, M. X.; Fielder, P. J.; Tibbitts, J. Modeling the efficacy of trastuzumab-DM1, an antibody drug conjugate, in mice. *J. Pharmacokinet. Pharmacodyn.* 2010, 37, 221–242.
12. Roberts, S. A.; Andrews, P. A.; Blanset, D.; Flagella, K. M.; Gorovits, B.; Lynch, C. M.; Martin, P. L.; Kramer-Stickland, K.; Thibault, S.; Warner, G. Considerations for the nonclinical safety evaluation of antibody drug conjugates for oncology. *Regul. Toxicol. Pharmacol.* 2013, 67, 382–391.
13. Dixit, R.; Iciek, L. A.; McKeever, K.; Ryan, P. C. Challenges of general safety evaluations of biologics compared to small molecule pharmaceuticals in animal models. *Expert Opin. Drug Discov.* 2010, 5, 79–94.
14. Poon, K. A.; Flagella, K.; Beyer, J.; Tibbitts, J.; Kaur, S.; Saad, O.; Yi, J.-H.; Girish, S.; Dybdal, N.; Reynolds, T. Preclinical safety profile of trastuzumab emtansine (T-DM1): mechanism of action of its cytotoxic component retained with improved tolerability. *Toxicol. Appl. Pharmacol.* 2013, 273, 298–313.
15. (a)Anderl, J.; Faulstich, H.; Hechler, T.; Kulke, M. Antibody–drug conjugate payloads. *Methods Mol. Biol.* 2013, 1045, 51–70;(b)Ducry, L. (ed.) (2013) Antibody–Drug Conjugates, Humana Press, New York, pp. 267–273.
16. Mathe, G.; Tran Ba, L. O. C.; Bernard, J. Effet sur la leucémie 1210 de la Souris d'une combinaison par diazotation d'A-méthoptérine et de y-globulines de hamsters porteurs de cette leucémie par hétérogreffe. *C. R. Hebd. Seances Acad. Sci.* 1958, 246, 1626–1628.
17. Dillman, R. O.; Johnson, D. E.; Shawler, D. L.; Koziol, J. A. Superiority of an acid-labile daunorubicin-monoclonal antibody immunoconjugate compared to free drug. *Cancer Res.* 1988, 48, 6097–6102.
18. Junutula, J. R.; Flagella, K. M.; Graham, R. A.; Parsons, K. L.; Ha, E.; Raab, H.; Bhakta, S.; Nguyen, T.; Dugger, D. L.; Li, G.; et al. Engineered

thio-trastuzumab-DM1 conjugate with an improved therapeutic index to target human epidermal growth factor receptor 2-positive breast cancer. *Clin. Cancer Res.* 2010, 16, 4769–4778.

19. McDonagh, C. F.; Turcott, E.; Westendorf, L.; Webster, J. B.; Alley, S. C.; Kim, K.; Andreyka, J.; Stone, I.; Hamblett, K. J.; Francisco, J. A.; et al. Engineered antibody–drug conjugates with defined sites and stoichiometries of drug attachment. *Protein Eng. Des. Sel.* 2006, 19, 299–307.

20. Strop, P.; Liu, S.-H.; Dorywalska, M.; Delaria, K.; Dushin, R. G.; Tran, T.-T.; Ho, W.-H.; Farias, S.; Casas, M. G.; Abdiche, Y.; et al. Location matters: site of conjugation modulates stability and pharmacokinetics of antibody drug conjugates. *Chem. Biol.* 2013, 20, 161–167.

21. McDonagh, C. F.; Kim, K. M.; Turcott, E.; Brown, L. L.; Westendorf, L.; Feist, T.; Sussman, D.; Stone, I.; Anderson, M.; Miyamoto, J.; et al. Engineered anti-CD70 antibody–drug conjugate with increased therapeutic index. *Mol. Cancer Ther.* 2008, 7, 2913–2923.

22. Boswell, C. A.; Mundo, E. E.; Zhang, C.; Bumbaca, D.; Valle, N. R.; Kozak, K. R.; Fourie, A.; Chuh, J.; Koppada, N.; Saad, O.; et al. Impact of drug conjugation on pharmacokinetics and tissue distribution of anti-STEAP1 antibody–drug conjugates in rats. *Bioconjug. Chem.* 2011, 22, 1994–2004.

23. Erickson, H. K.; Lambert, J. M. ADME of antibody–maytansinoid conjugates. *AAPS J.* 2012, 14, 799–805.

24. Drake, P. M.; Albers, A. E.; Baker, J.; Banas, S.; Barfield, R. M.; Bhat, A. S.; de Hart, G. W.; Garofalo, A. W.; Holder, P.; Jones, L. C.; et al. Aldehyde tag coupled with HIPS chemistry enables the production of ADCs conjugated site-specifically to different antibody regions with distinct *in vivo* efficacy and PK outcomes. *Bioconjug. Chem.* 2014, 25, 1331–1341.

25. Jackson, D.; Atkinson, J.; Guevara, C. I.; Zhang, C.; Kery, V.; Moon, S.-J.; Virata, C.; Yang, P.; Lowe, C.; Pinkstaff, J.; et al. *In vitro* and *in vivo* evaluation of cysteine and site specific conjugated herceptin antibody–drug conjugates. *PLoS ONE* 2014, 9, e83865.

26. Tian, F.; Lu, Y.; Manibusan, A.; Sellers, A.; Tran, H.; Sun, Y.; Phuong, T.; Barnett, R.; Hehli, B.; Song, F.; et al. A general approach to site-specific antibody drug conjugates. *Proc. Natl. Acad. Sci. U. S. A.* 2014, 111, 1766–1771.

27. Shen, B.-Q.; Xu, K.; Liu, L.; Raab, H.; Bhakta, S.; Kenrick, M.; Parsons-Reponte, K. L.; Tien, J.; Yu, S.-F.; Mai, E.; et al. Conjugation site modulates the *in vivo* stability and therapeutic activity of antibody–drug conjugates. *Nat. Biotechnol.* 2012, 30, 184–189.

28. Jeffrey, S. C.; Burke, P. J.; Lyon, R. P.; Meyer, D. W.; Sussman, D.; Anderson, M.; Hunter, J. H.; Leiske, C. I.; Miyamoto, J. B.; Nicholas, N. D.; et al. A potent anti-CD70 antibody–drug conjugate combining a dimeric pyrrolobenzodiazepine drug with site-specific conjugation technology. *Bioconjug. Chem.* 2013, 24, 1256–1263.

29. Wade, R.; Whisson, M. E.; Szekerke, M. Some serum protein nitrogen mustard complexes with high chemotherapeutic selectivity. *Nature* 1967, 215, 1303–1304.
30. Rowland, G. F.; O'Neill, G. J.; Davies, D. A. Suppression of tumour growth in mice by a drug–antibody conjugate using a novel approach to linkage. *Nature* 1975, 255, 487–488.
31. Hurwitz, E.; Levy, R.; Maron, R.; Wilchek, M.; Arnon, R.; Sela, M. The covalent binding of daunomycin and adriamycin to antibodies, with retention of both drug and antibody activities. *Cancer Res.* 1975, 35, 1175–1181.
32. Winkelhake, J. L. Effects of chemical modification of antibodies on their clearance from the circulation. Addition of simple aliphatic compounds by reductive alkylation and carbodiimide-promoted amide formation. *J. Biol. Chem.* 1977, 252, 1865–1868.
33. Kim, M. T.; Chen, Y.; Marhoul, J.; Jacobson, F. Statistical modeling of the drug load distribution on trastuzumab emtansine (Kadcyla), a lysine-linked antibody drug conjugate. *Bioconjug. Chem.* 2014, 25, 1223–1232.
34. Liu, H.; May, K. Disulfide bond structures of IgG molecules: structural variations, chemical modifications and possible impacts to stability and biological function. *mAbs* 2012, 4, 17–23.
35. (a) Konigsberg, W. Reduction of disulfide bonds in proteins with dithiothreitol. *Methods Enzymol.* 1972, 25, 185–188; (b) Hirs, C. H. W.; Timasheff, S. N. (eds) (1972) Enzyme Structure, Part B, Academic Press, New York, pp. 3–27.
36. Linford, J. H. 2,3,5-Tris-ethylenimino-1,4-benzoquinone (Trenimon): some chemical and biological properties. *Chem. Biol. Interact.* 1973, 6, 149–168.
37. Linford, J. H.; Froese, G.; Berczi, I.; Israels, L. G. An alkylating agent–globulin conjugate with both al-kylating and antibody activity. *J. Natl. Cancer Inst.* 1974, 52, 1665–1667.
38. Linford, J. H.; Froese, G. Comparisons of the chemical and biologic properties of triaziquone and triaziquone–protein conjugates. *J. Natl. Cancer Inst.* 1978, 60, 307–316.
39. Willner, D.; Trail, P. A.; Hofstead, S. J.; King, H. D.; Lasch, S. J.; Braslawsky, G. R.; Greenfield, R. S.; Kaneko, T.; Firestone, R. A. (6-Maleimidocaproyl) hydrazone of doxorubicin. A new derivative for the preparation of immunoconjugates of doxorubicin. *Bioconjug. Chem.* 1993, 4, 521–527.
40. Trail, P. A.; Willner, D.; Lasch, S. J.; Henderson, A. J.; Hofstead, S.; Casazza, A. M.; Firestone, R. A.; Hellstrom, I.; Hellstrom, K. E. Cure of xenografted human carcinomas by BR96–doxorubicin immunoconjugates. *Science* 1993, 261, 212–215.
41. Mosure, K. W.; Henderson, A. J.; Klunk, L. J.; Knipe, J. O. Disposition of conjugate-bound and free doxorubicin in tumor-bearing mice following administration of a BR96–doxorubicin immunoconjugate (BMS 182248). *Cancer Chemother. Pharmacol.* 1997, 40, 251–258.

42. Saleh, M. N.; Sugarman, S.; Murray, J.; Ostroff, J. B.; Healey, D.; Jones, D.; Daniel, C. R.; LeBherz, D.; Brewer, H.; Onetto, N.; et al. Phase I trial of the anti-Lewis Y drug immunoconjugate BR96–doxorubicin in patients with Lewis Y-expressing epithelial tumors. *J. Clin. Oncol.* 2000, 18, 2282–2292.
43. Tolcher, A. W.; Sugarman, S.; Gelmon, K. A.; Cohen, R.; Saleh, M.; Isaacs, C.; Young, L.; Healey, D.; Onetto, N.; Slichenmyer, W. Randomized phase II study of BR96–doxorubicin conjugate in patients with metastatic breast cancer. *J. Clin. Oncol.* 1999, 17, 478–484.
44. Singh, R.; Maloney, E. K. Labeling of antibodies by *in situ* modification of thiol groups generated from selenol-catalyzed reduction of native disulfide bonds. *Anal. Biochem.* 2002, 304, 147–156.
45. Francisco, J. A. cAC10-vcMMAE, an anti-CD30–monomethyl auristatin E conjugate with potent and selective antitumor activity. *Blood* 2003, 102, 1458–1465.
46. Hamblett, K. J.; Senter, P. D.; Chace, D. F.; Sun, M. M.; Lenox, J.; Cerveny, C. G.; Kissler, K. M.; Bernhardt, S. X.; Kopcha, A. K.; Zabinski, R. F.; et al. Effects of drug loading on the antitumor activity of a monoclonal antibody drug conjugate. *Clin. Cancer Res.* 2004, 10, 7063–7070.
47. Wolfe, C.; Hage, D. Studies on the rate and control of antibody oxidation by periodate. *Anal. Biochem.* 1995, 231, 123–130.
48. Hage, D. S.; Wolfe, C. A. C.; Oates, M. R. Development of a kinetic model to describe the effective rate of antibody oxidation by periodate. *Bioconjug. Chem.* 1997, 8, 914–920.
49. Starling, J. J.; Maciak, R. S.; Hinson, N. A.; Nichols, C. L.; Briggs, S. L.; Laguzza, B. C.; Smith, W.; Corvalan, J. R. F. *In vivo* antitumor activity of a panel of four monoclonal antibody–vinca alkaloid immunoconjugates which bind to three distinct epitopes of carcinoembryonic antigen. *Bioconjug. Chem.* 1992, 3, 315–322.
50. Abraham, R.; Moller, D.; Gabel, D.; Senter, P.; Hellström, I.; Hellström, K. E. The influence of periodate oxidation on monoclonal antibody avidity and immunoreactivity. *J. Immunol. Methods* 1991, 144, 77–86.
51. Rodwell, J. D.; Alvarez, V. L.; Lee, C.; Lopes, A. D.; Goers, J. W.; King, H. D.; Powsner, H. J.; McKearn, T. J. Site-specific covalent modification of monoclonal antibodies: *in vitro* and *in vivo* evaluations. *Proc. Natl. Acad. Sci. U. S. A.* 1986, 83, 2632–2636.
52. Ranadive, G.; Rosenzweig, H.; Epperly, M.; Seskey, T.; Bloomer, W. A new method of Tc-99m labeling of monoclonal-antibodies through sugar residues—a study with tag-72 specific Cc-49 antibody. *Nucl. Med. Biol.* 1993, 20, 719–726.
53. Apelgren, L. D.; Zimmerman, D. L.; Briggs, S. L.; Bumol, T. F. Antitumor activity of the monoclonal antibody–vinca alkaloid immunoconjugate LY203725 (KS1/4-4-desacetylvinblastine-3-carboxhydrazide) in a nude mouse model of human ovarian cancer. *Cancer Res.* 1990, 50, 3540–3544.

54. Apelgren, L. D.; Bailey, D. L.; Briggs, S. L.; Barton, R. L.; Guttman-Carlisle, D.; Koppel, G. A.; Nichols, C. L.; Scott, W. L.; Lindstrom, T. D. Chemoimmunoconjugate development for ovarian carcinoma therapy: preclinical studies with vinca alkaloid–monoclonal antibody constructs. *Bioconjug. Chem.* 1993, 4, 121–126.
55. Petersen, B.; Barrett, P.; Labus, J.; Woodworth, J.; Zimmermann, J.; Butler, F.; Dugan, W.; Schneck, D. Murine monoclonal antibody–vinca conjugate KS1/4-DAVLB hydrazide—phase 1 studies in patients with adenocarcinoma. *Antib. Immunoconjug. Radiopharm.* 1993, 6, 127–139.
56. Pearson, J. W.; Fogler, W. E.; Volker, K.; Usui, N.; Goldenberg, S. K.; Gruys, E.; Riggs, C. W.; Komschlies, K.; Wiltrout, R. H.; Tsuruo, T.; et al. Reversal of drug resistance in a human colon cancer xenograft expressing MDR1 complementary DNA by *in vivo* administration of MRK-16 monoclonal antibody. *J. Natl. Cancer Inst.* 1991, 83, 1386–1391.
57. Lindstrom, T. D.; Althaus, W. A.; Ruterbories, K. J.; Kau, D. Pharmacokinetics and disposition of the KS1/4 monoclonal antibody–desacetylvinblastine hydrazide conjugate LY203725 in rats and monkeys. *J. Pharmacol. Exp. Ther.* 1990, 252, 1117–1124.
58. Shih, L. B.; Goldenberg, D. M.; Xuan, H.; Lu, H. W.-Z.; Mattes, M. J.; Hall, T. C. Internalization of an intact doxorubicin immunoconjugate. *Cancer Immunol. Immunother.* 1994, 38, 92–98.
59. Zuberbühler, K.; Casi, G.; Bernardes, G. J. L.; Neri, D. Fucose-specific conjugation of hydrazide derivatives to a vascular-targeting monoclonal antibody in IgG format. *Chem. Commun.* 2012, 48, 7100.
60. Zhou, Q.; Stefano, J. E.; Manning, C.; Kyazike, J.; Chen, B.; Gianolio, D. A.; Park, A.; Busch, M.; Bird, J.; Zheng, X.; et al. Site-specific antibody–drug conjugation through glycoengineering. *Bioconjug. Chem.* 2014, 25, 510–520.
61. Junutula, J. R.; Raab, H.; Clark, S.; Bhakta, S.; Leipold, D. D.; Weir, S.; Chen, Y.; Simpson, M.; Tsai, S. P.; Dennis, M. S.; et al. Site-specific conjugation of a cytotoxic drug to an antibody improves the therapeutic index. *Nat. Biotechnol.* 2008, 26, 925–932.
62. Liu, C. C.; Schultz, P. G. Adding new chemistries to the genetic code. *Annu. Rev. Biochem.* 2010, 79, 413–444.
63. Axup, J. Y.; Bajjuri, K. M.; Ritland, M.; Hutchins, B. M.; Kim, C. H.; Kazane, S. A.; Halder, R.; Forsyth, J. S.; Santidrian, A. F.; Stafin, K.; et al. Synthesis of site-specific antibody–drug conjugates using unnatural amino acids. *Proc. Natl. Acad. Sci. U. S. A.* 2012, 109, 16101–16106.
64. Walsh, G.; Jefferis, R. Post-translational modifications in the context of therapeutic proteins. *Nat. Biotechnol.* 2006, 24, 1241–1252.
65. Walsh, C. T.; Garneau-Tsodikova, S.; Gatto, G. J. Protein posttranslational modifications: the chemistry of proteome diversifications. *Angew. Chem. Int. Ed.* 2005, 44, 7342–7372.

66. Schmaltz, R. M.; Hanson, S. R.; Wong, C.-H. Enzymes in the synthesis of glycoconjugates. *Chem. Rev.* 2011, 111, 4259–4307.
67. Okeley, N. M.; Toki, B. E.; Zhang, X.; Jeffrey, S. C.; Burke, P. J.; Alley, S. C.; Senter, P. D. Metabolic engineering of monoclonal antibody carbohydrates for antibody–drug conjugation. *Bioconjug. Chem.* 2013, 24, 1650–1655.
68. Schmidt, B.; Selmer, T.; Ingendoh, A.; von Figura, K. A novel amino acid modification in sulfatases that is defective in multiple sulfatase deficiency. *Cell* 1995, 82, 271–278.
69. Dierks, T.; Miech, C.; Hummerjohann, J.; Schmidt, B.; Kertesz, M. A.; von Figura, K. Posttranslational formation of formylglycine in prokaryotic sulfatases by modification of either cysteine or serine. *J. Biol. Chem.* 1998, 273, 25560–25564.
70. Dierks, T.; Schmidt, B.; Von Figura, K. Conversion of cysteine to formylglycine: a protein modification in the endoplasmic reticulum. *Proc. Natl. Acad. Sci. U. S. A.* 1997, 94, 11963–11968.
71. Dierks, T.; Lecca, M. R.; Schlotterhose, P.; Schmidt, B.; von Figura, K. Sequence determinants directing conversion of cysteine to formylglycine in eukaryotic sulfatases. *EMBO J.* 1999, 18, 2084–2091.
72. Rabuka, D.; Rush, J. S.; deHart, G. W.; Wu, P.; Bertozzi, C. R. Site-specific chemical protein conjugation using genetically encoded aldehyde tags. *Nat. Protoc.* 2012, 7, 1052–1067.
73. Palcic, M. M. Glycosyltransferases as biocatalysts. *Curr. Opin. Chem. Biol.* 2011, 15, 226–233.
74. Zhou, Q.; Stefano, J. E.; Harrahy, J.; Finn, P.; Avila, L.; Kyazike, J.; Wei, R.; Van Patten, S. M.; Gotschall, R.; Zheng, X.; et al. Strategies for neoglycan conjugation to human acid A–glucosidase. *Bioconjug. Chem.* 2011, 22, 741–751.
75. Pietersz, G. A.; Kanellos, J.; Smyth, M. J.; Zalcberg, J.; McKenzie, I. F. The use of monoclonal antibody conjugates for the diagnosis and treatment of cancer. *Immunol. Cell Biol.* 1987, 65 (Pt 2), 111–125.
76. Lang, K.; Chin, J. W. Cellular incorporation of unnatural amino acids and bioorthogonal labeling of proteins. *Chem. Rev.* 2014, 114, 4764–4806.
77. Stephanopoulos, N.; Francis, M. B. Choosing an effective protein bioconjugation strategy. *Nat. Chem. Biol.* 2011, 7, 876–884.
78. Patterson, D. M.; Nazarova, L. A.; Prescher, J. A. Finding the right (bioorthogonal) chemistry. *ACS Chem. Biol.* 2014, 9, 592–605.
79. Lang, K.; Chin, J. W. Bioorthogonal reactions for labeling proteins. *ACS Chem. Biol.* 2014, 9, 16–20.
80. Rabuka, D. Chemoenzymatic methods for site-specific protein modification. *Curr. Opin. Chem. Biol.* 2010, 14, 790–796.
81. Hermanson, G. T. (1996) 2—The chemistry of reactive groups. In: Bioconjugate Techniques (ed. G. T. Hermanson), Academic Press, San Diego, CA, pp. 137–166.

82. Timkovich, R. Polymerization side reactions during protein modifications with carbodiimide. *Biochem. Biophys. Res. Commun.* 1977, 74, 1463–1468.

83. Peters, K.; Richards, F. M. Chemical cross-linking: reagents and problems in studies of membrane structure. *Annu. Rev. Biochem.* 1977, 46, 523–551.

84. Levy, R.; Hurwitz, E.; Maron, R.; Arnon, R.; Sela, M. The specific cytotoxic effects of daunomycin conjugated to antitumor antibodies. *Cancer Res.* 1975, 35, 1182–1186.

85. Anjaneyulu, P. S. R.; Staros, J. V. Reactions of *N*-hydroxysulfosuccinimide active esters. *Int. J. Pept. Protein Res.* 1987, 30, 117–124.

86. Braslawsky, G. R.; Edson, M. A.; Pearce, W.; Kaneko, T.; Greenfield, R. S. Antitumor activity of adriamycin (hydrazone-linked) immunoconjugates compared with free adriamycin and specificity of tumor cell killing. *Cancer Res.* 1990, 50, 6608–6614.

87. Greenfield, R. S.; Kaneko, T.; Daues, A.; Edson, M. A.; Fitzgerald, K. A.; Olech, L. J.; Grattan, J. A.; Spitalny, G. L.; Braslawsky, G. R. Evaluation *in vitro* of adriamycin immunoconjugates synthesized using an acid-sensitive hydrazone linker. *Cancer Res.* 1990, 50, 6600–6607.

88. Smyth, M. J.; Pietersz, G. A.; McKenzie, I. F. C. Selective enhancement of antitumor activity of *N*-acetyl melphalan upon conjugation to monoclonal antibodies. *Cancer Res.* 1987, 47, 62–69.

89. Chari, R. V. J.; Martell, B. A.; Gross, J. L.; Cook, S. B.; Shah, S. A.; Blättler, W. A.; McKenzie, S. J.; Goldmacher, V. S. Immunoconjugates containing novel maytansinoids: promising anticancer drugs. *Cancer Res.* 1992, 52, 127–131.

90. Chih, H.-W.; Gikanga, B.; Yang, Y.; Zhang, B. Identification of amino acid residues responsible for the release of free drug from an antibody–drug conjugate utilizing lysine–succinimidyl ester chemistry. *J. Pharm. Sci.* 2011, 100, 2518–2525.

91. Wakankar, A. A.; Feeney, M. B.; Rivera, J.; Chen, Y.; Kim, M.; Sharma, V. K.; Wang, Y. J. Physicochemical stability of the antibody–drug conjugate trastuzumab-DM1: changes due to modification and conjugation processes. Bioconjug. Chem. 2010, 21, 1588–1595.

92. Miao, Z.; Hong, Y.; Zhu, T.; Chucholowski, A. W. Drug-conjugates, conjugation methods, and uses thereof. WOA1, November 21, 2013.

93. Lewis Phillips, G. D.; Li, G.; Dugger, D. L.; Crocker, L. M.; Parsons, K. L.; Mai, E.; Blattler, W. A.; Lambert, J. M.; Chari, R. V. J.; Lutz, R. J.; et al. Targeting HER2-positive breast cancer with trastuzumab-DM1, an antibody–cytotoxic drug conjugate. Cancer Res. 2008, 68, 9280–9290.

94. Erickson, H. K.; Park, P. U.; Widdison, W. C.; Kovtun, Y. V.; Garrett, L. M.; Hoffman, K.; Lutz, R. J.; Goldmacher, V. S.; Blättler, W. A. Antibody–maytansinoid conjugates are activated in targeted cancer cells by lysosomal degradation and linker-dependent intracellular processing. *Cancer Res.* 2006, 66, 4426–4433.

95. Lillo, A. M.; Sun, C.; Gao, C.; Ditzel, H.; Parrish, J.; Gauss, C.-M.; Moss, J.; Felding-Habermann, B.; Wirsching, P.; Boger, D. L.; et al. A human single-chain antibody specific for integrin α3β1 capable of cell internalization and delivery of antitumor agents. *Chem. Biol.* 2004, 11, 897–906.

96. Katz, J.; Janik, J. E.; Younes, A. Brentuximab vedotin (SGN-35). *Clin. Cancer Res.* 2011, 17, 6428–6436.

97. LoRusso, P. M.; Weiss, D.; Guardino, E.; Girish, S.; Sliwkowski, M. X. Trastuzumab emtansine: a unique antibody–drug conjugate in development for human epidermal growth factor receptor 2-positive cancer. *Clin. Cancer Res.* 2011, 17, 6437–6447.

98. Casi, G.; Huguenin-Dezot, N.; Zuberbühler, K.; Scheuermann, J.; Neri, D. Site-specific traceless coupling of potent cytotoxic drugs to recombinant antibodies for pharmacodelivery. *J. Am. Chem. Soc.* 2012, 134, 5887–5892.

99. Bernardes, G. J. L.; Steiner, M.; Hartmann, I.; Neri, D.; Casi, G. Site-specific chemical modification of antibody fragments using traceless cleavable linkers. *Nat. Protoc.* 2013, 8, 2079–2089.

100. Ren, H.; Xiao, F.; Zhan, K.; Kim, Y.-P.; Xie, H.; Xia, Z.; Rao, J. A biocompatible condensation reaction for the labeling of terminal cysteine residues on proteins. *Angew. Chem. Int. Ed.* 2009, 48, 9658–9662.

101. Yuan, Y.; Liang, G. A biocompatible, highly efficient click reaction and its applications. *Org. Biomol. Chem.* 2014, 12, 865–871.

102. Wang, P.; Zhang, C.-J.; Chen, G.; Na, Z.; Yao, S. Q.; Sun, H. Site-specific immobilization of biomolecules by a biocompatible reaction between terminal cysteine and 2-cyanobenzothiazole. *Chem. Commun.* 2013, 49, 8644–8646.

103. Wang, H.-C.; Yu, C.-C.; Liang, C.-F.; Huang, L.-D.; Hwu, J.-R.; Lin, C.-C. Site-selective protein immobilization through 2-cyanobenzothiazole–cysteine condensation. *ChemBioChem* 2014, 15, 829–835.

104. Jeon, J.; Shen, B.; Xiong, L.; Miao, Z.; Lee, K. H.; Rao, J.; Chin, F. T. Efficient method for site-specific ^{18}F-labeling of biomolecules using the rapid condensation reaction between 2-cyanobenzothiazole and cysteine. *Bioconjug. Chem.* 2012, 23, 1902–1908.

105. Godinat, A.; Park, H. M.; Miller, S. C.; Cheng, K.; Hanahan, D.; Sanman, L. E.; Bogyo, M.; Yu, A.; Nikitin, G. F.; Stahl, A.; et al. A biocompatible *in vivo* ligation reaction and its application for noninvasive bioluminescent imaging of protease activity in living mice. *ACS Chem. Biol.* 2013, 8, 987–999.

106. Kundu, R.; Ball, Z. T. Rhodium-catalyzed cysteine modification with diazo reagents. *Chem. Commun.* 2013, 49, 4166–4168.

107. Koniev, O.; Leriche, G.; Nothisen, M.; Remy, J.-S.; Strub, J.-M.; Schaeffer-Reiss, C.; Van Dorsselaer, A.; Baati, R.; Wagner, A. Selective irreversible chemical tagging of cysteine with 3-arylpropiolonitriles. *Bioconjug. Chem.* 2014, 25, 202–206.

108. Badescu, G.; Bryant, P.; Swierkosz, J.; Khayrzad, F.; Pawlisz, E.; Farys, M.; Cong, Y.; Muroni, M.; Rumpf, N.; Brocchini, S.; et al. A new reagent for stable thiol-specific conjugation. *Bioconjug. Chem.* 2014, 25, 460–469.

109. Tilley, S. D.; Francis, M. B. Tyrosine-selective protein alkylation using π-allylpalladium complexes. *J. Am. Chem. Soc.* 2006, 128, 1080–1081.

110. Joshi, N. S.; Whitaker, L. R.; Francis, M. B. A three-component mannich-type reaction for selective tyrosine bioconjugation. *J. Am. Chem. Soc.* 2004, 126, 15942–15943.

111. McFarland, J. M.; Joshi, N. S.; Francis, M. B. Characterization of a three-component coupling reaction on proteins by isotopic labeling and nuclear magnetic resonance spectroscopy. *J. Am. Chem. Soc.* 2008, 130, 7639–7644.

112. Romanini, D. W.; Francis, M. B. Attachment of peptide building blocks to proteins through tyrosine bioconjugation. *Bioconjug. Chem.* 2008, 19, 153–157.

113. Seim, K. L.; Obermeyer, A. C.; Francis, M. B. Oxidative modification of native protein residues using cerium(IV) ammonium nitrate. *J. Am. Chem. Soc.* 2011, 133, 16970–16976.

114. Antos, J. M.; Francis, M. B. Selective tryptophan modification with rhodium carbenoids in aqueous solution. *J. Am. Chem. Soc.* 2004, 126, 10256–10257.

115. Antos, J. M.; McFarland, J. M.; Iavarone, A. T.; Francis, M. B. Chemoselective tryptophan labeling with rhodium carbenoids at mild pH. *J. Am. Chem. Soc.* 2009, 131, 6301–6308.

116. Ban, H.; Gavrilyuk, J.; Barbas, C. F. III. Tyrosine bioconjugation through aqueous ene-type reactions: a click-like reaction for tyrosine. *J. Am. Chem. Soc.* 2010, 132, 1523–1525.

117. Ban, H.; Nagano, M.; Gavrilyuk, J.; Hakamata, W.; Inokuma, T.; Barbas, C. F. Facile and stabile linkages through tyrosine: bioconjugation strategies with the tyrosine-click reaction. *Bioconjug. Chem.* 2013, 24, 520–532.

118. Gilmore, J. M.; Scheck, R. A.; Esser-Kahn, A. P.; Joshi, N. S.; Francis, M. B. N-terminal protein modification through a biomimetic transamination reaction. *Angew. Chem. Int. Ed.* 2006, 45, 5307–5311.

119. Netirojjanakul, C.; Witus, L. S.; Behrens, C. R.; Weng, C.-H.; Iavarone, A. T.; Francis, M. B. Synthetically modified Fc domains as building blocks for immunotherapy applications. *Chem. Sci.* 2013, 4, 266.

120. Witus, L. S.; Netirojjanakul, C.; Palla, K. S.; Muehl, E. M.; Weng, C.-H.; Iavarone, A. T.; Francis, M. B. Site-specific protein transamination using *N*-methylpyridinium-4-carboxaldehyde. *J. Am. Chem. Soc.* 2013, 135, 17223–17229.

121. Klinguer-Hamour, C.; Strop, P.; Shah, D. K.; Ducry, L.; Xu, A.; Beck, A. World antibody–drug conjugate summit, October 15–16, 2013, San Francisco, CA. *mAbs* 2014, 6, 18–29.

122. Agarwal, P.; Kudirka, R.; Albers, A. E.; Barfield, R. M.; de Hart, G. W.; Drake, P. M.; Jones, L. C.; Rabuka, D. Hydrazino-Pictet-Spengler ligation as a biocompatible method for the generation of stable protein conjugates. *Bioconjug. Chem.* 2013, 24, 846–851.

123. Chalker, J. M.; Gunnoo, S. B.; Boutureira, O.; Gerstberger, S. C.; Fernández-González, M.; Bernardes, G. J. L.; Griffin, L.; Hailu, H.; Schofield, C. J.; Davis, B. G. Methods for converting cysteine to dehydroalanine on peptides and proteins. *Chem. Sci.* 2011, 2, 1666–1676.

124. You, Y. O.; Levengood, M. R.; Ihnken, L. A. F.; Knowlton, A. K.; van der Donk, W. A. Lacticin 481 synthetase as a general serine/threonine kinase. *ACS Chem. Biol.* 2009, 4, 379–385.

125. Strop, P. Versatility of microbial transglutaminase. *Bioconjug. Chem.* 2014, 25, 855–862.

126. Farias, S. E.; Strop, P.; Delaria, K.; Galindo Casas, M.; Dorywalska, M.; Shelton, D. L.; Pons, J.; Rajpal, A. Mass spectrometric characterization of transglutaminase based site-specific antibody–drug conjugates. *Bioconjug. Chem.* 2014, 25, 240–250.

127. Dennler, P.; Chiotellis, A.; Fischer, E.; Brégeon, D.; Belmant, C.; Gauthier, L.; Lhospice, F.; Romagne, F.; Schibli, R. Transglutaminase-based chemo-enzymatic conjugation approach yields homogeneous antibody–drug conjugates. *Bioconjug. Chem.* 2014, 25, 569–578.

128. Strop, P.; Dorywalska, M. G.; Rajpal, A.; Shelton, D.; Liu, S.-H.; Pons, J.; Dushin, R. Engineered polypeptide conjugates and methods for making thereof using transglutaminase. WO2012059882 (A3), November 22, 2012.

129. Geierstanger, B. H.; Grunewald, J.; Bursulaya, B. Site-specific labeling methods and molecules produced thereby. WOA1, March 6, 2014.

130. Antos, J. M.; Miller, G. M.; Grotenbreg, G. M.; Ploegh, H. L. Lipid modification of proteins through sortase-catalyzed transpeptidation. *J. Am. Chem. Soc.* 2008, 130, 16338–16343.

131. Yamamoto, T.; Nagamune, T. Expansion of the sortase-mediated labeling method for site-specific N-terminal labeling of cell surface proteins on living cells. *Chem. Commun.* 2009, 9, 1022–1024.

132. Kornberger, P.; Skerra, A. Sortase-catalyzed *in vitro* functionalization of a HER2-specific recombinant Fab for tumor targeting of the plant cytotoxin gelonin. *mAbs* 2013, 6, 354–366.

133. Madej, M. P.; Coia, G.; Williams, C. C.; Caine, J. M.; Pearce, L. A.; Attwood, R.; Bartone, N. A.; Dolezal, O.; Nisbet, R. M.; Nuttall, S. D.; et al. Engineering of an anti-epidermal growth factor receptor antibody to single chain format and labeling by sortase A-mediated protein ligation. *Biotechnol. Bioeng.* 2012, 109, 1461–1470.

134. Zhang, C.; Spokoyny, A. M.; Zou, Y.; Simon, M. D.; Pentelute, B. L. Enzymatic "click" ligation: selective cysteine modification in polypeptides enabled by promiscuous glutathione *S*-transferase. *Angew. Chem. Int. Ed.* 2013, 52, 14001–14005.

135. Zhang, C.; Dai, P.; Spokoyny, A. M.; Pentelute, B. L. Enzyme-catalyzed macrocyclization of long unprotected peptides. *Org. Lett.* 2014, 16, 3652–3655.

136. Francis, M. B.; Carrico, I. S. New frontiers in protein bioconjugation. *Curr. Opin. Chem. Biol.* 2010, 14, 771–773.

137. Szajewski, R. P.; Whitesides, G. M. Rate constants and equilibrium constants for thiol-disulfide interchange reactions involving oxidized glutathione. *J. Am. Chem. Soc.* 1980, 102, 2011–2026.

138. Kellogg, B. A.; Garrett, L.; Kovtun, Y.; Lai, K. C.; Leece, B.; Miller, M.; Payne, G.; Steeves, R.; Whiteman, K. R.; Widdison, W.; et al. Disulfide-linked antibody–maytansinoid conjugates: optimization of *in vivo* activity by varying the steric hindrance at carbon atoms adjacent to the disulfide linkage. *Bioconjug. Chem.* 2011, 22, 717–727.

139. Erickson, H. K.; Widdison, W. C.; Mayo, M. F.; Whiteman, K.; Audette, C.; Wilhelm, S. D.; Singh, R. Tumor delivery and *in vivo* processing of disulfide-linked and thioether-linked antibody–maytansinoid conjugates. *Bioconjug. Chem.* 2010, 21, 84–92.

140. Saito, G.; Swanson, J. A.; Lee, K.-D. Drug delivery strategy utilizing conjugation via reversible disulfide linkages: role and site of cellular reducing activities. *Adv. Drug Deliv. Rev.* 2003, 55, 199–215.

141. Jencks, W. P. Studies on the mechanism of oxime and semicarbazone formation. *J. Am. Chem. Soc.* 1959, 81, 475–481.

142. King, M.; Wagner, A. Developments in the field of bioorthogonal bond forming reactions—past and present trends. *Bioconjug. Chem.* 2014, 25, 825–839.

143. Rostovtsev, V. V.; Green, L. G.; Fokin, V. V.; Sharpless, K. B. A stepwise Huisgen cycloaddition process: copper(I)-catalyzed regioselective "ligation" of azides and terminal alkynes. *Angew. Chem. Int. Ed.* 2002, 41, 2596–2599.

144. Tornøe, C. W.; Christensen, C.; Meldal, M. Peptidotriazoles on solid phase: [1,2,3]-triazoles by regiospecific copper(I)-catalyzed 1,3-dipolar cycloadditions of terminal alkynes to azides. *J. Org. Chem.* 2002, 67, 3057–3064.

145. Agard, N. J.; Prescher, J. A.; Bertozzi, C. R. A strain-promoted [3 + 2] azide–alkyne cycloaddition for covalent modification of biomolecules in living systems. *J. Am. Chem. Soc.* 2004, 126, 15046–15047.

146. Zimmerman, E. S.; Heibeck, T. H.; Gill, A.; Li, X.; Murray, C. J.; Madlansacay, M. R.; Tran, C.; Uter, N. T.; Yin, G.; Rivers, P. J.; et al. Production of site-specific antibody–drug conjugates using optimized non-natural amino acids in a cell-free expression system. *Bioconjug. Chem.* 2014, 25, 351–361.

147. Hooker, J. M.; Esser-Kahn, A. P.; Francis, M. B. Modification of aniline containing proteins using an oxidative coupling strategy. *J. Am. Chem. Soc.* 2006, 128, 15558–15559.

148. Obermeyer, A. C.; Jarman, J. B.; Netirojjanakul, C.; El Muslemany, K.; Francis, M. B. Mild bioconjugation through the oxidative coupling of *ortho*-aminophenols and anilines with ferricyanide. *Angew. Chem. Int. Ed.* 2014, 53, 1057–1061.

149. Behrens, C. R.; Hooker, J. M.; Obermeyer, A. C.; Romanini, D. W.; Katz, E. M.; Francis, M. B. Rapid chemoselective bioconjugation through oxidative coupling of anilines and aminophenols. *J. Am. Chem. Soc.* 2011, 133, 16398–16401.

150. Carrico, Z. M.; Romanini, D. W.; Mehl, R. A.; Francis, M. B. Oxidative coupling of peptides to a virus capsid containing unnatural amino acids. *Chem. Commun.* 2008, 10, 1205–1207.

151. Obermeyer, A. C.; Capehart, S. L.; Jarman, J. B.; Francis, M. B. Multivalent viral capsids with internal cargo for fibrin imaging. *PLoS ONE* 2014, 9, e100678.

152. Stephanopoulos, N.; Tong, G. J.; Hsiao, S. C.; Francis, M. B. Dual-surface modified virus capsids for targeted delivery of photodynamic agents to cancer cells. *ACS Nano* 2010, 4, 6014–6020.

153. Farkas, M. E.; Aanei, I. L.; Behrens, C. R.; Tong, G. J.; Murphy, S. T.; O'Neil, J. P.; Francis, M. B. PET imaging and biodistribution of chemically modified bacteriophage MS2. *Mol. Pharm.* 2013, 10, 69–76.

154. Kodama, K.; Fukuzawa, S.; Nakayama, H.; Kigawa, T.; Sakamoto, K.; Yabuki, T.; Matsuda, N.; Shirouzu, M.; Takio, K.; Tachibana, K.; et al. Regioselective carbon–carbon bond formation in proteins with palladium catalysis; new protein chemistry by organometallic chemistry. *ChemBioChem* 2006, 7, 134–139.

155. Dumas, A.; Spicer, C. D.; Gao, Z.; Takehana, T.; Lin, Y. A.; Yasukohchi, T.; Davis, B. G. Self-liganded Suzuki-Miyaura coupling for site-selective protein PEGylation. *Angew. Chem. Int. Ed.* 2013, 52, 3916–3921.

156. Li, N.; Lim, R. K. V.; Edwardraja, S.; Lin, Q. Copper-free Sonogashira cross-coupling for functionalization of alkyne-encoded proteins in aqueous medium and in bacterial cells. *J. Am. Chem. Soc.* 2011, 133, 15316–15319.

157. King, H. D.; Yurgaitis, D.; Willner, D.; Firestone, R. A.; Yang, M. B.; Lasch, S. J.; Hellström, K. E.; Trail, P. A. Monoclonal antibody conjugates of doxorubicin prepared with branched linkers: a novel method for increasing the potency of doxorubicin immunoconjugates. *Bioconjug. Chem.* 1999, 10, 279–288.

158. Sutherland, M. S. K.; Sanderson, R. J.; Gordon, K. A.; Andreyka, J.; Cerveny, C. G.; Yu, C.; Lewis, T. S.; Meyer, D. L.; Zabinski, R. F.; Doronina, S. O.; et al. Lysosomal trafficking and cysteine protease metabolism confer target-specific cytotoxicity by peptide-linked anti-CD30–auristatin conjugates. *J. Biol. Chem.* 2006, 281, 10540–10547.

159. Beaumont, K.; Cole, S. M.; Gibson, K.; Gosset, J. R. (2010) Chapter 2. ADMET for the medicinal chemist. In: Metabolism, Pharmacokinetics and Toxicity of Functional Groups: Impact of Chemical Building Blocks on ADMET (ed. D. A. Smith), Royal Society of Chemistry, Cambridge, pp. 61–98.

160. Yoshitake, S.; Yamada, Y.; Ishikawa, E.; Masseyeff, R. Conjugation of glucose oxidase from *Aspergillus niger* and rabbit antibodies using *N*-hydroxysuccinimide ester of *N*-(4-carboxycyclohexylmethyl)-maleimide. *Eur. J. Biochem.* 1979, 101, 395–399.

161. Webster, R.; Didier, E.; Harris, P.; Siegel, N.; Stadler, J.; Tilbury, L.; Smith, D. PEGylated proteins: evaluation of their safety in the absence of definitive metabolism studies. *Drug Metab. Dispos.* 2007, 35, 9–16.

162. Harris, J. M.; Zalipsky, S. (eds) (1997) Poly(ethylene Glycol): Chemistry and Biological Applications, vol. 680, ACS Symposium Series; American Chemical Society, Washington, DC.

163. King, H. D.; Dubowchik, G. M.; Mastalerz, H.; Willner, D.; Hofstead, S. J.; Firestone, R. A.; Lasch, S. J.; Trail, P. A. Monoclonal antibody conjugates of doxorubicin prepared with branched peptide linkers: inhibition of aggregation by methoxytriethyleneglycol chains. *J. Med. Chem.* 2002, 45, 4336–4343.

164. Jeffrey, S. C.; Torgov, M. Y.; Andreyka, J. B.; Boddington, L.; Cerveny, C. G.; Denny, W. A.; Gordon, K. A.; Gustin, D.; Haugen, J.; Kline, T.; et al. Design, synthesis, and *in vitro* evaluation of dipeptide-based antibody minor groove binder conjugates. *J. Med. Chem.* 2005, 48, 1344–1358.

165. Hapuarachchige, S.; Zhu, W.; Kato, Y.; Artemov, D. Bioorthogonal, two-component delivery systems based on antibody and drug-loaded nanocarriers for enhanced internalization of nanotherapeutics. *Biomaterials* 2014, 35, 2346–2354.

166. Zhao, R. Y.; Wilhelm, S. D.; Audette, C.; Jones, G.; Leece, B. A.; Lazar, A. C.; Goldmacher, V. S.; Singh, R.; Kovtun, Y.; Widdison, W. C.; et al. Synthesis and evaluation of hydrophilic linkers for antibody–maytansinoid conjugates. *J. Med. Chem.* 2011, 54, 3606–3623.

167. Pasut, G.; Veronese, F. M. State of the art in PEGylation: the great versatility achieved after forty years of research. *J. Control. Release* 2012, 161, 461–472.

168. El-Faham, A.; Albericio, F. Peptide coupling reagents, more than a letter soup. *Chem. Rev.* 2011, 111, 6557–6602.

169. Brown, R. S.; Bennet, A. J.; Slebocka-Tilk, H. Recent perspectives concerning the mechanism of H_3O^+-and hydroxide-promoted amide hydrolysis. *Acc. Chem. Res.* 1992, 25, 481–488.

170. Hedstrom, L. Introduction: proteases. *Chem. Rev.* 2002, 102, 4429–4430.

171. Dubowchik, G. M.; Firestone, R. A.; Padilla, L.; Willner, D.; Hofstead, S. J.; Mosure, K.; Knipe, J. O.; Lasch, S. J.; Trail, P. A. Cathepsin B-labile dipeptide linkers for lysosomal release of doxorubicin from internalizing immunoconjugates: model studies of enzymatic drug release and antigen-specific *in vitro* anticancer activity. *Bioconjug. Chem.* 2002, 13, 855–869.

172. Doronina, S. O.; Mendelsohn, B. A.; Bovee, T. D.; Cerveny, C. G.; Alley, S. C.; Meyer, D. L.; Oflazoglu, E.; Toki, B. E.; Sanderson, R. J.; Zabinski, R. F.; et al. Enhanced activity of monomethylauristatin F through monoclonal antibody delivery: effects of linker technology on efficacy and toxicity. *Bioconjug. Chem.* 2006, 17, 114–124.

173. Asmus, K. D. Stabilization of oxidized sulfur centers in organic sulfides. Radical cations and odd-electron sulfur–sulfur bonds. *Acc. Chem. Res.* 1979, 12, 436–442.

174. Musker, W. K. Chemistry of aliphatic thioether cation radicals and dications. *Acc. Chem. Res.* 1980, 13, 200–206.

175. Coleman, B. R.; Glass, R. S.; Setzer, W. N.; Prabhu, U. D. G.; Wilson, G. S. (1982) Electrochemistry of aliphatic thioethers as models for biological electron transfer. In: Electrochemical and Spectrochemical Studies of Biological Redox Components, vol. 201, Advances in Chemistry; American Chemical Society, Washington, DC, pp. 417–441.

176. Vogt, W. Oxidation of methionyl residues in proteins: tools, targets, and reversal. *Free Radic. Biol. Med.* 1995, 18, 93–105.

177. Alley, S. C.; Benjamin, D. R.; Jeffrey, S. C.; Okeley, N. M.; Meyer, D. L.; Sanderson, R. J.; Senter, P. D. Contribution of linker stability to the activities of anticancer immunoconjugates. *Bioconjug. Chem.* 2008, 19, 759–765.

178. Guo, J.; Kumar, S.; Prashad, A.; Starkey, J.; Singh, S. K. Assessment of physical stability of an antibody drug conjugate by higher order structure analysis: impact of thiol–maleimide chemistry. *Pharm. Res.* 2014, 31, 1710–1723.

179. Lyon, R. P.; Setter, J. R.; Bovee, T. D.; Doronina, S. O.; Anderson, M. E.; Leiske, C. L.; Senter, P. D. Self-stabilizing ADCs: antibody–drug conjugates prepared with maleimido drug-linkers that catalyze their own thiosuccinimide ring hydrolysis. *Cancer Res.* 2013, 73, 4333.

180. Fishkin, N.; Maloney, E. K.; Chari, R. V. J.; Singh, R. A novel pathway for maytansinoid release from thioether linked antibody–drug conjugates (ADCs) under oxidative conditions. *Chem. Commun.* 2011, 47, 10752–10754.

181. Austin, C. D.; Wen, X.; Gazzard, L.; Nelson, C.; Scheller, R. H.; Scales, S. J. Oxidizing potential of endosomes and lysosomes limits intracellular cleavage of disulfide-based antibody–drug conjugates. *Proc. Natl. Acad. Sci. U. S. A.* 2005, 102, 17987–17992.

182. Dalvie, D.; Kang, P.; Loi, C.-M.; Goulet, L.; Nair, S. (2010) Chapter 7. Influence of heteroaromatic rings on ADME properties of drugs. In: Metabolism, Pharmacokinetics and Toxicity of Functional Groups: Impact of Chemical Building Blocks on ADMET (ed. D. A. Smith), RSC Publishing, Cambridge, pp. 275–327.

183. Egberink, H.; Van Heerden, C. The Mechanism of the formation and hydrolysis of cyclohexanone oxime in aqueous solutions. *Anal. Chim. Acta* 1980, 118, 359–368.

184. Cordes, E. H.; Jencks, W. P. The mechanism of hydrolysis of Schiff bases derived from aliphatic amines. *J. Am. Chem. Soc.* 1963, 85, 2843–2848.

185. Doronina, S. O.; Toki, B. E.; Torgov, M. Y.; Mendelsohn, B. A.; Cerveny, C. G.; Chace, D. F.; DeBlanc, R. L.; Gearing, R. P.; Bovee, T. D.; Siegall, C. B.; et al. Development of potent monoclonal antibody auristatin conjugates for cancer therapy. *Nat. Biotechnol.* 2003, 21, 778–784.

186. Chari, R. V. J.; Jackel, K. A.; Bourret, L. A.; Derr, S. M.; Tadayoni, B. M.; Mattocks, K. M.; Shah, S. A.; Liu, C.; Blättler, W. A.; Goldmacher, V. S. Enhancement of the selectivity and antitumor efficacy of a CC-1065 analogue through immunoconjugate formation. *Cancer Res.* 1995, 55, 4079–4084.

187. Braslawsky, G. R.; Kadow, K.; Knipe, J.; McGoff, K.; Edson, M.; Kaneko, T.; Greenfield, R. S. Adriamycin(hydrazone)–antibody conjugates require internalization and intracellular acid hydrolysis for antitumor activity. *Cancer Immunol. Immunother.* 1991, 33, 367–374.

188. Austin, C. D.; Mazière, A. M. D.; Pisacane, P. I.; van Dijk, S. M.; Eigenbrot, C.; Sliwkowski, M. X.; Klumperman, J.; Scheller, R. H. Endocytosis and sorting of ErbB2 and the site of action of cancer therapeutics trastuzumab and geldanamycin. *Mol. Biol. Cell* 2004, 15, 5268–5282.

189. Geisow, M. J.; Evans, W. H. pH in the endosome: measurements during pinocytosis and receptor-mediated endocytosis. *Exp. Cell Res.* 1984, 150, 36–46.

190. Froesch, B. A.; Stahel, R. A.; Zangemeister-Wittke, U. Preparation and functional evaluation of new doxorubicin immunoconjugates containing an acid-sensitive linker on small-cell lung cancer cells. *Cancer Immunol. Immunother.* 1996, 42, 55–63.

191. Blattler, W. A.; Kuenzi, B. S.; Lambert, J. M.; Senter, P. D. New heterobifunctional protein crosslinking reagent that forms an acid-labile link. *Biochemistry* 1985, 24, 1517–1524.

192. Dubowchik, G. M.; Padilla, L. M.; Firestone, R. A. An acid-cleavable linker stable at neutral pH that releases doxorubicin at lysosomal pH. *Bioorg. Med. Chem. Lett.* 1993, 3, 2843–2846.

193. Castañeda, L.; Maruani, A.; Schumacher, F. F.; Miranda, E.; Chudasama, V.; Chester, K. A.; Baker, J. R.; Smith, M. E. B.; Caddick, S. Acid-cleavable thiomaleamic acid linker for homogeneous antibody–drug conjugation. *Chem. Commun.* 2013, 49, 8187–8189.

194. Cheung, J. C.; Kim Chiaw, P.; Deber, C. M.; Bear, C. E. A novel method for monitoring the cytosolic delivery of peptide cargo. *J. Control. Release* 2009, 137, 2–7.

195. Hastings, K. T.; Cresswell, P. Disulfide reduction in the endocytic pathway: immunological functions of gamma-interferon-inducible lysosomal thiol reductase. *Antioxid. Redox Signal.* 2011, 15, 657–668.

196. Pillay, C.; Elliott, E.; Dennison, C. Endolysosomal proteolysis and its regulation. *Biochem. J.* 2002, 363, 417–429.

197. Hinman, L. M.; Hamann, P. R.; Wallace, R.; Menendez, A. T.; Durr, F. E.; Upeslacis, J. Preparation and characterization of monoclonal antibody conjugates of the calicheamicins: a novel and potent family of antitumor antibiotics. *Cancer Res.* 1993, 53, 3336–3342.

198. Trouet, A.; Masquelier, M.; Baurain, R.; Deprez-De Campeneere, D. A covalent linkage between daunorubicin and proteins that is stable in serum and reversible by lysosomal hydrolases, as required for a lysosomotropic drug-carrier conjugate: *in vitro* and *in vivo* studies. *Proc. Natl. Acad. Sci. U. S. A.* 1982, 79, 626–629.

199. Mohamed, M. M.; Sloane, B. F. Cysteine cathepsins: multifunctional enzymes in cancer. *Nat. Rev. Cancer* 2006, 6, 764–775.

200. Aronson, N. N.; Barrett, A. J. The specificity of cathepsin B. Hydrolysis of glucagon at the C-terminus by a peptidyldipeptidase mechanism. *Biochem. J.* 1978, 171, 759–765.

201. Dubowchik, G. M.; Firestone, R. A. Cathepsin B-sensitive dipeptide prodrugs. 1. A model study of structural requirements for efficient release of doxorubicin. *Bioorg. Med. Chem. Lett.* 1998, 8, 3341–3346.

202. Dubowchik, G. M.; Mosure, K.; Knipe, J. O.; Firestone, R. A. Cathepsin B-sensitive dipeptide prodrugs. 2. Models of anticancer drugs paclitaxel (Taxol®), mitomycin C and doxorubicin. *Bioorg. Med. Chem. Lett.* 1998, 8, 3347–3352.

203. Sanderson, R. J.; Hering, M. A.; James, S. F.; Sun, M. M.; Doronina, S. O.; Siadak, A. W.; Senter, P. D.; Wahl, A. F. *In vivo* drug-linker stability of an anti-CD30 dipeptide-linked auristatin immunoconjugate. *Clin. Cancer Res.* 2005, 11, 843–852.

204. Graaf, M.; Boven, E.; Scheeren, H.; Haisma, H.; Pinedo, H. Beta-glucuronidase-mediated drug release. *Curr. Pharm. Des.* 2002, 8, 1391–1403.

205. Albin, N.; Massaad, L.; Toussaint, C.; Mathieu, M.-C.; Morizet, J.; Parise, O.; Gouyette, A.; Chabot, G. G. Main drug-metabolizing enzyme systems in human breast tumors and peritumoral tissues. *Cancer Res.* 1993, 53, 3541–3546.

206. Desbène, S.; Van, H. D.; Michel, S.; Koch, M.; Tillequin, F.; Fournier, G.; Farjaudon, N.; Monneret, C. Doxorubicin prodrugs with reduced cytotoxicity suited for tumour-specific activation. *Anticancer Drug Des.* 1998, 13, 955–968.

207. Leenders, R. G. G.; Damen, E. W. P.; Bijsterveld, E. J. A.; Scheeren, H. W.; Houba, P. H. J.; van der Meulen-Muileman, I. H.; Boven, E.; Haisma, H. J. Novel anthracycline-spacer-B-glucuronide, -B-glucoside, and -B-galactoside prodrugs for application in selective chemotherapy. *Bioorg. Med. Chem.* 1999, 7, 1597–1610.

208. Jeffrey, S. C.; Nguyen, M. T.; Moser, R. F.; Meyer, D. L.; Miyamoto, J. B.; Senter, P. D. Minor groove binder antibody conjugates employing a water soluble B-glucuronide linker. *Bioorg. Med. Chem. Lett.* 2007, 17, 2278–2280.

209. Burke, P. J.; Senter, P. D.; Meyer, D. W.; Miyamoto, J. B.; Anderson, M.; Toki, B. E.; Manikumar, G.; Wani, M. C.; Kroll, D. J.; Jeffrey, S. C. Design, synthesis, and biological evaluation of antibody–drug conjugates comprised of potent camptothecin analogues. *Bioconjug. Chem.* 2009, 20, 1242–1250.

210. Jeffrey, S. C.; De Brabander, J.; Miyamoto, J.; Senter, P. D. Expanded utility of the B-glucuronide linker: ADCs that deliver phenolic cytotoxic agents. *ACS Med. Chem. Lett.* 2010, 1, 277–280.

211. Carl, P. L.; Chakravarty, P. K.; Katzenellenbogen, J. A. A novel connector linkage applicable in prodrug design. *J. Med. Chem.* 1981, 24, 479–480.

212. Greenwald, R. B.; Pendri, A.; Conover, C. D.; Zhao, H.; Choe, Y. H.; Martinez, A.; Shum, K.; Guan, S. Drug delivery systems employing 1,4- or 1,6-elimination: poly(ethylene glycol) prodrugs of amine-containing compounds. *J. Med. Chem.* 1999, 42, 3657–3667.

213. Niculescu-Duvaz, I.; Springer, C. J. Antibody-directed enzyme prodrug therapy (ADEPT): a review. *Adv. Drug Deliv. Rev.* 1997, 26, 151–172.

214. Saari, W. S.; Schwering, J. E.; Lyle, P. A.; Smith, S. J.; Engelhardt, E. L. Cyclization-activated prodrugs. Basic carbamates of 4-hydroxyanisole. *J. Med. Chem.* 1990, 33, 97–101.

Index

Biosimilars of Monoclonal Antibodies: A Practical Guide to Manufacturing, Preclinical, and Clinical Development, First Edition. Edited by Cheng Liu and K. John Morrow, Jr.

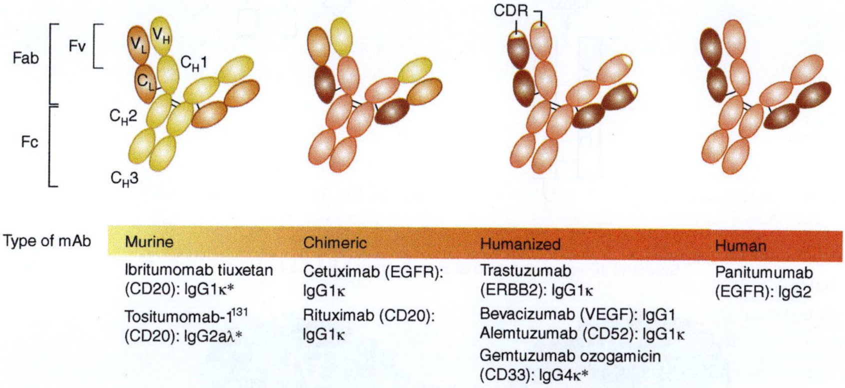

Figure 2.1 Classification of therapeutic mAb according to its increasing content of human sequence. ©2006 Nature Publishing Group. Nature Reviews/Cancer.

Biosimilars of Monoclonal Antibodies: A Practical Guide to Manufacturing, Preclinical, and Clinical Development, First Edition. Edited by Cheng Liu and K. John Morrow, Jr.

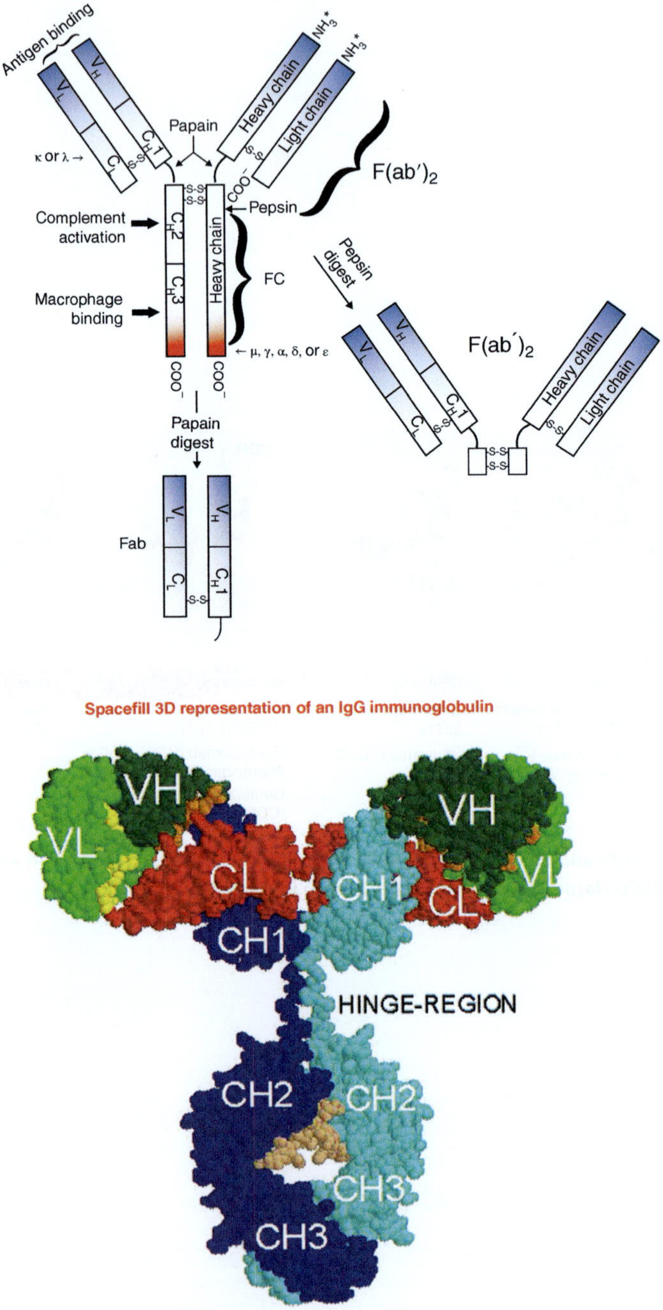

Figure 2.2 General structure of antibody.

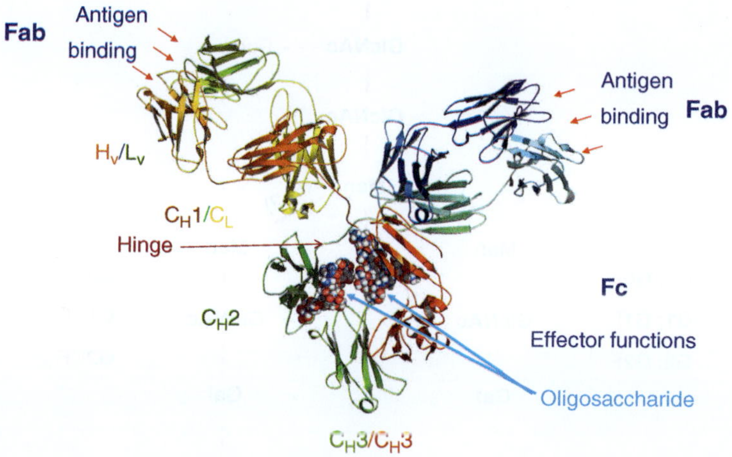

Figure 5.1 Domain structure of IgG. Courtesy of R. Emery and D. Fernandez, Ludger United Kingdom; based on the pdb 1IGY molecule.

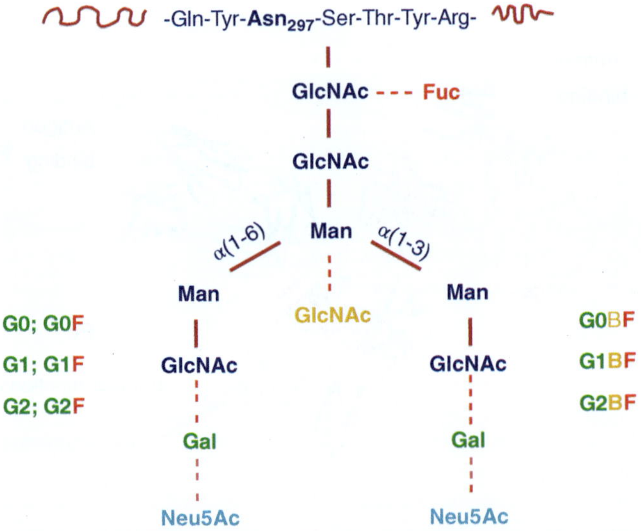

Figure 5.2 Representative IgG-Fc complex diantennary oligosaccharide. The "core" heptasaccharide has the composition: (GlcNAc)2Man3(GlcNAc)2, in blue [69, 72–75].

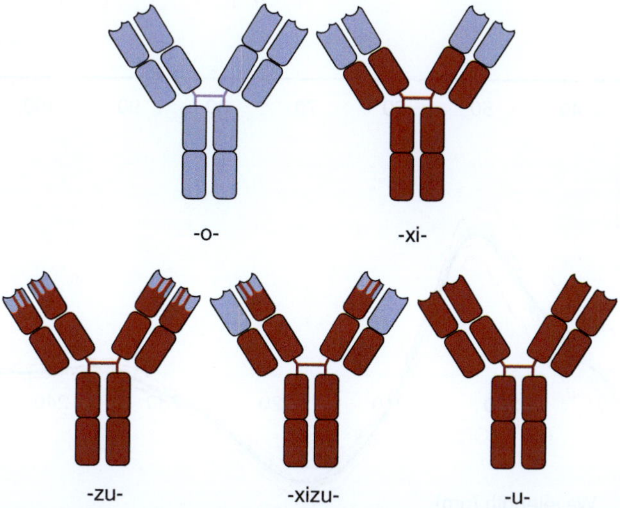

Figure 7.1 Nomenclature of different types of monoclonal antibodies. Comparison of monoclonal antibodies (dark brown, human; light blue, nonhuman or mouse). Top row: murine mAbs (name ends in "-omab"); ibritumomab tiuxetan. Top row: chimeric mAbs (name ends in "-ximab"); example: rituximab. Bottom row: humanized mAbs (name ends in "-zumab"); example: trastuzumab. Bottom row: chimeric/humanized mAbs (name ends in "-xizumab"); non are FDA approved. Bottom row: human mAbs (name ends in "-umab"); example: panitumumab. Note: Public domain image from Wikimedia Commons. Accessed on October 18, 2016, from https://en.wikipedia.org/wiki/Nomenclature_of_monoclonal_antibodies.

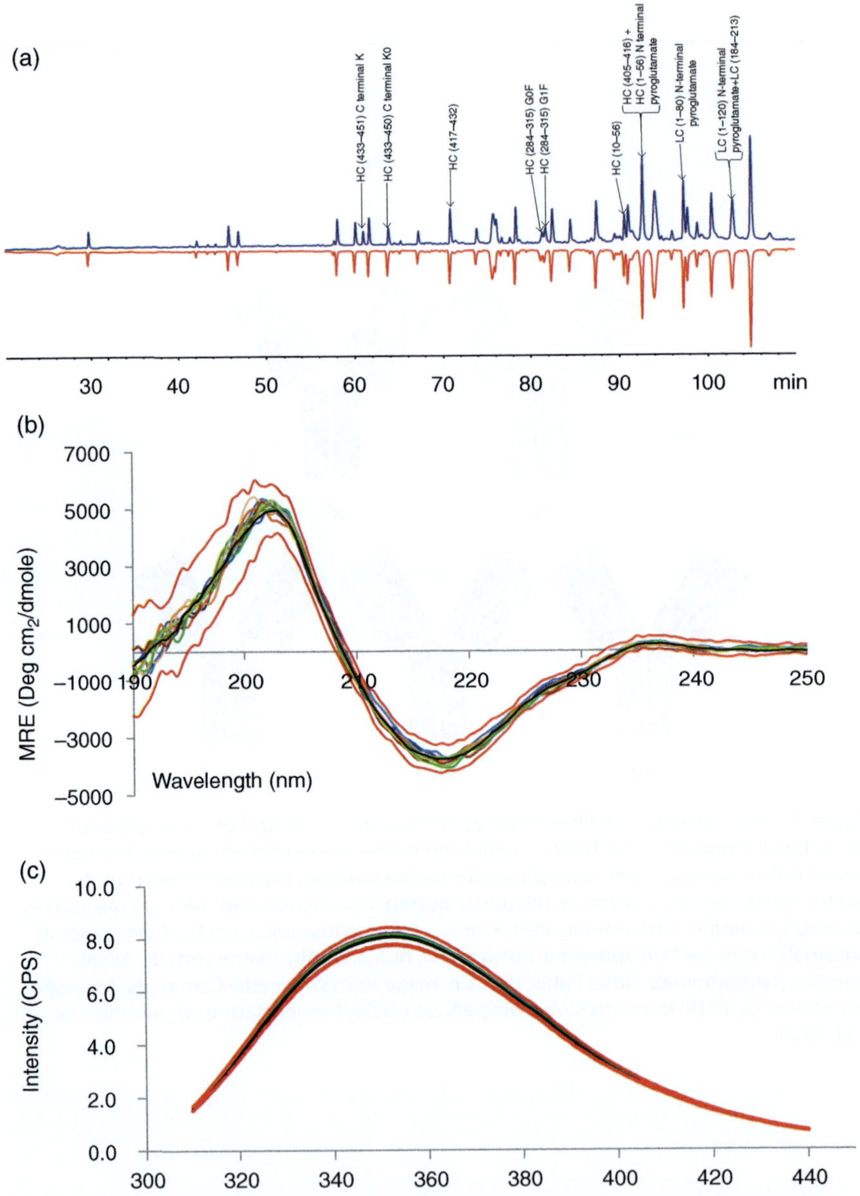

Figure 8.1 Structural similarity of Dr. Reddy's rituximab and originator product. Similar identity in amino acid sequence (a), secondary (b), and tertiary structure (c) for Reditux and originator products. Spectral profiles in red indicate method/batch to batch variability determined from Innovator product sourced in the European Union and the United States.

(d)

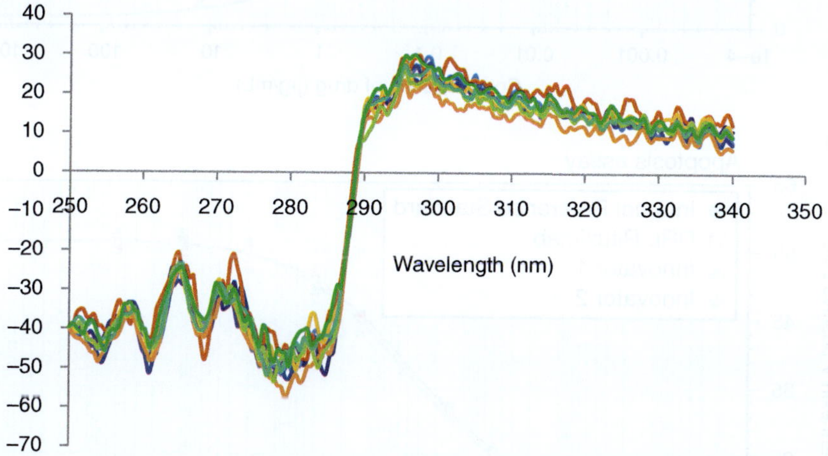

Figure 8.1 (Continued)

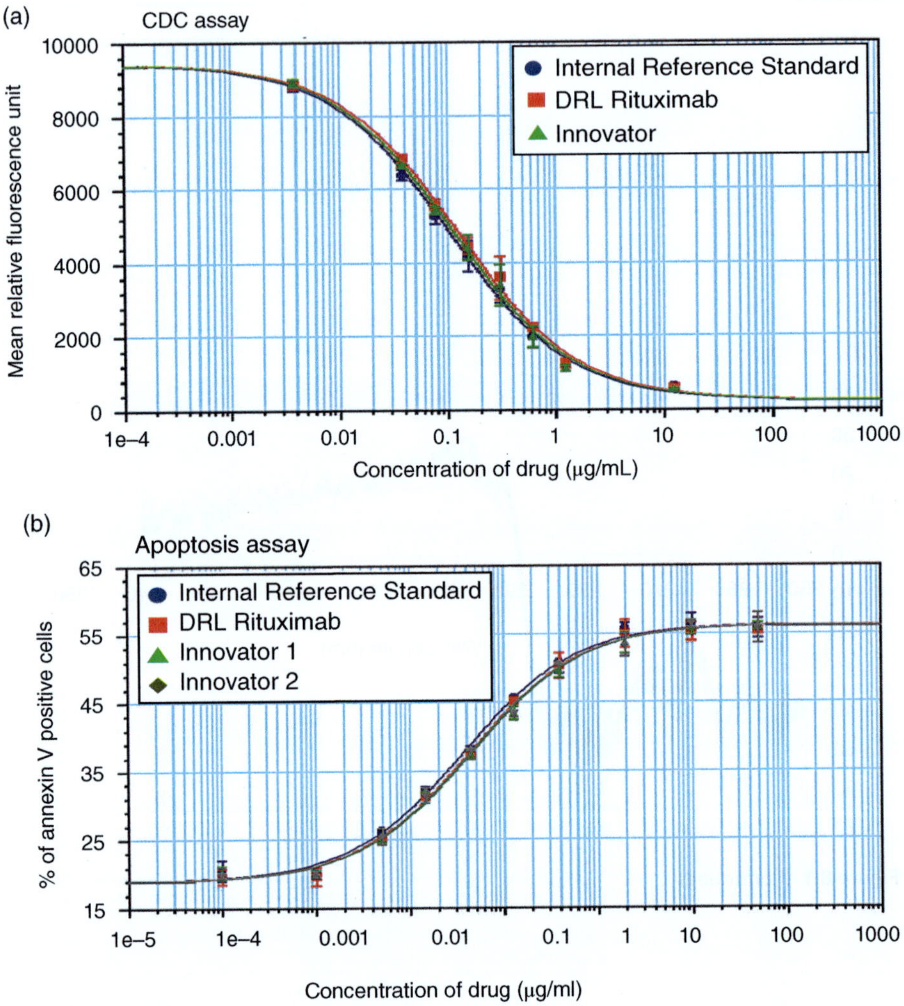

Figure 8.2 Functional similarity of Dr. Reddy's rituximab and originator product. Cell-based cytotoxicity induced by complement establishes CDC, cell-based cytotoxicity induced by PBMCs establishes ADCC, and cell-based assay for detection of annexin V-positive cells establishes apoptosis. Assays used for binding comparisons include cell-based competitive binding FACS assay for CD20 ligand binding; ELISA-based binding assay for C1q, FcγRIIa, and FcγRIIb binding; and binding constant determination by surface plasmon resonance for FcγRI, FcγRIIIa, and FcRn binding. The figure shows CDC (a) and apoptosis (b) assays and the FcγRIIIa binding sensorgrams (c).

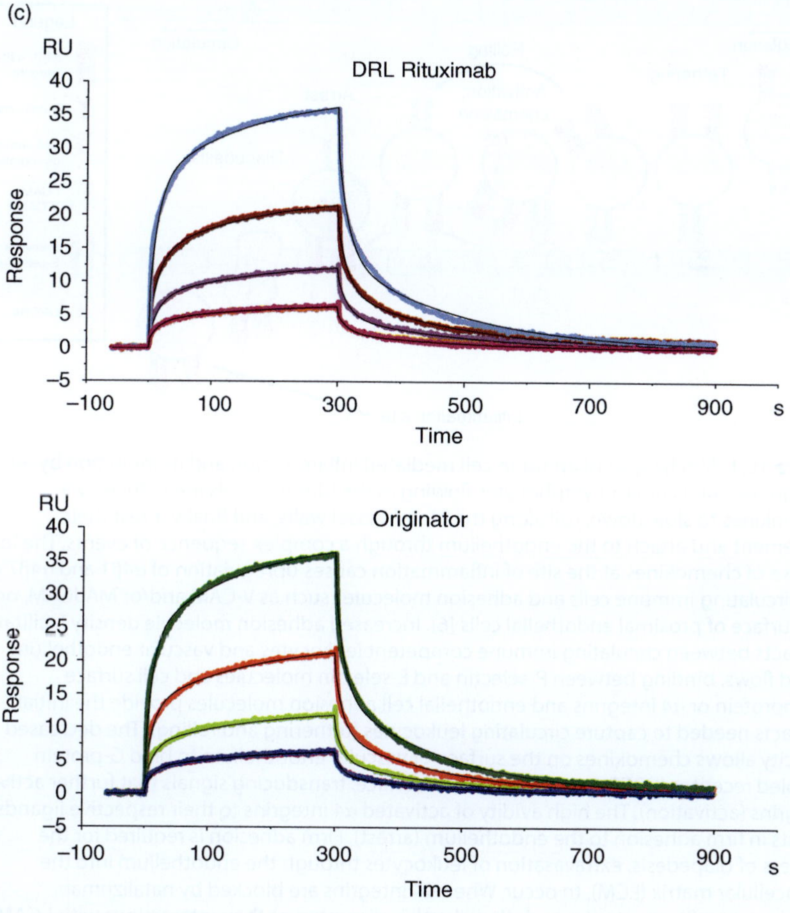

Figure 8.2 (Continued)

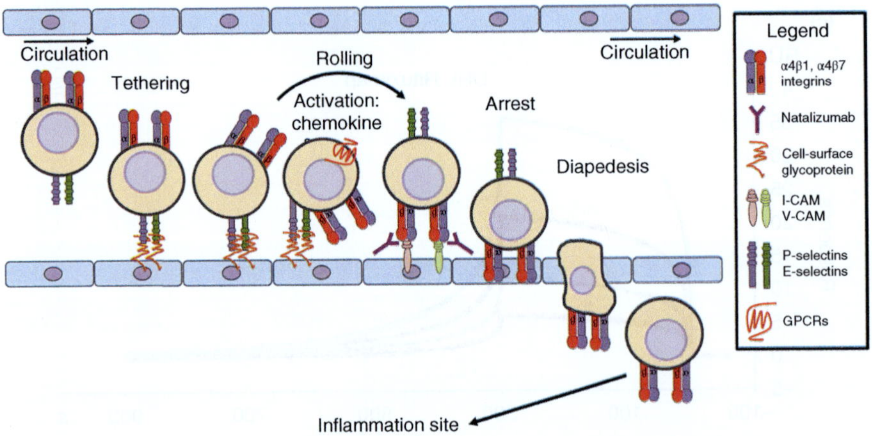

Figure 10.1 Mechanism of immune-cell mediated inflammation and its inhibition by α4 antagonist natalizumab. Lymphocytes flowing in the blood vessels are induced via chemokines to slow down, roll along the blood vessel walls, and finally arrest their movement and attach to the endothelium through a complex sequence of events. The local release of chemokines at the site of inflammation causes upregulation of α4β1 and α4β7 on the circulating immune cells and adhesion molecules such as V-CAM and/or MAdCAM, on the surface of proximal endothelial cells [6]. Increased adhesion molecule density facilitates contacts between circulating immune competent leukocytes and vascular endothelium. As blood flows, binding between P-selectin and E-selectin molecules and cell surface glycoprotein or α4 integrins and endothelial cell adhesion molecules provide the initial contacts needed to capture circulating leukocytes (tethering and rolling). The decreased velocity allows chemokines on the surface of vascular endothelium to bind G-protein coupled receptor (GPCRs) on the leukocyte surface, transducing signals that further activate integrins (activation). The high avidity of activated α4 integrins to their respective ligands results in firm adhesion to the endothelium (arrest). Firm adhesion is required for the process of diapedesis, extravasation of leukocytes through the endothelium into the extracellular matrix (ECM), to occur. When α4 integrins are blocked by natalizumab, lymphocyte adhesion to the endothelial wall is disrupted as their interactions with I-CAM and V-CAM are impaired.

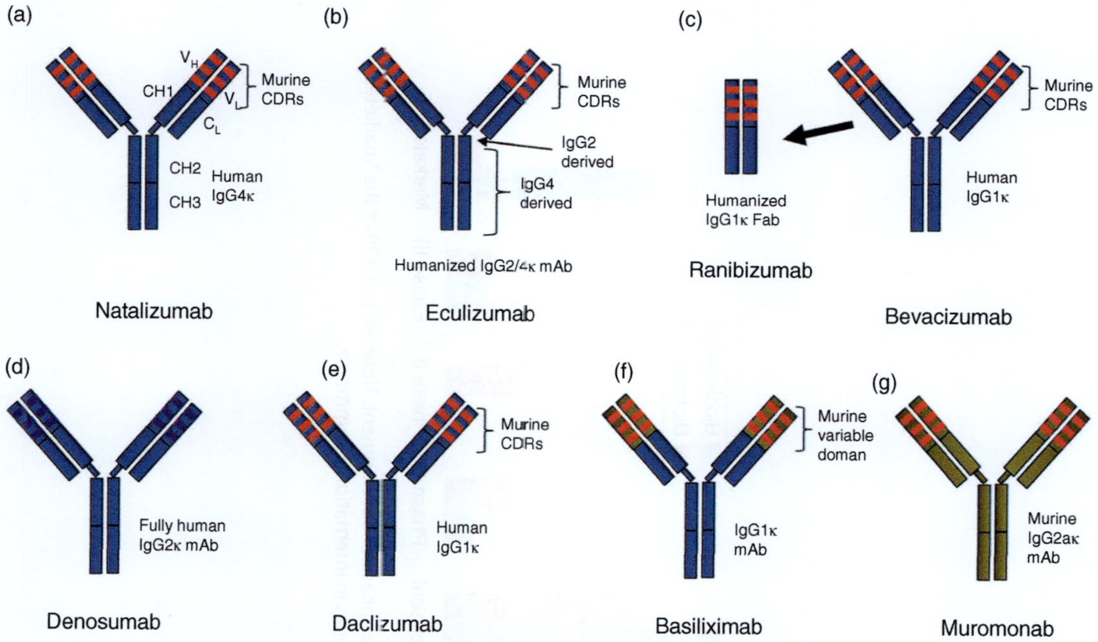

Figure 10.6 The schematic structures of studied antibodies in this chapter. Monoclonal antibodies covered in this chapter illustrate the evolution of monoclonal antibody therapeutics. Muromonab (g) is a murine antibody. Basiliximab (f) is a human/murine chimera with whole variable domain from murine origin. Natalizumab (a), eculizumab (b), bevacizumab (c), and daclizumab (e) are humanized monoclonal antibodies with only murine CDRs. Ranibizumab (c) is a humanized antibody fragment (Fab) derived from bevacizumab. Denosumab (d) is a fully human monoclonal antibody.

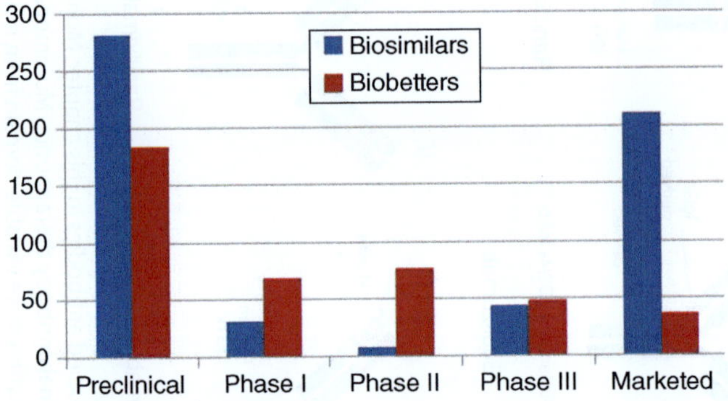

Figure 13.3 Pipelines by phase of development. Note: All but ≥20 of the "marketed" biosimilars are biogenerics in international commerce.

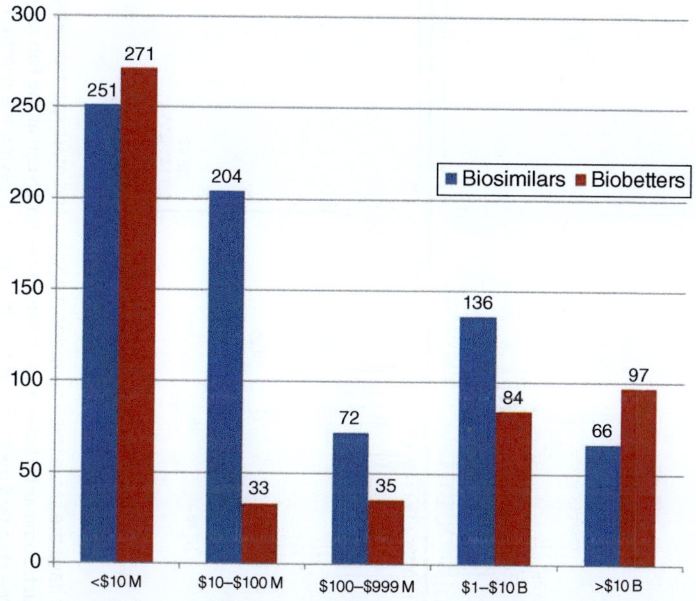

Figure 13.4 Number of companies in involved in biosimilars by size. The previous figure shows the numbers of companies known to be active in biosimilars development by company size. The table in the following shows the number of biosimilars in development by leading companies, in terms of number of biosimilars having been reported in development.

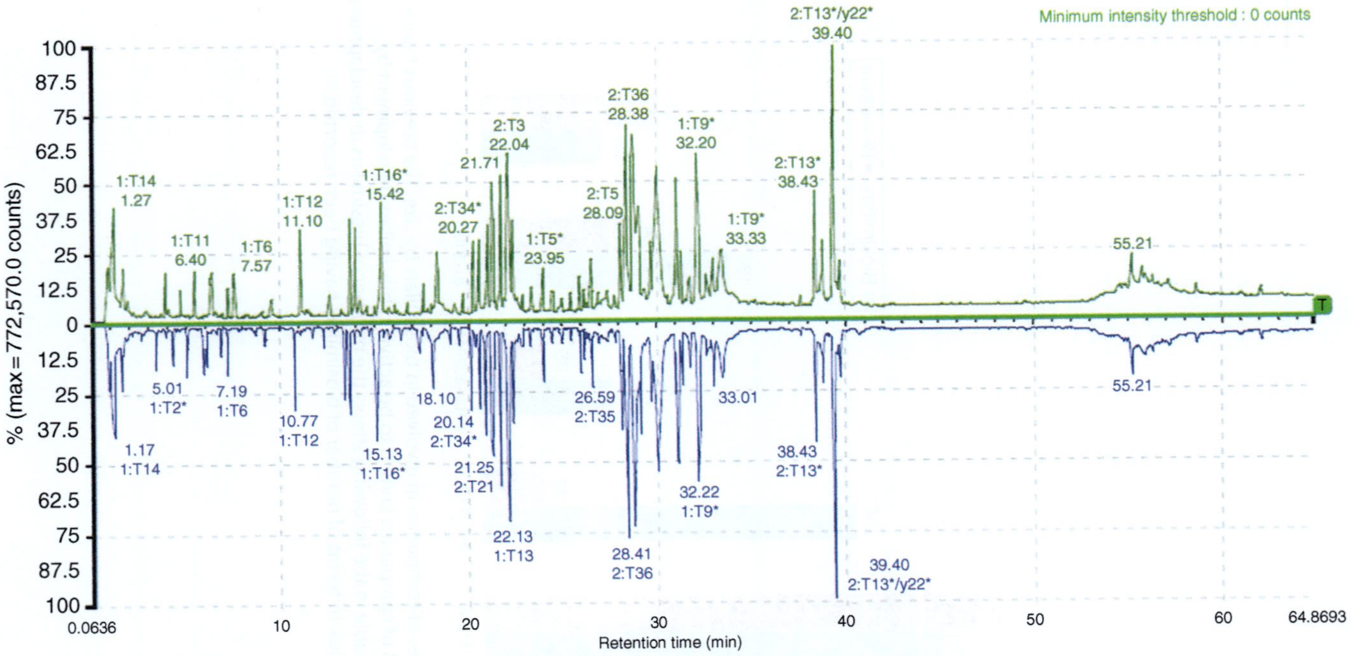

Figure 15.3 Biosimilar and originator products have the same tryptic peptide mapping and mass spectrometry sequences. Top panel: originator drug; bottom panel: biosimilar product. Some antibodies may contain an extra glycan at the Fab region in addition to the Fc site. The extra glycan has to be treated differently from the standard method due to difficulties in cleaving the glycans from the Fab region for analysis (see more details in the glycan analysis section).

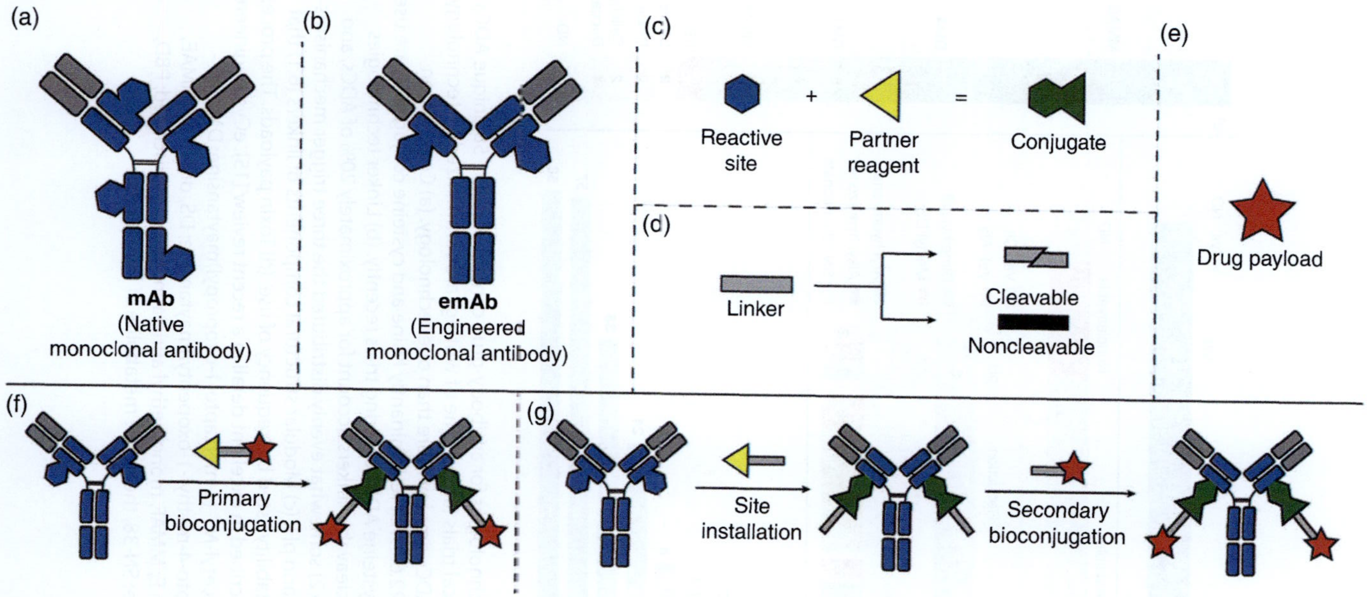

Figure 22.1 Modular components of an antibody-drug conjugate. A monoclonal antibody (mAb) (a) can be engineered (b) to contain reactive sites (c). These sites include amino acids (AAs) and posttranslational modifications (PTMs). Partner reagents are added to form a conjugate with the reactive site (c). A linker (d) is attached to a payload (e), which can include a release trigger. Synthesis of an ADC involves addition of an elaborated drug–linker to an antibody containing a reactive site (f and g).

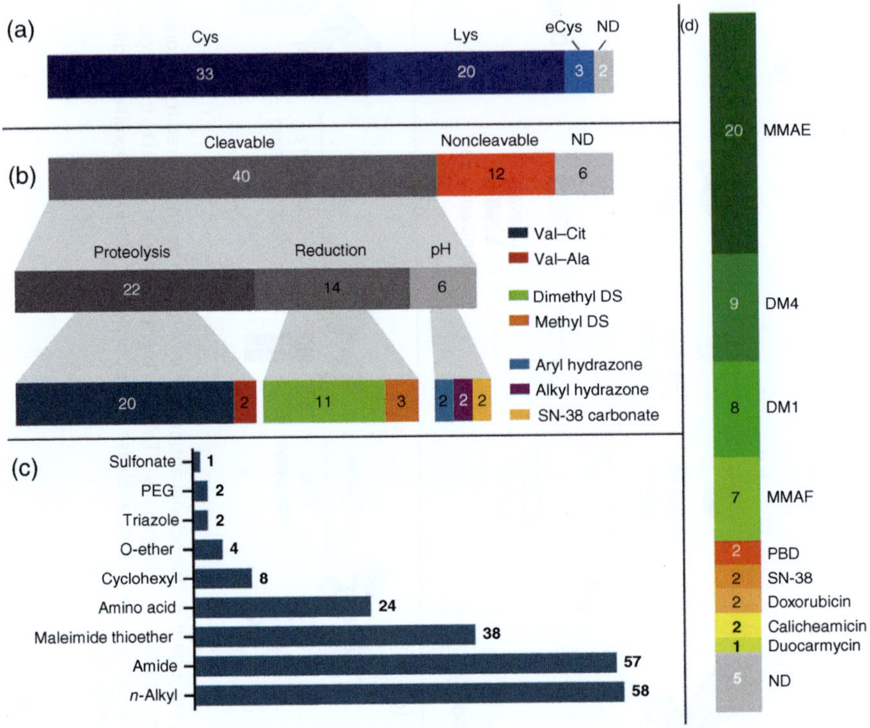

Figure 22.2 Proven technologies for antibody-drug conjugation. The 58 unique ADCs that have advanced to clinical trials (at the time of writing) are summarized by technology used. Note that most ADCs employ more than one technology. (a) Conjugation technologies (Section 22.6.2.4). Predominantly lysine and cysteine conjugation are used, with two engineered cysteine ADCs entering trials recently. (b) Linker technologies (Section 22.7.4.4). Noncleavable linkers account for approximately 20% of ADCs, and cleavable constructs (6.2) somewhat evenly distributed the three trigger mechanisms—proteolysis, reduction, and pH. (c) Modular structural components of linkers (6.1) that impart solubility and stability listed by frequency of use. (d) Toxin payloads. The process of choosing a toxin is described in excellent detail in a recent review [15]. eCys, engineered cysteine; DM1, $N^{2'}$-deacetyl-$N^{2'}$-(3-mercapto-1-oxopropyl)maytansine; DM4, $N^{2'}$-deacetyl-$N^{2'}$-(4-mercapto-4-methyl-1-oxopentyl)maytansine; DS, disulfide; MMAE, monomethyl auristatin E; MMAF, monomethyl auristatin F; ND, not disclosed; PBD, pyrrolobenzodiazepine; SN-38, the active metabolite of irinotecan.

Printed and bound by CPI Group (UK) Ltd, Croydon, CR0 4YY
30/09/2026
14973385-0001